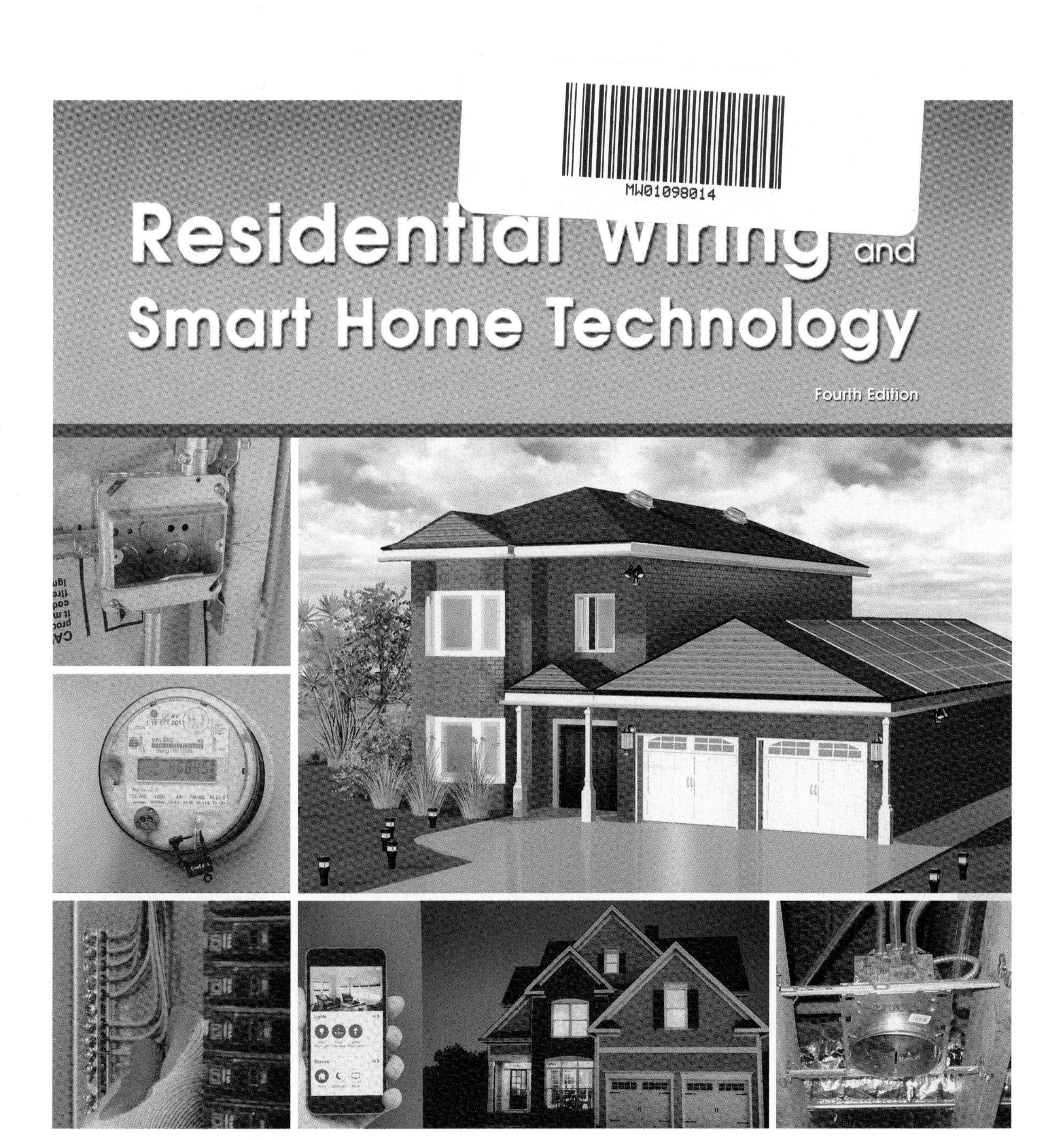

atp AMERICAN TECHNICAL PUBLISHERS
Orland Park, Illinois

Gary J. Rockis

American Technical Publishers Editorial Staff

Editor in Chief:
Peter A. Zurlis
Assistant Production Manager:
Nicole D. Bigos
Technical Editor:
William A. Hendricks
Kyle D. Wathen
Scott C. Bloom
Supervising Copy Editor:
Catherine A. Mini
Copy Editor:
Dane K. Hamann
Editorial Assistant:
Erin M. Magee

Cover Design:
Bethany J. Fisher
Art Supervisor:
Sarah E. Kaducak
Illustration/Layout:
Bethany J. Fisher
Richard O. Davis
Nicholas W. Basham
Nicholas G. Doornbos
Steven E. Gibbs
Kevin L. Cuasay
Melanie G. Doornbos
Digital Media Manager:
Adam T. Schuldt

4 5 6 7 8 9 – 19 – 9 8 7 6 5 4 3 2 1

Printed in the United States of America

ISBN 978-0-8269-1833-8

 This book is printed on recycled paper.

Acknowledgments

In memory of

Robert "Bob" Narvick

mentor, fellow traveler, and wonderful friend

The author and publisher are grateful for the photographs, technical information, and assistance provided by the following companies and organizations:

AEMC® Instruments
Briggs & Stratton Corporation
Coleman Cable Systems, Inc.
Cooper Wiring Devices
DOE/NREL, Altair Energy
East Penn Manufacturing Co., Inc.
Energy Policy Initiatives Center
Fluke Corporation
Fluke Networks
GE Security
Honeywell International, Inc.
IDEAL Industries, Inc.
Kennedy Mfg. Co.
Kidde
Klein Tools, Inc.
Lab Safety Supply, Inc.
Legrand®
Leviton Manufacturing Co., Inc.
Lew Electric Fittings Co®
Milwaukee Tool Corporation
Nutone, Division of Scovill
Omni Training
Panduit Corp.
Potter Electric Signal Co.
Raco Inc.
RIGID®
Salisbury by Honeywell
SMA Technologie AG
Square D/Schneider Electric
The Stanley Works
The Wadsworth Electric Mfg. Co., Inc.
Weller®
www.ftc.gov
U.S. Navy

Contents

Contents

LEARNER RESOURCES

- Quick Quizzes™
- Illustrated Glossary
- Flash Cards
- Interactive Wiring Activities
- Smart Home Technology
- Media Library
- Internet Resources

Features

Residential Wiring and Smart Home Technology focuses on the principles, installation, and operation of wired and wireless residential electrical and electronic systems. This edition provides expanded material on utility power generation and distribution, electrical safety, and NEC® guidelines.

New topics include smart home infrastructure, security and fire alarm systems, and energy management applications supported by the smart grid. A lifestyle applications chapter covers improvements to convenience and comfort provided by smart home technology.

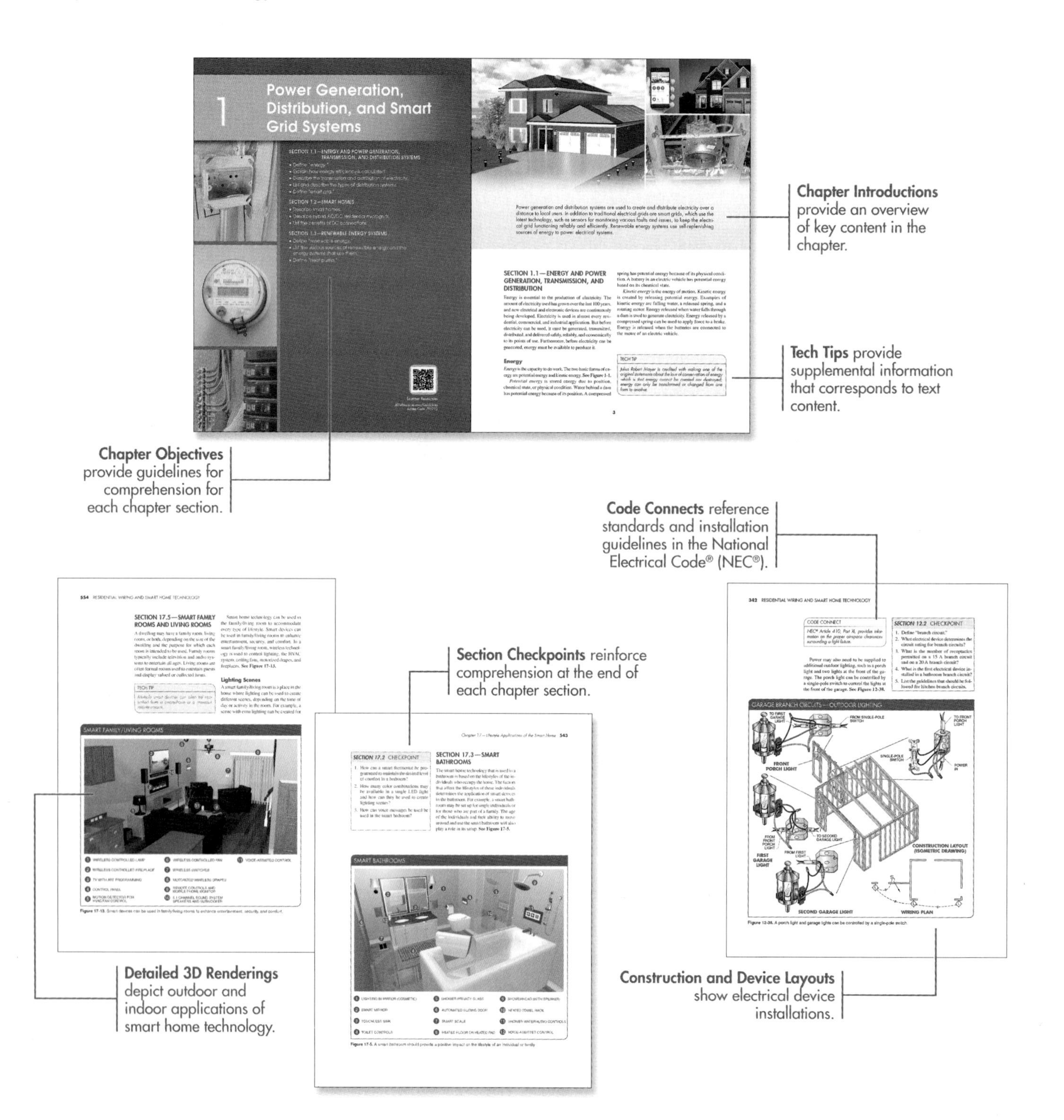

Learner Resources

Residential Wiring and Smart Home Technology online learner resources are self-study tools that reinforce the content covered in the book. These online learner resources can be accessed using either of the following methods:

- Key ATPeResources.com/QuickLinks into a web browser and enter QuickLinks™ Access code 791712.
- Use a quick response (QR) reader app to scan the QR code with a mobile device.

The online learner resources include the following:

- **Quick Quizzes™** that provide interactive questions for each section, with embedded links to highlighted content within the textbook and to the Illustrated Glossary
- **Illustrated Glossary** that serves as a helpful reference to commonly used terms, with selected terms linked to textbook illustrations
- **Flash Cards** that provide a self-study/review of common terms and their definitions
- **Interactive Wiring Activities** that reinforce residential wiring fundamentals
- **Smart Home Technology** that shows how security, convenience, and comfort can be improved for people of different lifestyles
- **Media Library** that consists of videos and animations that reinforce textbook content
- **Internet Resources** that provide access to additional online resources to support continued learning

1 Power Generation, Distribution, and Smart Grid Systems

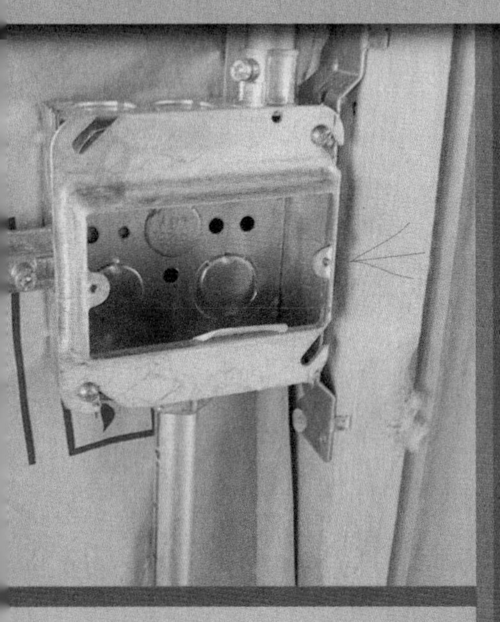

SECTION 1.1—ENERGY AND POWER GENERATION, TRANSMISSION, AND DISTRIBUTION SYSTEMS

- Define "energy."
- Explain how energy efficiency is calculated.
- Describe the transmission and distribution of electricity.
- List and describe the types of distribution systems.
- Define "smart grid."

SECTION 1.2—SMART HOMES

- Describe smart homes.
- Describe hybrid AC/DC residential microgrids.
- List the benefits of DC connections.

SECTION 1.3—RENEWABLE ENERGY SYSTEMS

- Define "renewable energy."
- List the various sources of renewable energy and the energy systems that use them.
- Define "heat pump."

Learner Resources

ATPeResources.com/QuickLinks
Access Code: 791712

Power generation and distribution systems are used to create and distribute electricity over a distance to local users. In addition to traditional electrical grids are smart grids, which use the latest technology, such as sensors for monitoring various faults and issues, to keep the electrical grid functioning reliably and efficiently. Renewable energy systems use self-replenishing sources of energy to power electrical systems.

SECTION 1.1—ENERGY AND POWER GENERATION, TRANSMISSION, AND DISTRIBUTION SYSTEMS

Energy is essential to the production of electricity. The amount of electricity used has grown over the last 100 years, and new electrical and electronic devices are continuously being developed. Electricity is used in almost every residential, commercial, and industrial application. But before electricity can be used, it must be generated, transmitted, distributed, and delivered safely, reliably, and economically to its points of use. Furthermore, before electricity can be generated, energy must be available to produce it.

Energy

Energy is the capacity to do work. The two basic forms of energy are potential energy and kinetic energy. **See Figure 1-1.**

Potential energy is stored energy due to position, chemical state, or physical condition. Water behind a dam has potential energy because of its position. A compressed spring has potential energy because of its physical condition. A battery in an electric vehicle has potential energy based on its chemical state.

Kinetic energy is the energy of motion. Kinetic energy is created by releasing potential energy. Examples of kinetic energy are falling water, a released spring, and a rotating motor. Energy released when water falls through a dam is used to generate electricity. Energy released by a compressed spring can be used to apply force to a brake. Energy is released when the batteries are connected to the motor of an electric vehicle.

> TECH TIP
>
> *Julius Robert Mayer is credited with making one of the original statements about the law of conservation of energy which is that energy cannot be created nor destroyed; energy can only be transformed or changed from one form to another.*

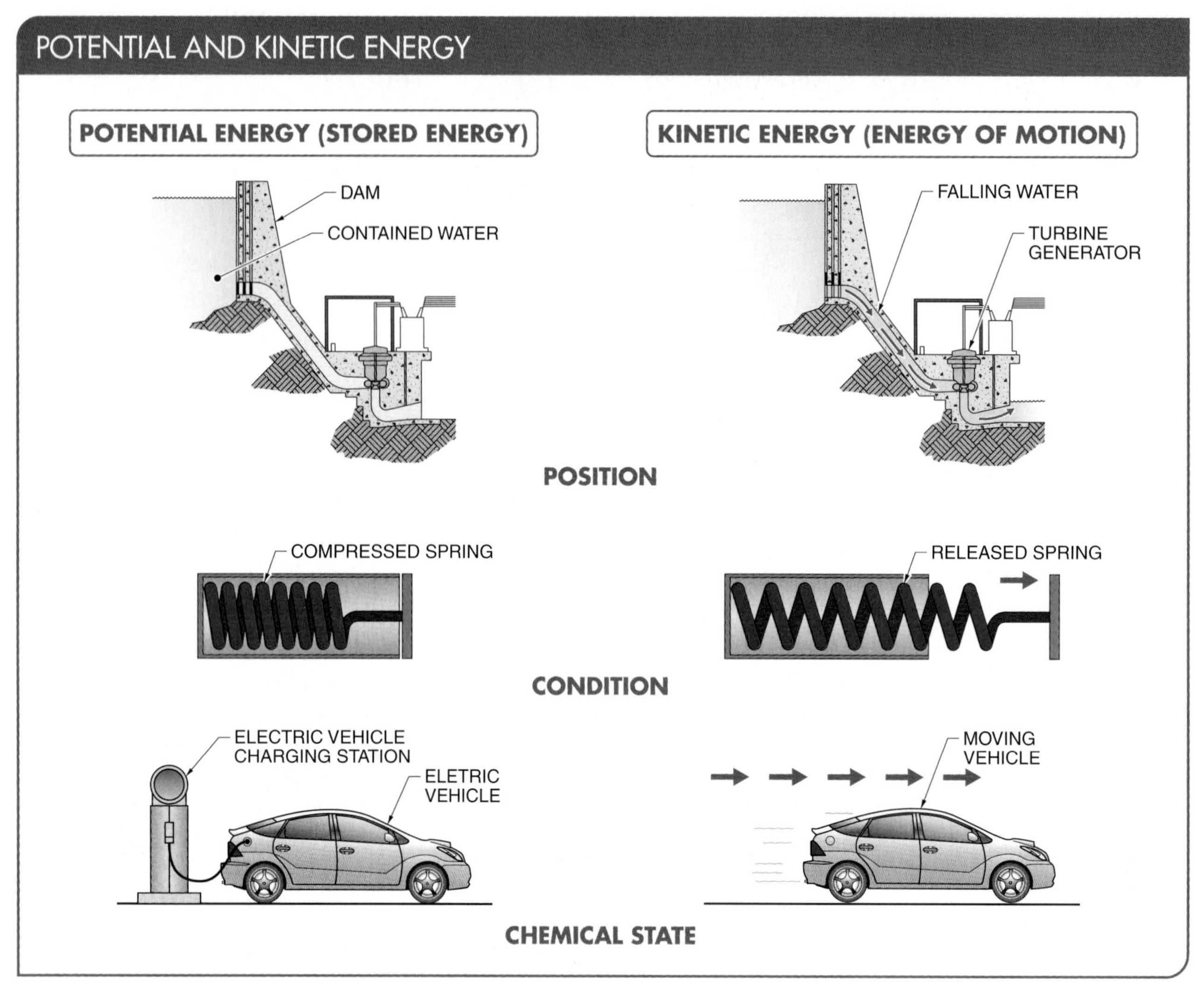

Figure 1-1. The two forms of energy are potential energy and kinetic energy; when potential energy is transformed into a moving position, a dynamic condition, or a chemical state, it becomes kinetic energy.

Energy Usage. Electrical energy is converted into motion, light, heat, sound, and visual outputs. Approximately 62% of all electricity generated is converted into rotating motion by motors. **See Figure 1-2.** Although three-phase (3ϕ) motors use the largest amount of electricity of all motors, they are also the most energy efficient. Unfortunately, most residential structures do not have 3ϕ power lines. However, 240 V, single-phase (1ϕ) power is available and is used for equipment that requires higher current to operate efficiently, such as electric ovens, electric clothes dryers, and heating, ventilating, and air conditioning (HVAC) systems. For the effective and efficient use of energy, electrical equipment must be designed and controlled properly for the maximum output.

Energy Efficiency. According to the law of conservation of energy, total output energy must equal total input energy. However, some input energy does not contribute to output energy because it is lost through friction and heat. For example, energy is lost through the heat and friction of an operating electric motor. **See Figure 1-3.** The use of advanced sensors and controls along with high-efficiency motors can significantly increase energy efficiency.

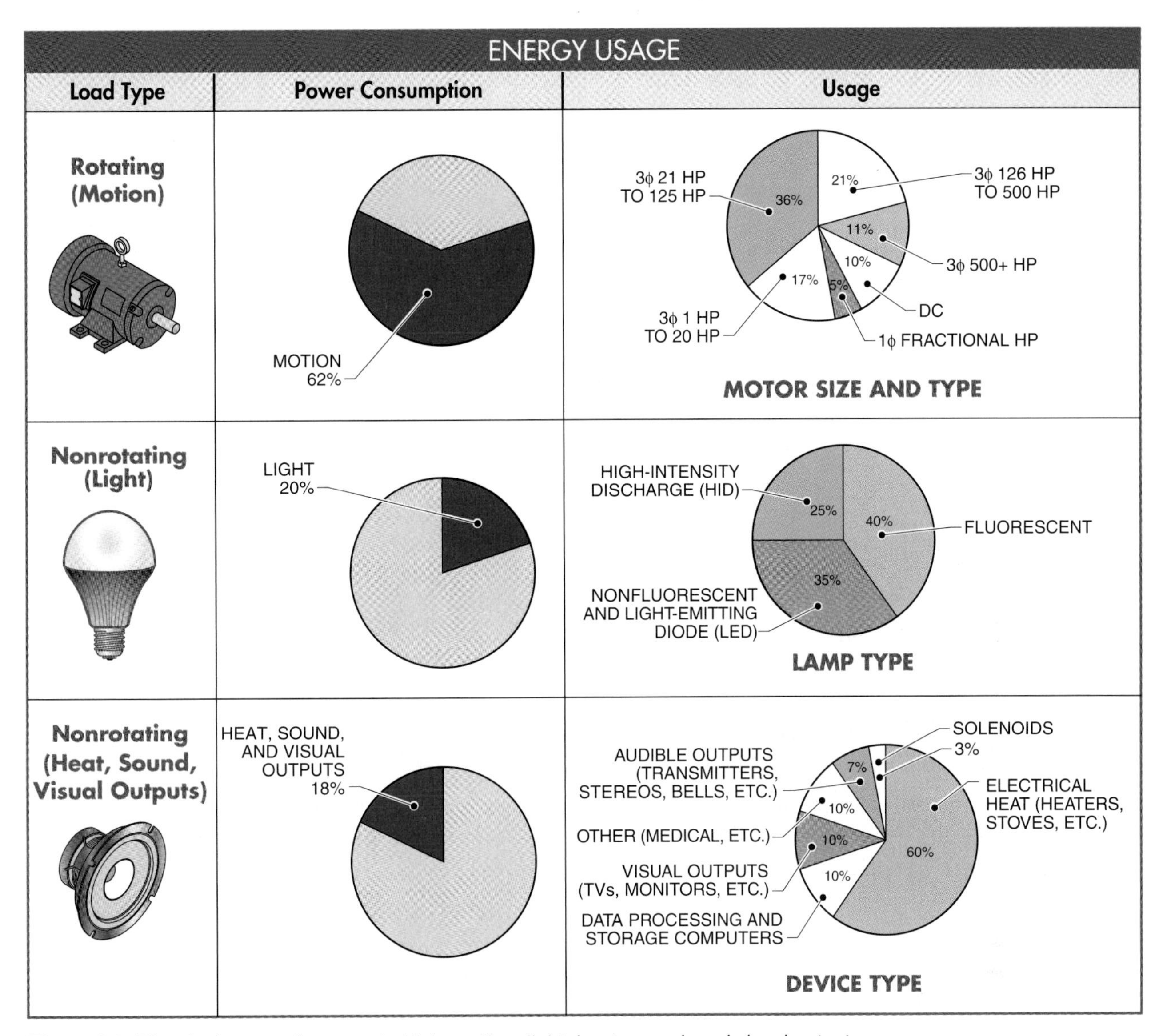

Figure 1-2. Electrical energy is converted into motion, light, heat, sound, and visual outputs.

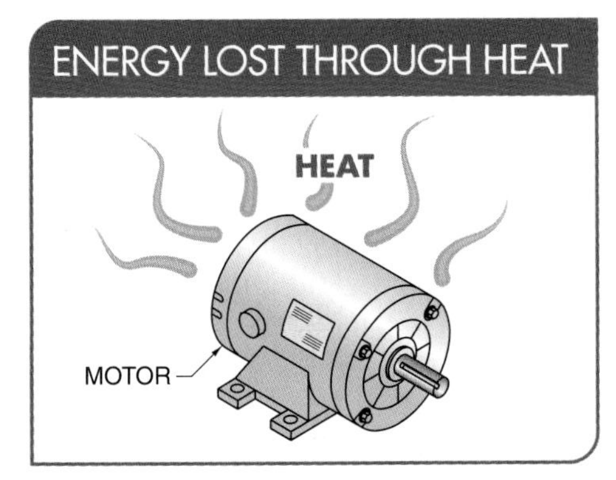

Figure 1-3. No machine is 100% efficient, as some motor efficiency is lost due to heat from the coils and friction from the bearings.

The cost of electricity to operate a machine is affected by the machine's energy efficiency. The energy efficiency of a machine indicates how well its input energy is converted to useful output energy or work. Energy efficiency (η) can be determined through the following formula:

$$\eta = \frac{E_o}{E_i} \times 100\%$$

where

η = energy efficiency (in %)

E_o = energy output (in W)

E_i = energy input (in W)

For example, a home appliance that is provided with 120 W of power produces 90 W of power. What is the energy efficiency of the appliance?

$$\eta = \frac{E_o}{E_i} \times 100\%$$

$$\eta = \frac{90\text{ W}}{120\text{ W}} \times 100\%$$

$$\eta = 0.75 \times 100$$

$$\eta = \mathbf{75\%}$$

Energy Star. Energy Star® is a US government program that mandates limits on the power consumption of electric equipment. Energy Star approval ratings are found on electrical appliances and are used to help consumers make well-informed decisions related to the energy efficiency of the appliances. In addition to the Energy Star label for appliances, residential homes can also earn the Energy Star label. To earn the Energy Star label, a residential home must be at least 15% more efficient than those built to code. Also, it must include additional energy-saving features to provide a performance advantage of up to 30% compared to a typical new home.

Electricity

Electricity is a form of energy where electrons move from the outer shell of one atom to the outer shell of another atom when an electromotive force (EMF) is applied to a material. **See Figure 1-4.** Conductive material, such as a copper conductor (wire), allows electrons to move easily. Insulating material, such as plastic, does not allow electrons to pass because it has a high resistance to electron movement.

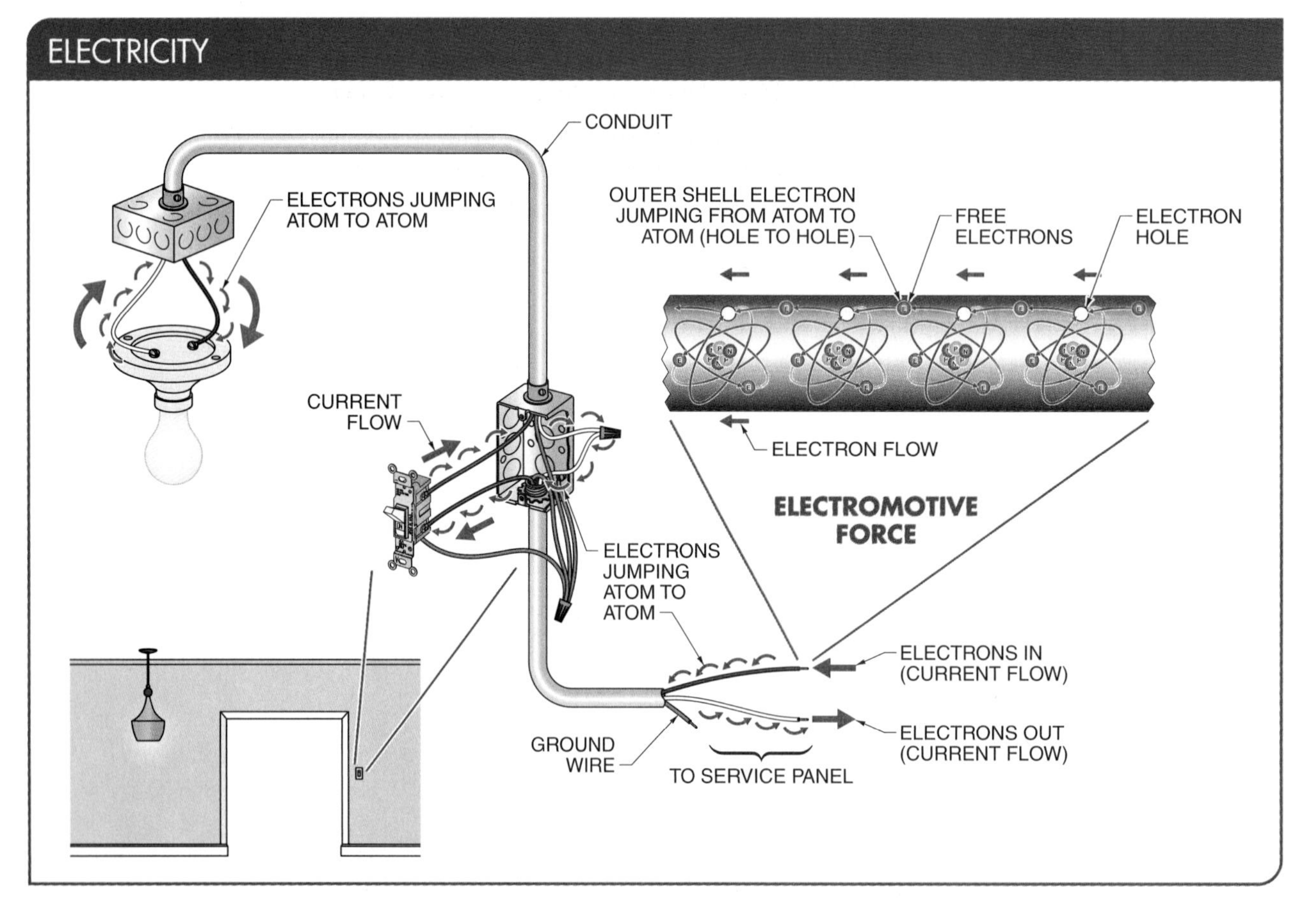

Figure 1-4. Electricity is the movement of electrons from atom to atom.

Power Transmission

Electricity that has been generated must be transmitted from its generation source, such as a power plant, over long distances to transmission stations and distribution substations before finally reaching residential areas. Generated electricity is transmitted through transmission lines. A *transmission line* is a conductor that carries large amounts of electrical power at high voltages over long distances. Typically, aerial transmission lines must be spaced far enough apart and elevated in order to be safe, since they are typically uninsulated. **See Figure 1-5.**

Figure 1-5. Transmission lines safely carry large amounts of electrical power at high voltages over long distances.

The transmission voltage level varies depending on the required transmission distance and amount of power carried. The longer the distance or higher the transmitted power, the higher the transmitted voltage must be. Electricity is typically generated and transmitted as alternating current (AC). However, new technology has allowed for advances in the transmission of high-voltage direct current (DC).

The transmission and distribution of electricity typically results in losses of as much as 6% to 7% of the generated electrical power annually, mostly attributed to heat. Transmission line voltages can vary from a few kilovolts (kV) to hundreds of kilovolts. Transmission-line conductor sizes are based on the amount of current they can safely carry without overheating. When transmission voltage increases, the amount of current decreases. Therefore, when power is transmitted at high voltages, the required size and weight of the conductors is reduced. Higher transmitted voltages allow for reduced conductor size, allow more power to be transmitted, and result in lower material cost. Increasing the transmitted voltage also reduces the power losses between the utility and the consumer. Power loss can be reduced up to 75% when transmitted voltage is doubled.

Alternating Current (AC) Transmission. Alternating current (AC) electricity has long been the primary source of generated electricity that is transmitted through transmission lines. *Alternating current (AC)* is current that reverses or alternates its direction of flow at regular intervals. AC is typically transmitted at high voltages to allow for transmission at long distances.

High-Voltage Direct Current (HVDC) Transmission. New technology has allowed utilities to transmit high-voltage direct current (HVDC) over long distances by improving the efficiency of transmission. *Direct current (DC)* is current that flows in only one direction. In an HVDC transmission system, electric power is taken from one point in a 3ϕ AC network, converted to DC in a converter station, transmitted to the receiving point by an overhead transmission line or cable, and then converted back to AC in another converter station and injected into the receiving AC network. **See Figure 1-6.**

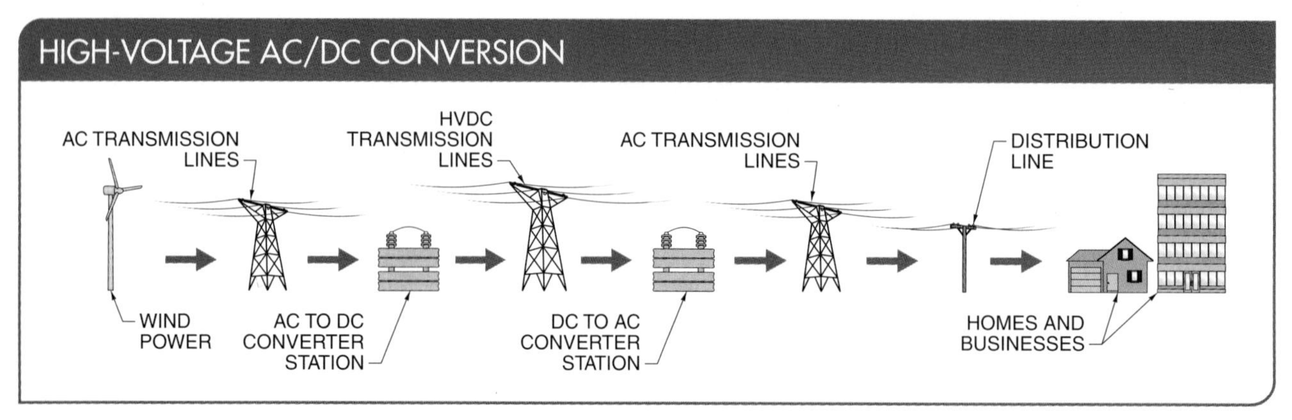

Figure 1-6. In an HVDC system, electric power is taken from one point in a 3ϕ AC network, converted to DC in a converter station, transmitted to another receiving point by an overhead line or cable, and then converted back to AC through another converter station and injected into the receiving AC network.

TECH TIP

Thomas Edison developed the first commercial central power plant and patented the first electrical power (DC) distribution system.

HVDC transmission lines can economically transfer significantly more power with lower losses over longer distances than comparable AC transmission lines. HVDC transmission lines require fewer conductors, smaller rights-of-way, and a smaller tower footprint than comparable AC transmission infrastructure. **See Figure 1-7.**

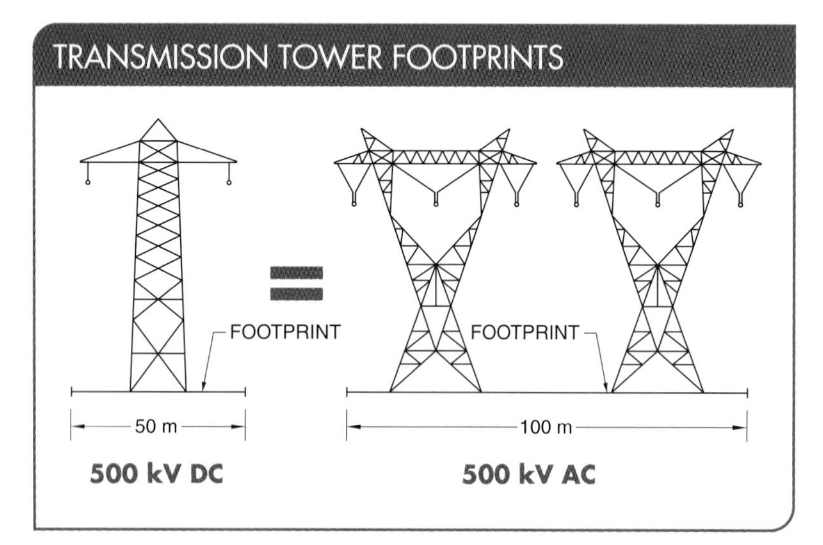

Figure 1-7. HVDC transmission line requires fewer conductors, smaller rights-of-way, and a smaller footprint than comparable AC transmission infrastructure.

HVDC transmission lines act as firewalls and are not overloaded by unrelated power outages. The amount of power delivered is strictly limited by the HVDC converters at each end of the HVDC transmission lines, reducing the likelihood that power outages will move from one part of the power grid to another. HVDC technology gives the operators direct control of energy flow, which makes HVDC particularly well-suited for managing the injection of variable wind generation. HVDC power lines can also be safely placed under water bodies for offshore applications.

Transformers

To efficiently transmit electricity throughout the power grid, transformers are used to increase the voltage of generated electricity. A *transformer* is an electric device that uses electromagnetism to change (step up or step down) AC voltage from one level to another. Transformers are often used at power plants to step up the voltage from the generators at 22,000 V to approximately 240,000 V to more easily distribute electricity across power lines to step-down transformer substations. **See Figure 1-8.**

Transformer substations then step down the electrical voltage from 240,000 V to 12,000 V for local distribution. The local distribution voltage is further stepped down at pole-mounted transformers from 12,000 V to 600 V, 480 V, 240 V, or 120 V for consumer use.

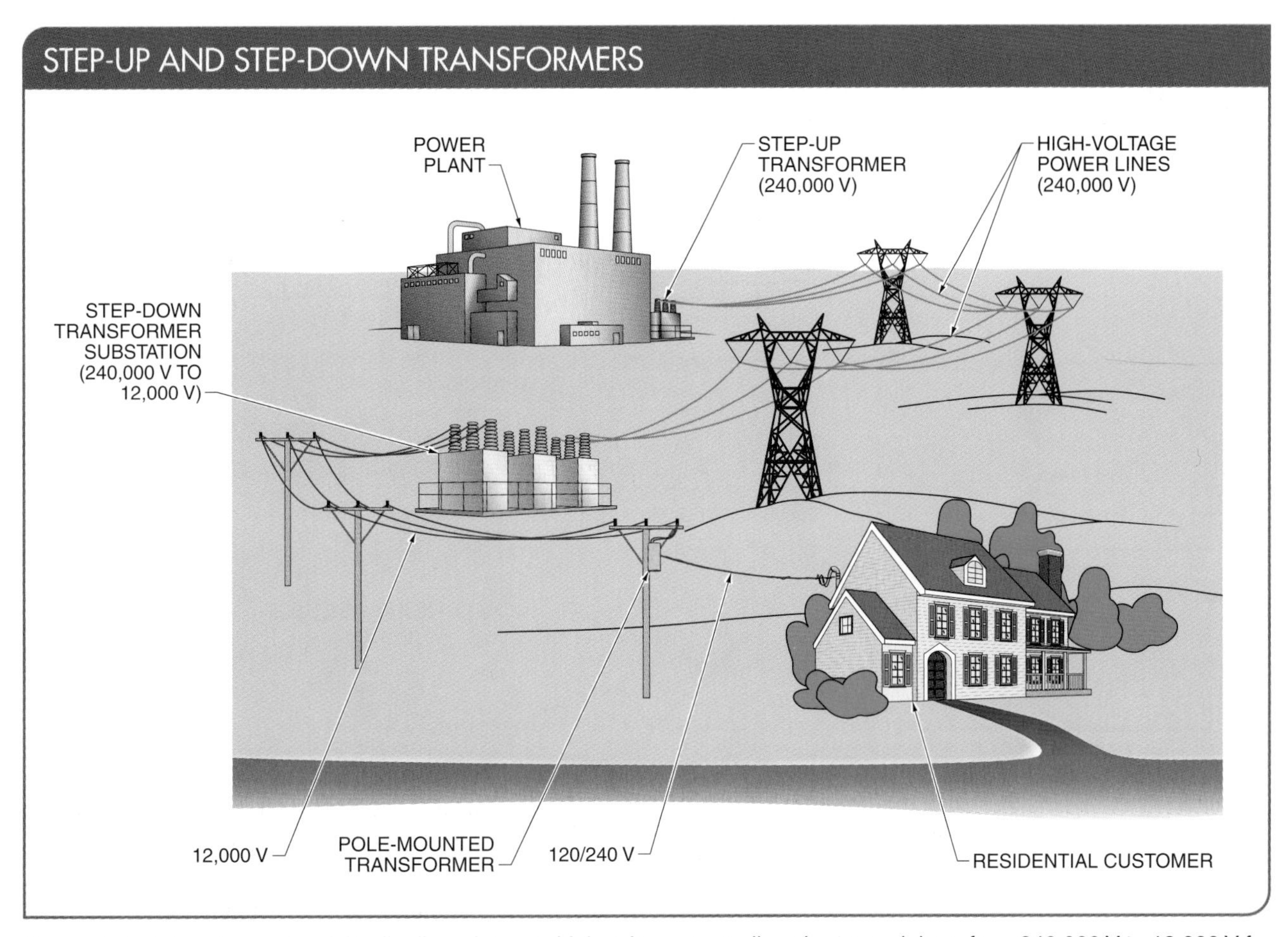

Figure 1-8. High-voltage electricity distributed across high-voltage power lines is stepped down from 240,000 V to 12,000 V for local distribution. The local voltage distribution level of 12,000 V is further stepped down by transformers to voltages of 480 V, 240 V, or 120 V for consumer use.

Transformer Connections. Transformers may be connected in various configurations depending on the application. Configurations consist of 1ϕ and 3ϕ connections. Single-phase connections are typically found in residential applications, while 3ϕ connections are found in commercial and industrial applications.

Electricity is used in residential applications (one-family, two-family, and multifamily dwellings) to provide energy for lighting, heating, cooling, cooking, etc. The electrical service to dwellings is normally 1ϕ, 120/240 V. The low voltage (120 V) is used for general-purpose receptacles and general lighting. The high voltage (240 V) is used for heating, cooling, cooking, and other large electrical uses.

Residential electrical services are typically connected to the secondary side of distribution transformers through overhead or underground connections (service entrances). A *service entrance* is the wires installed between the meter fitting (socket) and the disconnecting means (main breaker, fuse box, or breaker panel) inside a dwelling. **See Figure 1-9.** The disconnecting means also serves as the panel for distributing the electrical service to the residence (service distribution panel). Many electrical services contain service entrance conductors (wires) that are run through the air from the utility pole to the building. Some electrical services contain service entrance conductors (wires) that are run underground from the utility system to the service point.

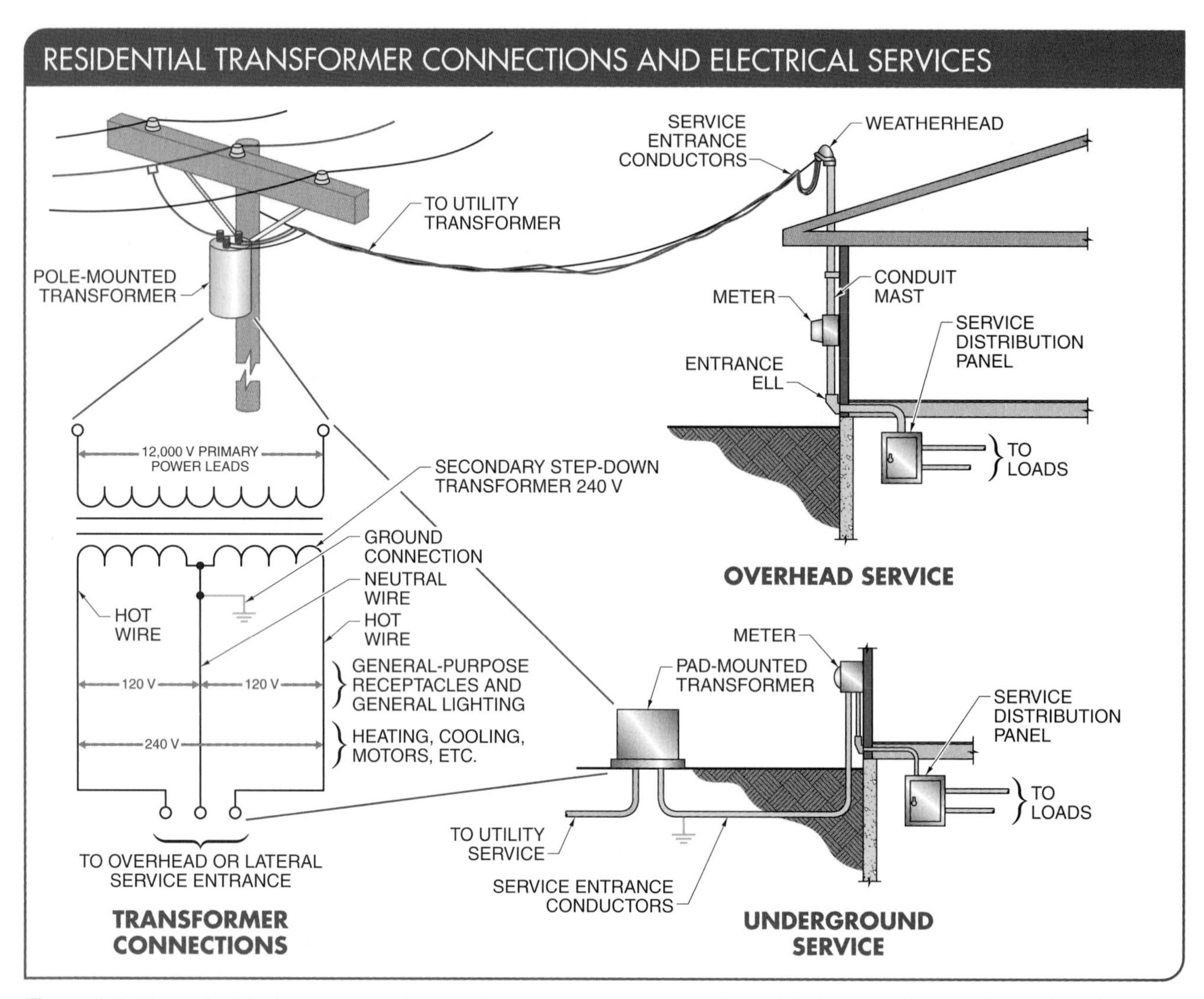

Figure 1-9. Many electrical services contain service entrance conductors (wires) that are run through the air from the utility pole to the building, while other electrical services contain service entrance conductors (wires) that are run underground from the utility system to the service point.

Power Distribution

Before transmitted electricity can be distributed to residential areas, the voltage must be stepped-down by transformers at a distribution station. Distribution lines distribute power at a relatively lower voltage than transmission lines, which allows for smaller equipment and a smaller footprint to distribute power throughout a residential area. Distribution systems typically include traditional power distribution, distributed power generation, and interactive distributed generation systems.

Traditional Power Distribution Systems. Currently, most electrical power is generated at large, centralized power-generating stations such as nuclear power plants, fossil fuel power stations, and hydroelectric powers stations. The electricity is then transmitted to transmission substations, distributed through a network of distribution substations (transformers) and distribution lines (conductors), and finally delivered to residential consumers. This system of power distribution is referred to as "the electrical grid." **See Figure 1-10.** The organization that produces and/or distributes electricity through the electrical grid to customers is often referred to as the utility.

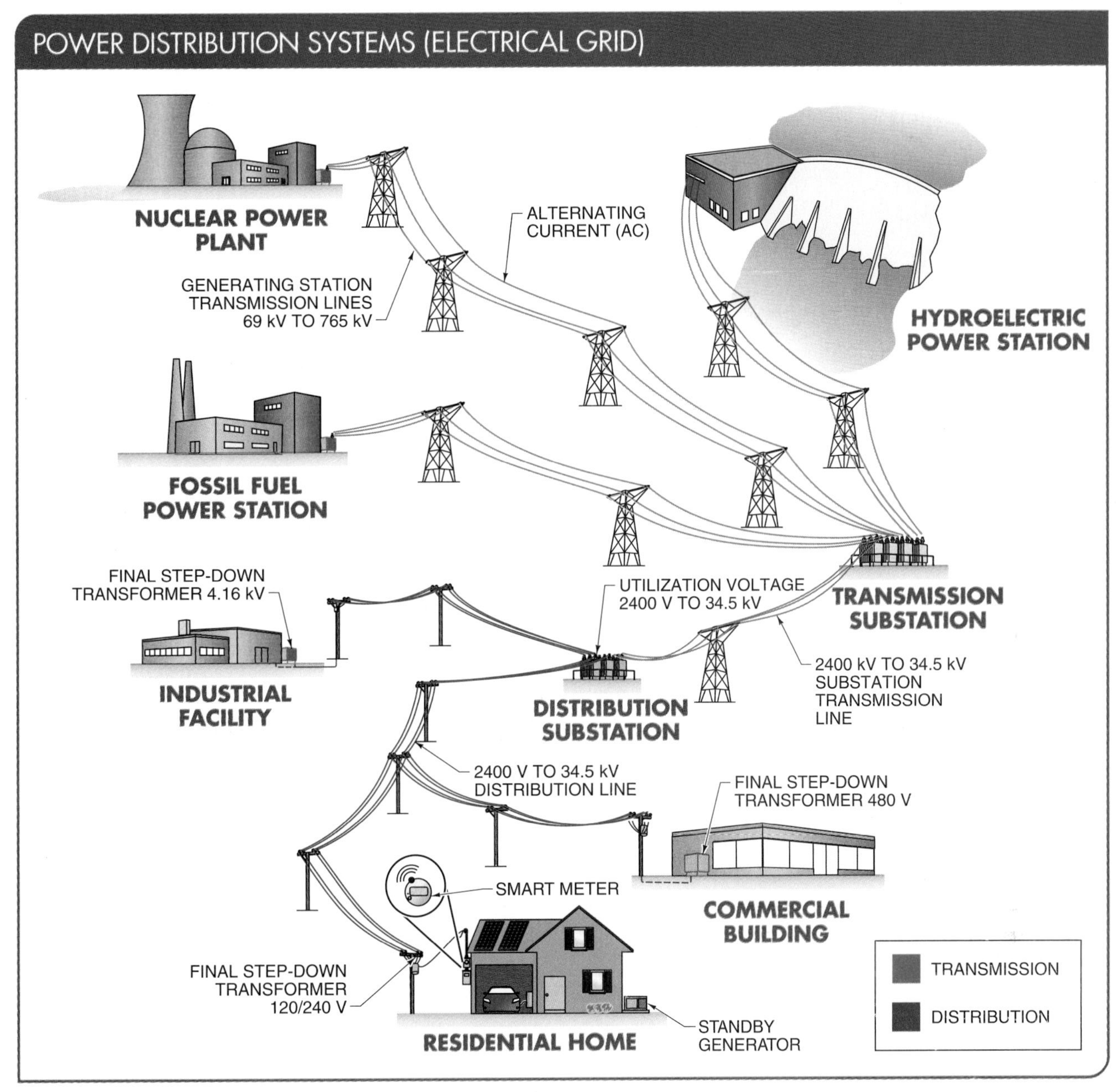

Figure 1-10. An electric utility generally produces electricity at a large, centralized power-generating station and distributes it to consumers through power lines, substations, and transformers.

Large, centralized power-generating stations are often located near abundant energy sources such as coal, oil, and natural gas. Large, centralized power-generating stations may also be located near other natural resources, such as water. For example, nuclear power plants require large sources of water to readily cool the reactor. The need to be located near abundant energy sources and natural resources limits the location of large, centralized power-generating stations.

Although devices that use electricity continue to increase in number and complexity, the means of generating and distributing electrical power has not changed much over time. However, as the amount of available electrical power reaches its maximum and costs increase, the way of producing and distributing power must change.

New sources of power generation include photovoltaic (PV) arrays, wind farms, and fuel cells. To help integrate and control these sources of power generation, the older, traditional distribution systems are being upgraded through the addition of HVDC transmission lines and new electronic monitoring equipment. These innovations and electronic devices, coupled with alternative sources of energy, are all part of a smart distribution system or smart grid.

Distributed Power Generation Systems. *Distributed power generation* is the generation of power on site or in close proximity to where the power is used. **See Figure 1-11.** Distributed power generation systems typically use renewable energy sources for power generation. Renewable energy sources may include photovoltaic (PV) arrays, wind turbines, biodiesel generators, fuel cells, and other relatively small-scale power systems.

Distributed power generation systems may be set up as microgrids. A *microgrid* is a small local electrical network or grid that can operate completely separate from the main electrical grid. A microgrid is essentially a smaller version of a traditional electrical grid that can be disconnected from the traditional electrical grid and operate as a stand-alone local electrical grid or network.

Interactive Distributed Generation Systems. *Interactive distributed generation* is on-site power generation where switchgear can activate additional generation sources on demand. Interactive distributed generation systems can be used for back-up power and peak demand power generation. These systems are interactive because they are connected to switchgear that can sense a power disruption from the utility and automatically switch to the back-up power source. Interactive distributed generation systems may be connected to a

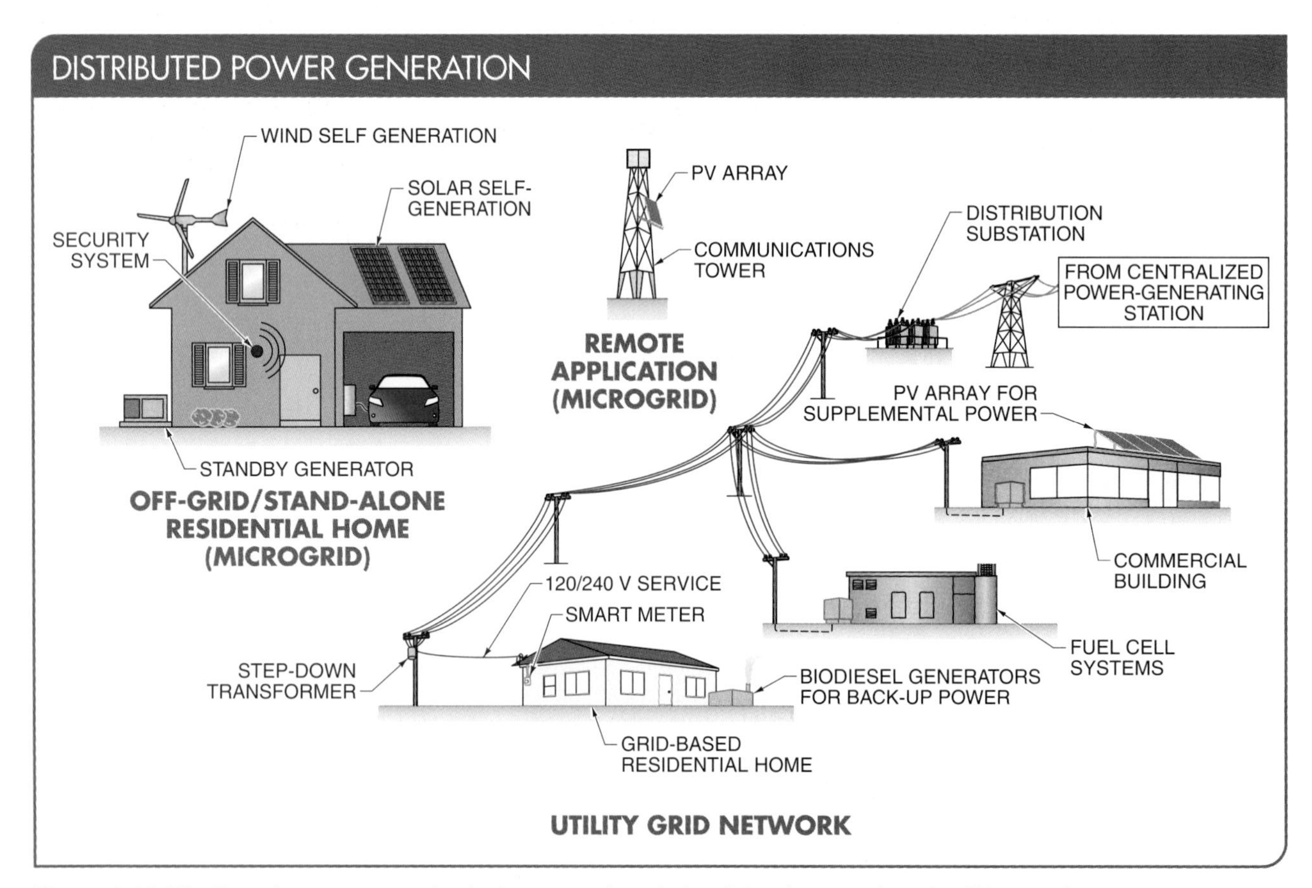

Figure 1-11. Distributed power generation is the generation of electricity close to where it will be used.

utility distribution grid and supply power when the centralized power-generating station fails or needs additional power to meet peak demand. **See Figure 1-12.**

For customers with critical power needs, these generation systems can provide continuous power to on-site loads or back up their stand-alone systems in the event of a utility power outage. For example, back-up generators are placed in a fixed location, such as a hospital, and automatically connect to that location's power distribution system through an electric transfer switch. The electric transfer switch detects when utility power has been lost, disconnects the utility distribution system from the loads, and connects the generator to the loads. Back-up generators normally supply power only to critical loads that must have power during a power outage.

Distribution transformers are used to reduce high transmitted voltage to lower levels that can be used for residential applications.

Interactive distributed generation of electrical power is becoming increasingly common as a supplement to traditional centralized power generation. Interactive distributed generation systems increase the diversity and security of the electrical energy supply and benefit both consumers and electric utilities.

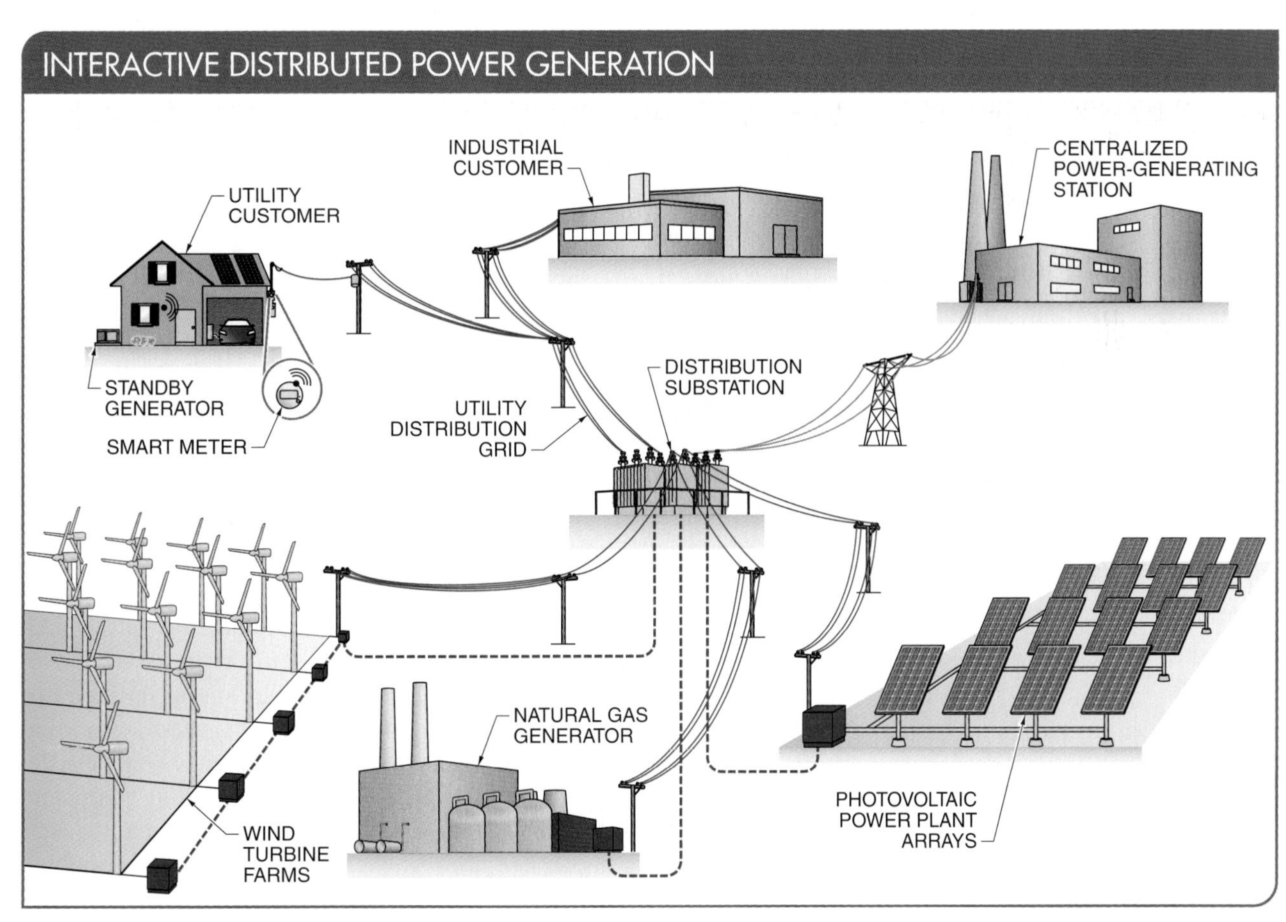

Figure 1-12. Interactive distributed generation is on-site power generation where switchgear can activate additional generation sources on demand.

Traditional Electrical Grids

An *electrical grid* is the entire network of power generation stations, transmission stations, distribution systems, and corresponding power lines. The traditional electrical grid produces the most electricity when it is either very cold or very hot outside, since heating and cooling systems require large amounts of energy. Electrical demand hits its highest point during the summer due to the widespread use of electric air conditioners. The electrical grid works the hardest during the afternoons of the hottest summer days. These time periods are referred to as "peak demand." **See Figure 1-13.**

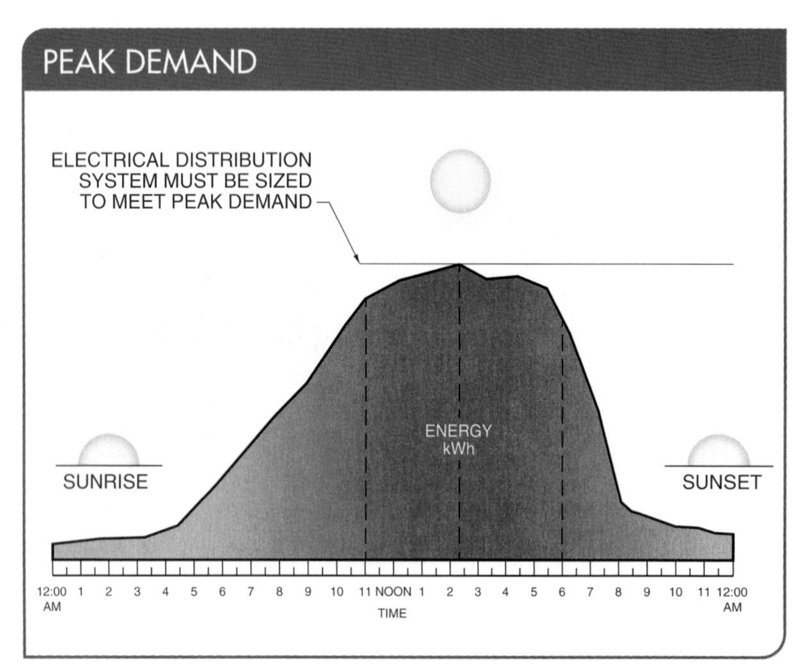

Figure 1-13. Peak demand is the greatest demand for electrical power over a specific time period, especially in the afternoon during the summer.

Since energy cannot be easily stored on a large scale, utilities have met increases in demand with more generation. Most of the electricity needed to meet peak demand comes from peaker power plants or reserve storage battery banks. A *peaker power plant* is a back-up power plant that is reserved to run only during times of increased demand. Peaker power plants are much more expensive to operate, resulting in a wide range of generation costs not only throughout the year but also each day. For example, it may cost many times more to generate electricity during a hot summer afternoon compared to a cool summer night. Failure to meet peak demand can result in power outages.

Smart Grids

The word "traditional" has been used to help differentiate between a typical electrical grid and the newer innovation of smart grids. A *smart grid* is an electrical grid in which electricity is delivered and monitored from a power source to an end user. Smart grid systems are being implemented across the United States, from electrical generation to the point of use. **See Figure 1-14.**

A smart grid may use renewable energy sources, such as wind, sun, and water, to reduce harmful emissions during the production of electricity. The benefits of smart grids include higher power quality, diagnostics and self-healing, reduced vulnerability, and improved efficiency.

Power Quality. *Power quality* is the condition of incoming power to a load. Power quality includes voltage, frequency, and reliability. Through the use of smart grid technology, power disturbances can be minimized and corrected in fractions of a second, reducing or eliminating power quality problems. High power quality is required by consumers as they are becoming more and more dependent on computers and environmental controls, such as digital thermostats, in their residences.

Diagnostics and Self-Healing. A smart grid system allows operators of the electrical grid to monitor the condition of the electrical grid at any given moment. Through the use of advanced sensors and high-speed computing, an operator can detect threats and identify failing equipment before a failure occurs. In many cases, through the use of advanced programming and resettable circuit protectors, a smart grid section can reroute power to restore electricity without human intervention.

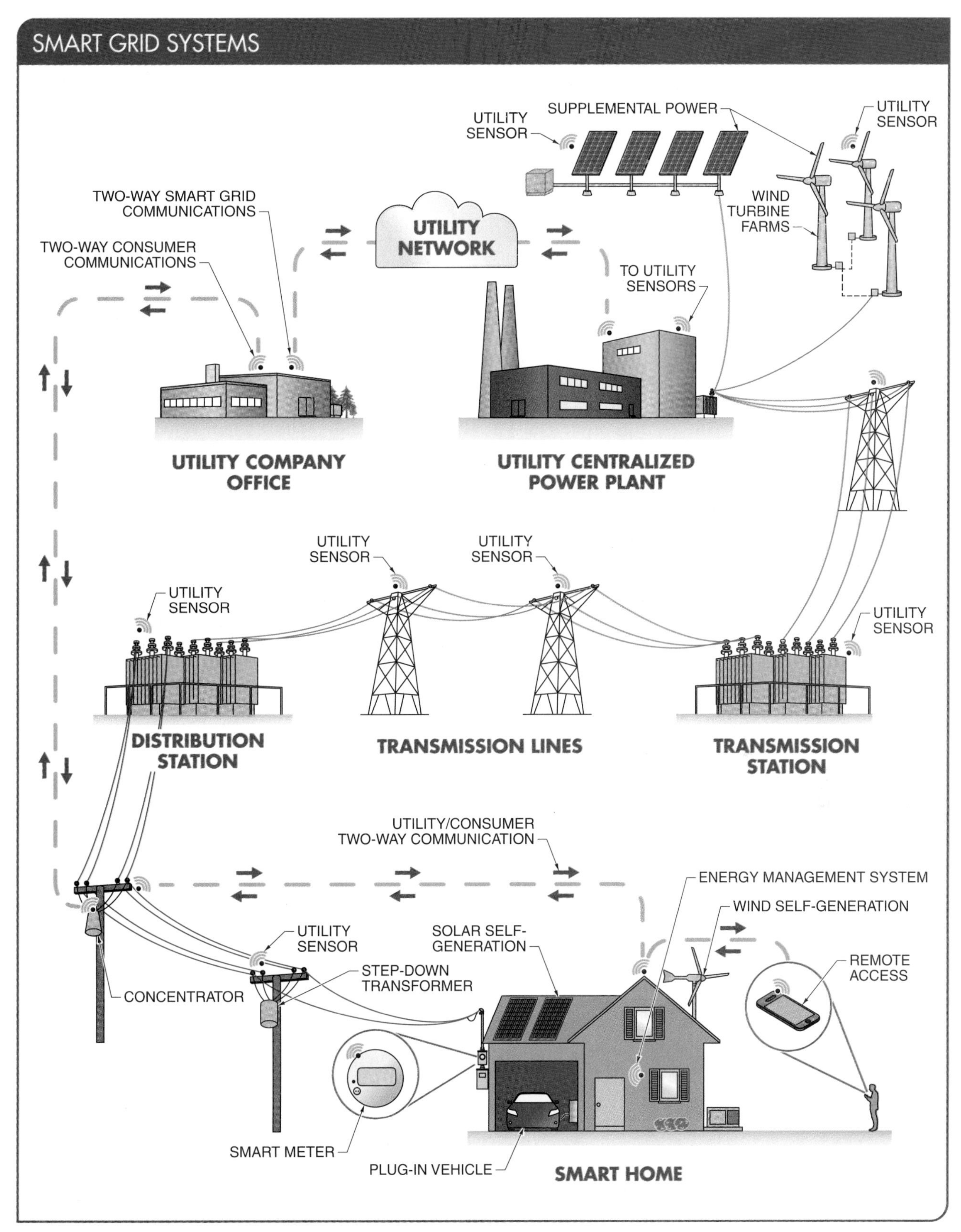

Figure 1-14. A smart grid is an electric grid (network of transmission lines, substations, transformers, and electrical devices) that is used to monitor and deliver electricity from a power source to an end user.

TECH TIP

Smart meters have improved power outage detection and notification over analog meters. Smart meters can immediately notify the electrical utility of power outages or power problems at a residence.

Reduced Vulnerability. Two-way communication and intelligent control of smart grid systems allow system operators to detect, minimize, and even stop potential threats to their systems. With these new capabilities, the electrical grid is able to support a wide range of options for power generation. This capability reduces the effects of a fault or power outage in one part of a system by using its network to respond to the emergency with a variety of options.

Improved Efficiency. The smart grid can provide consumers and utilities with information that allows both to operate more efficiently and help reduce the cost of electrical production and consumption. Utilities receive information that can allow them to manage and maintain the assets in the electrical grid more efficiently. The smart grid can provide consumers with the information and feedback needed for them to make well-informed choices regarding electrical consumption. New devices, such as light-emitting diode (LED) lighting and variable speed drives, also help reduce electrical consumption. Smart grid technology can also be used to gather information from remote sensors and use energy harvesting to power these sensors remotely. **See Figure 1-15.**

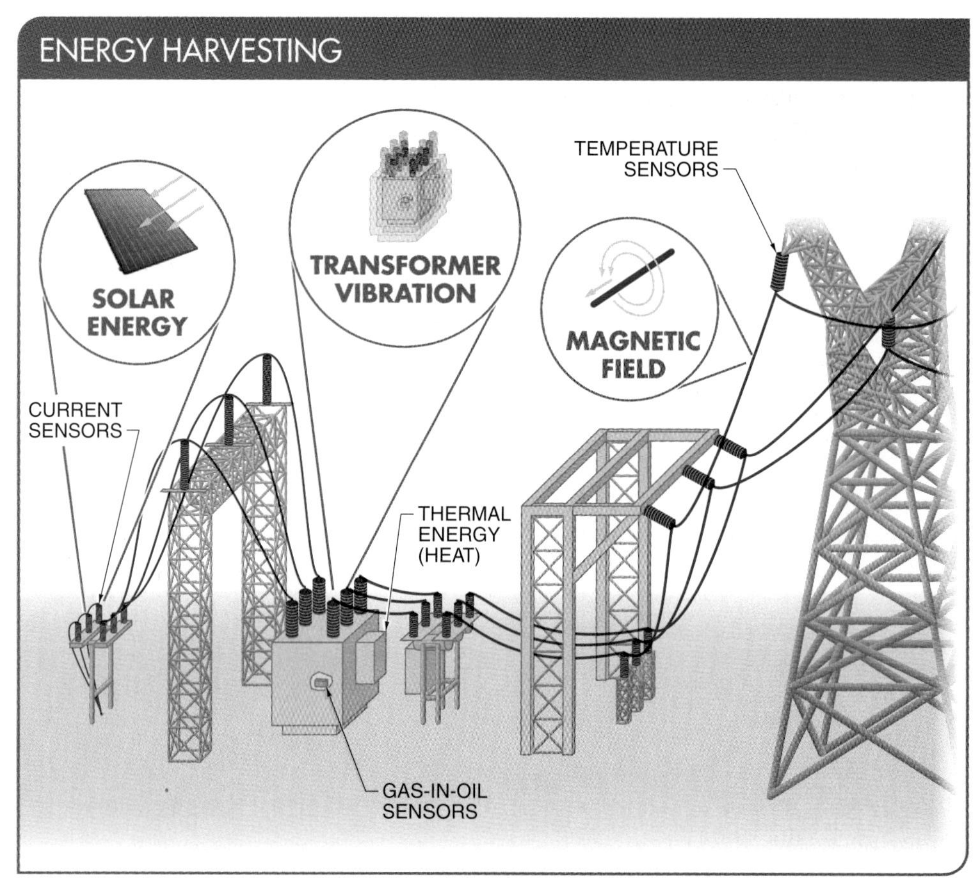

Figure 1-15. Energy harvesting is the process of obtaining power from the surrounding environment and using it to power sensors.

SECTION 1.1 CHECKPOINT

1. What are the two basic forms of energy?
2. What is the formula for determining energy efficiency?
3. What is a transformer?
4. What is distributed power generation?
5. What is interactive distributed generation?
6. What is a microgrid?
7. What is a smart grid?

SECTION 1.2—SMART HOMES

Traditionally, residences have only been wired for AC. The increasing use of electronic devices such as computers, cell phones, digital TVs, and electric vehicle (EV) charging stations in a residence is making total AC systems inefficient. Electronic devices use DC and must convert the available AC into DC through adapters, resulting in losses of as much as 20%.

A *smart home* is a residence that contains equipment, appliances, and software that are designed to provide security, energy management, and convenience. **See Figure 1-16.** In a smart home, sensors and controllers connected to a touchscreen graphic management display and the internet provide local and remote control, as well as the capability to monitor the residence. Sensors can automatically dim lights and controllers enable homeowners to change heating/cooling setpoints, helping to reduce energy costs. A smart home also enables control of devices used for entertainment, such as a home theater system.

Smart home technology also enables off-site supervision. Remote access allows an owner of a smart home to view security video of the residence on a laptop or other mobile devices. In addition, the off-site supervision enabled through smart home technology can help people care for elderly parents, children, and/or individuals with disabilities.

Legrand®

USB receptacles can be installed in smart homes to eliminate the need for external AC to DC adapters.

Energy Losses in AC Systems

An environmentally responsible and energy-efficient residence may include the addition of solar photovoltaic (PV) panels or a wind turbine system. However, the addition of an alternative energy source to a residence wired for AC may not be as energy efficient as possible. For example, a solar array may not provide the desired energy efficiency due to the AC/DC conversion losses. **See Figure 1-17.**

In a system that includes a solar array, the solar array charges a battery bank that feeds a DC-to-AC converter to deliver AC at a receptacle. However, there is a 20% energy loss due to the DC-to-AC conversion process. In addition, at the receptacle, an AC-to-DC adapter (converter) is needed to convert the AC into DC to power electronic devices. This results in an additional 20% energy loss for a potential loss of 25% or more. The more times AC and DC are converted, the more energy is lost. For this reason, to increase the overall energy efficiency of a residence, the number of times AC and DC is converted must be decreased.

Figure 1-16. Smart homes incorporate the use of sensors and controllers connected to a network and the internet to provide security, energy management, and convenience to a residence.

Hybrid AC/DC Microgrids

A *hybrid AC/DC microgrid* is a localized electrical grid or network that operates on AC and DC and can be disconnected from the traditional electrical grid or smart grid to operate on its own power source. Commercial businesses and cities have been developing local microgrids for years. The major reason for homeowners to consider a hybrid AC/DC microgrid is for increased energy efficiency. **See Figure 1-18.**

The typical 40% energy loss when converting from AC to DC and back to AC can be reduced by using a hybrid AC/DC microgrid. The use of a DC power system reduces the loss of DC-to-AC conversion. In addition, a DC system can provide a variety of voltage levels. If a solar array has an output of 48 V, stepping the voltage down to 5 V, 12 V, or 24 V to meet the needs of a variety of electronic devices is more efficient than multiple DC to AC conversions.

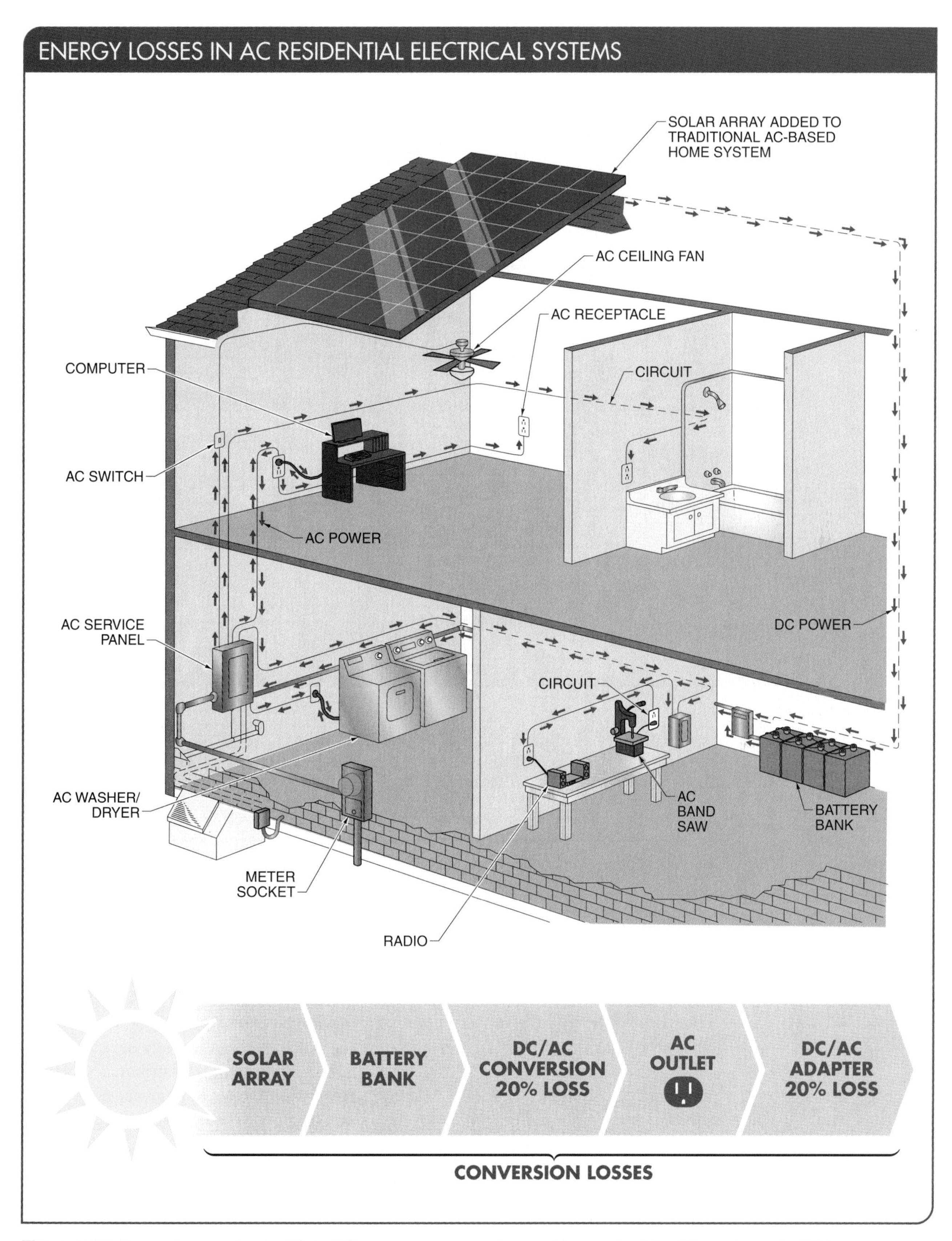

Figure 1-17. Energy losses due to AC-to-DC power conversion in a residence wired for AC can be up to 40%.

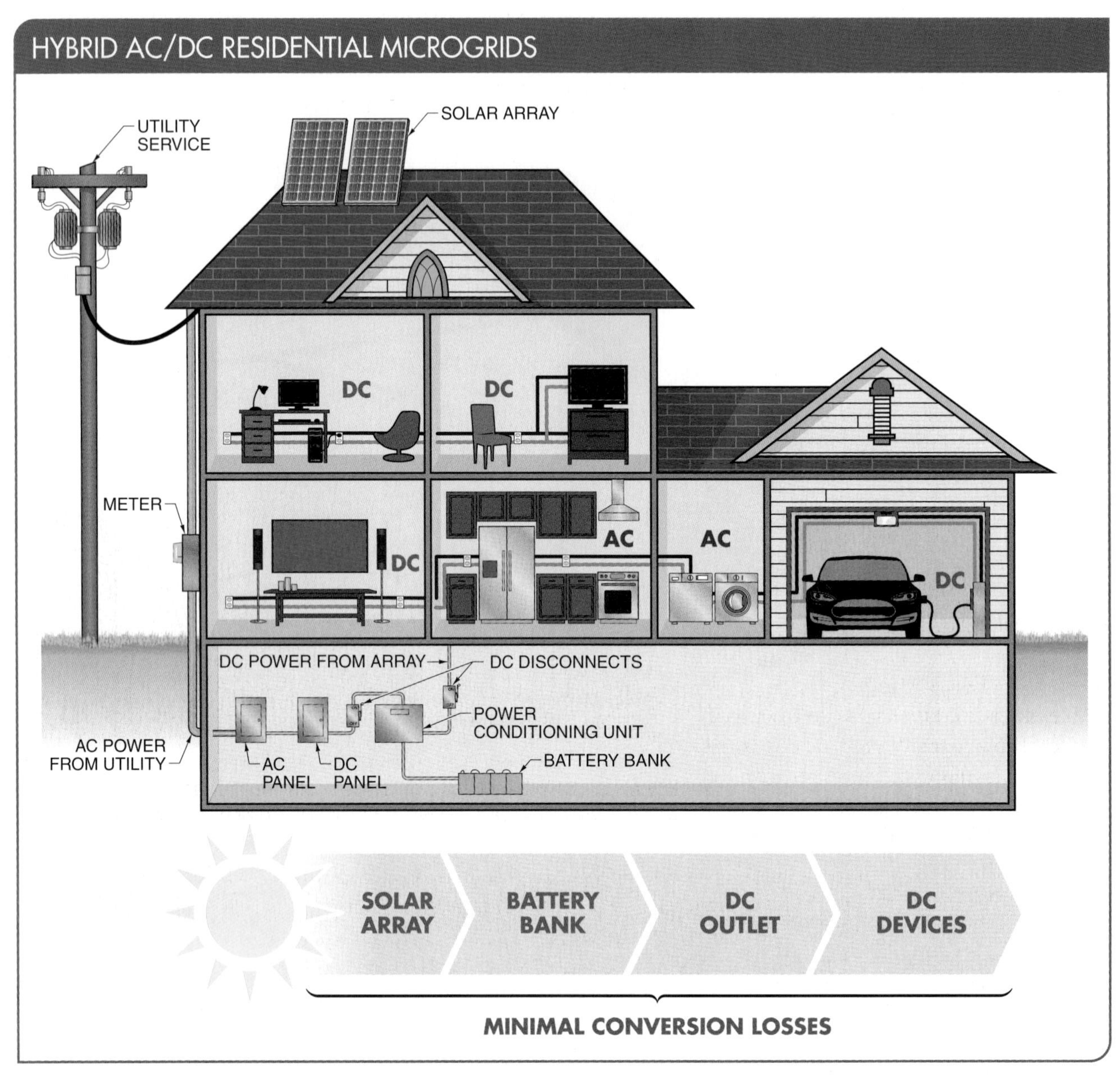

Figure 1-18. In hybrid AC/DC microgrids, AC-to-DC conversion energy losses are reduced by connecting devices directly to a DC-based wiring system.

A DC power system maximizes not only efficiency but the return on investment (ROI) of rooftop solar arrays. A DC power system could provide power directly to consumer electronics, LED lights, and EV charging stations with minimum energy loss. This type of system also provides the homeowner with a choice of either storing excess power or selling it back to the utility. Over time, improvements in the design of DC power systems will help increase the efficiency of the hybrid AC/DC microgrid systems.

AC/DC Receptacles with USB Connector Ports. Developments in universal serial bus (USB) technology has resulted in a standard system to distribute DC. High-power USB cables can transfer up to 100 W of DC power. While AC is convenient to step up and step down voltage using transformers, direct DC is more efficient for powering electronic devices because an adapter (converter) is not required. USB ports in receptacles are used for charging electronic devices. **See Figure 1-19.**

AC receptacles with USB ports are becoming widely adopted in existing residences and new construction.

USB connector ports are able to power an ever-increasing number of electronic devices without the need for external adapters. As DC power becomes more acceptable in a residence, manufacturers will develop a greater number of DC-powered appliances. A direct DC connection to devices eliminates the need for a power adapter for electronic products. For example, since LED lights use low-voltage DC power, a direct DC connection eliminates the energy loss from DC-to-AC and AC-to-DC conversions. **See Figure 1-20.**

An appropriately sized rechargeable battery bank can keep DC-powered devices running during a power outage. DC could be fed directly into a residence without the need for an inverter or adapter.

A downside for direct DC connections is that not all digital or electronic control devices are 100% efficient. Most electronic devices allow a small amount of leakage current to pass through. Even when a device is off, the small leakage current reduces the overall efficiency of the system. Some devices have internal switching capability to prevent the small leakage current. This leakage current is often referred to as phantom power.

In addition to cost, one of the major road blocks for direct DC power in residences is the lack of standardization. Standards for the quality and safety of AC have been around for a long time, but the standards for DC are still being developed.

DC System Electrical Safety. Although low-voltage systems have a reduced potential for electrical shock, hazards still exist when working with DC systems. For example, battery storage units can deliver hundreds of amps of stored energy. In addition, some batteries require proper ventilation to vent toxic or even explosive gases. Wind turbine systems and solar PV systems must be installed per all National Electrical Code (NEC®) requirements and must be equipped with proper disconnects.

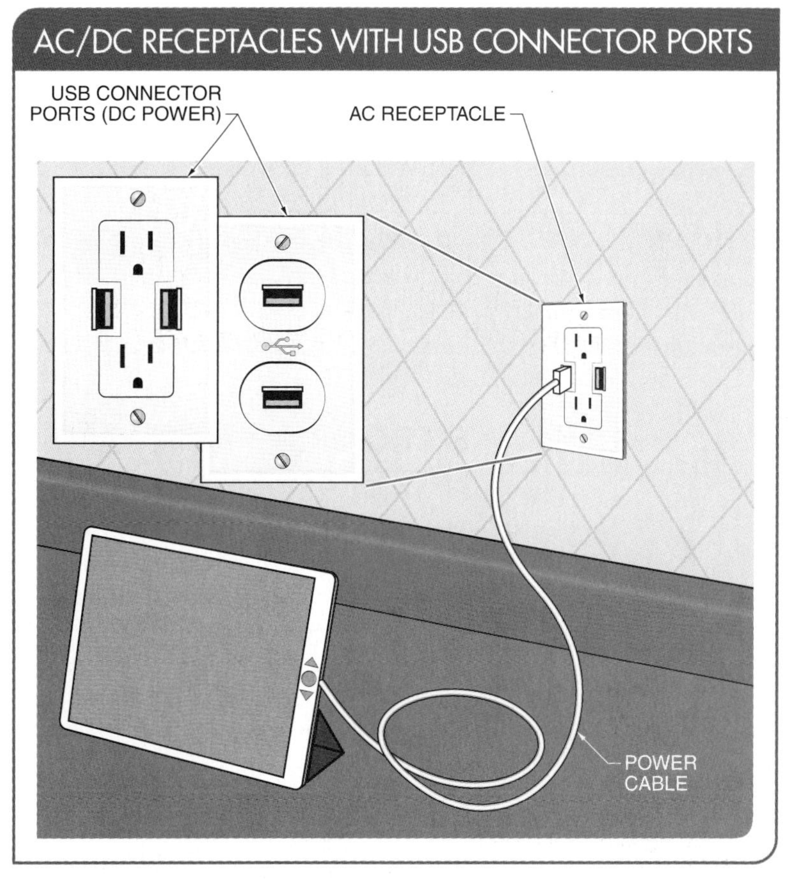

Figure 1-19. USB connector ports in receptacles are used for charging electronic devices and can deliver up to 100 W of power through a USB cable.

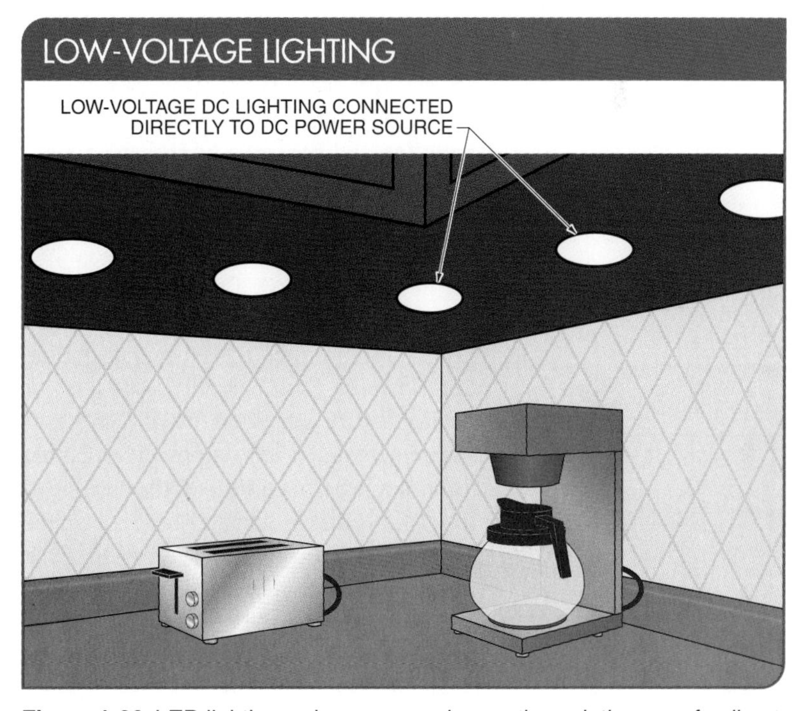

Figure 1-20. LED lighting reduces energy losses through the use of a direct connection to a DC power source.

SECTION 1.2 CHECKPOINT

1. Traditionally, residences have only been wired for what type of current?
2. What is a smart home?
3. What is a hybrid AC/DC microgrid?
4. What are the benefits of direct DC connection?

SECTION 1.3—RENEWABLE ENERGY SYSTEMS

Renewable energy is energy from self-replenishing sources. The main sources of renewable energy include the sun, wind, and water. In addition to those main sources, there are other sources such as solar thermal and geothermal energy. Renewable energy systems use these sources to generate electricity.

Solar Energy Systems

Solar energy from the sun is typically captured through the use of photovoltaics. *Photovoltaics* is solar energy technology that uses the unique properties of semiconductors to convert solar radiation into electricity. **See Figure 1-21.** Photovoltaics are environmentally friendly, or "green," technology that produce energy without pollution and conserve nonrenewable energy sources. Photovoltaic (PV) systems use silicon wafers, called "cells," that are sensitive to light and produce a small direct current when exposed to light. When PV cells are combined into large systems (arrays), they produce a significant amount of electricity.

Electricity supplied by PV systems supplements electricity supplied from other technologies. Concerns over the burning of fossil fuels and a demand for renewable, clean energy are making PV systems an increasingly attractive source of energy.

PV systems are flexible and can be adapted to many different applications. The modular nature of PV system components makes them easy to expand for increased capacity. PV systems can be used in almost any application where electricity is needed and the sun is readily available. Most PV systems also include a battery storage system.

Wind Energy Systems

Wind energy is harnessed by mechanically rotating wind turbines. A *wind turbine* is a power generation system that converts the kinetic energy of wind into mechanical energy, which is used to rotate a generator that produces electrical energy. Wind turbines can be used to produce power where no utility lines exist or they can be connected to the utility electrical grid to supplement grid power. Wind turbine output ranges from a few kilowatts (kW) to several megawatts (MW). A commercial wind turbine primarily consists of a hub, rotor, nacelle, gearbox, generator, power converter/controller, tower, and a number of blades. **See Figure 1-22.**

Wind turbines must be located where a sustained wind is present. Wind turbines can be grouped in parallel to construct a wind farm that produces large amounts of power. In standard generating power plants, a large percentage of the power generated is lost as line losses and heat before it reaches the consumer. Adding wind turbines to a utility system increases system efficiency locally and reduces the need to transport additional fuel to consume at a utility generating station. Since wind energy can be generated locally, long-distance distribution lines can be shortened, thus reducing transmission losses.

Wind turbine systems can be grid-connected (connected to an electrical utility grid) systems or off-grid (stand-alone) systems. A grid-connected wind turbine system is usually a large set of wind turbines or a wind farm that is used to supplement utility-supplied electricity. If utility electrical grid cannot produce the required amount of electricity, additional electricity is available from the wind farm. A small off-grid wind turbine system is usually not connected to the electrical utility grid. Off-grid wind turbine systems are usually designed to provide adequate power for local applications.

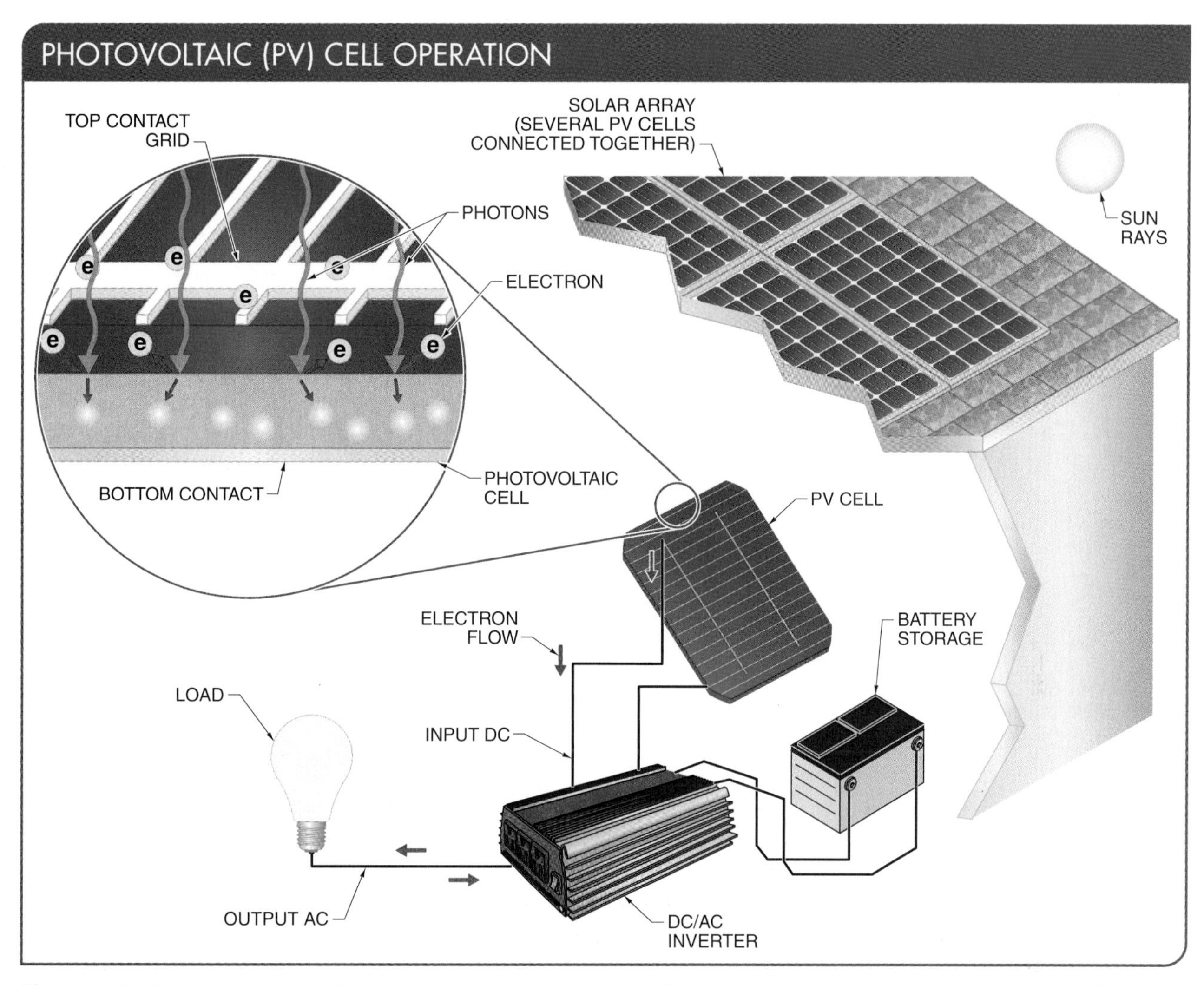

Figure 1-21. PV cells can be combined into arrays to produce a significant amount of power. The DC produced can be stored for use later or converted into AC in a DC/AC converter for usage.

Residential Wind Turbine System Efficiency. Wind turbines are normally used where sunlight is not in constant supply. However, the biggest problem with small residential wind turbines is efficiency. There are many conversions that take place in a typical stand-alone wind turbine. Because of these energy conversions, there is some energy loss. Energy losses include the friction in the turbine generator, heat losses, and losses due to chemical impurities in the wet cell batteries, and the AC to DC conversion and the DC to AC conversion. The conversion losses also take place in the electronic components of the converters.

DOE/NREL, Altair Energy

Some types of dwellings use solar panels to supplement power from the utility and reduce energy bills or provide extra power during times of peak demand.

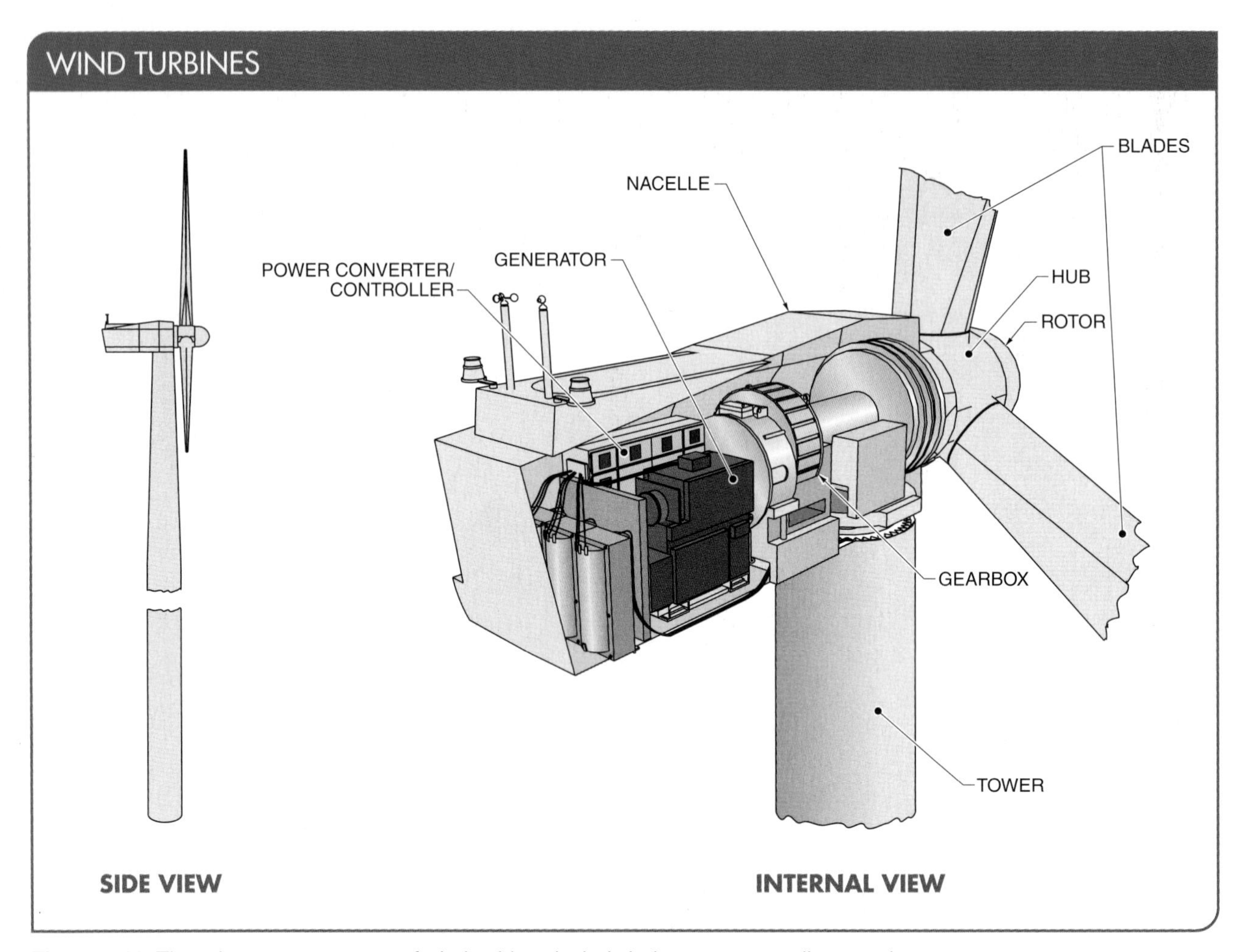

Figure 1-22. The primary components of wind turbines include hubs, rotors, nacelles, gearboxes, generators, power converters, controllers, towers, and blades.

Water (Hydroelectric) Energy Systems

Energy from the flow of water is typically harnessed through the use of hydroelectric power plants. A hydroelectric power plant is an electricity-generating plant that uses falling or running water to mechanically rotate turbines and convert the kinetic energy of the water into electricity. The generators of hydroelectric plants then convert the mechanical energy from the turbines into electrical energy. **See Figure 1-23.** Although dams are the primary installation of hydroelectric power plants, some small plants are installed in fast-moving rivers and streams.

Construction of Hydroelectric Power Plants. Hydroelectric power plants have four major components: a dam, a turbine, a generator, and transmission lines. The dam raises the water level of a body of water such as a river to create falling water. The turbine uses the force of falling water against the turbine blades to rotate the turbine, which converts the kinetic energy of the water into mechanical energy. The generator is connected to the turbine, which causes the generator to rotate and convert the mechanical energy from the turbine into electric energy. Transmission lines conduct electricity from the hydroelectric power plant to the electrical grid.

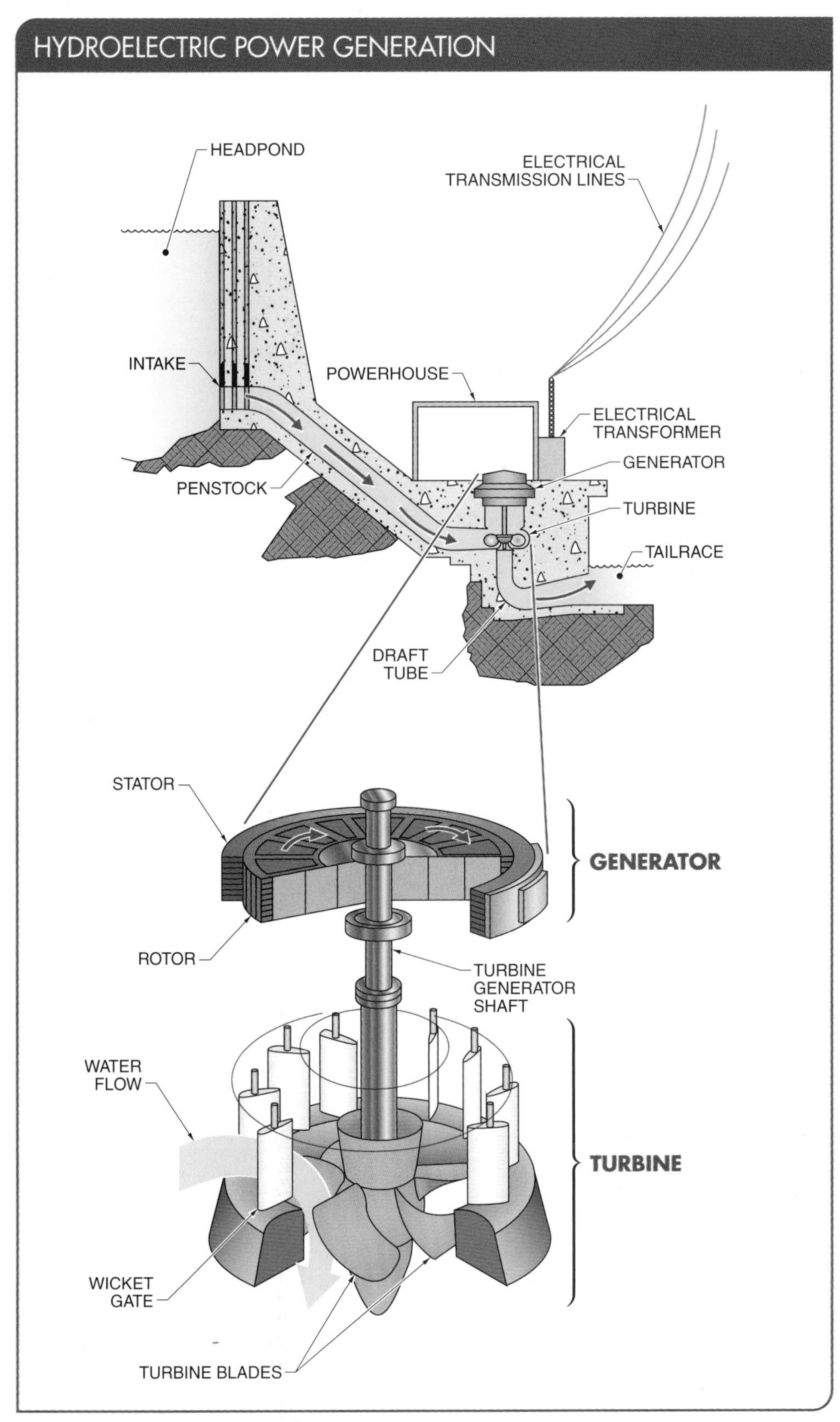

Figure 1-23. Hydroelectric power plants use the energy from falling or moving water to generate electricity.

Solar Thermal Energy Systems

A *solar thermal energy system* is a renewable energy system that collects and stores solar energy and is used to heat air and water in a residential structure. Well-designed solar thermal energy systems maximize the amount of thermal energy received. **See Figure 1-24.** Solar thermal energy systems decrease the demand for electricity for temperature control and hot water heating. *Note:* Properly located windows and skylights can increase direct solar heat in living spaces and decrease the amount of illumination required by lighting systems.

Geothermal Energy Systems

Geothermal energy is renewable energy that is derived from heat contained within the earth. Heat contained within the earth is maintained at a constant temperature (50°F to 70°F) in the soil and rocks below the frost line. Geothermal energy systems decrease the demand for electricity for temperature control because they use ground-source heat pumps to pull heat energy from the earth and convey it through a residence to provide heat during cold weather and cooling during warm weather. **See Figure 1-25.**

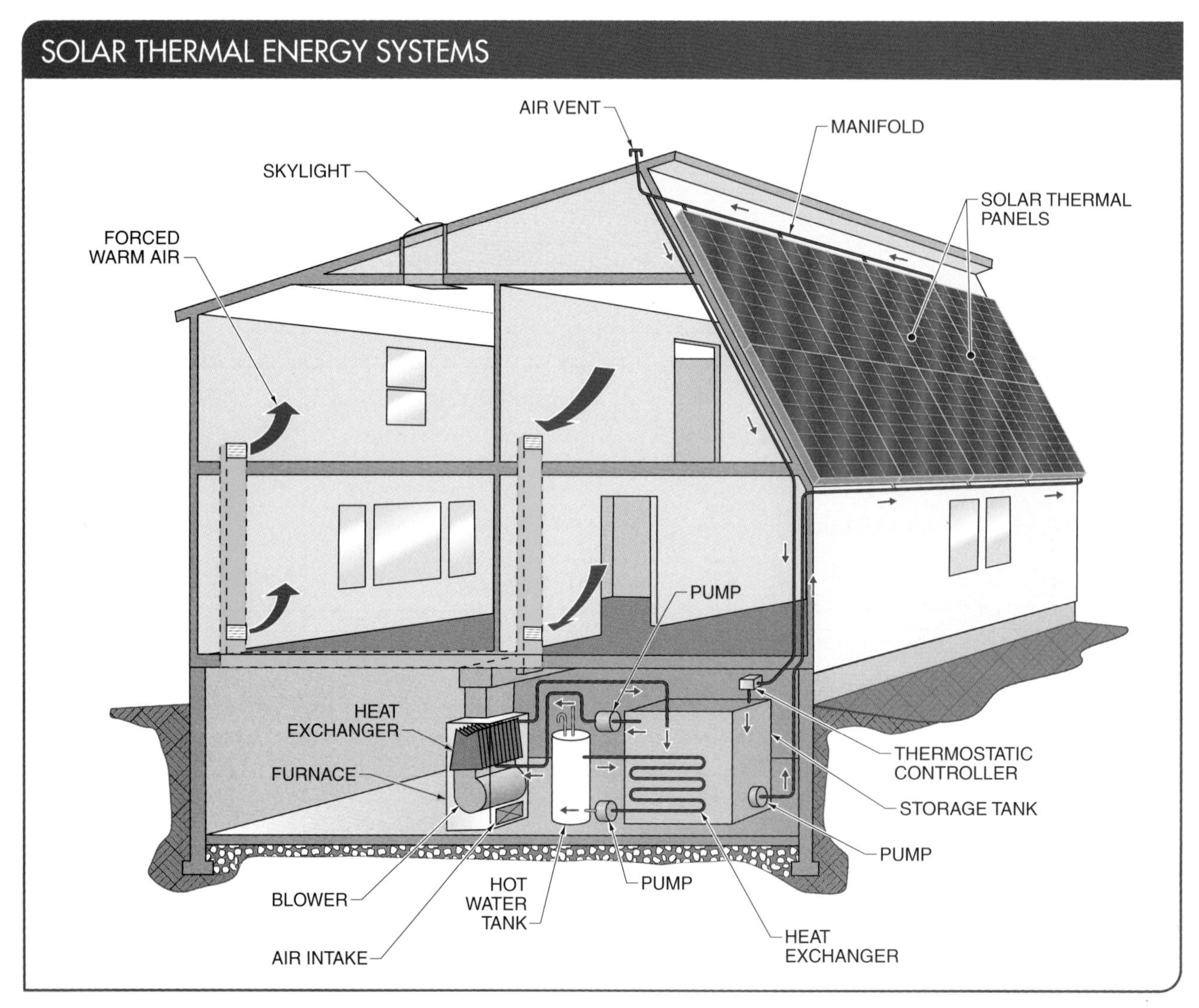

Figure 1-24. A solar thermal energy system is a renewable energy system that collects and stores solar energy and is used to heat air and water in a residential structure.

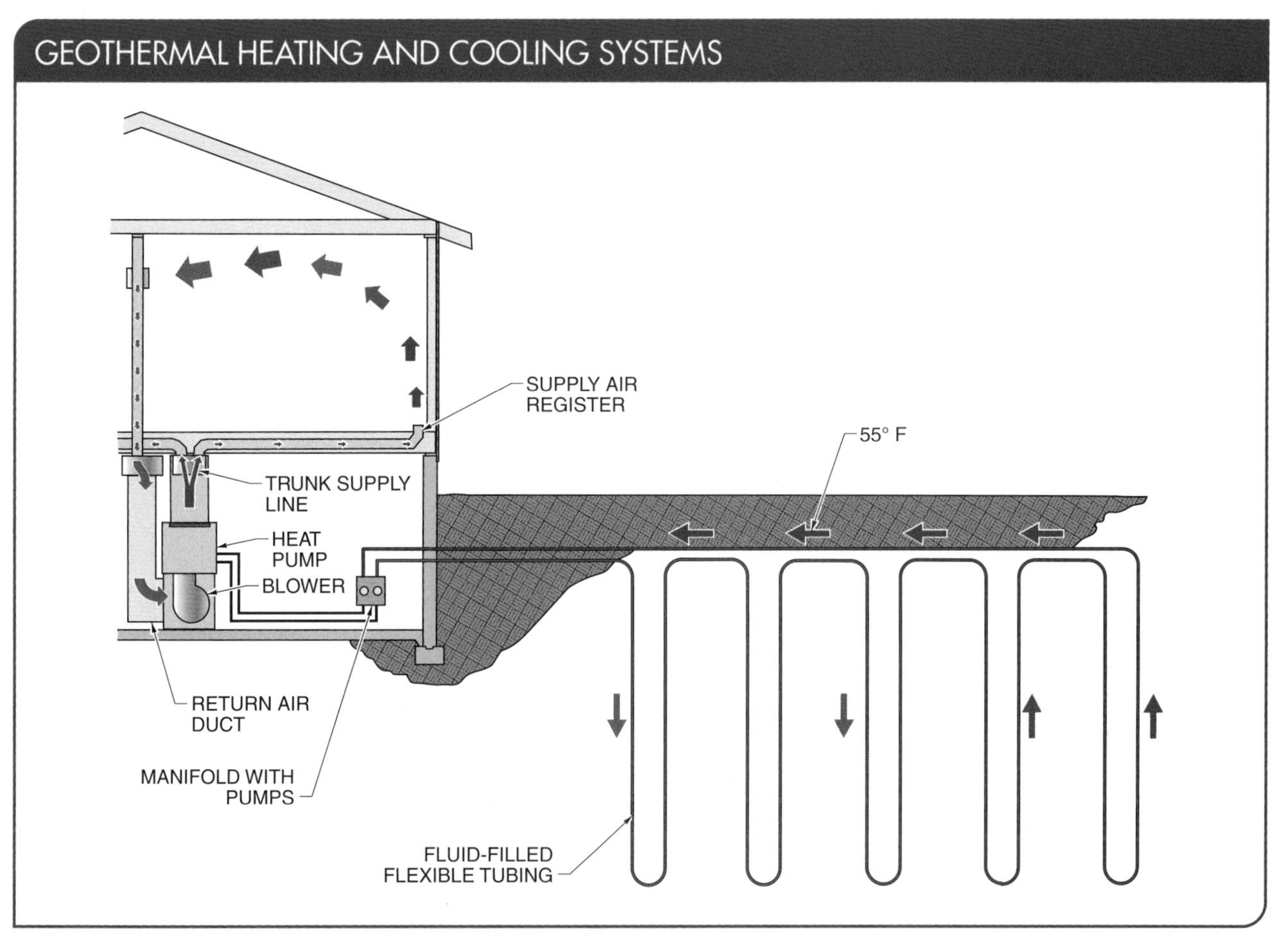

Figure 1-25. Geothermal systems use the constant temperature of the earth to efficiently heat and cool a residence.

TECH TIP

On average, geothermal energy systems are 48% more efficient than gas furnaces and are 75% more efficient than oil furnaces.

Heat Pumps

A *heat pump* is a mechanical compression refrigeration system containing devices and controls that reverse the flow of refrigerant to move heat from one area to another. **See Figure 1-26.** In a heat pump system, reversing the flow of refrigerant switches the function of the indoor and outdoor units. During the heating operation, the indoor unit of a heat pump functions as a condenser and the outdoor unit functions as an evaporator. During the cooling operation, the flow of refrigerant is reversed and the indoor unit functions as an evaporator and the outdoor unit functions as a condenser. A reversing valve is used to reverse the flow of refrigerant through the system.

A heat pump in the cooling mode moves heat from inside a building to the air outside the building. When a heat pump is in the cooling mode, the indoor unit is the evaporator and the outdoor unit is the condenser. A heat pump in the heating mode moves heat from outside a building to the air inside a building. When a heat pump is in the heating mode, the indoor unit is the condenser and the outdoor unit is the evaporator. Solar thermal systems, geothermal systems, and heat pumps are all more efficient at heating a residence than using electricity.

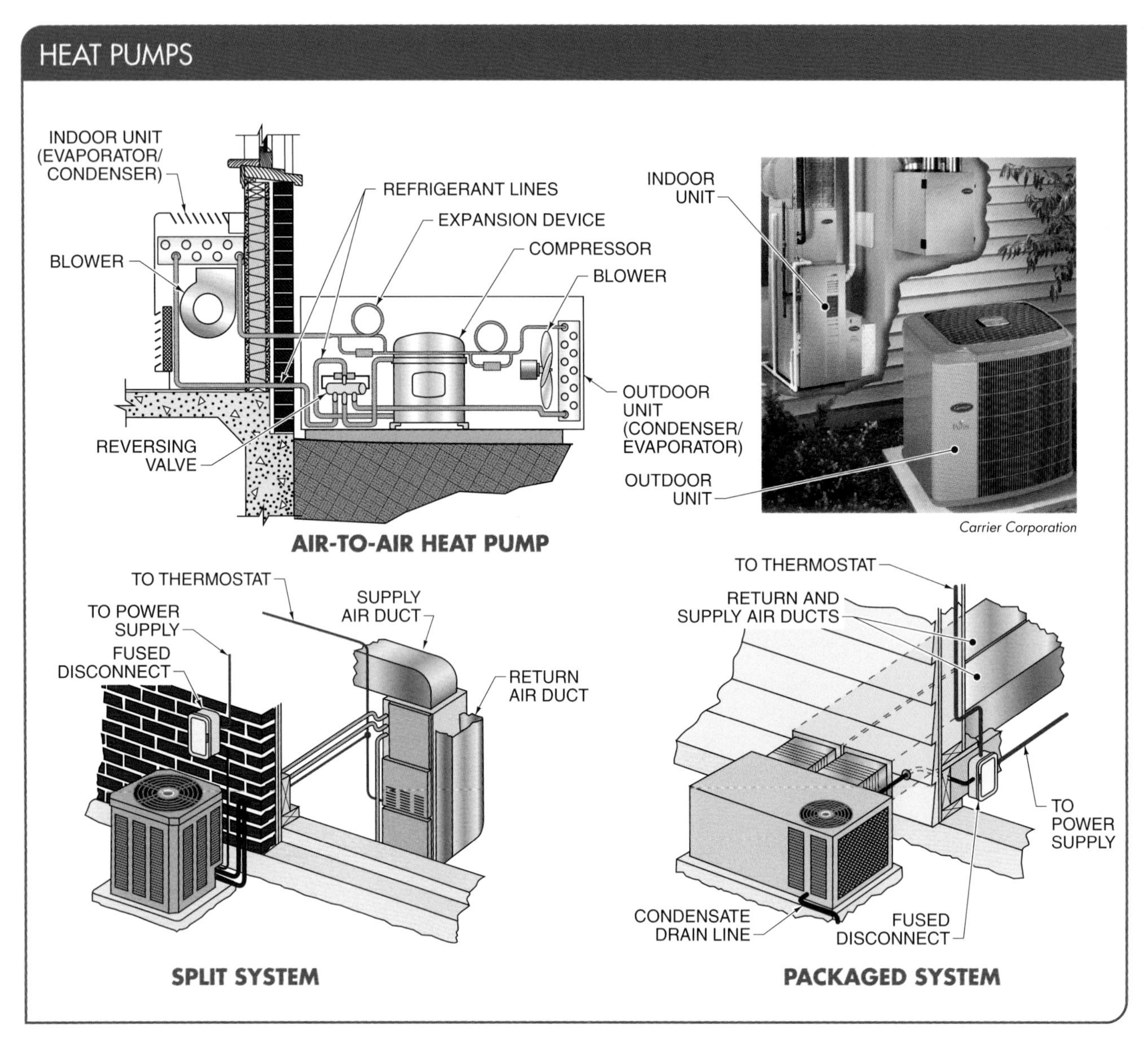

Figure 1-26. The most common type of heat pump used in residential applications is an air-to-air heat pump, which is available as a split system or a packaged system.

TECH TIP

Heat pumps can be an energy-efficient alternative for heating and cooling a dwelling in seasonal climates. A heat pump system can provide the equivalent operation of a conventional heating and cooling system in as little as a quarter of the cost. Geothermal heat pump systems are one of the most efficient systems and can effectively reduce energy use by up to 60%.

SECTION 1.3 CHECKPOINT

1. What are the three main sources of renewable energy?
2. What are the other sources of renewable energy?
3. Where does geothermal energy come from?
4. What is a heat pump?

Chapter Activities

1 Power Generation, Distribution, and Smart Grid Systems

Name ______________________________ Date ____________

Energy Usage

Identify the load types from the percentage of power consumption.

__________ **1.** Heat, sound, and visual outputs

__________ **2.** Light

__________ **3.** Motion

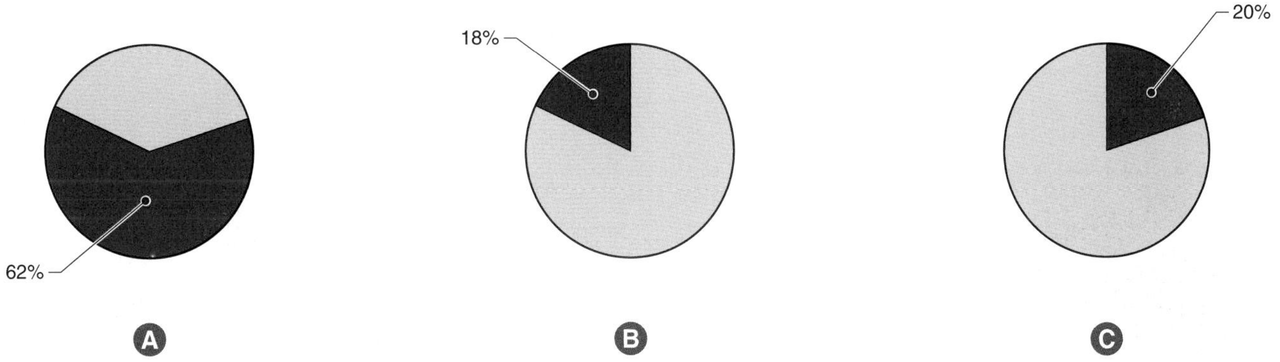

62%

A

18%

B

20%

C

Energy Efficiency

Calculate energy efficiency.

1. A home appliance is provided 120 W of power and uses 100 W. What is the energy efficiency of the appliance?

2. A washing machine is provided 500 W of power and uses 460 W. What is the energy efficiency of the washing machine?

3. A sump pump motor draws 1200 W from the power lines and delivers 900 W of power. What is the energy efficiency of the sump pump?

Potential and Kinetic Energy

Identify each as potential or kinetic energy.

____________________ **1.** Energy at A is ___.

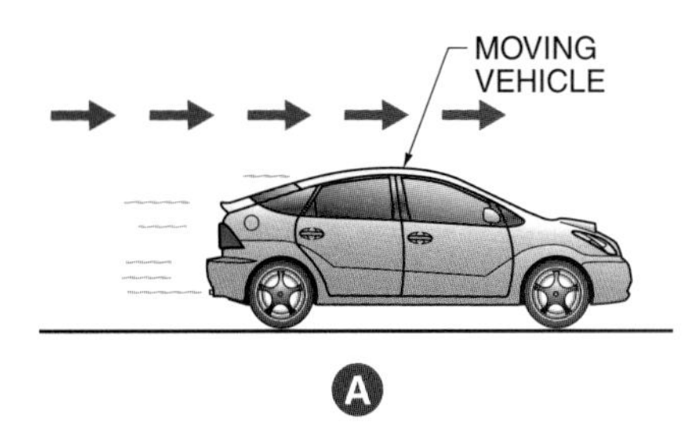

____________________ **2.** Energy at B is ___.

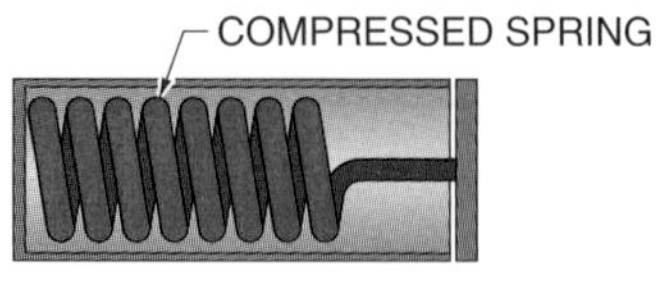

____________________ **3.** Energy at C is ___.

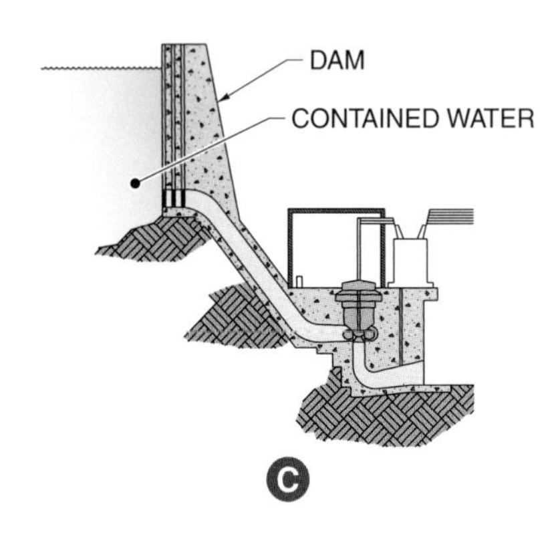

____________________ **4.** Energy at D is ___.

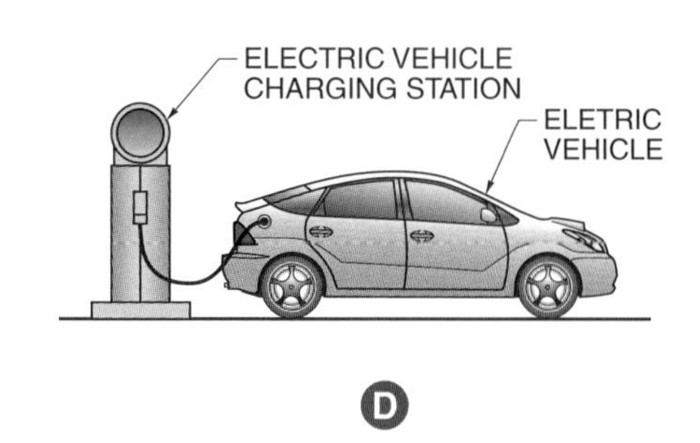

____________________ **5.** Energy at E is ___.

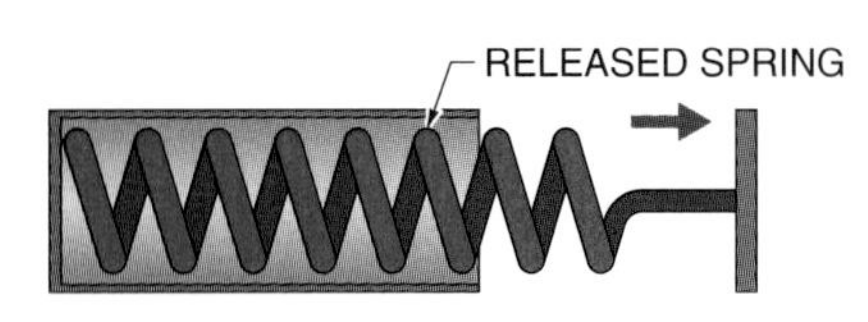

____________________ **6.** Energy at F is ___.

FALLING WATER
TURBINE
GENERATOR
F

2 Electrical Quantities, Voltage Sources, and Generators

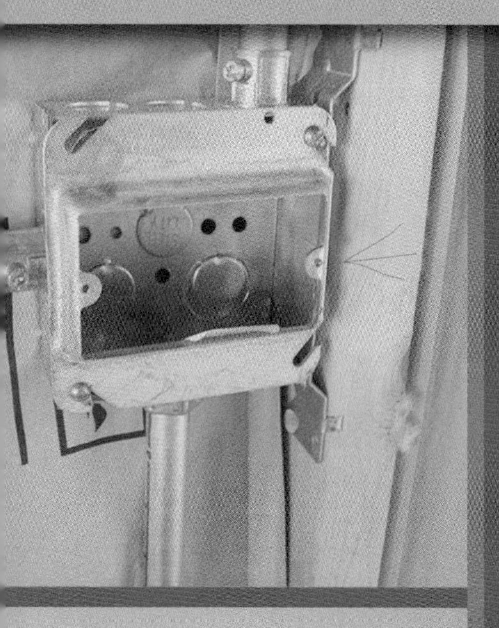

SECTION 2.1—ELECTRICAL QUANTITIES AND ABBREVIATIONS

- Define "voltage."
- Explain current.
- Explain resistance.
- Describe Ohm's law.

SECTION 2.2—VOLTAGE SOURCES

- Explain voltage sources.
- Describe electrochemical cells.
- Describe battery storage systems.
- Explain cell and battery capacity.
- Describe fuel cells.

SECTION 2.3—GENERATORS

- Explain how generators work.
- Compare and contrast DC and AC generators.
- Compare and contrast 1ϕ and 3ϕ generators.

Learner Resources

ATPeResources.com/QuickLinks
Access Code: 791712

Electrical abbreviations, prefixes, and keywords used in residential wiring must be understood to know how electricity is generated, how various aspects of electricity are represented by units of measure, how electricity operates, and how electricity is measured. Voltage sources, including electrochemical cells, batteries, and fuel cells, are used to supply electrical power to a load or store power until it is needed to power a load.

SECTION 2.1—ELECTRICAL QUANTITIES AND ABBREVIATIONS

Units of measure are used to define and quantify the various aspects of electricity. Electrical abbreviations and prefixes are used to simplify the expression of common electrical quantities. **See Figure 2-1.**

An *abbreviation* is a letter or combination of letters that represents a word. The exact abbreviation used may be a letter or a symbol. For example, voltage may be abbreviated using capital letters *E* or *V*. The letter used may not always be the first letter of the word being abbreviated. For voltage, both abbreviations *E* and *V* are often used interchangeably and are equally acceptable. The abbreviation for an "ohm" is an exception because it uses the omega symbol (Ω) as the abbreviation.

Prefixes are used to avoid long numbers when referring to amounts of voltage (V), current (A), resistance (Ω), etc. **See Figure 2-2.** For example, instead of "1000 volts" or "1 kilovolt" being written out, the abbreviation "1 kV" can be used.

COMMON ELECTRICAL TERMS		
Name	**Name Abbreviation**	**Unit of Measure Abbreviation**
Voltage	E	Volt—V
Current	I	Ampere—A
Resistance	R	Ohm—Ω
Power	P	Watt—W
Power (apparent)	PA	Volt-ampere—VA
Capacitance	C	Farad—F
Inductance	L	Henry—H
Impedance	Z	Ohm—Ω
Frequency	f	Hertz—Hz
Period	T	Second—s

Figure 2-1. Electrical abbreviations and prefixes are used to simplify the expression of common electrical terms and amounts.

COMMON ELECTRICAL QUANTITIES AND PREFIXES	
Name	**Unit of Measure Abbreviations**
1000 volts	1 kilovolt—1kV
One thousandth (1/1000) of an amp	1 milliamp—1 mA
47 thousand ohms	1 kiloohms—47 kΩ
1 millionth (1/1,000,000) of a farad	1 microfarad—1 μF
1000 watts	1 kilowatt—1 kW
One million (1,000,000) hertz	1 mega hertz—1 MHz

Figure 2-2. Prefixes are used to prevent long numbers when referring to amounts of voltage (V), current (A), and resistance (Ω). Instead of 1000 volts or 1 kilovolt being written out, the abbreviation "1 kV" can be used.

Voltage

All electrical circuits must have a source of power to produce work. The source of power used depends on the application and the amount of power required by that application. All sources of power produce a set voltage level or a voltage range. *Voltage* is the force or electrical potential between two points.

> TECH TIP
>
> *A person making direct contact with a source of AC or DC voltage can experience increased muscle contraction and ventricular fibrillation, which is a detrimental irregular heart rhythm that can eventually cause death if the person is not removed from the source voltage.*

Voltage can be understood by comparing it to the water pressure at a closed hose spigot or valve. Water pressure (voltage) is largest when the valve is closed and there is no water flow (current). When the valve is opened, the pressure (voltage) causes the water to flow (current) through the hose.

Voltage, as well as current, is produced when electrons are freed from atoms by electromagnetism (generators), chemicals (batteries), sunlight (PV cells), heat (thermocouples), pressure/vibrations (microphones), and friction (static electricity). **See Figure 2-3.** Voltage can be either direct current (DC) or alternating current (AC).

AC Voltage. *AC voltage* is voltage that causes current to reverse its direction of flow at regular intervals or cycles. An *AC cycle* is the complete positive and negative alternation of a wave form. An *alternation* is one half of an AC cycle. A complete AC cycle has one positive alternation and one negative alternation per cycle. **See Figure 2-4.** AC voltage is either 1ϕ or 3ϕ. Single-phase (1ϕ) AC voltage contains only one alternating voltage waveform.

AC and DC motors are both rated in horsepower (HP); 1 HP is equivalent to 746 W of power.

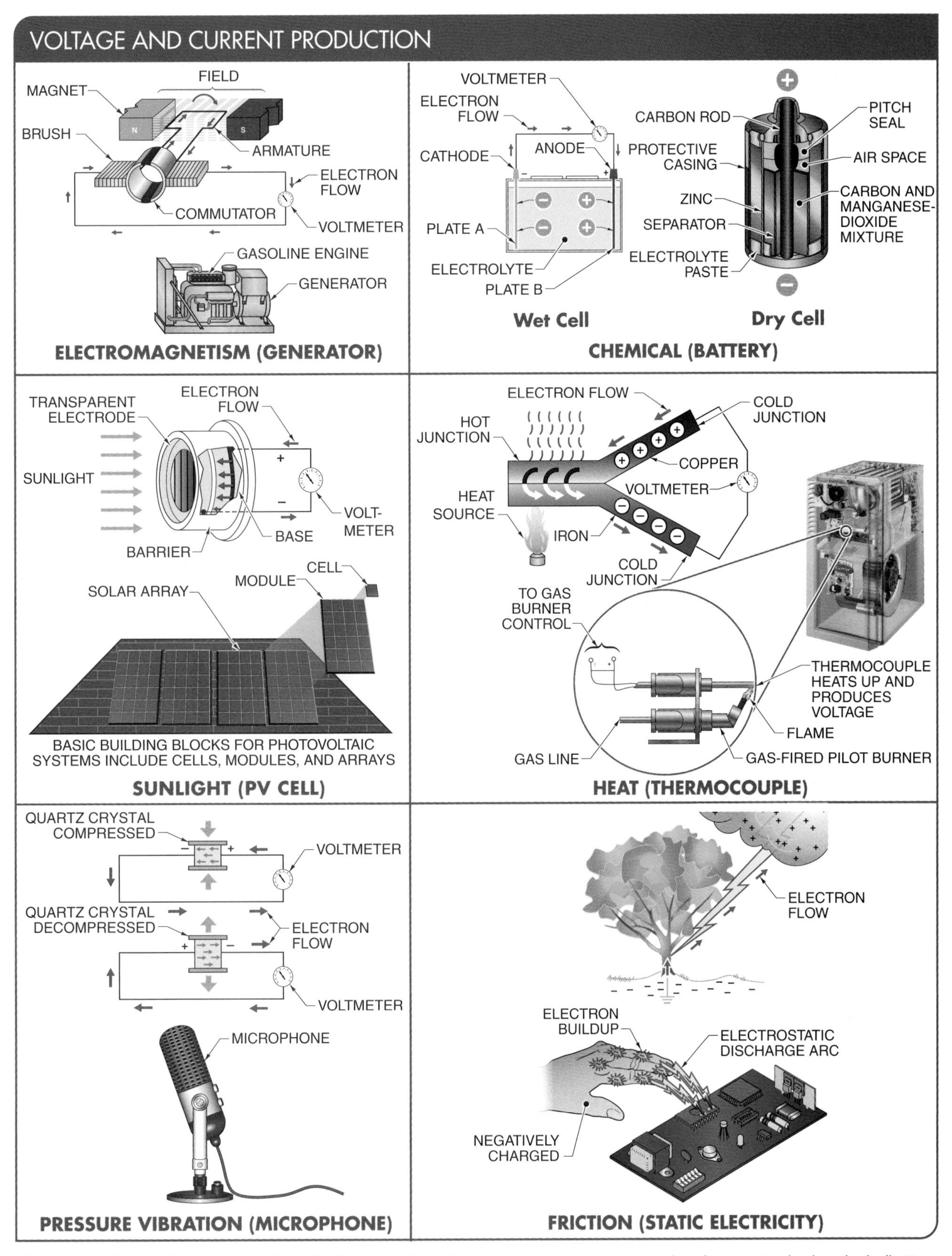

Figure 2-3. Voltage is produced when electrons are freed from atoms by electromagnetism (generators), chemicals (batteries), sunlight (PV cells), heat (thermocouples), pressure/vibrations (microphones), and friction (static electricity).

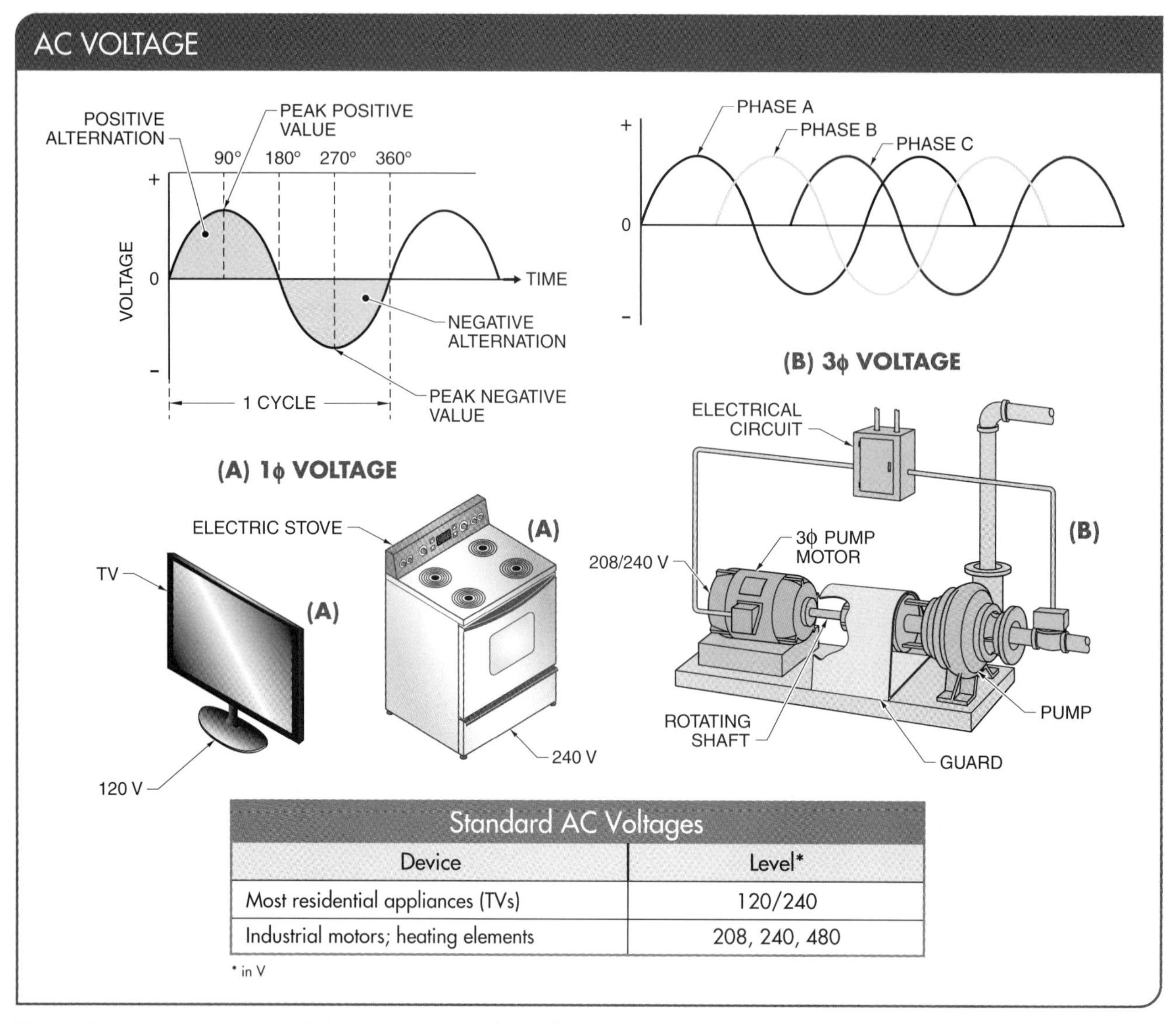

Standard AC Voltages	
Device	Level*
Most residential appliances (TVs)	120/240
Industrial motors; heating elements	208, 240, 480

* in V

Figure 2-4. A complete AC cycle has one positive alternation and one negative alternation per cycle. Single-phase (1ϕ) AC voltage contains only one alternating voltage waveform; 3ϕ AC voltage is a combination of three alternating voltage waveforms, each displaced 120 electrical degrees (one-third of a cycle) apart.

Single-phase (1ϕ) AC voltage is the most commonly used voltage in a residence. Typically, a test light can be used to determine whether 120 VAC is present at a typical residential outlet. **See Figure 2-5.** Test lights can also be used for checking 220 V outlets, such as those used for electric ranges. With 220 V, the tester will glow more brightly.

Three-phase (3ϕ) AC voltage is a combination of three alternating voltage waveforms, each displaced 120 electrical degrees (one-third of a cycle) apart. Three-phase (3ϕ) AC voltage is used for commercial and industrial applications where large loads are present. Three-phase (3ϕ) AC voltage is often used for heavy-duty equipment driven by 3ϕ motors.

DC Voltage. *DC voltage* is voltage that causes current to flow in one direction only. All DC voltage sources have a positive terminal and a negative terminal. The positive and negative terminals establish polarity in a circuit. *Polarity* is the positive (+) or negative (–) state of an object. All points in a DC circuit have polarity.

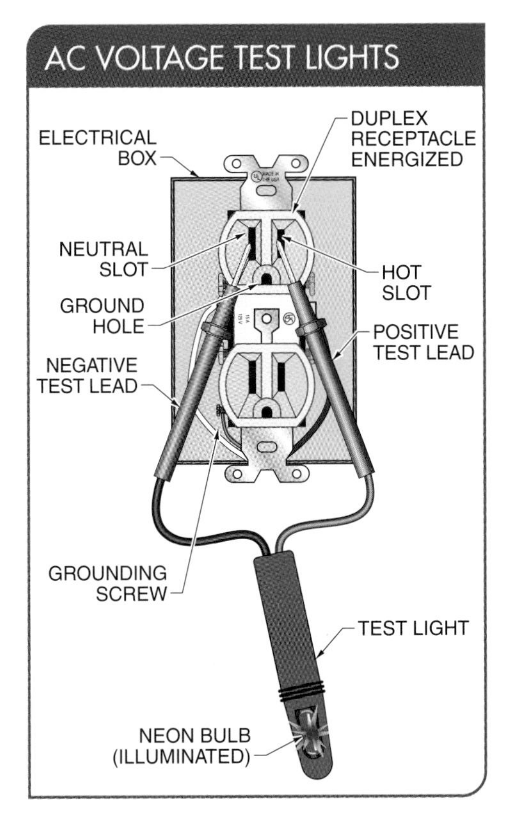

Figure 2-5. Single-phase AC voltage is the most common voltage used in a residence. A test light illuminates when a standard receptacle is energized.

Batteries, which are the most common power source that provide DC voltage in a residence, have positive and negative terminals that are often easily identified. These terminals can also be used to measure DC voltage, typically through the use of a voltage tester or digital multimeter (DMM). **See Figure 2-6.**

Besides batteries, DC voltage can also be obtained from photocells and rectified AC voltage supplies. DC voltage is obtained any time an AC voltage is passed through a rectifier. A *rectifier* is a device that converts AC voltage to DC voltage by allowing the voltage and current to move in only one direction. DC voltage obtained from a rectified AC voltage supply varies from almost pure DC voltage to half-wave DC voltage. AC-to-DC adapters, as seen in laptops, radios, and cell phone chargers, are examples of low-voltage rectifiers. **See Figure 2-7.**

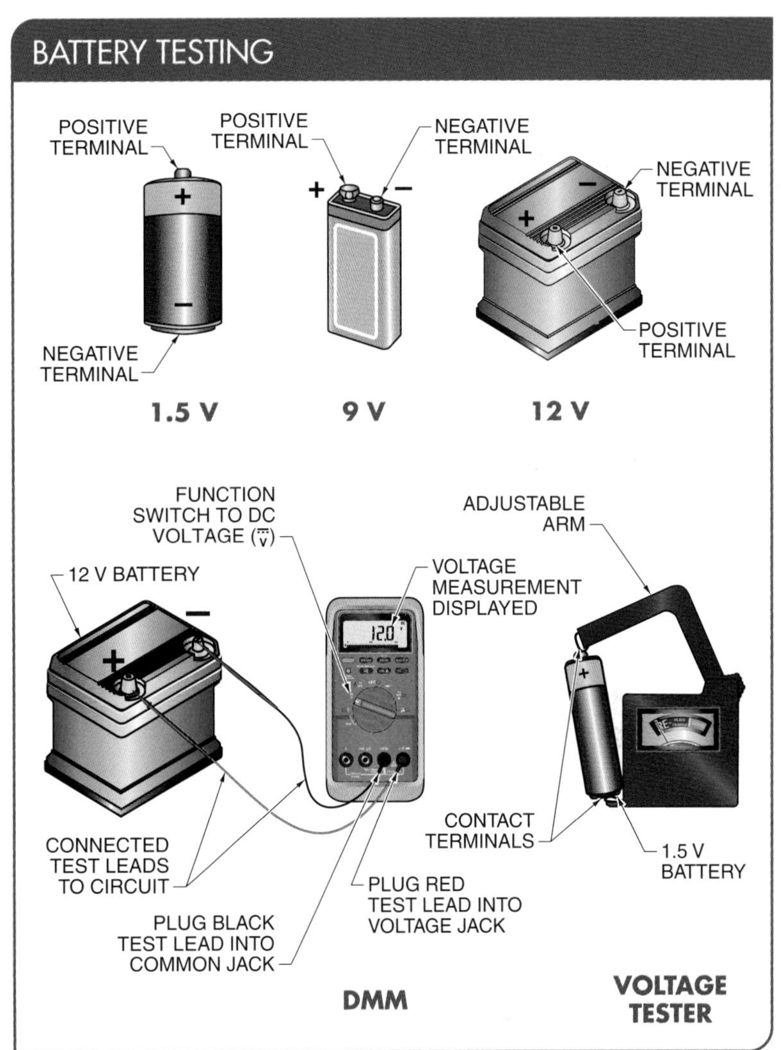

Figure 2-6. Batteries are the most common power source and directly provide DC voltage around a residence. The typical methods of measuring DC voltage include voltage testers and DMMs.

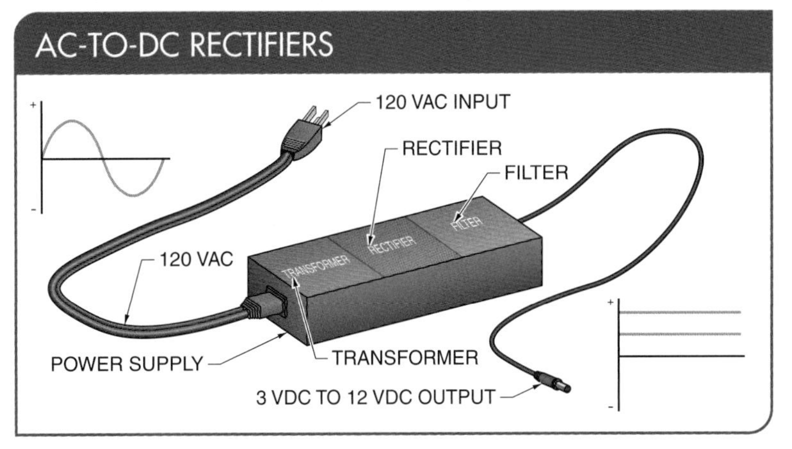

Figure 2-7. DC voltage can also be obtained from a rectified AC-to-DC power supply, which uses a rectifier made up of diodes to convert AC to DC.

Current

Current flows through a circuit when a voltage source is connected to a device that uses electricity. *Current* is the measure of the flow of electrons through an electrical circuit. **See Figure 2-8.** Current is abbreviated with the letter *I* and is measured in amperes. An *ampere (A)* is the number of electrons passing a given point in one second. The more power a load requires, the larger the amount of current it must draw. For example, a 10 horsepower (HP) motor draws approximately 28 A when wired for 230 V. A 20 HP motor draws approximately 56 A when wired for 230 V.

Current may be direct current (DC) or alternating current (AC). *Direct current (DC)* is current that flows in only one direction. *Alternating current (AC)* is current that reverses or alternates its direction of flow at regular intervals. AC is found in typical residential buildings.

Current is usually measured by trained electricians since they are more aware of the safety needs surrounding these measurements. The most common method of safely measuring current is to use a clamp-on ammeter. **See Figure 2-9.** Since current measurement requires working with live circuits, all safety precautions, especially the use of protective gloves, should be taken.

Resistance

Resistance is the opposition that a material offers to the flow of electric current. This means that the larger the resistance, the smaller the current flow; and, conversely, the smaller the resistance, the larger the current flow. Resistance is represented by the capital letter *R*. The standard unit of resistance is the ohm, sometimes written out as a word or, more commonly, symbolized by the omega symbol (Ω).

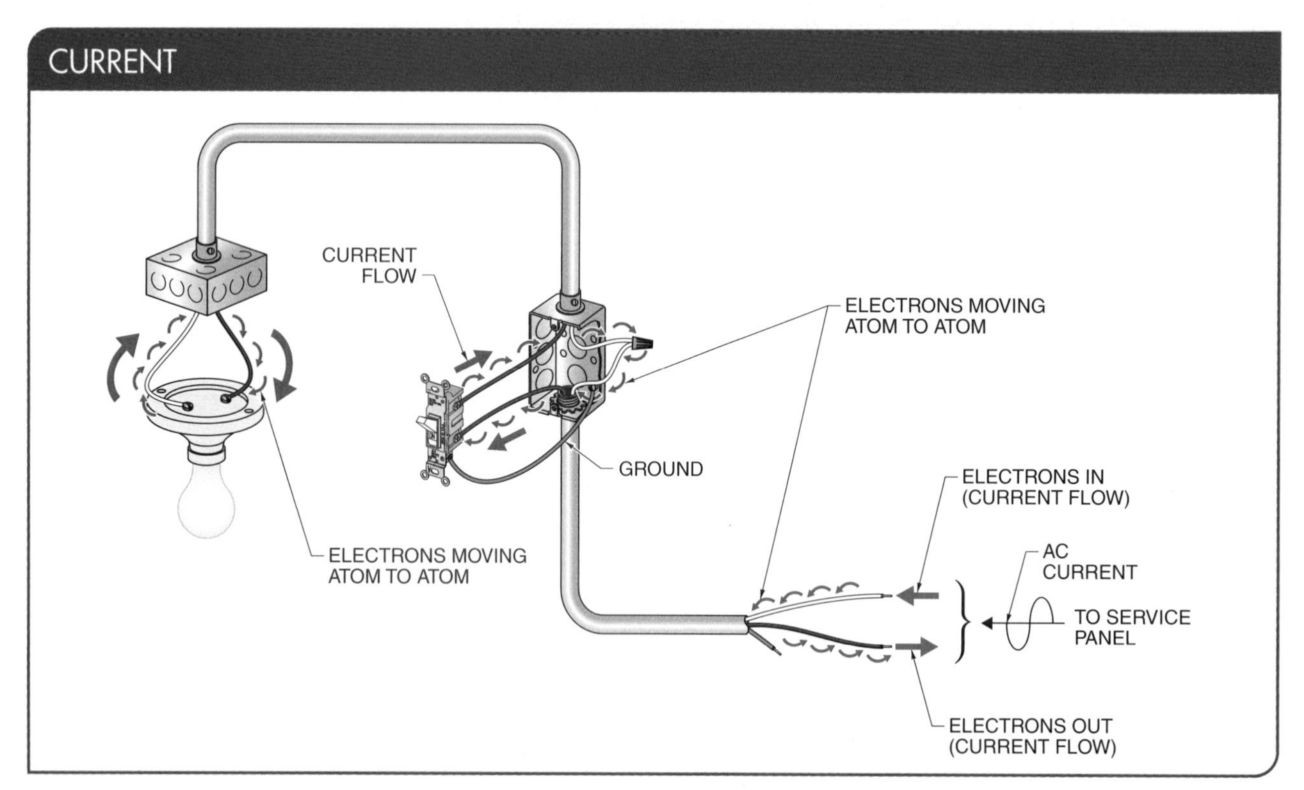

Figure 2-8. Current is the movement of electrons from atom to atom in a conductor. Current flows through an electrical circuit when a voltage source is connected to a device, or load, that uses electricity.

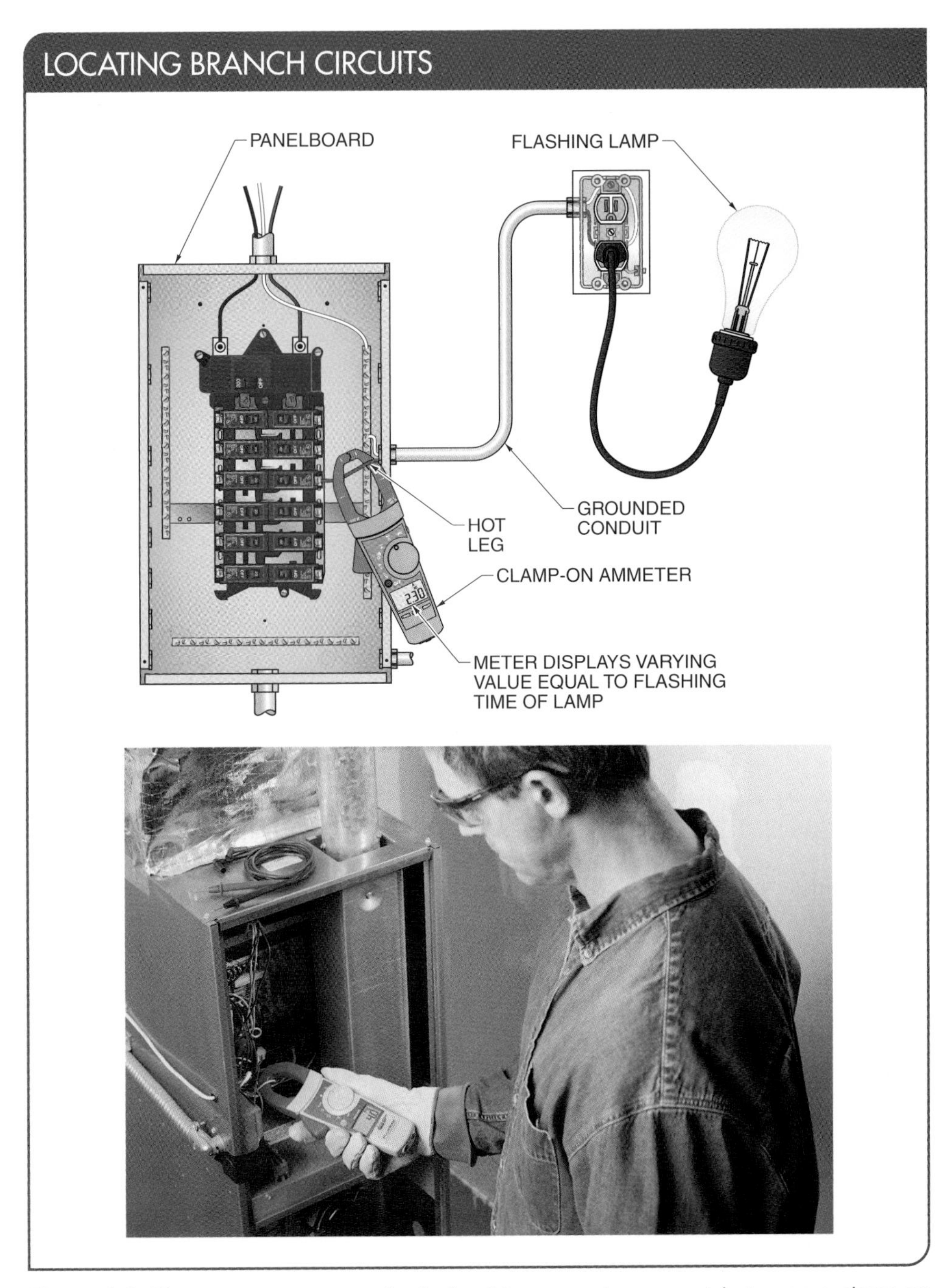

Figure 2-9. The most common method of safely measuring current is to use a clamp-on ammeter; however, since measuring current requires working with a live circuit, all safety precautions should still be followed.

Ohm's Law

Ohm's law is the relationship between voltage, current, and resistance properties in an electrical circuit. The relationship between voltage, current, and resistance is typically visualized as a pie chart of Ohm's law. **See Figure 2-10.** Ohm's law is important because it helps explain concepts such as why current is high when voltage is high, but it also shows why current can also be high when resistance is low. Since most things related to voltage, current, and resistance can be easily measured, these values do not need to be calculated very often.

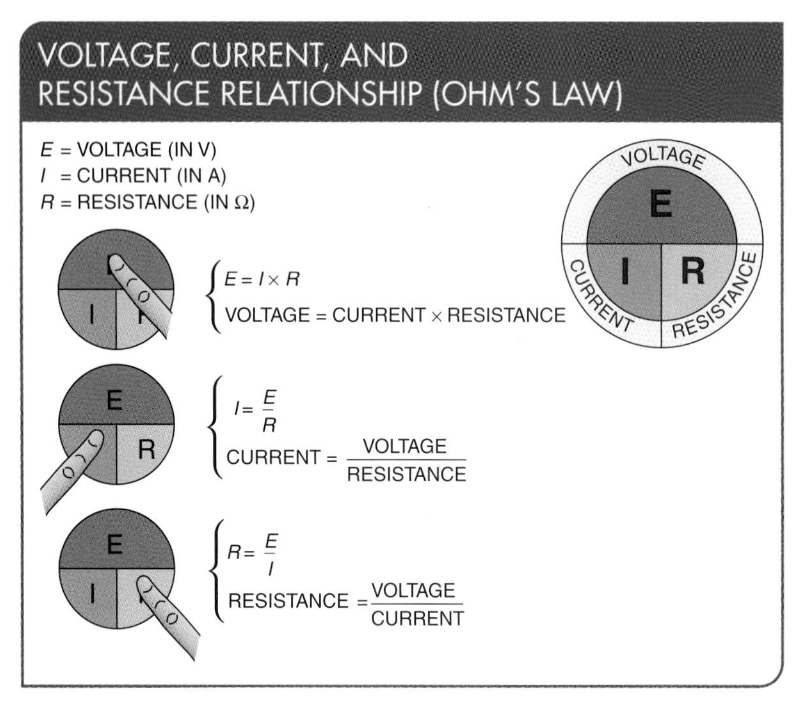

Figure 2-10. Ohm's law is the relationship between voltage, current, and resistance properties in an electrical circuit.

SECTION 2.1 CHECKPOINT

1. What is voltage?
2. What is AC voltage?
3. What is DC voltage?
4. What is current?
5. What is resistance?

SECTION 2.2—VOLTAGE SOURCES

The two types of electrical power are DC and AC, which are generated, distributed, and used to operate both DC and AC loads. DC power can be converted into AC using an inverter and AC power can be converted into DC using a rectifier (converter). Most residences use a combination of both DC and AC circuits and components.

DC and AC voltage sources include electrochemical cells, battery storage systems, primary and secondary cells, and fuel cells. Each type of source has advantages and disadvantages depending upon the application.

Electrochemical Cells

An *electrochemical power source* is a device that uses chemical reactions to produce or store electrical energy. The most basic type of electrochemical power source is a single electrochemical cell.

A *battery* is a group of connected electrochemical cells. The amount of energy available from a battery is dependent on the amount of chemical reactant stored in the battery. Once the chemical reactants are consumed (i.e., the battery is discharged), the battery will cease to produce electrical energy.

Electrochemical cells are classified as either dry cells or wet cells. A dry cell uses an electrolyte paste. For example, the dry cells in flashlights use ammonium chloride as an electrolyte paste. A wet cell uses a liquid electrolyte to create a chemical reaction. For example, the wet cells in automobile batteries use sulfuric acid as a liquid electrolyte.

Electrochemical cells have electrodes that connect to terminals outside of electrolyte. A cell has two terminals: one positive (+) and one negative (–). With one electrode positively charged and the other electrode negatively charged, a voltage (or difference of potential) exists between the electrodes.

> TECH TIP
>
> *Most electrical devices use either nickel-cadmium or lithium-ion batteries. Lithium-ion batteries have a higher energy density than nickel-cadmium batteries. One lithium-ion cell can provide 3.6 V, whereas three nickel-cadmium batteries are required to produce the same voltage.*

Dry Cells. A *dry cell* is two electrodes in an electrolyte paste that are used to create a chemical reaction. **See Figure 2-11.** A dry cell is typically made of a carbon rod, which acts as a positive electrode, surrounded by an electrolyte paste that is enclosed in a zinc case, which acts as a negative electrode. A dry cell also includes an expansion space and a metal cap that is attached to the positive electrode (carbon rod).

Wet Cells. A *wet cell* is two electrodes in a liquid electrolyte solution that are used to create a chemical reaction. **See Figure 2-12.** The liquid electrolyte is usually a solution of acid in water. Through an electrochemical reaction, the electrolyte breaks down particles in the cell and forms ions, which are groups of electrically charged atoms. Certain ions have excess electrons and are negatively charged. Other ions have a shortage of electrons and are positively charged. The negative ions and the positive ions float in the electrolyte. The positive ions form a positive electrode while the negative ions form a negative electrode.

Current flows from one electrode to the other when a closed circuit is formed between each electrode of a wet cell. For example, a closed circuit may consist of a lamp connected to the negative and positive electrodes using conductors to form a closed loop or circuit. The electrodes and electrolyte chemically change while supplying current to the circuit. After a period of time, the electrodes change to the point where the cell is no longer able to supply current. When this happens, the cells must be replaced or recharged and used again, depending on their type.

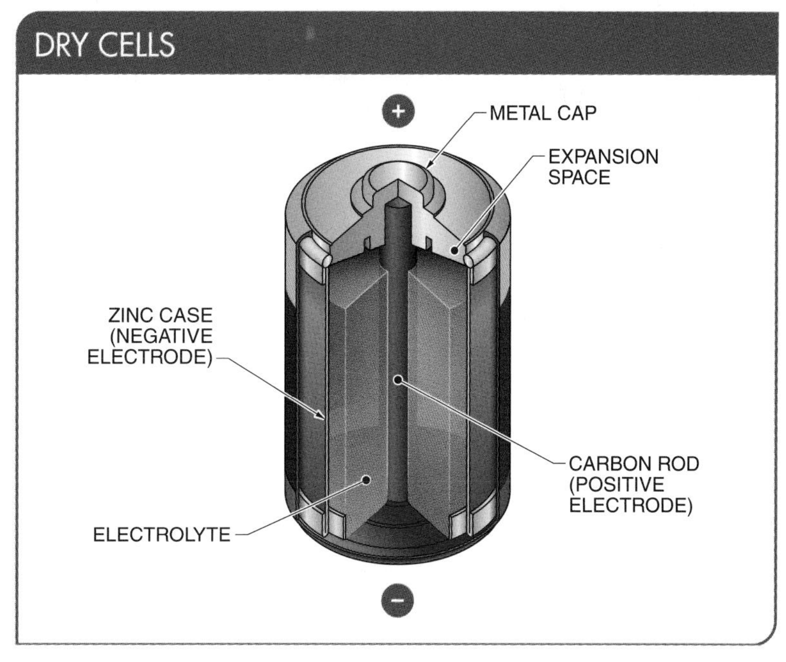

Figure 2-11. A dry cell consists of two electrodes in an electrolyte paste.

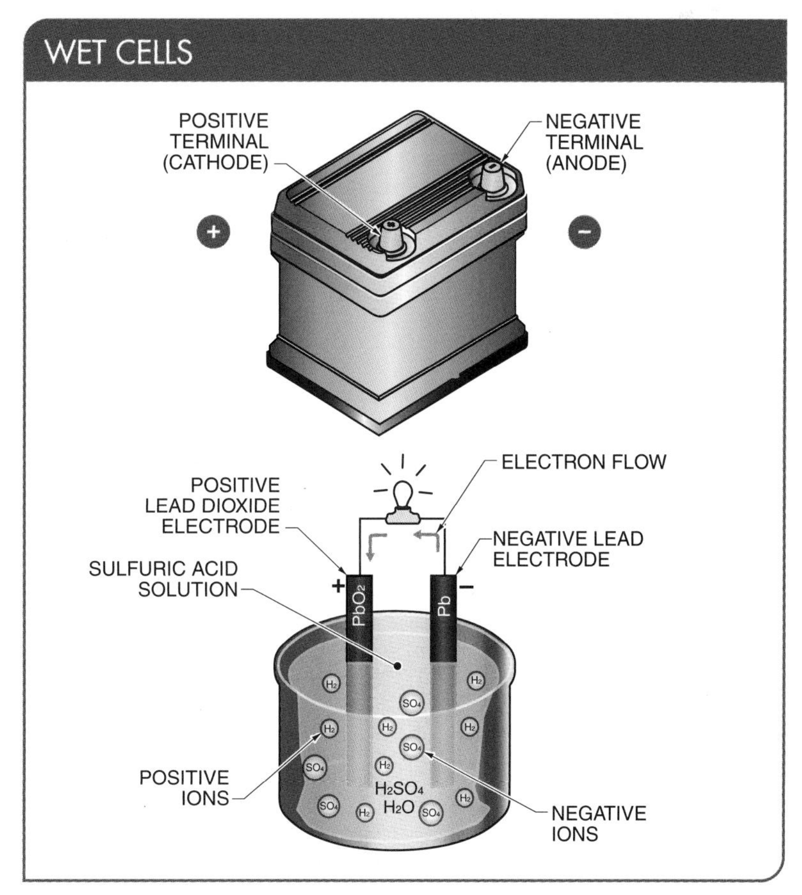

Figure 2-12. A wet cell consists of two electrodes in a liquid electrolyte solution that breaks down particles in the cell through an electrochemical reaction to form ions.

Battery Storage Systems

A *battery storage system* is a network of connected batteries that store electricity, typically generated from renewable energy sources, for use at a later time. Utilities and homeowners may use battery storage systems. **See Figure 2-13.**

Batteries connected to an inverter system can be set to regulate the amount of current used by a homeowner or fed into the grid without adversely affecting the performance of the batteries. Installing many small battery storage systems could offset the need to build a large grid storage facility. The small storage systems can also serve as a backup during power outages.

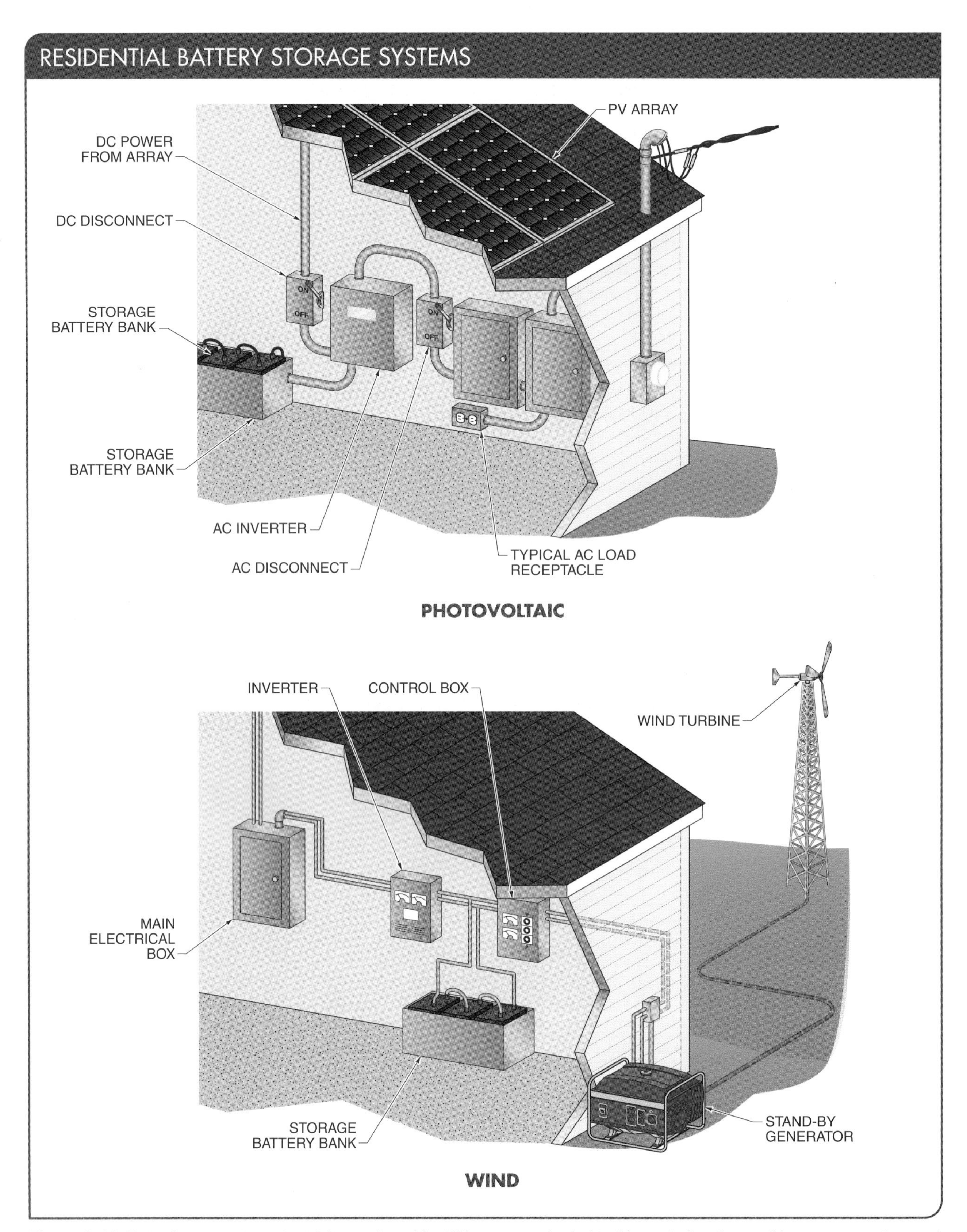

Figure 2-13. Batteries can store electricity produced by PV arrays and wind turbines during times of low consumption and discharge it during periods of high consumption or peak demand.

During times of low consumption, batteries can store electricity produced by renewable energy sources. Then during times of high consumption or peak demand, the batteries can discharge the electricity. This reduces the need for expanding the electrical grid, which also saves money. Although battery prices are currently high, increases in manufacturing capacity will make batteries more affordable.

Batteries deliver power for only a fixed amount of operating time. Once the battery power is used, the battery must be replaced or recharged. The operating time of a battery depends on the load that it is powering. The more power the load requires, the less operating time. The less power the load requires, the greater the operating time. The amount of current that a battery can supply depends largely on the size of the cell or battery. **See Figure 2-14.**

Battery Monitoring. The current in a battery bank is an issue that always needs to be considered and is one of the most misunderstood. Regardless of whether the batteries are old or new, they can be worn out due to improper charging or maintenance and will not be able to handle any load beyond a few amperes. Powering a small load of a few amperes (1 A to 5 A) is usually possible from a dead battery bank for many years after the batteries have passed their useful lifetime. However, if a large load is applied, the batteries will be unable to supply power to the heavier load. A battery monitor should be installed to monitor the battery bank. Also, a load meter should be used to regularly test the bank.

Battery Maintenance and Storage. General storage batteries, which are used to provide back-up power for large power systems, must be properly maintained and serviced. There are many problems that may occur due to improper or poor battery maintenance, including corrosion, dirt contamination, missing vent caps, low electrolyte levels, cracked cases, cracked cell covers, and frayed or broken cables. **See Figure 2-15.** Most of these problems can be avoided with preventive maintenance.

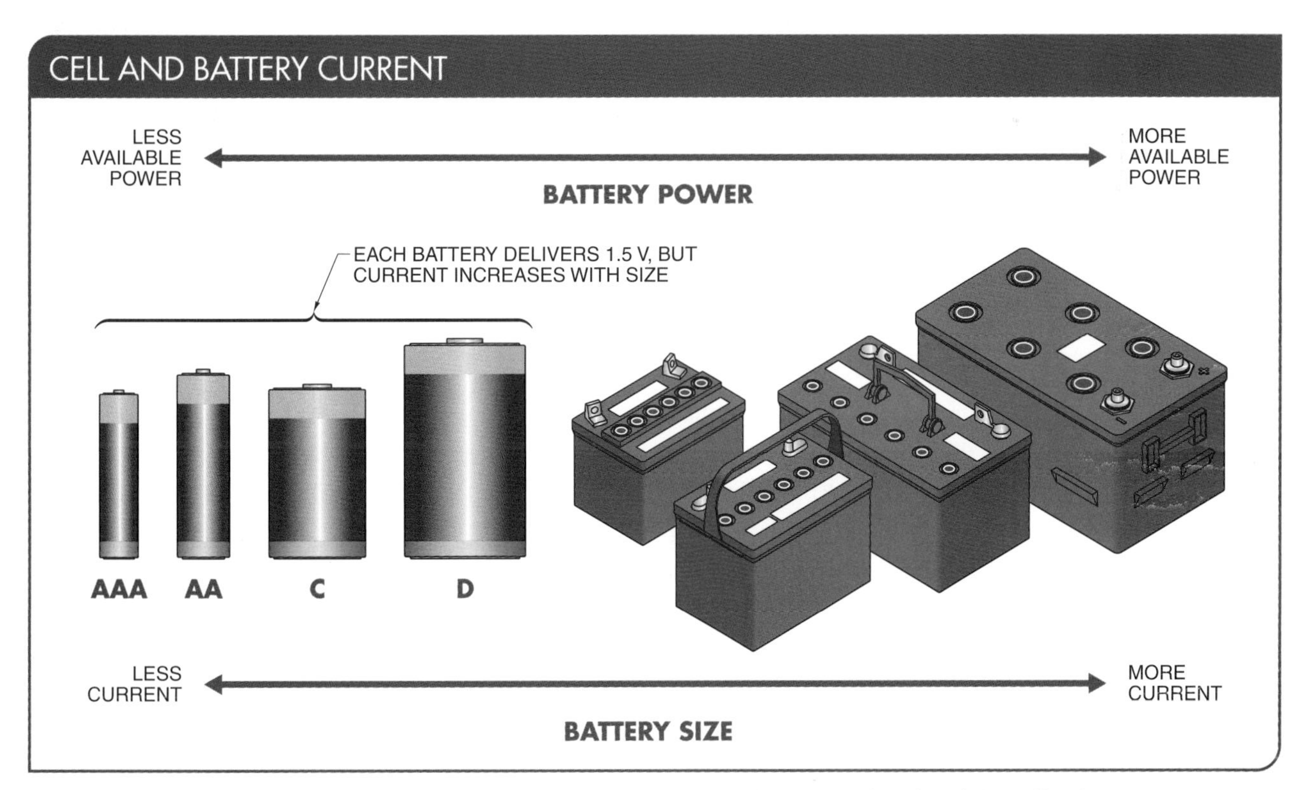

Figure 2-14. The amount of current that a battery can supply depends largely on the size of the cell or battery.

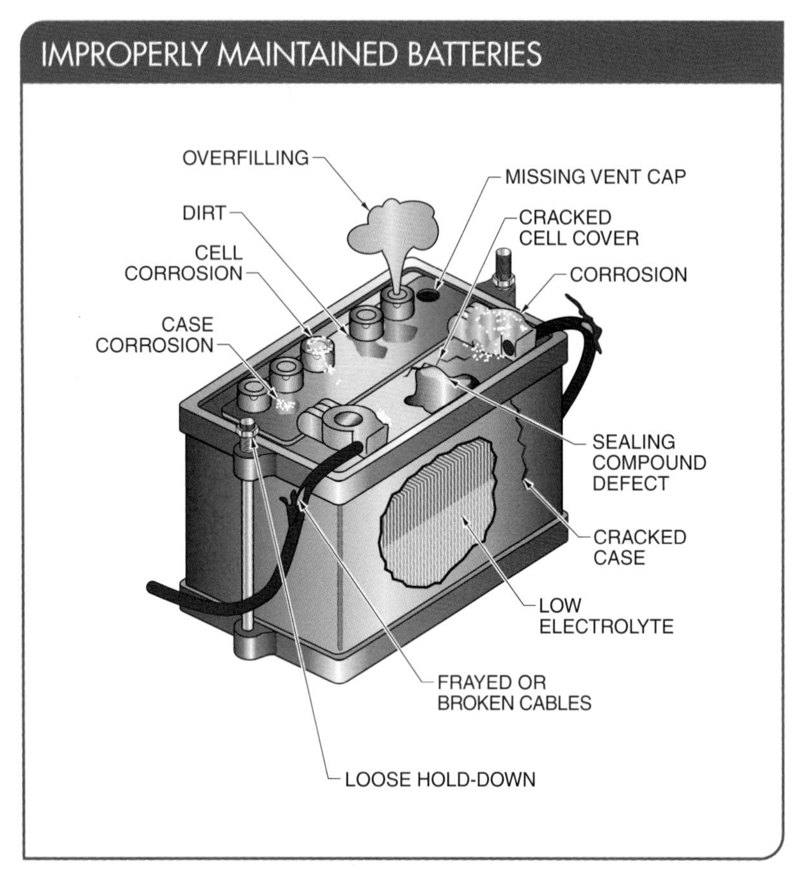

Figure 2-15. There are many problems that may occur due to improper or poor battery maintenance, most of which can be avoided with preventive maintenance.

Battery Safety. Batteries can produce toxic and explosive mixtures of gases. Ventilation of the battery enclosure is sometimes required. Wet cell batteries can explode from gases generated when batteries are being jumped or charged due to sparks from the cable connections. Battery acid is corrosive and can burn the skin. If battery acid comes in contact with the skin, the area should be immediately washed with plenty of soap and water, and anyone injured should seek medical attention.

Cell and Battery Capacity

The capacity of a storage battery is stated in ampere-hours (Ah). One ampere-hour is the maximum current a 1 A battery can continuously deliver until the battery is completely discharged after a period of 1 hr. The capacity for other batteries can be found by simply multiplying the number of amperes of current carried by the number of hours the current flows. **See Figure 2-16.** For example, a battery capable of providing 2 A of current continuously for a period of 10 hr has a capacity of 20 Ah ($2 \times 10 = 20$). Likewise, a battery that can deliver a current of 10 A for 2 hr also has a capacity of 20 Ah (10 A × 2 hr = 20 Ah).

Cell and Battery Connections. A battery is made up of a group of cells connected together. Cells and batteries are connected in series to produce a higher voltage than the voltage that is available from just one cell or battery. Cells and batteries are connected in parallel to make a greater current available than the current that can be provided by a single cell or battery. **See Figure 2-17.**

Cell and Battery Disposal. A large number of batteries are being used worldwide in computers and other electronics. Therefore, it is important to dispose of batteries properly and safely to minimize their impact on the environment.

Primary and Secondary Cells. A *primary cell* is a cell that uses up its electrodes and electrolyte to the point where recharging is not practical. Primary cells are also known as dry cells and single-use or nonrechargeable cells. A *secondary cell* is a cell that can be recharged. These cells are recharged by passing current through them in the direction opposite to that of the load current.

Fuel Cells

A *fuel cell* is an electrochemical device that combines hydrogen and oxygen to produce electricity. When fuel cells produce electricity, they create by-products in the form of both water and useful heat. There are many different types of fuel cells, each with their own unique operating characteristics. Many fuel cells are powered with hydrogen gas, but they can also be powered with many other sources. Fuel cells have twice the efficiency of traditional combustion technologies.

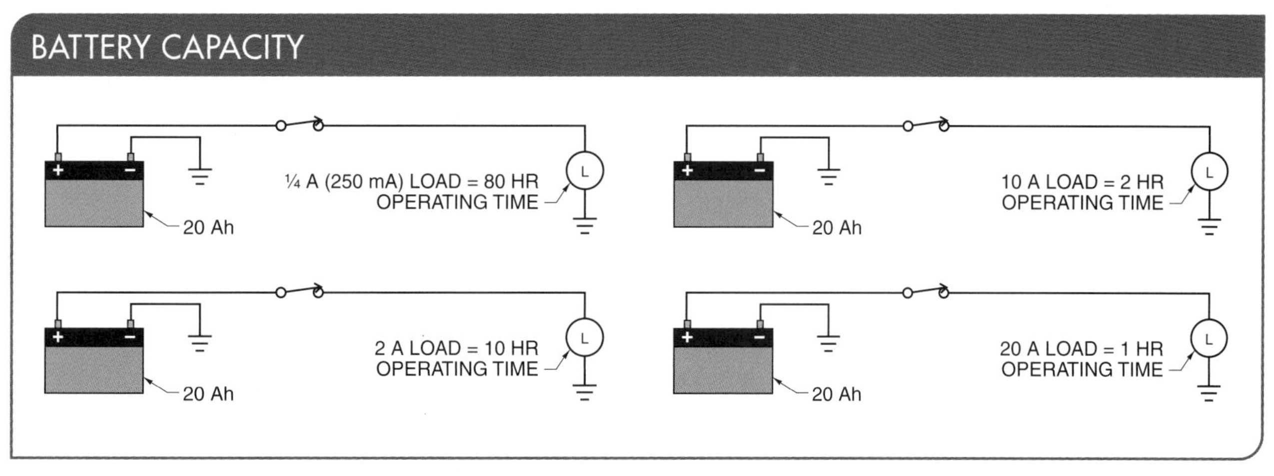

Figure 2-16. The capacity of a storage battery is stated in ampere-hours (Ah). One ampere-hour is the maximum current that a 1 A battery can continuously deliver until the battery is completely discharged after a period of 1 hr.

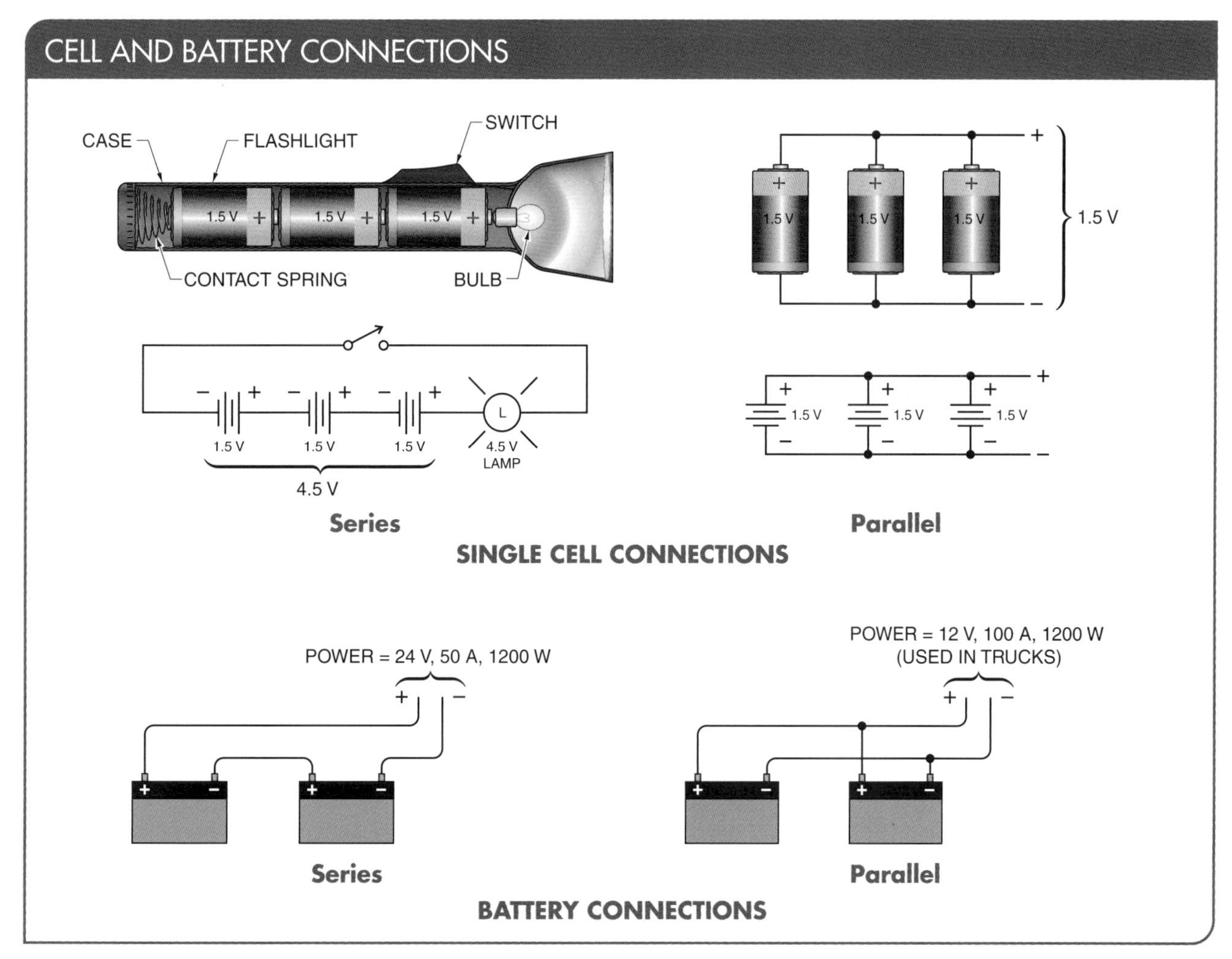

Figure 2-17. Cells and batteries are connected in series to produce a higher voltage than the voltage that is available from just one cell or battery. They are connected in parallel to make a greater current available than the current that can be provided by a single cell or battery.

Fuel Cell Operation. Fuel cells operate by electrochemical reaction. Fuel cells are different from batteries in that they produce but do not store energy. A fuel cell will continue to provide electrical power as long as it is supplied with fuel. The basic structure of a single fuel cell consists of an electrolyte layer in contact with a porous anode on one side and a porous cathode on the other side. **See Figure 2-18.** A single fuel cell produces a small amount of power.

Fuel Cell Capacity. The power produced by a fuel cell depends on several factors, including the fuel cell type, size, temperature at which it operates, and pressure at which gases are supplied. Commercial fuel cells are actually a collection or "stack" of individual fuel cells.

Depending on the application, the construction of a fuel cell stack may contain only a few cells or as many as 100 or more individual cells layered together. This scalability makes fuel cells ideal for a wide variety of applications, such as laptops, residences, transportation, and central power generation.

Utility companies have begun to deploy megawatt-scale fuel cell stacks at electric power stations to generate and deliver electricity to the power grid. **See Figure 2-19.** Adding fuel cell power generation capacity helps energy utility companies meet clean energy requirements required by government regulations. In this way, fuel cells have become complementary to, rather than competitive with, other electricity generation technologies, including renewable energy technology.

Advantages of Fuel Cells. Fuel cells have flexible fuel sources and can operate on conventional fuels or renewable fuels. Fuel cells typically provide high-quality, reliable power with exceptionally low emissions. They are very quiet, rugged, and compatible with solar energy, wind energy, and other renewable energy technologies. They also reduce the long-term costs of maintaining traditional transmission and distribution grids.

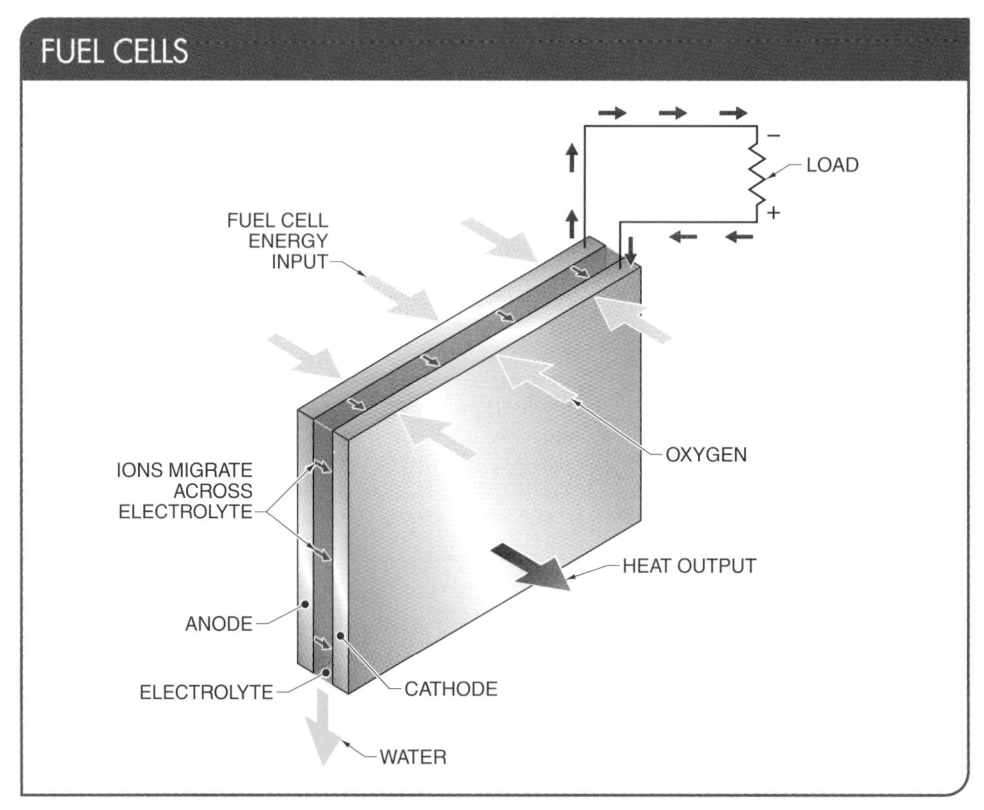

Figure 2-18. Fuel cells, which produce but do not store energy, consist of an electrolyte layer in contact with a porous anode on one side and a porous cathode on the other side.

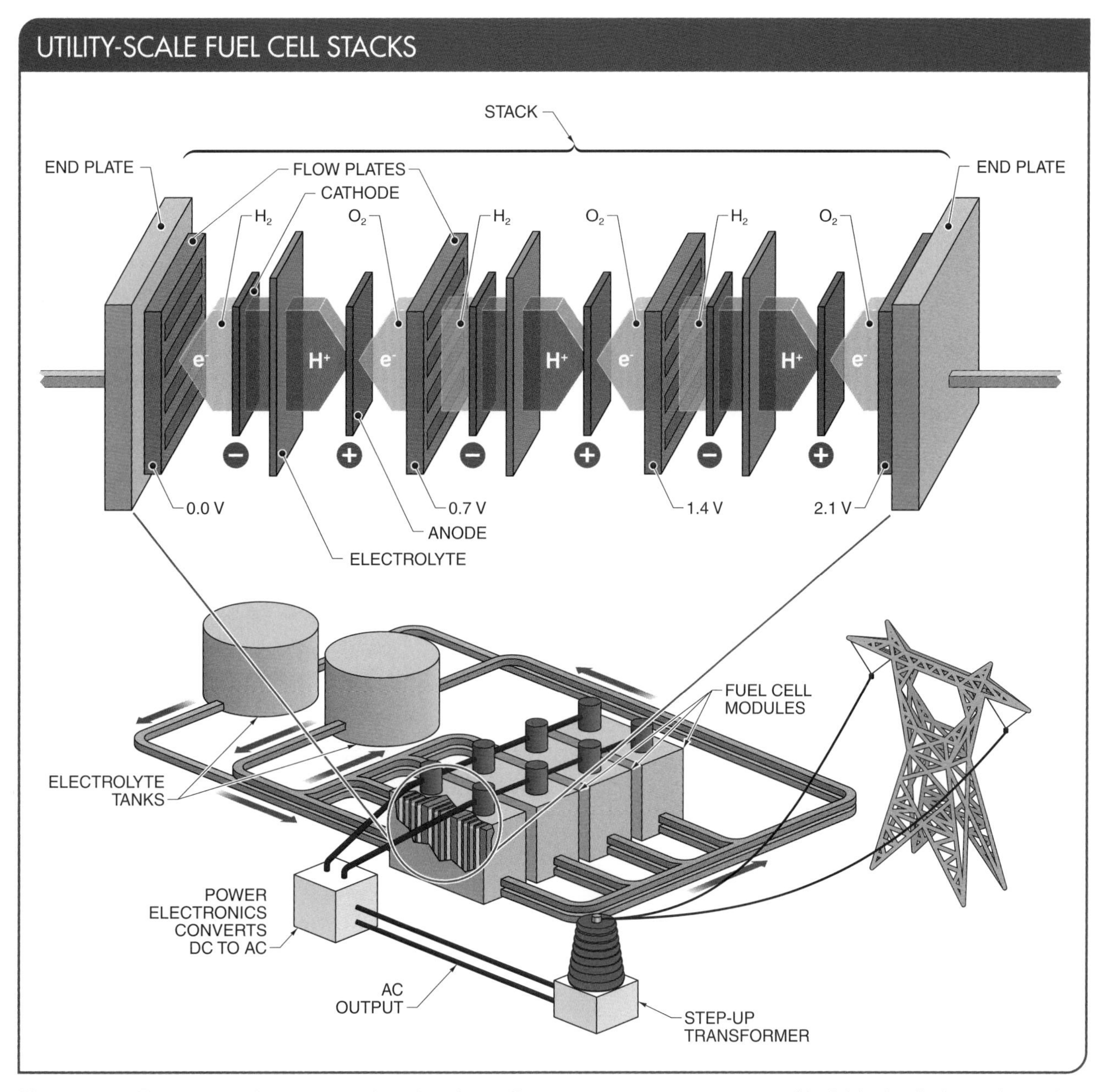

Figure 2-19. Fuel cell stacks may contain only a few cells or as many as 100 or more of individual cells layered together. Utility companies have begun to deploy large-scale fuel cells at electric power stations to generate and deliver electricity to the power grid.

TECH TIP

Residential hybrid solar energy systems use a battery bank to store energy during operation. A homeowner can benefit from the reduced energy costs and power-grid consumption of these energy systems, as well as the ability to still have power during blackouts.

SECTION 2.2 CHECKPOINT

1. What is an electrochemical power source?
2. What is a battery?
3. What is a dry cell?
4. What is a wet cell?

SECTION 2.3—GENERATORS

A *generator* is a machine that converts mechanical energy into electrical energy by means of electromagnetic induction. Generators can either be DC or AC. Typically, generators are used to power motors, battery banks, and arc welding machines. *Electromagnetic induction* is the ability of a coil of wire to induce a voltage in another circuit or conductor. **See Figure 2-20.**

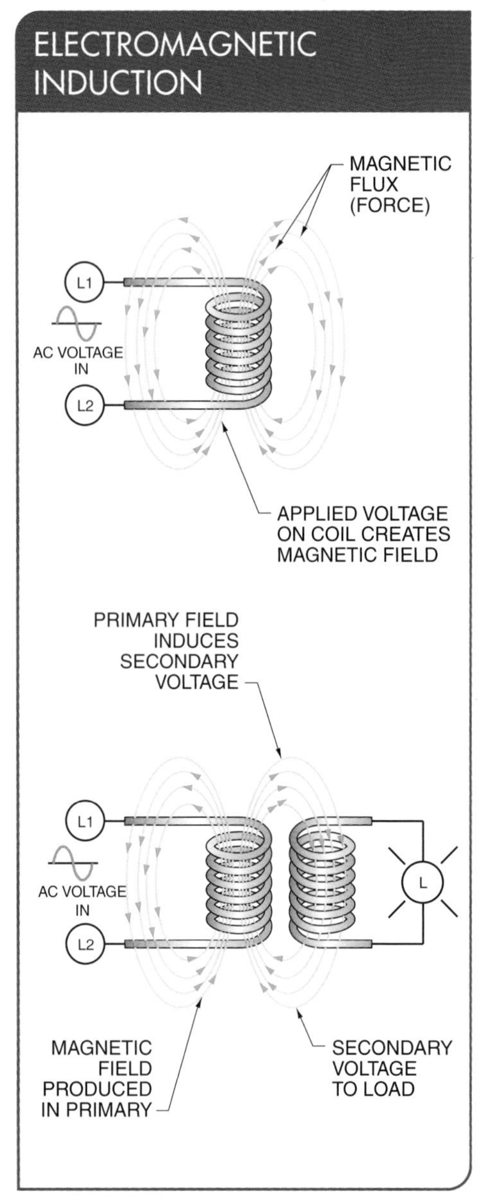

Figure 2-20. Electromagnetic induction is the process where a coil of wire induces voltage in another coil of wire.

DC Generators

A *DC generator* is a machine that converts mechanical energy into DC electricity. DC generators operate on the principle that when a coil of wire is rotated in a magnetic field, a voltage is induced in the coil. The amount of voltage induced in the coil is determined by the rate at which the coil is rotated in the magnetic field. DC generators consist of field windings, an armature, a commutator, and brushes. **See Figure 2-21.** Brushes instead of slip rings are used so that electricity generated can travel in only one direction.

The three types of DC generators are series-wound, shunt-wound, and compound-wound generators. The difference between the types of DC generators is based on the way in which the field windings are wired to the external circuit or load.

Series-Wound Generators. A *series-wound generator* is a generator that has its field windings connected in series with the armature and the external circuit (load). The ability of a generator to have a constant voltage output under varying load conditions is referred to as the generator's voltage regulation. Series-wound generators have poor voltage regulation. Because of their poor voltage regulation, series-wound DC generators are not frequently used.

Shunt-Wound Generators. A *shunt-wound generator* is a generator that has its field windings connected in parallel (shunt) with the armature and the external circuit (load). A shunt-wound generator is used where the load is constant, such as for lighting fixtures.

CODE CONNECT

Per NEC® Section 445.11, a generator shall have a nameplate providing the manufacturer's name, the rated frequency, the number of phases (if AC), the rating in kW or kVA, the power factor, and the normal volts and amperes corresponding to the power rating, the rated ambient temperature, and the rated temperature rise.

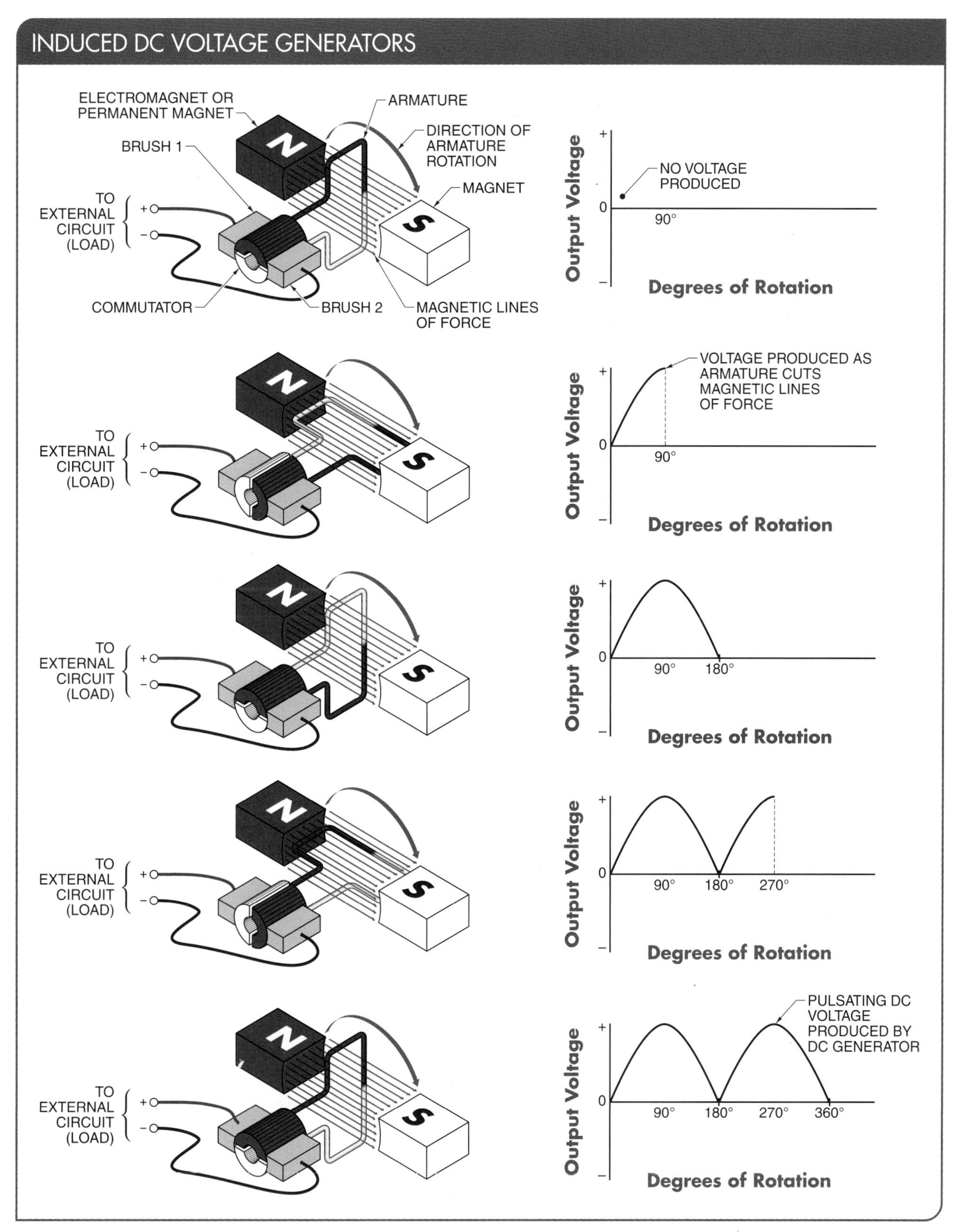

Figure 2-21. DC generators operate on the principle that when a coil of wire is rotated in a magnetic field, a voltage is induced in the coil.

Compound-Wound Generators. A *compound-wound generator* is a generator that includes series and shunt field windings. Compound-wound generators are used with larger DC power loads, such as for charging a battery bank, driving a DC motor, or supplying electrical energy for use in arc welding machines.

Briggs & Stratton Corporation

Some dwellings use 1ϕ generators as emergency back-up systems to provide power during a blackout.

In a compound-wound generator, electric current from the wire coil is transferred to the load through the metal slip rings. The voltage produced by a generator depends on the strength of the magnetic field and the rotational speed of the rotor. The stronger the magnetic lines of force and the faster the rotational speed, the higher the voltage that is produced. The output of a generator may be connected directly to the load (such as in a portable generator located on a construction site), connected to transformers, or connected to a rectifier (such as in an automobile alternator).

AC Generators

An *AC generator* is a machine that converts mechanical energy into AC electricity. Electricity is produced by converting potential energy directly or indirectly into electricity. For example, PV cells convert solar energy directly into electricity. The majority of most electricity, however, is produced indirectly by converting potential energy into electricity using a generator.

Single-Phase Generators. Magnetic lines of force are produced by the magnetic field of a permanent magnet or electromagnet. As the rotor of a generator moves through the magnetic field, electric current flow is produced through the wire coil(s) of the rotor. **See Figure 2-22.** An AC generator that has only one rotating coil produces a 1ϕ output. Single-phase generators are used for small power demands, but are not practical or economical for producing large amounts of power.

CODE CONNECT

Per NEC® Section 445.13, the ampacity of the conductors from the generator output terminals to the first distribution device(s) containing overcurrent protection shall not be less than 115% of the nameplate current rating of the generator.

Three-Phase Generators. To produce large amounts of power, three coils are coupled in a generator to produce 3ϕ power. **See Figure 2-23.** The three separate coils are spaced 120 electrical degrees apart. The individual AC voltage outputs are phase A, phase B, and phase C. Three-phase generators produce power more efficiently than single-phase generators and are typically used by utility companies.

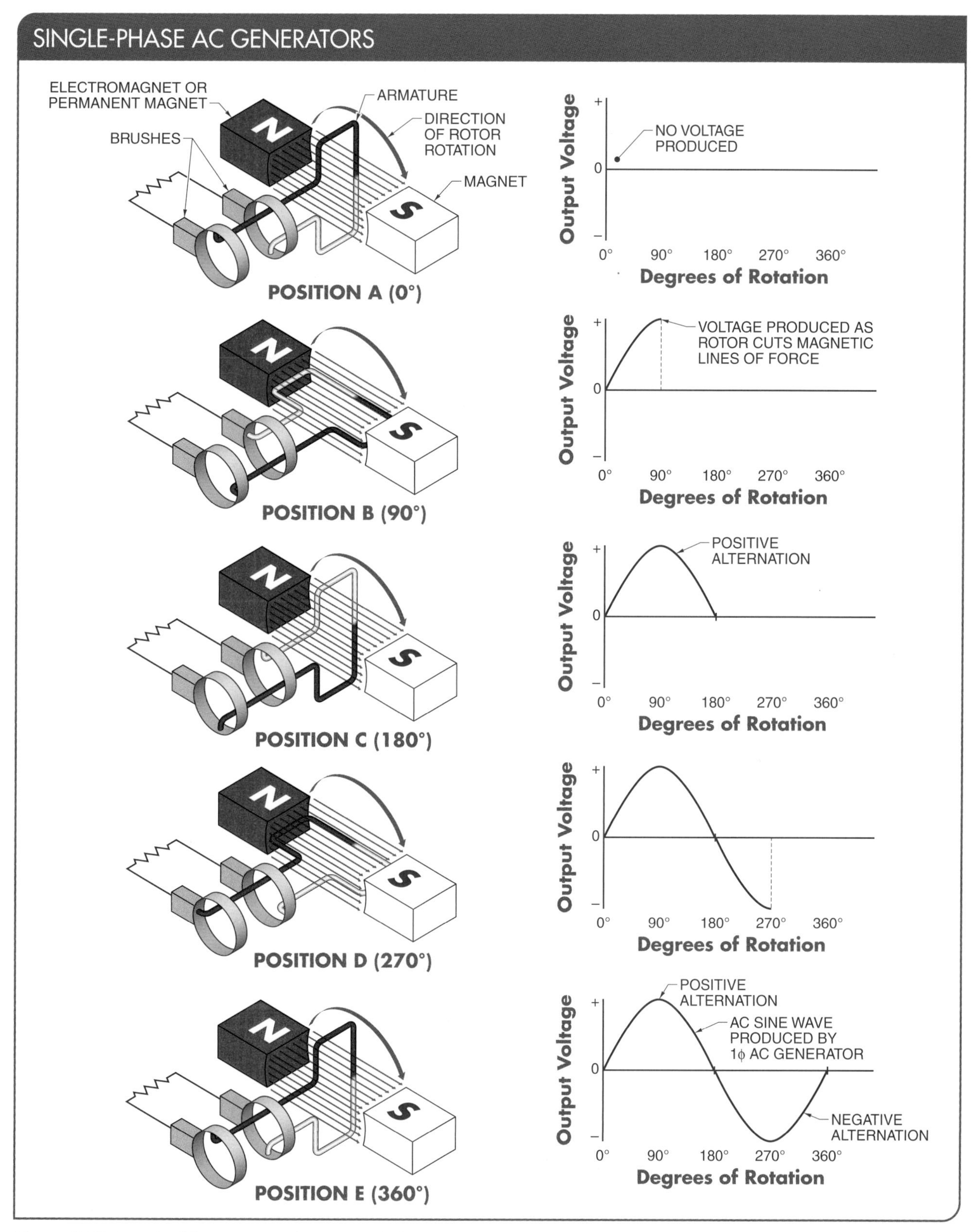

Figure 2-22. In a 1ϕ AC generator, the magnetic lines of force are produced by the magnetic field of a permanent magnet or an electromagnet. As the rotor moves through the magnetic field, electric current flow is produced through the wire coil(s) of the rotor.

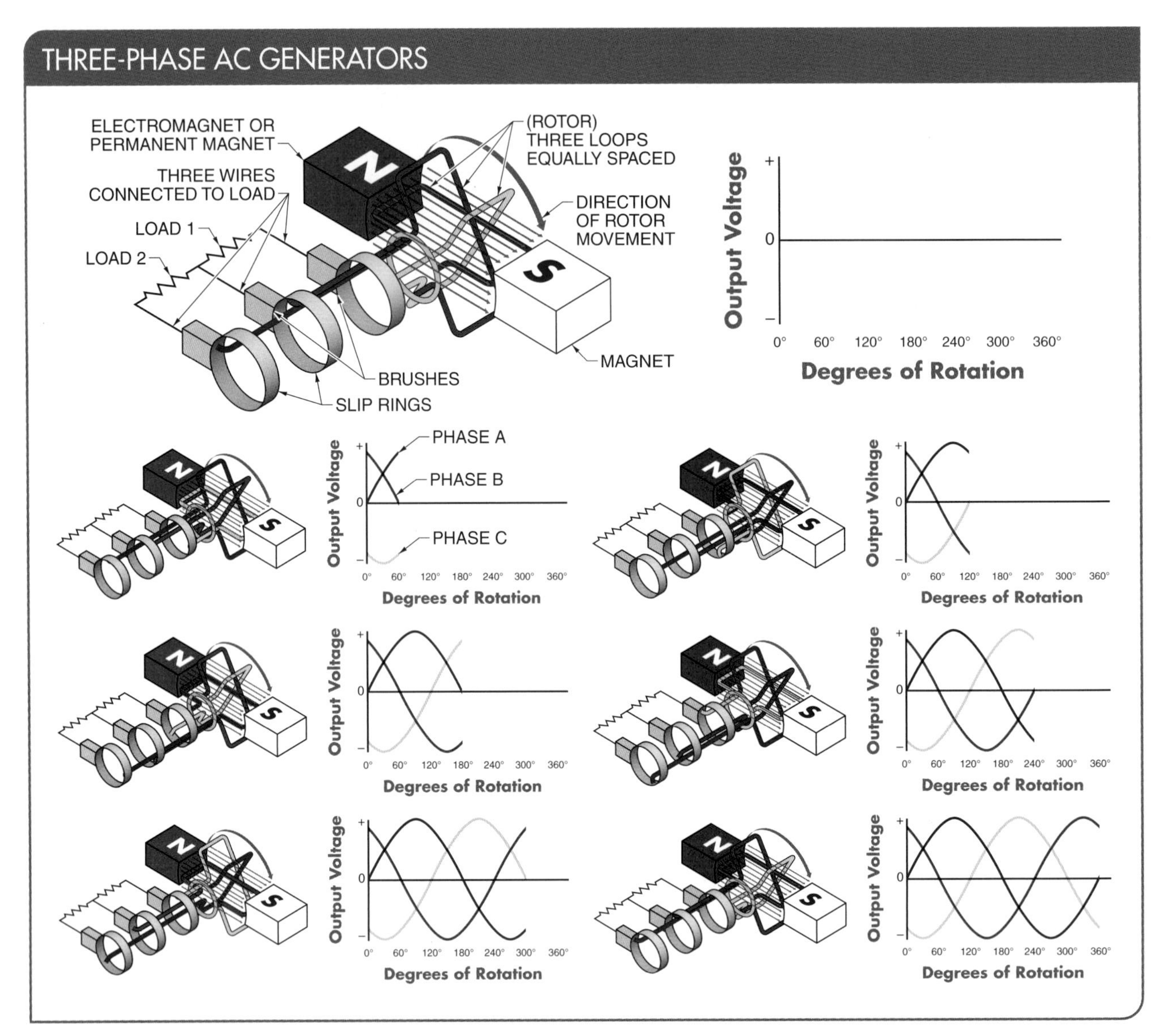

Figure 2-23. To produce large amounts of power, three coils are used to produce 3ϕ power. The three separate coils are spaced 120 electrical degrees apart.

CODE CONNECT

Per NEC® Section 445.20, receptacle outlets that are part of a 15 kW or smaller portable generator shall have listed ground-fault circuit interrupter (GFCI) protection for personnel integral to the generator or receptacle for either unbonded (floating neutral) generators or bonded neutral generators.

SECTION 2.3 CHECKPOINT

1. What is a generator?
2. What is electromagnetic induction?
3. What is a DC generator?
4. What happens when a coil of wire is rotated in a magnetic field?
5. What is an AC generator?

Chapter Activities

2 Electrical Quantities, Voltage Sources, and Generators

Name ______________________________ Date ______________

Common Electrical Terms

Identify the unit of measure and its abbreviation for the common electrical terms.

______________ **1.** Capacitance

______________ **2.** Resistance

______________ **3.** Power

______________ **4.** Voltage

______________ **5.** Period

______________ **6.** Frequency

______________ **7.** Inductance

______________ **8.** Current

______________ **9.** Power (apparent)

______________ **10.** Impedance

Ohm's Law

Calculate for voltage, current, or resistance using Ohm's law.

____________ **1.** A circuit has a current of 4 A and a resistance of 6 Ω. What is its voltage?

____________ **2.** A circuit has a voltage of 120 V and a resistance of 60 Ω. What is its amperage?

____________ **3.** A circuit has a voltage of 12 V and a current of 1 A. What is its resistance?

____________ **4.** A circuit has a current of 40 A and a resistance of 60 Ω. What is its voltage?

____________ **5.** A circuit has a voltage of 6 V and a resistance of 12 Ω. What is its amperage?

3 Service Entrances and Smart Meter Installation

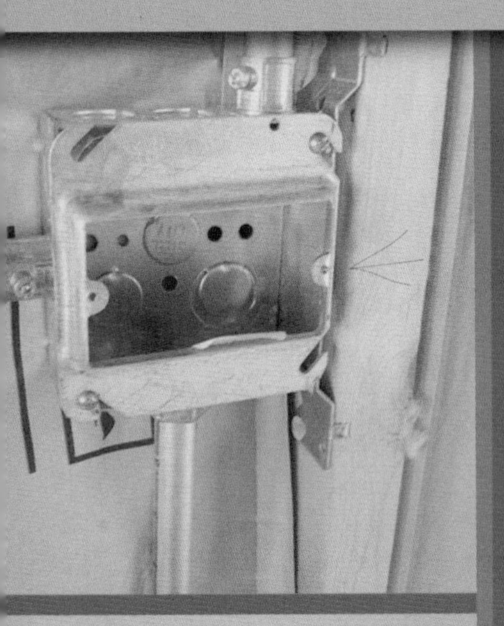

SECTION 3.1—RESIDENTIAL SERVICE ENTRANCES

- Define "service entrance" and "service point."
- Describe service drop entrances.
- Describe service lateral entrances.
- Explain the placement of service entrances.
- Describe how a service entrance is sized.

SECTION 3.2—ROUGHING-IN AND TRIMMING SERVICE ENTRANCES

- Describe service heads.
- Define "service entrance cables."
- Describe service panels.
- Explain the use of subpanels.
- Compare and contrast mobile home electrical service and residential electrical service.
- Explain service hookup and meter installation.

SECTION 3.3—DIGITAL SMART METERS

- Describe smart meters and how they work.
- Describe advanced metering infrastructure (AMI).
- Explain smart meter security.

SECTION 3.4—POWER QUALITY

- Describe power quality.
- Identify the three types of power interruptions.
- Differentiate between power outages, surges, and transient voltages.
- Describe whole-house protection plans that include panel protectors, surge-protector breakers, and power strip plug-in protection.
- Describe uninterruptible power supply (UPS) systems and standby UPSs.

SECTION 3.5—WHOLE-HOUSE GENERATORS

- Describe whole-house generators.
- Describe automatic transfer switches (ATSs).
- Identify generator hazards.
- Explain whole-house generator battery installation safety.

Learner Resources

ATPeResources.com/QuickLinks
Access Code: 791712

A utility provides electrical service to residential homes by connecting to the customer's service entrance. A homeowner should contact the utility and the local governing body before and after the installation of a service entrance. Smart meters allow the utility and the homeowner to remotely monitor power consumption. Remote monitoring can help the homeowner save money and enables improved service by automatically sending meter readings that result in quicker response time during power outages.

SECTION 3.1 — RESIDENTIAL SERVICE ENTRANCES

The two methods used to provide electrical service to the meter of a residence are wires that run overhead (service drop) or wires that are buried in the ground (service lateral). A *service entrance* is the wires installed between the meter fitting (socket) and the disconnecting means (main breaker, fuse box, or breaker panel) inside a dwelling. **See Figure 3-1.**

A *service point* is the connection between the electric utility and the premises wiring. The service point is where the utility's maintenance responsibility ends and the homeowner's responsibility begins. The location of the service point can change depending on the design and configuration of the distribution system. Prior to installation, the exact location of the service point should be determined. **See Figure 3-2.**

Service Drop Entrances

A *service drop entrance* is a service entrance that has wires running from a utility pole to a service head. The service head is connected to a riser pipe that is secured to the exterior of the dwelling and connected to the meter socket box. The meter socket box is connected to the service panel inside the dwelling. **See Figure 3-3.**

Service Lateral Entrances

A *service lateral entrance* is a service entrance where wires are buried underground. The four types of service laterals used for residences can originate from the following locations:

- from a utility hole in the street to the service panel disconnect
- from a utility sidewalk handhole to the service panel disconnect
- from a pole riser to the service panel disconnect
- from a transformer pad to the service panel disconnect

> CODE CONNECT
>
> *Per NEC® Section 230.41, service entrance conductors on the exterior of a building or entering a building shall be insulated.*

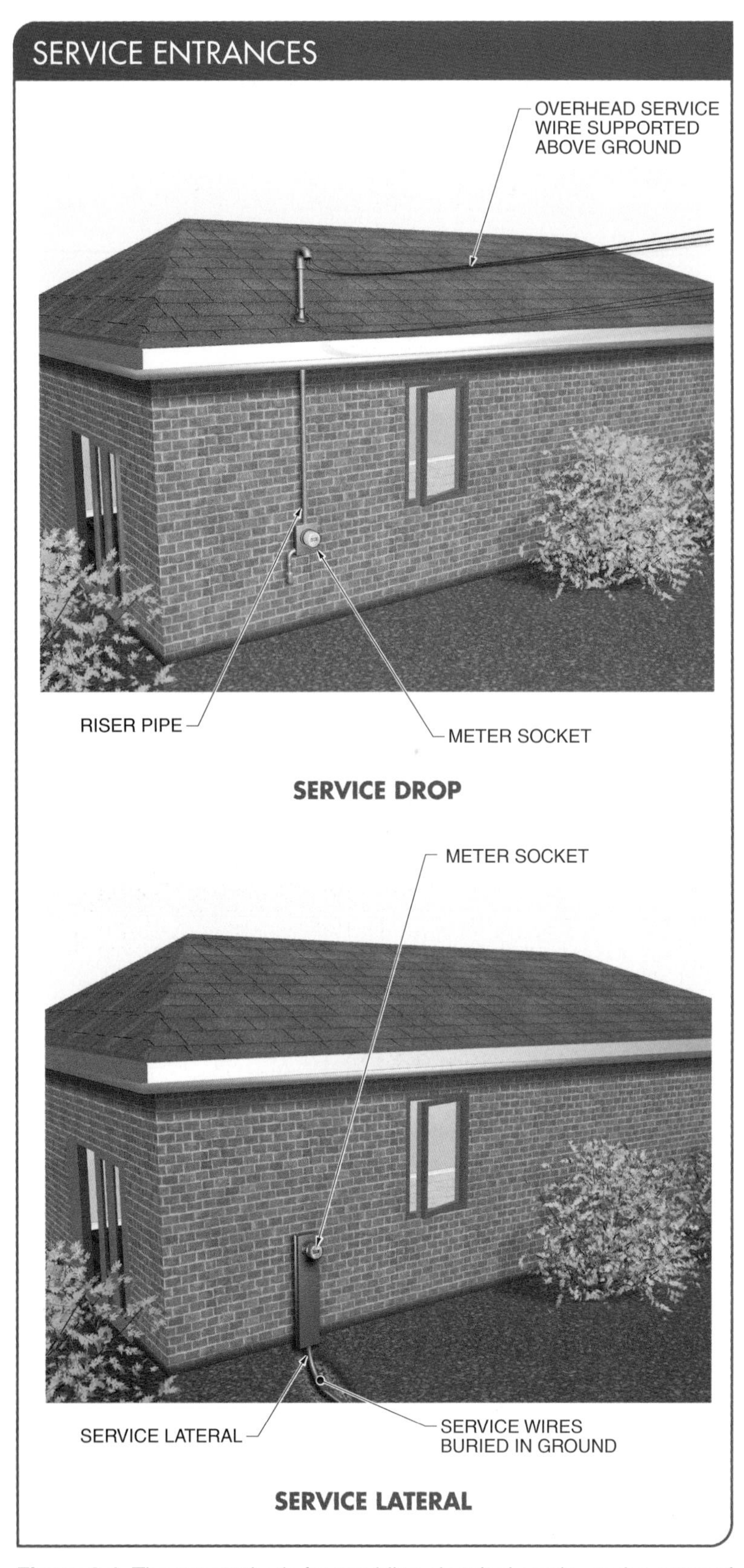

Figure 3-1. The two methods for providing electrical service to the meter of a residence are the service drop and service lateral.

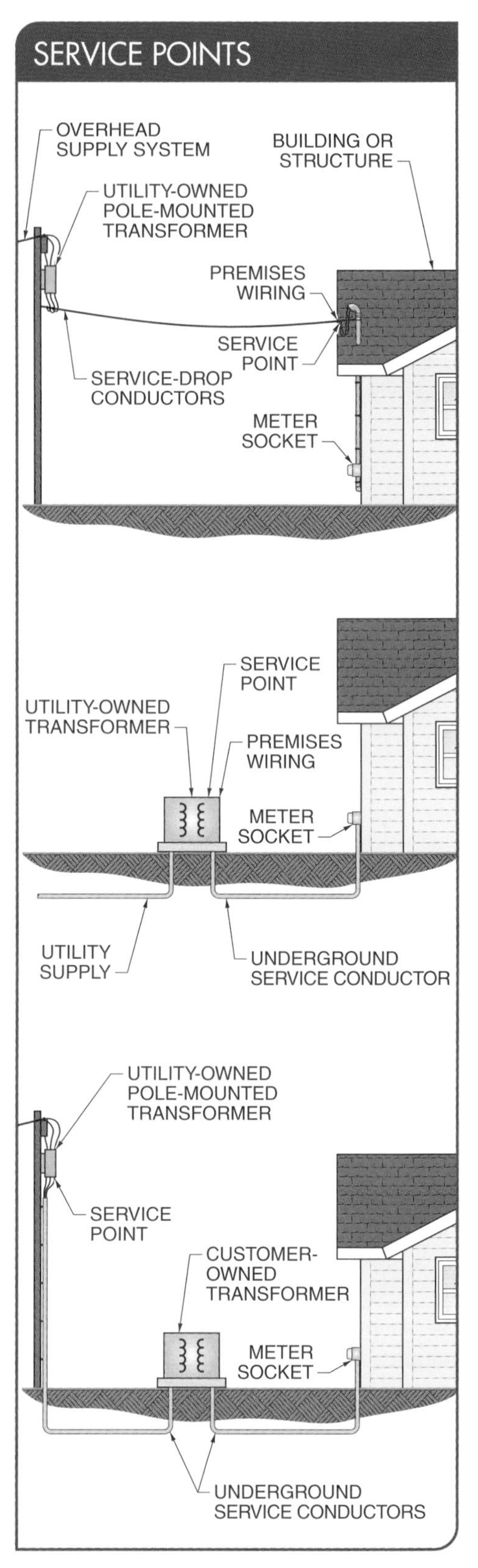

Figure 3-2. The service point is the point of connection between the electric utility and the premises wiring of the dwelling.

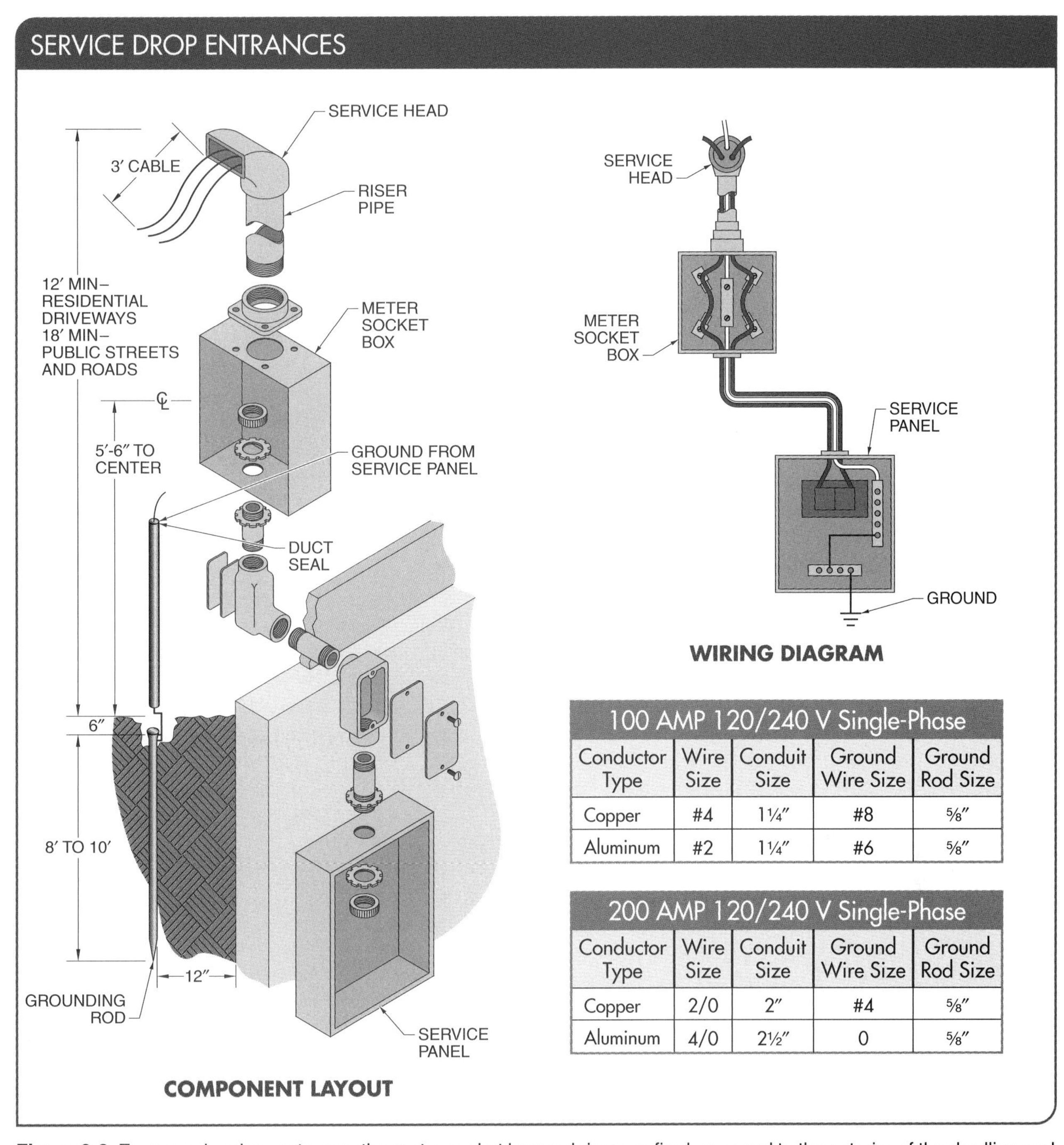

100 AMP 120/240 V Single-Phase				
Conductor Type	Wire Size	Conduit Size	Ground Wire Size	Ground Rod Size
Copper	#4	1¼″	#8	⅝″
Aluminum	#2	1¼″	#6	⅝″

200 AMP 120/240 V Single-Phase				
Conductor Type	Wire Size	Conduit Size	Ground Wire Size	Ground Rod Size
Copper	2/0	2″	#4	⅝″
Aluminum	4/0	2½″	0	⅝″

Figure 3-3. For a service drop entrance, the meter socket box and riser are firmly secured to the exterior of the dwelling and the service panel is mounted inside the dwelling.

When the service lateral originates in a utility hole in the street, the work is usually performed by the utility. The requirements for entering a utility hole are strict because it is a potentially dangerous space and should only be accessed by utility workers.

When the service lateral originates in a sidewalk handhole, electricians are typically permitted to run conduit and wires into the handhole. The utility makes the appropriate connections in the handhole. **See Figure 3-4.**

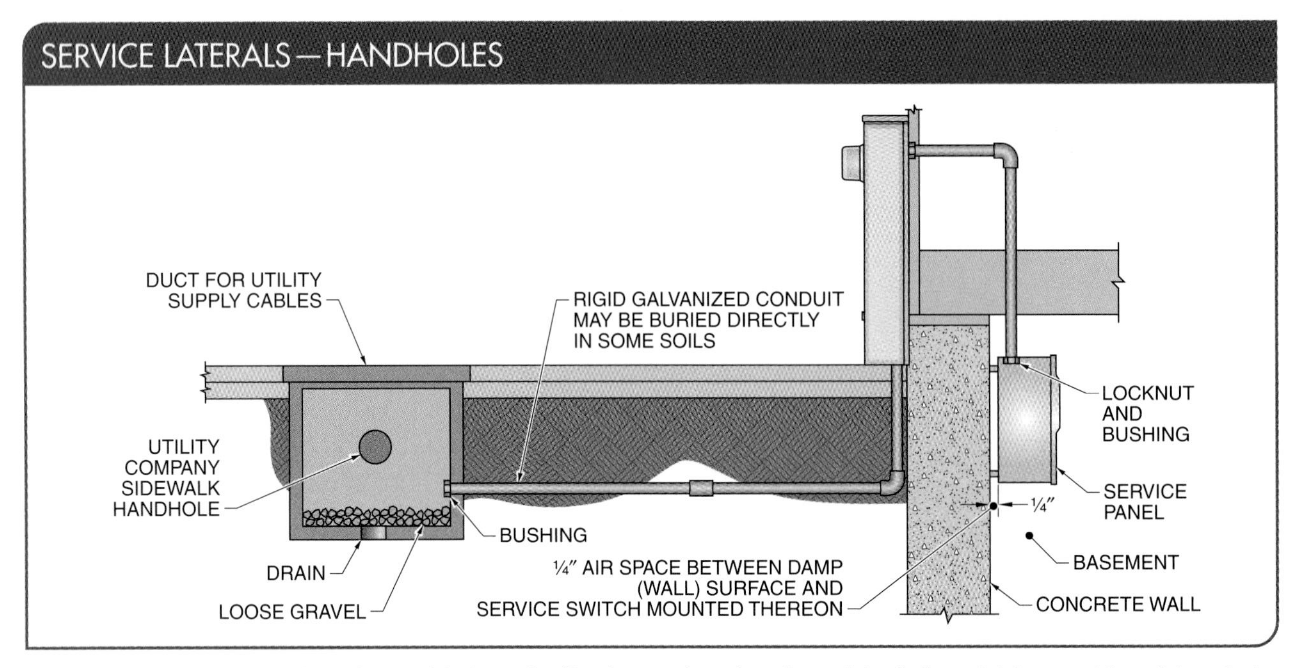

Figure 3-4. A service lateral can be provided to a dwelling by running wires through buried conduit from a sidewalk handhole.

When the service lateral originates from a pole riser, the service installation is shared by the homeowner and utility. In many instances, electricians install conduit to a point at least 8′ above grade and pull enough wire to reach the top of the pole. The utility extends the protection for the drop cables with approved molding to the final connection. In some cases, a short piece of conduit with a service head is used to terminate the run. When the utility extends protection, the homeowner furnishes the conduit and service head. In other cases, homeowners must furnish conduit all the way with a service head at the end. **See Figure 3-5.**

When a service lateral entrance originates from a transformer pad, special direct-burial cable is used to make the installation. Service lateral from a pad-mounted transformer may require the homeowner to furnish a transformer pad in addition to the regular electrical service equipment. The utility typically creates the trench from the transformer pad to the dwelling, buries the cable, and makes all final connections. **See Figure 3-6.**

Service Lateral Entrances Using Cabinets. A service lateral entrance may use a cabinet to enclose the service wires and to provide a location for the meter socket and meter. The entire enclosure is then covered with a metal cover that is typically locked. A cabinet should be neat and is used to enable the easy routing of wires. **See Figure 3-7.**

A service lateral entrance may use a cabinet to enclose the service wires and meter socket.

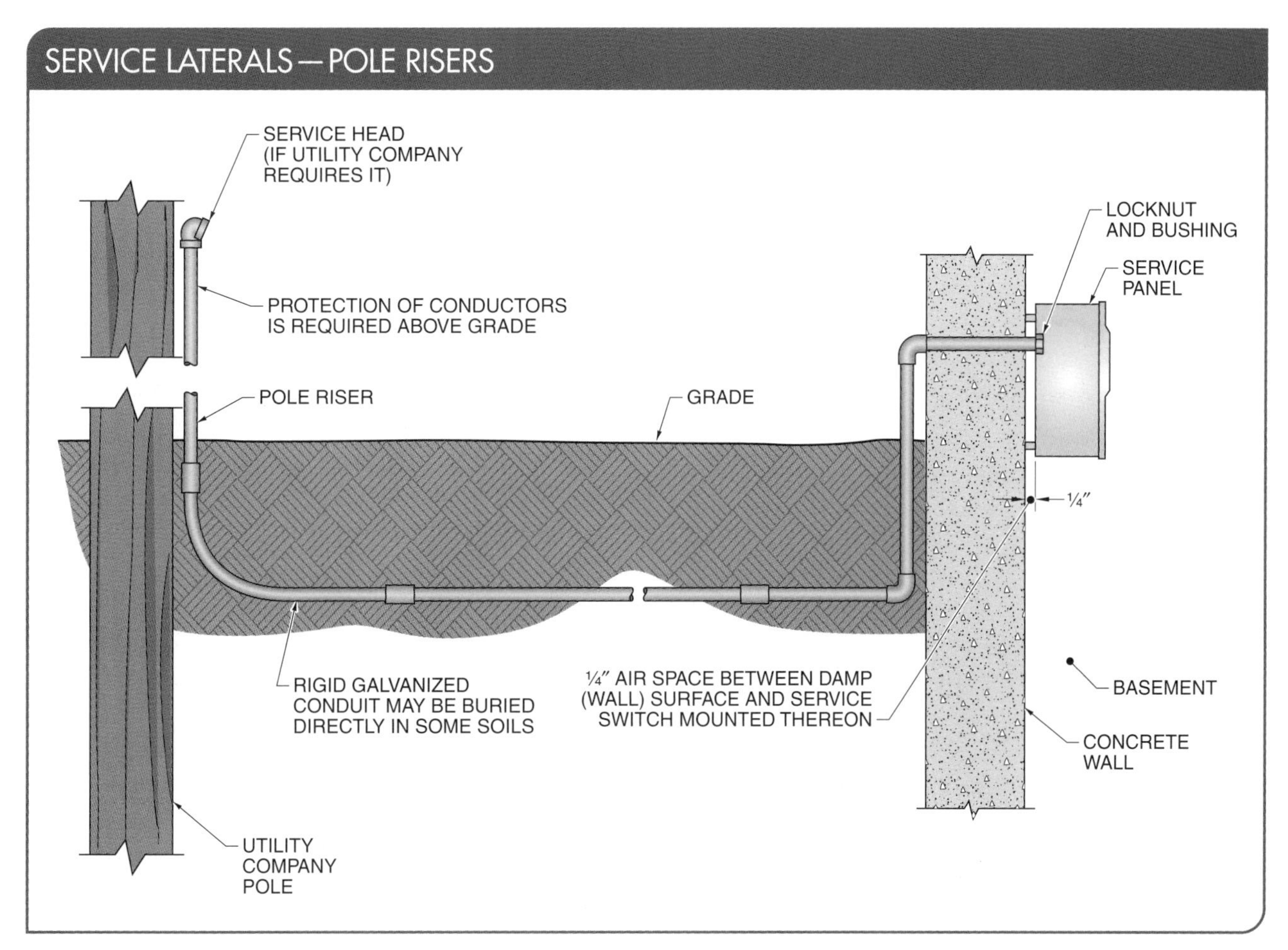

Figure 3-5. A service lateral can originate from a pole riser.

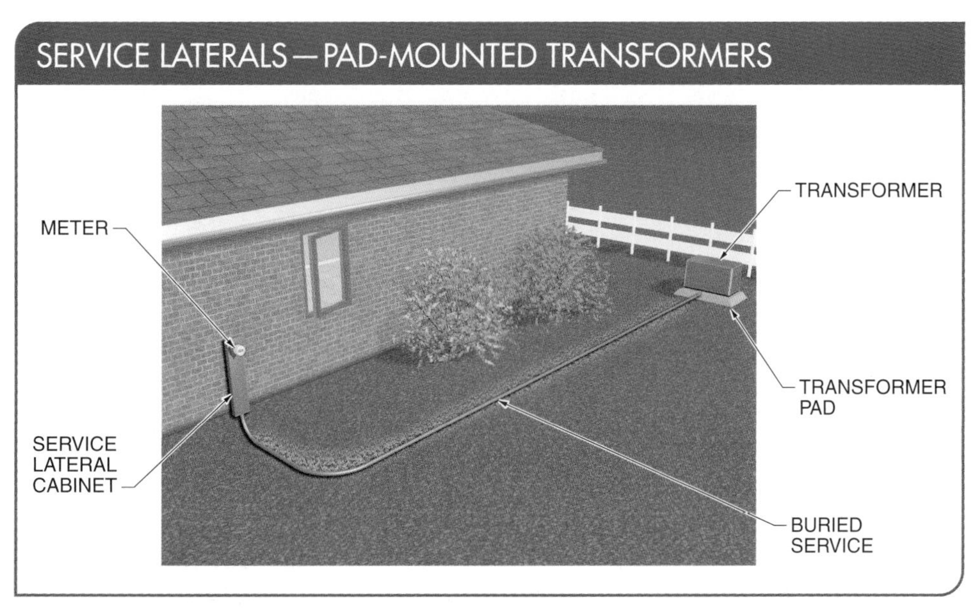

Figure 3-6. A service lateral can originate from a pad-mounted transformer.

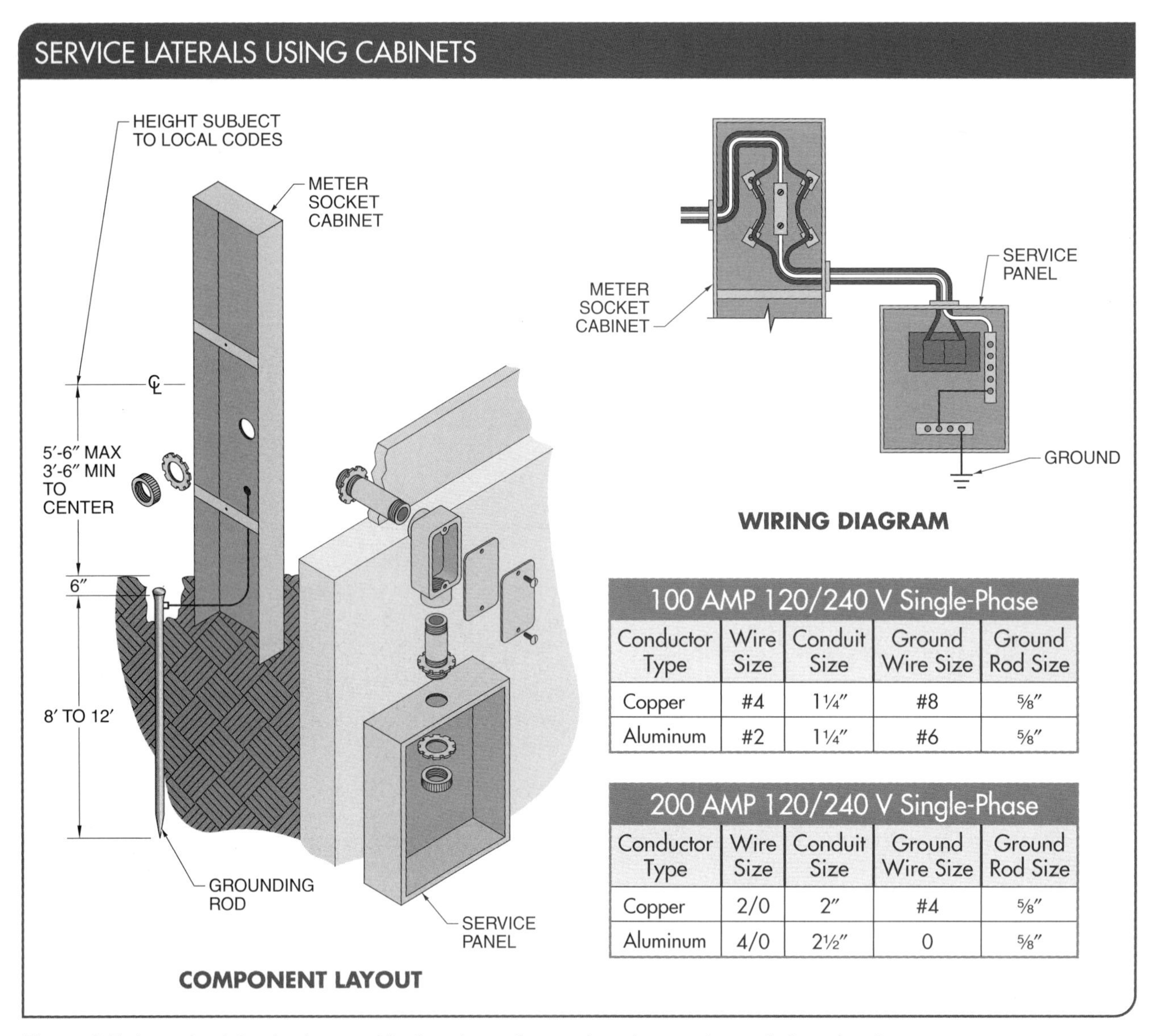

100 AMP 120/240 V Single-Phase				
Conductor Type	Wire Size	Conduit Size	Ground Wire Size	Ground Rod Size
Copper	#4	1¼″	#8	⅝″
Aluminum	#2	1¼″	#6	⅝″

200 AMP 120/240 V Single-Phase				
Conductor Type	Wire Size	Conduit Size	Ground Wire Size	Ground Rod Size
Copper	2/0	2″	#4	⅝″
Aluminum	4/0	2½″	0	⅝″

Figure 3-7. A service lateral using a cabinet encloses the service wires, meter socket, and meter.

Service Lateral Entrances Using Conduit. A service lateral entrance may use conduit that includes all service wires and is buried underground up to the meter socket box. Service lateral entrances in conduit are protected from weather and subjected to less damage than buried cables. **See Figure 3-8.**

Placement of Service Entrances

Cost, convenience, and safety must be taken into consideration when a service entrance is placed. Installers must ensure that the service entrance to a dwelling is as accessible as possible using the minimum of materials and incorporating all safety rules and regulations. The following are three useful guidelines for determining the placement of a service entrance:

- Keep the service wires as short as possible.
- Locate the service panel as close to the kitchen as possible to avoid costly wire runs to major appliances.
- Avoid placing the meter socket on the outside of a bedroom wall.

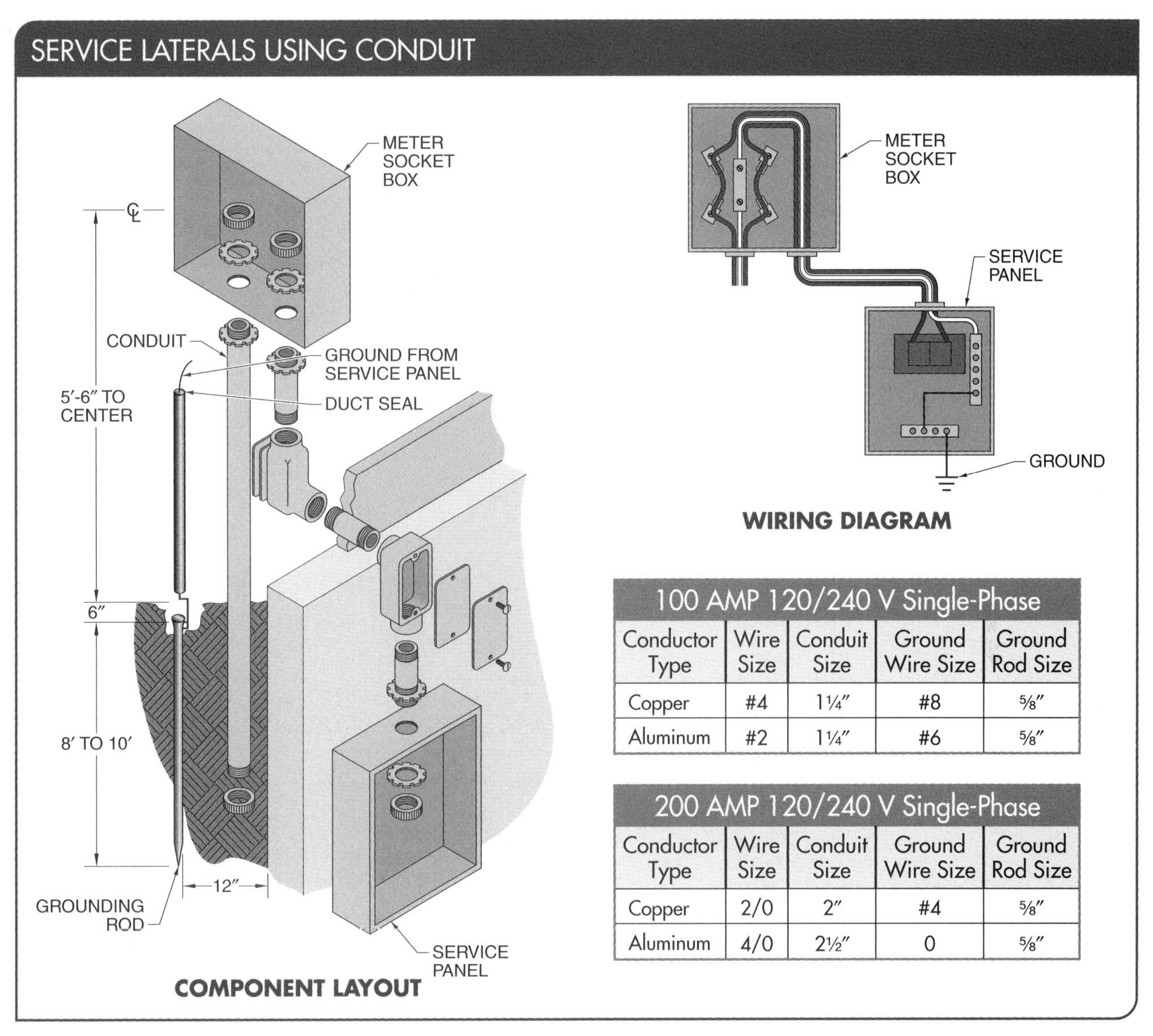

100 AMP 120/240 V Single-Phase				
Conductor Type	Wire Size	Conduit Size	Ground Wire Size	Ground Rod Size
Copper	#4	1¼″	#8	⅝″
Aluminum	#2	1¼″	#6	⅝″

200 AMP 120/240 V Single-Phase				
Conductor Type	Wire Size	Conduit Size	Ground Wire Size	Ground Rod Size
Copper	2/0	2″	#4	⅝″
Aluminum	4/0	2½″	0	⅝″

Figure 3-8. A service lateral using conduit has all wiring buried underground in conduit.

The homeowner should consult with the utility concerning future home additions, such as a pool, fence, deck, or shed, before the trench is dug for a service lateral. Any home additions after the underground service is installed requires that the service lateral be located and marked with flags or paint to avoid digging into it.

Sizing Service Entrances

A service entrance for a single-family dwelling requires a minimum 100 A capability. For larger residences and structures using electric heat, a 200 A service entrance may be required. Residential service entrances are typically rated in amperes and the voltage is typically at a standard 120 V/240 V. Further information concerning service entrances can be obtained from Article 230 of the NEC® or the local building inspector.

> TECH TIP
>
> *New residential developments typically have the water lines, sewer, and easements installed before underground electric service is installed.*

SECTION 3.1 CHECKPOINT

1. What are the two methods to provide electric service?
2. What is a service entrance?
3. What is a service point?
4. What is a service drop entrance?
5. What is a service lateral entrance?
6. What are three useful guidelines for determining the placement of a service entrance?
7. Describe how a service entrance is sized for single-family dwellings and for larger residences and structures using electric heat.

SECTION 3.2—ROUGHING-IN AND TRIMMING SERVICE ENTRANCES

Licensed electricians install the service entrance prior to the utility making the final connection. The NEC®, local codes, and the electric utility set the standards for how service entrances are to be installed. To make roughing-in a service entrance installation easier, manufacturers provide service entrance installation kits that include service heads, service entrance caps, sill plates, oval service cable connectors, insulator brackets, conduit supports, roof flashing, and pipe straps. **See Figure 3-9.**

Trimming out or finishing a service entrance requires the installation of internal service entrance cables, service panels, and a service disconnect. Wires must be bent to fit some service heads. Similar bends are required when the meter socket and service panel are wired.

Service entrance caps and associated hardware are shaped to accept oval service cables. The sill plate covers the cable and the cable opening where the service enters the dwelling. Sealing compound or caulking is applied to make the sill plate watertight. Service entrance cable connectors are available as standard box connectors or as watertight connectors when used on top of the meter socket. Mounting straps are used to secure service cables in place.

A service drop entrance terminates at the service head.

Service Heads

A *service head* (weatherhead) is a weatherproof service point where overhead service drop conductors enter the service entrance riser pipe going to the meter socket. **See Figure 3-10.** The NEC® includes standards relating to the use of service heads. Careful consideration of each NEC® standard during installation ensures a permanent weatherproof connection.

To comply with NEC® standards, a service head must be at least 10′ above and 3′ to the side of any platform. Service wires must be at least 3′ from the bottom and sides of a window. Conductors that run above a window are considered out of reach from the window. NEC® standards governing the elevation of service wires specify a 3′ clearance over adjacent buildings. Various elevations are also required for sidewalks, private drives, alleys, and streets. **See Figure 3-11.**

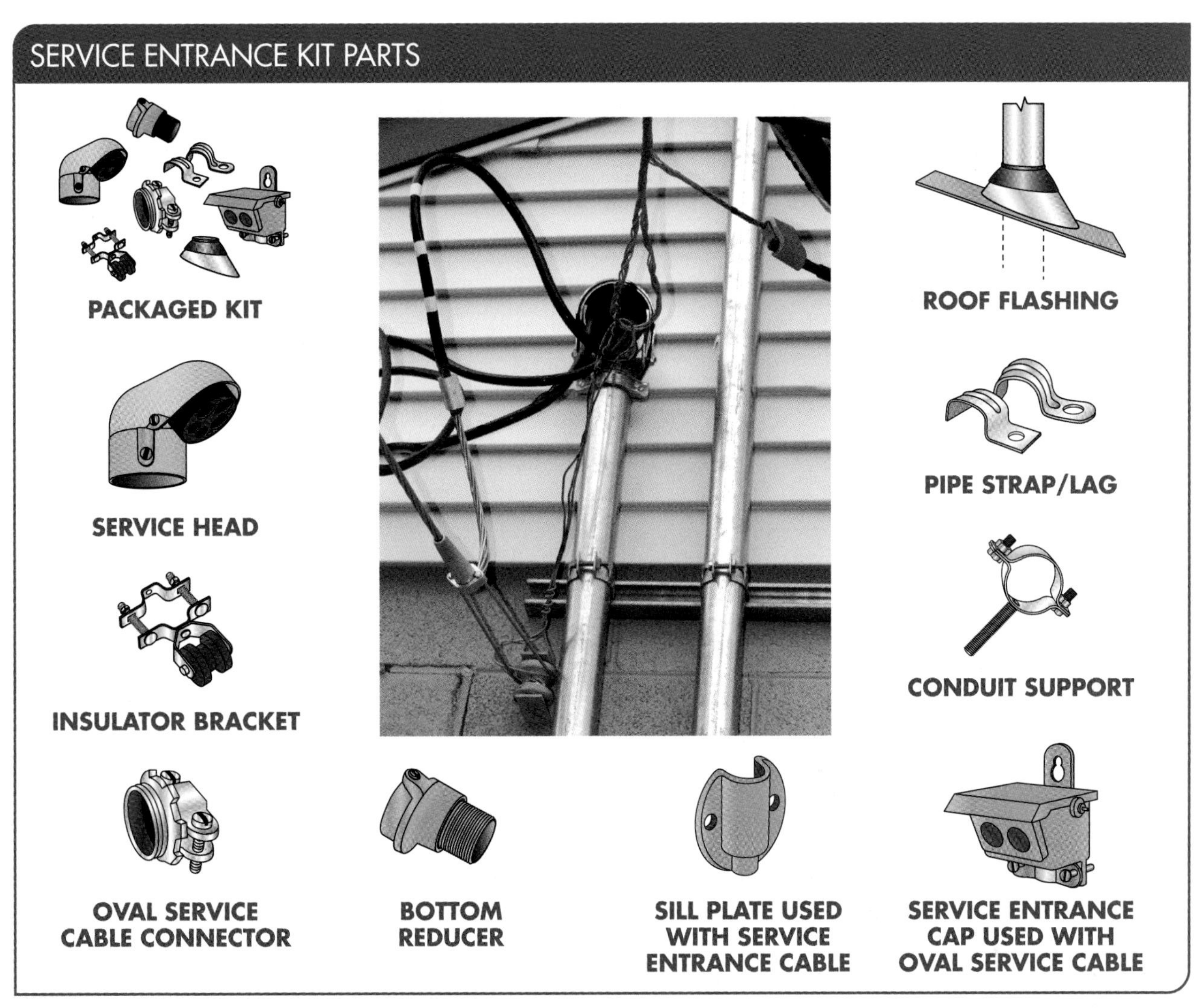

Figure 3-9. To make service entrance installation to a residence easier, manufacturers sell service entrance installation kits.

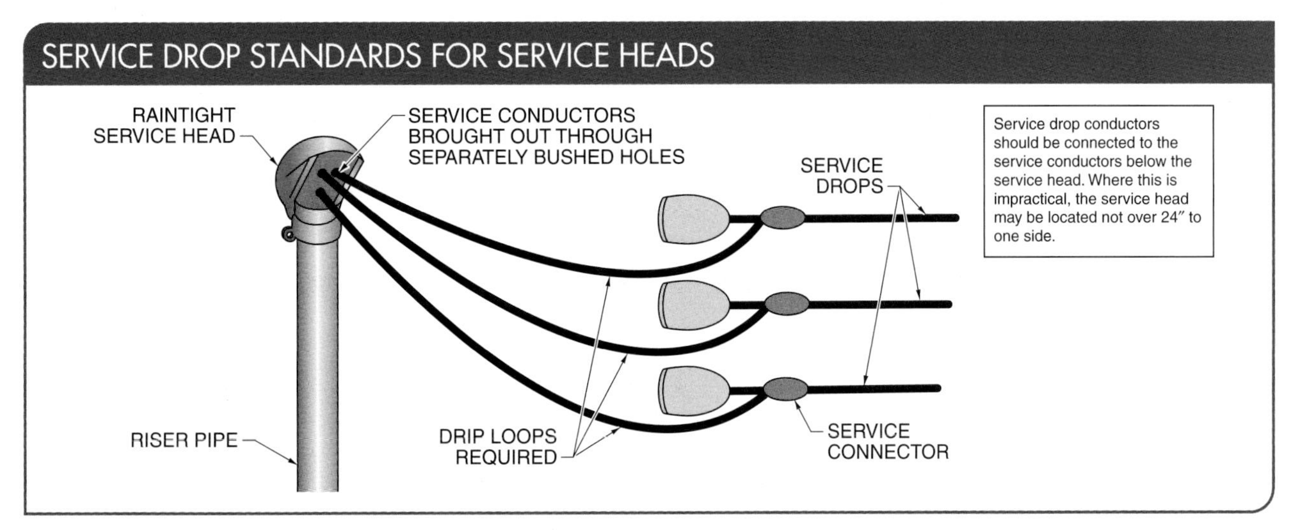

Figure 3-10. Service drop conductors are typically connected to the service conductors of a residence by means of a service head.

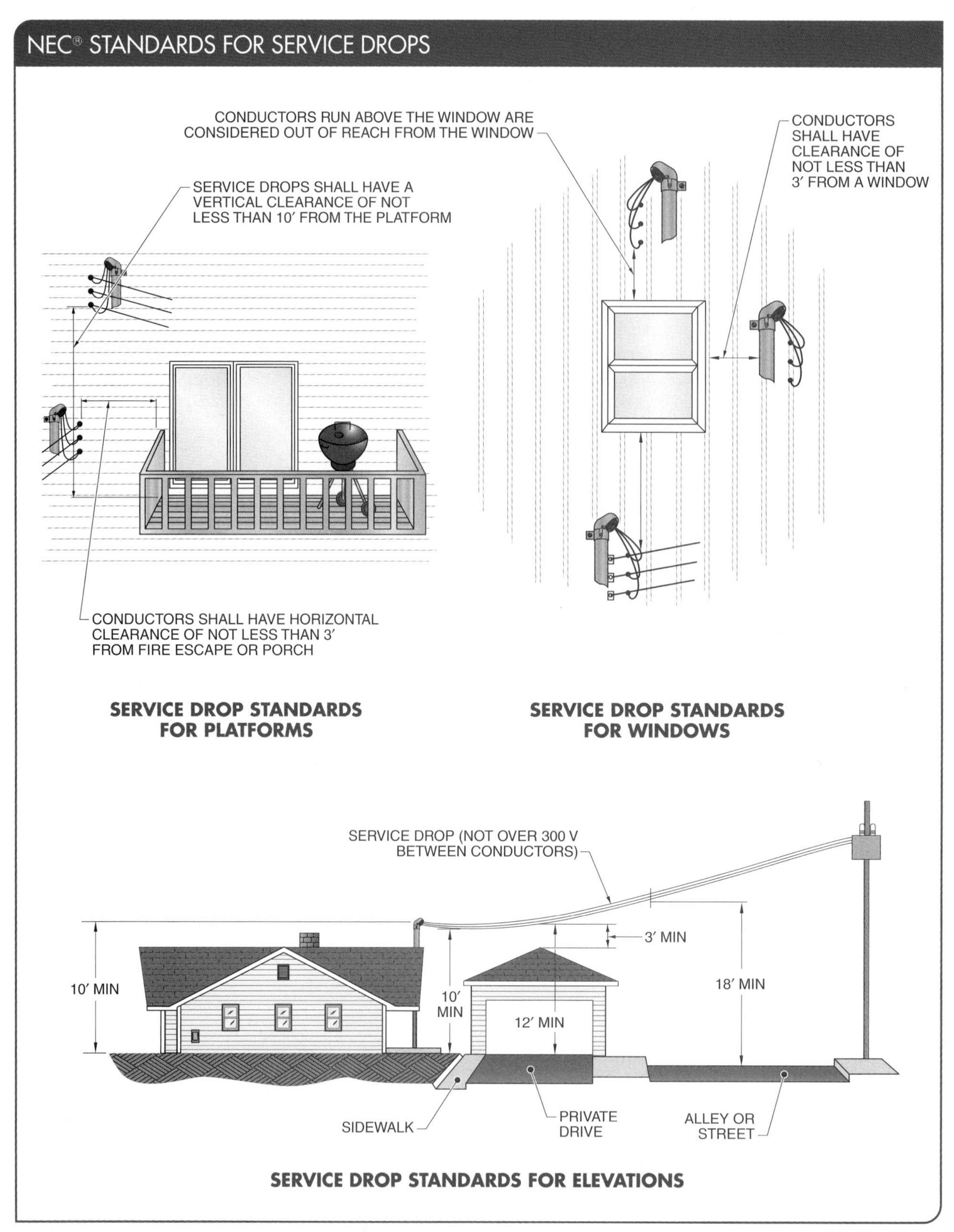

Figure 3-11. To comply with NEC® standards, a service head must be at least 10′ above and 3′ to the side of any platform, at least 3′ from the bottom and sides of a window, and at various elevations for sidewalks, private drives, alleys, and streets.

Service Entrance Cables

Service entrance cables are often used instead of individual wires pulled through conduit. A *service entrance cable* is a cable that has a bare conductor wound around the insulated conductors and is often used for service drops. The spiral winding prevents tampering with the conductors ahead of the meter and serves as the neutral conductor. **See Figure 3-12.**

Service Panels

A *service panel* is an electrical box designed to contain fuses or circuit breakers for protecting the individual circuits of a dwelling from the distribution system. Service panels are mounted on the wall in basements, utility rooms, or attached garages. Wherever a service panel is mounted, the panel must be easily accessible to the occupants of the dwelling.

The electric power brought to a dwelling is typically a three-wire 120 V/240 V, 1ϕ system. Red, black, and white service panel wires enter the dwelling through a service head. The red and black wires are used for the hot or live wires, and the white wire is used as the neutral wire. Two black wires can also be used for the hot wires. As with most residential services, 240 V can be obtained between the hot wires and 120 V can be obtained between either of the hot wires and the neutral wire. The 120 V and 240 V voltages are constant throughout the meter socket and into the service panel.

Inside the service panel, where the service wires are terminated, metal conductors called busbars are used to distribute the power to fuses and circuit breakers in an organized manner. Service panel busbars convey power to fuses and single-pole and double-pole circuit breakers to form 120 V and 240 V circuits.

A single-pole circuit breaker only furnishes power to one 120 V circuit. A double-pole circuit breaker is used for a 240 V circuit. Manufacturers make 240 V circuit breakers with handles that are tied together so that a short on any one phase will open both phases. **See Figure 3-13.** Circuit breakers are snapped into position when installed into the busbars of a service panel.

After all the circuit breakers have been properly installed and each circuit tested, electricians must prepare the trim cover to the service panel. The proper technique must be used to remove the circuit breaker knockouts from the service panel trim cover. Only the knockouts necessary for the installed circuit breakers should be removed. **See Figure 3-14.**

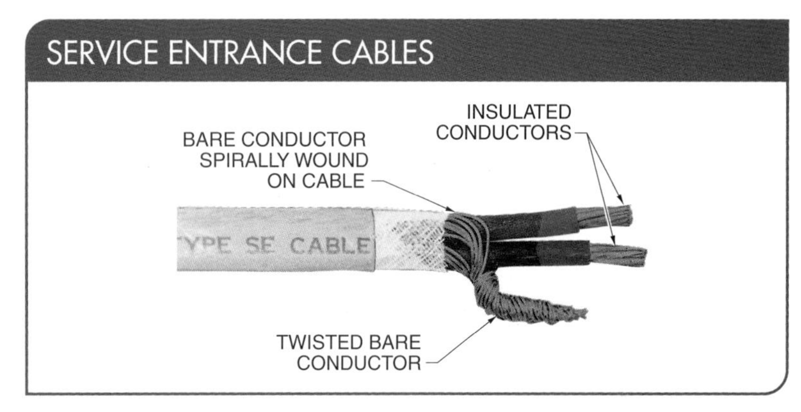

Figure 3-12. A twisted bare conductor in a service entrance cable wraps around and protects the current-carrying wires, serves as the neutral wire, and is installed using various hardware.

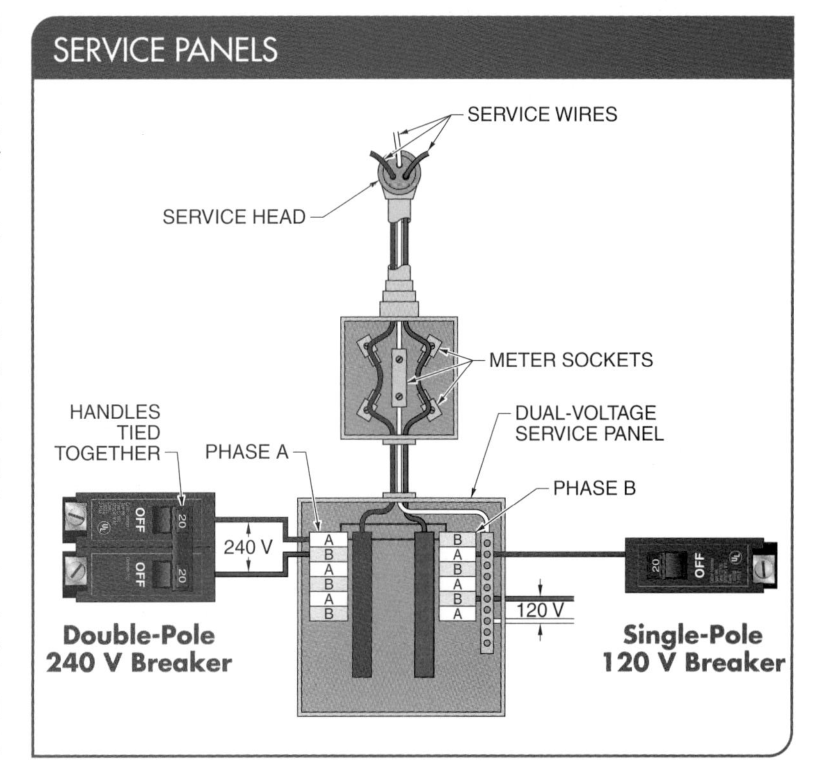

Figure 3-13. A standard 120 V/240 V system consists of two incoming hot (red and black) wires and a neutral (white) wire. Busbars alternatively supply circuit breakers with phase A and phase B power.

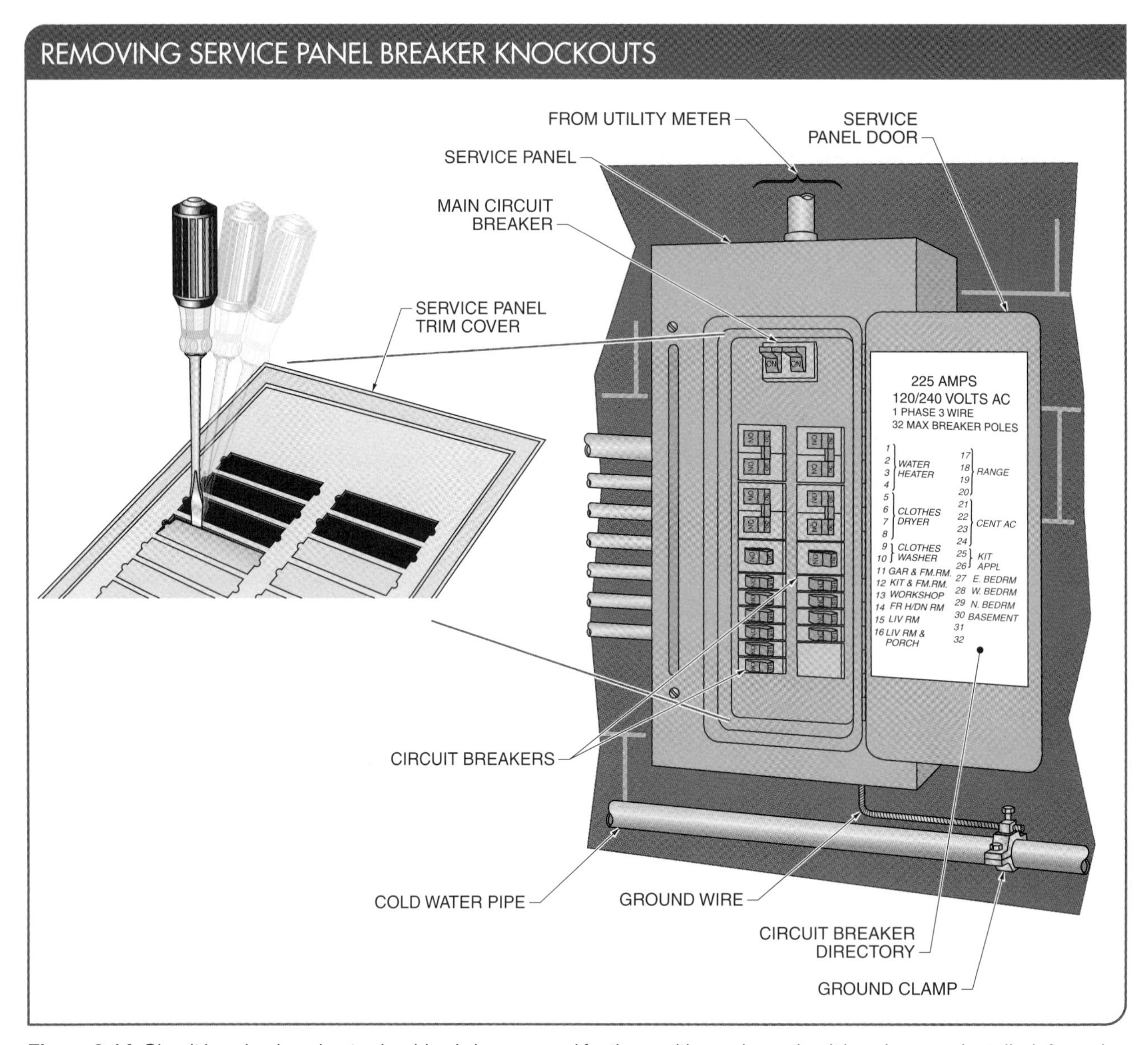

Figure 3-14. Circuit breaker knockouts should only be removed for the positions where circuit breakers are installed. A service panel is finished when the cover is installed, and each breaker is labeled according to the location and use of each circuit.

The service panel cover is installed to finish trimming out a service panel. The cover panel has a directory on the inside of the door. The directory can be filled out to indicate the location and use of each circuit.

Securing Service Equipment and Service Panels

The utility service wires are not connected by the utility until the service panel inside the residence is properly secured. Meter sockets, service panels, and connecting conduit must all be firmly secured with mechanical fasteners before any wires or cables are pulled into place.

> TECH TIP
>
> *Service panels should be easily accessible, must not be mounted in small closets or near obstructions, and must have a minimum clearance of 36″.*

Whenever possible, large knockouts should be removed before the service panel is mounted. Removing the knockouts before mounting reduces any chance of loosening the panel. When service equipment is mounted to masonry, drills and special bolts with anchors are used. Plastic anchors support a considerable amount of weight when properly installed. When service equipment is mounted on wood, wood screws or lag bolts are used to hold the equipment in place. **See Figure 3-15.**

Subpanels

A *subpanel* is a panel supplied by the main service panel that distributes circuits to a specific area of the dwelling or outbuilding. The main service panel must have a circuit breaker large enough to meet the demands of the subpanel (30 A, 50 A, or 60 A). The service panel circuit breaker protects the subpanel from overloading. The specifications from the subpanel manufacturer must be checked to determine the size of the circuit breaker that must be used in the service panel.

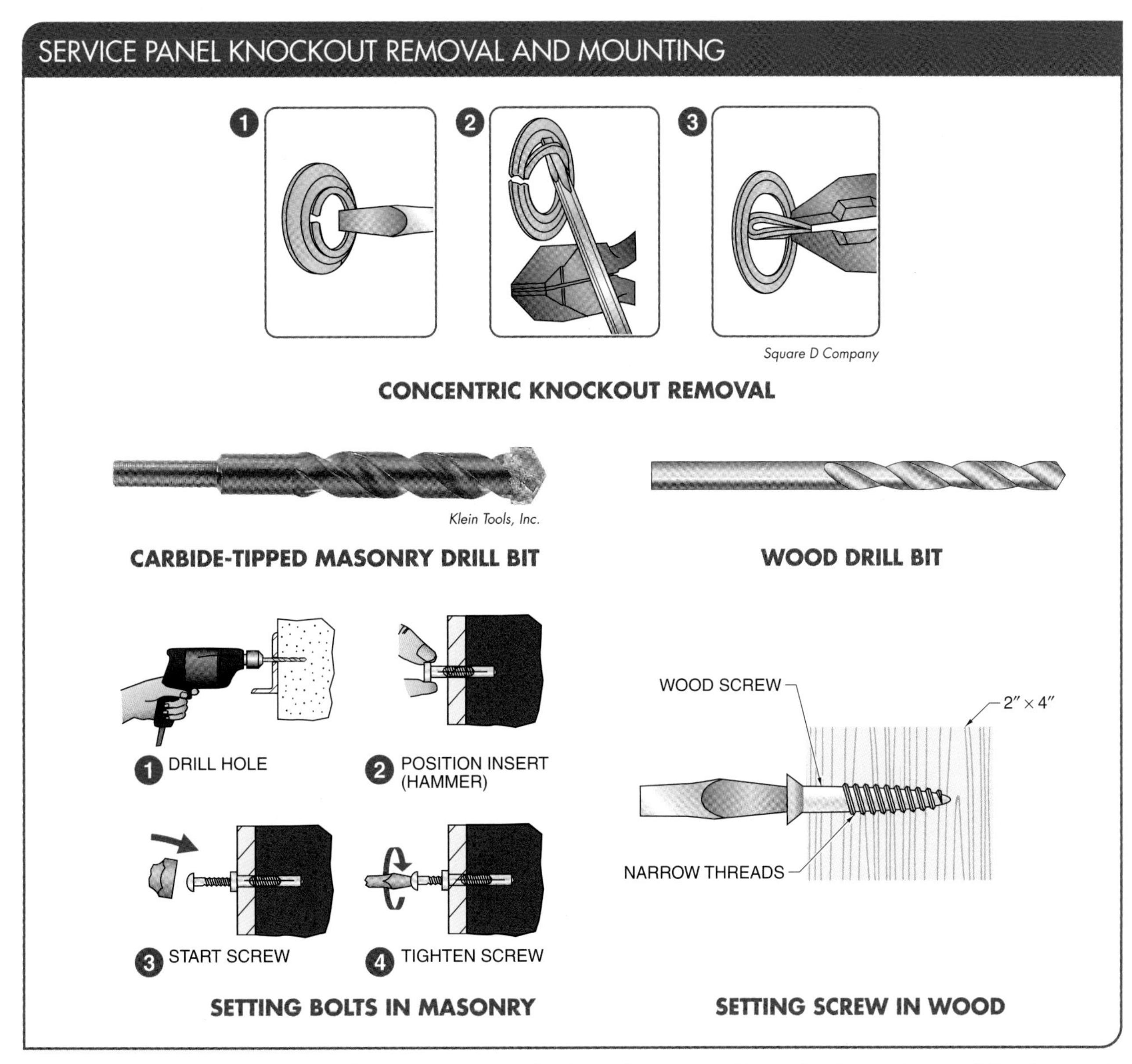

Figure 3-15. Removing knockouts before mounting a panel reduces the chance of loosening the service panel. Service panels are mounted to masonry, using drills and special bolts with anchors, and wood surfaces, using woood screws.

Subpanels may be in another part of the dwelling or may extend out to a separate building such as a detached garage or workshop. **See Figure 3-16.** In many cases, the subpanel is connected by a direct-burial cable that protects the underground cables and conductors from damage. The service conductors not under the exclusive control of the utility must be buried 18″ below ground and must have their location identified by a warning ribbon placed in the trench above the underground cable installation.

Mobile Home Electrical Service

Mobile home service entrances are not typically attached directly to the mobile home. Instead, the service is permanently mounted near the mobile home and a cable is run to the mobile home. The overhead and service lateral methods are used to connect electrical service to mobile homes. **See Figure 3-17.**

A conduit from the meter runs to a disconnect box, and then a conduit or buried cable from the disconnect box runs to the service panel of the mobile home. The service panel

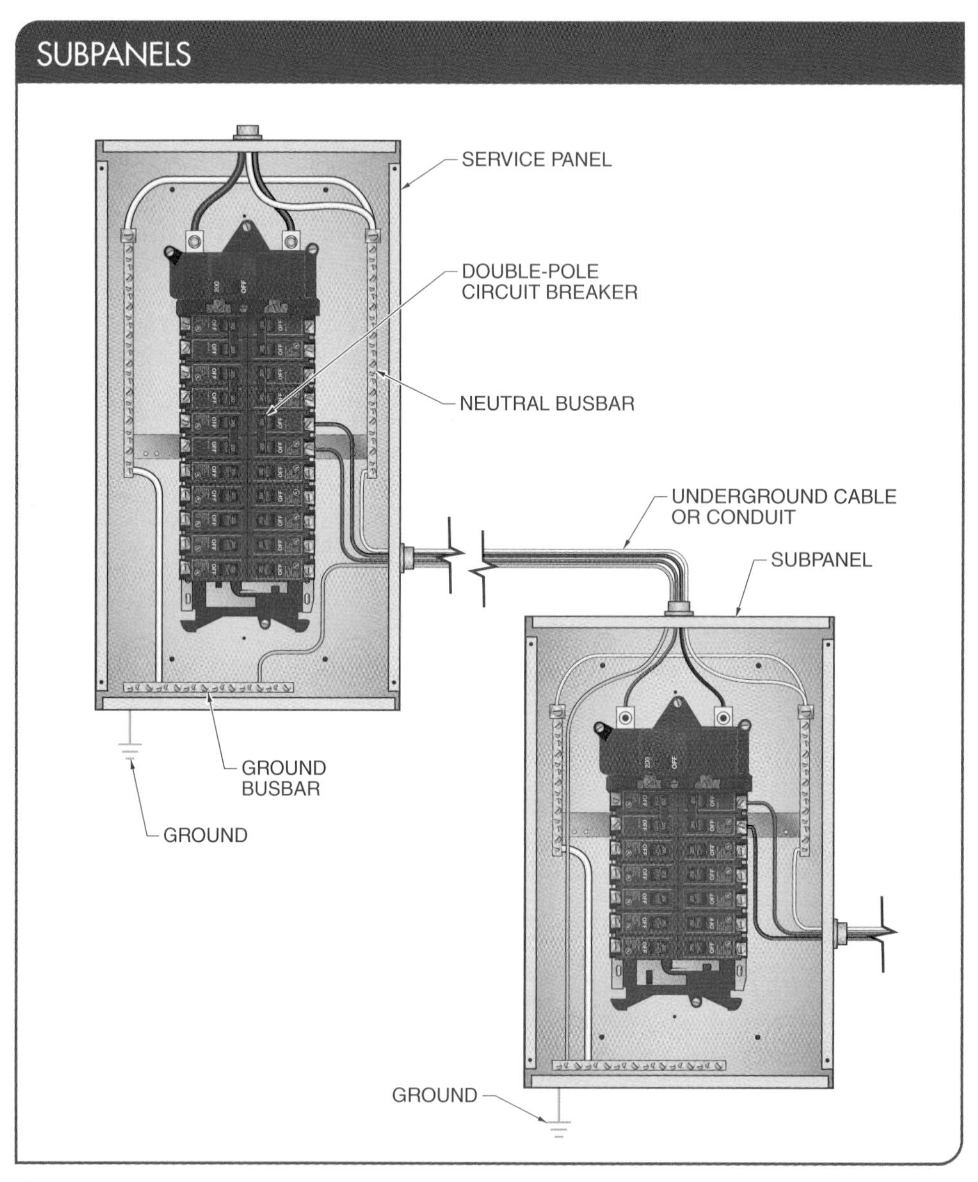

Figure 3-16. Subpanels distribute circuits to a specific area of the dwelling such as a room addition or outbuilding. The neutral is never connected to ground in a subpanel.

distributes electricity throughout the mobile home much like other residential distribution panels. Since the cable is run outside from the disconnect box, GFCI receptacles may be installed for additional safety.

CODE CONNECT

Per NEC® Subsection 230.42(2), the minimum service entrance conductor ampacity shall not be less than the maximum load to be served.

Service Connection and Meter Installation

After the service entrance has been completed and inspected, the utility company makes the final connection and installs a meter in the meter socket. Almost all new construction will have a smart meter incorporated into the new electrical system for billing the consumer and monitoring grid performance. To prevent an arc flash, all circuit breakers are turned off so there is no load when service wires are connected.

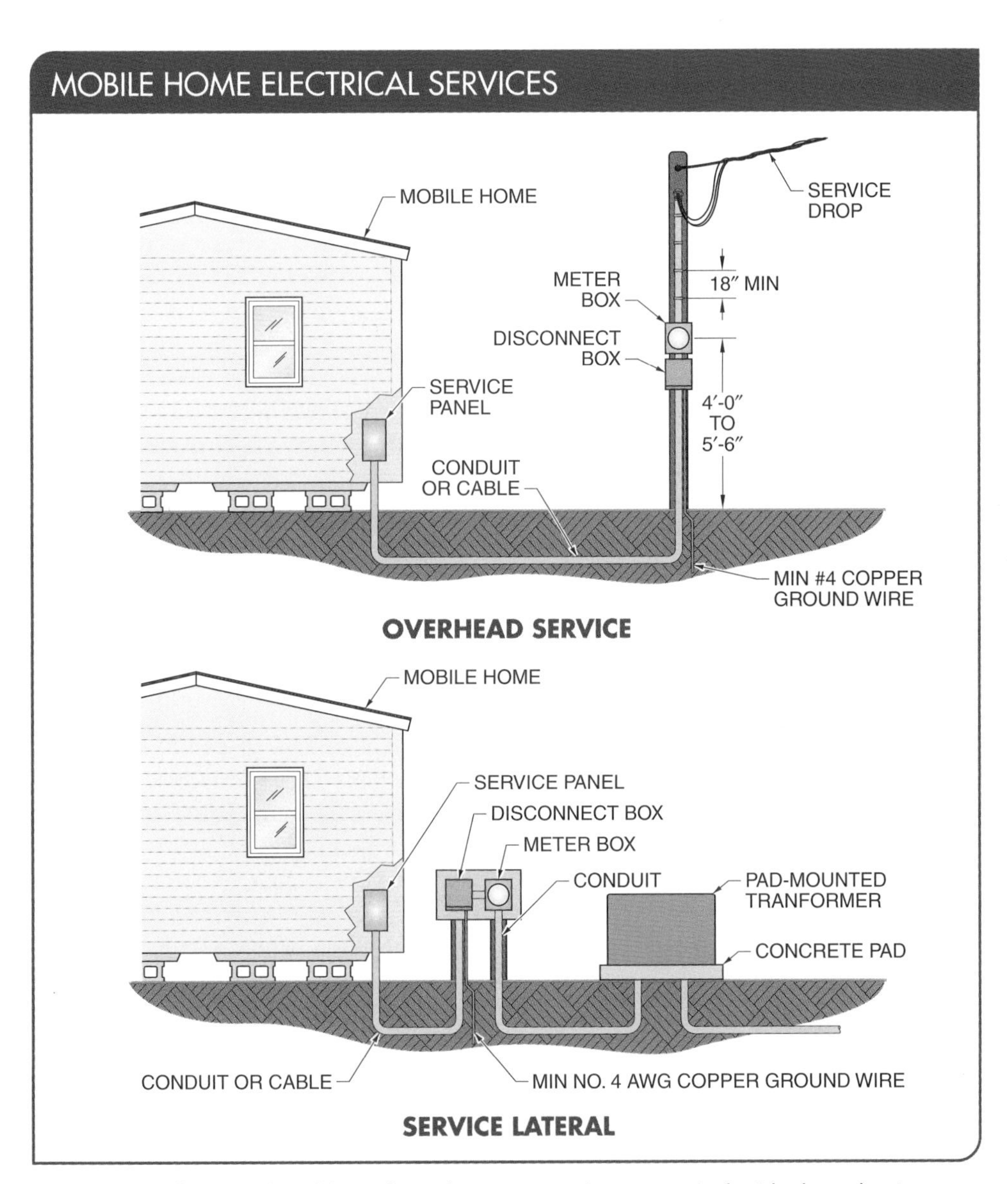

Figure 3-17. Overhead and lateral services are used to connect electrical service to a mobile home. The electric service connects to a disconnect box before being connected to the service panel.

SECTION 3.2 CHECKPOINT

1. Define "service entrance cable."
2. Define "service panel."
3. How is service equipment secured?
4. What are subpanels used for?
5. Describe a mobile home electrical service.

SECTION 3.3—DIGITAL SMART METERS

The two types of electric meters in use are analog meters and digital electric meters, commonly referred to as smart meters. A *smart meter* is a meter that uses two-way digital communication to enable a utility to securely collect information regarding the electrical consumption (usage) of a consumer. Analog meters are being replaced by smart meters. **See Figure 3-18.**

DIGITAL (SMART) METERS

Figure 3-18. Digital electric meters, commonly referred to as smart meters, contain transceivers (radios with both transmit and receive functions) that enable information to be communicated between the utility and consumer.

TECH TIP

Upgrading an analog meter to a digital smart meter may require replacing the old meter socket and enclosure.

Electrical consumption (usage) is the total amount of electrical energy used during a billing period. Electrical consumption is measured in kilowatt-hours (kWh). A *kilowatt-hour (kWh)* is the amount of electricity used in an hour. Smart meters also collect information from the utility grid to monitor and detect problems, and they enable a utility to turn on and off appliances during peak demand if allowed by the consumer.

A smart meter consists of a digital display, a radio frequency (RF) transceiver, and control software. The RF transceiver consists of two low-power radios. One low-power radio is a transmitter that sends information to the utility for billing and grid monitoring purposes. The other low-power radio is a receiver that allows the consumer to receive information which can be displayed on the digital screen of a home energy management system (HEMS). **See Figure 3-19.**

Smart meters transmit radio waves (RF waves) to communicate information between the consumer and the utility. The Federal Communications Commission (FCC) created maximum permissible exposure (MPE) guidelines for RF waves from stationary devices that operate at least 8″ away from human contact. The FCC document detailing how to measure and/or calculate RF levels titled "Evaluating Compliance with FCC Guidelines for Human Exposure to Radiofrequency Electromagnetic Fields."

All RF wave exposure is based on the transmitter power and the distance from the source. In general, doubling the distance cuts the power density by a factor of four. RF waves from smart meters 10′ away from a person are only about 1⁄1000 as much as a typical cell phone. **See Figure 3-20.** In addition, the metal meter box and outside wall materials block some of the RF waves from entering the dwelling.

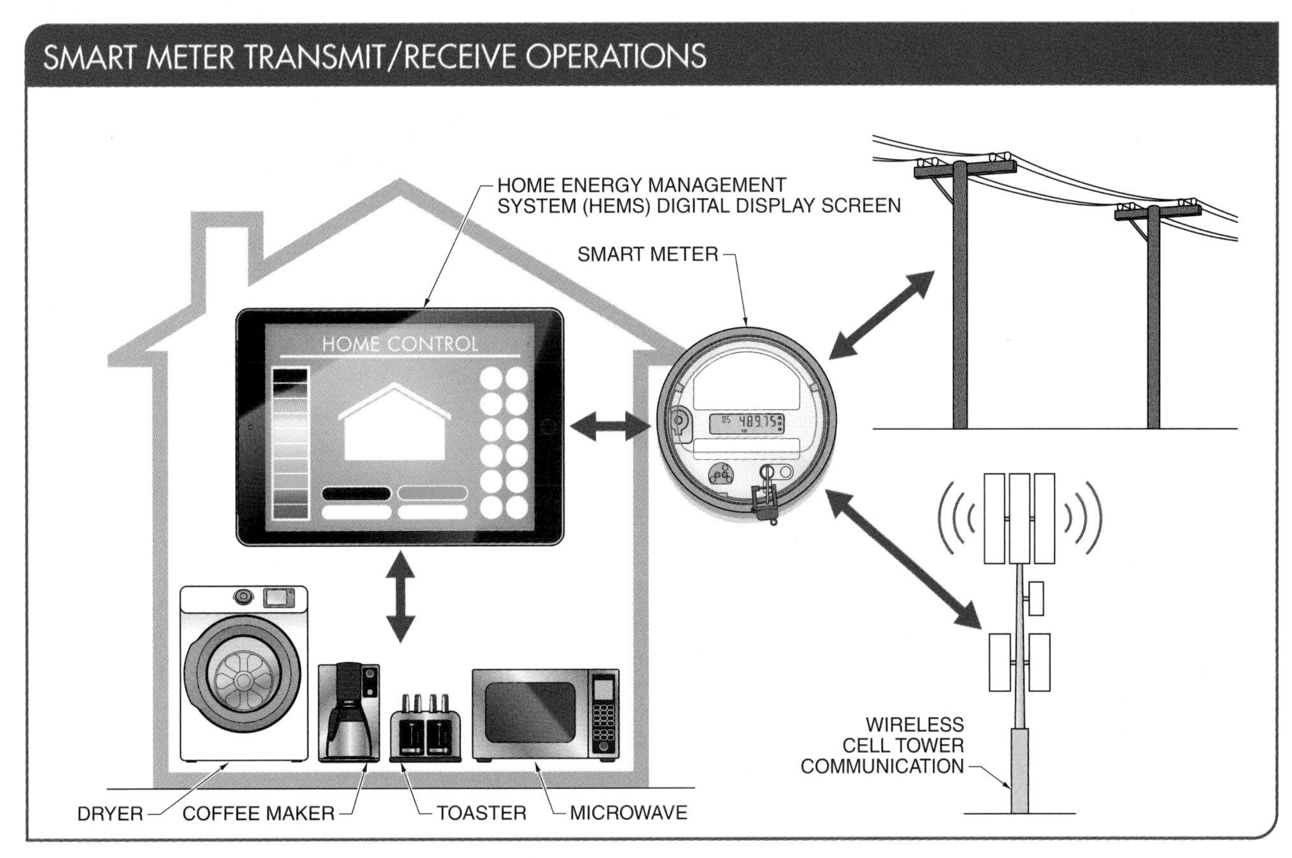

Figure 3-19. Smart meters can send information to the digital display screen of a home energy management system (HEMS).

Advanced Metering Infrastructure (AMI)

A smart meter is part of an advanced metering infrastructure (AMI) system. An AMI system provides a utility with automatic notification of outages, which pinpoints outages quickly and reduces false alerts. An AMI system can also send messages to consumers through the meter network when power is restored. This is usually done by a text message, but it can also show up on an in-home display (IHD) of a home energy management system (HEMS).

AMI implementation also reduces the chances of inaccurate meter readings since readings are taken electronically. In addition, AMI systems can identify patterns and suggest theft or tampering by automatically analyzing real-time data and comparing it with historical trends from the same smart meter. The system then automatically generates work reports for field managers to investigate. An AMI system is linked with a supervisory control and data acquisition (SCADA) management system, maintaining grid security.

Consumers benefit from an AMI system by receiving faster notifications of power outages, power restorations, and information on billing and power consumption. Consumers can create energy profiles and view the power consumption of equipment and appliances on the graphic displays of personal computers, cell phones, or home energy management systems (HEMSs).

Besides seeing a reduction in meter tampering and energy theft, utilities also benefit from an AMI system by using real-time data to balance loads and reduce brownouts, blackouts, and rolling blackouts. An AMI system improves efficiency and quality by providing information to outage management systems that detect and locate outages. Systems may also reroute outages to alternate power resources.

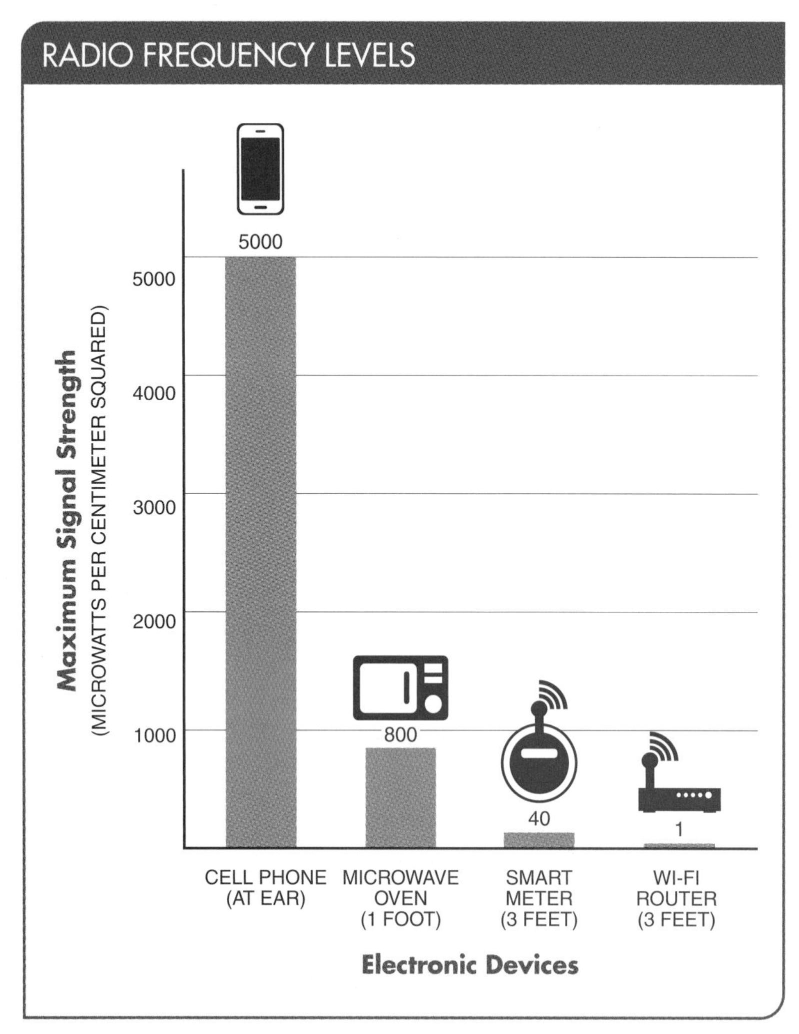

Figure 3-20. Consumer exposure to the radio frequency (RF) levels of smart meters is much less compared to other electronic devices.

An AMI system consists of three interconnected networks that include home area networks (HANs), neighborhood area networks (NANs), and wide area networks (WANs). NANs are sometimes referred to as metropolitan area networks (MANs) in large cities. **See Figure 3-21.**

HANs. A *home area network (HAN)* is a computer network that facilitates communication among devices within the close vicinity of a dwelling. The two types of HANs include one type provided by a consumer for personal use in a residence and another type provided by the utility through a smart meter.

A HAN provided and supported by a utility is a secure wireless network that works on the same principle as a personal home wireless network. A utility-provided HAN allows a smart meter to collect information for billing purposes. A utility-provided HAN can also monitor appliance usage, if a consumer chooses to participate, and can network appliances. A HAN also provides information for use by the consumer through the in-home display (IHD) found in a home energy management system (HEMS). **See Figure 3-22.** The key components of a HAN system include the following:

- a HEMS, which is a combination of a HAN with hardware that has autopricing response capabilities, demand response (DR) load control, and home automation controls
- an IHD, which can show pricing signals, meter-based consumption, and bill-to-date information
- smart home appliances (or appliances connected through smart plugs), which can be remotely controlled either through an IHD or web portal

HANs enable consumers to create schedules for switching off devices during peak demand to conserve energy and reduce electric bills. In addition, homeowners can view the consumption patterns of different devices and make decisions regarding energy efficiency, demand response, monthly billing, etc. Finally, hourly energy consumption patterns help homeowners analyze equipment performance, which can help them develop household energy efficiency programs.

NANs and MANs. A *neighborhood area network (NAN)*, or metropolitan area network (MAN) for a large city, is the utility's outdoor network that collects data from multiple smart meters in a specific community or city location. A smart meter area network data collector is also referred to as a concentrator. A *concentrator* is a device that primarily functions to serve as the collection point for several smart meters within a NAN. **See Figure 3-23.**

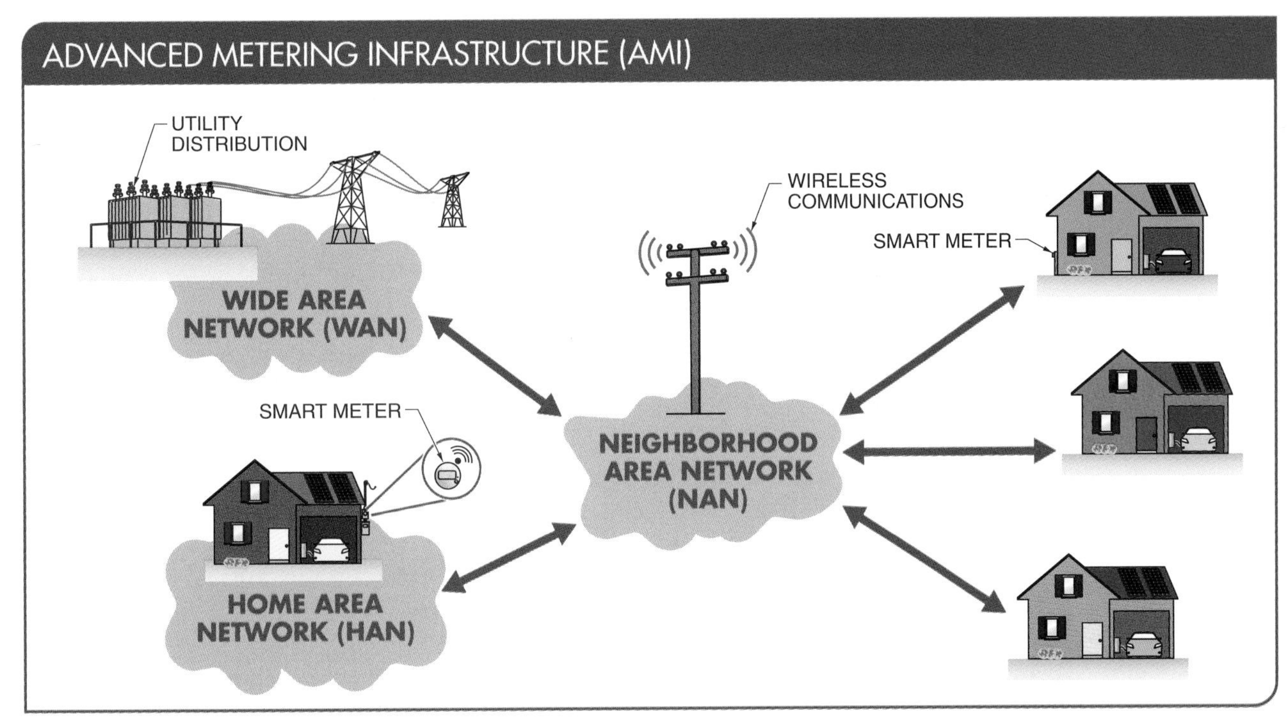

Figure 3-21. An AMI system consists of three main interconnected networks: HANs, NANs (sometimes referred to as MANs in large cities), and WANs.

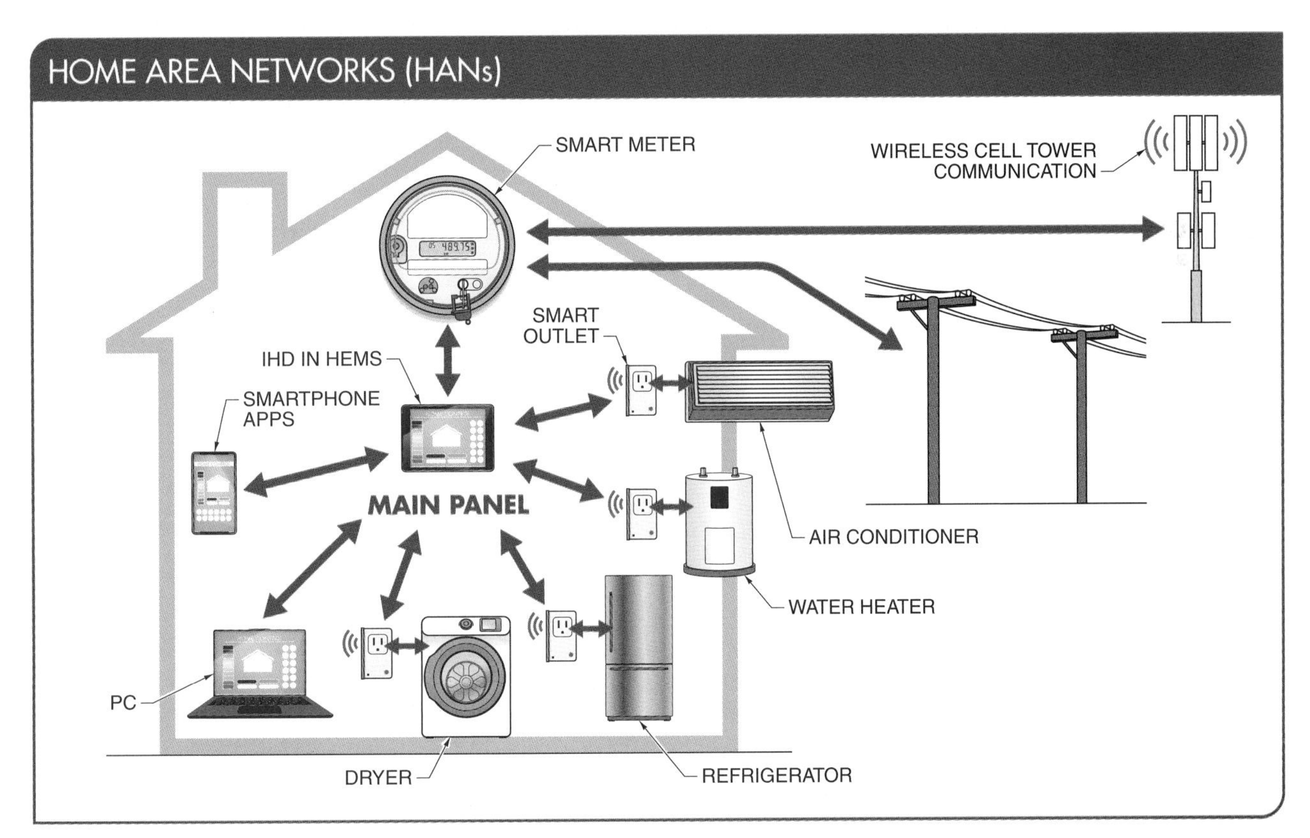

Figure 3-22. HANs and HEMSs allow access to energy consumption and billing information from inside residences or remotely through smartphone apps.

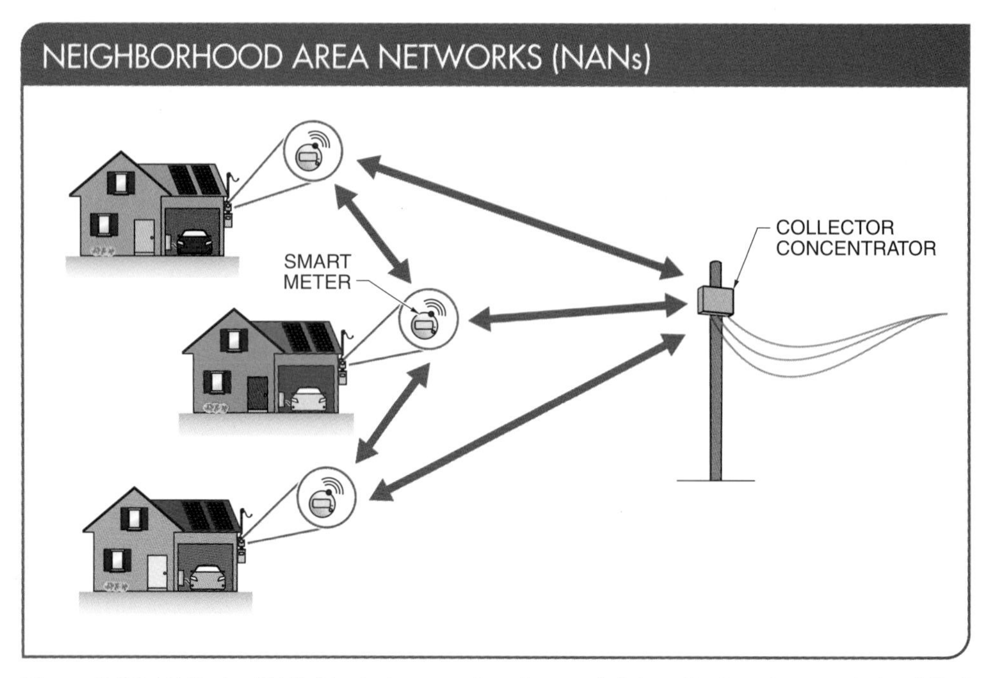

Figure 3-23. NANs (or MANs) includes smart meters and data collectors (concentrators) that collect information from several smart meters.

WANs. A *wide area network (WAN)* is the final communication link between the consumer and the utility. A WAN is similar to the networks used with cell phones to send and receive information. In certain areas of the United States, some utilities contract with local cell phone carriers for WAN service. This level of wireless networking allows smart meter information to be communicated securely to the utility. **See Figure 3-24.**

Mesh Networks. Most of the AMI networks use some form of mesh technology in their communications systems. A *mesh network* is a network that has no centralized access point but relies on all nodes to communicate with each other and form an interconnected network. One important benefit of a mesh network is that the communication path can be continued even if one or more of the smart meters are out of service. By using other smart meters in a mesh network as repeaters, new communication paths can be created. Mesh networks can also extend the range of a network by using repeaters.

CODE CONNECT

Per NEC® Subsection 250.142(B), Exception No. 2, some conditions allow the grounding of meter enclosures by connection to the grounded circuit conductor on the load side of the service disconnect.

Smart Meter Security and Privacy Laws

Since smart meters handle private consumer information, the utility is responsible for establishing and maintaining a secure network to guard against information interception, information tampering, and unauthorized access. Utilities encrypt all consumer-related, energy-use data transmitted by smart meters using methods similar to those used for online banking and ATM machines.

Currently, there are many state and national laws in place to protect consumers. These laws restrict utilities from disclosing or selling any personally identifiable energy use information without approval from the consumer.

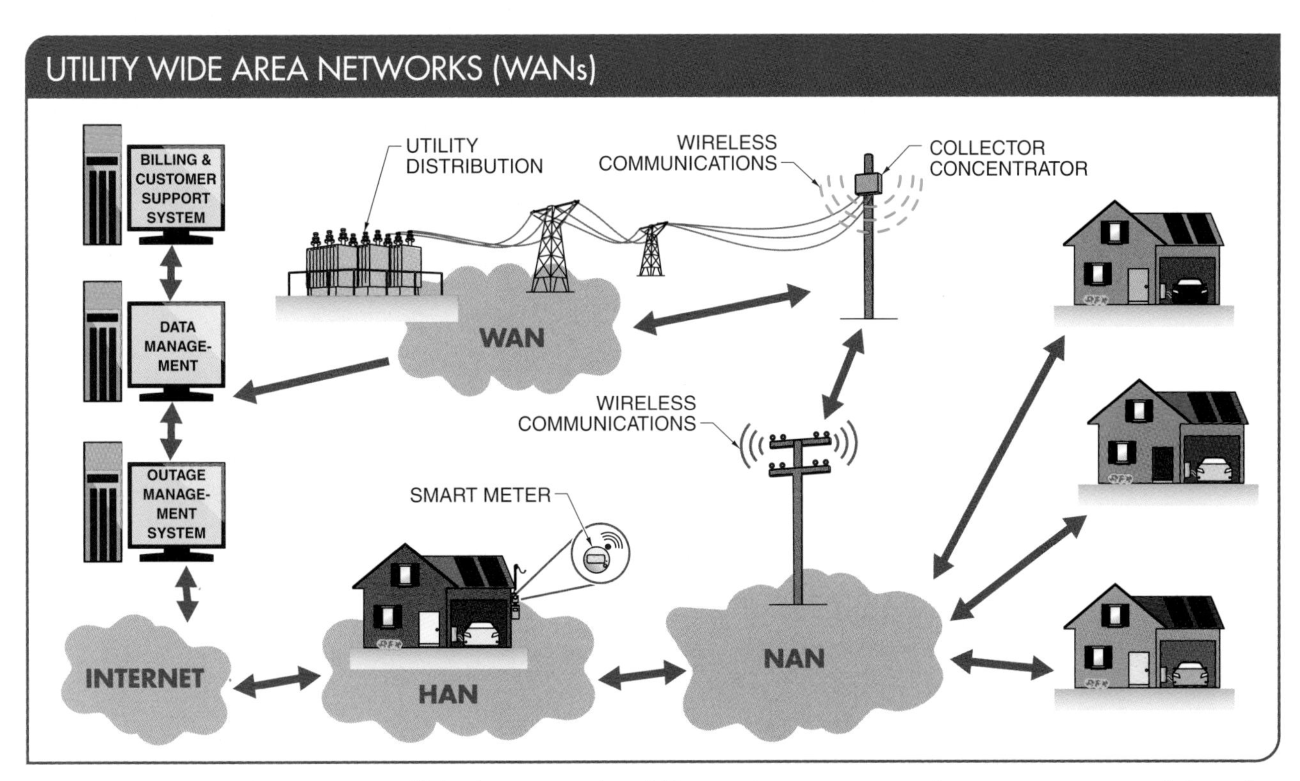

Figure 3-24. WANs allow utilities to efficiently manage data, billing, and power outages. Customers can access information through the utility WAN or the internet.

SECTION 3.3 CHECKPOINT

1. What is a kilowatt-hour (kWh)?
2. How many radios are contained in a smart meter and what do they do?
3. What does "AMI" stand for?
4. What are the three interconnected networks that make up an AMI system?
5. What are the key components of a HAN system?
6. What is an important benefit of using a mesh network?

SECTION 3.4— POWER QUALITY

Power quality refers to the consistency of the electric voltage. Voltage outside the proper range can damage electrical equipment. Poor power quality occurs with power interruptions, power outages, power surges, and transient voltages. Equipment can be protected against poor power quality with properly rated surge protection devices.

Power Interruptions

A *power interruption* is a loss of electric power for less than a second to a few minutes. Power interruptions include momentary power interruptions, temporary power interruptions, and sustained power interruptions. **See Figure 3-25.**

Momentary Power Interruptions. A *momentary power interruption* is a decrease to 0 V on one or more power lines lasting from 0.5 cycles up to 3 sec. All power distribution systems have momentary power interruptions during normal operation. Momentary power interruptions can be caused by lightning strikes, utility grid switching during a short on a line, or during open circuit transition switching. *Open circuit transition switching* is the process in which power is momentarily disconnected when switching a circuit from one voltage supply to another.

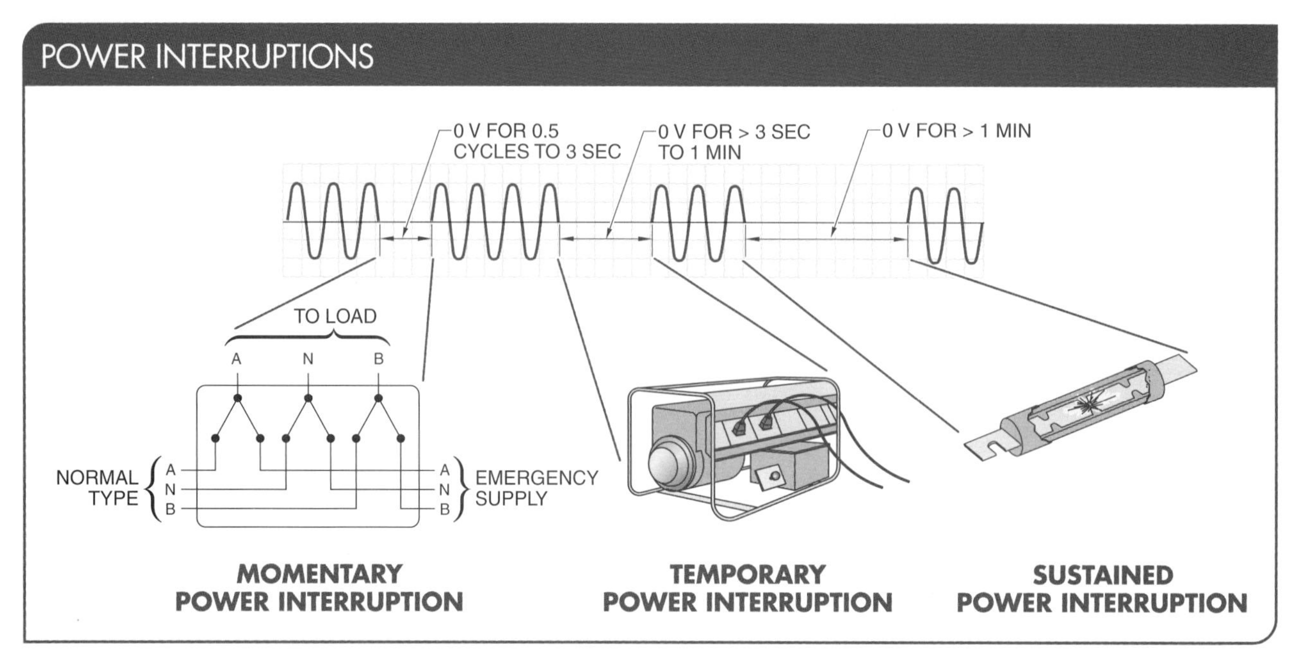

Figure 3-25. Power interruptions in an electrical system can be categorized as momentary, temporary, or sustained.

Temporary Power Interruptions. A *temporary power interruption* is a decrease to 0 V on one or more power lines lasting between 3 sec and 1 min. Automatic circuit breakers and other circuit protection equipment protect all power distribution systems. Circuit protection equipment is designed to remove faults and restore power. An automatic circuit breaker takes 20 cycles to 5 sec to close.

Sustained Power Interruptions. A *sustained power interruption* is a decrease to 0 V on all power lines for a period of more than 1 min. All power distribution systems have a complete loss of power at some time. Sustained power interruptions (power outages) are commonly the result of storms, tripped circuit breakers, blown fuses, and/or damaged equipment.

Power Outages

A *power outage* is a short- or long-term sustained loss of electric power. The three types of power outages are blackouts (complete loss of power), brownouts (intentional reduction in power during peak times), and rolling blackouts (rotating power outages to minimize blackouts).

Power Surges

A *power surge* is a major increase in the electrical energy at any point in a power line. Power surges can increase voltage in the power line to a point where the initial surge can produce increases in voltage at a receptacle. Lightning strikes are a common source of power line surges. When lightning strikes near a power line, the electrical energy from the lightning can increase electrical voltage by thousands of volts. This extremely large surge can damage utility transformers, switching gear, and consumer equipment unless some type of protective device is in place.

Transient Voltages

A *transient voltage* is a temporary, undesirable voltage in an electrical system. A transient voltage lasts for only a few milliseconds to a few microseconds. Transient voltages differ from power surges by being higher in voltage, shorter in duration, and erratic.

High-voltage transients can permanently damage sensitive electrical circuits or electronic equipment. Computers, electronic circuits, and specialized electrical equipment require protection against transient voltages.

Protection methods commonly include proper wiring, grounding, shielding of the power lines, and use of transient surge suppressors.

Transients can be created when high-current draw electrical devices such as air conditioners and refrigerators are turned on and off. This switching creates sudden and brief demands for power that upset the steady flow of the electrical system. Other sources of transients include faulty wiring in a dwelling, problems with utility equipment, and surges in the power lines.

Whole-House Protection Plans

A *whole-house protection plan* is the use of proper grounding methods and devices designed to suppress power surges and transient voltages to protect electrical and electronic circuits. Transient voltages and surges can enter electronic equipment (such as routers, computers, and televisions) through telephone lines, television cables, or satellite dish connections. Most surge protective devices (SPDs) designed for AC power lines do not prevent telephone or communication line transients from damaging equipment.

Equipment Grounding Protection. Proper grounding of the electrical system is needed to protect electrical equipment. Sensitive electronic equipment should be placed on its own dedicated grounded circuit (isolated ground) and should only use three-prong outlets.

Whole-House Panel Protectors. A *whole-house panel protector* is a surge suppressor connected to the main service panel that diverts heavy electrical surges into the grounding system. This permits the point-of-use devices to serve as filters to shut down electrical noise on the line and stop any small remaining line surges. A subpanel that is 20′ or more from the main service panel should have a second whole-house panel protector installed to protect it and its equipment. **See Figure 3-26.**

Low-cost surge suppressors that plug directly into a power strip or an outlet do not provide adequate protection against power surges. Total protection must be accomplished on two levels. The first level of protection is at the main panel where the line surge can be prevented from entering the house wiring. The second level of protection is at the point-of-use where any remaining surges or transients can be removed just before they enter an appliance or other electrical device.

Surge-Protector Breakers. Surge-protector breakers, which mount inside of the main electrical panel box, protect circuits and can be used as regular circuit breakers. Surge-protector breakers usually have a red light indicating the surge protector is functional. Once the device is installed, all the devices plugged into this circuit are automatically protected. **See Figure 3-27.**

Power Strip Plug-In Protection. A power strip surge suppressor provides protection by limiting the level of voltage change that can pass through the surge suppressor. Surge suppressors can react within one billionth of a second (nanosecond) to divert excess voltage to the electrical grounding system.

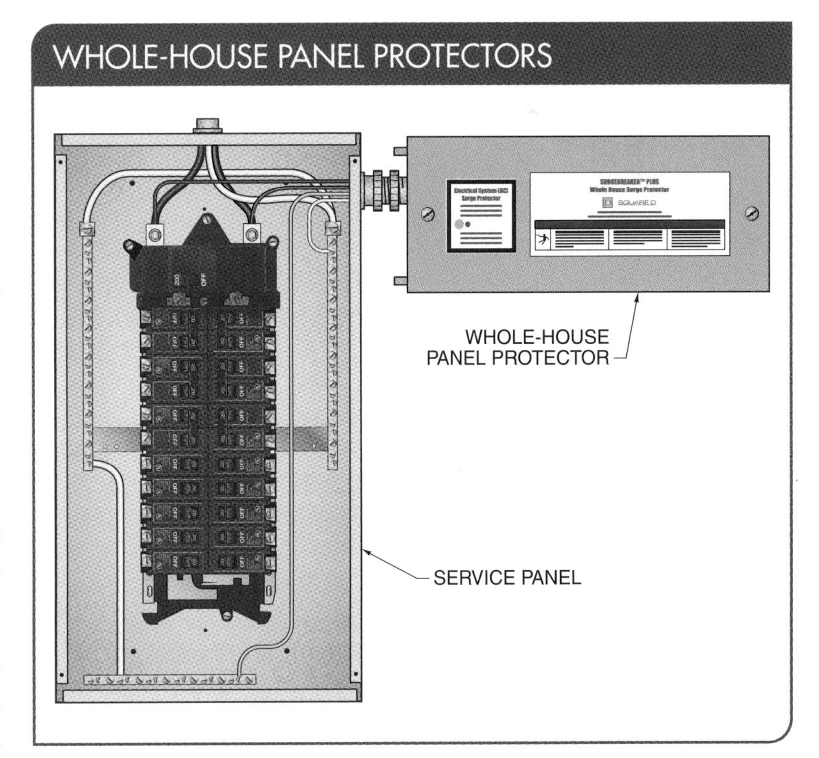

Figure 3-26. The first level of protection against surge and transient voltages is to install a whole-house panel protector.

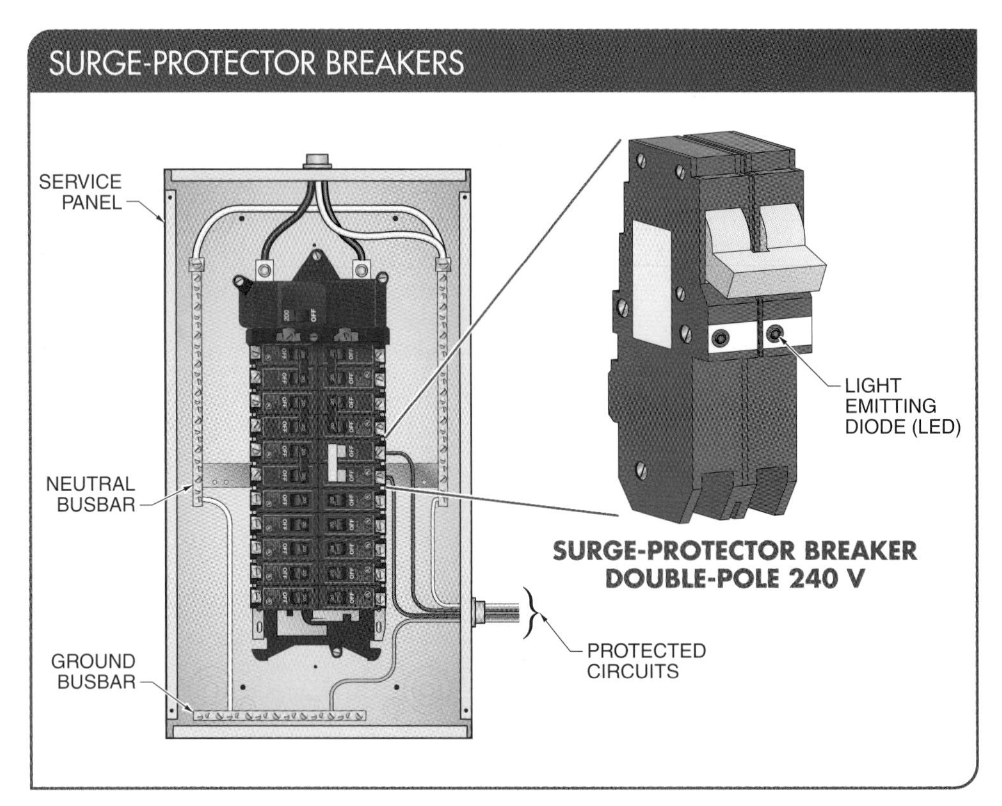

Figure 3-27. Surge-protector breakers automatically protect all devices plugged into the circuit.

Most electronic equipment needs the protection of plug-in surge suppressors. Plug-in surge suppressors are an effective way to prevent problems with electronic appliances. Surge suppressors are installed between the appliance and the wall outlet. A power strip surge suppressor allows several electrical or electronic devices to be plugged into one source. If voltage surges or transients in the line feeding the power strip are above an acceptable level, the protective circuitry in the power strip diverts the unwanted electricity into the power strip grounding wire. **See Figure 3-28.**

Line-Conditioning Systems. Some surge protectors include a line-conditioning system for filtering electrical noise on the line, which is small fluctuations in electrical current. Basic line-conditioning includes passing the hot wire through a wound wire coil. The wound wire coil is constructed as a small electromagnet. As the changing current passes through the electromagnet, the changing electromagnetic forces smooth the small increases and decreases in current.

Uninterruptible Power Supply (UPS) Systems. An *uninterruptible power supply (UPS)* is an electrical storage device that provides stable and reliable power during fluctuations or failures of the primary power source. UPSs are used to maintain the operation of critical loads such as computer systems, medical equipment, and sensitive electronics during short interruptions in the primary power source. During normal operations, power from the utility is supplied to a UPS. When this power is unstable, the UPS supplies stable output power to critical loads. A UPS is a standalone device installed at the point of use. **See Figure 3-29.**

UPSs use a battery back-up system to ensure quality power during interruptions protecting both data and hardware. Most UPSs provide 10 min to 20 min of back-up time, which, for example, provides enough time for an individual to close files and exit programs. Most UPSs provide surge suppression as well.

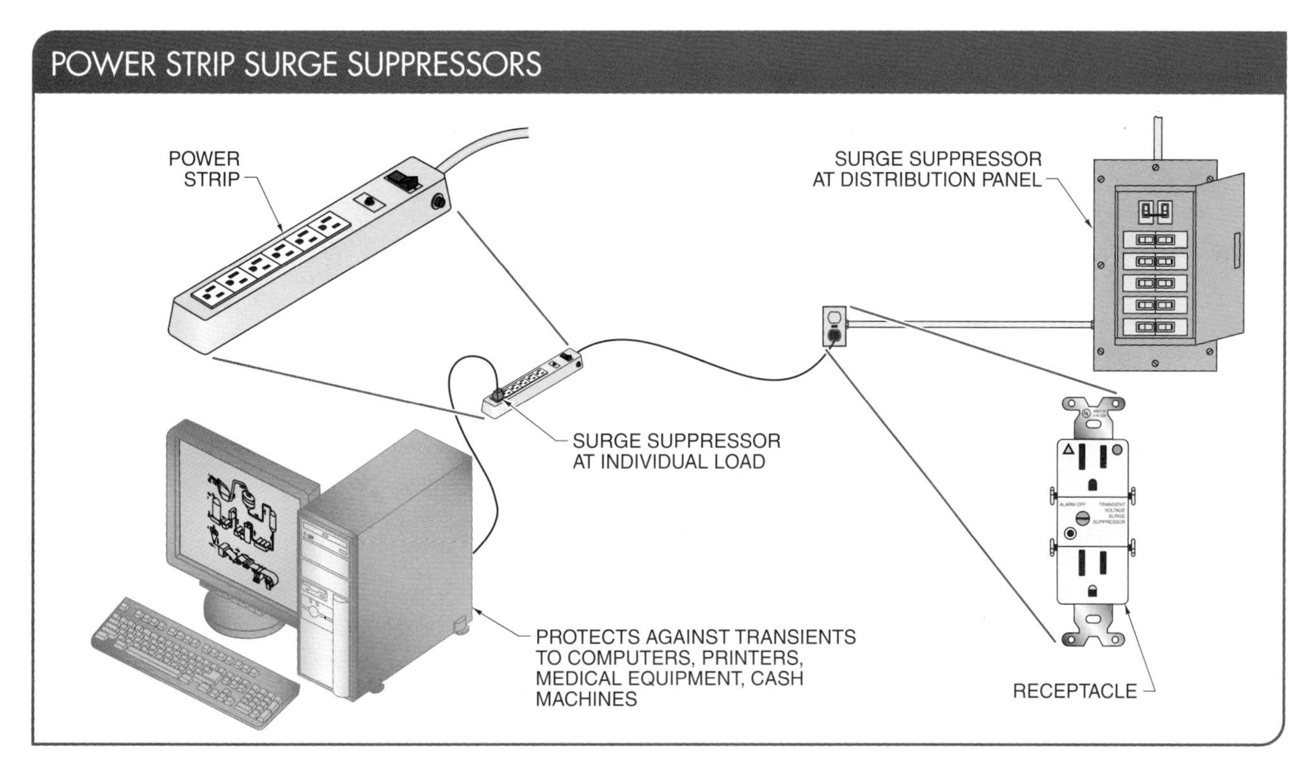

Figure 3-28. A power strip surge suppressor allows several electrical or electronic devices to be protected at one source.

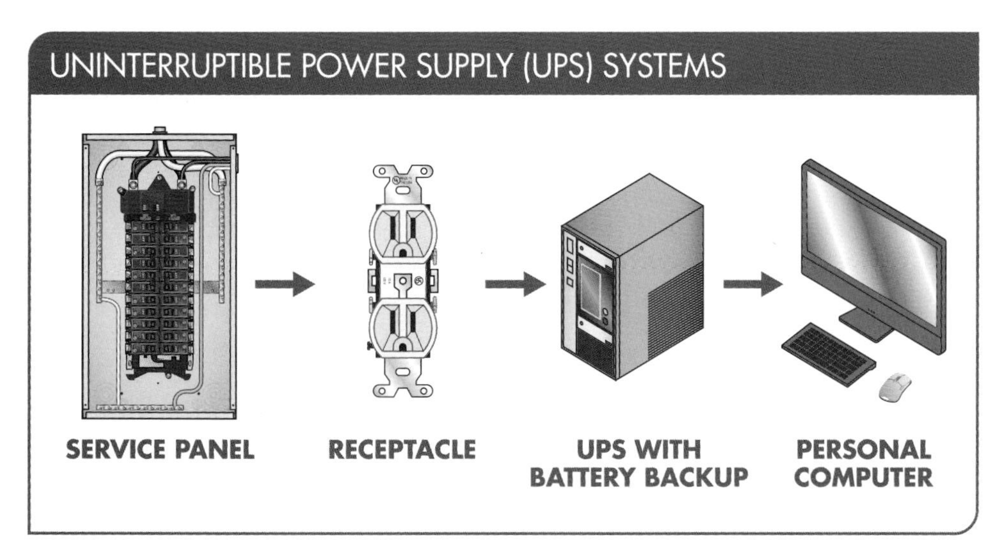

Figure 3-29. UPS systems are used to maintain the operation of critical loads such as computer systems, medical equipment, and sensitive electronics during short interruptions in primary power.

A UPS must be UL® listed. A UL® listing means that UL® has tested the product and determined that it meets UL® requirements. These requirements are based primarily on UL® published and nationally recognized standards for safety.

TECH TIP

Outdoor UPS systems should include a battery heater mat for extreme cold conditions and a fan or air conditioning for extreme heat conditions.

When utility power is interrupted or the power quality is poor, current is drawn from the back-up power source to continue supplying adequate power from the output section of a UPS. A UPS includes an inverter to convert DC power to AC power. When acceptable utility power returns, the UPS input returns to using utility power to provide power to the load while also charging the back-up power source (batteries).

TECH TIP

UPS batteries should all be replaced at the same time because old batteries discharge and recharge faster than new batteries.

Standby UPSs. Standby UPSs are typically used for small applications (less than 600 VA) and are commonly used to protect personal computers. Standby UPSs are also known as "off-line UPSs" because under normal conditions they pass utility power through to the output side and the inverter does not operate. In this case, the back-up power system inverter is in standby or off-line mode. A standby UPS provides some power conditioning to suppress surges or filter noise, but it only uses back-up power if the utility power falls below minimum requirements. Then, a transfer switch changes completely over to the back-up power source.

SECTION 3.4 CHECKPOINT

1. What is power quality?
2. Name the three types of power interruptions.
3. What are the three types of power outages?
4. What are power surges?
5. What is a transient voltage?
6. What is a whole-house panel protector?
7. What are surge-protector breakers?
8. What is an uninterruptible power supply (UPS)?
9. What is a standby UPS used for?

SECTION 3.5—WHOLE-HOUSE GENERATORS

A *whole-house generator* is a permanently installed standby generator designed to supply temporary power to a dwelling during a power outage. During a power outage, a whole-house generator allows the continued use of lighting and appliances. A whole-house generator can start automatically when it is properly connected to the service panel.

Whole-house generators are usually installed in new construction and placed outside on a solid surface such as packed gravel or a concrete pad. Residential whole-house generators are usually powered by natural gas or liquefied petroleum gas (LPG) and have a capacity between 6500 W to 40,000 W. To prevent exhaust gases from entering the dwelling, the generator should be placed in an area away from doors and windows. **See Figure 3-30.**

Selecting Whole-House Generators

A whole-house generator is selected based on the number of lighting and appliance circuits that must be powered during a power outage, as well as the extra power required to start those appliances. A whole-house generator is selected based on the number of lights and appliances that must be powered during a power outage, as well as the extra power required to start appliances with motors. Whole-house generators are rated based on their power output capacity expressed in watts.

A *watt* is a unit of power used to quantify the rate of energy transfer produced by a current of one ampere across a potential difference of one volt. The formula for calculating power in watts is as follows:

$$P = V \times A$$

where

P = power (in watts)

V = voltage (in volts)

A = amperage (in amps)

Figure 3-30. Whole-house generators are permanently installed standby generators designed to supply temporary power to a dwelling during a power outage.

For example, how many watts of power will an appliance plugged into a 120 V outlet that draws 5 A consume?

$P = V \times A$

$P = 120 \times 5$

$P = \mathbf{600\ W}$

Appliance nameplates often list voltage and amperage information that can be used to calculate wattage. **See Figure 3-31.** When the total wattage needed from a generator is calculated, both the starting wattage and the running wattage of each motor or pump driven appliance must be accounted for. For example, most HVAC compressors and motors need two to three times more power to start compared to when they are running. A 2 HP pump needs 6000 W to start pumping but only 2000 W to remain pumping. The best generators have built-in surge current capabilities for when a motor goes on-line.

Another method that can be used to determine power is to measure amperage and calculate the wattage. A clamp-on ammeter can measure starting amperage and running amperage, which can then be used to calculate start wattage and running wattage. **See Figure 3-32.**

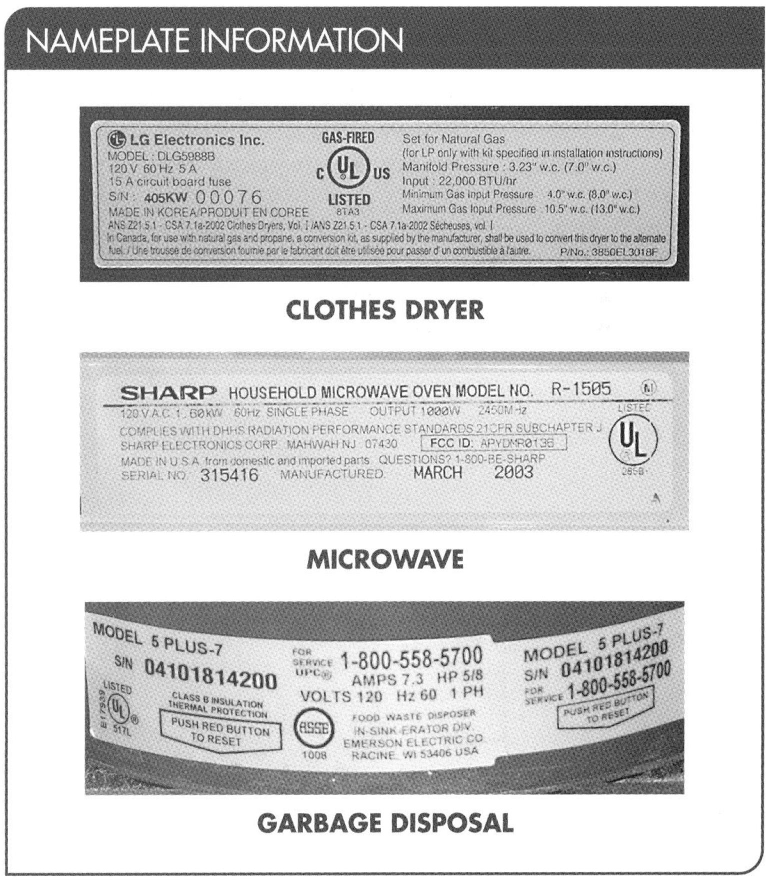

Figure 3-31. Appliance nameplates often list voltage and amperage information that can be used to calculate wattage.

APPLIANCE STARTING AND RUNNING WATTAGES		
Appliance	**Wattage (Starting)**	**Wattage (Running)**
Refrigerator	2800	700
Freezer	2500	500
Well Pump –2 HP	6000	2000
Well Pump –3 HP	9000	3000
Gas Furnace Fan	500 to 2350	300 to 875

TYPICAL APPLIANCE WATTAGES	
Appliance	**Typical Wattage**
Radio	40 to 225
Toaster	200
Computer Monitor	200 to 800
Fan	275
Television	300 to 800
Printer	400 to 800
Microwave	600 to 1200
Dishwasher	700
Computer	700 to 1000
Electric Fry Pan	1000
Coffee Maker	1100
Washing Machine	1200
Hair Dryer	1500
Electric Cooktop Top – 8″	2100
Lights	4500
Electric Clothes Dryer	6000
Electric Oven	6500
Electric Water Heater	10,000

Figure 3-32. To calculate the total wattage of a dwelling, both the starting wattage and the running wattage of each motor and pump driven appliance must be accounted for.

A properly maintained generator can provide over 20 years of operation when operated as an emergency standby generator. Air-cooled generators do not provide continuous power. Liquid-cooled generators are designed for longer operation but not for continuous duty year-round. The expected life for liquid-cooled generators is 10,000 hr to 12,000 hr with proper maintenance.

Whole-House Generator Connections

A licensed electrician should make all electrical connections to a whole-house generator. The installation may need approval by the utility and a permit may be required if propane gas is part of the system. A whole-house generator should be UL® listed, be properly sized for the calculated load, be properly located to minimize generated noise, and be properly housed against weather exposure. Other considerations include the type of fuel, the battery, an automatic safety shutoff for low oil pressure or high temperatures, and a seven-day exerciser programmed to run the system for several minutes each week to maintain top running condition.

Automatic Transfer Switches (ATSs)

An *automatic transfer switch (ATS)* is a device that switches a load between two different sources of electric power. An ATS continuously monitors utility power for fluctuations or outages and can trigger a command to start a generator and bring it online. With an ATS, power failures are detected immediately and the transfer from utility power to generator power is instantaneous. **See Figure 3-33.**

ATSs operate using a process called open-transition transfer (OTT). In an OTT system, one power source must be completely disconnected from the electrical load before the other power source is connected. This is also known as a "break-before-make" transfer. This process prevents the power from back-feeding into the utility grid. Normally, an ATS has three positions: LINE, OFF, and GEN. When switching between LINE and GEN, an ATS always passes through the OFF position. This break prevents arcing or short circuits during the transition.

Back-feeding. *Back-feeding* (islanding) is the flow of electrical energy in the reverse direction from its normal direction of flow. Back-feeding could energize the line transformer serving a residence and pose an electrocution hazard for utility workers who may not know the voltage is present in the line. In addition, back-feeding a home generator with neighboring loads damages the generator.

Voltage Sensing. The main function of the control panel of an ATS is to detect a voltage drop or a complete failure of the utility grid. The failure is determined as a voltage drop below a preset setting on any phase. Voltage information is provided by the sensors in the control panel. Only a small percent of voltage drop must be reached before the dwelling is switched to generator power.

ATSs with AC power control modules allow loads to be prioritized when a voltage drop is detected. The power control module monitors the load current in the circuits and temporarily turns off lower priority loads to allow major appliances, such as an air conditioner, to turn on.

Time Delays. ATSs come with a time delay feature. A time delay is necessary for an ATS because of momentary outages. The time delay feature overrides any momentary outages that can cause a false alert. The time delay is between 1 sec and 6 sec, with 1 sec being the most common setting.

ATS Safety. An ATS is installed indoors on a firm and sturdy supporting structure. To prevent switch distortion, the switch should be level. An ATS should never be installed where water or any corrosive substance might drip onto the enclosure. ATSs must also be protected from excessive moisture, dust, dirt, and lint.

CODE CONNECT

Per NEC® Section 445.10, generators must be suitable for the location where they are installed. A suitable location for a generator prevents generator exhaust from entering adjacent buildings.

Generator Hazards

Only qualified persons should install, operate, and maintain generator equipment. A generator contains rotating parts and may be hot during operation. Installation must always comply with applicable codes, standards, laws, and regulations. In addition, a running generator emits carbon monoxide (CO). Inhaling CO can cause headaches, fatigue, dizziness, nausea, vomiting, confusion, fainting, seizures, and death.

Grounding Hazards. The NEC® may require the frame and external electrically conductive parts of a generator to be connected to an approved earth ground. Local electrical codes may also require proper grounding of the generator electrical system. Grounding requirements should be verified with the local authority having jurisdiction (AHJ) before a generator is installed or used.

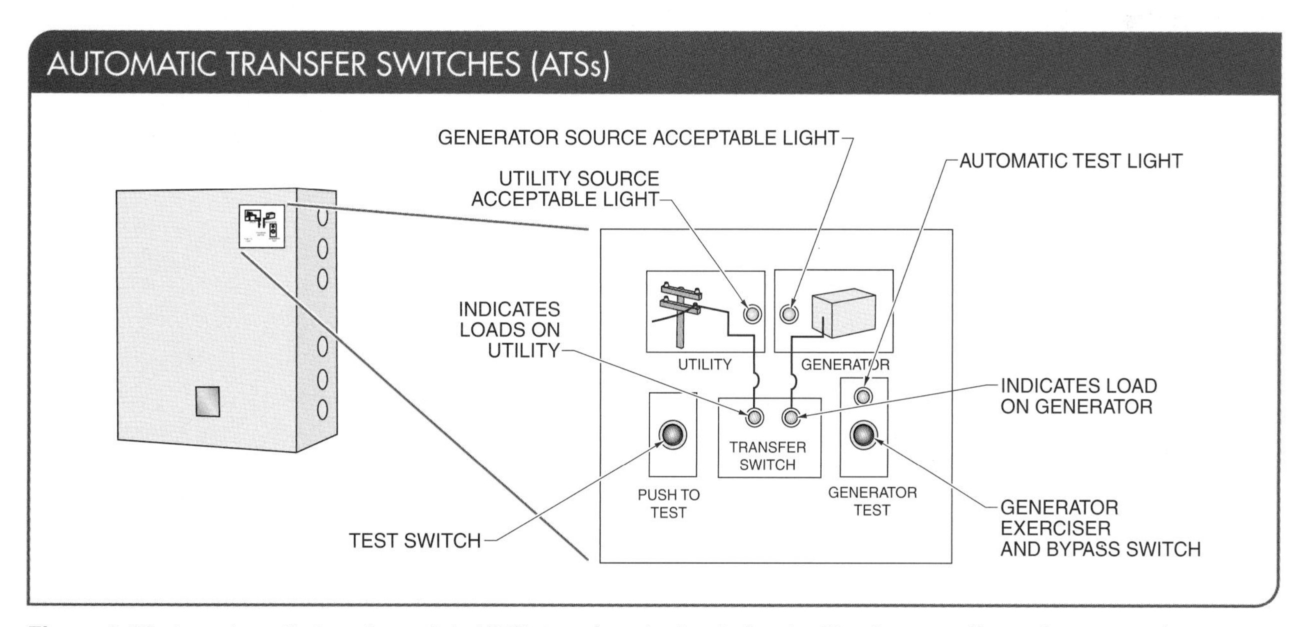

Figure 3-33. An automatic transfer switch (ATS) transfers the load of a dwelling from a utility to the output of a standby generator during a power failure.

Explosion Hazards. Flammable fluids, such as natural gas and liquid propane, are extremely explosive. To avoid the possibility of an explosion, the fuel supply system must be installed according to applicable fuel-gas codes. Before a whole-house generator is placed into service, fuel system lines must be properly purged and leak-tested according to applicable codes. After installation, the fuel system should be inspected periodically for leaks. Smoking, flames, combustible material, or debris should not be permitted near a generator, as a fire or explosion may result. Fuel and oil spills should be cleaned up immediately.

Whole-House Generator Battery Installation Safety

Prior to the installation of a battery in a generator, the battery must be filled with the proper electrolyte fluid, if necessary, and be fully charged. Before the battery is installed and connected, the following procedure is applied:

1. Set the auto/off/manual switch of the generator to the OFF position.
2. Turn off the utility power supply to the transfer switch.
3. Remove the fuse from the generator control panel. *Note:* Battery cables are factory connected at the generator.
4. Connect the red battery cable (from the starter contactor) to the battery post indicated by a positive identification mark (+).
5. Connect the black battery cable (from frame ground) to the battery post indicated by a negative identification mark (–).

Cable leads and connectors must be inspected for fraying, looseness, and burning of the contact areas. **See Figure 3-34.** Cable leads and connectors should be repaired or replaced as needed. Batteries can fail or malfunction for multiple reasons. Common causes for battery failure include corrosion, damaged cables, cracked cases, freezing, and undercharging.

If the auto/off/manual switch is not set to the OFF position, the generator can crank and start as soon as the battery cables are connected. If the utility power supply is not turned off, sparking can occur at the battery posts and cause an explosion. To prevent an explosion, the following safety precautions must be followed:

- Do not smoke near the battery.
- Do not have open flames or sparks in the battery area.
- Be sure the auto/off/manual switch is set to the OFF position before connecting the battery cables. If the switch is not set to the OFF position, the generator can crank and start as soon as the battery cables are connected.
- Do not dispose of the battery in a fire. The battery can explode.

A battery presents the risk of electrical shock from high short-circuit current. To prevent electrical shock when handling a battery, the following safety precautions must be followed:

- Remove the fuse from the generator control panel.
- Remove jewelry or other metal objects from the body before working on or near the generator.
- Only use tools with insulated handles.
- Wear rubber gloves with leather insulators and shoes with electrical hazard (EH) rated soles.
- Do not lay tools or metal parts on top of the battery.
- Disconnect the charging source prior to connecting or disconnecting battery terminals.

Many whole-house generators use lead-acid batteries that contain electrolyte. Electrolyte is a solution of sulfuric acid and water that is harmful to the skin and eyes. The two types of lead-acid batteries are maintenance free and conventional. Maintenance-free batteries are sealed and do not need maintenance. Conventional batteries have individual caps for filling and testing electrolyte in each cell. To ensure conventional batteries operate at full capacity, each cell should be regularly checked and topped off using distilled water if necessary. This prevents the cells from being damaged due to drying out.

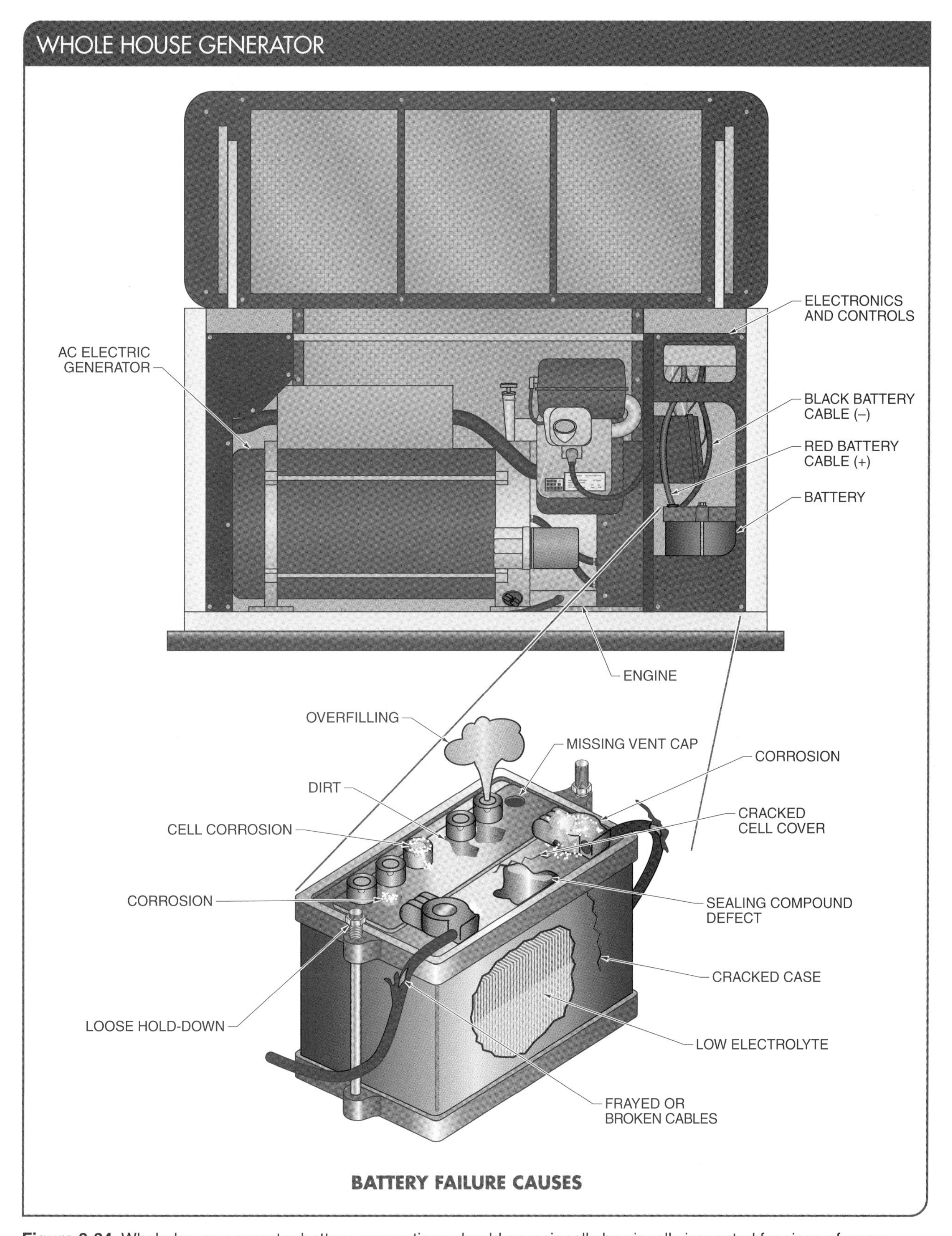

Figure 3-34. Whole-house generator battery connections should occasionally be visually inspected for signs of wear.

TECH TIP

Always turn off the generator power and wear the appropriate PPE before handling generator batteries.

Note: Always wear full eye protection and protective clothing when handling electrolyte. If electrolyte contacts the skin, flush with water for 10 min to 15 min. If electrolyte contacts the eyes, flush immediately with water for 10 min to 15 min and get medical attention.

SECTION 3.5 CHECKPOINT

1. What is a whole-house generator?
2. Describe automatic transfer switches (ATSs).
3. What is back-feeding?
4. What is the main function of the ATS control panel?
5. What is the use for the time delay feature of an ATS?
6. What are two generator hazards?
7. What five steps are required before a generator battery is installed and connected?

3 Service Entrances and Smart Meter Installation

Chapter Activities

Name ______________________ Date ______________

Service Drop Entrances

Identify the different components shown.

______________ **1.** Meter socket box

______________ **2.** Service panel

______________ **3.** Service head

______________ **4.** Ground

______________ **5.** Service wires

A

B

C

D

E

Service Entrance Parts

Identify the service entrance parts shown.

______________ **1.** Conduit support

______________ **2.** Pipe strap/lag

______________ **3.** Sill plate used with service entrance cable

______________ **4.** Service head

______________ **5.** Insulator bracket

______________ **6.** Roof flashing

A B

C D

E F

Whole-House Panel Protectors

1. Draw lines to connect the whole-house panel protector to the proper termination points on the service panel.

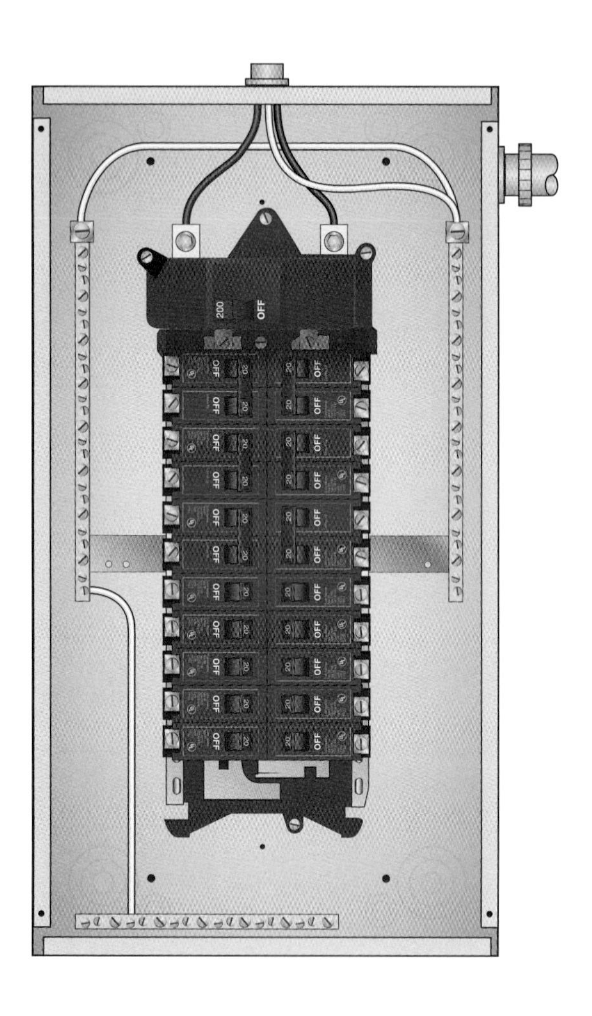

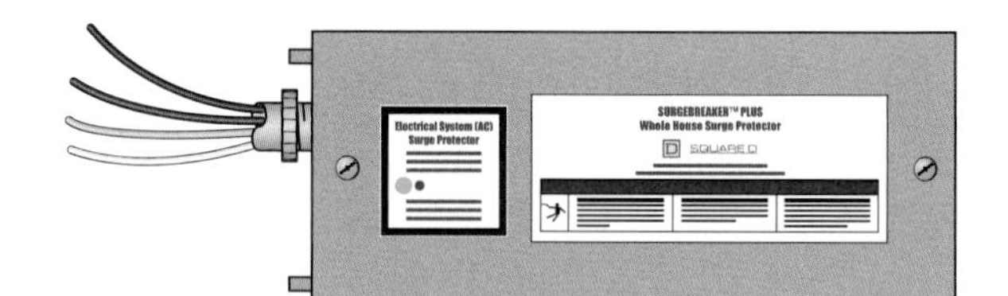

4 Electrical Grounding and Overcurrent Protection

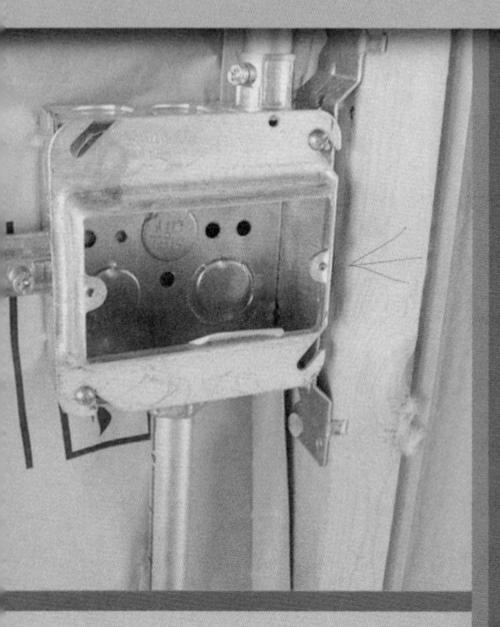

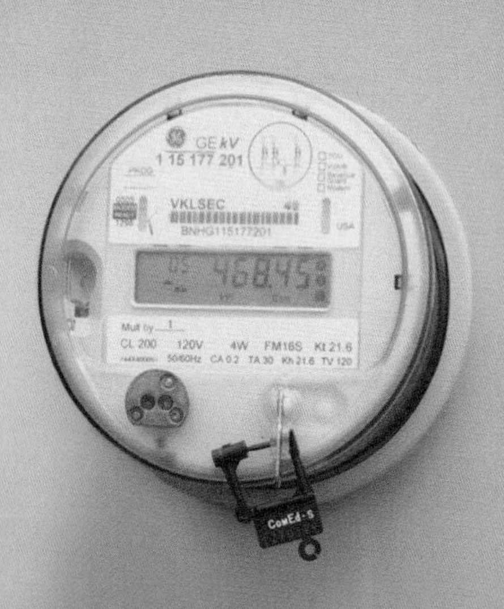

SECTION 4.1—GROUNDING

- Explain grounding.
- Compare and contrast system and equipment grounding.
- Describe grounding for small appliances.

SECTION 4.2—OVERCURRENT PROTECTIVE DEVICES

- Define "overcurrent protective device (OCPD)."
- Describe fuses and the three categories of fuses.
- Describe circuit breakers and the three types of circuit breakers.
- Define "DC circuit breaker."
- Explain DC and AC arc suppression.

SECTION 4.3—GROUND-FAULT AND ARC-FAULT CIRCUIT INTERRUPTERS

- Explain electrical shock.
- Describe ground-fault circuit interrupters (GFCIs), as well as their operation and installation locations.
- Describe arc-fault circuit interrupters (AFCIs), as well as their operation and installation locations.

Learner Resources

ATPeResources.com/QuickLinks
Access Code: 791712

Residential electrical systems consist of safety devices that act as overcurrent protection to prevent damage to electrical equipment, the residence, and occupants. Grounding, overcurrent protective devices (OCPDs), ground-fault circuit interrupters (GFCIs), and arc-fault circuit interrupters (AFCIs) are used to protect the residence and its occupants from any electrical faults, overcurrents, and unintended arcing.

SECTION 4.1—GROUNDING

Grounding is an integral part of any properly operating electrical system. In a residence, grounding protects the occupants by providing a safe pathway for unwanted electricity that might otherwise create a hazard. Electricity always takes the easiest flow path to earth. A *ground* is a low-resistance conducting connection between electrical circuits, equipment, and the earth.

The purpose of grounding is to provide a safe path for a fault current to flow. A complete ground path must be maintained when installing switches, light fixtures, appliances, and receptacles. In a properly grounded system, the unwanted current flow blows fuses or trips circuit breakers. Once a fuse is blown or a circuit breaker is tripped, the circuit is open and no additional current will flow.

Grounding is typically established at two levels: system grounding and equipment grounding. A *system ground* is a special circuit designed to protect the entire distribution system of a residence. An *equipment ground* is a circuit designed to protect individual components of an electrical system. Grounded conductors are used to provide a path to ground for system and equipment grounds. A *grounded conductor* is a conductor that has been intentionally grounded. Grounded conductors are typically identified with green or green and yellow markings and may be installed as bare conductors.

System Grounding

The primary function of system grounding is to protect the service entrance wiring and the circuits connected to it. There are several methods of grounding an entire system. The two most popular methods used for grounding an electrical system are electrode grounding and water pipe grounding. **See Figure 4-1.** Other grounding methods use a concrete-encased electrode or a ground ring, both of which are less common in residential wiring systems.

Electrode Grounding. An *electrode* is a long metal rod used to make contact with the earth for grounding purposes. When no satisfactory grounding electrode is readily available, the common practice is to drive one or more metal rods (connected in parallel) into the ground. The electrode and circuit must provide a flow path to the earth with less than 25 Ω of resistance.

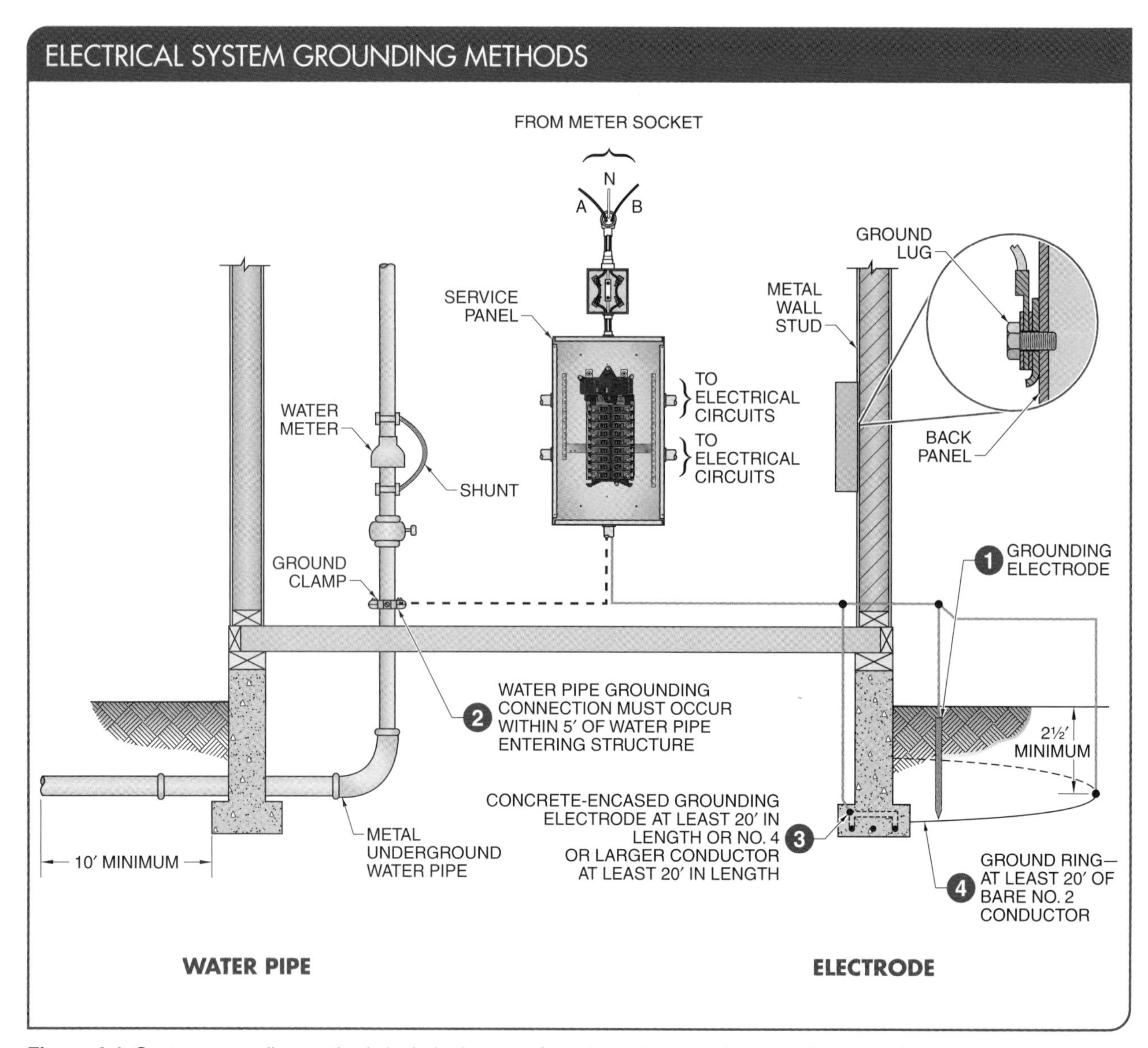

Figure 4-1. System grounding methods include the use of an electrode ground, water pipe ground, concrete-encased electrode, or ground ring.

Water Pipe Grounding. A water pipe ground uses the underground metal pipe that supplies a residence with water and is typically the best electrical ground for a residential electrical system. Water pipes work well as grounds because a large surface area of the pipe is in contact with the earth, as it connects the municipal water main to the water distribution system in the residence. This large surface area reduces resistance and allows any unwanted electricity to easily pass through the pipe to the earth.

When a water pipe is used for grounding, the water pipe run must never be interrupted by a plastic fitting or have an open section of plumbing. Water meters are a source of an open ground circuit when removed. To provide protection when a water meter is removed, a shunt (or meter bonding wire), must be permanently installed. A *shunt* is a permanent conductor placed across a water meter to provide a continuous flow path for ground current. **See Figure 4-2.**

All internal piping systems capable of becoming energized must be bonded and connected. A *bonding conductor* is a reliable conductor that ensures the electrical conductivity between two metal parts that must be connected electrically. The term "grounding conductor" no longer appears by itself in the NEC®. Instead, conductors are referred to by their function, such as "grounding electrode conductor," "bonding conductor," or "equipment grounding conductor."

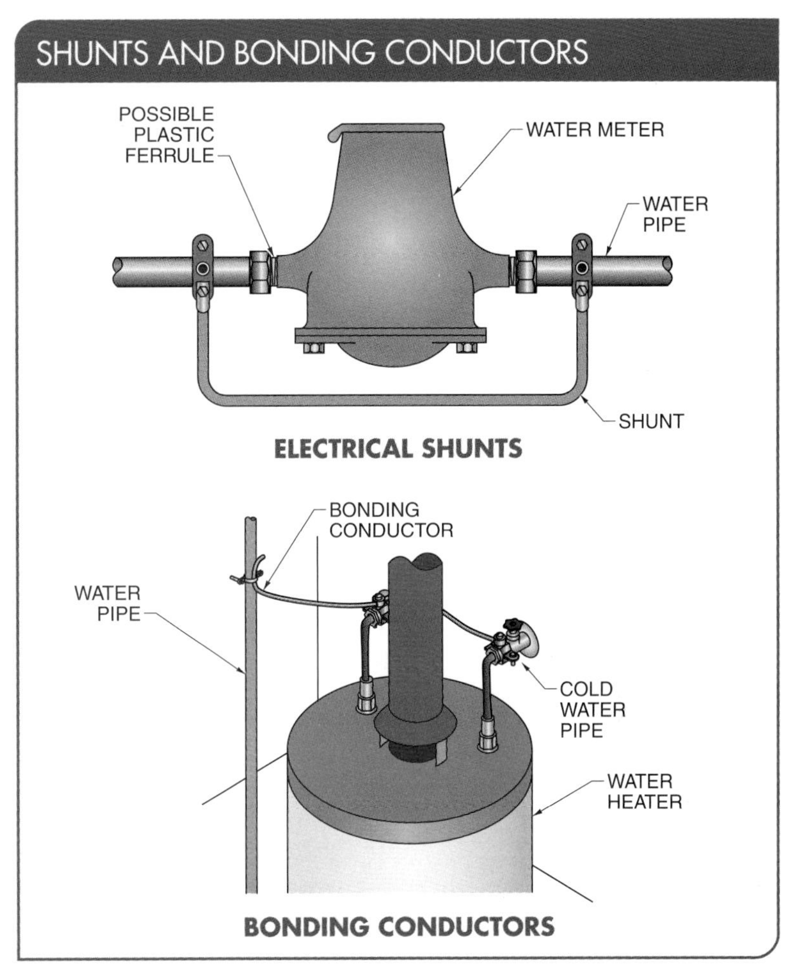

Figure 4-2. A shunt is a permanent conductor placed across a water meter to provide a continuous flow path for ground current, while bonding conductors are used to bond and connect all internal piping systems capable of becoming energized.

Ground clamps are used to properly secure ground conductors to water pipes.

Equipment Grounding

The primary function of equipment grounding is to protect individual electrical devices. Equipment grounding safely grounds any devices or appliances attached to an electrical system or plugged into receptacles inside a home. For example, when a refrigerator has not been properly grounded, electrical current caused by a short will seek the easiest path to earth. Unfortunately, the human body is an electrical conductor and allows current to reach earth by traveling through the body (electric shock). Proper equipment grounding protects the body by harmlessly conducting unwanted electricity to ground. **See Figure 4-3.**

Grounding Small Appliances. Small appliances are easily incorporated into a grounded system. Most small electrical appliances are designed with three-prong grounded plugs that match a standard three-prong grounded receptacle. **See Figure 4-4.** The U-shaped blade of the plug and the U-shaped hole in the receptacle are the ground connections. The U-shaped blade of a plug is longer than the current-carrying blades. The added length ensures a strong ground connection while the plug is being inserted or removed from a receptacle. The ground wire is connected to all receptacles and metal boxes to provide a continuous pathway for short-circuit current. The ground wire may be connected to each box using a pigtail, screw, or ground clip.

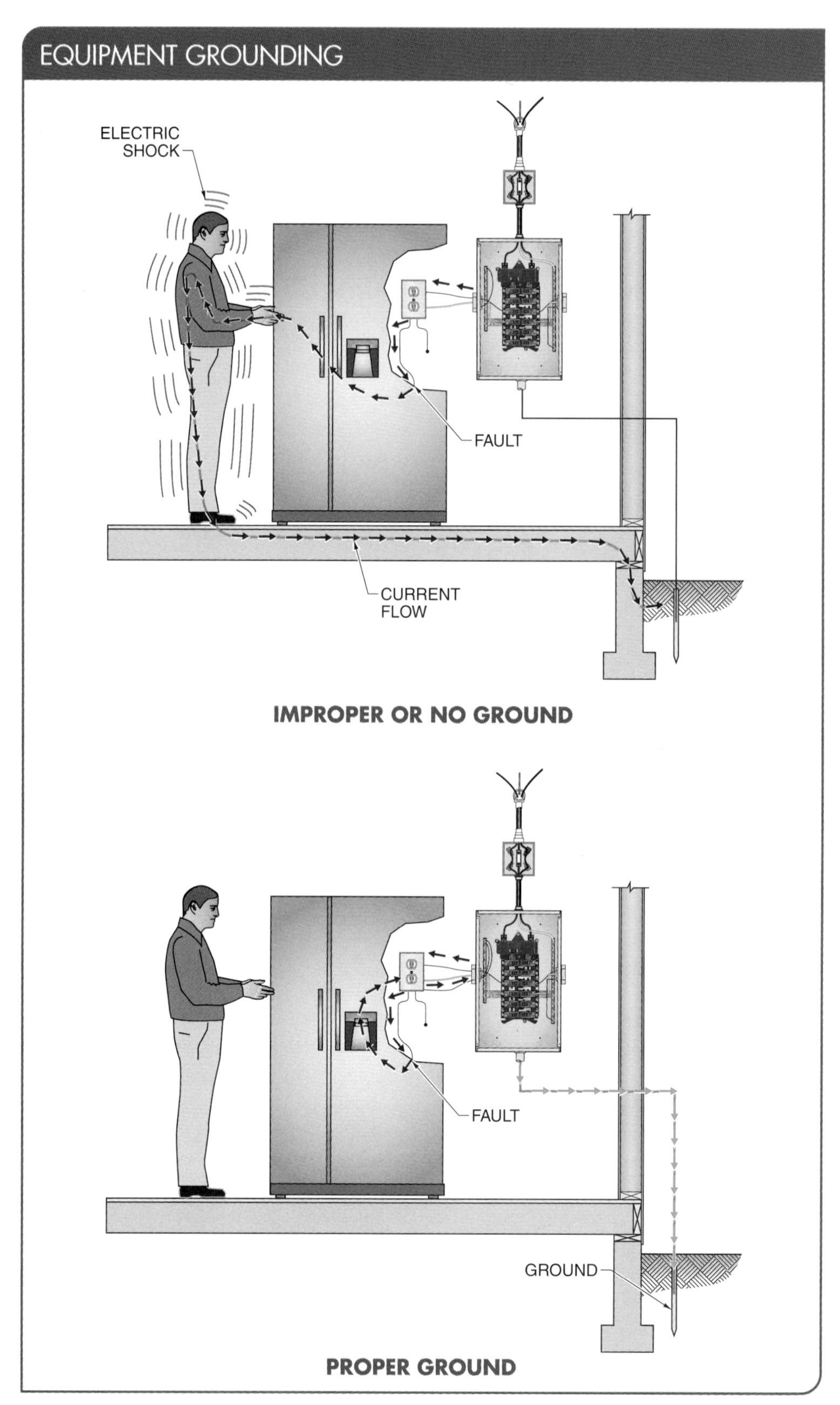

Figure 4-3. The primary function of equipment grounding is to protect individual electrical devices.

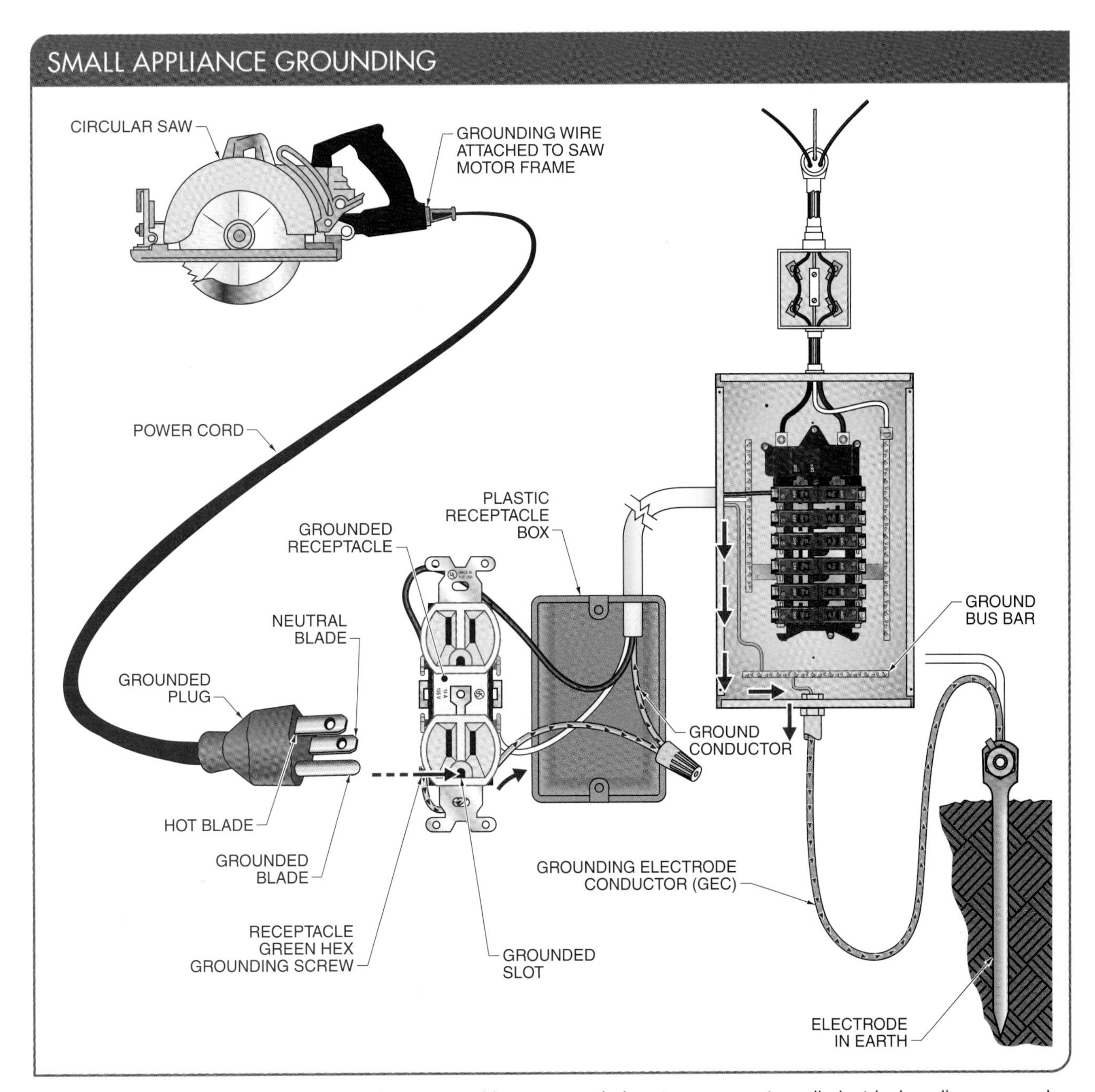

Figure 4-4. Small appliances are easily incorporated into a grounded system, as most small electrical appliances are designed with three-prong grounded plugs that match standard three-prong grounded receptacles.

CODE CONNECT

According to the NEC®, an equipment grounding conductor (EGC) provides a ground-fault current path and connects the non-current-carrying metal parts of the equipment together and to the system ground and/or grounding electrode in order to establish a direct path to earth.

SECTION 4.1 CHECKPOINT

1. What is a system ground?
2. What is an equipment ground?
3. What is the purpose of grounding in a residence?
4. What is an electrode?

SECTION 4.2—OVERCURRENT PROTECTIVE DEVICES

An *overcurrent protective device (OCPD)* is electrical equipment that is used to protect service, feeder, and branch circuits and equipment from excess current by interrupting the flow of current. OCPDs include fuses, electromechanical circuit breakers, and DC circuit breakers.

Fuses

A *fuse* is an OCPD used to limit the rate of current flow in a circuit. Fuses typically contain a piece of soft metal that melts, opening the circuit, when the fuse is overloaded. Three categories of fuses include ferrule cartridge, blade cartridge, and plug fuses. **See Figure 4-5.**

A typical residence generally requires 15 A or 20 A plug fuses and 30 A or 40 A cartridge fuses. Occasionally, a service panel blade cartridge fuse may need to be replaced. Blade cartridge fuses typically have 60 A or 100 A ratings and are found in the main disconnect of the service panel or subpanel. For each category of fuse, several specific types of fuses can be found for specific applications. **See Figure 4-6.**

Standard Plug Fuses. A *standard plug fuse* is a screw-in OCPD that contains a metal conducting element designed to melt when the current through the fuse exceeds the rated value. **See Figure 4-7.** During a serious overload condition, the melting of the conductor strip is almost instantaneous.

Time-Delay Plug Fuses. A *time-delay plug fuse* is a screw-in OCPD with an internal dual element. The first element

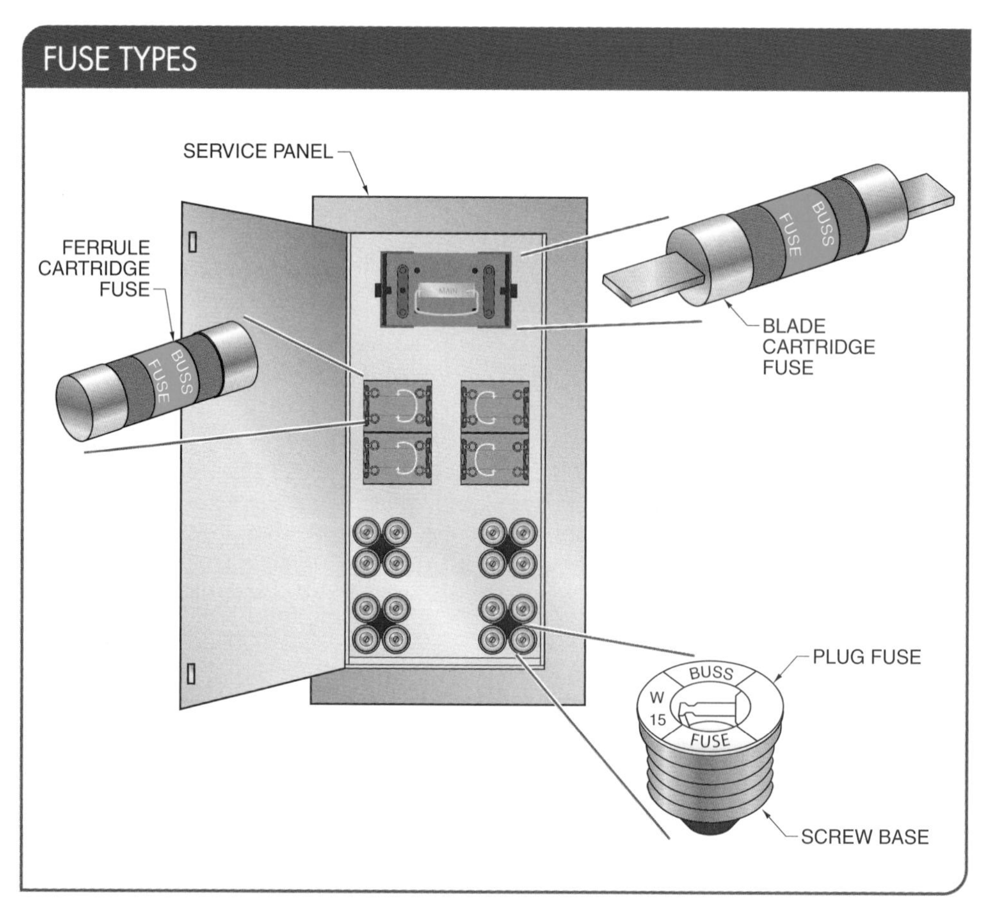

Figure 4-5. Three categories of fuses include ferrule cartridge, blade cartridge, and plug fuses.

provides the protection of a Type T standard plug fuse for short circuits. The second element protects against heating due to light overloads. **See Figure 4-8.** The second element is useful in preventing nuisance tripping caused from the startup of motor-driven appliances, such as refrigerators and air conditioners. Manufacturers make time-delay plug fuses with various responses to shorts and overloads.

A Type S fuse is a screw-in OCPD that has all the operating characteristics of a time-delay plug fuse as well as the added advantage of being "nontamperable." A Type S fuse is considered nontamperable because the fuse cannot be installed into a base unless the fuse matches the size of the base. Each Type S base adapter is sized for a particular size fuse. For example, a nontamperable 20 A Type S fuse will not fit into a 15 A Type S base.

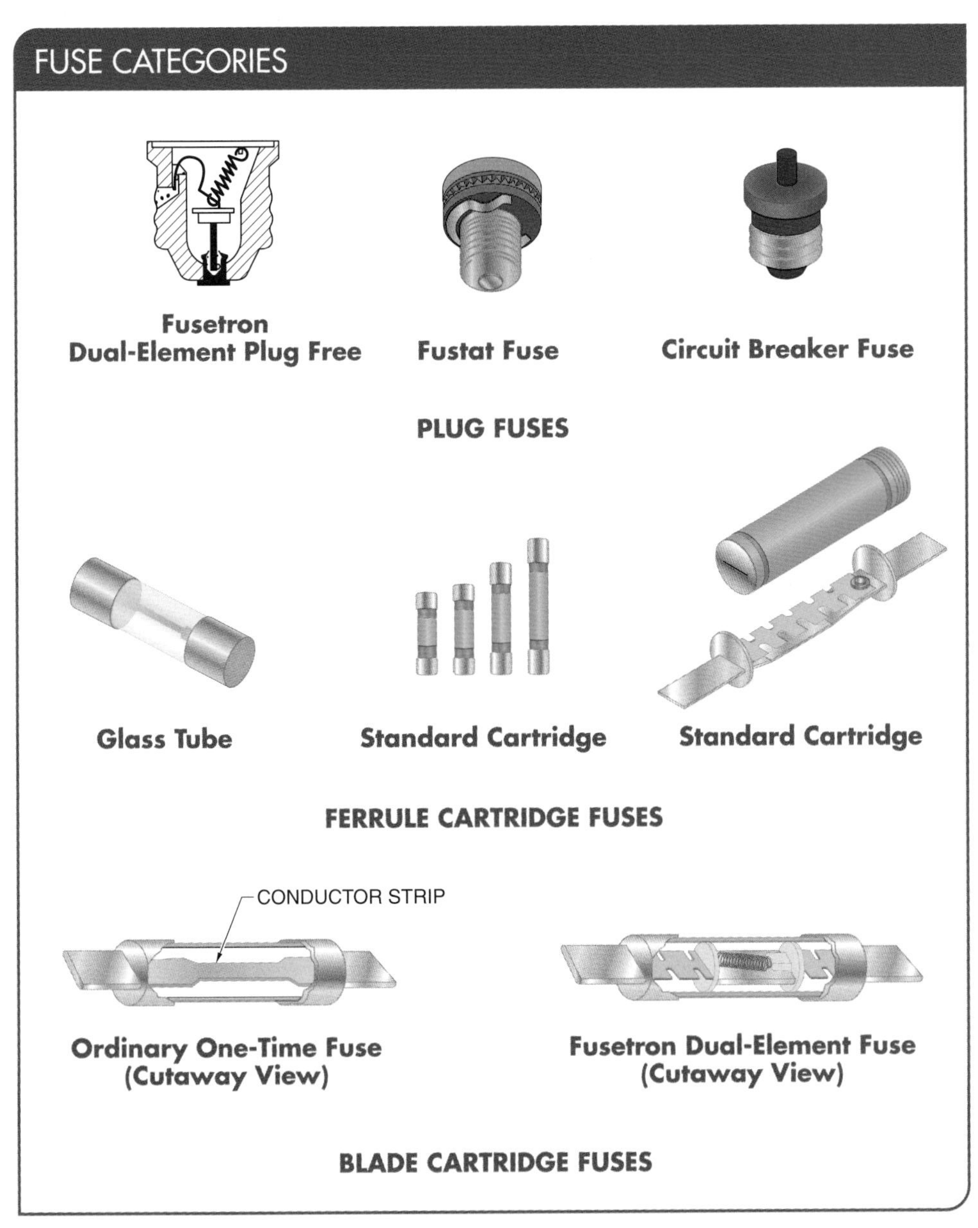

Figure 4-6. For each category of fuse, several specific types of fuses can be found for specific applications.

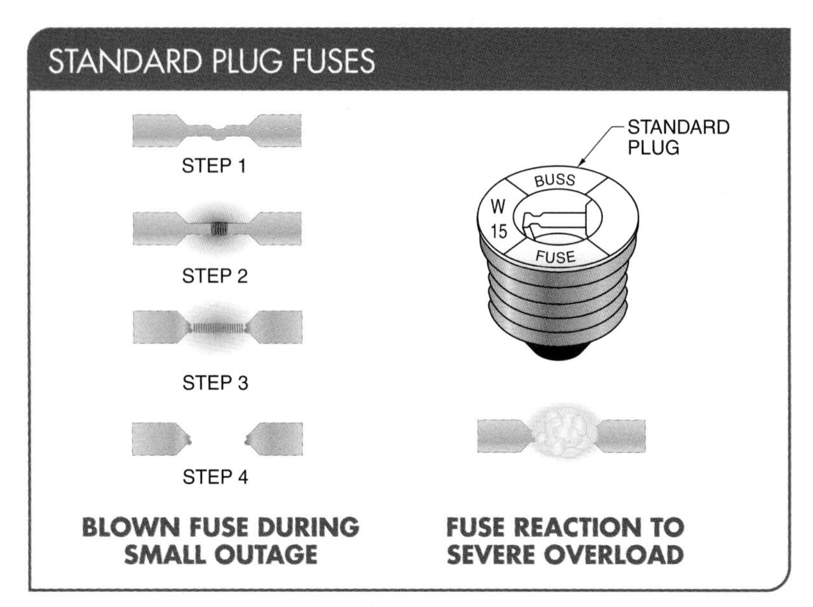

Figure 4-7. A standard plug fuse is a screw-in OCPD that contains a metal conducting element designed to melt when the current through the fuse exceeds the rated value.

Cartridge Fuses. A *cartridge fuse* is a snap-in OCPD that operates on the same basic heating principle as a plug fuse. However, instead of a screw base, cartridge fuses are secured by clips. Cartridge fuses include both blade and ferrule types. **See Figure 4-9.**

A *blade cartridge fuse* is a snap-in OCPD with blade-like connectors at each end of the fuse that operates based on the heating effect of an element. Blade cartridge fuses are typically used for applications with high amperage ratings. A *ferrule cartridge fuse* is a snap-in OCPD that has conductive ferrules at each end of the fuse. Many cartridge fuses have dual elements that absorb temporary overloads (such as when starting up a motor) without blowing, while still protecting against dangerous overloads and short circuits.

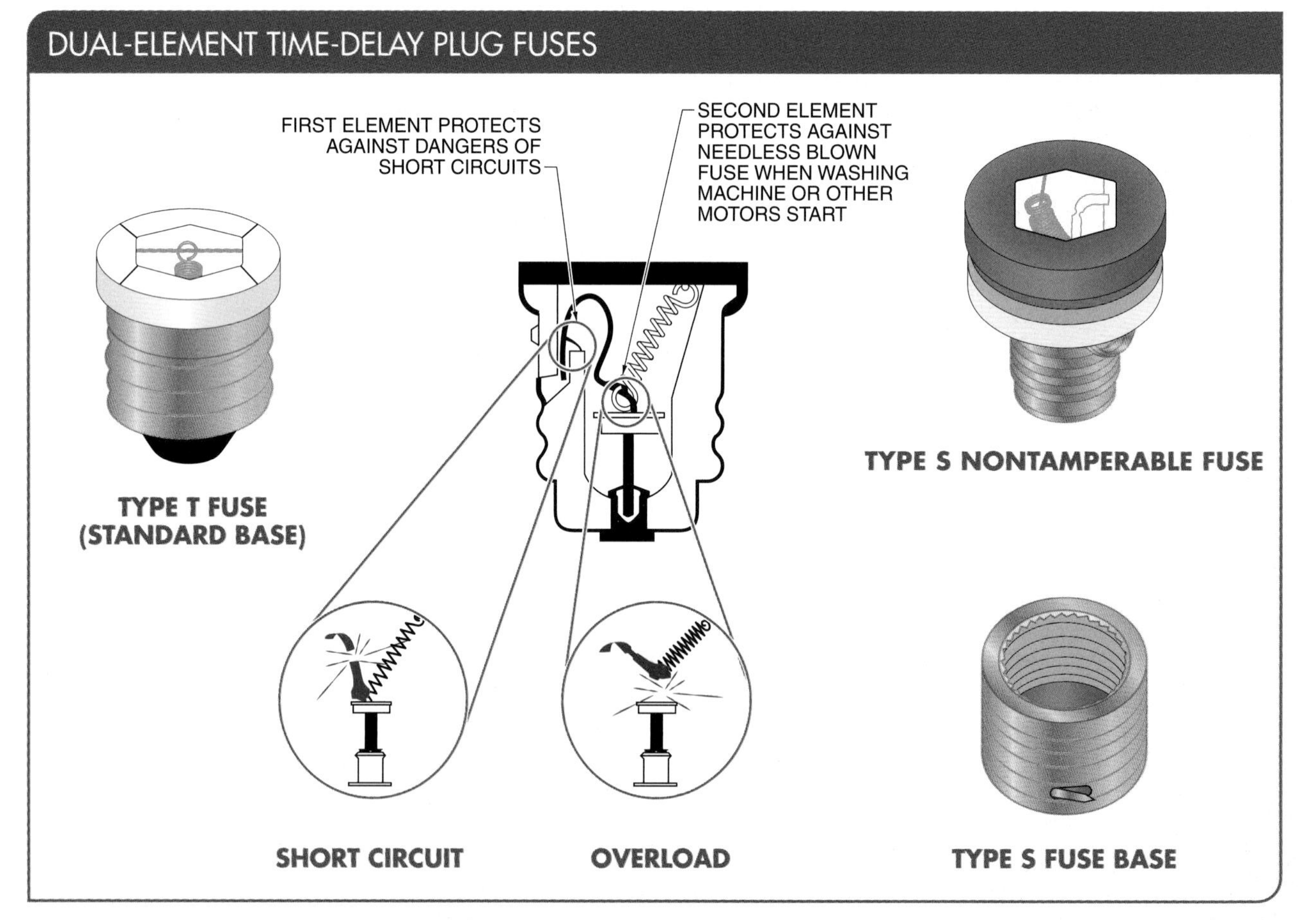

Figure 4-8. A time-delay plug fuse is a screw-in OCPD with an internal dual element.

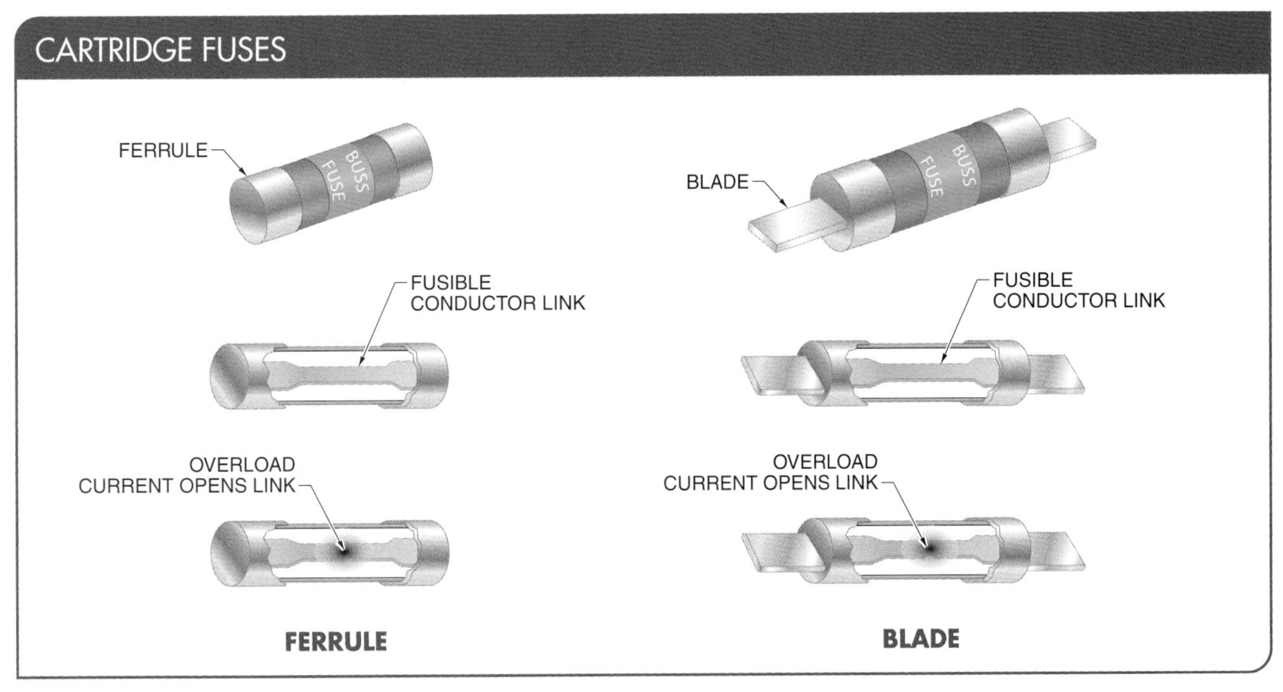

Figure 4-9. A cartridge fuse is a snap-in OCPD that operates on the same basic heating principle as a plug fuse, but it is secured by clips instead of a screw base.

Cartridge fuses can be purchased as one-time-use fuses or as renewable-element fuses. Renewable-element cartridge fuses have replaceable elements. Cartridge fuses should be removed and installed using fuse pullers. **See Figure 4-10.**

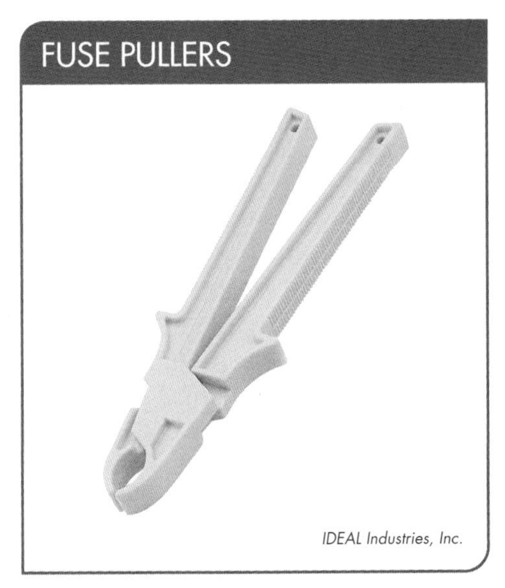

Figure 4-10. Cartridge fuses should be removed and installed using fuse pullers.

Circuit Breakers

A *circuit breaker* is an OCPD designed to protect electrical devices and individuals from overcurrent conditions. Unlike most fuses, circuit breakers can be reset, which makes them a popular choice for overcurrent protection. Circuit breakers operate based on the principle of electromagnetism.

Electromagnetism is magnetism that is produced when electric current passes through a conductor. A magnetic field is created around an electrical conductor when electric current flows through the conductor. **See Figure 4-11.**

If a conductor is wound into multiple loops (a coil), the magnetic lines of force combine and increase the magnetic effect. Thus, the magnetic force of a coil with multiple turns is stronger than the magnetic force of a coil with a single loop. The magnetic effect can be increased even more by inserting an iron core into the electromagnetic coil. An iron core, such as an iron plunger, inserted into a coil increases the magnetic effect to a point where the core can be used as a tool that can do useful work.

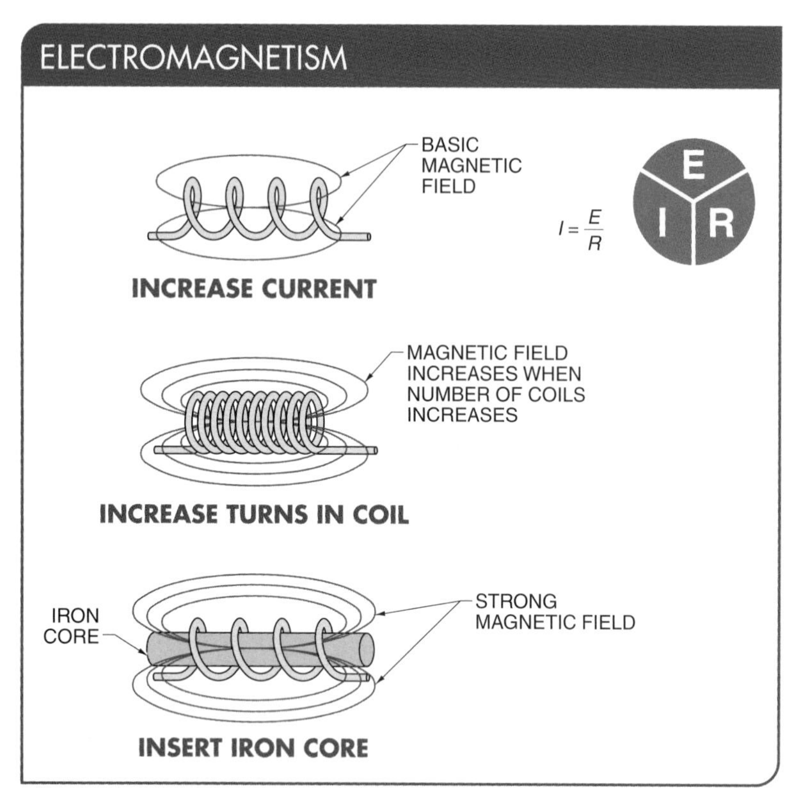

Figure 4-11. An electromagnet can be strengthened by increasing the amount of current by increasing the voltage, increasing the number of turns in the coil, and inserting an iron core through the coil.

An example of electromagnetism performing work is in the operation of an electromagnetic solenoid. A *solenoid* is an electric actuator that consists of an iron plunger surrounded by an encased coil of wire. When a set of contacts are attached to an iron plunger, moving the iron plunger through the coil can open or close contacts in the solenoid. **See Figure 4-12.**

The position of the contacts in a circuit breaker is represented by the trip lever handle. When the circuit breaker is on, the contacts are closed; and when the breaker is tripped off, the contacts are open. A circuit breaker may be reset by moving the trip lever handle to the full OFF position and then returning the handle to the ON position. Individuals must ensure the source of an overload is cleared before attempting to reset a breaker. Three types of circuit breakers are magnetic, thermal, and thermal-magnetic circuit breakers, which are differentiated by their internal mechanisms for tripping.

Regardless of which internal mechanism a circuit breaker uses, most circuit breakers look the same externally, with the exception of the circuit breaker fuse. A *circuit breaker fuse* is a screw-in OCPD that has the operating characteristics of a circuit breaker. The advantage of a circuit breaker fuse is that the fuse can be reset after an overload. Circuit breakers are available in a variety of amperages, but the voltage is typically rated as 110 V for single-pole residential breakers or 220 V for double-pole residential breakers. To gain access to the circuit breaker connections in a service panel, the cover of the panel must be removed.

Circuit breakers are available in a number of configurations, including single-pole and double-pole breakers.

Magnetic Circuit Breakers. A *magnetic circuit breaker* is an OCPD that operates by using miniature electromagnets to open and close contacts. In a magnetic circuit breaker, electric current passed through the coil causes the contacts attached to the iron core to be pulled toward the coil. A solenoid in a magnetic circuit breaker opens the circuit based on current limit of the breaker. When the current through the coil exceeds the rated value of the breaker, the magnetic attraction becomes strong enough to activate the trip lever handle and open the circuit. **See Figure 4-13.** Once the overload is removed, the trip lever handle can be reset to the original position, reactivating the circuit.

TECH TIP

A properly operating circuit breaker will not reset until the problem is corrected.

Thermal Circuit Breakers. A *thermal circuit breaker* is an OCPD that operates with a bimetallic strip that warps when overheated. The bimetallic strip is two pieces of dissimilar metals that are permanently joined together. Because metals expand and contract at different rates, heating the bimetallic strip causes the strip to warp or curve. The warping effect of the bimetallic strip is utilized as the tripping mechanism for the thermal breaker. **See Figure 4-14.**

Thermal-Magnetic Circuit Breakers. A *thermal-magnetic circuit breaker* is an OCPD that combines the heating effect of a bimetallic strip with the pulling strength of a magnet to move a trip bar. The magnetic portion consists of an electromagnet in series with the bimetallic strip. Thermal-magnetic circuit breakers have the fastest response times to serious overloads of any type of circuit breaker.

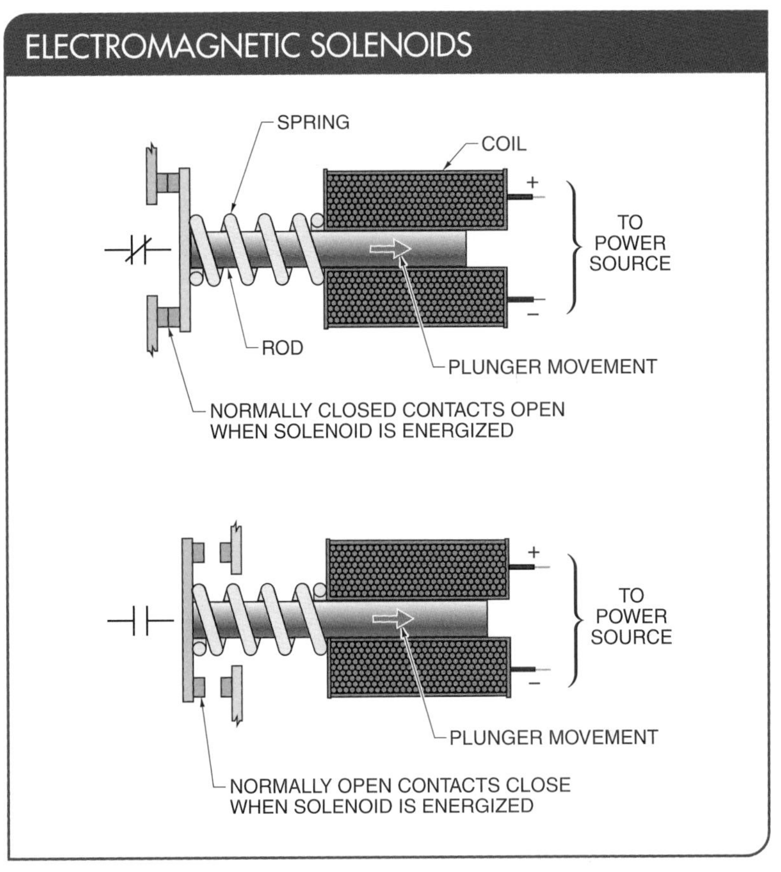

Figure 4-12. Electromagnetic solenoids are an example of using electromagnetism to do work.

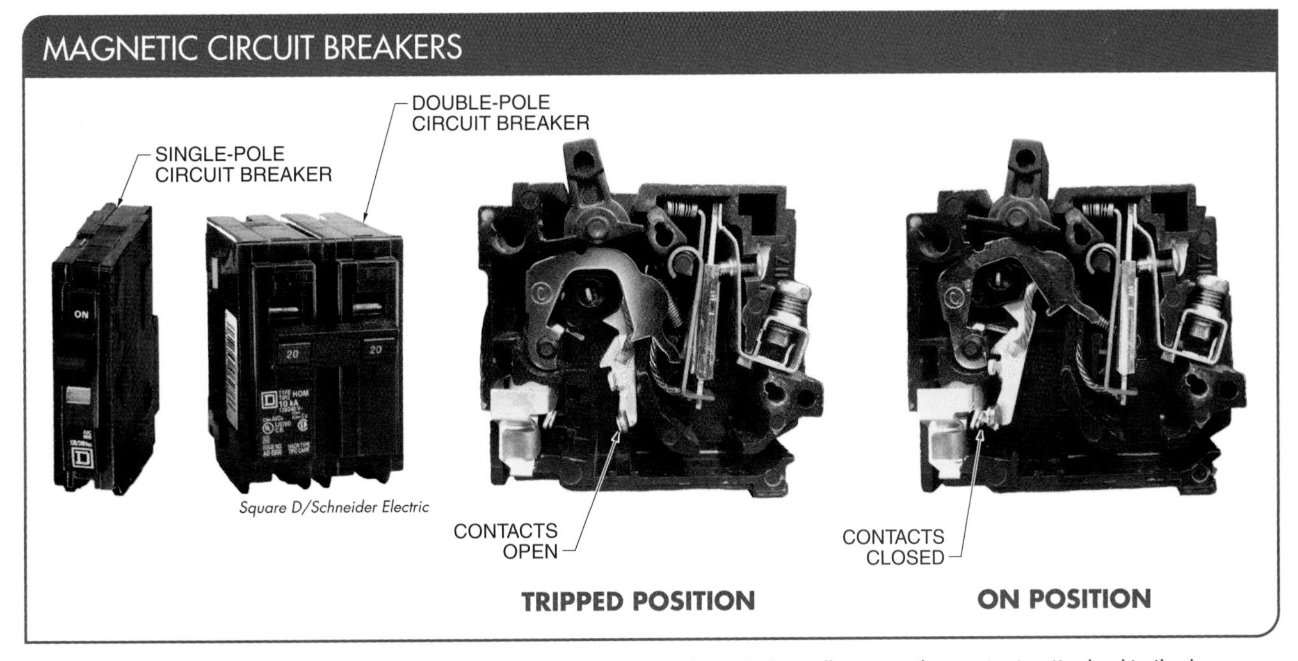

Figure 4-13. In a magnetic circuit breaker, passing electric current through the coil causes the contacts attached to the iron core to be pulled toward the coil. The solenoid in a magnetic circuit breaker opens and closes the contacts based on current flow.

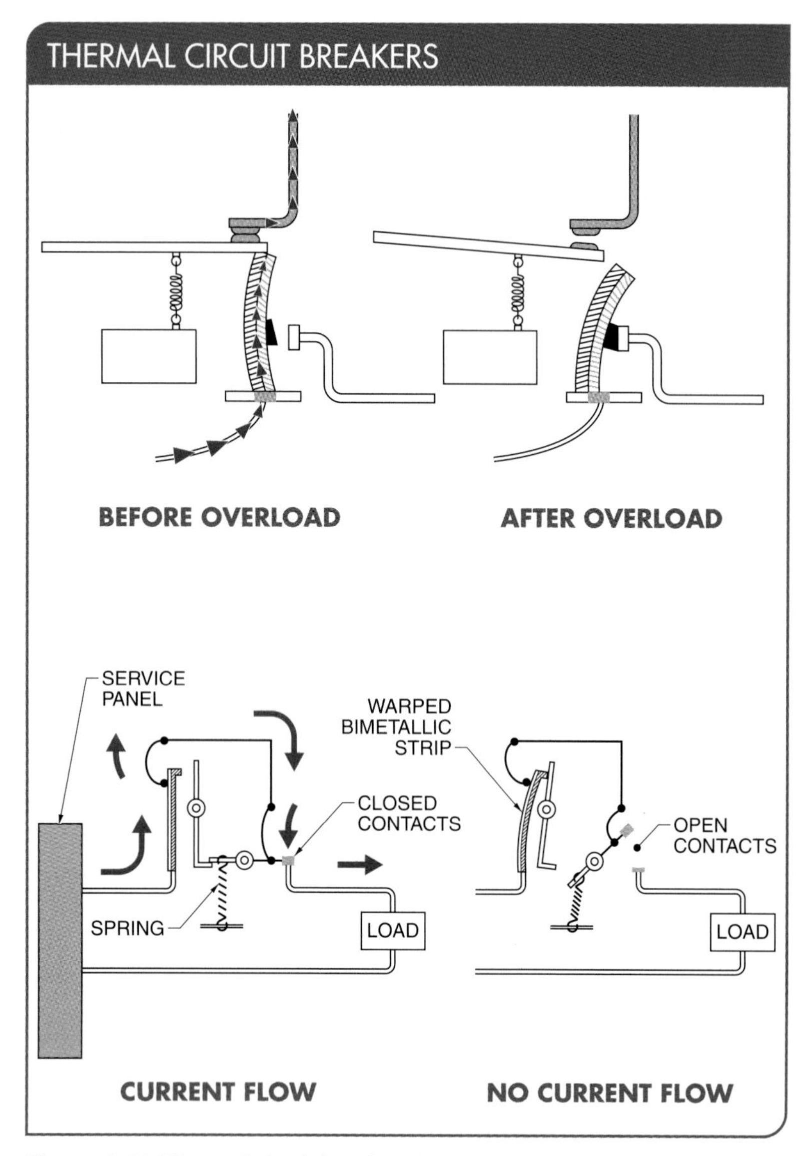

Figure 4-14. Thermal circuit breakers have bimetallic strips that warp with increases in heat caused by excessive current flow.

DC Circuit Breakers

A *DC circuit breaker* is an OCPD that protects electrical devices operating with DC and contains additional arc-extinguishing measures. DC circuit breakers are a relatively new technology to most homeowners since most devices used in the home work with AC and AC circuit breakers. Some manufacturers produce circuit breakers that are dual rated for both AC/DC from 48 VDC to 125 VDC.

> **TECH TIP**
>
> *DC circuit breakers are used with 24 VDC to 48 VDC programmable logic controllers (PLCs) and in wind power applications.*

Though AC and DC breakers appear similar in form and function, they operate very different internally. During an overload, the internal contacts of both AC and DC circuit breakers separate to protect the circuit. However, as the contacts pull apart from each other, an arc will form as the current jumps across the air gap created. *Contact arcing* is an electrical arc that occurs when opening and closing circuit breakers. **See Figure 4-15.** As the arc continues to jump across the air gap, current will continue to flow through the circuit. These arcs must be extinguished quickly.

The ways in which AC and DC breakers are designed to extinguish the arc are very different and are why AC and DC breakers are not interchangeable. Only breakers that are labeled as DC rated should be used for DC applications. An AC-rated breaker should never be used in a DC circuit. AC circuit breakers are not designed to handle the problems of arcing associated with DC. DC circuit breakers include additional arc-extinguishing measures to elongate and dissipate the electrical arc when opening and closing.

DC Arc Suppression. DC arcs are considered the most difficult to extinguish because the continuous DC supply causes current to flow constantly and with great stability across a much wider gap than an AC supply of equal voltage. To reduce arcing in DC circuits, the switching mechanism must be such that the contacts separate rapidly and with enough of an air gap to extinguish the arc as soon as possible when opening. When DC contacts are being closed, it is necessary that the contacts move together as quickly as possible to prevent some of the same problems encountered in opening them. If a circuit breaker is DC rated, it will be indicated on the breaker by the manufacturers.

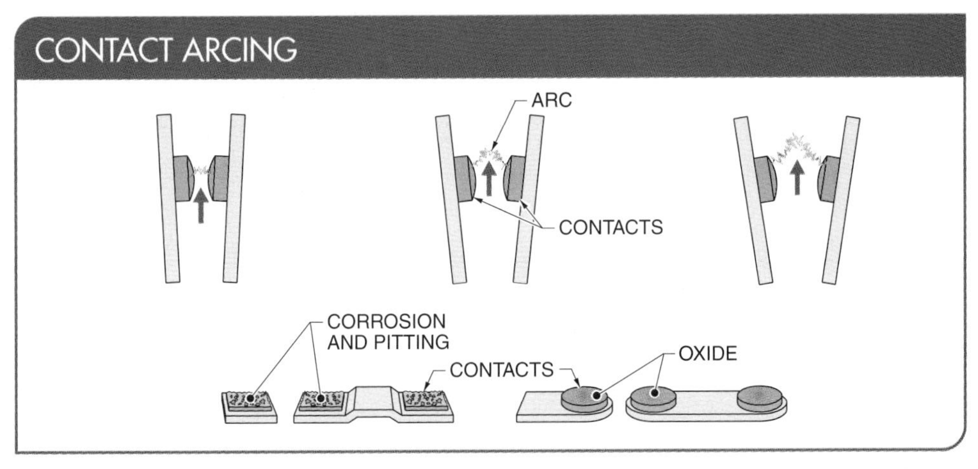

Figure 4-15. Contact arcing is an electrical arc that occurs when opening and closing circuit breakers.

Note: It is permissible to use a circuit breaker that has a dual AC/DC rating. This information will be stated on the manufacturer's label.

AC Arc Suppression. An AC arc self-extinguishes when the set of contacts opens. An AC supply has a voltage that reverses its polarity 120 times a second when operated on a 60 Hz line frequency. The alternation allows the arc to have a maximum duration of no more than a half-cycle. During any cycle, AC current reaches zero 60 times each second. **See Figure 4-16.** When AC reaches zero, no current flows and therefore the arc is extinguished.

Square D/Schneider Electric

Some types of circuit breakers are rated AC/DC for use with either type of application.

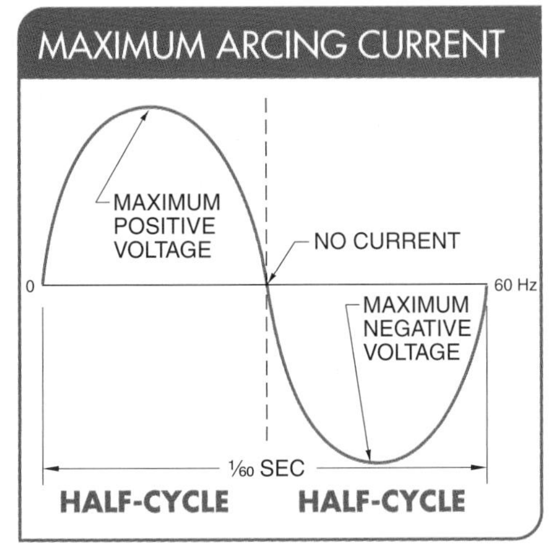

Figure 4-16. When AC current reaches zero, no current flows and therefore the arc is extinguished.

TECH TIP

When a short circuit occurs across the terminals of a DC circuit, the current increases from the operating current to the short-circuit current depending on the resistance and the inductance of the short-circuited loop.

SECTION 4.2 CHECKPOINT

1. What is an overcurrent protective device (OCPD)?
2. What is a fuse?
3. What is a circuit breaker?
4. What is a DC circuit breaker?

SECTION 4.3—GROUND-FAULT AND ARC-FAULT CIRCUIT INTERRUPTERS

Fuses and circuit breakers protect the entire circuit from overload. These types of OCPDs respond to typical currents of 15 A to 20 A, which are more than enough to electrocute a person or start a fire. Ground-fault circuit interrupters (GFCIs) and arc-fault circuit interrupters (AFCIs) were developed in response to the US Consumer Product Safety Commission (CPSC) and other prominent organizations such as the National Fire Protection Association® (NFPA®) to reduce the possibility of fires and electrical accidents. Electricity is the number one cause of fires. More than 1700 people die and/or are injured in electrical fires each year in the United States.

The primary difference between a GFCI and an AFCI is their intended purpose. GFCIs are designed to protect a person from being shocked or electrocuted when they come in contact with a device that is accidentally touching a circuit. AFCIs, on the other hand, are designed to protect a home from catching fire due to arcing in defective conductors or appliances. **See Figure 4-17.**

AFCI receptacles are easily identified by the AFCI marking on the face of the receptacle.

> TECH TIP
>
> *An AFCI is designed to detect a loose connection that causes an arc.*

Electrical Shock

An *electrical shock* is a shock that results anytime a body becomes part of an electrical circuit. Electrical shock kills over 1000 people a year in the United States according to the National Safety Council (NSC). Electrocution can take place when any electrical current entering the human body is strong enough to stop the natural rhythm of the heart. Water with impurities is an excellent conductor of electricity, and the human body is 70% water. When a person touches an energized bare electrical wire or a faulty appliance, the electricity will use the body as the shortest path to the ground. **See Figure 4-18.**

The severity of an electrical shock depends on the amount of electric current (in mA) that flows through the body, the length of time the body is exposed to the current flow, and the path the current takes through the body. The amount of current that passes through a circuit depends on the voltage and resistance of the circuit. During an electrical shock, a person's body becomes part of the electrical circuit. The resistance a person's body offers to the flow of current can vary. Sweaty hands have less resistance than dry hands. A wet floor has less resistance than a dry floor. The lower the resistance, the greater the current flow. The greater the current flow, the greater the severity of shock.

GFCIs have a trip level of 6 mA, which is the amount of electrical current a human body can withstand without serious harm. Physical effects will vary from person to person depending on whether they are male or female, and an adult or a child. Typically, at about 10 mA, a person may not be able to let go if they touch an energized conductor. Current flow as low as 30 mA may cause breathing difficulty and heart problems in small children. A current over 2000 mA (2 A) can cause heart paralysis and tissue burning.

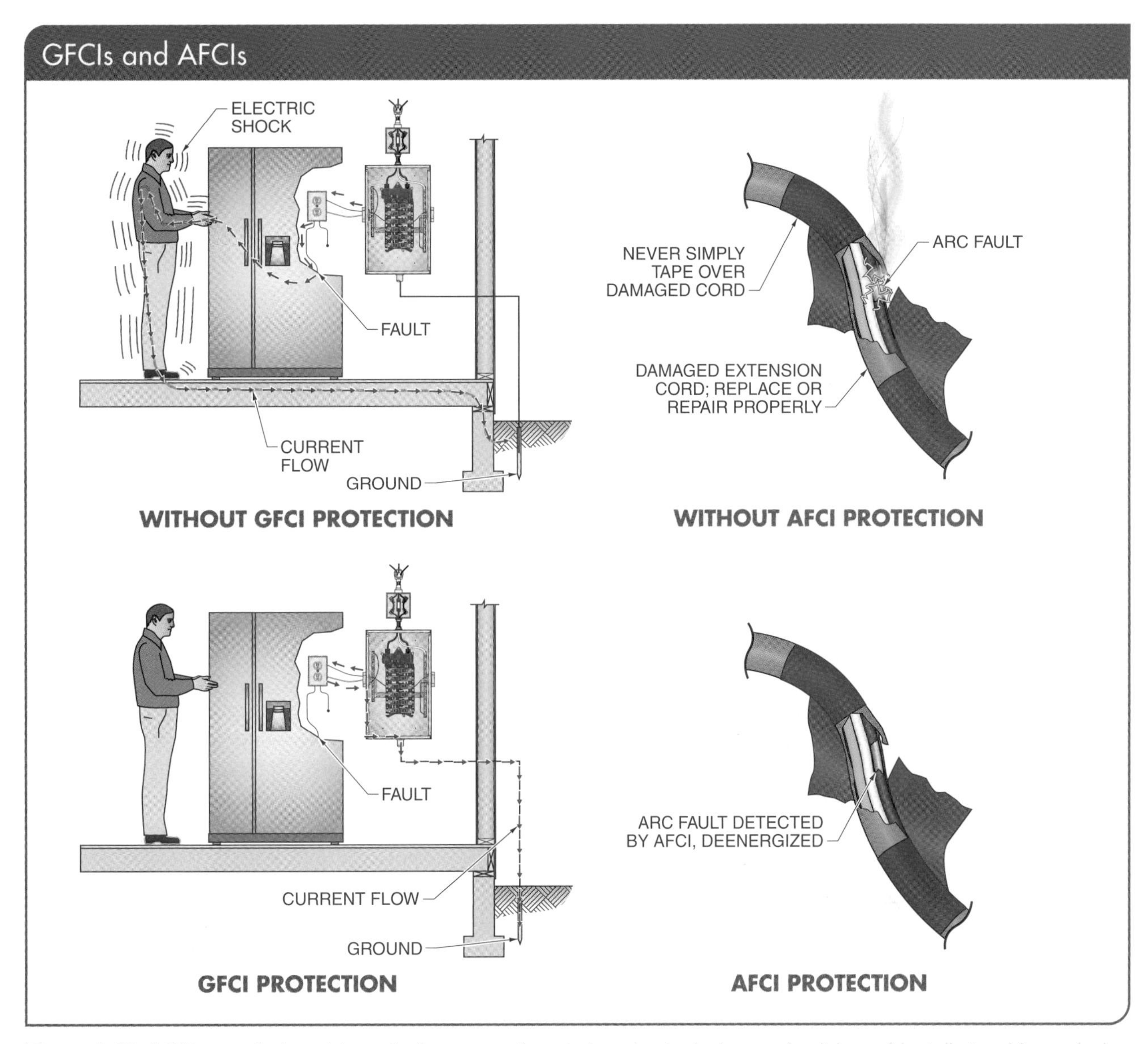

Figure 4-17. GFCIs are designed to protect a person from being shocked when a circuit is accidentally touching a device that a person comes in contact with. AFCIs are designed to protect a home from catching fire due to arcing in defective conductors or appliances.

GFCIs

A *ground-fault circuit interrupter (GFCI)* is an electric device that protects personnel by detecting ground faults and quickly disconnecting power from the circuit. **See Figure 4-19.** GFCIs were created due to the limitations of commonly used circuit breakers, which were designed to trip only when large currents are present (short or overload). Individuals have been electrocuted by equipment in electrical systems where the fault current was not great enough to trip a standard circuit breaker. Currents present from deteriorated insulation and minor equipment damage do not produce enough current flow to open a standard circuit breaker. GFCIs were specifically designed with such leakage currents in mind and can react to current as small as 5⁄1000 (0.005) of an amp in a fraction of a second.

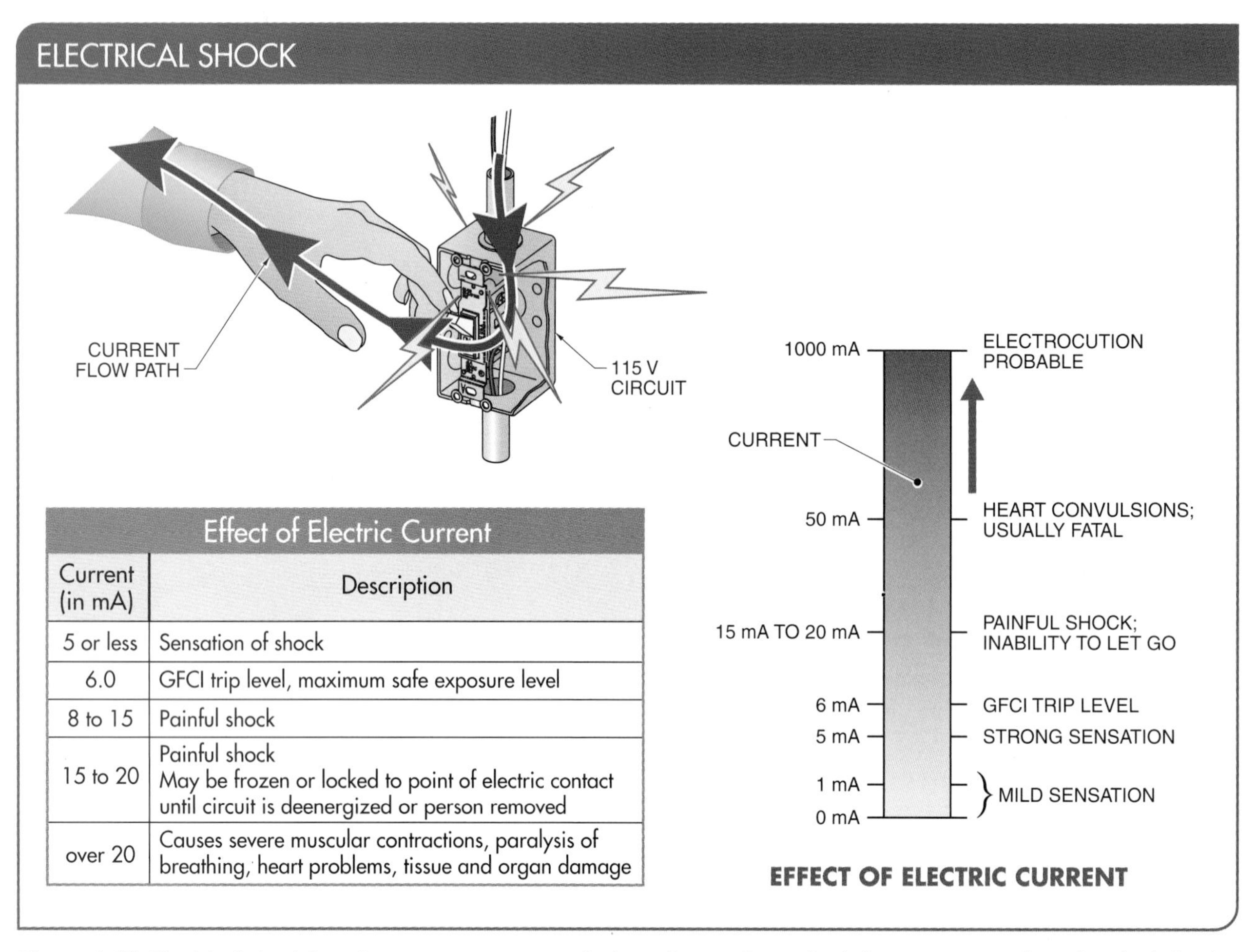

Effect of Electric Current

Current (in mA)	Description
5 or less	Sensation of shock
6.0	GFCI trip level, maximum safe exposure level
8 to 15	Painful shock
15 to 20	Painful shock May be frozen or locked to point of electric contact until circuit is deenergized or person removed
over 20	Causes severe muscular contractions, paralysis of breathing, heart problems, tissue and organ damage

Figure 4-18. Electrical shock is a dangerous occurrence that results any time a body becomes part of an electrical circuit.

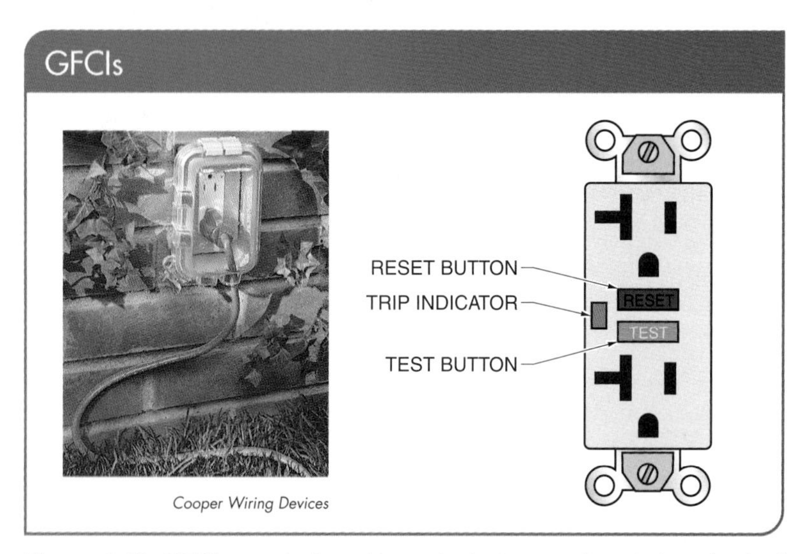

Figure 4-19. GFCIs are designed to protect a person from being shocked or electrocuted when they come in contact with a device that is accidentally touching a circuit. GFCIs can react to current as small as 5/1000 (0.005) of an ampere in a fraction of a second.

GFCI Operation. A sequence of events takes place when a fault current is detected by a GFCI. Typically, a load (for example, an electric shaver, drill, or garden tool) has the same amount of current flowing to it through the black or red (hot) wire as flowing away from it through the white (neutral) wire. However, in the event of a ground fault, some of the current, which normally returns to the power source through the white wire, is diverted to ground. **See Figure 4-20.**

Because a current imbalance exists between the hot and neutral wires, the sensing device detects the current difference and signals an amplifier. When the amplified signal is large enough, the amplifier activates the interrupting device. Once the interrupting device is activated, the circuit is opened and current to the load is shut off.

After the ground fault has been discovered and repaired, the GFCI is reset and is ready to begin the process again. GFCIs have a test circuit built into the unit so that the GFCI can be tested on a monthly basis without a ground fault condition.

GFCI Required Locations. The National Electric Code® (NEC®) requires ground-fault circuit interrupters for protection in many areas. It is always a good idea to check the NEC® requirements as well as local building codes. Required residential GFCI locations include outdoor receptacles, bathrooms, residential garages, kitchens (countertop areas), unfinished basements, and crawl spaces. **See Figure 4-21.**

Additional locations that require GFCI receptacles include construction sites as well as the outdoor receptacles and bathrooms of mobile homes and mobile home parks. GFCIs must also be installed nearby swimming pools and fountains. These locations include receptacles near pools, lighting fixtures and lighting outlets near pools, underwater lighting fixtures over 15 V, electrical equipment used with storable pools, fountain equipment operating at over 15 V, and cord- and plug-connected equipment for fountains.

TECH TIP

Installed GFCIs should be tested for proper operation with a GFCI receptacle tester.

GFCI Installation. GFCIs are installed in standard receptacle boxes indoors and in weatherproof boxes outdoors. GFCI receptacles are typically installed individually. GFCIs may also be installed to protect several standard receptacles in one continuous circuit. GFCIs are considered easy to install and most manufacturers supply an installation kit to aid in installation. In addition to being available as receptacles, GFCIs are also available as circuit breakers to protect entire circuits.

Plug-in GFCIs. Plug-in GFCIs are the most convenient type of GFCI device to use. Through the simple use of a plug-in GFCI in any grounded receptacle, any tool plugged into the GFCI device and its operator are protected. Individuals should typically keep a plug-in GFCI unit in a toolbox, and use the device whenever working with power tools outdoors, especially when conditions are wet. **See Figure 4-22.**

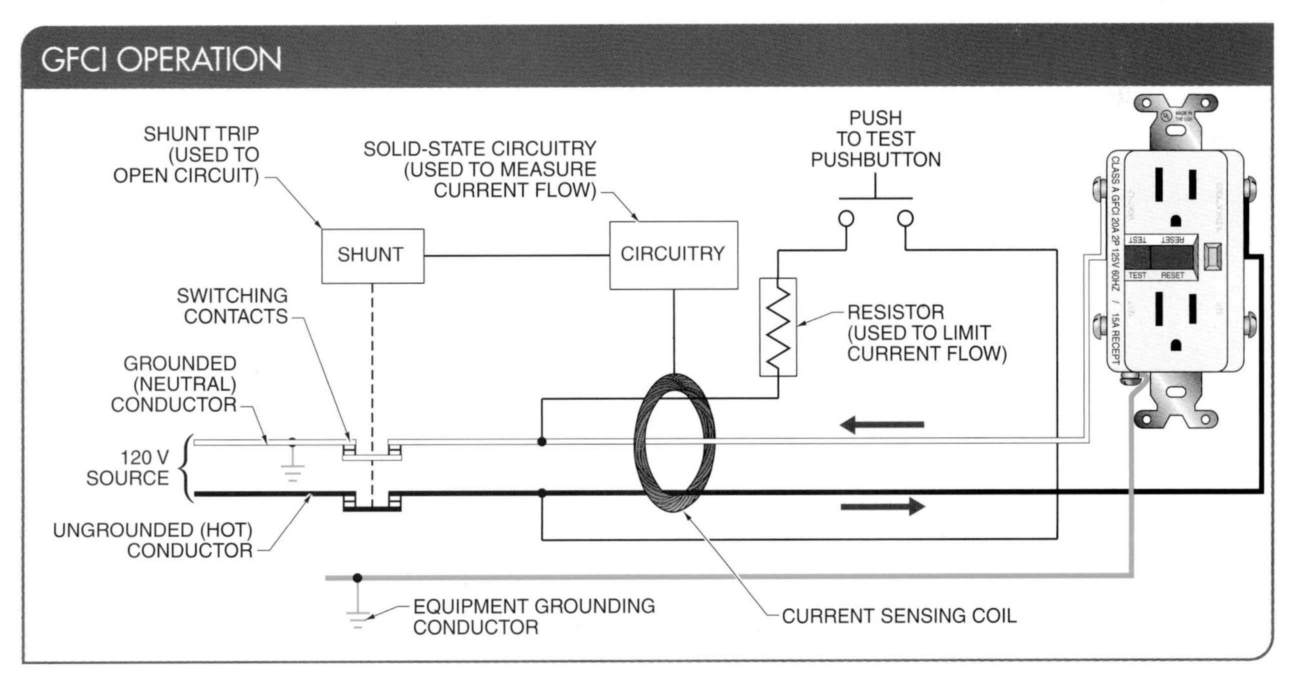

Figure 4-20. GFCIs operate by detecting a current imbalance between the hot and neutral wires and signaling an amplifier. When the amplified signal is large enough, the amplifier activates the interrupting device.

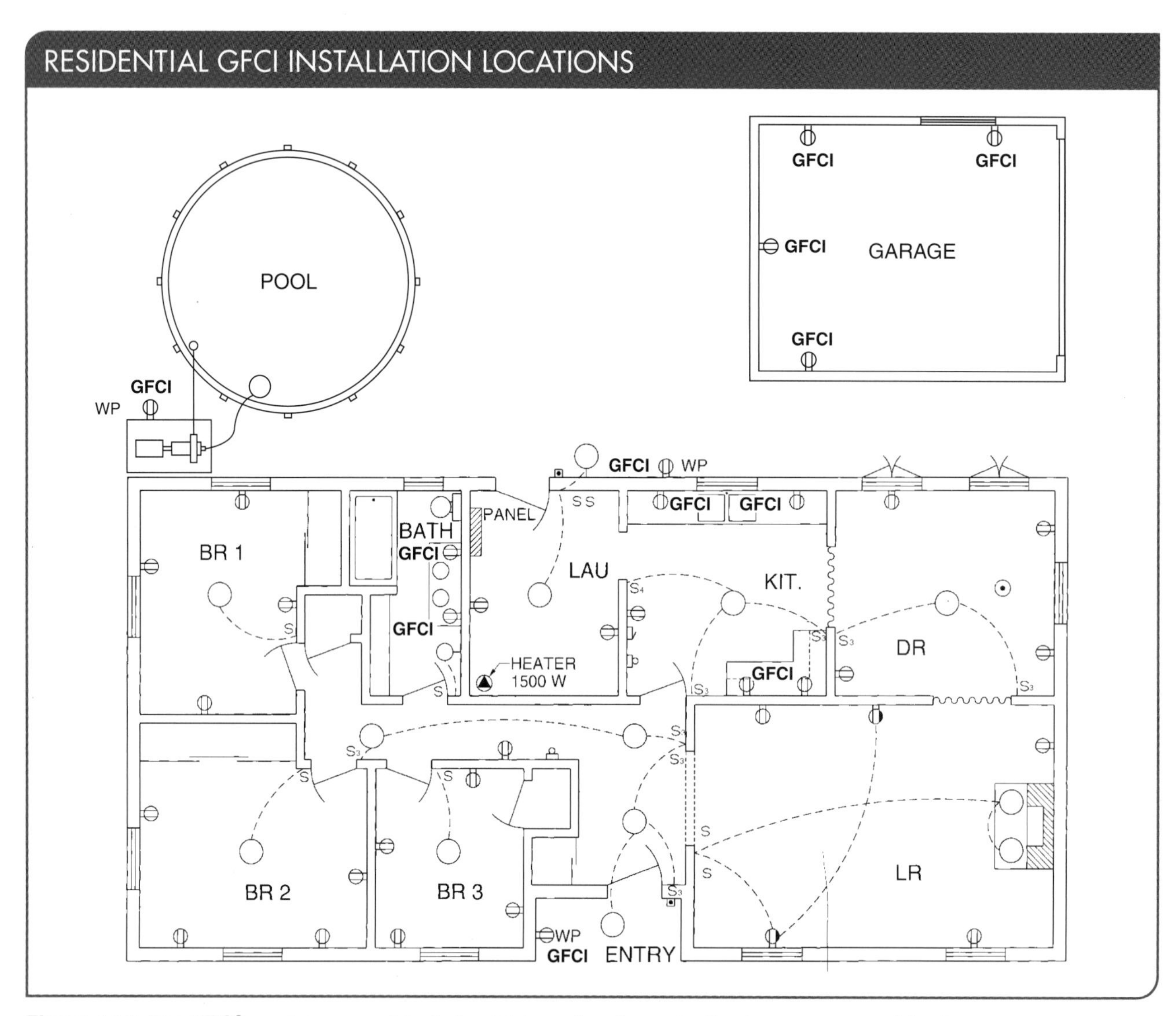

Figure 4-21. The NEC® requires ground-fault circuit interrupters for protection in many areas of the home.

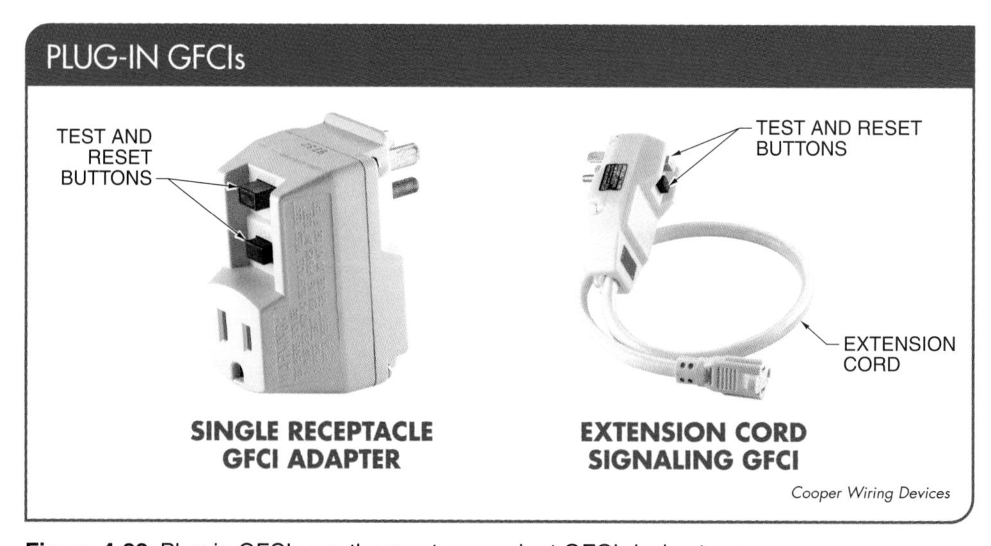

Figure 4-22. Plug-in GFCIs are the most convenient GFCI device to use.

AFCIs

An *arc-fault circuit interrupter (AFCI)* is a current-sensing device designed to detect a wide range of arcing electrical faults. An arc fault is an unintended arc created by current flowing through an unplanned path. **See Figure 4-23.**

Arcing creates high-intensity heating at the point of the arc where burning particles may easily ignite surrounding material, such as wood framing, carpeting, or furniture fabric. The temperatures of these arcs can range from 2000°F to 10,000°F. AFCI outlets and circuit breakers are needed because arc faults are often unseen and not detected by conventional OCPDs. The objective of an AFCI is to protect the circuit in a manner that will reduce the chances that the circuit becomes a source of an electrical fire.

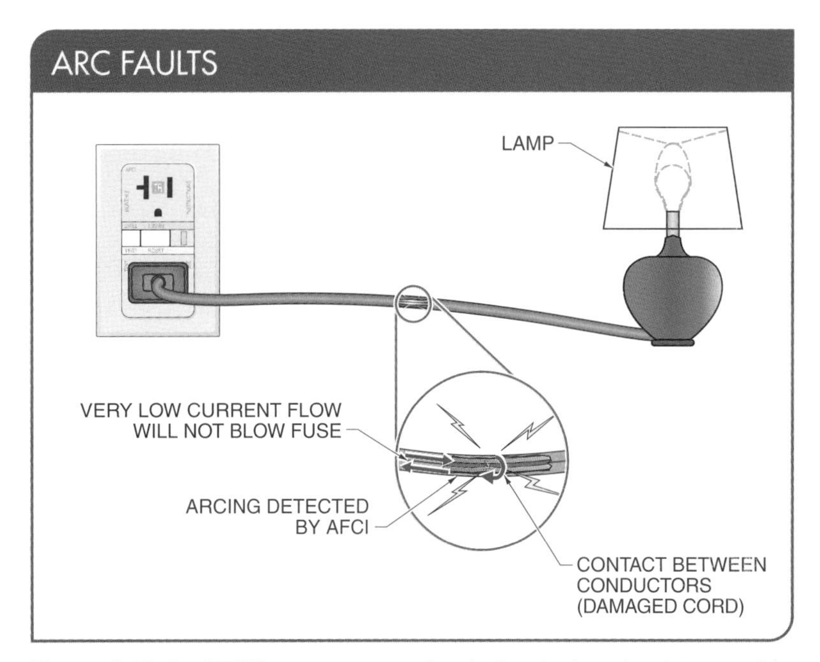

Figure 4-23. An AFCI is a current-sensing device designed to detect a wide range of arcing electrical faults.

ACFI Operation. AFCIs are very sophisticated electronic devices. An AFCI must be able to determine when a hazardous arc is taking place compared to an operational arc. A hazardous arc can lead to an electrical fire. An operational arc is one that occurs normally when switches are turned on or off or motors are started or stopped.

AFCI manufacturers test for the hundreds of possible operating conditions and then program their devices to monitor constantly for both the normal and hazardous arcing conditions. An AFCI contains a microprocessor (computer on a chip) that constantly compares the current voltage patterns in the protected circuit to those found in a normal circuit. When the microprocessor detects a pattern of a hazardous arc, the AFCI trips open the circuit to stop the flow of electricity. The analysis of voltage patterns is one method of determining when the arc is hazardous or normal and compares a normal AC sine wave to one created by a hazardous arc fault. **See Figure 4-24.**

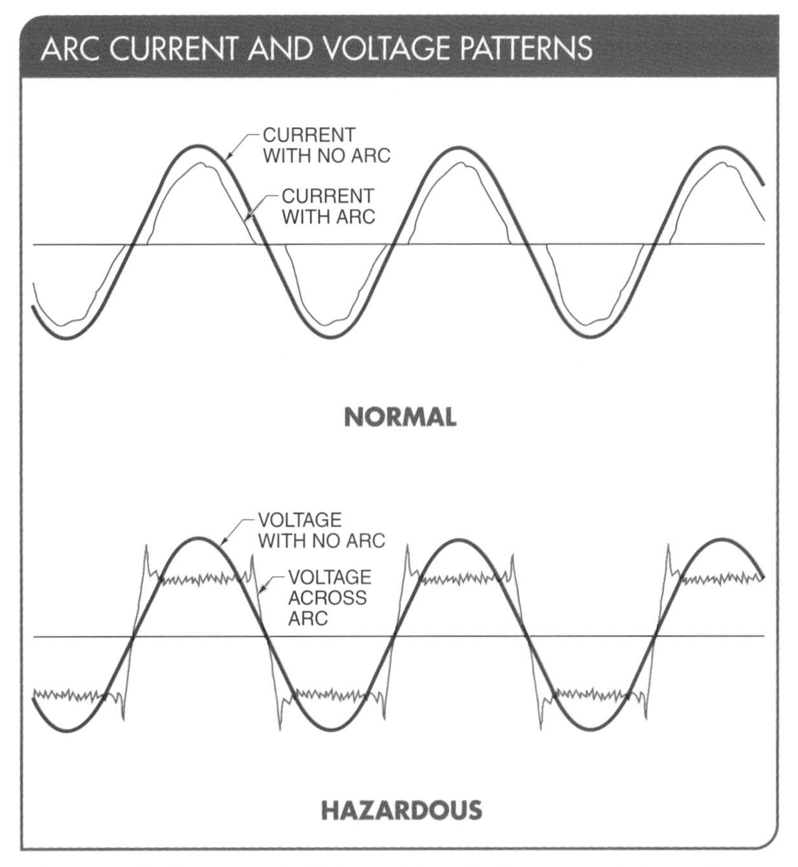

Figure 4-24. One method of determining whether an arc is hazardous or normal is by comparing a normal AC sine wave to that of one created by a hazardous arc fault.

TECH TIP

AFCI receptacles are similar in cost to GFCI receptacles with the same rating.

AFCI Required Locations. The NEC® requires arc-fault protection of all 120 V, 15 A, and 20 A branch circuits in dwelling unit family rooms, dining rooms, living rooms, parlors, libraries, dens, bedrooms, sunrooms, recreation rooms, closets, hallways, or any other similar rooms or areas. An AFCI is most often used in bedroom branch circuits.

AFCI Installation. An AFCI is wired similar to a GFCI in that the AFCI breaker is wired to the hot and neutral of a branch circuit with a pigtail from the AFCI to connect the neutral bus. **See Figure 4-25.** As with a GFCI, an AFCI has a test button to check its operation. Some AFCI manufacturers color-code the test button, making it a different color than the test button on a GFCI.

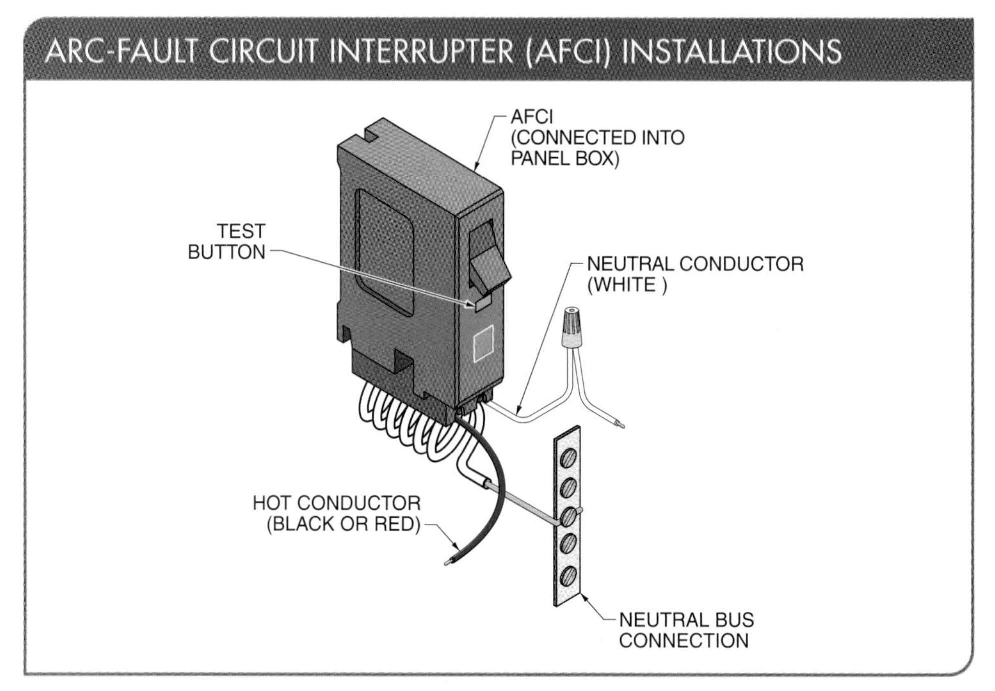

Figure 4-25. AFCIs are wired similar to GFCIs in that the AFCI breaker is wired to the hot and neutral of the branch circuit with a pigtail from the AFCI to connect the neutral bus. As with a GFCI, an AFCI has a test button to check its operation.

TECH TIP

A report by the National Association of State Fire Marshals (NASFM) states that of 73,500 electrical fires from 1994 to 1998 (which caused 591 deaths, 2247 injuries, and an estimated $1,047,900,000 in damage), 60,900 or 82% were caused by arcing, rather than overloads or short circuits.

SECTION 4.3 CHECKPOINT

1. What is an electrical shock?
2. What is a ground-fault circuit interrupter (GFCI)?
3. What is an arc-fault circuit interrupter (AFCI)?
4. What is an arc fault?

4 Electrical Grounding and Overcurrent Protection

Chapter Activities

Name ______________________________ Date ______________

Electrical System Grounding Methods

Identify the electrical system grounding methods shown.

______________ **1.** Grounding electrode

______________ **2.** Water pipe grounding

______________ **3.** Concrete-encased grounding

______________ **4.** Ground ring

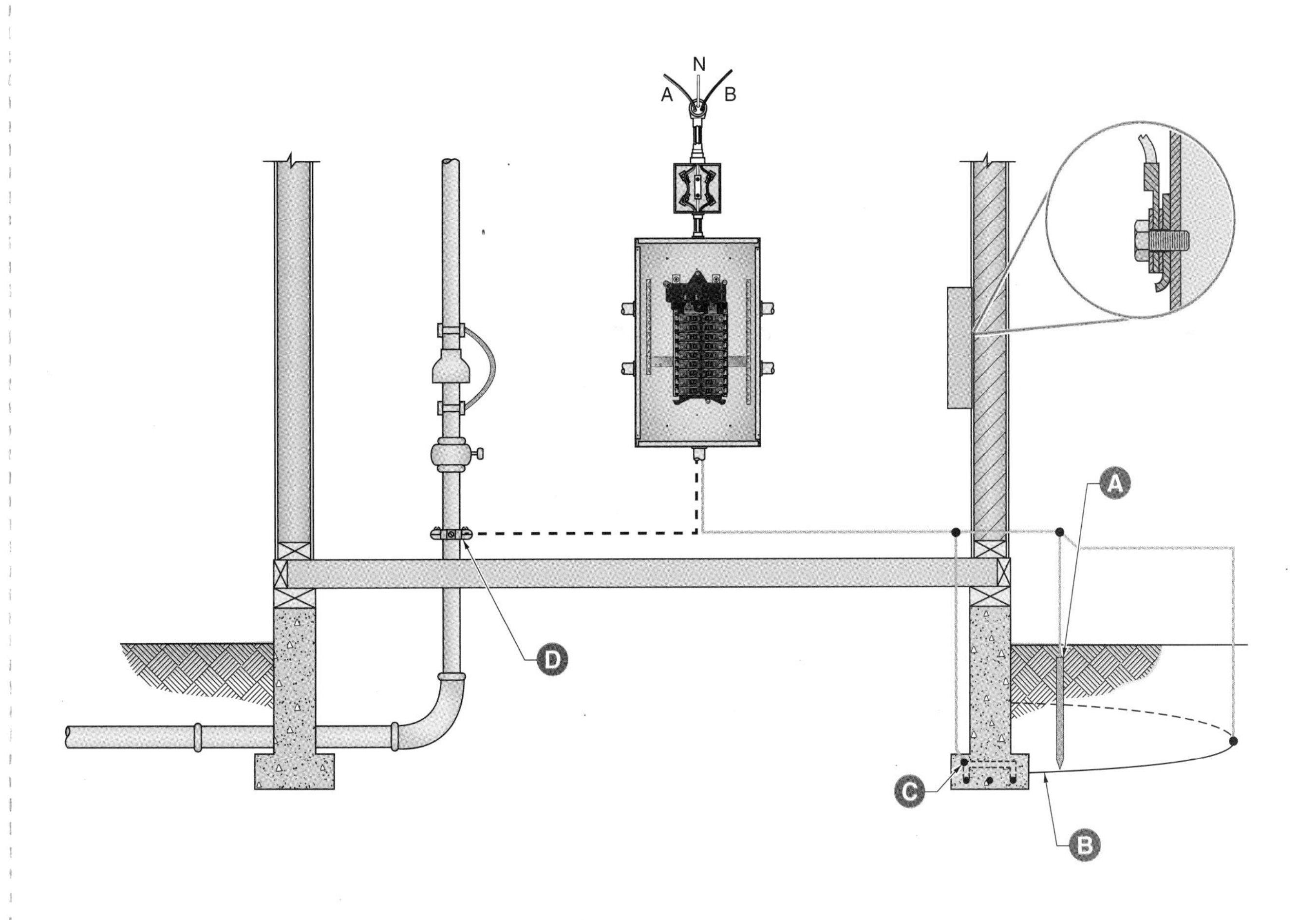

Small Appliance Grounding

1. Draw lines to connect the circular saw to earth ground.

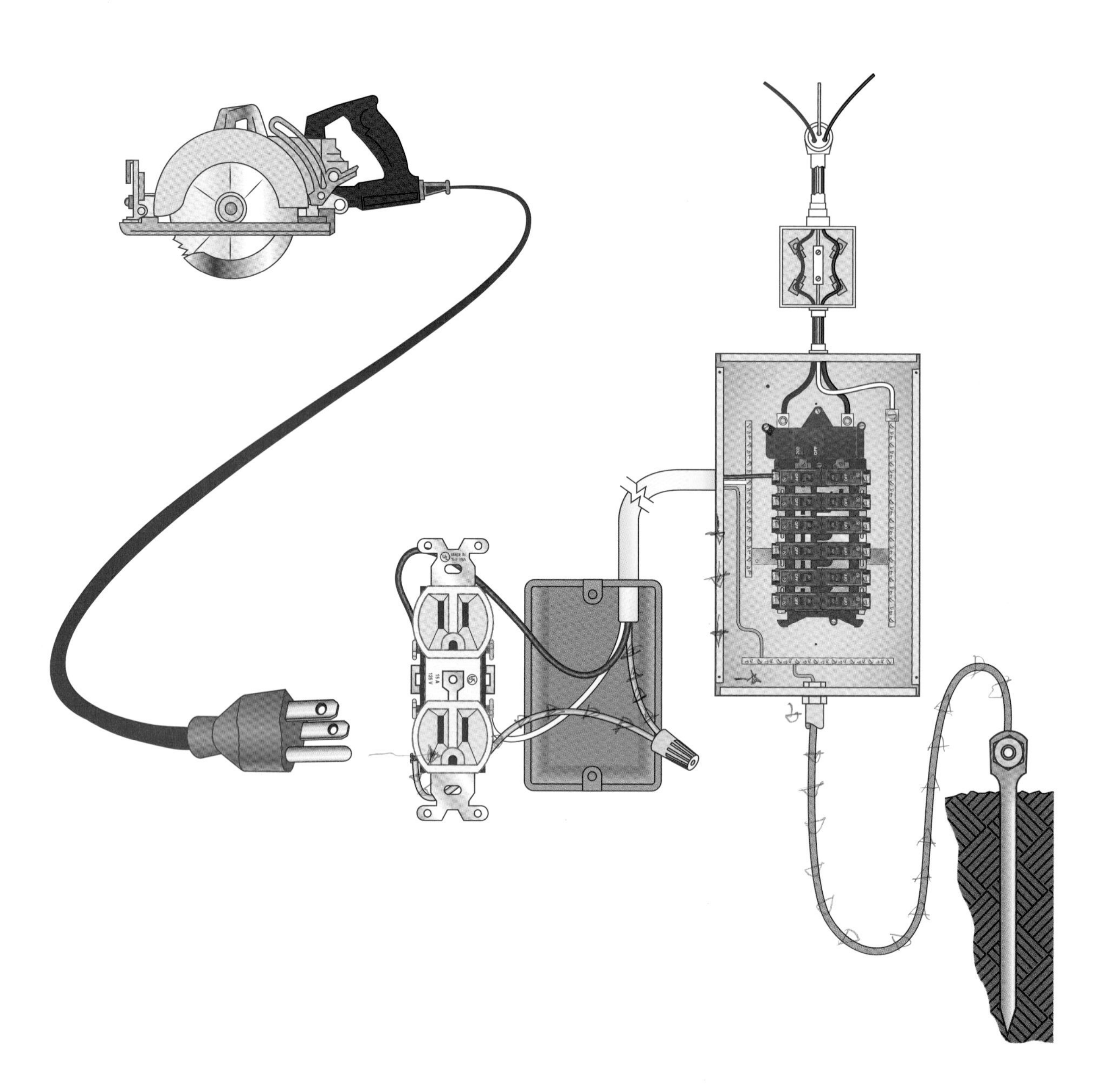

5 Electrical Safety, Tools, and Test Instruments

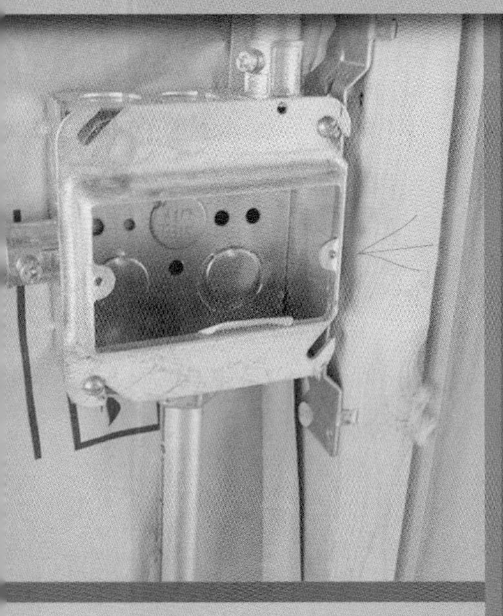

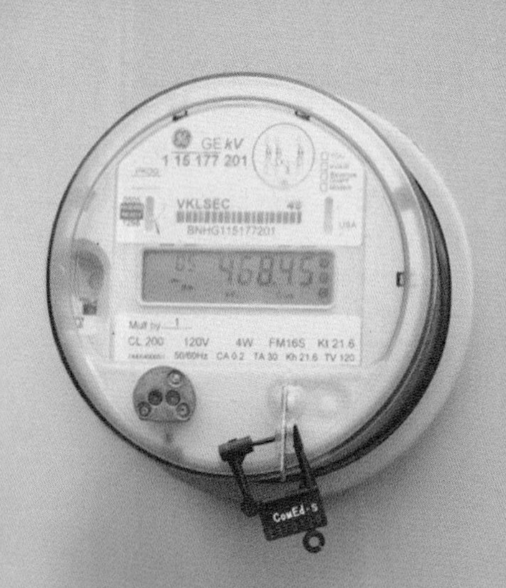

SECTION 5.1—ELECTRICAL SAFETY

- Describe the types of code and standard organizations.
- Define and describe "electrical shock."
- List and describe the shock protection boundaries associated with electrical systems.
- Identify preventive safety measures used in electrical environments.
- List the types of personal protective equipment (PPE) used by electricians.

SECTION 5.2—ELECTRICAL TOOLS

- List types of hand tools and power tools.
- Describe how to safely use hand tools and power tools.
- Explain how hand tools and power tools can be organized.

SECTION 5.3—TEST INSTRUMENTS

- Identify how test instruments are categorized by rating.
- Identify three different types of voltage test instruments.
- Explain the procedures used for digital multimeter (DMM) measurements and tests.
- Describe the use of thermal imagers.
- Explain how to safely operate test instruments.

Learner Resources

ATPeResources.com/QuickLinks
Access Code: 791712

Electrical personnel perform a number of work tasks on residential electrical systems. Work tasks involve the use of hand tools and power tools for installations and the use of test instruments for maintenance, testing, and inspecting operations. Testing operations may cause hazardous situations if not performed properly. Individuals can avoid serious injury by knowing the hazards associated with electricity and the safety measures that should be put in place to reduce the chance of electrical shock. Proper personal protective equipment (PPE) is worn to protect personnel from potential workplace hazards.

SECTION 5.1 — ELECTRICAL SAFETY

Individuals working on electrical circuits must use proper safety measures and personal protective equipment. An individual must thoroughly understand basic electrical principles, hazards, and the codes and standards associated with electrical circuit installations and maintenance tasks. Individuals who perform electrical tasks without regard to safety disregard their personal safety, the safety of any persons in the vicinity, and the equipment being worked on.

Code and Standard Organizations

The safety of an individual working with electricity in a home is the most important concern for any type of installation or maintenance task. National, state, and local codes and standards are in place to create a safe working environment and to protect people and property from electrical hazards. A *code* is a regulation or minimum requirement. A *standard* is an accepted reference or practice.

Some code, standard, and certification organizations include the Occupational Safety and Health Administration (OSHA), National Electrical Manufacturers Association (NEMA), International Electrotechnical Commission (IEC), UL® (formerly Underwriters Laboratories), Canadian Standards Association (CSA), and National Fire Protection Association® (NFPA®). **See Figure 5-1.**

Occupational Safety and Health Administration (OSHA). The *Occupational Safety and Health Administration (OSHA)* is a federal agency that requires all employers to provide a safe environment for their employees. Safe environments are work areas free from hazards that are likely to cause serious injury. OSHA standards are enforced through federal inspections.

In electrical environments, OSHA uses *NFPA 70®: National Electrical Code® (NEC®)* to help ensure safe installations. OSHA also uses *NFPA 70E®: Standard for Electrical Safety in the Workplace®* to prove that employers have taken necessary steps to protect individuals working with electricity in certain situations.

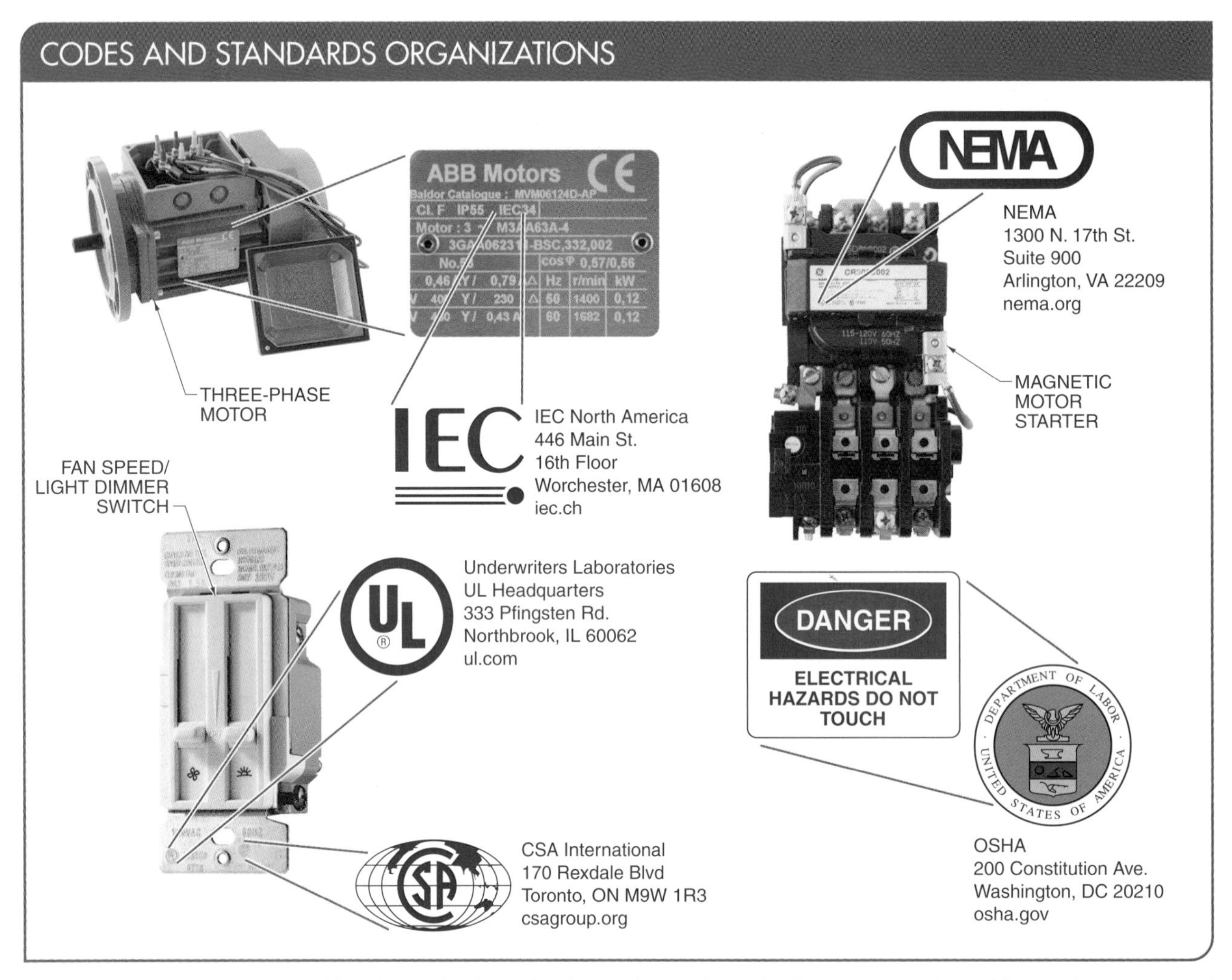

Figure 5-1. National, state, and local organizations develop codes and standards to create safe working environments and to protect people and property from electrical hazards.

National Electrical Manufacturers Association (NEMA). The *National Electrical Manufacturers Association (NEMA)* is a national organization that assists with information and standards concerning proper selection, ratings, construction, testing, and performance of electrical equipment. NEMA develops specific standards in the construction, testing, performance, and manufacture of electrical motor control equipment.

International Electrotechnical Commission (IEC). The *International Electrotechnical Commission (IEC)* is an organization that develops international safety standards for electrical equipment. Like NEMA, the IEC develops standards for motor control equipment. The IEC also issues recommendations on electrical terms, ratings, test methods, and dimensional requirements for equipment designs.

The UL®. UL®, formerly Underwriters Laboratories, is an independent organization that tests equipment and products to see if they conform to national codes and standards. Electrical equipment in homes are tested and approved by the UL for compliance and regulatory issues. UL-approved equipment and products are listed in their annual publication.

TECH TIP

Underwriters Laboratories® evaluates more than 19,450 types of products and 21 billion UL® marks appear in the marketplace.

Canadian Standards Association (CSA). The *Canadian Standards Association (CSA)* is a Canadian organization similar to the UL® that tests equipment and devices for conformity to meet national standards.

National Fire Protection Association® (NFPA®). The *National Fire Protection Association® (NFPA®)* is a national organization that provides guidance in assessing the hazards of products of combustion. The NFPA® sponsors the development of *NFPA 70®: National Electrical Code® (NEC®)* and *NFPA 70E®: Standard for Electrical Safety in the Workplace®*. The *National Electrical Code® (NEC®)* is a code book of electrical standards that indicate how electrical systems must be installed and how work must be performed. **See Figure 5-2.** The purpose of the NEC® is to protect electrical workers and equipment from electrical hazards. NFPA 70E® is a voluntary standard for electrical safety-related work practices. NFPA 70E® covers specific work practices required by electrical workers, both qualified and unqualified.

The NEC® establishes the installation standards for all residential, commercial, and industrial electrical work and is amended every three years to stay current with new safety issues, technology, electrical products, and procedures. In most areas, the NEC® and local code requirements are the same, but the NEC® sets only minimum requirements. Local codes, particularly in large cities, may be stricter than those in the NEC®. In the case of stricter local codes, the local code must be followed. When more specific requirements (codes) are established by local city councils, the local codes are obtained from a local building department. Individuals working around a home should become aware of the local and state codes regarding safe installation and maintenance of electrical devices and equipment, and systems.

NATIONAL ELECTRICAL CODE® (NEC®)

Section	Description
Article 90	Introduction
Chapter 1	General
Chapter 2	Wiring and Protection
Chapter 3	Wiring Methods and Materials
Chapter 4	Equipment for General Use
Chapter 5	Special Occupancies
Chapter 6	Special Equipment
Chapter 7	Special Conditions
Chapter 8	Communication Systems
Chapter 9	Tables

Figure 5-2. The NFPA® sponsors the development of the NEC®, which indicates how electrical systems must be installed and how work must be performed.

Electrical Hazards

Per NFPA 70E®, "Only qualified persons shall perform testing on or near live parts operating at 50 V or more." A *qualified person* is an individual with the necessary education and training who is familiar with the construction and operation of electrical equipment and devices. Individuals become qualified persons when they are certified by passing local or state electrical licensing exams. Certified electricians/technicians are the least likely to cause, but most likely to prevent, an electrical hazard when working on electrical circuits.

Equipment, devices, tools, and test instruments must be suited for the operating voltage being used or worked on. If not, an electrician can be exposed to potential electrical hazards such as electrical shock, arc flash, and arc blasts. In most cases, preventive safety measures are put in place to safeguard electricians from potential electrical hazards. The following safety guidelines must be used to prevent electrical hazards around a home:

- Always plug major appliances such as refrigerators, stoves, washers, and dryers into grounded receptacles.
- Never use an extension cord with a major appliance—extension cords are for temporary use only and must be properly sized for the intended use.
- Never overload wall outlets.
- Use light bulbs that match the recommended wattage on the lamp or fixture.
- Use extension cords for temporary purposes only.
- Confine long hair near power tools.
- Avoid putting cords where they can be damaged or pinched, like under a carpet or rug.
- Avoid wearing jewelry such as watches, rings, and neck pendants. Jewelry is capable of conducting electricity and causing electrical shock.

Electrical Shock. An *electrical shock* is a shock that results any time a body becomes part of an electrical circuit. Electrical shock effects can vary from a mild sensation to paralysis or death. The severity of electrical shock depends on the amount of current, in milliamps (mA), that flows through the body, the length of time the body is exposed to the current flow, and the path the current takes. **See Figure 5-3.** Improper wiring

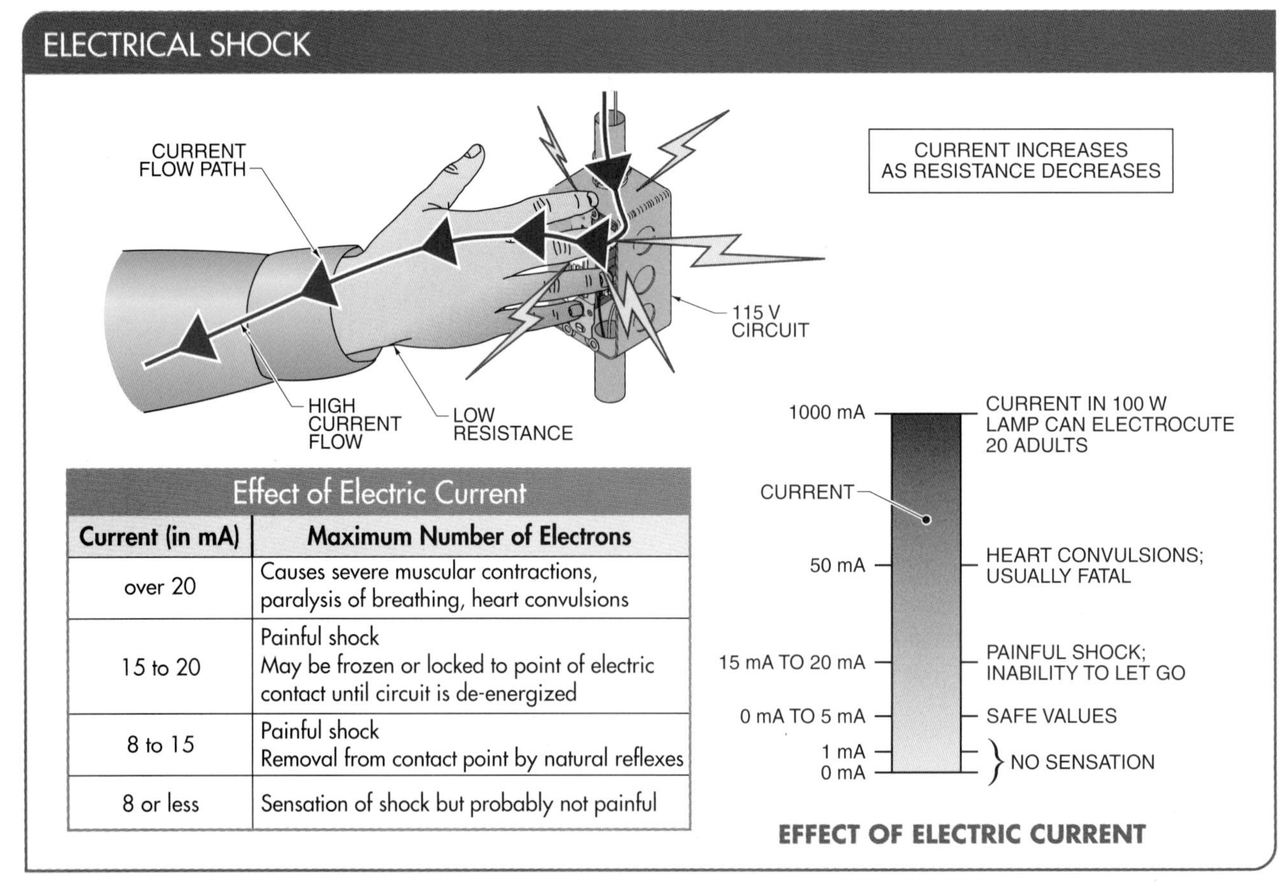

Effect of Electric Current

Current (in mA)	Maximum Number of Electrons
over 20	Causes severe muscular contractions, paralysis of breathing, heart convulsions
15 to 20	Painful shock May be frozen or locked to point of electric contact until circuit is de-energized
8 to 15	Painful shock Removal from contact point by natural reflexes
8 or less	Sensation of shock but probably not painful

Figure 5-3. Electrical shock results any time a body becomes part of an electrical circuit, and the effects can vary from a mild sensation to paralysis or death.

of equipment, circuit-faults, defective appliances, arc flash, or arc blasts can cause electrical shock in homes.

During electrical shock, a person's body becomes part of the electrical circuit. The resistance of the person's body offers the path for current to flow. The lower the resistance, the higher the current flow. As current increases, the severity of electrical shock increases. Electrical burns can also occur where current enters the body. When a person is receiving an electrical shock, the following procedures should be applied to securely remove them from the electrical circuit and treat them:

1. Shut the power source off immediately. If power cannot be removed, remove the person from the electrical circuit by using insulated devices or protective equipment such as rubber gloves, blankets, wood poles, or plastic pipes. Never touch any part of a victim's body in contact with the circuit.
2. After the person is detached from the circuit, call for help immediately and determine whether the person is breathing and has a pulse. If the person is incapable of breathing, perform cardiopulmonary resuscitation (CPR) if trained to do so. Always get medical attention for a person after an electrical shock.
3. When the person is capable of breathing and has a pulse, check for cuts or burns. Burns are caused by contact with the circuit, and are found at the point of contact where the electrical current entered and exited the body. Treat the burns as thermal burns and get medical help immediately.

Shock Protection Boundaries. Shock protection boundaries establish the conditions for working safely within an established area containing energized conductors or circuit parts. NFPA 70E identifies two shock protection boundaries to protect against electrical shock when working on energized equipment. These boundaries are the limited approach boundary and the restricted approach boundary. **See Figure 5-4.**

The size of each boundary is based on the phase-to-phase nominal system voltage of the energized conductor or circuit part. NFPA 70E Table 130.4(D)(a) and 130.4(D)(b) can be used to determine approach boundaries. **See Figure 5-5.**

Limited Approach Boundary. The *limited approach boundary* is an approach boundary that establishes the distance around exposed energized parts that a qualified person may enter. Unqualified persons may not enter unless escorted by a qualified person. Unqualified persons may enter the limited approach space to perform a minor task like inspections. However, entry is permitted only after the person has been trained and briefed on the possible hazards, is wearing the appropriate protective clothing, and remains under the direct supervision of a qualified person.

Where there is poor lighting or obstructed views, an individual must not reach blindly into areas that may contain exposed energized conductors or devices. If an individual is unable to see clearly because of an obstruction or lack of light, the limited approach boundary may not be crossed to perform work on energized conductors and devices. Adequate lighting must be provided before an individual can enter spaces containing energized conductors or devices. Adequate lighting means enough lighting so that all potential electrical hazards can be clearly seen and identified.

Restricted Approach Boundary. The *restricted approach boundary* is an approach boundary that establishes the distance around exposed energized parts that may only be crossed by qualified persons using the correct shock protection techniques and equipment. There is a high risk of potential contact with an exposed, energized part within this boundary. Personnel are required to use appropriate arc-rated clothing, PPE, and approved tools.

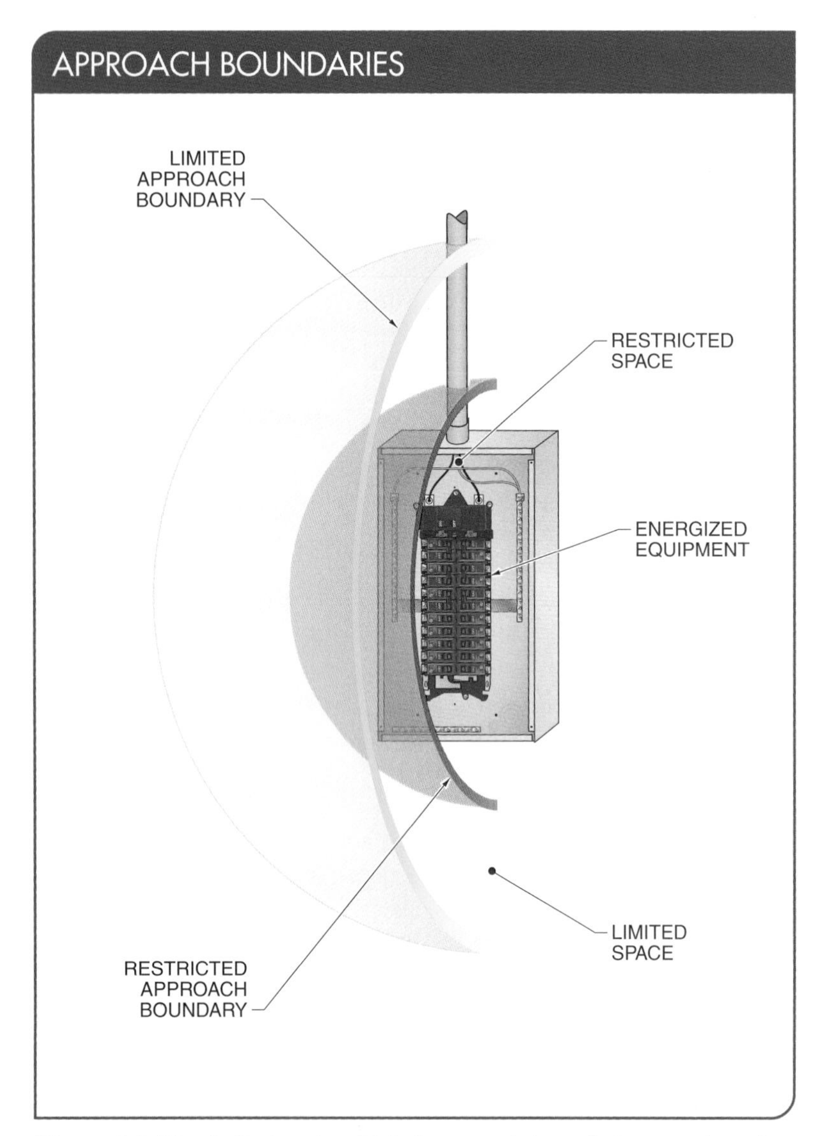

Figure 5-4. The limited and restricted approach boundaries are viewed as spherical areas, extending 360° around the exposed electrical equipment, conductor, or circuit.

NFPA 70E Tables 130.7(C)(15)(a) and (c) can be used for the selection of the appropriate PPE for a specific task.

Preventive Safety Measures

In order to prevent any type of electrical hazard accident, preventive safety measures should be understood and followed to keep all personnel and equipment safe. Types of preventive safety measures include lockout and tagout processes with associated devices, fire safety, and basic first aid.

Lockout/Tagout. Electrical power must be shut off before any inspection, repair, or maintenance is performed. To ensure the safety of an electrician, the equipment must then be locked out and tagged out. **See Figure 5-6.** *Lockout* is the process of removing the source of electrical power and installing a lock that prevents the power from being turned on. *Tagout* is the process of placing a danger tag on the source of electrical power. Tagouts indicate that the equipment may not be operated until the danger tag is removed.

APPROACH BOUNDARIES TO ENERGIZED EQUIPMENT FOR SHOCK PREVENTION

Nominal System (Voltage, Range, Phase to Phase*)	Limited Approach Boundary		Restricted Approach Boundary (Allowing for Accidental Movement)
	Exposed Movable Conductor	Exposed Fixed-Circuit Part	
less than 50	N/A	N/A	N/A
50 to 150	10′-0″	3′-6″	Avoid contact
151 to 750	10′-0″	3′-6″	1′-0″
751 to 15,000	10′-0″	5′-0″	2′-2″

* in V

Figure 5-5. The distance for each approach boundary is based on the nominal voltage of the energized conductor or circuit equipment.

Figure 5-6. Lockout and tagout is the process of removing the source of electrical power and indicating that the equipment may not be operated until the danger tag is removed.

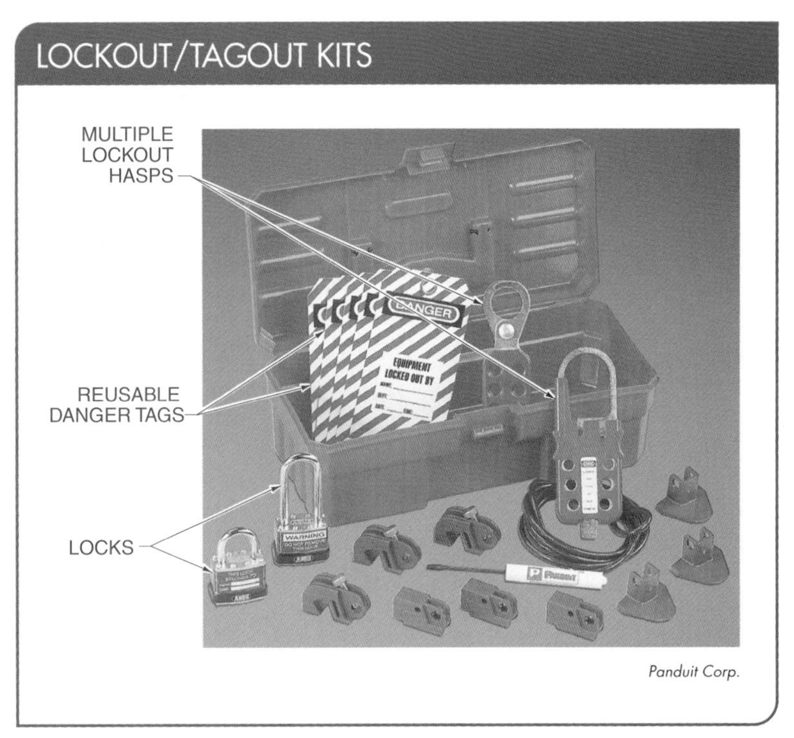

Figure 5-7. Lockout/tagout kits contain usable danger tags, multiple lockouts, locks, tag ties, and information regarding tagout procedures.

Individuals should comply with OSHA 29 Code of Federal Regulations (CFR) 1910.147—The Control of Hazardous Energy (Lockout/Tagout) for guidelines on applying lockout/tagout. Lockout/tagout kits are available for electricians. Lockout/tagout kits contain reusable danger tags, multiple lockouts, locks, cable ties, and information regarding tagout procedures. **See Figure 5-7.**

Lockout Devices. A *lockout device* is a lightweight enclosure that allows the lockout of standard control devices. Lockout devices are available in colors as well as various shapes and sizes that allow for the lockout of electrical plugs, switches, circuit breakers, and service panels. Lockout devices resist chemicals, cracking, abrasion, and temperature changes. Lockout devices are sized to fit standard residential control devices. **See Figure 5-8.**

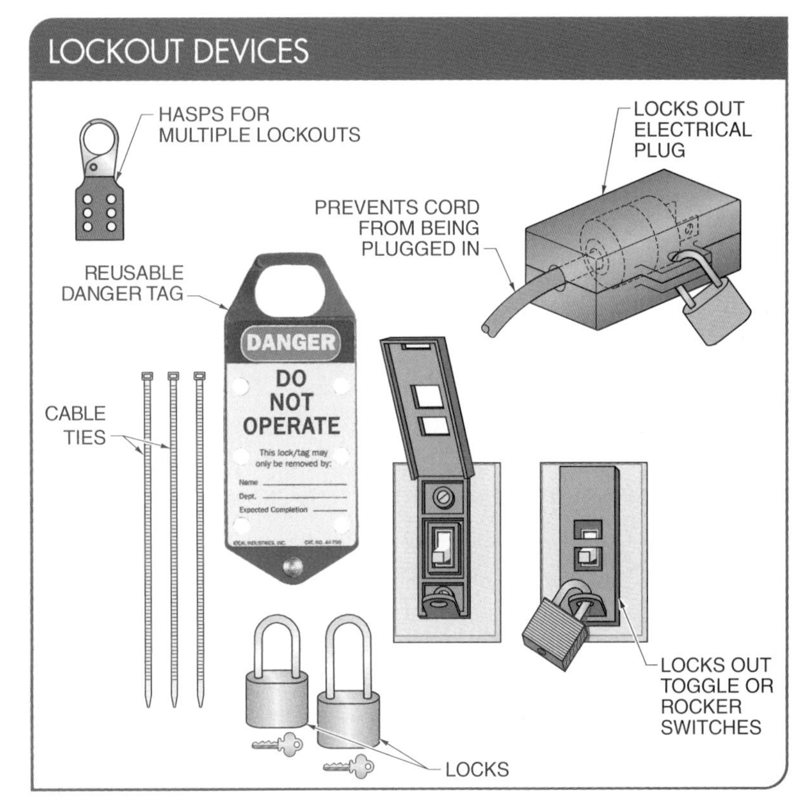

Figure 5-8. Lockout devices are lightweight enclosures that allow the lockout of standard control devices and are available in various shapes and sizes.

Danger tags provide additional lockout and warning information. Various danger tags are available. Danger tags may include warnings such as "Do Not Start" or "Do Not Operate," or they may provide space to enter information on the worker, date, and lockout reason. Tag ties must be strong enough to prevent accidental removal and must be self-locking and non-reusable.

Salisbury by Honeywell

Arc flash temperatures can exceed 35,000°F. If possible, electrical equipment should be deenergized and locked out before any work is performed.

Fire Safety. Fire safety requires procedures be put in place to reduce the conditions that could cause a fire. All residences should have a fire safety plan that establishes procedures for individuals to follow when a fire occurs. The chance of a fire is greatly reduced by good housekeeping. This can be done by keeping debris and liquids in designated containers. Rags containing oil, gasoline, or other solvents must be properly stored in designated containers.

If a fire should occur, the first thing an individual must do is alert everyone in the home or work area and call the fire department. During the time before the fire department arrives, a reasonable effort should be made to contain the fire. In the case of a small fire, portable fire extinguishers should be used to extinguish the fire. **See Figure 5-9.** The stream from fire extinguishers must be directed to the base of fires. When firefighters arrive, an individual must be prepared to direct the firefighters to the location of the fire. Also, individuals should inform the firefighters of any special problems or conditions that exist, such as burned electrical wires or chemical leaks from equipment.

FIRE EXTINGUISHERS

Kidde

Figure 5-9. Portable fire extinguishers should be used to extinguish or contain fires before the fire department arrives.

Fire extinguishers are labeled with operating instructions for extinguishing fires and provide class symbols for each type of fire. Individuals must be aware of how to identify which fire extinguisher to use on what type of fire. The five classes of fires are Class A, Class B, Class C, Class D, and Class K fires. Each class is denoted by a letter that informs the user of the type of materials that started the fire and how to extinguish them. **See Figure 5-10.**

FIRE EXTINGUISHER CLASSES

Class A Fires

Class A fires occur in wood, clothing, paper, rubbish, and other such items. Class A fires are typically controlled with water.

ORDINARY COMBUSTIBLES

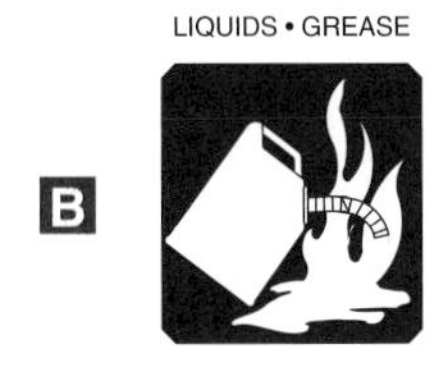

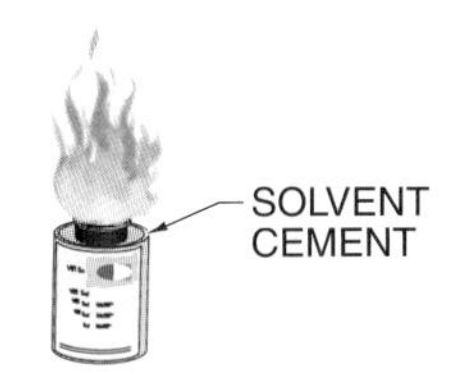

Class B Fires

Class B fires occur in flammable liquids such as gasoline, fuel oil, lube oil, grease, thinner, and paints. The agents required for extinguishing Class B fires dilute or eliminate the air to the fire by blanketing the surface of the fire. Chemicals such as foam CO_2, dry chemical, and Halon are used on Class B fires.

FLAMMABLE LIQUIDS

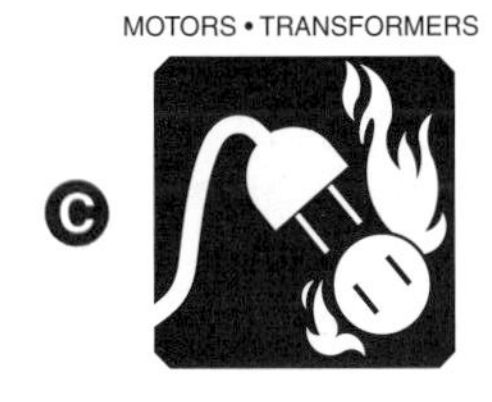

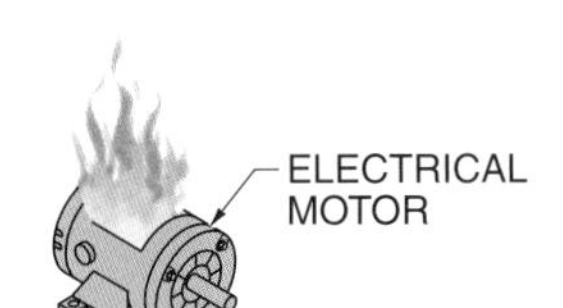

Class C Fires

Class C fires occur in facilities near or in electrical equipment. The extinguishing agent for Class C fires must be a nonconductor of electricity and provide a smothering effect. CO_2, dry chemical, and halon gas extinguishers are typically used.

ELECTRICAL EQUIPMENT

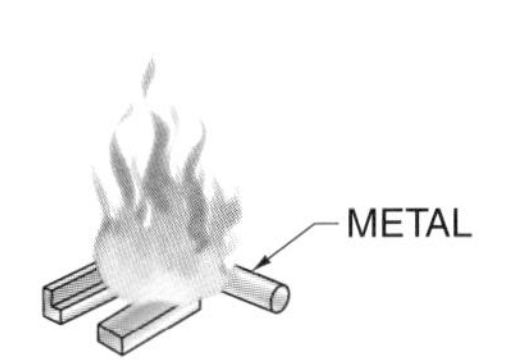

Class D Fires

Class D fires occur in combustible metals such as magnesium, potassium, powdered aluminum, zinc, sodium, titanium, zirconium, and lithium. The extinguishing agent for Class D fires is a dry powdered compound. The powdered compound creates a smothering effect that is somewhat effective.

COMBUSTIBLE METALS

Class K Fires

Class K fires occur in kitchens with grease. The extinguishing agent for Class K fires is a wet chemical (potassium acetate).

COMMERCIAL COOKING GREASE

Figure 5-10. Fire extinguishers are labeled with operating instructions and symbols to extinguish Class A, B, C, D, or K fires. Each class is denoted by a letter that informs the user of the type of materials used to start the fire and how each class is extinguished.

Basic First Aid. Accidents that cause injury to individuals can happen at any time and in any location. Immediate medical care is required for the victim, regardless of the extent of the injury. Often, first aid given immediately at the scene of an accident can improve the victim's chances of survival and recovery. *First aid* is help for a victim immediately after an injury and before professional medical help arrives. First aid includes procedures performed to help save a person's life before professional help arrives. The following steps should be taken to keep an injured person as safe as possible until professional help arrives:

- Remain calm.
- Call 911 or the workplace emergency number immediately if an individual is severely injured.
- Never move an injured person unless a fire or explosives are involved. Moving an injured person may make the injury worse.
- Assess the injured person carefully and perform basic first aid procedures.
- Maintain first aid procedures until professional medical help arrives.
- Report all injuries to the supervisor.

A conscious victim of an accident must provide consent before care can be administered. To obtain consent, the victim must be asked if help can be provided. Once the victim gives consent, the appropriate care can be provided. If the victim does not give consent, care should not be given. In such cases, 911 or the workplace emergency number should be called. However, consent is implied in cases where a victim is unconscious, confused, or severely injured. Implied consent allows an individual to provide care to a victim because the victim would likely agree to the care if possible.

Personal Protective Equipment

Personal protective equipment (PPE) is clothing and/or equipment worn by individuals to reduce the possibility of injury in the work area. The use of PPE is required whenever work may occur on or near energized exposed electrical circuits. For maximum safety, PPE and safety requirements must be followed as specified in NFPA 70E®, OSHA 29 CFR 1910 Subpart I—Personal Protective Equipment (1910.132 through 1910.138), and other applicable safety mandates.

Electrical PPE includes arc-rated clothing, head protection, eye protection, ear protection, hand protection, foot protection, knee protection, and rubber insulating matting. However, not all electrical PPE is used for residential purposes. Residential electrical PPE includes protective helmets (hard hats), arc-rated (AR) clothing, safety glasses, and electrical gloves. **See Figure 5-11.** When work is performed on an electrical circuit in a residence, the following PPE should be worn:

- arc-rated protective clothing
- safety glasses or goggles, depending on the task
- thick-soled rubber work shoes for protection against sharp objects and insulation against electrical shock, or boots with steel toes as required by the task
- rubber boots in damp locations
- an approved protective helmet (hard hat)
- electrical gloves for protection against electrical shock when taking measurements on energized circuits

> TECH TIP
>
> *Personal protective equipment (PPE) with good ergonomic design can help ensure worker safety by not impairing the wearer's ability to work or causing discomfort.*

Arc-Rated (AR) Clothing. Approved arc-rated (AR) clothing must be worn for protection from electrical arcs when performing certain operations on or near energized equipment or circuits. AR clothing must be kept as clean and sanitary as practical and must be inspected prior to any electrical task. Defective clothing must be removed from service immediately and replaced. Defective AR clothing must be tagged "unsafe" and returned to a supervisor.

RESIDENTIAL PERSONAL PROTECTIVE EQUIPMENT (PPE)

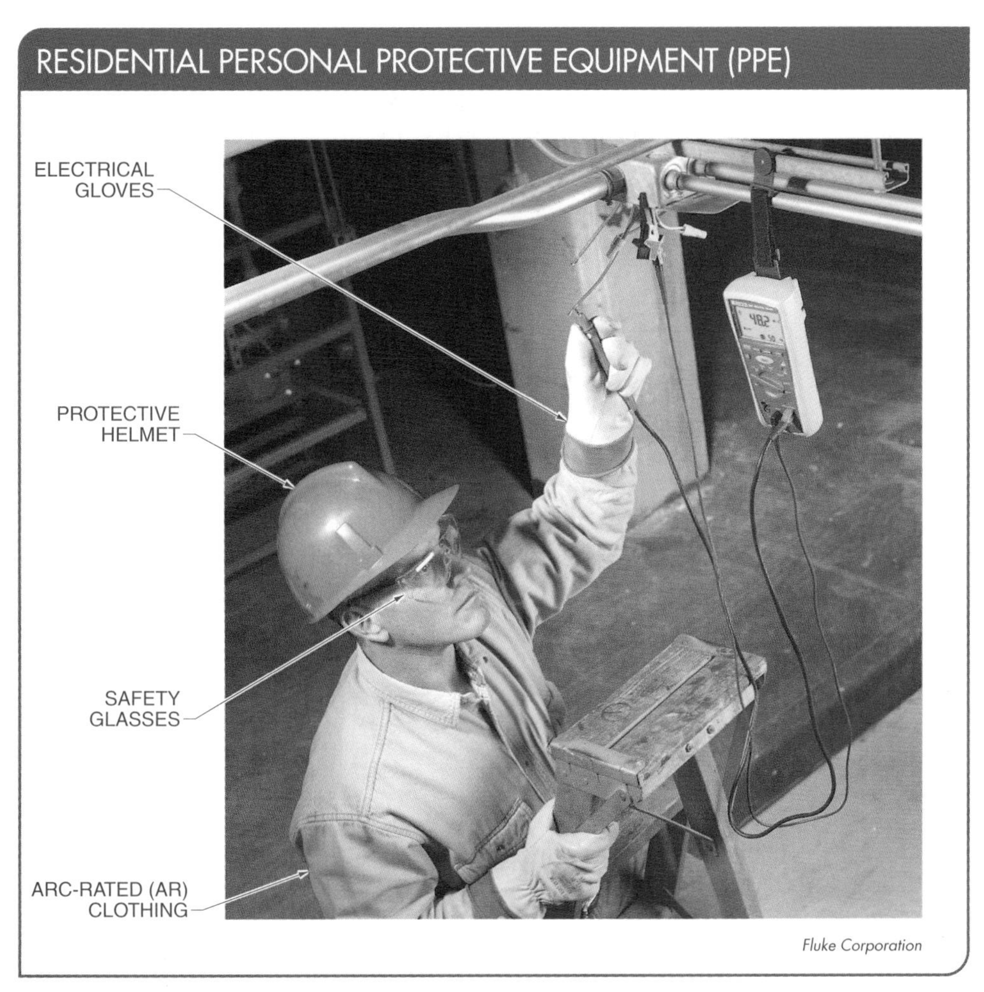

Figure 5-11. Residential electrical PPE includes protective helmets, arc-rated (AR) clothing, safety glasses, and electrical gloves.

Head Protection. Head protection requires the use of a protective helmet (hard hat). A *protective helmet* is a hard hat that is used in the workplace to prevent injury from the impact of falling and flying objects and from electrical shock. Protective helmets resist penetration and absorb impact force. Protective helmet shells are made of durable, lightweight materials. A shock-absorbing lining that consists of crown straps and a headband keeps the shell away from the head to provide ventilation.

Protective helmets are identified by class of protection against specific hazardous conditions. Class A, B, and C helmets are used for construction applications. Class A protective helmets protect against low-voltage shocks and burns and impact hazards and are commonly used in construction and manufacturing facilities. Class B protective helmets protect against high-voltage shocks and burns, impact hazards, and penetration by falling or flying objects. Class C protective helmets are manufactured with light materials, yet provide adequate impact protection. **See Figure 5-12.**

Eye Protection. Eye protection must be worn to prevent eye or face injuries caused by flying particles, contact arcing, and radiant energy. Eye protection must comply with OSHA 29 CFR 1910.133—Eye and Face Protection. Eye protection includes safety glasses or arc-rated face shields. **See Figure 5-13.**

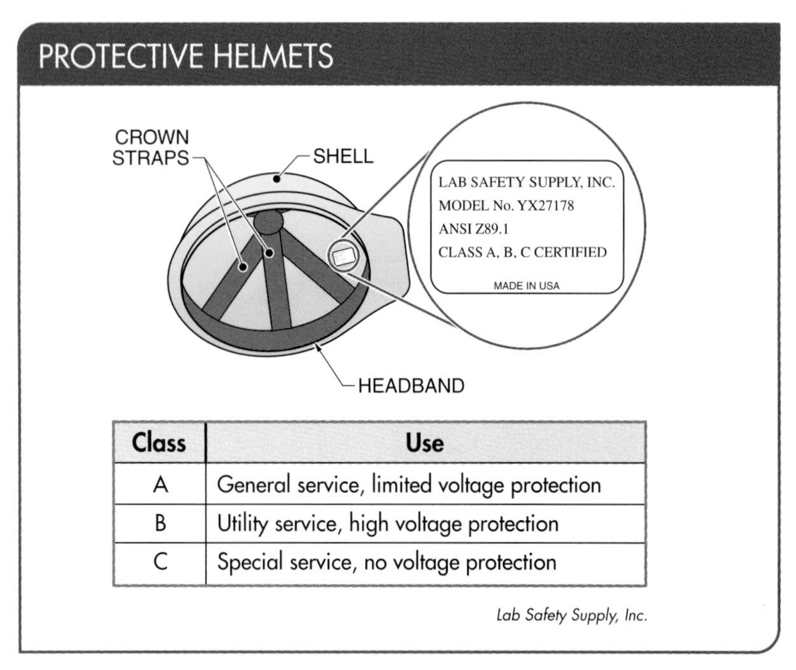

Class	Use
A	General service, limited voltage protection
B	Utility service, high voltage protection
C	Special service, no voltage protection

Figure 5-12. Protective helmets are identified as Class A, B, or C, each protecting against specific hazardous conditions.

Figure 5-13. Eye protection, such as safety glasses or arc-rated face shields, must be worn to prevent eye or face injuries.

Safety glasses are an eye protection device with special impact-resistant glass or plastic lenses, a reinforced frame, and side shields. Plastic frames are designed to keep the lenses secured in the frame if an impact occurs and to minimize the shock hazard when working with electrical equipment. Side shields provide additional protection from flying objects. Tinted-lens safety glasses protect against low-voltage arc hazards.

An *arc-rated face shield* is an eye and face protection device that covers the entire face with a plastic shield. Face shields are used to protect against electrical hazards and flying objects. Tinted face shields protect against low-voltage arc hazards.

> **TECH TIP**
>
> *Arc-rated face shields protect against arc flashes. Deenergizing systems protects against arc blasts.*

Ear Protection. Ear protection devices are worn to limit the noise entering the ear and include earplugs and earmuffs. An *earplug* is an ear protection device made of moldable rubber, foam, or plastic and inserted into the ear canal. An *earmuff* is an ear protection device worn over the ears. A tight seal around an earmuff is required for proper protection. **See Figure 5-14.**

Power tools and equipment can produce excessive noise levels. Electricians subjected to excessive noise levels may develop hearing loss over time. The severity of hearing loss depends on the intensity and duration of exposure. Noise intensity is expressed in decibels. A *decibel (dB)* is a unit of measure used to express the relative intensity of sound. Ear protection is worn to prevent hearing loss.

Ear protection devices are assigned a noise reduction rating (NRR) number based on the noise level reduced. For example, an NRR of 27 means that the noise level is reduced by 27 dB when the device is tested at the factory. To determine approximate noise reduction in the field, 7 dB is subtracted from the NRR. For example, an NRR of 27 provides a noise reduction of approximately 20 dB in the field.

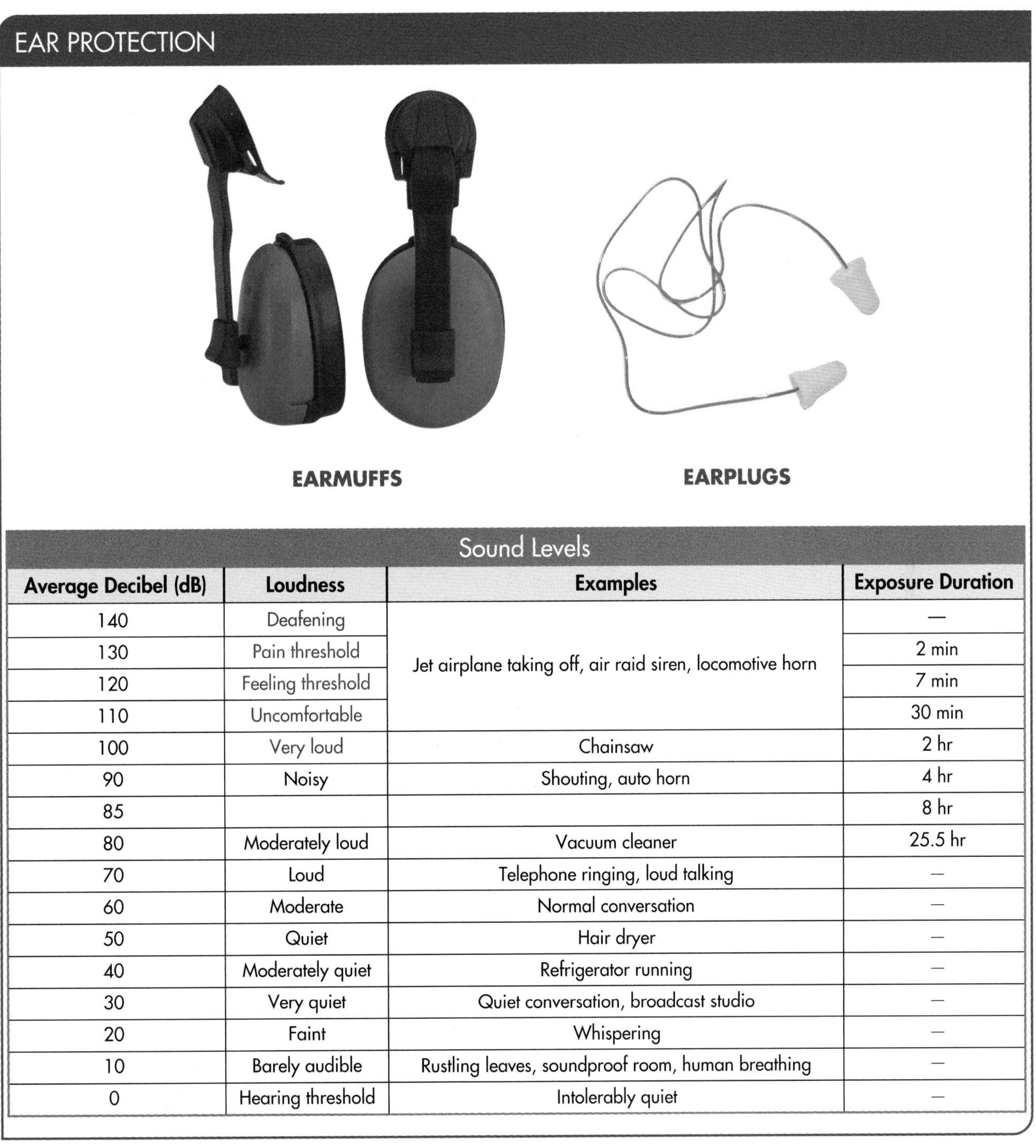

Sound Levels			
Average Decibel (dB)	**Loudness**	**Examples**	**Exposure Duration**
140	Deafening	Jet airplane taking off, air raid siren, locomotive horn	—
130	Pain threshold		2 min
120	Feeling threshold		7 min
110	Uncomfortable		30 min
100	Very loud	Chainsaw	2 hr
90	Noisy	Shouting, auto horn	4 hr
85			8 hr
80	Moderately loud	Vacuum cleaner	25.5 hr
70	Loud	Telephone ringing, loud talking	—
60	Moderate	Normal conversation	—
50	Quiet	Hair dryer	—
40	Moderately quiet	Refrigerator running	—
30	Very quiet	Quiet conversation, broadcast studio	—
20	Faint	Whispering	—
10	Barely audible	Rustling leaves, soundproof room, human breathing	—
0	Hearing threshold	Intolerably quiet	—

Figure 5-14. Ear protection devices, such as earplugs and earmuffs, are worn to limit the noise entering the ears.

Hand Protection. Hand protection consists of rubber insulating gloves and leather protectors worn to prevent injuries to hands caused by cuts or electrical shock. The primary purpose of rubber insulating gloves and leather protectors is to insulate hands and lower arms from possible contact with live conductors. Rubber insulating gloves offer a high resistance to current flow to help prevent an electrical shock, and the leather protectors protect the rubber glove and add additional insulation. Rubber insulating gloves are rated, labeled, and color coded to show the maximum voltage allowed. **See Figure 5-15.**

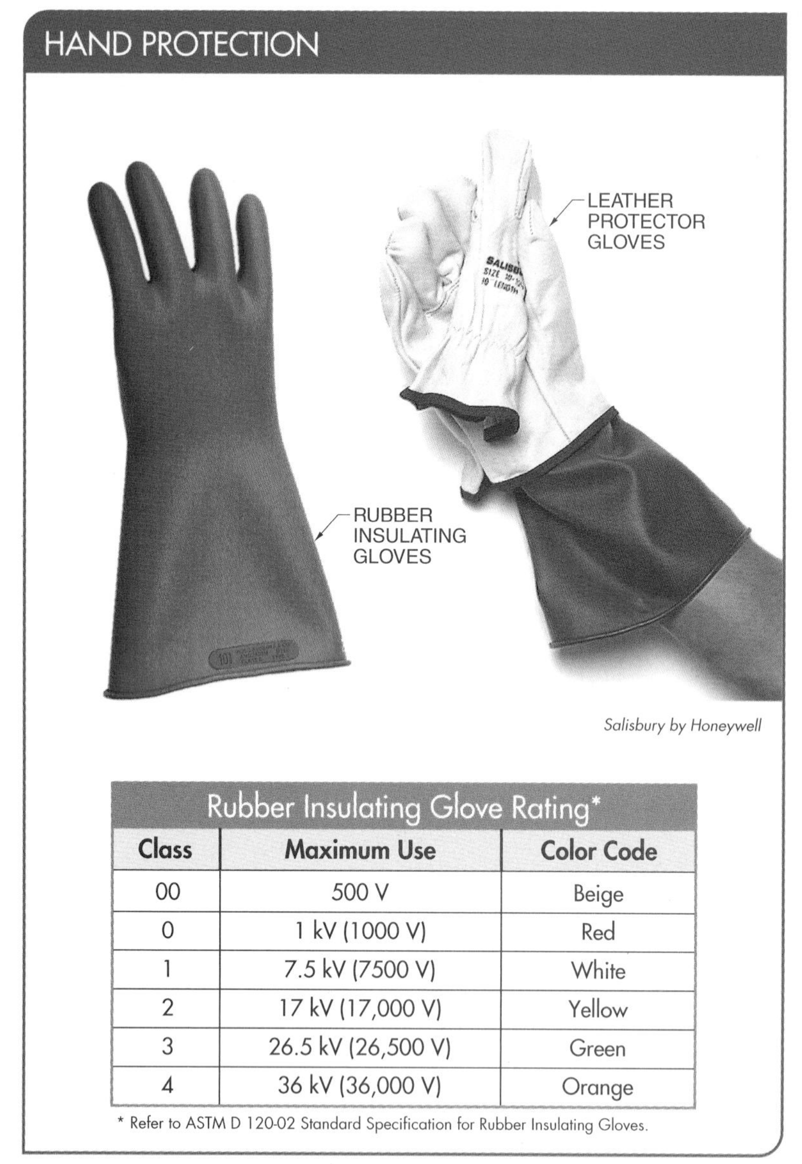

Rubber Insulating Glove Rating*		
Class	**Maximum Use**	**Color Code**
00	500 V	Beige
0	1 kV (1000 V)	Red
1	7.5 kV (7500 V)	White
2	17 kV (17,000 V)	Yellow
3	26.5 kV (26,500 V)	Green
4	36 kV (36,000 V)	Orange

* Refer to ASTM D 120-02 Standard Specification for Rubber Insulating Gloves.

Figure 5-15. Hand protection consists of rubber insulating gloves and leather protectors, which are worn to prevent injuries to hands caused by cuts or electrical shock.

WARNING: Rubber insulating gloves are designed for specific applications. Leather protector gloves are required for protecting rubber insulating gloves. Rubber insulating gloves must not be used alone. Rubber insulating gloves offer the highest resistance and greatest insulation. Serious injury or death can result from improper use of rubber insulating gloves or from using outdated and/or the wrong type of rubber insulating gloves for an electrical application.

Foot Protection. Foot protection consists of shoes or boots worn to prevent foot injuries that are typically caused by objects falling less than 4′ and having an average weight of less than 65 lb. Safety shoes/boots with reinforced steel toes protect against injuries caused by compression and impact. Insulated rubber-soled shoes are commonly worn during electrical work to prevent electrical shock.

Knee Protection. A *knee pad* is a rubber, leather, or plastic pad strapped onto the knees for protection. A knee pad is worn by personnel who work in close areas and must kneel for proper access to electrical equipment. Buckle straps or Velcro® closures secure knee pads in position. **See Figure 5-16.**

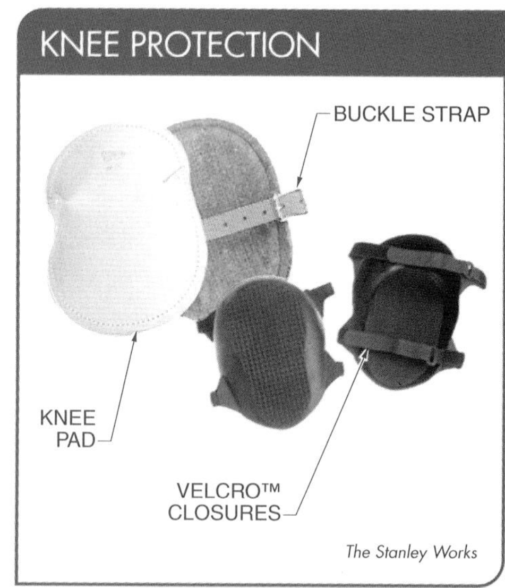

Figure 5-16. Knee pads are worn by electricians who work in close areas and must kneel for proper access to electrical equipment.

Rubber Insulating Matting. *Rubber insulating matting* is a floor covering that provides individuals protection from electrical shock when working on live electrical circuits. Rubber insulating matting is specifically designed for use in front of open cabinets or high-voltage equipment to protect electricians when voltages are over 50 V. Two types of matting that differ in chemical and physical characteristics are designated as Type I natural rubber and Type II elastomeric compound matting. **See Figure 5-17.**

RUBBER INSULATING MATTING RATINGS					
Safety Standard	Material Thickness		Material Width (in in.)	Test Voltage	Maximum Working Voltage
	Inches	Millimeters			
BS921*	0.236	6	36	11,000	450
BS921*	0.236	6	48	11,000	450
BS921*	0.354	9	36	15,000	650
BS921*	0.354	9	48	15,000	650
VDE0680†	0.118	3	39	10,000	1000
ASTM D178‡	0.236	6	24	25,000	17,000
ASTM D178‡	0.236	6	30	25,000	17,000
ASTM D178‡	0.236	6	36	25,000	17,000
ASTM D178‡	0.236	6	48	25,000	17,000

* BSI–British Standards Institute
† VDE–Verband Deutscher Elektrotechniker Testing and Certification Institute
‡ ASTM International–American Society for Testing and Materials

Figure 5-17. Rubber insulating matting is a floor covering that protects from electrical shock when working on live electrical circuits.

SECTION 5.1 CHECKPOINT

1. What is the National Electrical Code® (NEC®) and what standard organization sponsors its development?
2. What are three guidelines that can be used to prevent electrical hazards around a home?
3. What are the three approach boundaries established by NFPA 70E® to protect individuals?
4. What is lockout/tagout?
5. What kinds of PPE should be worn when working in a residence?

SECTION 5.2— ELECTRICAL TOOLS

Electrical work requires various types of hand and power tools. In a residence, the main types of electrical work include maintenance, testing, and installation of electrical equipment. Different hand and power tools are required to complete these types of electrical work. While wiring installation is performed only on deenergized circuits, maintenance and testing operations are often performed around or near energized equipment and conductors. Proper use of hand and power tools is required for safe and efficient electrical tasks.

Individuals can become more productive with experience in the proper use, maintenance, and storage of tools. When tools are used correctly, an individual avoids minimal to severe personal injury and maintains the condition of the tools. As more tools are acquired for use, organized tools systems allow for safe and efficient storage.

Hand Tools

A *hand tool* is a tool that uses no external power other than that supplied by the person using the tool. Individuals use hand tools for twisting, turning, bending, cutting, stripping, attaching, pulling, and securing operations. Hand tools used for noncutting operations include fish tapes, fuse pullers, conduit benders, hammers, pliers, screwdrivers, and wrenches. Hand tools used for cutting operations include wire strippers, knives, and saws. Other miscellaneous hand tools include tapes rules, torpedo levels, cable tie guns, PVC pipe cutters, conduit screwdrivers, and cable strippers.

Fish Tapes. A *fish tape* is a retractable tape used to pull conductors through conduit, through inaccessible spaces, or around obstructions in walls. Fish tapes can range in length from 25′ to 200′ depending on the application. Flexible wire conductors can be pulled through conduit when they are attached to the trailing end of fish tape. **See Figure 5-18.**

FISH TAPES

CONDUIT

FISH TAPE

Figure 5-18. Fish tapes pull conductors through conduit and are constructed of nylon, steel, fiberglass, or multistrand steel.

Fish tapes are constructed of nylon, tempered steel, stainless steel, fiberglass, or multistrand steel. Steel fish tapes may be coated with nylon to prevent corrosion and reduce friction and jamming. Nylon-coated and fiberglass fish tapes are recommended for pulling conductors through conduit that might come into contact with energized wires or equipment. However, any attempt to run fish tapes through energized panels and conduit should be avoided.

TECH TIP

Pulling wire is easier if two people are involved—one person to push the wire and another to pull the fish tape.

Fuse Pullers. A *fuse puller* is a nonmetallic device that is used to safely remove fuses from service panels and subpanels. **See Figure 5-19.** Fuse pullers have solid-grip jaws and are constructed of nonmetallic materials such as nylon, laminated fiber, or hard plastic. Fuse pullers can be used to remove nonoperational (blown) fuses or install operational cartridge-type fuses. Before any installation or removal of fuses, the circuit must always be deenergized.

FUSE PULLERS

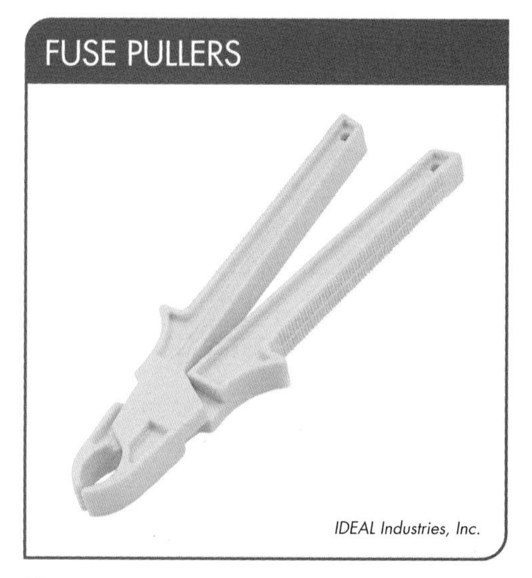

IDEAL Industries, Inc.

Figure 5-19. Fuse pullers have solid-grip jaws and are constructed of nonmetallic materials for removing fuses from service panels.

Conduit Benders. A *conduit bender* is a device used to bend certain types of conduit in order to clear obstructions. **See Figure 5-20.** Conduit benders have bending shoes with supporting sidewalls to prevent flattening or kinking of conduit and are designed for quick, efficient bending. Individuals use handheld conduit benders to install and bend small conduit ranging from ½″ to 1¼″. Conduit benders are normally made from heat-treated aluminum or iron and are available with flared handles that can be used as a conduit-straightening tool.

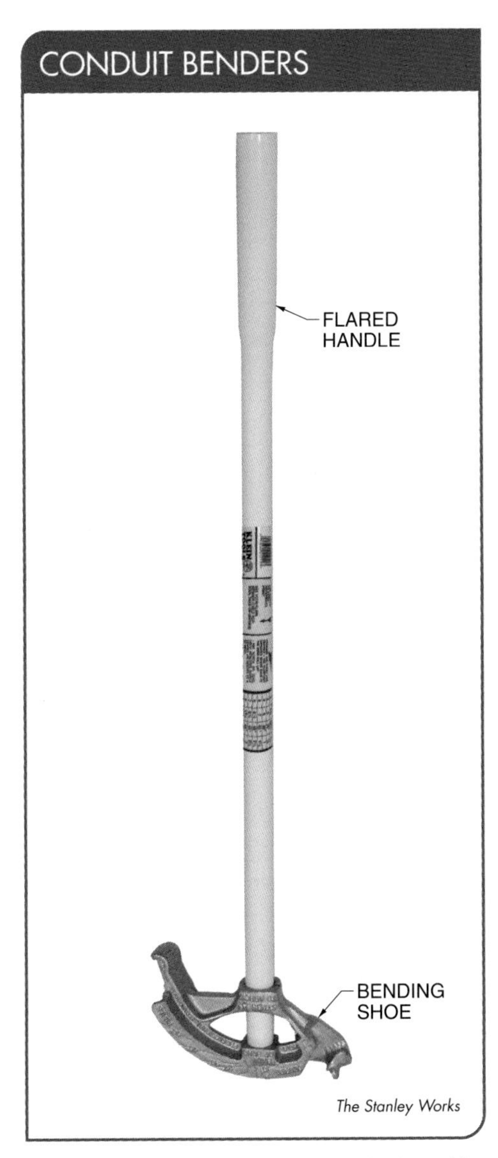

Figure 5-20. Conduit benders are designed for quick and efficient bends in conduit in order to clear obstructions.

Hammers. A *hammer* is a striking hand tool with a hardened head fastened perpendicularly to a handle. Common hammers used for installations include an electrician's hammer and ball-peen hammer. **See Figure 5-21.** An electrician's hammer is used to mount electrical boxes and drive nails. Because electrician's hammers are designed to be 12″ in length from head to end, they can be useful in determining the height of most receptacle boxes from the floor. Ball-peen hammers are designed for striking chisels and punches. Ball-peen hammers can be useful for riveting, shaping, and straightening unhardened metal.

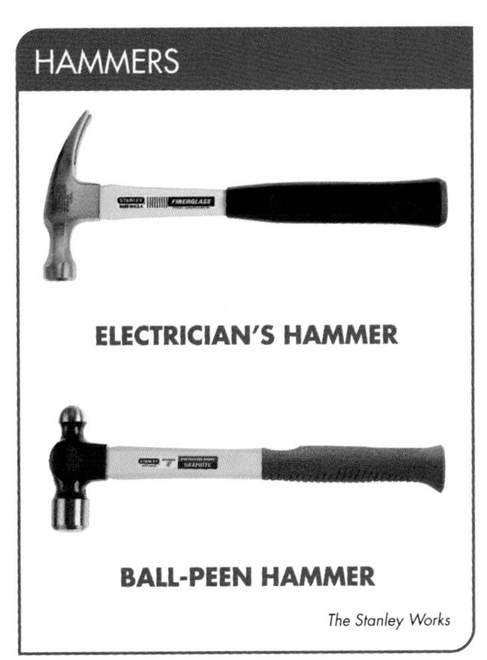

Figure 5-21. Hammers commonly used for installing electrical equipment are the electrician's hammer and the ball-peen hammer.

Pliers. Pliers are a hand tool with opposing jaws for gripping and/or cutting. Individuals use pliers for installations containing various gripping, turning, cutting, positioning, and bending operations. Common pliers used include tongue-and-groove, long-nose, diagonal-cutting, and side-cutting pliers. **See Figure 5-22.**

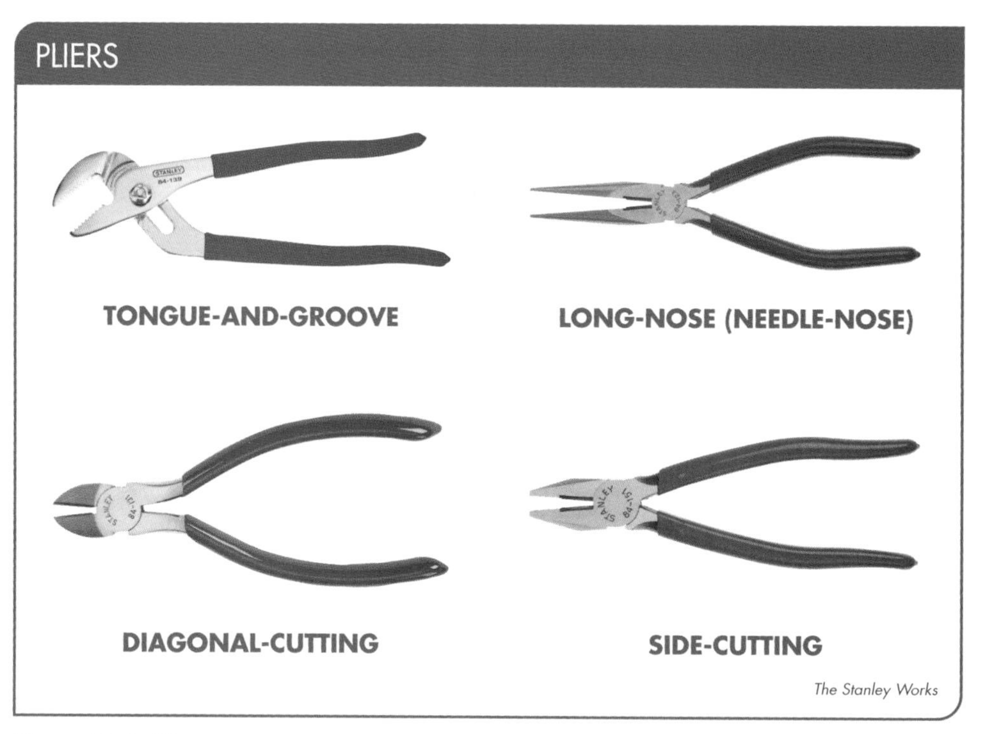

Figure 5-22. Pliers commonly used by electricians to install equipment include tongue-and-groove, long-nose (needle-nose), diagonal-cutting, and side-cutting pliers.

Tongue-and-groove pliers are used for a wide range of applications involving gripping, turning, and bending. The adjustable jaw of tongue-and-groove pliers allows them to be used on a wide range of nut sizes, such as for tightening box connectors, locknuts, and small-size conduit couplings. Long-nose (needle-nose) pliers are used for bending or cutting wires and for positioning small components. Diagonal-cutting pliers are used for cutting cables and wires that are difficult to cut with side-cutting pliers. Side-cutting pliers are used to cut cable, remove knockouts, and twist wire.

Screwdrivers. A *screwdriver* is a hand tool with a tip designed to fit into a screw head for fastening operations. Individuals use screwdrivers for installation or maintenance purposes to secure or remove various threaded fasteners. The two main types of screwdrivers are flathead and Phillips screwdrivers. These two types of screwdrivers are available in standard, offset, torque, and screw-starting designs. **See Figure 5-23.**

Standard screwdrivers are used for the installation and removal of threaded fasteners. Standard screwdrivers come in many sizes and lengths. Some are available with square shanks to which a wrench can be applied for removing screws that are jammed or rusted in place. Offset screwdrivers are designed to reach screws located in areas that cannot be reached with a standard screwdriver. Torque screwdrivers are used to tighten screws to a specified torque value. An internal mechanism in the screwdriver automatically releases pressure when the specified torque value has been reached. A screw-starting screwdriver is used to hold screws in place when working in tight spots.

Wrenches. A *wrench* is a hand tool with jaws at one or both ends that is designed to turn bolts, nuts, or other types of fasteners. Common wrenches used to install equipment include socket, adjustable, hex key, and combination wrenches. **See Figure 5-24.**

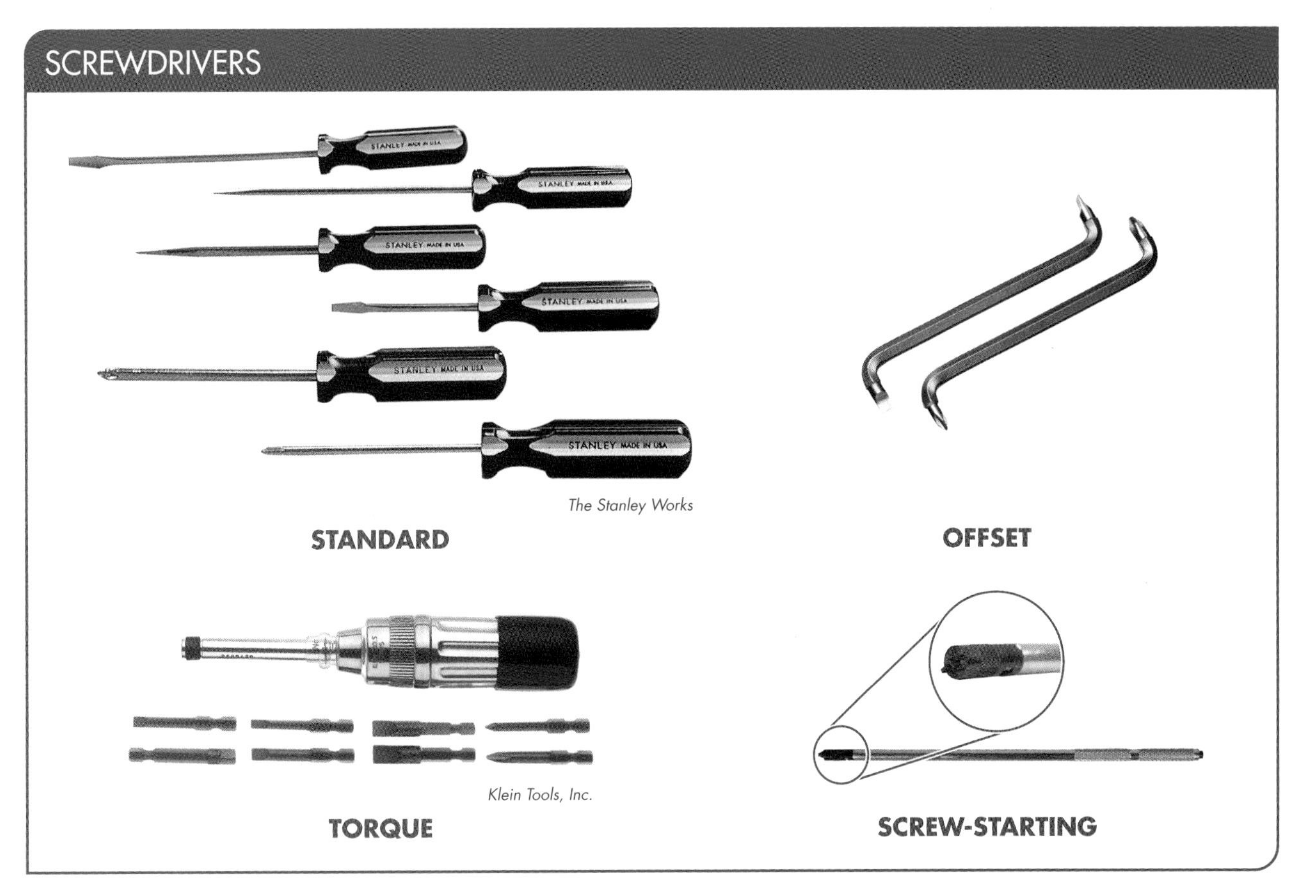

Figure 5-23. Individuals use screwdrivers for installation and maintenance purposes to secure or remove various types of electrical equipment.

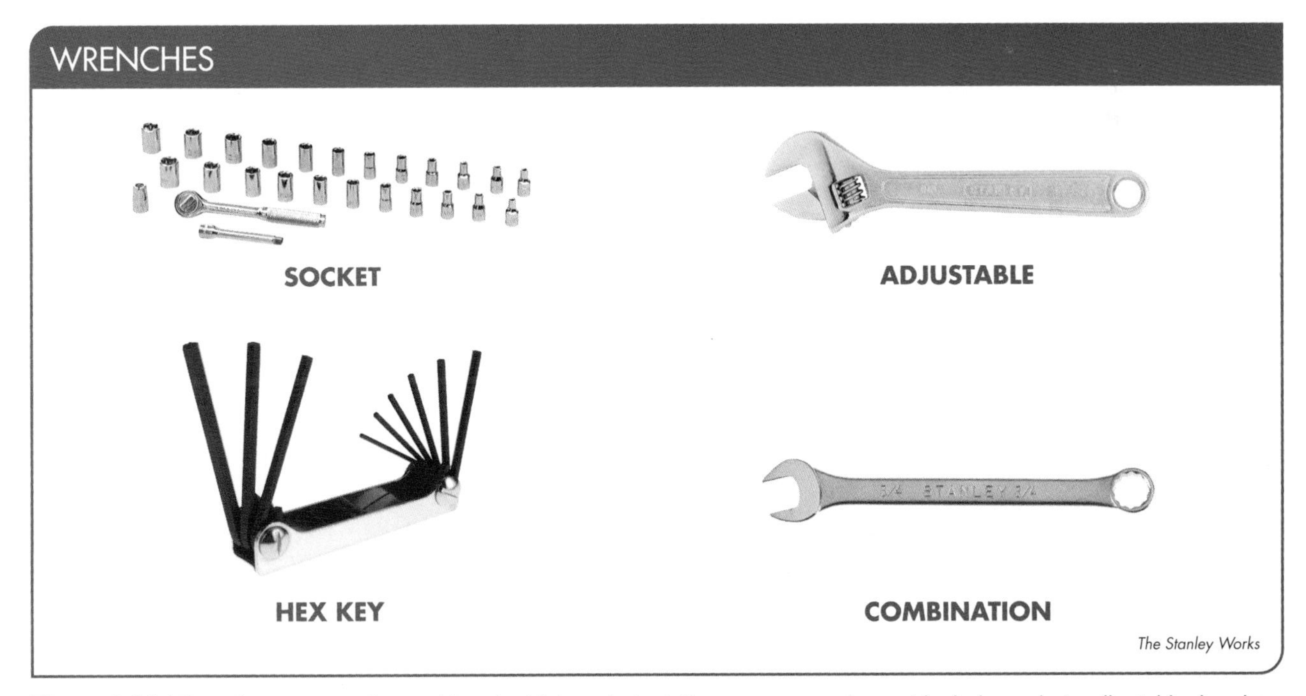

Figure 5-24. Wrenches commonly used by electricians to install or remove equipment include socket, adjustable, hex key, and combination wrenches.

Socket wrenches are used to tighten a variety of fasteners such as hex-head lag screws, bolts, and various electrical connectors. Adjustable wrenches are used to tighten items such as hex-head lag screws, bolts, and large conduit couplings. Hex key wrenches are used for tightening hex-head bolts. A combination wrench has an open jaw on one end and a closed box-end wrench on the other.

Wire Strippers. A *wire stripper* is a hand tool that is designed to remove insulation from small-gauge wires. Most types of wire strippers can remove insulation from stranded conductors that are sizes 22 AWG (American Wire Gauge) to 10 AWG and solid conductors that are sizes from 18 AWG to 8 AWG. Some wire strippers are designed with extra functions to crimp connections and cut wires. A *wire stripper/crimper/cutter* is a hand tool used to strip conductor insulation, crimp conductor terminals, and cut conductors. A wire stripper/crimper/cutter is used for conductors that are sizes 22 AWG to 10 AWG. **See Figure 5-25.**

TECH TIP

When wire is being stripped, care must be taken to cut through the insulation without nicking the conductor. A nick in a conductor creates a weak spot that may later result in a break.

Some wire strippers are designed with a wire cutter in the middle and a small-diameter bolt cutting section. Most wire stripper/crimper/cutters can be used to shear bolts ranging from sizes 4-40 (0.1120″) to 10-24 (0.1900″). A wire stripper without the crimper and bolt cutting functions are easier to handle when removing insulation. **See Figure 5-26.** To remove insulation from a wire using a wire stripper/crimper/cutter, the following procedure is applied:

1. Insert the wire into the correct wire gauge hole size.
2. Squeeze and rotate the stripper around the wire (or back and forth) to make the cut through the insulation.
3. Strip the insulation off the end of the wire.

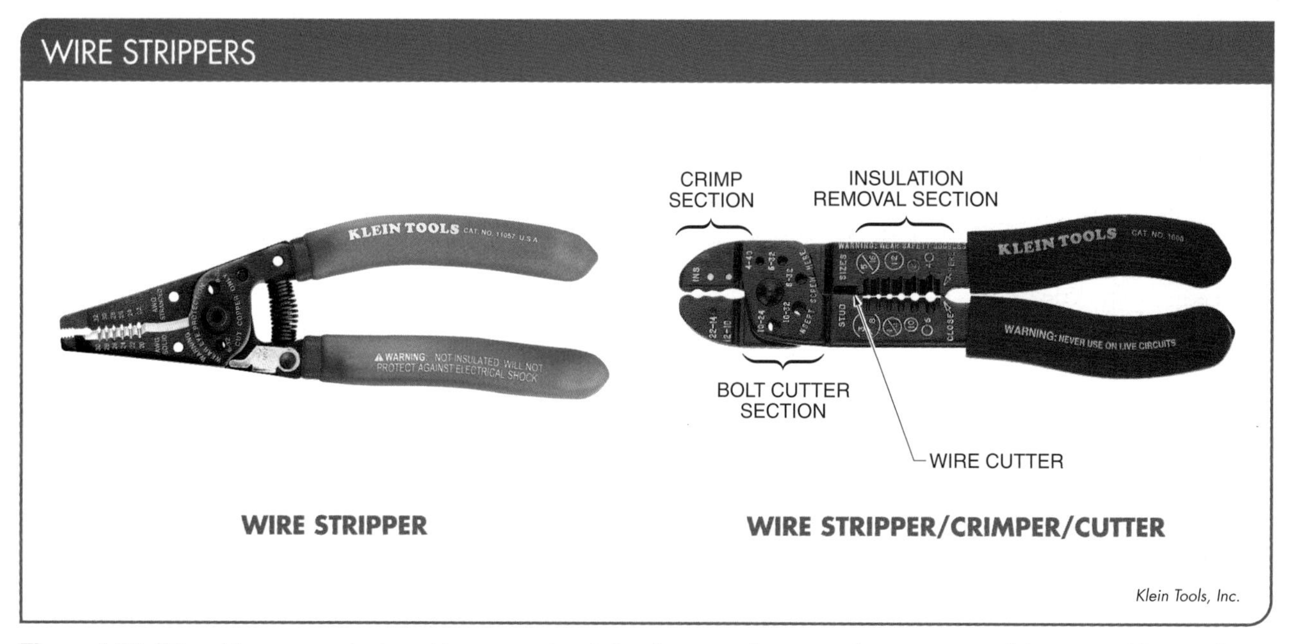

Figure 5-25. Wire strippers are designed to remove insulation from small-gauge wires; some models come with extra functions to crimp connections and cut wires.

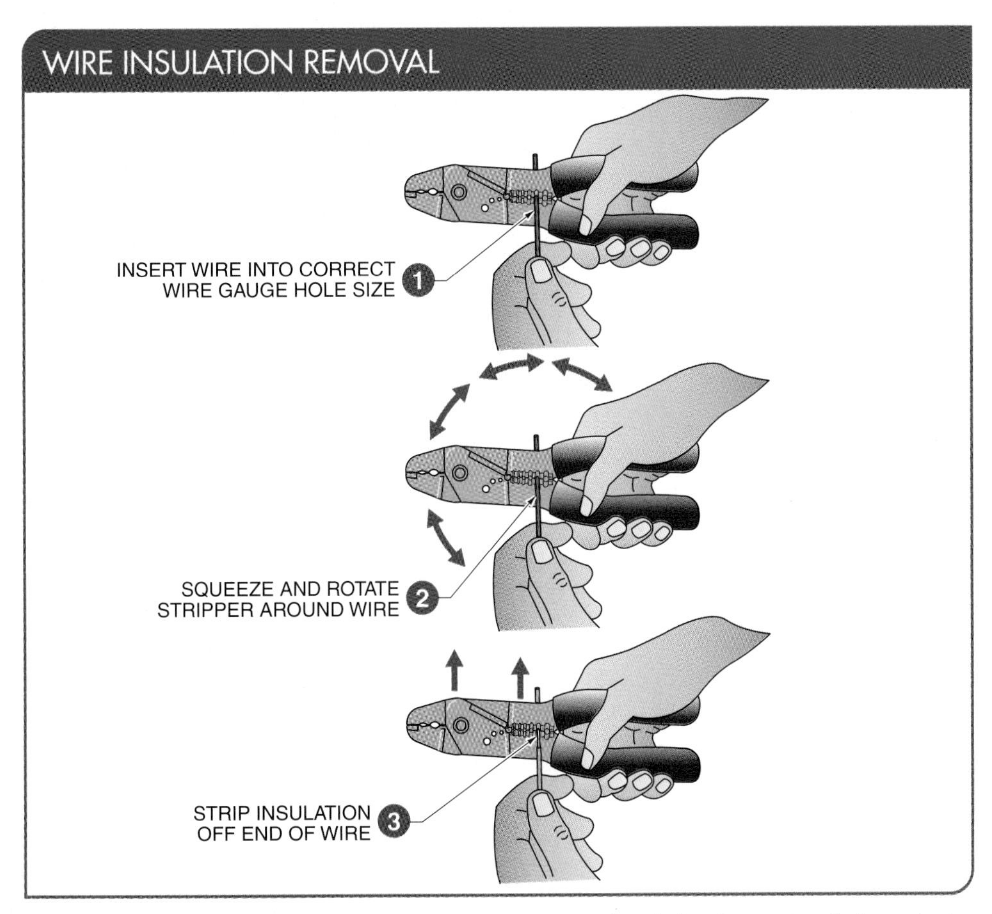

Figure 5-26. A wire stripper/crimper/cutter can remove insulation from stranded conductor sizes 22 AWG to 10 AWG and solid conductor sizes from 18 AWG to 8 AWG.

Knives. Knives are frequently used for cutting insulation or other construction materials. Three common types of knives used by electricians include utility, lineman's, and electrician's knives. **See Figure 5-27.**

> TECH TIP
>
> *When using a knife to skin insulation off a conductor, always cut away from the body to prevent injury.*

Figure 5-27. Three knives commonly used by electricians to cut insulation include utility, lineman's, and electrician's knives.

A *utility knife,* also known as a box-cutting knife, is a knife with a blade that can be retracted into the knife body when not in use. Additional blades are often stored in the handle. A *lineman's knife* is a knife that is used for skinning and scraping insulation off of conductors and cables. An *electrician's knife* is a knife that is similar in design to a pocket knife and is used for cutting and scraping insulation from conductors. An electrician's knife is normally used on smaller conductors and cords.

Handsaws. A *handsaw* is a saw operated with one hand and used to cut and trim materials to their proper dimensions. Three common types of handsaws used include compass saws, drywall saws, and hacksaws. **See Figure 5-28.**

HANDSAWS

The Stanley Works

COMPASS SAW

The Stanley Works

DRYWALL SAW

Klein Tools, Inc.

HACKSAW

Figure 5-28. Three handsaws commonly used in electrical work include compass saws, drywall saws, and hacksaws.

A *compass saw* is a handsaw that is used to saw curves, holes, and other internal openings. It can be used to cut in tight spaces where a regular saw will not fit. A *drywall saw* is a handsaw that is used to saw cutouts in drywall for electrical boxes. A *hacksaw* is a handsaw that has an adjustable steel frame designed to hold various lengths and types of blades for cutting different materials. Hacksaws are used to cut workpieces such as metal pipe, PVC pipe, conduit, rigid plastics, bolts, and nails.

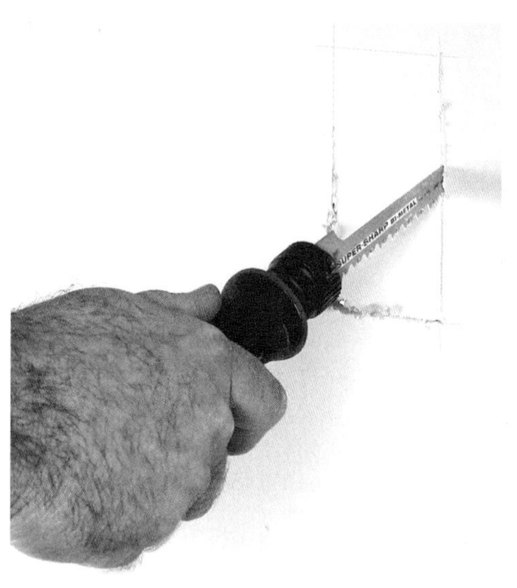

Milwaukee Tool Corporation

Fixed blade or retractable blade drywall saws can be used to cut drywall for the installation of electrical boxes.

Miscellaneous Hand Tools. There are many other hand tools that can be used for residential electrical work. Additional miscellaneous hand tools used include tape rules, torpedo levels, cable tie guns, cable strippers, PVC pipe cutters, and conduit screwdrivers. **See Figure 5-29.**

A *tape rule,* or tape measure, is a measuring hand tool that consists of a long, continuous strip of fabric, plastic, or metal graduated in regular increments. Tape rules are used often by electricians. Tape rules are available in various lengths, such as 12′, 16′, 20′, 25′, 30′, and 35′.

Figure 5-29. Miscellaneous hand tools that are used in electrical work include tape rules, torpedo levels, cable tie guns, cable strippers, PVC pipe cutters, and conduit screwdrivers.

A *torpedo level,* or pocket level, is a small plumb and leveling tool. Torpedo levels are usually 9″ to 10″ long. Torpedo levels include 45° bubble vials to check 45° offsets, as well as vertical and horizontal bubble vials.

A *cable tie gun* is a handheld device that is used to tighten plastic or steel cable ties around a bundle of wires or small-diameter cables. Wire and cable can be quickly bound together in bundles up to 2″ in diameter and in multiple locations. A *cable stripper* is a tool used to strip insulation from cables with outside diameters from ¼″ to 1½″.

A *PVC pipe cutter* is a handheld tool designed to quickly and accurately cut plastic pipe (PVC or PE) and rigid rubber hose up to 2″ diameter without the use of a vise. The pipe is held in place by a hooked jaw. Alloy steel blades are ratcheted through the pipe being cut.

A *conduit screwdriver* is a hand tool designed with a hooded blade to fasten conduit screws. A hooded blade keeps the tip of the screwdriver from slipping when tightening overhead conduit fittings.

Hand Tool Safety

Safety is the primary concern when using hand tools for specific tasks. Hand tool safety not only requires that the proper type of tool be chosen, but also that the correct size of tool be chosen. Individuals should only use hand tools in the manner for which they were designed to be used.

Tools should be kept in a safe and designated location. Any hand tool can pose a danger when left in the wrong location. Many accidents are caused by tools falling off ladders, shelves, and scaffolds that are being used or moved. The following proper safety guidelines should be observed when using hand tools:

- Do not force hand tools beyond their rated capacity.
- Always inspect hand tools for worn, chipped, or damaged surfaces prior to use.
- Do not carry hand tools in clothes pockets unless the pocket is designed for the tool.
- Never use a hand tool that is in poor, faulty condition.
- Never use dull tools. The extra force exerted when using dull tools often results in losing control of the tool.
- Always carry sharp-edged hand tools pointed down and away from the body.
- Inspect tools for dangerous conditions, and if found, repair or replace the tool immediately.

For maximum safety in electrical maintenance or testing tasks, insulated hand tools should be used. **See Figure 5-30.** Insulated hand tools are rated for a maximum use of 1000 V and are designed to meet NFPA 70E® and OSHA requirements for personnel exposed to electrical hazards. Insulated tools are covered with a double-bonded, flame- and impact-resistant insulation material that reduces the chance of an arc flash or an electric shock given to personnel working on or near energized electrical equipment and circuits. Certain tools, such as pliers, screwdrivers, wrenches, hex keys, and wire strippers, are available as insulated hand tools.

INSULATED HAND TOOLS

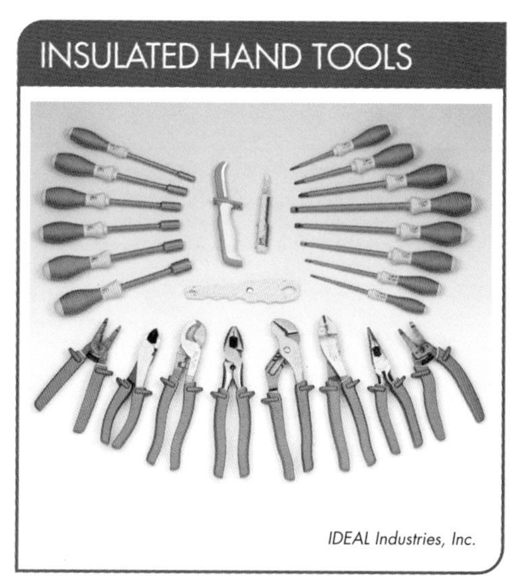

IDEAL Industries, Inc.

Figure 5-30. Insulated hand tools are rated for a maximum use of 1000 V and are designed to meet NFPA® 70E and OSHA requirements for personnel exposed to electrical hazards.

Proper Hand Tool Use and Maintenance

Individuals must understand how to properly use and maintain each hand tool in order to work safely and efficiently. Tools should only be used for the function for which they were designed. Tools must not be forced beyond their normal function. Potential injury can happen when individuals use hand tools not fitted for use or used inappropriately. The proper tools fitted for use will help an individual perform the work task safely and efficiently. **See Figure 5-31.**

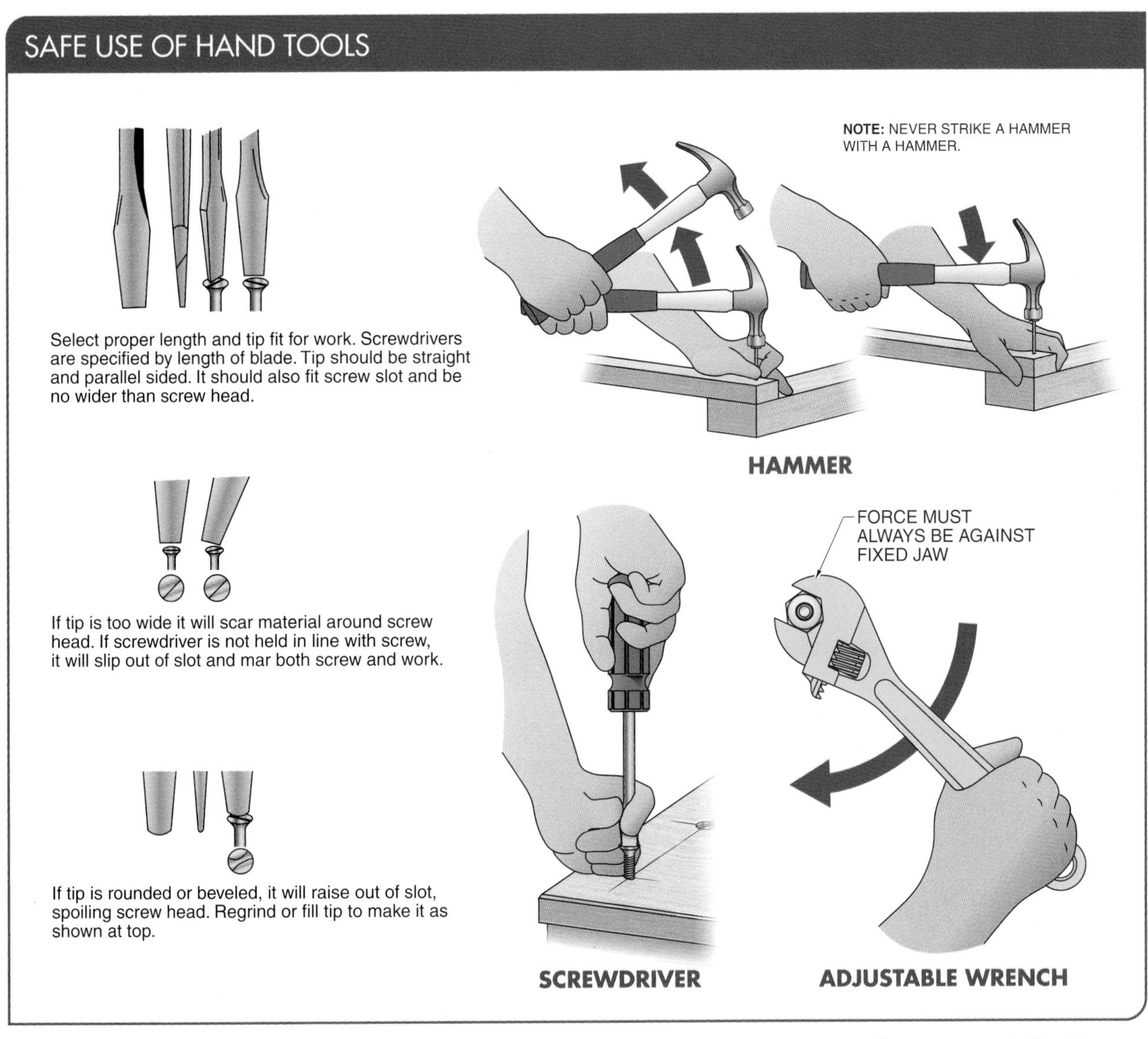

Figure 5-31. When individuals use hand tools not fitted for use, there is a potential for injury. The proper tools fitted for use will help an electrician perform the work task safely and efficiently.

The maintenance of hand tools is important for prolonging the life of the tools. Periodic checks of hand tools will aid in keeping them in good condition. The following maintenance guidelines should be observed to prolong the life of tools:

- Always inspect a tool before use. Tool handles must be free of cracks and splinters and be fastened securely to the working part of the tool.
- Do not use a tool that is in poor or faulty condition.
- Keep hand tools for cutting operations sharp and clean.
- Do not place hand tools in areas with solvents, prolonged moisture, or excessive heat. These conditions can permanently damage the tool.
- Always clean hand tools of dirt, oil, and debris to reduce tool slippage and prevent injury.
- Place hand tools in designated locations such as a toolbox, chest, or cabinet.

Power Tools

While much of the work performed by an individual can be accomplished with hand tools, many work tasks can also be performed quickly and efficiently with the use of power tools. Like hand tools, power tools perform drilling, driving, and cutting operations. Types of power tools include power drills, impact drivers, impact wrenches, cable cutters, and reciprocating saws.

Power Drills. Power drills are used for their ease of use and efficiency when installing, securing, or removing electrical equipment. Power drills can be either corded (AC powered) or cordless (battery-powered) and are available in D-handle or pistol-grip handle designs. **See Figure 5-32.**

Figure 5-32. Power drills are used for ease of use and efficiency for installing, securing, or removing electrical equipment.

Corded drills are attached to an AC power source through an electrical cord that is plugged into an outlet. Corded drills are designed with auxiliary handles to reinforce high-torque operations, such as drilling into hard surfaces.

A cordless drill receives power from a removable battery, which is located at the bottom of the handle. Batteries are recharged between intervals based on the amount of usage. Power drill sizes range from ¼″ to 1¾″ and are based on the diameter of the large bit shank that can fit into the chuck of the drill. The adjustable jaws of the chuck are used to hold drill bits.

A *drill bit,* or bit, is a rotary-end cutting tool that fits into the chuck of a power drill. **See Figure 5-33.** Drill bits have two or more cutting edges and two or more spiral or straight flutes used for the removal of chips and the admission of cutting fluid. A *flute* is a spiral groove that runs along the length of a drill bit. The flutes are shaped to help form the proper cutting edges on the cone-shaped point. The three major parts of a drill bit are the point, the body, and the shank. The point is the entire cone-shaped cutting end of the drill bit. The body of the drill bit extends from the point to the shank. The shank is the portion of the drill bit that fits into the drill.

Types of drill bits include twist drill bits, auger drill bits, carbide-tipped masonry drill bits, spade drill bits, step drill bits, and hole saws. A twist drill bit is a general-purpose drill bit used to drill holes up to 1″ in diameter in wood or light-gauge metal. Twist drill bits can have a square or a reduced shank. An auger bit is a drill bit used to bore deep holes in wood up to 1″ in diameter. Auger drill bits with screw points are used most often for wood-framed structures. A carbide-tipped masonry drill bit has a carbide cutting tip used to bore holes in concrete, plaster, slate, stone, brick, and other types of masonry. A spade drill bit has flat cutting tips at the outside edges and a central projecting point used to guide the bit into wood, plastic, or composition materials.

A step drill bit is a drill bit that is used to drill holes with multiple sizes in wood, metal, or plastic. A hole saw is a rotating saw blade with a pilot twist drill mounted in the center that is used to cut larger holes than can be cut with drill bits. Hole saws are used to cut wood, fiberglass, fiberboard, ceramic, tile, and soft metals, and can have carbide-tipped teeth.

TECH TIP

AC power drills should not be used if the cord or plug is damaged. When a large hole must be drilled, a small pilot hole should be drilled first to prevent the drill bit getting stuck in the hole.

Impact Drivers. An *impact driver* is a cordless power screwdriver that uses both rotational force and a hammer mechanism to drive fasteners. **See Figure 5-34.** Impact drivers typically deliver more torque and rotational speed than cordless drills. Impact drivers usually have a ¼″ hex chuck and accept either square-drive bits or hex-shaft screwdriver bits.

The hammer mechanism that produces the torque also creates some forward pressure, which reduces cam-out when using Phillips head bits. Cam-out is a condition that occurs when the screwdriver bit slips out of the screw head when exerting pressure to turn the screw. Cam-out often results in damaging the screw head, screwdriver bit, or the work surface.

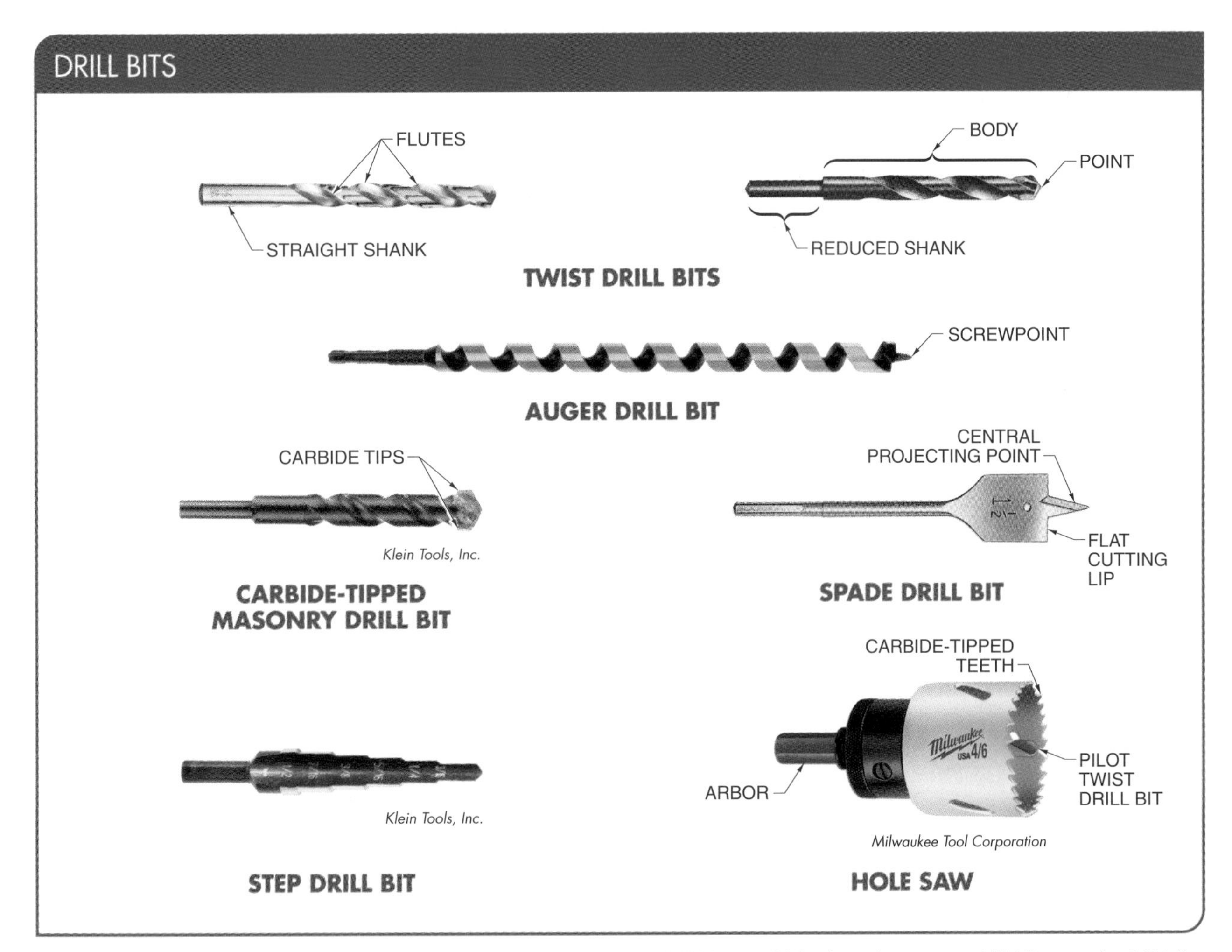

Figure 5-33. Types of drill bits used include twist drill bits, auger drill bits, carbide-tipped masonry drill bits, spade drill bits, step drill bits, and hole saws.

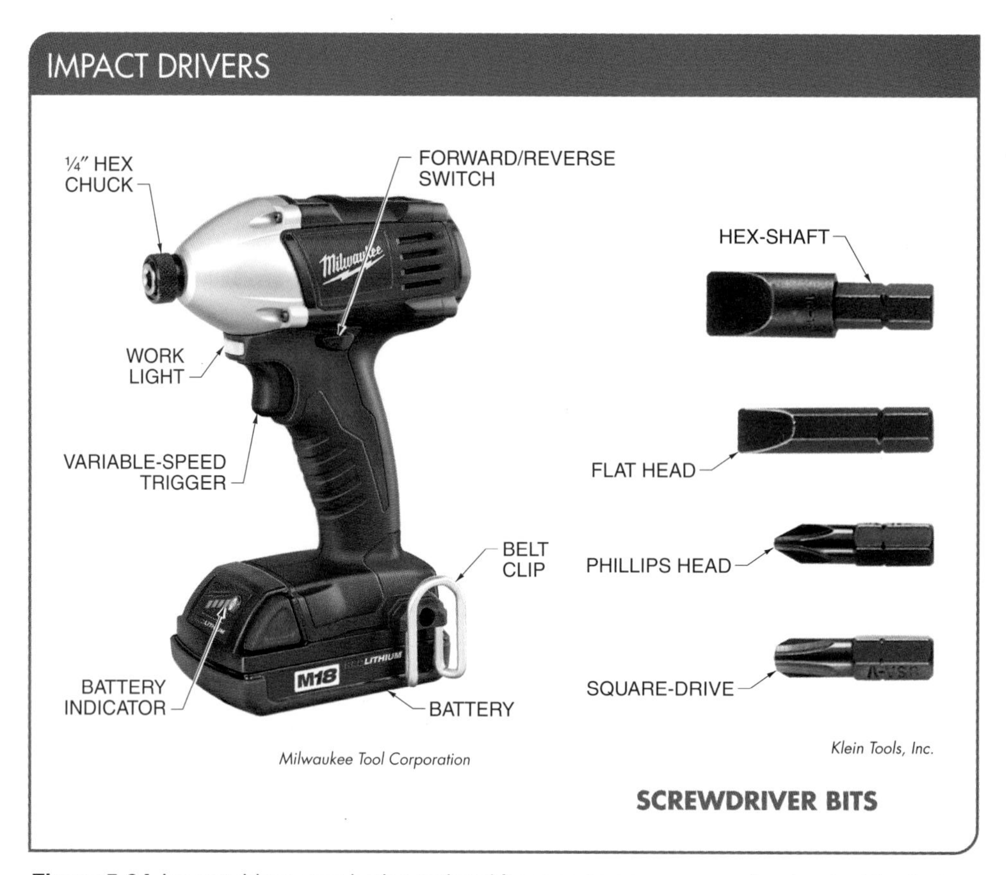

Figure 5-34. Impact drivers use both rotational force and a hammer mechanism to drive fasteners and typically deliver more torque and rotational speed than cordless drills.

Milwaukee Tool Corporation

Cordless power screwdrivers can be used for installing and removing electrical components.

Impact Wrenches. An *impact wrench* is a corded or cordless wrench used to supply short, rapid impulses to sockets. Impact wrenches are used for a variety of applications, including installing and removing nuts and bolts and driving lag bolts. **See Figure 5-35.** Only impact sockets should be used with an impact wrench. Impact wrenches typically have a ½″ or ¾″ square drive tang.

Battery-Powered Cable Cutters. A *battery-powered cable cutter* is a power tool designed to cut various diameters of electrical cables. **See Figure 5-36.** Battery-powered cable cutters are ideal for large projects that require numerous cuts. They allow electricians to reduce fatigue on the body and can increase productivity. Battery-powered cable cutters can cut aluminum and soft copper cables up to 2″ in diameter and hard copper cables up to about 1¼″ in diameter.

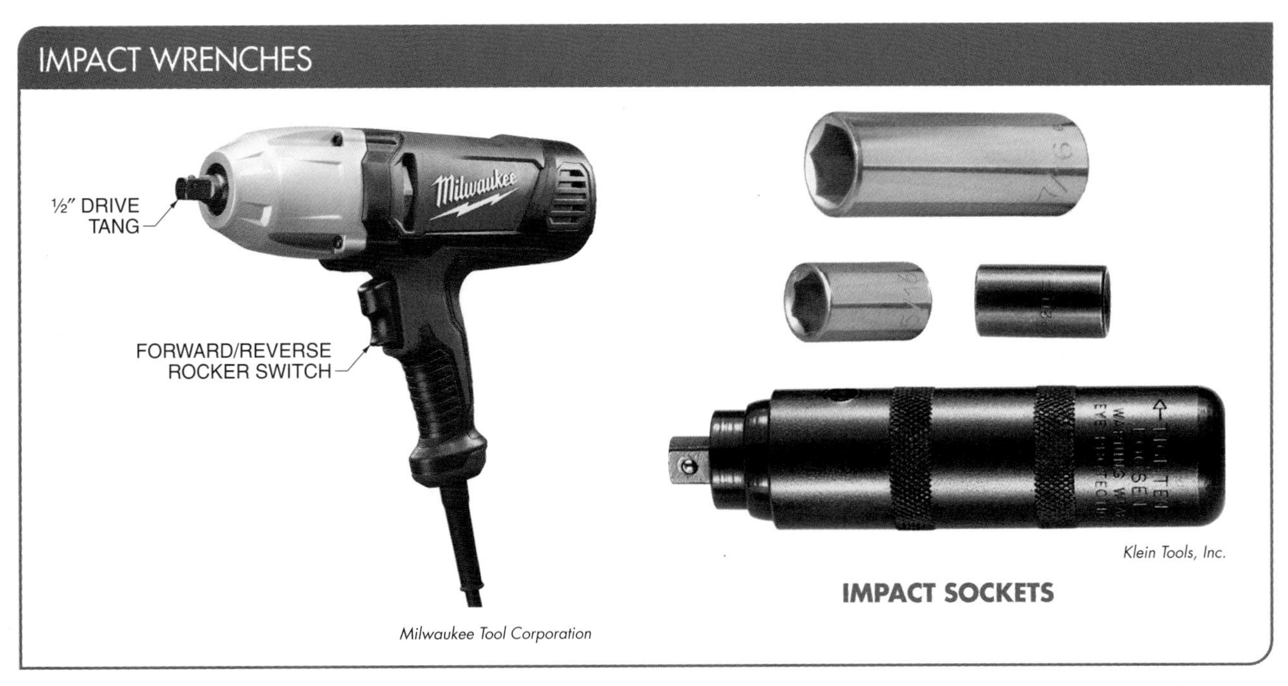

Figure 5-35. Impact wrenches are used to supply short, rapid impulses to sockets and can be used for a variety of applications, including installing and removing nuts and bolts and driving lag bolts.

Figure 5-36. Battery-powered cable cutters are designed to cut various diameters of electrical cables and are ideal for large projects that require numerous cuts.

Reciprocating Saws. A *reciprocating saw* is a multipurpose cutting power tool in which the blade reciprocates to create a cutting action. Reciprocating saw blades are plunged directly into walls, floors, and ceilings, and other durable materials. **See Figure 5-37.** Reciprocating saws are used by electricians to cut holes in drywall, plywood, and hard-surface flooring in order to install electrical panels, equipment, and conduit. Reciprocating saws are typically used after framing is completed and additional cuts need to be made.

Figure 5-37. Reciprocating saws are plunged directly into walls, floors, ceilings, and other durable materials to cut holes in order to install electrical panels, equipment, and conduit.

Power Tool Safety

Power tools should only be used after gaining knowledge of their operation, methods of use, and safety precautions. The advantage of power tools is that electrical work tasks are performed quicker and are easier to perform than hand tool operations. However, because power tools can deliver a great amount of energy in a short time, they can easily cause severe injury. The following safety guidelines must be observed when operating power tools:

- Read the owner's manual and manufacturer safety recommendations.
- Allow qualified repair technicians to inspect and service the power tools at regular intervals as specified by the manufacturer or OSHA.
- Inspect power cords to verify they are in good condition.
- Verify that all safety guards are in place and in proper working order. Do not remove, displace, or jam guards or any other safety devices.
- Make all adjustments, blade changes, and inspections with the power off.
- Ensure that the on/off switch is in the OFF position before plugging the power cord into a receptacle.
- Wear safety goggles and a dust mask where needed.
- Ensure that the material to be worked on is free of obstructions and securely clamped.
- Be attentive to the task at hand at all times.
- Investigate any changes in the sound of power tools immediately.
- Shut off the power when work is completed.
- Wait until the operation of the power tool stops before leaving or laying down the tool.
- Alert others when a power tool is defective, tag it, and remove from service.
- Avoid operating power tools in locations where sparks could ignite flammable materials.
- Unplug power tools by pulling on the plugs; never pull on the cord.

Grounding Power Tools. All corded power tools must have an acceptable means of grounding or use a ground-fault circuit interrupter (GFCI) to ensure worker safety from electrical shock. A GFCI protects against electrical shock by detecting the imbalance of current in the normal conductor pathway and then quickly opening the circuit. A common type of GFCI used on the job site is a portable GFCI. **See Figure 5-38.**

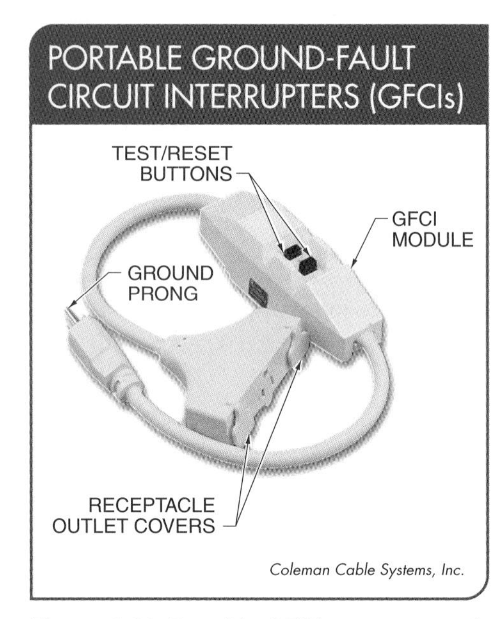

Figure 5-38. Portable GFCIs are commonly used on job sites.

If a portable GFCI is not available, an acceptable means of grounding must still be used to prevent electrical shock. OSHA recognizes an equipment grounding conductor program that requires periodic inspection and documentation of power tools and extension cords. In order to be grounded, power tools must have a three-conductor cord with a three-blade plug that connects to a grounded receptacle (outlet). **See Figure 5-39.** Two types of grounded receptacles include nonlocking and locking. Nonlocking receptacles are used for 120 V power tools, whereas locking receptacles can either be used for 120 V or 230 V power tools and equipment. *Note:* Always refer to and follow the NEC® and local codes for the proper grounding standards and procedures.

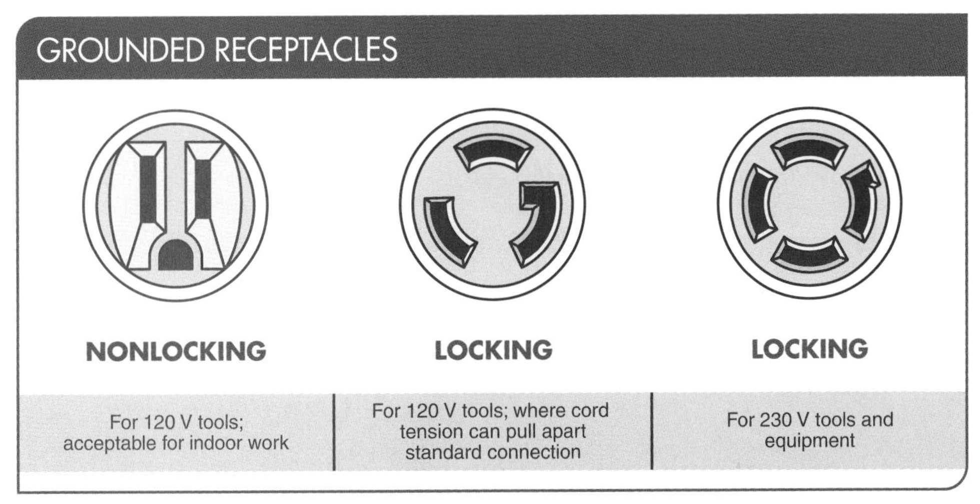

Figure 5-39. Power tools must have a three-conductor cord with a three-blade plug that connects to a grounded receptacle (outlet) such as nonlocking or locking receptacles.

Some power tools that are double-insulated have two prongs and a notation on the specification plate that the tool is double-insulated. Electrical parts in the motor of a double-insulated tool are surrounded by extra insulation to help prevent shock. Therefore, the tool does not have to be grounded. Double-insulated tools are relatively safe, but grounded power tools are typically used on job sites.

When a faulty power tool causes an electrical short, a proper grounding system will ensure a line-to-ground fault through a grounded outlet. **See Figure 5-40.** Starting at the ground blade, a fault current will take the path of least resistance through the grounding slot on the receptacle and beyond into the ground conductor. From there, the fault current will travel through the grounding conductors back to the service panel's ground busbar. The fault current will reach a copper grounding electrode conductor (GEC) and end in the earth.

> TECH TIP
>
> *Nonlocking receptacles should be installed properly. If they are installed vertically, the ground socket should be on the bottom. If they are installed horizontally, the smaller hot socket should be on top.*

Tool Organization Systems

Electrical tools must be available when needed for electrical work to be performed effectively. An tool organization system can provide both a central location for the retrieval of electrical tools and a means of storage for protection. Electrical tools can be organized in several ways depending on where and how frequently the tools are used. Three types of electrical tool organization systems used by electricians include tool pouches, toolboxes, and pegboards. Whichever system an individual chooses, proper organization will ensure that the tools are kept clean, dry, and ready when needed.

Tool Pouches. When electrical tools need to be close at hand, a leather or fabric tool pouch is typically used. A *tool pouch* is a small, open tool container (pouch) for storing electrical tools. **See Figure 5-41.** A tool pouch is typically made of heavy-duty leather or fabric and designed to be used with a belt that holds it in place around the waist. Tool pouches can be simple in design and hold only a few tools, or they can be relatively large and designed to hold a wide selection of tools. The type of pouch chosen depends on the type of work an electrician is planning to perform.

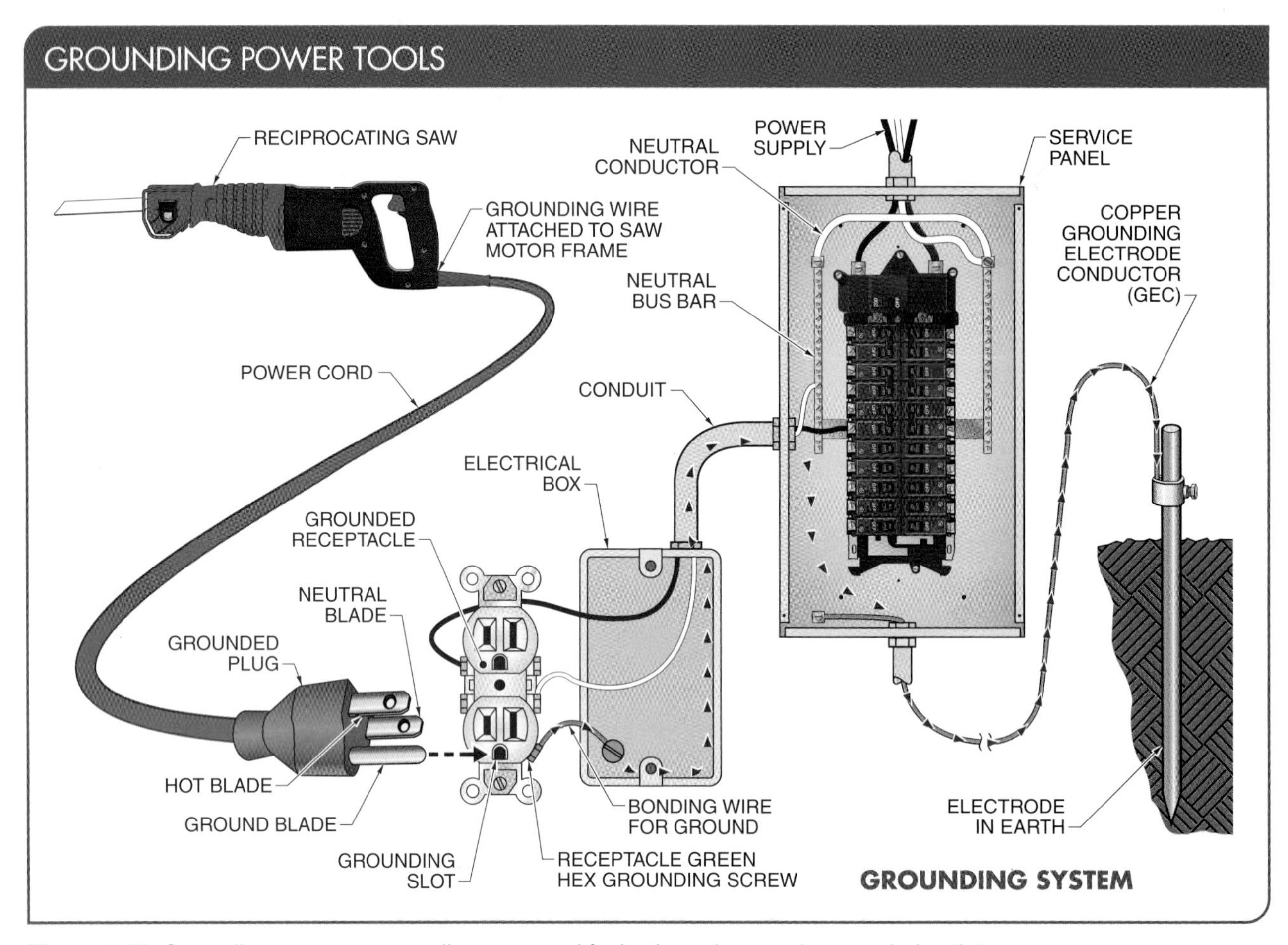

Figure 5-40. Grounding systems ensure line-to-ground faults through properly grounded outlets.

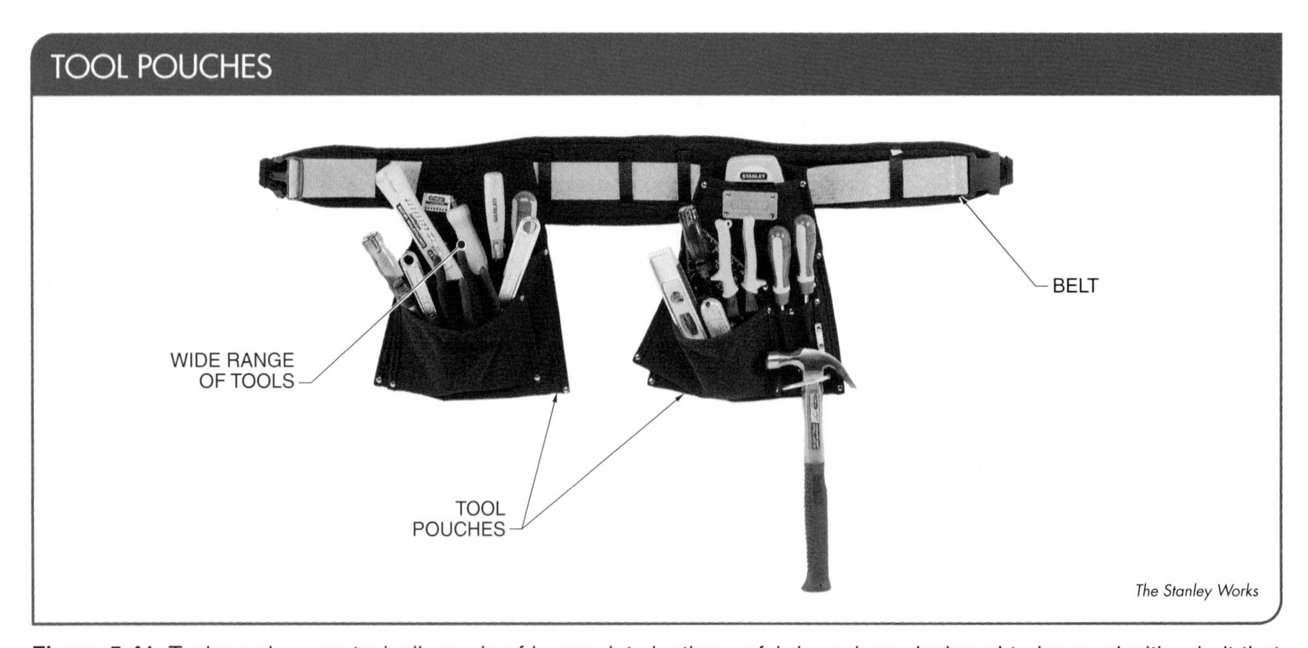

Figure 5-41. Tool pouches are typically made of heavy-duty leather or fabric and are designed to be used with a belt that holds positioning of the pouches for easy tool access.

Toolboxes. When tools need to be transported to another job site, a toolbox may be necessary. Some electrical installers also prefer to store tools in a toolbox. **See Figure 5-42.** A well-designed toolbox can be locked and will keep tools clean and dry. Common types of toolboxes include cantilever, handy carry, and fixed toolboxes. In addition, toolboxes, whether fixed or portable, provide a place where all tools can be collected and stored.

Pegboards. When tools are used at a repair bench in a residence, a pegboard may be appropriate. A *pegboard* is a hardboard that usually has a laminated finish on the front and is perforated with equally spaced holes for accepting hooks. When a pegboard is mounted, outlines of tools can be made to fit with the proper hooks for inventory purposes. **See Figure 5-43.** Pegboards are typically available in 4′ × 8′ sheets.

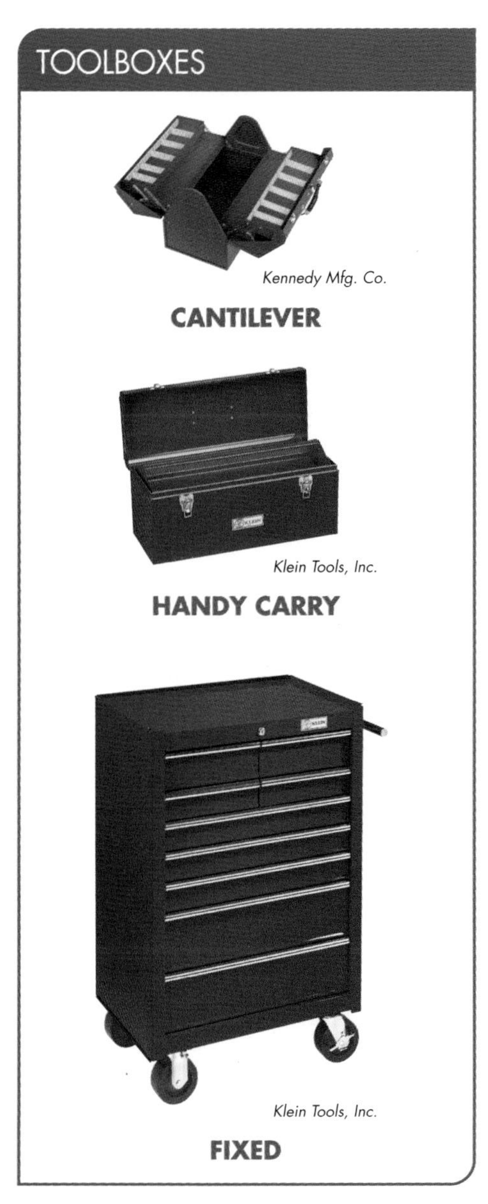

Figure 5-42. Types of toolboxes include cantilever, handy carry, and fixed toolboxes, which provide a place where all electrical tools can be collected and stored.

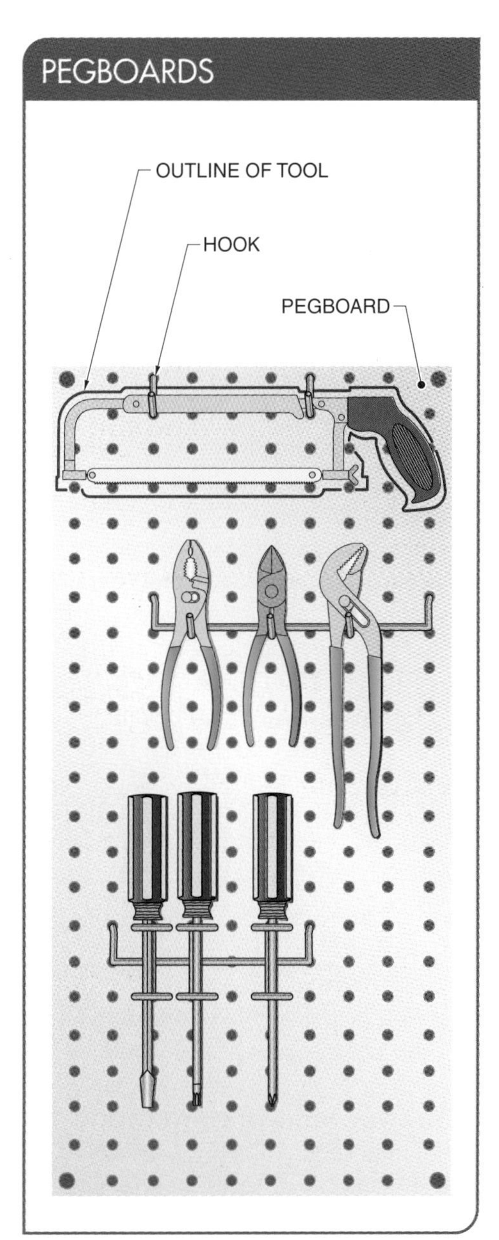

Figure 5-43. Pegboards are typically perforated with equally spaced holes to be fit with hooks for the inventory of electrical tools.

SECTION 5.2 CHECKPOINT

1. What type of hand tool is used to pull conductors through conduit, through inaccessible spaces, or around obstructions in walls?
2. Describe how to remove insulation from a wire using a wire stripper.
3. What type of hand tools provide maximum safety for performing maintenance or testing tasks?
4. Describe arc-rated PPE.
5. What are the three types of electrical tool organization systems?

SECTION 5.3—TEST INSTRUMENTS

When installation, inspection, or maintenance tasks are performed on electrical circuits in a residence, test instruments are used to verify the operation of the equipment, test circuit faults, and measure electrical quantities. Individuals must apply proper procedures when testing equipment or circuits and taking measurements. Failure to follow proper procedures can create an unsafe condition and can be damaging to test instruments and equipment.

There is a wide variety of test instrument types. The types of test instruments used include test lights, voltage indicators, voltage testers, branch circuit identifiers, digital multimeters (DMM), and thermal imagers. Some test instruments are meant for one purpose, whereas others have more capabilities.

To take a measurement, the test instrument must be set to the correct function setting and properly connected to the equipment or circuit being tested. When these steps are done correctly, the test instrument should display measurements, indicate power is present, or provide fault information. Test instruments set to the incorrect function setting or not properly connected to a circuit can allow inaccurate measurements, damage to the test instrument, and can warrant electrical shock. The following guidelines should be observed when conducting tests or measurements:

- Practice all electrical safety procedures.
- Wear appropriate PPE.
- Perform only authorized installation and maintenance tasks.
- Follow all manufacturer recommendations and procedures for test instruments and equipment.

TECH TIP

A test instrument should only be used within its rating limit on a circuit function it was designed to test. Using a meter to test circuit voltage above its rating could result in an arc flash. If a test instrument is set to an improper function setting, then the fuse inside the test instrument can blow.

CAT Ratings

The IEC creates standards to prevent electrical hazards that can occur in situations when using various test instruments. Certain risks, such as a voltage surge in a residential system, can cause electrical shock to individuals conducting tests or measurements or, in the worst case, an arc flash/blast. A type of voltage surge is a transient voltage. A *transient voltage* is a temporary, undesirable voltage in an electrical system. Transient voltages are caused by lightning strikes, unfiltered electrical equipment, and switching high-current loads (motors) on and off.

In order to reduce electrical hazards associated with transient voltages, the IEC created Standard 61010, *Safety Requirements for Electrical Equipment for Measurement, Control, and Laboratory Use.* This standard classifies the applications in which test instruments can be used into four overvoltage installation categories, or CAT ratings. **See Figure 5-44.**

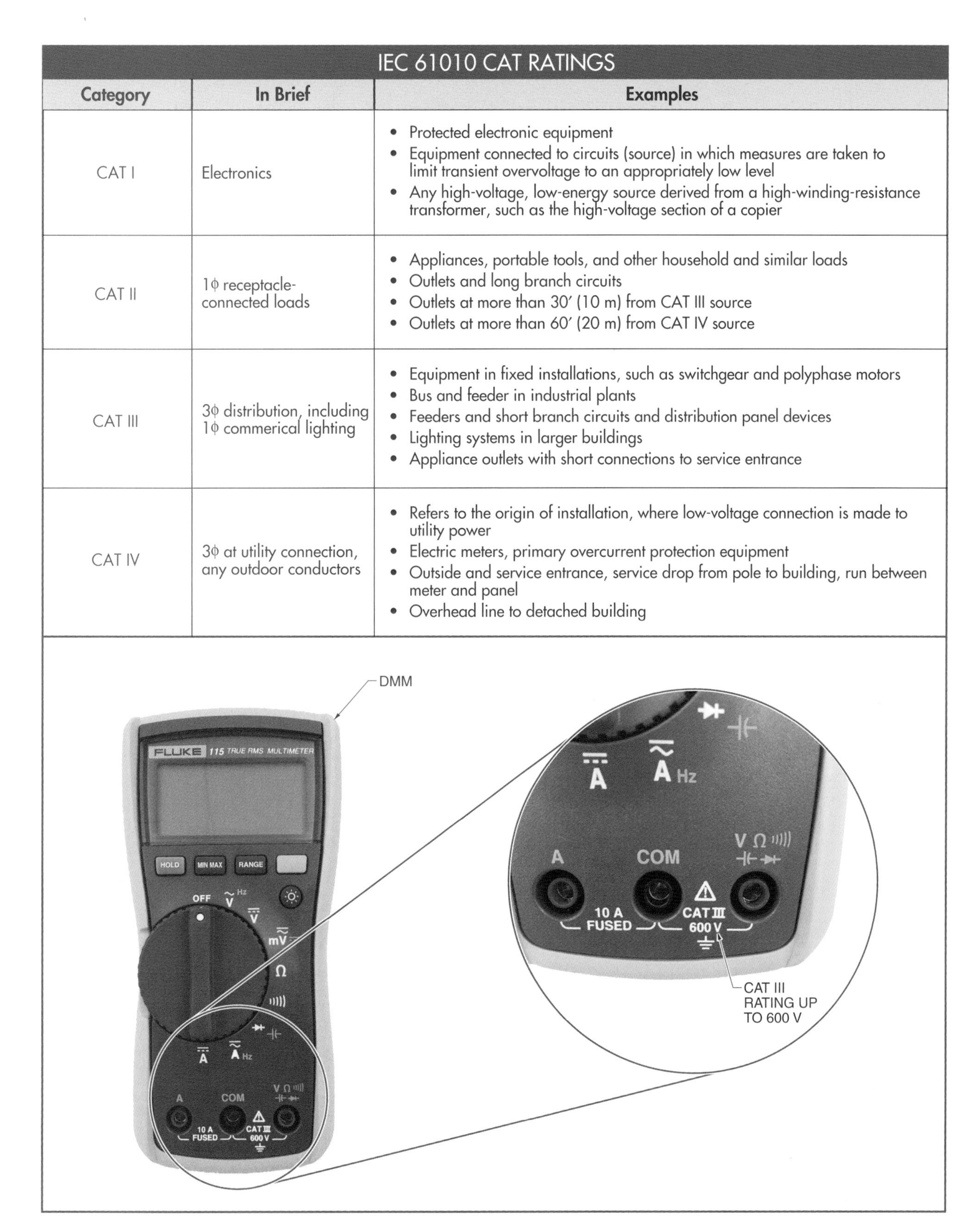

IEC 61010 CAT RATINGS		
Category	**In Brief**	**Examples**
CAT I	Electronics	• Protected electronic equipment • Equipment connected to circuits (source) in which measures are taken to limit transient overvoltage to an appropriately low level • Any high-voltage, low-energy source derived from a high-winding-resistance transformer, such as the high-voltage section of a copier
CAT II	1ϕ receptacle-connected loads	• Appliances, portable tools, and other household and similar loads • Outlets and long branch circuits • Outlets at more than 30′ (10 m) from CAT III source • Outlets at more than 60′ (20 m) from CAT IV source
CAT III	3ϕ distribution, including 1ϕ commerical lighting	• Equipment in fixed installations, such as switchgear and polyphase motors • Bus and feeder in industrial plants • Feeders and short branch circuits and distribution panel devices • Lighting systems in larger buildings • Appliance outlets with short connections to service entrance
CAT IV	3ϕ at utility connection, any outdoor conductors	• Refers to the origin of installation, where low-voltage connection is made to utility power • Electric meters, primary overcurrent protection equipment • Outside and service entrance, service drop from pole to building, run between meter and panel • Overhead line to detached building

Figure 5-44. IEC Standard 61010 establishes CAT ratings to determine the magnitude of voltage a test instrument can withstand when testing electrical equipment.

The four categories (ratings) are abbreviated as CAT I, CAT II, CAT III, and CAT IV. CAT ratings determine the magnitude of transient voltage that a test instrument can withstand when used in testing circuits or equipment. Test instruments designed to these standards offer high levels of protection to individuals working on residential systems. When a transient occurs, a test instrument with the appropriate CAT rating will be damaged but can save the electrician from electrical shock since it is rated to withstand the voltage surge. *Note:* CAT II-rated test instruments are sufficient for testing residential electrical equipment.

Voltage Indicators

A *voltage indicator* is an electrical device that indicates the presence of voltage when the test tip touches, or is near, an energized hot conductor or energized metal part. **See Figure 5-45.** When the test tip is near an energized wire or receptacle, the area near the tip glows. Some types of voltage indicators also create a sound when voltage is present. Voltage indicators are used to test receptacles, fuses, circuit breakers, wires, and other equipment where the presence of voltage must be detected. Voltage indicators are available in various types, with an assortment of voltage ranges for testing a variety of circuits and equipment.

The advantages of voltage indicators are that they are inexpensive, small enough to carry in a pocket, easy to use, nonconductive, and show the presence of voltage without touching live parts of a circuit, even through conductor insulation. A disadvantage of voltage indicators is that they only indicate when voltage is present and not the actual amount of voltage. Also, voltage indicators may not indicate a presence of voltage from shielded cables (insulated wires enclosed in a conductive material).

Before any circuit or equipment test, an electrician must ensure that the voltage indicator has the appropriate CAT rating. All safety guidelines must be followed before and after testing for voltage. **See Figure 5-46.** To test for voltage presence using a voltage indicator, the following procedure is applied:

1. Verify that the voltage indicator has a CAT rating higher than the highest potential voltage in the circuit being tested. Residential branch circuits typically provide 120 VAC or 240 VAC. When circuit voltage is unknown, slowly bring the voltage indicator near the wire or equipment being tested. The voltage indicator will glow and/ or sound when voltage is present.
2. Place the tip of the voltage indicator on or near the wire or equipment being tested. When testing an extension cord for a break, test several points along the wire. Expect the voltage tester to glow

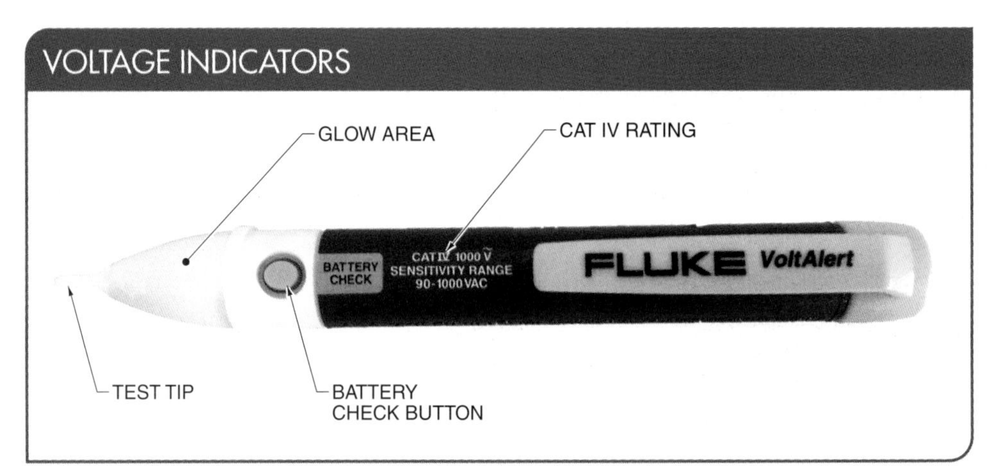

Figure 5-45. Voltage indicators detect the presence of voltage when the test tip touches or is near an energized hot conductor or energized metal part.

intermittently when moved along a cord that has twisted wires, because the hot wire will change position along the cord. The neutral wire will not indicate any voltage.

3. Remove the voltage indicator from the wire or equipment being tested.

When a voltage indicator does not indicate the presence of voltage by glowing or sounding, it should not be assumed that no voltage is present and that work can start on exposed wires or equipment. A second test instrument (voltage test light) should always be used to measure for the presence of voltage before working around or on exposed wires or electrical equipment. Also, if a 120 V receptacle is being tested, a lamp that is known to be operating properly could be plugged in to demonstrate whether voltage is present.

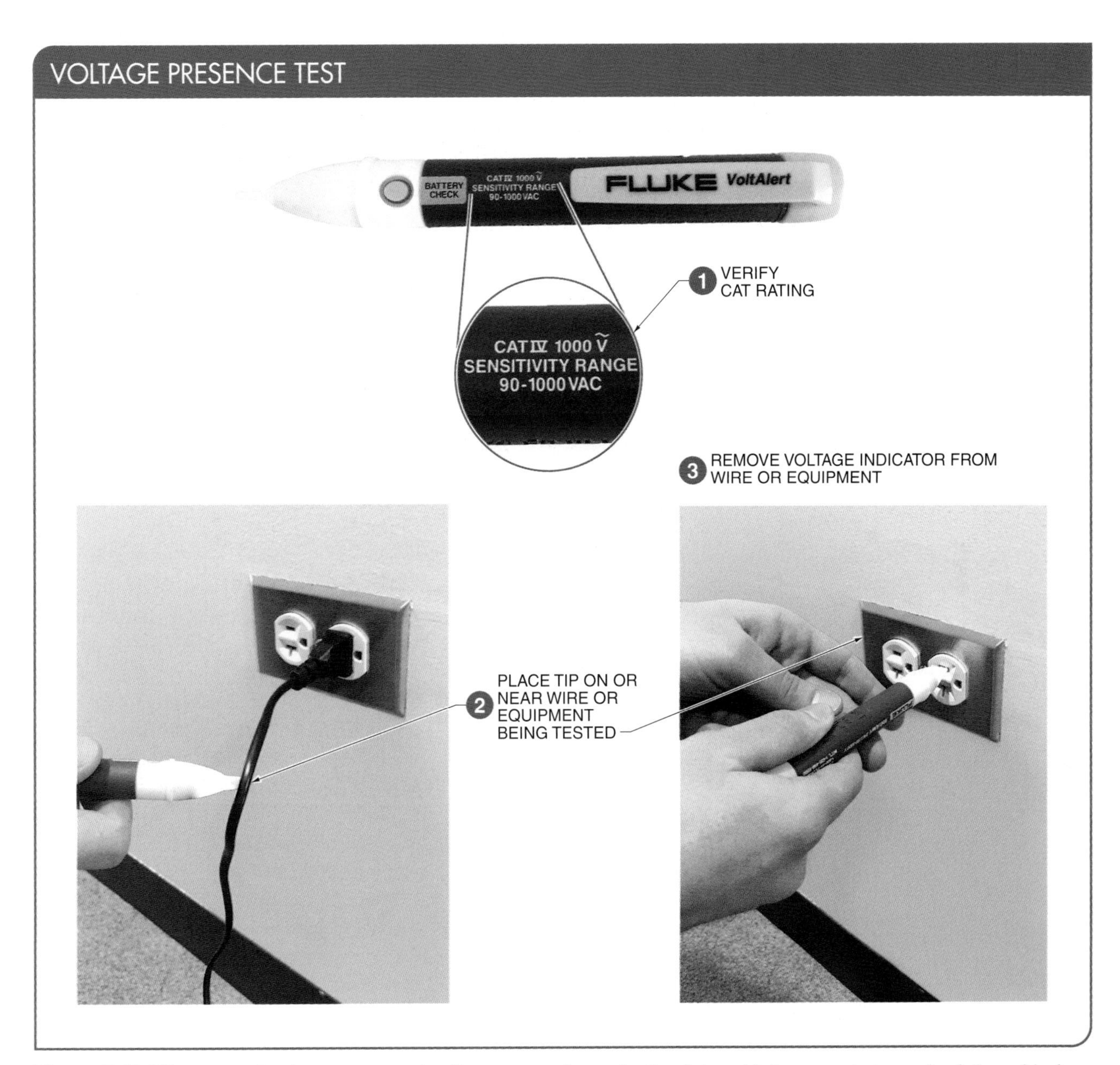

Figure 5-46. When tests for the presence of voltage are performed, all safety guidelines must always be followed before and after each test.

TECH TIP

Voltage indicators may read ghost voltages caused by a bad circuit ground. In this case, a circuit appears to be energized when it is not.

Voltage Test Lights

Electrical circuits can be tested safely and inexpensively with a voltage test light. **See Figure 5-47.** A *voltage test light*, commonly referred to as a neon tester, is an electrical test instrument that is designed to illuminate in the presence of 120 V, 240 V, 277 V, or 480 V AC power. Neon testers have two test leads: a common (black) test lead and a voltage (red) test lead. When a voltage test light is connected to a voltage source, a soft glow of the illuminated neon bulb indicates a lower voltage and a brighter glow indicates a higher voltage. For example, a soft glow on the AC 240 V bulb indicates a 220 V presence.

VOLTAGE TEST LIGHTS (NEON TESTER)

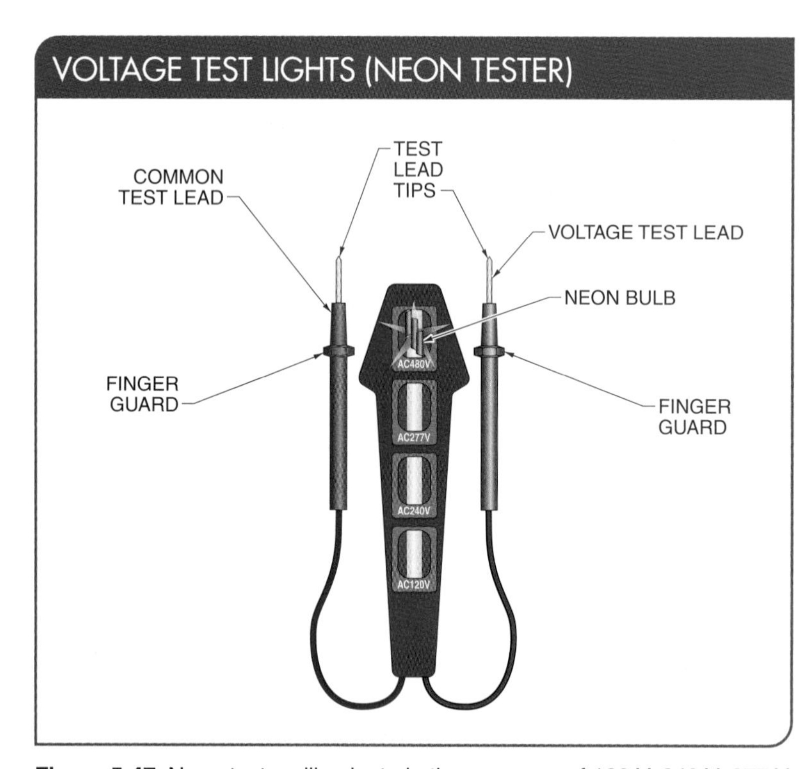

Figure 5-47. Neon testers illuminate in the presence of 120 V, 240 V, 277 V, or 480 V AC power and have a common (black) test lead and a voltage (red) test lead.

The advantages of using neon testers are that they are inexpensive, small enough to carry in a pocket, and easy to use. Low-cost neon testers are available at most hardware stores. Also, neon testers have a very high resistance, so the neon bulb draws very little current when taking measurements. The low current draw allows neon bulbs to have a longer life expectancy than other test lights. The disadvantages of neon testers are that they have a limited voltage range and usually determine only an approximate amount of voltage present in a circuit rather than the exact voltage. Also, neon test lights cannot be used to test GFCIs.

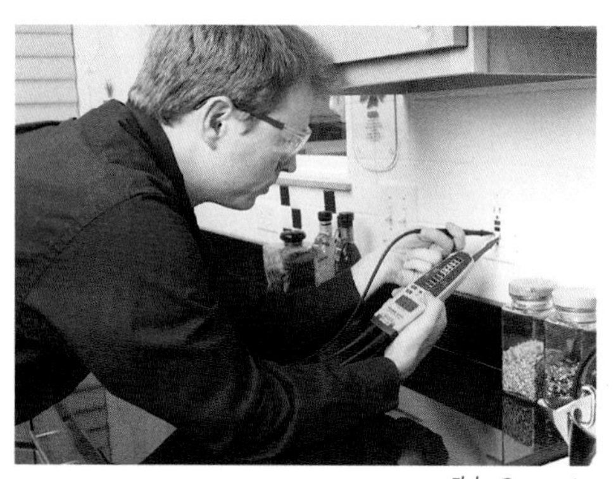

Fluke Corporation

Voltage testers are used to verify that a circuit is safe to work on and to troubleshoot wiring problems.

Voltage Testers

A *voltage tester* is an electrical test instrument that indicates an approximate amount of voltage and the voltage type (AC or DC) in an electrical circuit. Voltage testers are available as standard or advanced types, with one common (black) test lead and one voltage (red) test lead. **See Figure 5-48.**

Both standard and advanced voltage testers give an approximate level of either low-voltage (12 V, 24 V, or 48 V) or high-voltage (120 V, 208 V, 240 V, 277 V, 480 V, or 600 V) when testing branch circuits. They both also include continuity and GFCI testing modes. However, advanced voltage testers include more functions than the standard types, such as a resistance test mode, hold mode, and a digital display that indicates exact low- or high-voltage values.

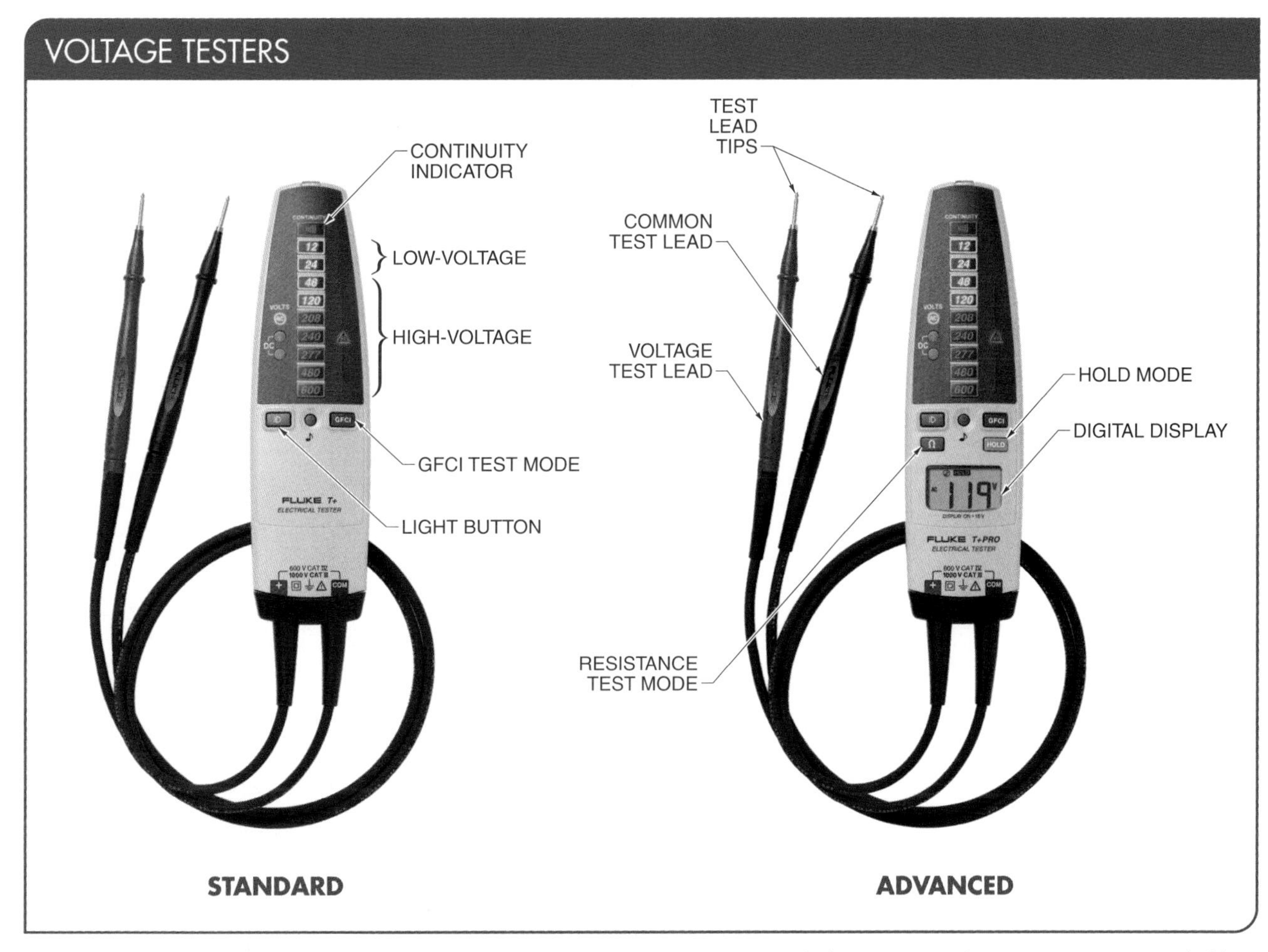

Figure 5-48. Voltage testers are available as standard or advanced types and give an approximate measurement of either low-voltage (12 V, 24 V, or 48 V) or high-voltage (120 V, 208 V, 240 V, 277 V, 480 V, or 600 V).

Voltage testers are primarily used to determine whether voltage is present and to show an approximate amount of voltage in a circuit. Before any measurements are taken, the voltage tester should always be tested on a known energized circuit to ensure that the voltage tester is operating correctly. Voltage tests are commonly conducted on receptacles to see if the receptacle is properly wired and energized. **See Figure 5-49.** To measure voltage using a voltage tester, the following procedure is applied:

1. Verify that the voltage tester has a CAT rating higher than the highest potential voltage in the circuit. Care must be taken to guarantee that the exposed voltage tester's lead tips do not touch fingers or any metal parts not being tested.
2. Insert the black test lead into the neutral side (long blade) slot of the receptacle. Always connect the black test lead to the neutral or ground side of the receptacle first.
3. Insert the red test lead into the hot side (short blade) slot of the receptacle. Always connect the red test lead to the hot side of the receptacle second.
4. Verify whether the voltage tester illuminates to determine that voltage is present at the receptacle. If the voltage tester does not illuminate, then the receptacle is not receiving power.
5. First remove the red test lead, followed by the black test lead, from the receptacle slots.

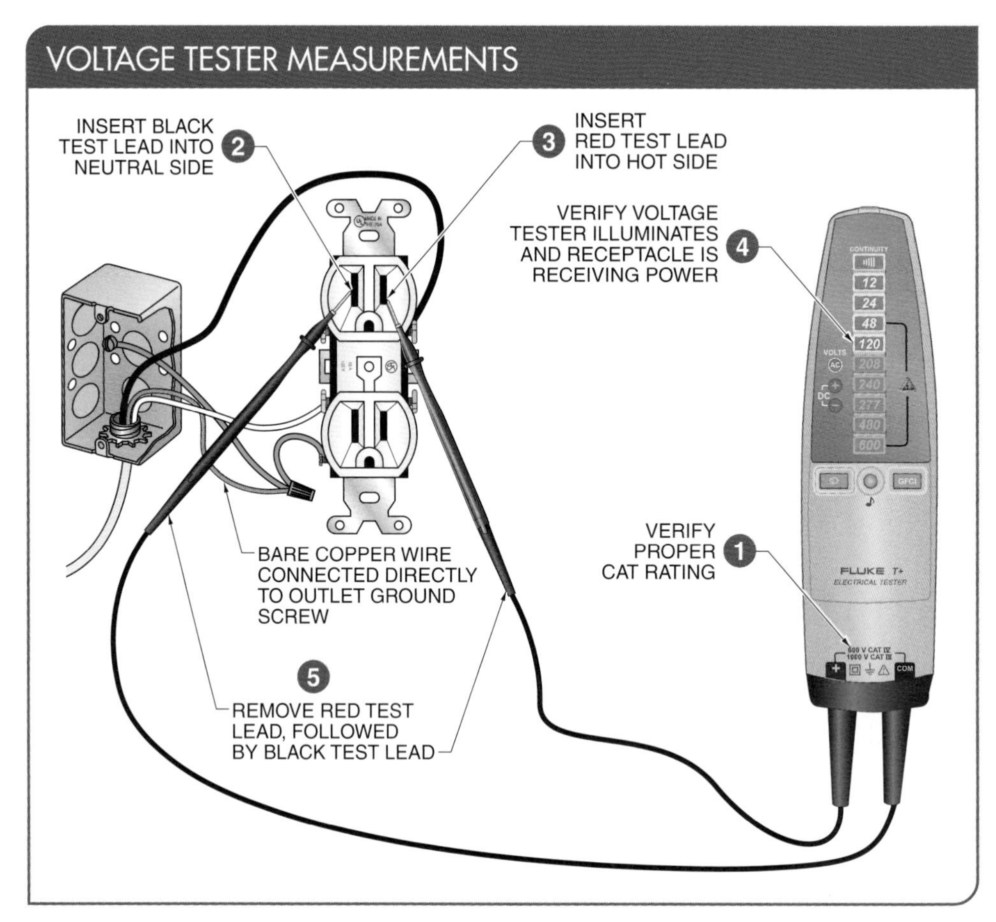

Figure 5-49. Voltage tests are commonly conducted on receptacles to see whether the receptacle is properly wired and energized.

Branch Circuit Identifiers

A *branch circuit identifier* is a two-piece test instrument consisting of a transmitter that is plugged into a receptacle and a receiver that provides an audible indication when located near the circuit to which the transmitter is connected. **See Figure 5-50.** A branch circuit identifier is used to identify a particular circuit breaker.

Before work is performed on any electrical circuit, the circuit must be deenergized. Normally, branch circuits are deenergized by turning off the circuit breaker for that circuit and using a lockout/tagout device. Often, the circuit breaker that is part of the circuit to be deenergized is not clearly marked or identifiable. The breaker box should be examined carefully before turning off breakers. Turning off an incorrect circuit breaker may unnecessarily require loads to be reset.

AEMC® Instruments

Figure 5-50. Branch circuit identifiers are used to identify a particular circuit breaker in electrical circuits.

Before any test using a branch circuit identifier is performed, it must be ensured that the branch circuit identifier is designed for the circuit to be tested. The manufacturer's operating manual for the test instrument should be consulted for any measuring precautions, limitations, and procedures. The necessary PPE should always be worn, and all safety rules should be followed when taking the measurement. **See Figure 5-51.** To identify the circuit breaker corresponding to a circuit using a branch circuit identifier, the following procedure is applied:

1. Turn on the transmitter and receiver of the branch circuit identifier.
2. Plug the transmitter of the branch circuit identifier into the receptacle (outlet) that is to be identified.
3. Verify that the receiver of the branch circuit identifier is operational by testing it at the same receptacle to which the transmitter is connected.
4. Turn the circuit breaker to the OFF position and lockout/tagout the circuit breaker.
5. Use a voltage tester to verify that the power is off before working on the identified circuit.
6. Use the receiver to identify the circuit breaker that controls the circuit being tested.
7. After work has been performed on the circuit, reset the breaker to the ON position.
8. Verify with a voltage tester that power is restored to the branch circuit receptacle.

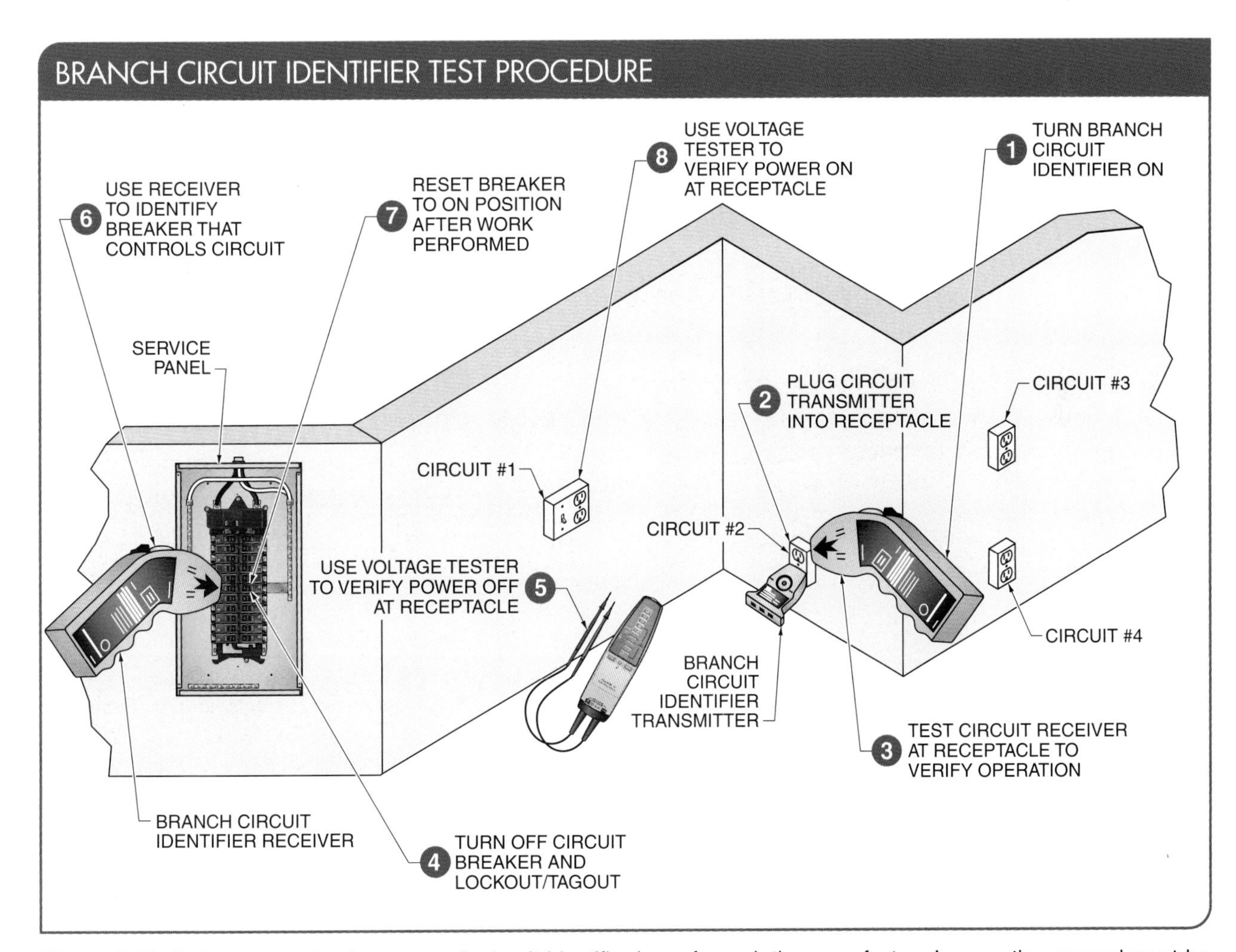

Figure 5-51. Before any test using a branch circuit identifier is performed, the manufacturer's operating manual must be consulted for any measuring precautions, limitations, and procedures.

TECH TIP

Branch circuit identifier features may include the ability to trace deenergized/energized circuits behind walls or underground.

Digital Multimeters (DMMs)

A *digital multimeter (DMM)* is an electrical test instrument that can measure more than one electrical quantity and display the measured quantities as numerical values. **See Figure 5-52.** DMMs have the capability to measure AC and DC voltage and current, resistance, capacitance, test continuity, and diodes. DMMs are designed with special functions such as recording, logging, and comparing stored measurements. The digital display of a DMM allows measured values to be easily read, and the function switch allows the easy selection of the quantity to be measured.

DIGITAL MULTIMETERS (DMMs)

Figure 5-52. DMMs can measure AC and DC voltage and current, resistance, capacitance, test continuity, and diodes and are designed with special functions such as recording, logging, and comparing stored measurements.

Digital multimeters are used on residential wiring installations, temporary wiring installations, maintenance tasks, and renovations to previously energized conductors and equipment. Exact measurements of certain quantities are required to determine whether fuses, circuit breakers, switches, receptacles, transformers, and motors are operational or to verify if a circuit is 12 V, 24 V, 120 V, 240 V, or any other voltage.

Fluke Corporation

Test leads with minimal exposed metal at the tip should be used to protect against electrical shorting.

Standard abbreviations and symbols are used on the digital display of a DMM to represent a quantity and its unit of measure. **See Figure 5-53.** For example, quantities such as voltage and current are identified with the abbreviations V (voltage) and A (current). Voltage and current can be measured in the base unit of measure or in the milli- (m) and micro- (μ) units of measure. Resistance is identified by the symbol Ω (ohms). Resistance can be measured in the base unit of measure or in the kilo- (k) and mega- (M) units of measure.

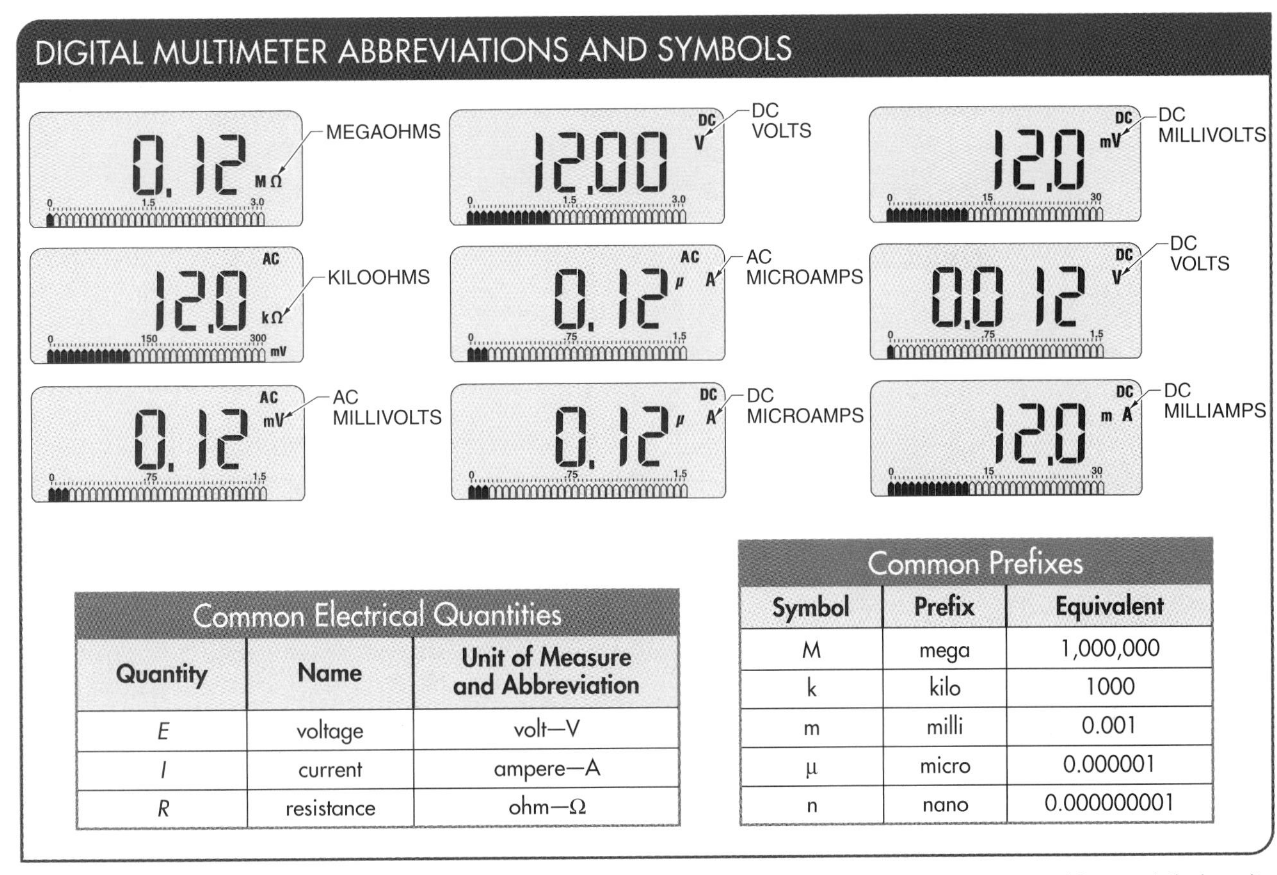

Common Electrical Quantities		
Quantity	Name	Unit of Measure and Abbreviation
E	voltage	volt—V
I	current	ampere—A
R	resistance	ohm—Ω

Common Prefixes		
Symbol	Prefix	Equivalent
M	mega	1,000,000
k	kilo	1000
m	milli	0.001
μ	micro	0.000001
n	nano	0.000000001

Figure 5-53. Standard abbreviations and symbols are used on DMM displays to represent quantities and their units of measure.

Voltage Measurements. Before a DMM is used, it should always be tested on a known energized source within the voltage rating of the meter to verify proper operation. This will ensure that the DMM will take proper measurements on the circuit or equipment being tested. The manufacturer's operating manual should be referred to for all measuring procedures, limitations, and precautions. The necessary PPE should always be worn, and all safety guidelines should be followed when taking measurements.

Typical voltage measurements can be made on small transformers, such as those that control doorbell circuits, which drop voltage from 120 V to 24 V. **See Figure 5-54.** To measure AC voltage at a transformer with a DMM, the following procedure is applied:

1. Verify that the voltage tester has a CAT rating higher than the highest potential voltage in the circuit.
2. Plug the black test lead into the common jack and the red test lead into the voltage jack.
3. Set the function switch on the DMM to AC voltage.
4. Connect the black test lead onto the neutral (white) terminal and the red test lead onto the positive (black) terminal of the transformer's primary side.
5. Read the voltage measurement displayed to verify 120 V for the transformer's primary side.
6. Connect the black test lead onto the neutral (white) terminal and the red test lead onto the positive (black) terminal of the transformer's secondary side.
7. Read the voltage measurement displayed to verify 24 V for the transformer's secondary side.
8. Remove the test leads from the transformer terminals and shut off the DMM.

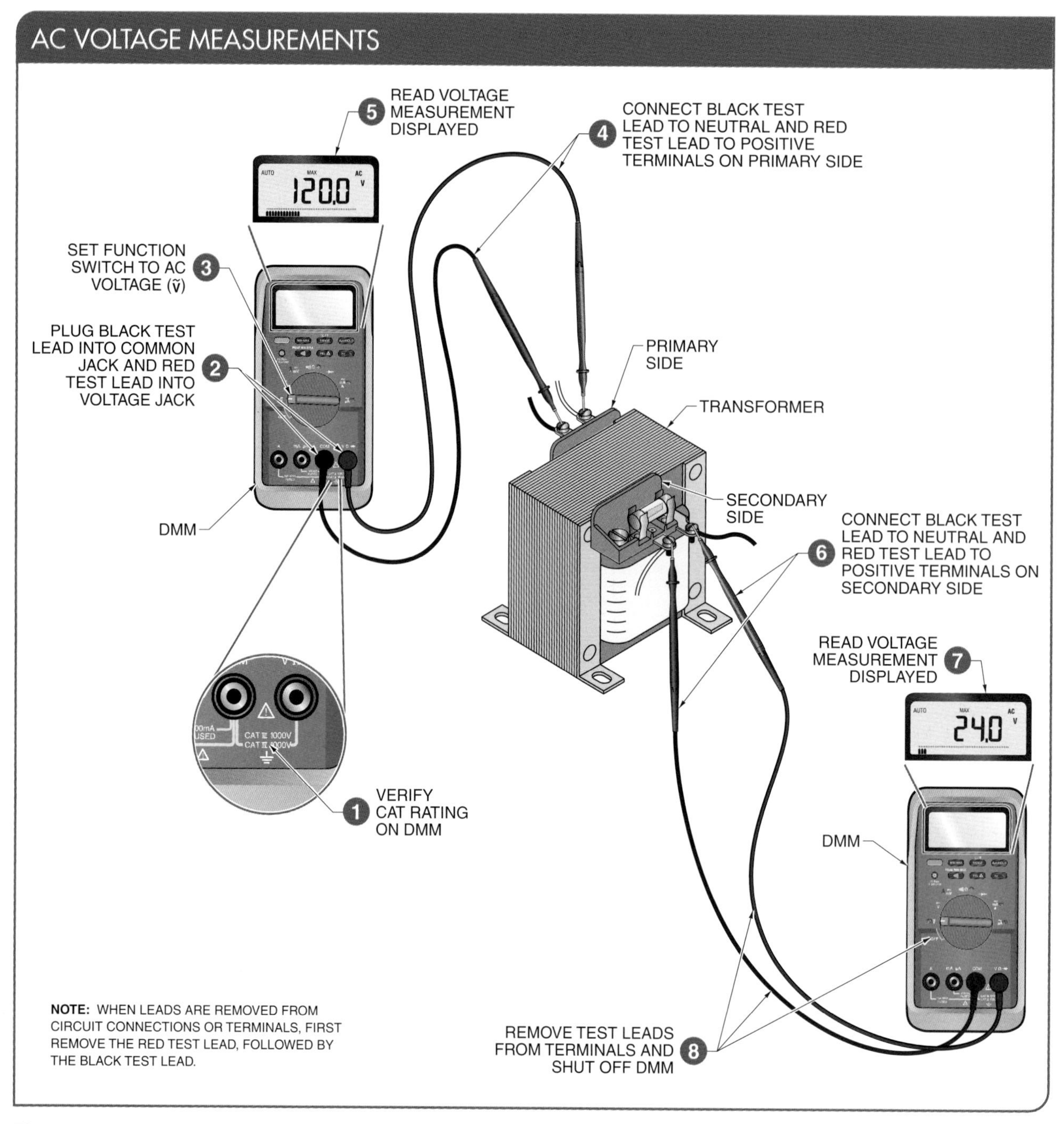

Figure 5-54. Typical voltage measurements can be made on small transformers that drop voltage from 120 V to 24 V to control doorbell circuits.

Measuring DC voltage is similar to measuring AC voltage with a DMM. **See Figure 5-55.** DC voltage measurements can be made on batteries or DC circuit equipment around a residence. Caution must be taken when taking measurements from DC voltage sources exceeding 60 V. **WARNING:** A battery can be shorted by contact with a metal object between the positive and negative battery terminals. To measure DC voltage at a battery with a DMM, the following procedure is applied:

1. Set the function switch to DC voltage. If the DMM includes more than one DC setting and the circuit voltage is unknown, select the highest setting.

2. Plug the black test lead into the common jack and the red test lead into the voltage jack.
3. Connect the black test lead to the negative terminal and the red test lead to the positive terminal.
4. Read the voltage measurement displayed to verify the 12 V source for the battery.
5. First remove the red test lead from the positive terminal, followed by the black test lead from the negative terminal.
6. Shut off the DMM.

TECH TIP

A DMM with a clamp-on ammeter adapter enables current measurements without having to disconnect the circuit for in-line measurements.

Current Measurements. A DMM can be used as an in-line ammeter or it can be equipped with a clamp-on ammeter adapter accessory to measure current (amperage). **See Figure 5-56.** A *clamp-on ammeter adapter* is a digital multimeter accessory that measures current in a circuit by measuring the strength of a magnetic field around a conductor. The jaws of a clamp-on ammeter adapter are placed around the conductor to measure current. Clamp-on ammeter adapters can measure either AC or DC voltage sources. The advantage of a clamp-on ammeter adapter is that current can be measured without opening or interrupting the circuit. Manufacturer recommendations and instructions should always be followed when measuring current with a clamp-on ammeter adapter.

A DMM connected as an in-line ammeter measures current in a circuit when the DMM is inserted in series with the equipment being tested. In-line current measurements are made with the circuit energized and open, which can increase the danger of electrical shock. The proper PPE must always be worn to prevent any electrical hazard when making in-line current measurements. In-line current measurements can be made on both AC and DC sources.

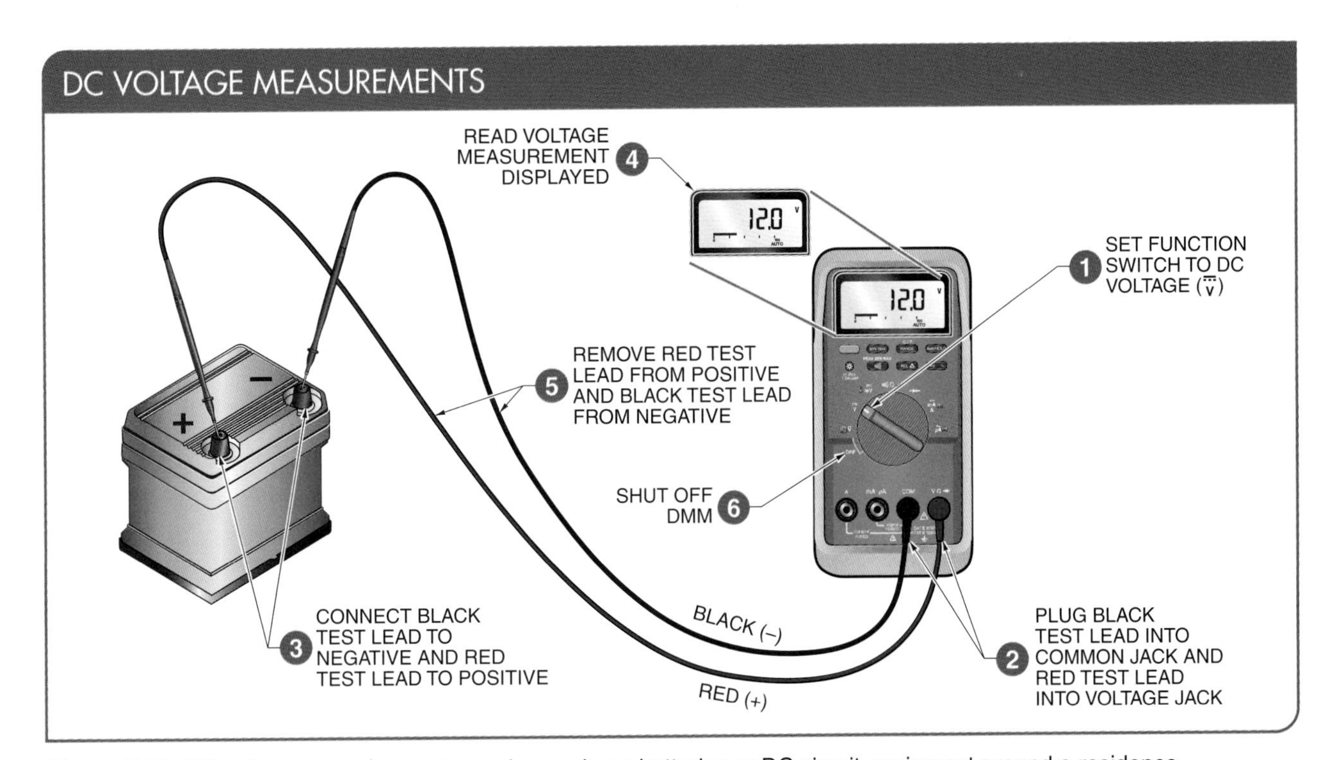

Figure 5-55. DC voltage measurements can be made on batteries or DC circuit equipment around a residence.

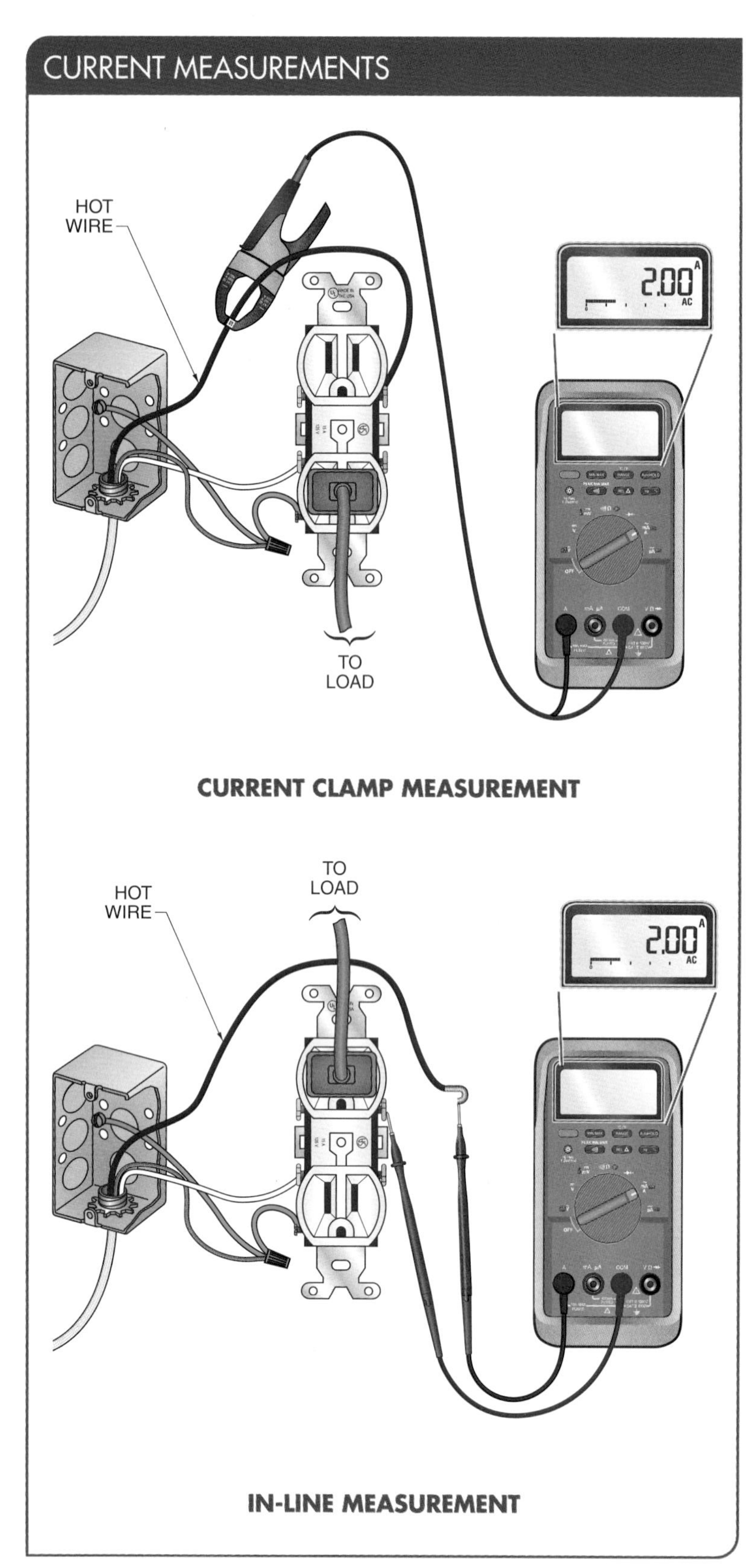

Figure 5-56. A DMM can be used as an in-line ammeter or it can be equipped with a clamp-on ammeter adapter accessory to measure current (amperage).

Resistance Measurements. A DMM can measure the resistance of a circuit or the resistance of a component removed from a circuit. **See Figure 5-57.** Circuits with heating elements can be measured for resistance. Other equipment such as fuses and resistors can be measured to verify operation. **WARNING:** Voltage applied to a DMM that is set to measure resistance will damage the DMM, even if the DMM has internal protection. *Note:* Check that all power is off to the circuit under test. Always remove any device under test from the circuit. To measure resistance with a DMM, the following procedure is applied:

1. Set the function switch to resistance mode on the DMM. The DMM should display "OL" and the "Ω" symbol when the DMM is in the resistance mode.
2. Plug the black test lead into the common jack and the red test lead into the resistance jack.
3. Connect the leads across the resistor under test. Ensure that there is a good connection between the test leads and the resistor leads.
4. Read the resistance displayed.
5. Turn off the DMM to prevent battery drain.

Fuse Tests. A fuse that is suspected of being defective must be tested for operation. Fuses can be used in dual-voltage subpanels (disconnects) that control motor circuits. DMMs that become part of the current measuring circuit also have a fuse built into them. **See Figure 5-58.**

With the DMM in resistance mode, the leads are placed on each side of the fuse. A good fuse should display a very low resistance of 0.0 Ω to 0.5 Ω. An open or bad fuse would display "OL." To verify the operation of the fuse inside the DMM, the DMM is set to resistance mode and the red lead tip is plugged into the current (A) terminal. A very low resistance of 0.0 Ω to 0.5 Ω will verify whether the DMM fuse is good. When the red test lead is plugged into the low-current (mA) terminal, a resistance of 0.995 kΩ to 1.005 kΩ would also indicate a good fuse.

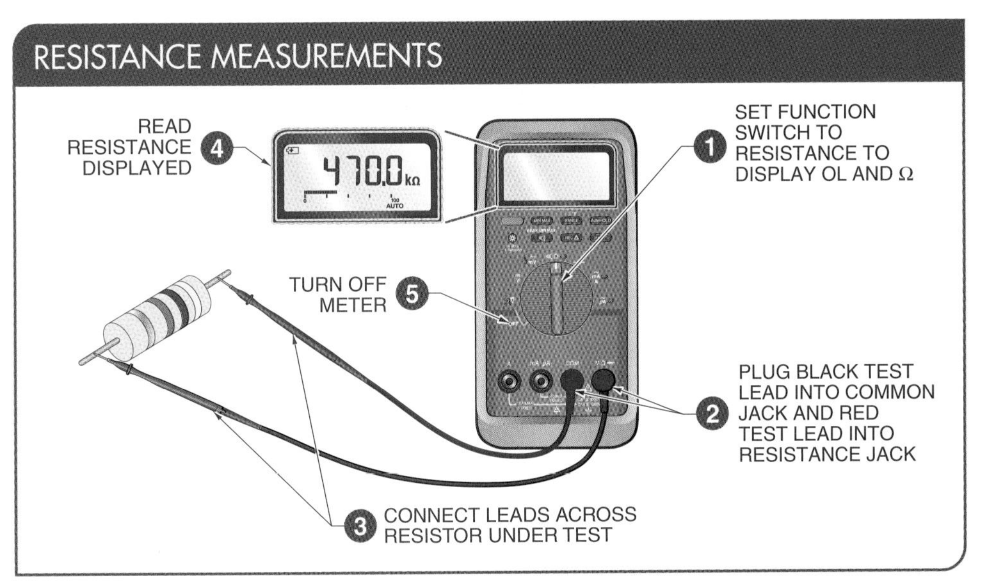

Figure 5-57. Electrical equipment such as fuses and resistors can be measured for resistance to verify their operation.

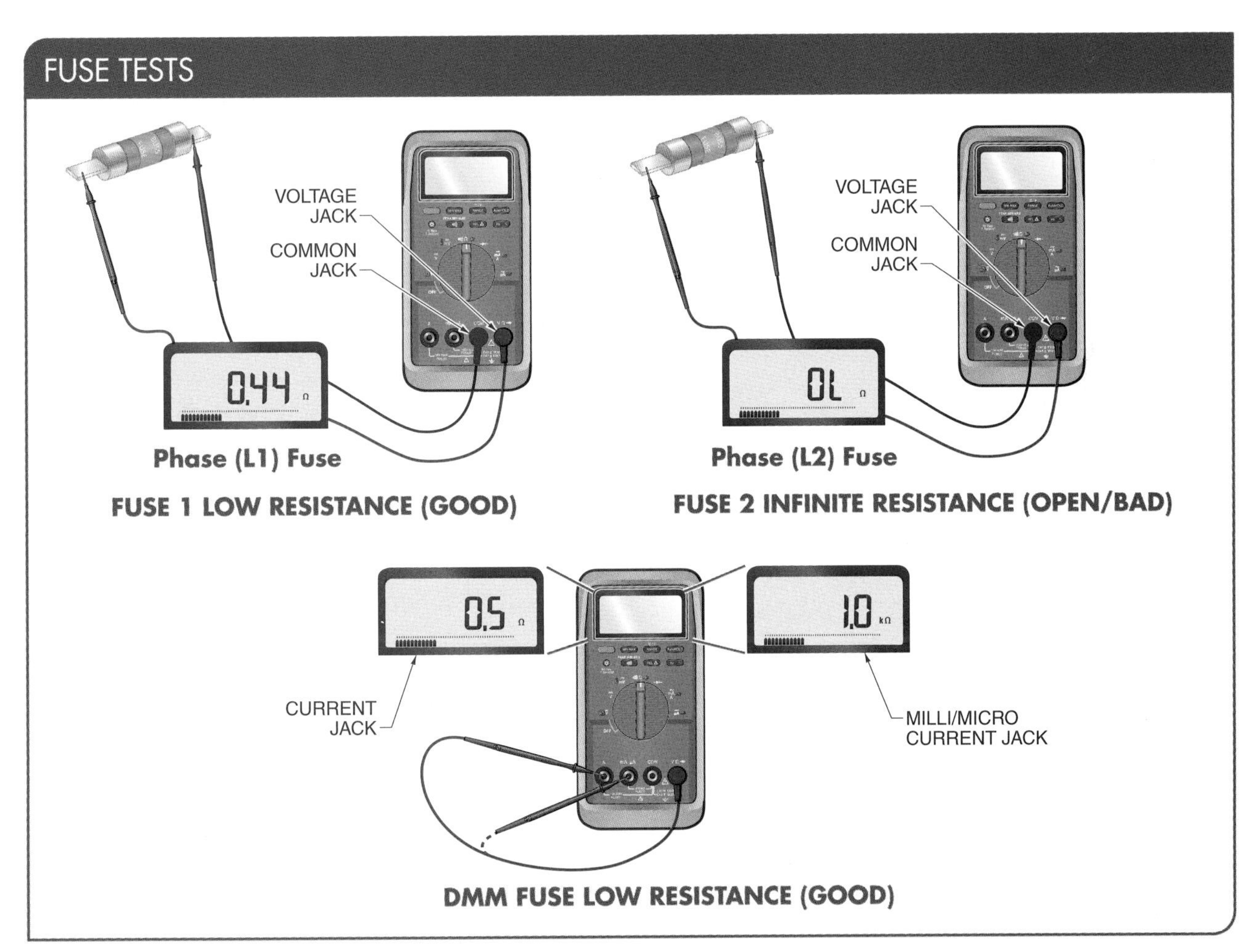

Figure 5-58. Fuses used in dual-voltage subpanels or built into a DMM that are suspected of being defective must be tested for operation.

Thermal Imagers

A *thermal imager* is a device that detects heat patterns in the infrared-wavelength spectrum without making direct contact with equipment. **See Figure 5-59.** Thermal imagers are advanced test instruments that display an electronically processed image of a target with different colors representing the various temperatures of the heat patterns. Individuals can use thermal imagers when conducting inspections to detect electrical hazards, such as faults or arcs, behind walls or under floorboards.

THERMAL IMAGERS

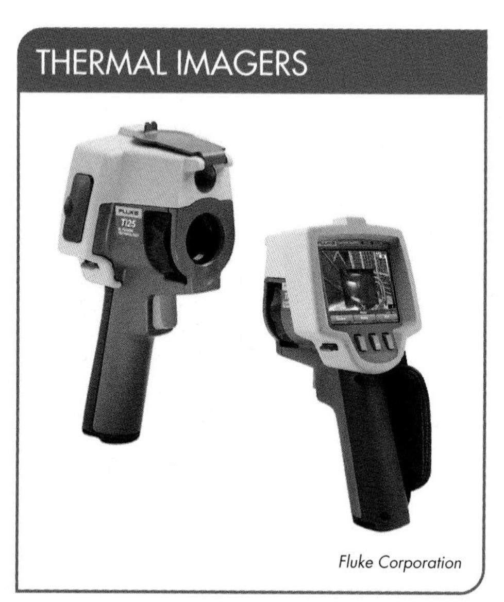

Fluke Corporation

Figure 5-59. Thermal imagers display an electronically processed image of a target with different colors representing the various temperatures of the heat signatures.

Test Instrument Safety

Proper test instrument procedures are required when taking measurements and testing electrical equipment. When any measurements are taken, the correct function setting and proper connection to the circuit are required to display correct measurements. Test instruments set to the incorrect function or incorrect range will be misread and will increase the likelihood of an improper or unsafe condition.

The following safety guidelines must be observed when using test instruments:

- Choose the correct test instrument for an application based on the CAT rating.
- Do not use damaged test instruments (cracked, missing parts, or worn insulation).
- Inspect the test leads of a test instrument on a regular basis and after each use.
- Keep the battery door of test instruments closed during operation.
- Test the fuses of a test instrument on a regular basis.
- Keep fingers behind the finger guards on the test leads when taking measurements.
- Do not use test instruments in hazardous locations (explosive gas, vapor, or dust), unless they are specifically designed for use in hazardous locations.
- Verify that the test leads of a test instrument are in the proper jacks (the jacks that correspond to the setting of the function switch) before taking any measurements.
- Verify the operation of a test instrument by measuring a known voltage source before performing tests on equipment or circuits.
- Verify that the function switch of a test instrument matches the desired measurement and the connections of the test leads to the circuit.
- Always connect the common (negative/black) test lead of a test instrument before connecting the voltage (positive/red) test lead. To disconnect, first disconnect the positive test lead, followed by the negative test lead.

SECTION 5.3 CHECKPOINT

1. What are the four overvoltage installation categories and what do they determine?
2. What should be done before taking measurements with a voltage tester?
3. What is a digital multimeter and what can it measure?
4. What is displayed on a DMM when a fuse is tested to determine whether the fuse is good, open, or bad?
5. What is a thermal imager and how can it be used?

Chapter Activities

5 Electrical Safety, Tools, and Test Instruments

Name ______________________________ Date ______________

Grounding Systems

1. Draw arrows to indicate the ground path from the power tool to earth ground.

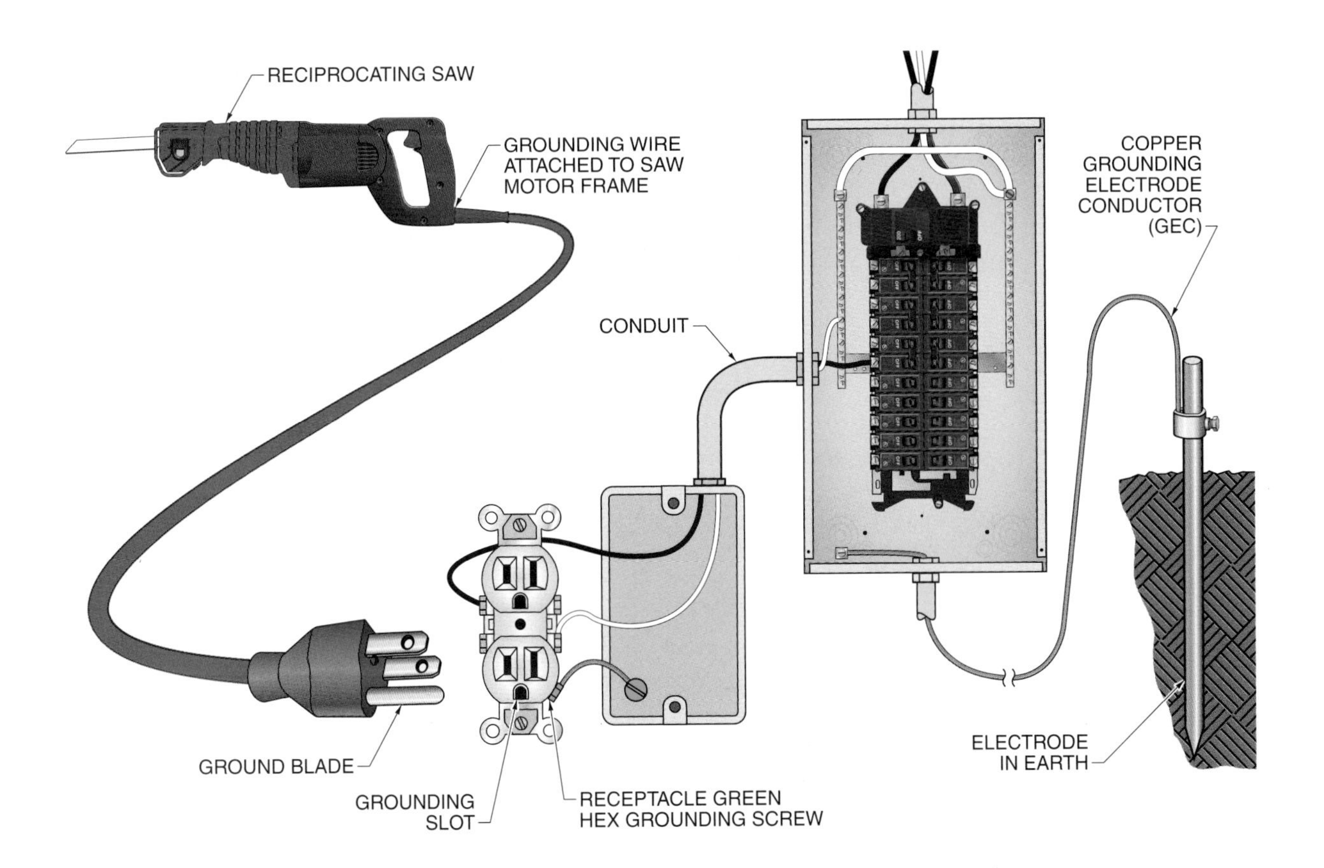

Approach Boundaries

Identify the different approach boundaries shown.

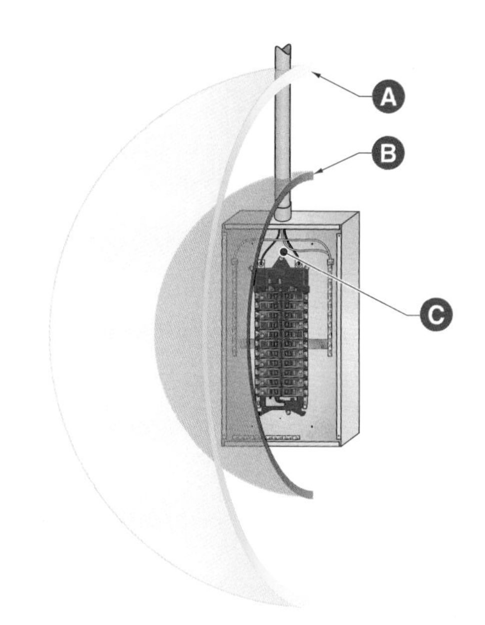

____________________ **1.** Restricted approach boundary

____________________ **2.** Restricted space

____________________ **3.** Limited approach boundary

Test Instrument Identification

Identify the test instruments shown.

____________________ **1.** Voltage tester—tests for high and low AC and DC voltages

____________________ **2.** Voltage test light—tests for 120 VAC, 240 VAC, 277 VAC, and 480 VAC

____________________ **3.** Voltage indicator—tests for the presence of AC voltage

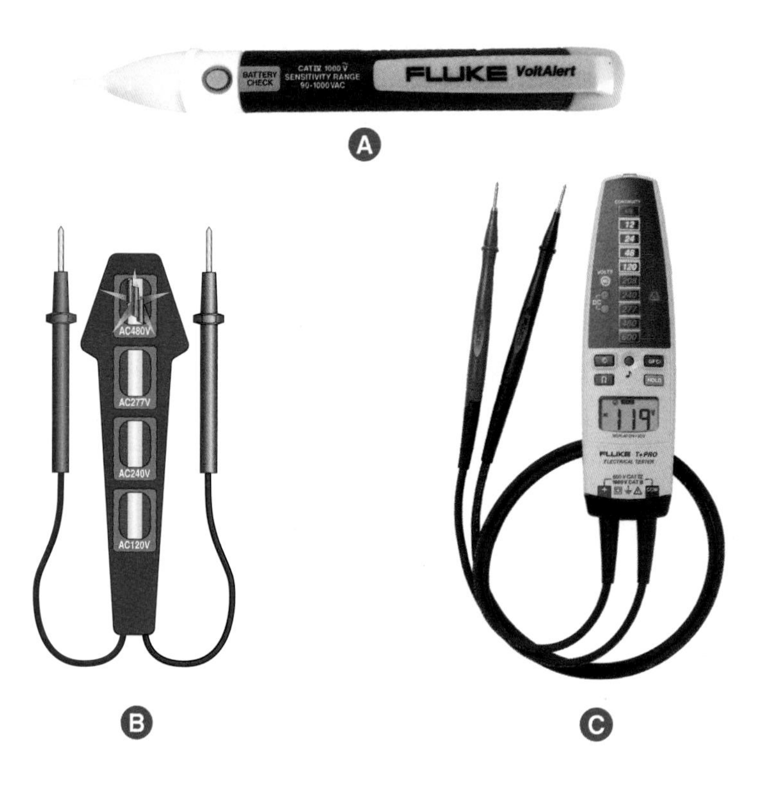

6 Electrical Prints and Diagrams

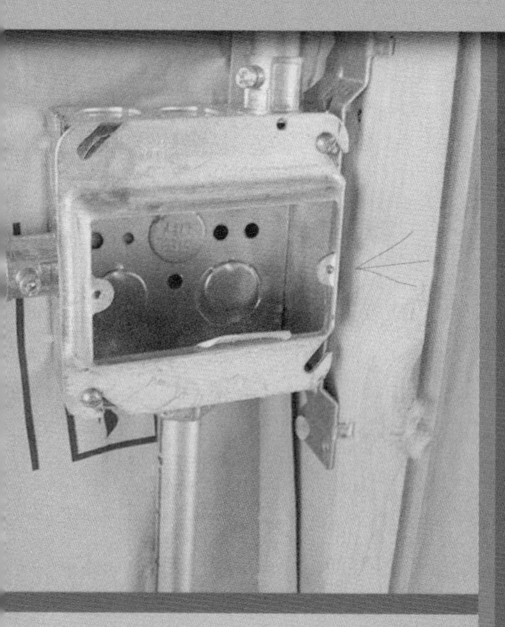

SECTION 6.1—ELECTRICAL PRINTS

- Explain the use of electrical prints in residential construction.
- Define "electrical floor plan."
- Differentiate between component plans and wiring plans.
- Identify device abbreviations and symbols used on electrical prints.

SECTION 6.2—ELECTRICAL DIAGRAMS

- Identify the five main parts of an electrical circuit.
- Describe a pictorial drawing.
- Identify device abbreviations and symbols used on electrical diagrams.
- Describe the use of line (ladder), wiring, schematic, and one-line electrical diagrams.

Learner Resources

ATPeResources.com/QuickLinks
Access Code: 791712

The electrical trades use a specific language on electrical prints and diagrams in order to communicate information that is used for installation, maintenance, or testing procedures. Electrical prints and diagrams include abbreviations and symbols that convey this specific language across the trade. Electrical prints are used during the construction of a residential system to see where and what kind of equipment is installed. Electrical diagrams are used to provide information for maintaining or testing electrical equipment and devices found throughout a residential electrical system. Individuals must understand how to read electrical prints and diagrams in order to install equipment and troubleshoot various problems.

SECTION 6.1 — ELECTRICAL PRINTS

Electrical prints are used to determine the location, type, and number of required devices installed in a residence. Electrical prints show the position of service panels, outlets, permanently connected loads, circuit numbers, and special installation information. Typically, architectural electrical symbols are used on prints to show the location of lights, receptacles, switches, power, and signaling devices found throughout a residence. Electrical prints that are used to convey electrical information are known as electrical floor plans.

Electrical Floor Plans

A *floor plan* is a print that provides a plan view of a floor of a residence. A floor plan shows the locations of equipment, doors, plumbing fixtures, rooms, stairways, walls, and windows on any floor. A description and/or room number is given to each room, including entryways and other locations.

An *electrical floor plan* is a print that shows all of the power circuits for a specific floor of a residence. **See Figure 6-1.** Electrical floor plans show all electrical devices and all directly connected loads, such as lighting, appliances, and other equipment.

Typical residential construction projects, such as one-family dwellings, provide only one electrical floor plan for each floor level. An electrical floor plan for a one-family dwelling may include the locations of service panels, receptacles, light fixtures, switches, appliances, telephones, television and data outlets, plumbing fixtures, fire alarms, smoke/carbon monoxide detectors, and locations of thermostats for temperature control. It also shows interior and exterior building dimensions, doors and windows, and floor and wall coverings.

In most residential construction projects, receptacles (outlets) are mounted at the same height, with the location of each identified on the electrical floor plan. Information on the specific dimensions of outlet and switch heights is found in the specifications part of electrical floor plans. Ceiling light fixtures are typically centered in a room. Mounting dimensions are only shown for light fixtures, outlets, and switches that require specific installation instructions.

ELECTRICAL FLOOR PLANS

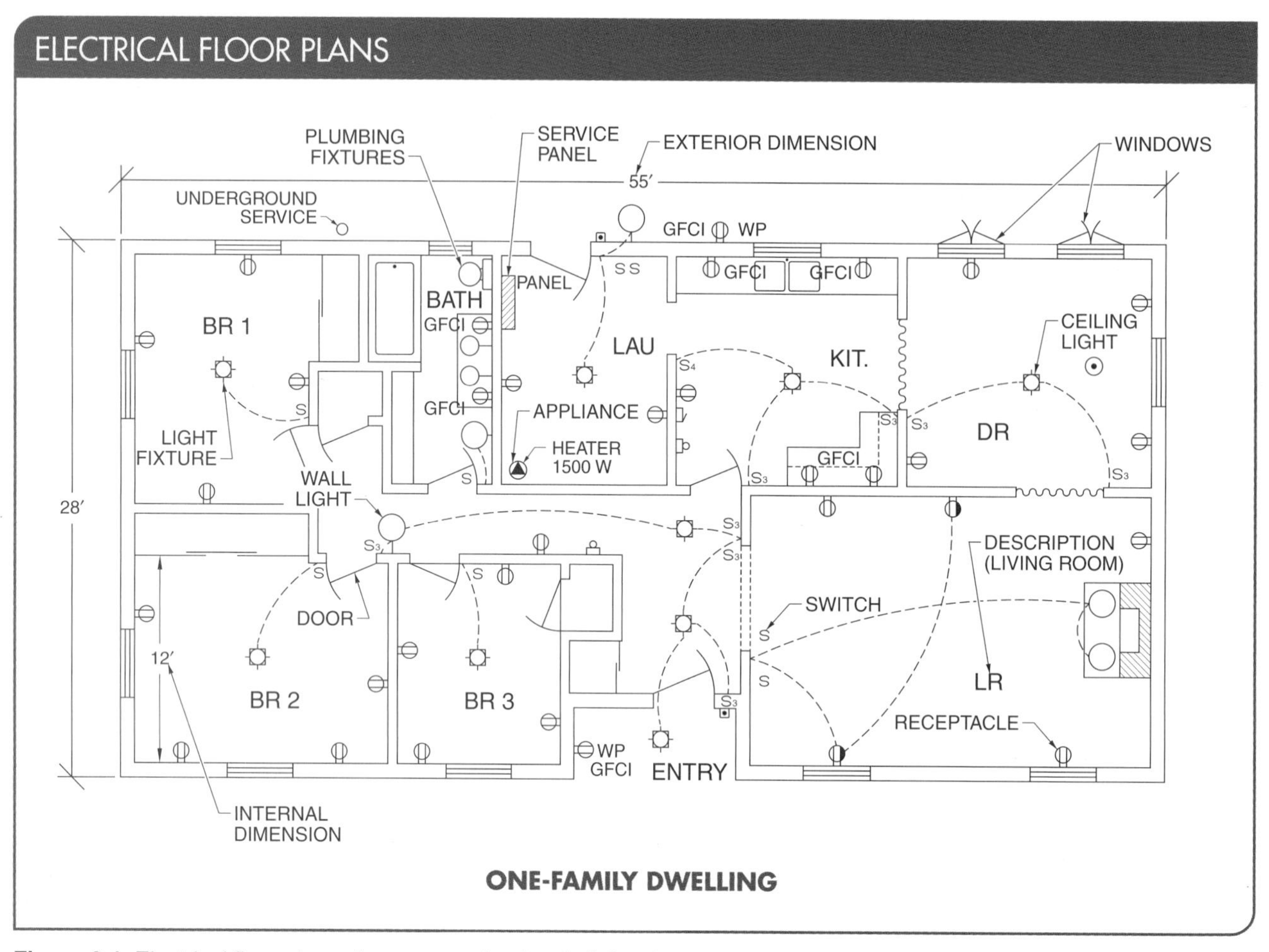

Figure 6-1. Electrical floor plans show power circuits of all the electrical devices and connected loads in a residence.

Individuals use electrical floor plans to interpret the type of installation that must be performed. Individuals can efficiently perform installation tasks by analyzing electrical floor plans to determine where the electrical devices are to be installed and how the devices are to be used and wired. This will also help determine the number of openings in walls and ceilings and estimate the amount of cable or conduit required for each installation.

Electrical floor plans must provide enough information to indicate the proper way to complete a job and must follow NEC® standards for local and state compliances. **See Figure 6-2.** A complete electrical floor plan, or set of plans, can have additional sets of information such as a component plan and a wiring plan.

Component Plans. A *component plan* is a group of electrical component schedules that state the required locations for receptacles, lights, and switches. **See Figure 6-3.** To determine which electrical devices are required for a certain room, electrical component schedules are used for each area of a residence such as bedrooms, living/family room, dining room, and bathrooms. An *electrical component schedule* is a list of electrical equipment that indicates manufacturer specifications and how many of each electrical device are required for a room or area in a component plan. To be effective, a component plan must include devices that satisfy all the electrical requirements for light, heat, and power to that room or area.

NEC® COMPLIANCE FOR ONE-FAMILY DWELLINGS

Electrical Plan—NEC®

A 210.52(A). All receptacles installed on 15 A and 20 A circuits shall be located in accordance with 210.52(A)(1) through (A)(4).

210.52(B). For small appliance load, including refrigeration equipment, in kitchen, pantry, and breakfast room, two or more 20 A appliance circuits shall be provided. Such circuits shall have no other outlets. Countertop outlets shall be supplied at each wall counter that is 12″ or wider.

B 210.8(A). All 125 V, 1ϕ receptacles shall have GFCI protection when installed in: (1) bathrooms, (2) garages, (3) outdoors where there is direct grade access, (4) crawl spaces below grade level and unfinished basements, (5) unfinished areas, (6) kitchen countertops, (7) sinks, (8) boathouses, (9) bathtubs or shower stalls, (10) laundry areas.

C 210.23. Individual branch circuits can supply any load for which they are rated.

D 210.63. A 125 V, 1ϕ, 15 A or 20 A-rated receptacle is installed for servicing heating, air-conditioning, or refrigeration equipment.

E 210.52(E). All receptacles installed outdoors must be of GFCI type and readily accessible.

F 210.52(H). Hallways 10′ in length or more shall have a receptacle provided.

G 220.12. Unit lighting load for dwelling occupancies shall not be less than 3 VA per sq ft. See Table 220.12 in determining load on "VA per sq ft" basis, outside dimensions of the building shall be used, excluding open porches, garages, unused spaces, unless adaptable for future use.

H 210.11(C)(2). At least one 20 A circuit shall be provided for a laundry receptacle.

I 220.14(J)(1). In dwelling units, all general-use receptacle outlets 20 A or less are included in general-lighting load calculations.

J 220.52(A). Branch circuit load for two small appliances circuits to be 3000 VA (1500 VA for each two-wire circuit).

220.52(B). Each laundry branch circuit load calculation to be at least 1500 VA.

K 230.42(A). Service-entrance conductors ampacity per 220 and 310.15.

L 410.16. Luminaires in Clothes Closets.

M 422.11(E). Branch circuit to a single non-motor appliance not to exceed OCPD rating. If an appliance is rated at 13.3 A or less, the next standard OCPD is required, not to exceed 150% of the appliance rating.

N 725. Remote control, signaling, and Power Limited circuits.

Electrical Key Plan

Letter	NEC®	Subject
A	210.52(A)	Receptacle Location
B	210.8(B)	GFCI Location
C	210.23	Receptacles
D	210.63	Heating Outlet
E	210.52(E)	Receptacle Location (Outdoors)
F	210.52(H)	Receptacle Location (Hallways)
G	220.12	Lighting Load
H	210.11(C)(2)	Laundry Receptacle
I	220.4(D)	Circuits (General Purpose)
J	220.52	Circuits (Small Appliance and Laundry)
K	230.42	Service Entrance Conductors Ampacity
L	410.2	Fixtures (Clothes Closet)
M	422.11(E)	Non-Motor-Operated Appliance Circuits
N	725	Signaling and Power Limited Circuits

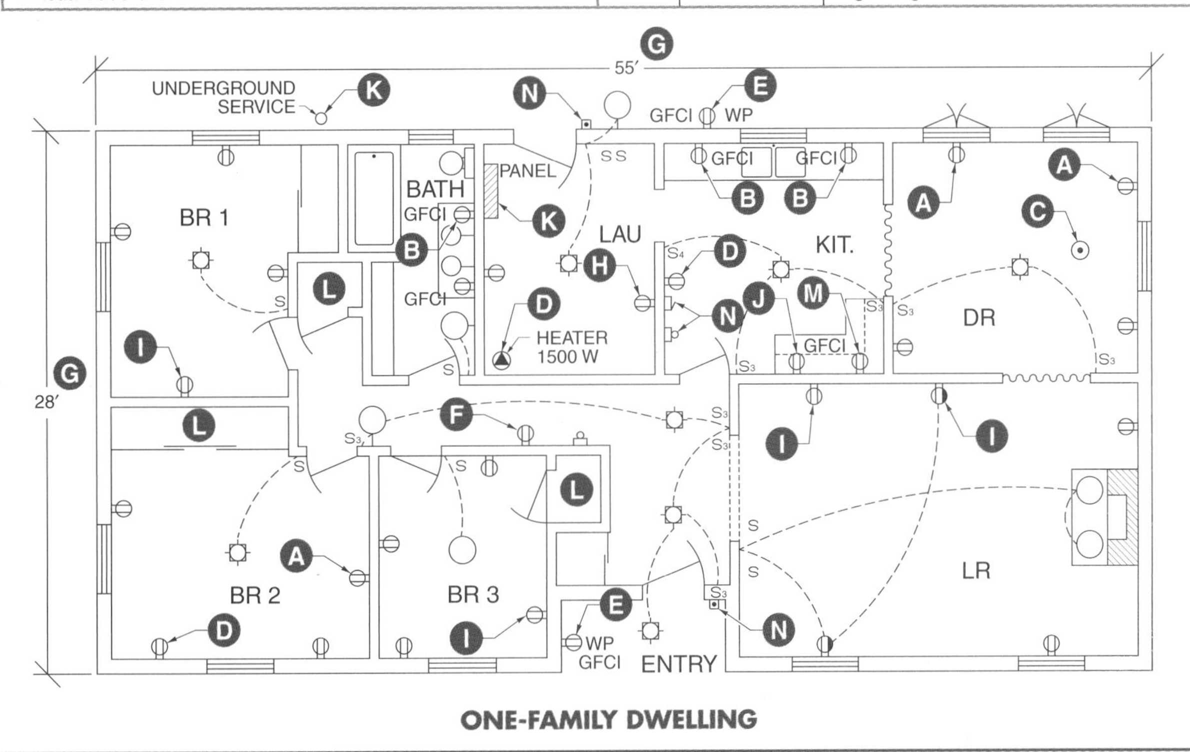

ONE-FAMILY DWELLING

Figure 6-2. Electrical floor plans must provide the proper information to complete a job and must be NEC® compliant.

COMPONENT PLANS

<table>
<tr><th>Area</th><th>Convenience Receptacles</th><th>Special-Purpose Outlets</th><th>General Lighting</th><th>Major Appliances</th><th>General Switching for All Areas</th></tr>
<tr><td>Bedroom(s) Schedule</td><td>No space along a wall should be more than 6′ from a receptacle outlet. Any wall space 2′ or larger should have a minimum of 1 receptacle. NEC® 210.52(A).</td><td>TV outlet, intercom, speakers (music), and telephone jack</td><td>Ceiling, wall or valence light, lamp switched at receptacle</td><td>Room air conditioner, electric baseboard heating</td><td rowspan="4">Switches are typically placed opposite the hinged side of door. When there are two or more entrances to a room, multiple switching should be used.

Door switch can be used for closet light so that the switch turns the light on and off as the door opens and closes.

Switches with pilot lights should be used on lights, fans, and other electrical devices which are in locations not readily observable.</td></tr>
<tr><td>Living Room Schedule</td><td>No space along a wall should be more than 6′ from a receptacle outlet. Any wall space 2′ or larger should have a minimum of 1 receptacle. NEC® 210.52(A).</td><td>TV outlet, intercom, speakers (music)</td><td>Ceiling fixture, recessed lighting, valence light, lamp switched at receptacle, possible dimmer</td><td>Room air conditioner, built-in stereo system, electric baseboard heating</td></tr>
<tr><td>Family Room Schedule</td><td>No space along a wall should be more than 6′ from a receptacle outlet. Any wall space 2′ or larger should have a minimum of 1 receptacle. NEC® 210.52(A).</td><td>TV outlet, intercom, speakers (music), bar area (ice maker, blenders, small refrigerator, hot plate), telephone jack, thermostat</td><td>Ceiling fixture, recessed lighting, valence lights, studio spot lights, lamp switched at receptacle, fluorescent light, possible dimmer</td><td>Room air conditioner, built-in stereo system, electric baseboard heating</td></tr>
<tr><td>Dining Room Schedule</td><td>No space along a wall should be more than 6′ from a receptacle outlet. Any wall space 2′ or larger should have a minimum of 1 receptacle. NEC® 210.52(A).</td><td>Elevated receptacles (48″) for buffet tables, speakers (music)</td><td>Ceiling chandelier, recessed lighting, valence lighting for china hutch</td><td>Room air conditioner, electric baseboard heating</td></tr>
</table>

Figure 6-3. Component plans determine which electrical devices are required and located in each room on a floor plan.

Component schedules can aid in developing a checklist for work tasks to be performed. Additional information for installations that were not initially recognized as being required from the electrical floor plan can be obtained from component plans.

Wiring Plans. Once the locations of receptacles, lights, and switches have been determined through the use of a component plan, a wiring plan can be used to wire circuits with the necessary devices in each room. **See Figure 6-4.** A *wiring plan* is an electrical floor plan that indicates the type, placement, and connection of all electrical devices required to wire an electrical circuit. The location or layout of all devices on a wiring plan should be an accurate representation of how the electrical circuit is connected for a particular room in a dwelling.

Wiring plans are essential when performing wiring installations because they indicate the type of switch, cable/wire, and receptacle used for each circuit in a room. Typical switches used to control electrical circuits include single-pole, three-way, and four-way combinations. Wiring plans help determine the correct number of devices per circuit (loads) and the proper placement to run cable. All connecting wires are shown going from one device to another with slash marks to indicate the number of wires in a cable.

Typically, the layout of a wiring plan uses designated symbols to represent each device and cable. An arrow symbol at the end of the circuit indicates the cable connection back to the service panel. Wiring plans can aid in efficient installations during construction, as well as reduce the amount of materials required per installation task.

When an individual installs an electrical appliance such as washing machine, dryer, or electric oven, it is important to follow the installation instructions and/or specifications provided by the manufacturer. Manufacturer specifications indicate the size of wire, type of receptacle, and overload protection that must be provided for appliance circuits. The specifications set by manufacturers must agree with the NEC®. A circuit breaker in the service panel (breaker box), at which a branch circuit originates, must identify that circuit in order to aid in future maintenance or to help anticipate problems during troubleshooting tasks. **See Figure 6-5.**

> TECH TIP
>
> *Wiring plans for a bedroom may show one circuit run to the service panel. A kitchen may show three or more circuits run to the service panel for separate appliance loads.*

Device Abbreviations and Symbols for Prints

Understanding electrical device abbreviations and symbols is the key to interpreting how electrical devices are installed and wired on electrical floor plans. In construction trades, architects use symbols and abbreviations when drafting electrical floor plans. A *symbol* is a graphic representation of a device, component, or object on a print. An *abbreviation* is a letter or combination of letters that represents a word.

Individuals must understand electrical device abbreviations and symbols to properly wire circuits and install devices. When interpreting electrical floor plans, individuals must identify unfamiliar symbols before any work is started. Most electrical plans used in new home construction today are drawn to provide the same types of information from one home to another. Because electrical symbols are easy to draw, they are used in place of actual pictorial drawings of electrical devices on prints. Real-world devices such as lighting, receptacles, switches, power, and signaling devices are represented by symbols.

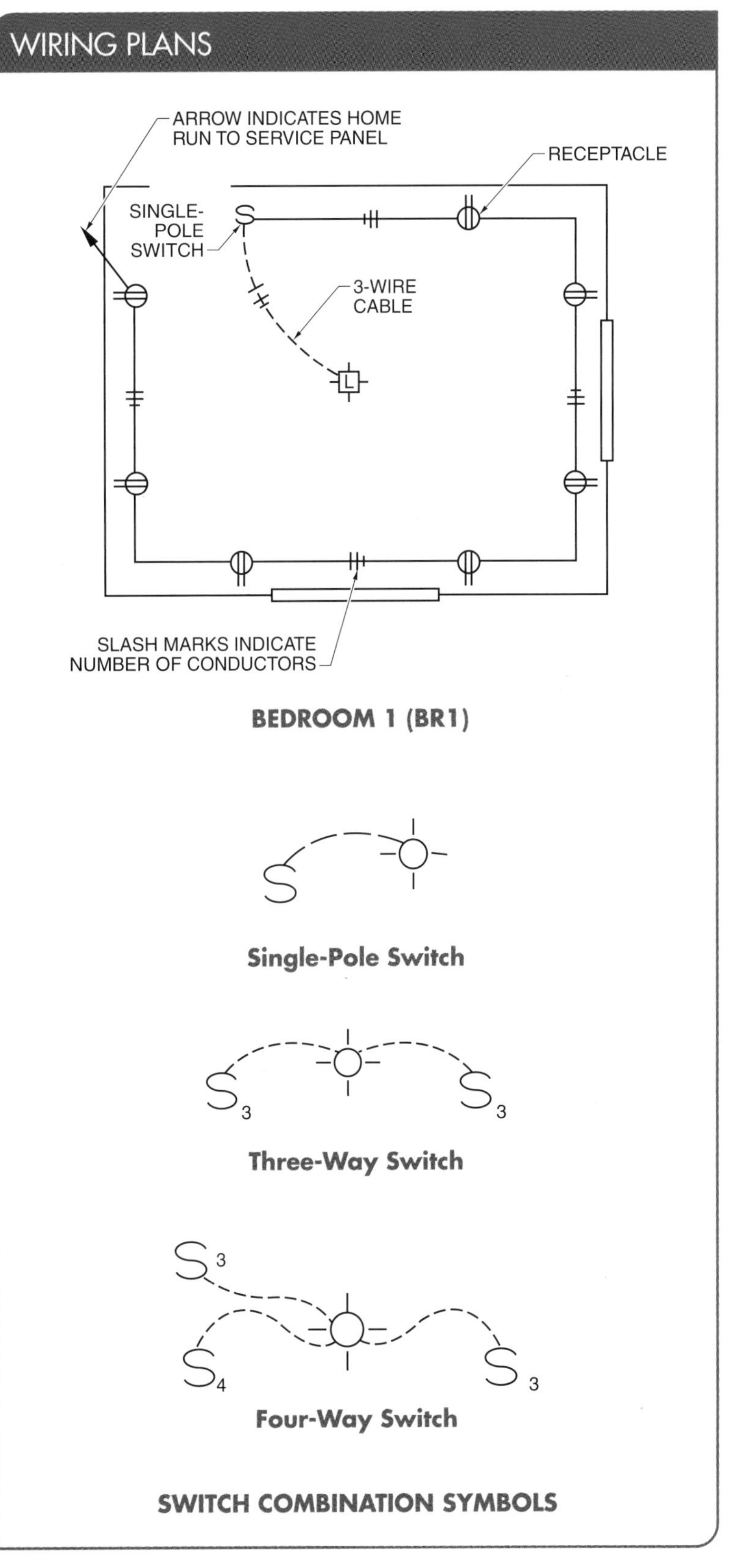

Figure 6-4. Wiring plans provide the layout of all electrical devices connected in an electrical circuit.

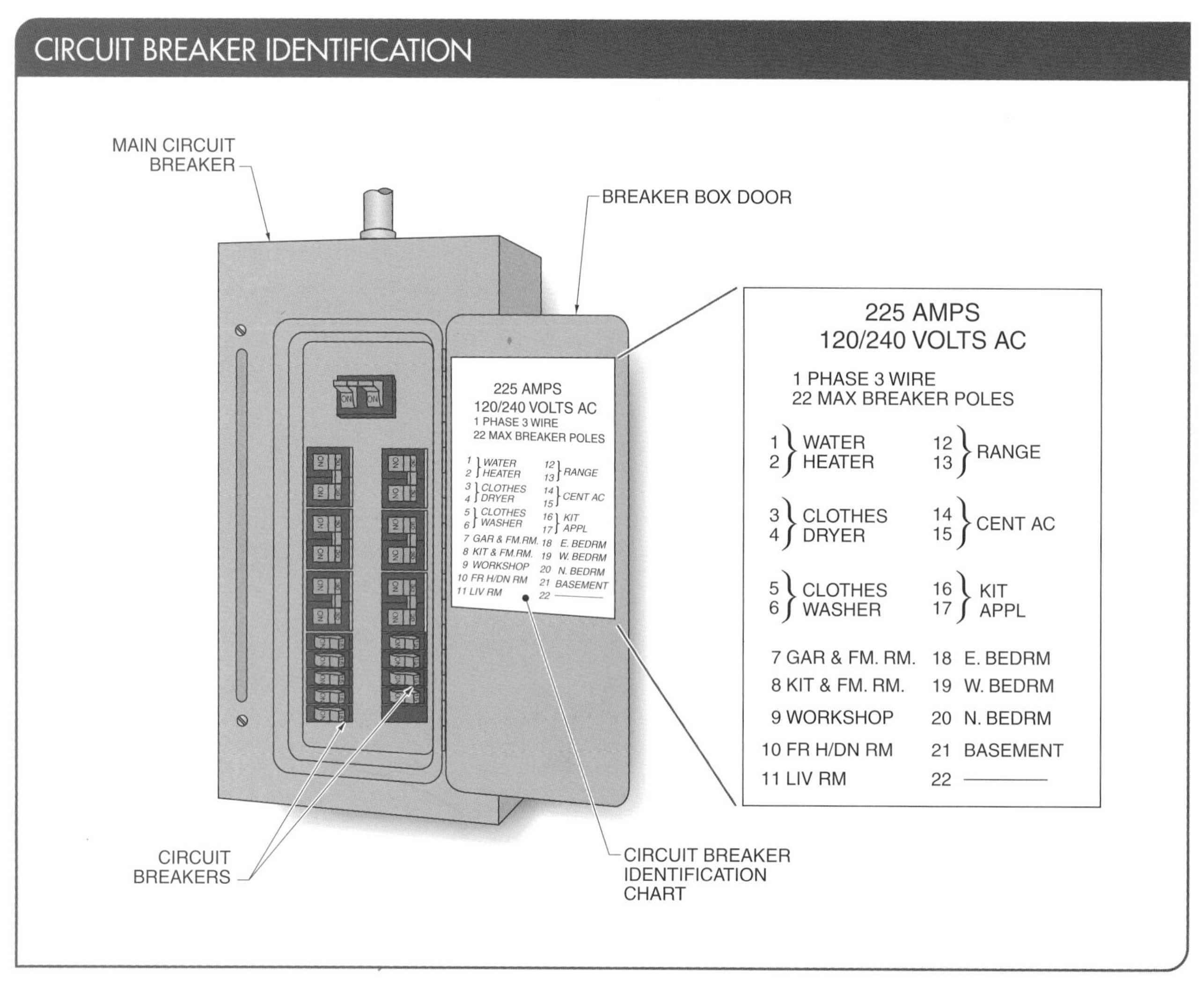

Figure 6-5. Circuit breakers that control each branch circuit must be identified for maintenance or troubleshooting tasks.

Locations for switch and device boxes are located on electrical wiring plans to enable correct installation.

Lighting. Lighting is the most common type of installation and application used in residential electricity distribution. Lighting symbols are used on electrical floor plans to show the location and style of fixture (recessed or surface), as well as the required type of light-fixture mounting (ceiling or wall). **See Figure 6-6.**

To indicate on an electrical floor plan that a light is installed onto a wall, a line is drawn from the wall to the light symbol. Different types of lighting symbols represent wall lights, ceiling lights, recessed lights, fluorescent lights, lamp holders, lamp holders with pull switches, track lighting, fan outlets, and junction boxes.

LIGHTING ABBREVIATIONS AND SYMBOLS			
Description	Device	Abbrev	Symbol
Wall Light	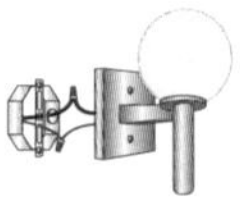	L	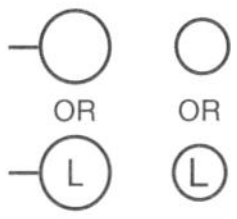 OR OR
Ceiling Light	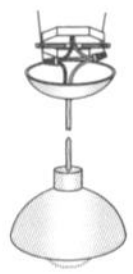	L	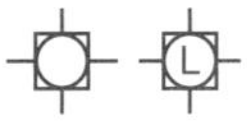
Recessed Ceiling Fixture Light	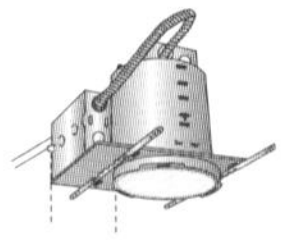	LAC	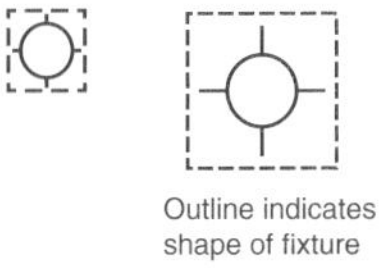 Outline indicates shape of fixture
Fluorescent Light	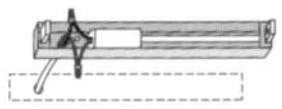	FLUOR	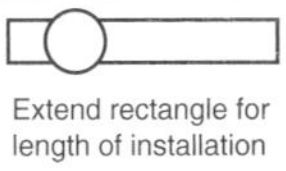 Extend rectangle for length of installation
Light with Lamp Holder		L	L
Light with Lamp Holder and Pull Switch		LPS	 L PS
Track Lighting	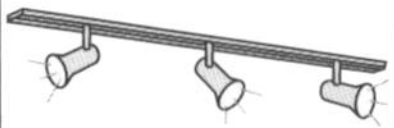	LTL	
Fan Outlet	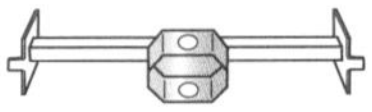	F	 F Box listed as acceptable for fan support
Junction Box	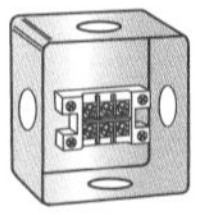	J	J

Figure 6-6. Lighting symbols are used on electrical floor plans to indicate the type and location of light fixtures.

Receptacles. Standard receptacles (outlets) are used throughout a residence to provide access to power for electrical loads. Typical residential electrical loads include free-standing lamps, computer equipment, televisions, and small or large appliances. Typical receptacles shown on electrical floor plans include standard, isolated-ground (IG), special-purpose, and ground-fault circuit interrupter (GFCI) receptacles. **See Figure 6-7.**

Standard receptacles are available in a variety of general-purpose types. These receptacles include single, duplex, split-wired duplex, triplex, and single floor outlets.

Isolated-ground receptacles look like standard duplex receptacles, but they can be distinguished by an orange triangle on the front of the receptacle. Isolated-ground receptacles are available as duplex receptacles with isolated ground and weatherproof (WP) for outdoor outlets.

RECEPTACLE OUTLET ABBREVIATIONS AND SYMBOLS			
Description	**Device**	**Abbrev**	**Symbol**
Single Receptacle (Outlet)	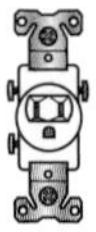	RCPT OR OUT	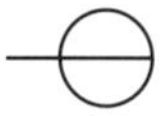
Duplex Receptacle		DX RCPT	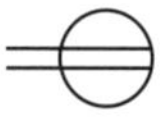
Split-Wired Duplex Receptacle	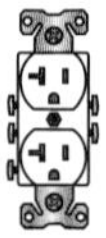	SPW RCPT	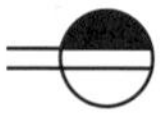
Triplex Receptacle		TRI RCPT	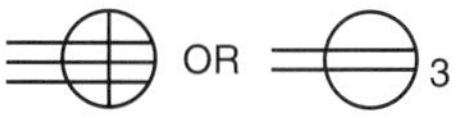 OR 3
Single Floor Outlet	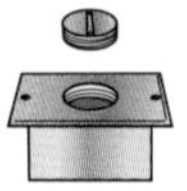	SF OUT	 OR
Duplex Receptacle with Isolated Ground	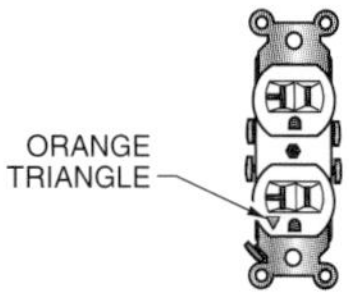 ORANGE TRIANGLE	DX IG	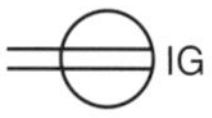 IG
Weatherproof Receptacle		IQ WP	 IG WP
Special-Purpose Outlet		OUT DW (DISHWASHER) OUT CD (CLOTHES DRYER)	 DW
Range Outlet		RNG OUT	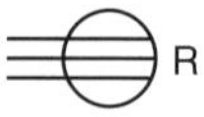 R
Combination Switch with Receptacle		SW & RCPT	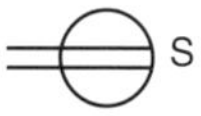 S
Ground-Fault Circuit Interuptor (GFCI)		GFCI	 GFCI

Figure 6-7. Receptacle symbols are used on electrical floor plans to show the type and location of each receptacle.

Special-purpose outlets, such as receptacles used for appliances, use particular letters to designate a variation of standard equipment on an electrical floor plan. Typical abbreviated letters used represent appliances include dishwashers (DW) or clothes dryers (CD). Other types of special-purpose receptacles seen on prints can include range (R) outlets and combination switches with receptacles. Typically, any variation of a special-purpose receptacle should be explained in the specifications.

Switches. Most residential electrical circuits contain control devices such as switches. On electrical prints, switches are used to show the combination of wiring to control an electrical load or circuit. The most common types of switches used to control lighting circuits are single-pole, double-pole, three-way, and four-way switches. Other electrical switches, such as dimmer switches, switches with pilot lights, or weatherproof switches, are used in certain locations throughout a dwelling. Dimmer switches control the brightness of lamps. **See Figure 6-8.**

SWITCH ABBREVIATIONS AND SYMBOLS			
Description	**Device**	**Abbrev**	**Symbol**
Single-Pole Switch		SPST	$S - S_1$
Double-Pole Switch		DPST	S_2
Three-Way Switch		SPDT	S_3
Four-Way Switch		DPDT	S_4
Dimmer Switch		DMR SW	S_D
Switched Pilot Light		SPST	S_P
Weatherproof Switch		WP SW	S_{WP}

Figure 6-8. Switch symbols are used on electrical floor plans to indicate the type, location, and wiring combination of lighting circuits.

Power. In order to deliver power to all lighting, receptacles, or loads, electrical power must be distributed throughout a residence. Typically, power is brought in from a utility company step-down transformer and travels through the service entrance to terminate at the service panel. A service panel in a residence includes a main circuit breaker or fuses to protect the entire electrical system.

Various types of power symbols are used to indicate devices and/or cables on electrical prints. Power symbols typically represent service panel and motor devices; 2-, 3-, or 4-wire cables; and other line symbols that show exposed or hidden cables in walls, ground connections, or home runs. **See Figure 6-9.** A *home run* is a line with an arrow that shows where circuit wiring should return to the service panel.

Power symbols are standardized, but some variations may be seen on electrical prints. Typically, hash marks on home-run lines indicate the number of hot, neutral, and ground wires in cable. Full-hash marks indicate hot and half-marks indicate neutral and ground wires. Dots may be used to indicate ground wires.

TECH TIP

Electrical workers often write "HR" on studs, joists, and cable to identify the home run to the service panel.

Signaling. In addition to lighting and power circuits, a residence can include specialized circuits consisting of signaling devices. These may include doorbell, fire alarm, sound system, and other signaling circuits, such as those for information technology. Signaling devices provide a visual and/or audible warning signal. Along with the symbols for lighting and power circuits, signaling device symbols are also shown on electrical prints. **See Figure 6-10.**

The types of devices that produce audible signals include pushbuttons, buzzers, bells, combination bell-buzzers, chimes, annunciators, and smoke/carbon monoxide detectors. Some other types of signaling devices are thermostats, television outlets, computer data outlets, and sound systems. Audible devices are typically used in applications designed to draw attention when abnormal conditions are presented.

A kitchen floor plan may include electrical symbols for lights, light switches, convenience receptacles, and appliance receptacles.

POWER ABBREVIATIONS AND SYMBOLS			
Description	**Device**	**Abbrev**	**Symbol**
Service Panel	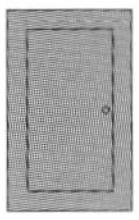	SRV PNL OR PWR PNL	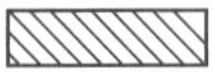
Switch Wire Concealed in Wall, Ceiling, or Floor	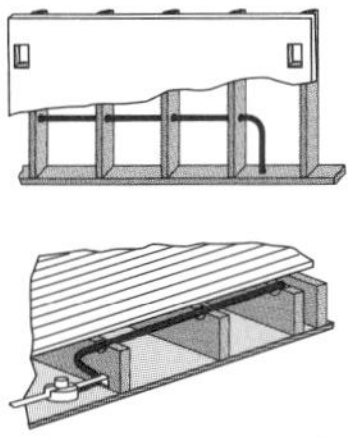		
Motor	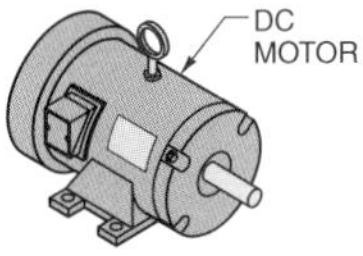 DC MOTOR	MTR	(M)
Ground Connection	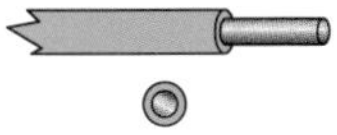	GR	
2-Wire Cable	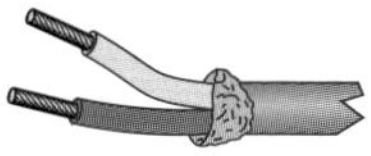		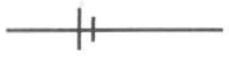
3-Wire Cable	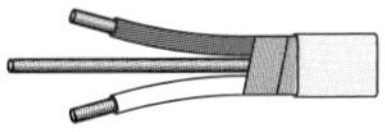		
4-Wire Cable	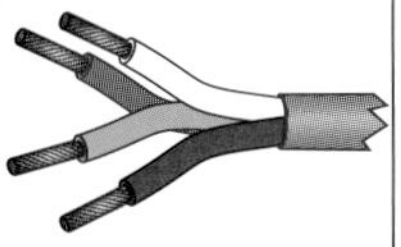		
Cable Return to Service Panel	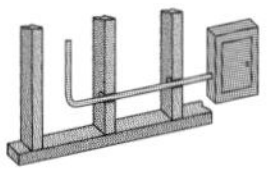	HR	
Wire Turned Up in Wall	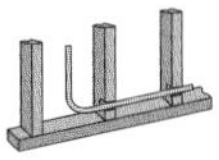		
Wire Turned Down in Wall	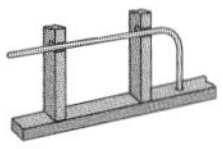		
Exposed Wire			

Figure 6-9. Power symbols typically represent service panel and motor devices; 2-, 3-, or 4-wire cables; and other line symbols that show exposed or hidden cables turning up or down in walls, ground connections, or home runs.

SIGNALING ABBREVIATIONS AND SYMBOLS			
Description	Device	Abbrev	Symbol
Pushbutton	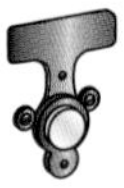	PB	
Buzzer		BZR	
Bell	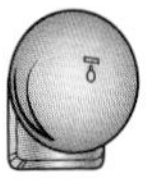	BEL	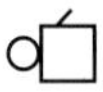
Combination Bell-Buzzer	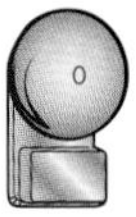	BEL & BZR	
Chime	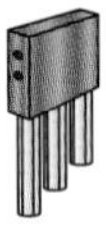	CHIM	CH
Annunciator	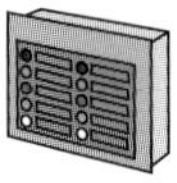	ANNT	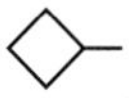
Smoke/Carbon Monoxide Detector	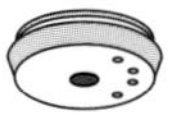	SD	 SD OR SD
Thermostat	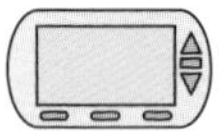	T	T
Television Outlet		TV	TV
Computer Data Outlet	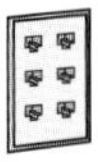	CMTR	
Sound System	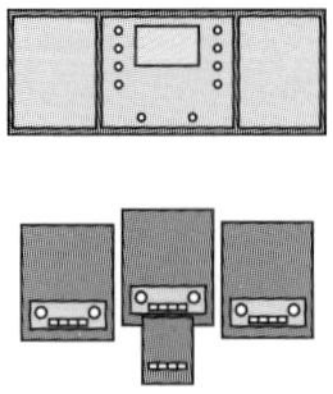	SND SYS	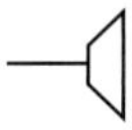

Figure 6-10. Signaling symbols represent signaling devices that provide a visual and/or audible warning signal, as well as other types of information technology.

SECTION 6.1 CHECKPOINT

1. What kind of devices are shown on electrical floor plans for one-family dwellings?
2. Why do electricians use electrical floor plans?
3. Why are wiring plans essential for installations?
4. What are the typical receptacles shown on electrical floor plans?
5. What are the typical power symbols shown on electrical floor plans?

SECTION 6.2—ELECTRICAL DIAGRAMS

In the electrical trades, individuals use a certain type of language to transfer information efficiently and to understand electrical circuit operations. This language includes abbreviations and symbols that are used in drawings and diagrams to convey the function of electrical circuits, components, and devices. Pictorial drawings provide a clear representation of electrical circuits, and electrical diagrams are used to interpret information. Types of electrical diagrams include line (ladder), wiring, schematic, and single-line diagrams.

Individuals should have a basic understanding of residential electrical circuits. An *electrical circuit* is an assembly of conductors (wires), electrical devices (switches and receptacles), and electrical components/loads (lights and motors) through which current flows. **See Figure 6-11.** Typical electrical circuits have five main parts, which are represented on electrical diagrams and include the following:

- a source of electricity (service panel)
- a load(s) that converts electrical energy into some other usable form of energy such as light, heat, or motion
- a method of controlling the flow of electricity (switch)
- circuit protection (fuses or circuit breakers) to ensure that the circuit operates safely within electrical limits
- a way of connection by conductors/cables to attach individual devices and loads together

Pictorial Drawings

A *pictorial drawing* is a drawing that shows the length, height, and depth of an object in one view. Pictorial drawings indicate physical details of an object or an electrical drawing as seen by the eye. For example, a simple pictorial drawing can display a lighting circuit, which would include a switch, lamp, and the type of conductors used to connect the devices. **See Figure 6-12.**

The pictorial diagrams for residential electrical circuits can also be quite complicated. For example, a complex pictorial drawing can display a construction layout that shows where cable is run for an electrical circuit in a bedroom of a dwelling. **See Figure 6-13.**

Voice, data, and video (VDV) systems are becoming increasingly common on residential electrical prints and diagrams.

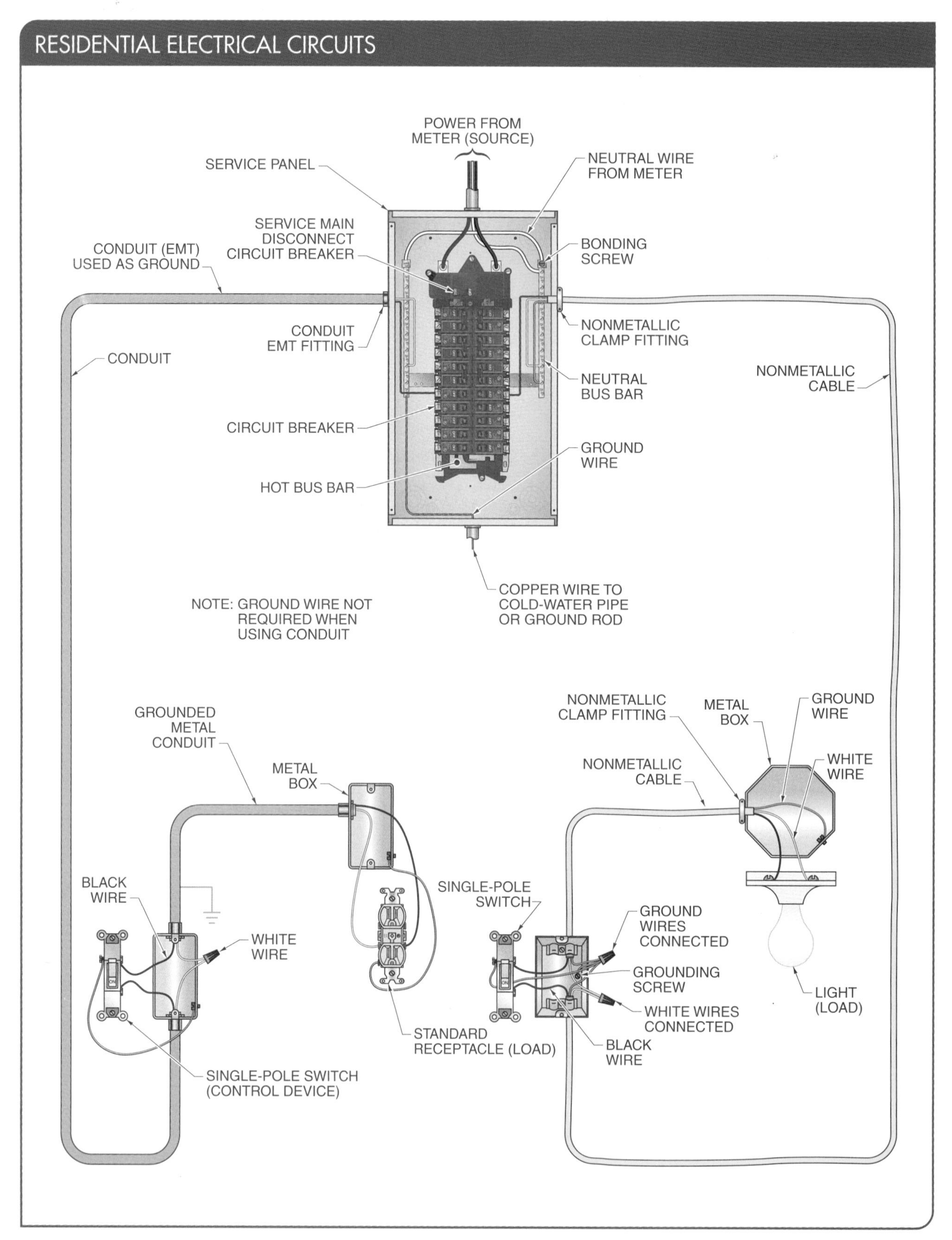

Figure 6-11. Typical residential electrical circuits have five main parts: a source of electricity, a load(s), a method of control, circuit protection, and a way of connection.

TECH TIP

The hot conductor from a switch to a light fixture should be a different color than the source conductor to the switch.

Device Abbreviations and Symbols for Diagrams

In electrical trades, abbreviations and symbols are used to represent electrical components and devices on electrical diagrams. Most graphic symbols can be referenced and used on the different types of electrical diagrams, such as line (ladder), wiring, schematic, and single-line diagrams. Different types of abbreviations and symbols are used on diagrams to represent electrical conductors and power sources, circuit protection and control devices, and electronic devices and loads.

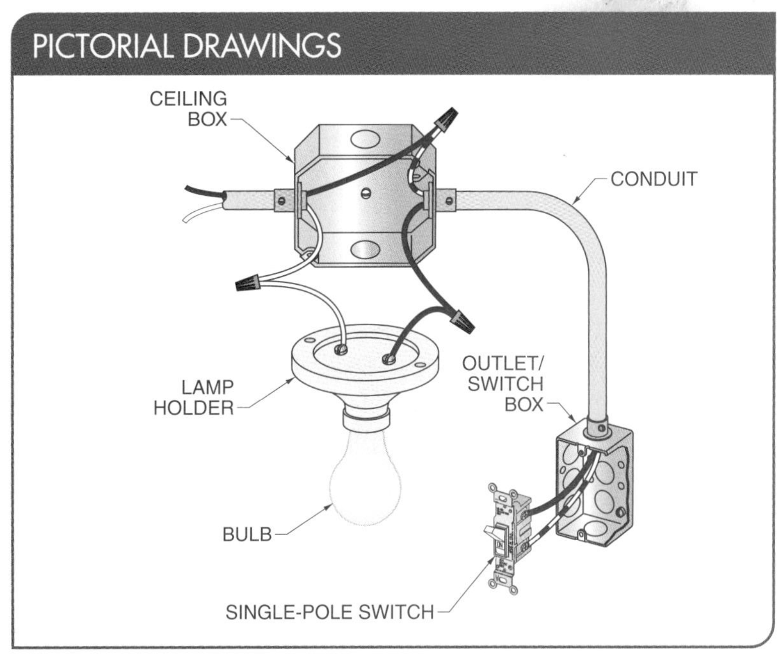

Figure 6-12. Pictorial drawings indicate the physical details of an object or an electrical drawing, such a lighting circuit, as seen by the eye.

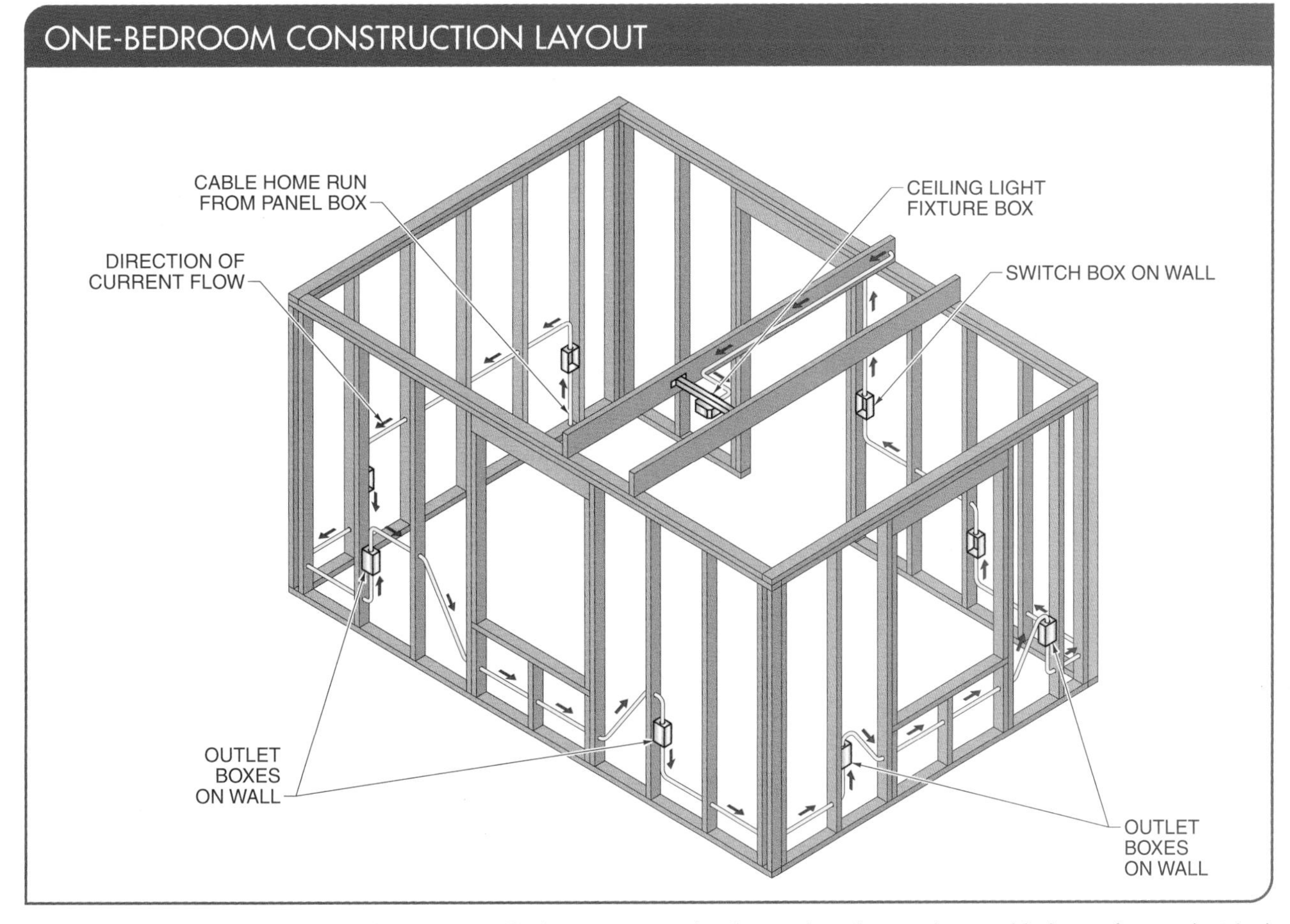

Figure 6-13. A complex pictorial drawing can display a construction layout that shows where cable is run for an electrical circuit in a bedroom.

Electrical Conductor and Power Sources. In a typical residence, a 120/240 V, single-phase, three-wire service is used to supply power to all electrical circuits. However, contemporary homes are starting to use many other types of power sources, such as single-cell or multiple-cell battery circuits or DC circuits powered by battery banks or photovoltaic (PV) cells. In electrical diagrams, these types of power sources are all given a standard symbol. **See Figure 6-14.**

Conductors (wires) are represented by lines that connect or do not connect. Connected lines are given a connection node and represent a connected wire or termination. Line weight on a diagram determines whether the line is a power wire or a control wire. A ground connection is shown as a vertical line connected with one perpendicular line followed by two smaller lines below. Cables can be shown as shielded or sheathed with each wire inside a dashed circle. The symbol for plugs and receptacles consists of a female part and a male part. The plug is described as the male part and the receptacle is described as the female part.

Electrical Circuit Protection and Control Devices. Most circuit protection in a modern home is located at the service panel. Circuit breakers are the typical overcurrent protective device (OCPD) used for circuit protection. In older dwellings, rather than circuit breakers, fuses may still be used to protect branch circuits.

The most common control device used in a residence is a switch. Switch types can consist of single-pole, three-way, and four-way switches; dimmer switches; temperature switches; and float switches. Other types of contacts can be either mechanical or solid-state connected components. Circuit protection and control devices are represented by standard symbols. **See Figure 6-15.**

Electronic Devices and Loads. Electronic devices and loads found in residential circuits include lamps, motors, and certain alarms. Typically, doorbells or other signaling circuits can be found in dwellings. Alarms usually consist of a bell, horn, or buzzer. Electronic devices are usually a part of an electrical device or appliance and consist of diodes, light-emitting diodes (LEDs), and photodiodes. These types of electronic devices and loads represented by standard symbols on electrical diagrams to convey information to electricians. **See Figure 6-16.**

TECH TIP

Transformers used for doorbells can be found in crawl spaces, attics, basements, or junction boxes.

Line (Ladder) Diagrams

A *line (ladder) diagram* is a diagram that shows the logic of an electrical circuit or system using standard electrical symbols. A line diagram is used to show the relationship between a circuit and its components, but not the actual location of the components. Line diagrams provide a fast, easy understanding of the connections and use of electrical components.

The arrangement of a line diagram should promote clarity. Graphic symbols, abbreviations, and device designations are drawn per electrical standards. The circuit should be shown in the most direct path and logical sequence.

Lines between symbols can be horizontal or vertical and should be drawn to minimize line crossing. The voltage level of a circuit is typically indicated at the top of a line diagram. Circuit voltages are nominal in a residence and are typically 120 VAC. Connection nodes are used on line diagrams to indicate an electrical conductor connection (junction). When a connection node is not present, the wires cross paths and are not electrically connected. Line diagrams are read from left (L1) to right (L2). Line diagrams show the function of manually controlled or automatically controlled circuits.

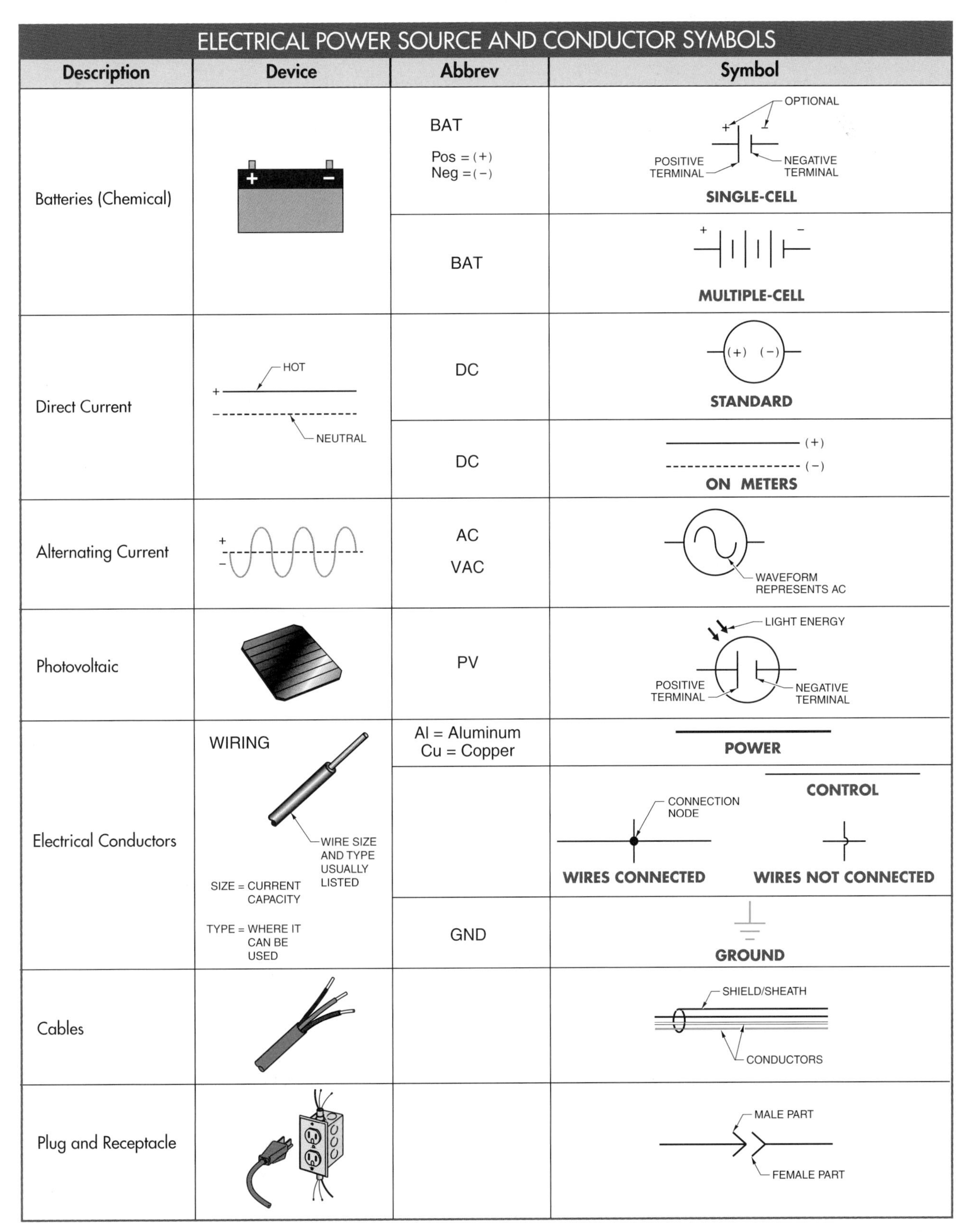

ELECTRICAL POWER SOURCE AND CONDUCTOR SYMBOLS			
Description	**Device**	**Abbrev**	**Symbol**
Batteries (Chemical)	+ −	BAT Pos = (+) Neg = (−)	OPTIONAL + POSITIVE TERMINAL NEGATIVE TERMINAL **SINGLE-CELL**
		BAT	+ − **MULTIPLE-CELL**
Direct Current	HOT + NEUTRAL	DC	(+) (−) **STANDARD**
		DC	(+) (−) **ON METERS**
Alternating Current	+ −	AC VAC	WAVEFORM REPRESENTS AC
Photovoltaic		PV	LIGHT ENERGY POSITIVE TERMINAL NEGATIVE TERMINAL
Electrical Conductors	WIRING WIRE SIZE AND TYPE USUALLY LISTED SIZE = CURRENT CAPACITY TYPE = WHERE IT CAN BE USED	Al = Aluminum Cu = Copper	**POWER**
			CONTROL CONNECTION NODE **WIRES CONNECTED** **WIRES NOT CONNECTED**
		GND	**GROUND**
Cables			SHIELD/SHEATH CONDUCTORS
Plug and Receptacle			MALE PART FEMALE PART

Figure 6-14. Electrical diagrams use different types of standard power symbols to provide information for installation tasks.

CIRCUIT PROTECTION AND CONTROL DEVICE SYMBOLS			
Description	Device	Abbrev	Symbol
Fuses		FU	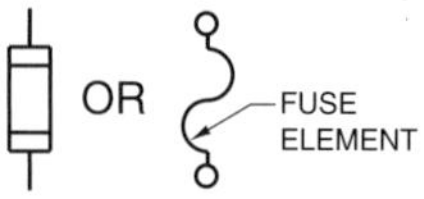 OR / FUSE ELEMENT / SINGLE FUSE
Single-Pole Circuit Breaker	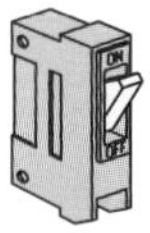	SPCB	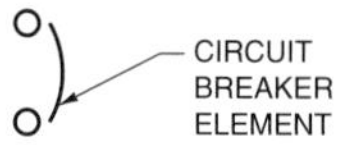 CIRCUIT BREAKER ELEMENT / SINGLE-POLE CIRCUIT BREAKER
Double-Pole Circuit Breaker	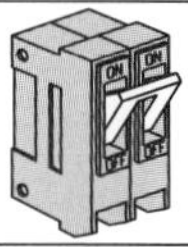	DPCB	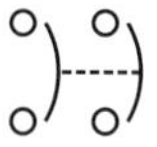 DOUBLE-POLE CIRCUIT BREAKER
Single-Pole Switch	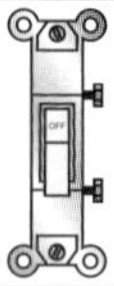	SPST	SINGLE-POLE SINGLE-THROW
Three-Way Switch	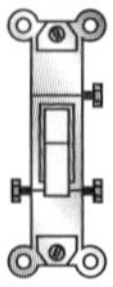	SPDT	SINGLE-POLE DOUBLE-THROW
Four-Way Switch	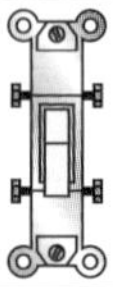	DPDT	DOUBLE-POLE DOUBLE-THROW
	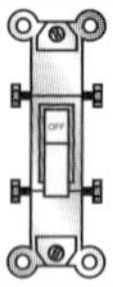	DPST	DOUBLE-POLE SINGLE-THROW
Dimmer Switch		DS	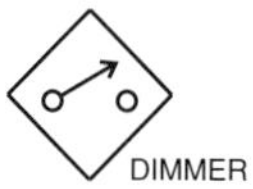 DIMMER
Float Switch		FS	 LEVEL OPERATOR / NORMALLY OPEN (NO) / NORMALLY CLOSED (NC)
Temperature Switch		TEMP SW	 TEMPERATURE OPERATOR / NO / NC
Contacts	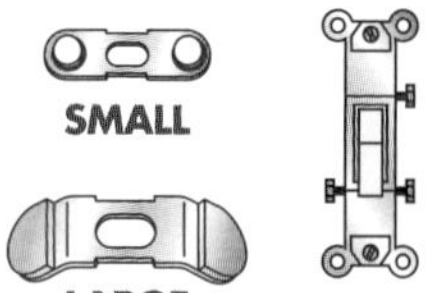 SMALL / LARGE / SWITCH	NO	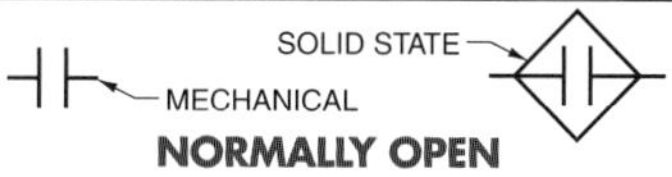 SOLID STATE / MECHANICAL / NORMALLY OPEN
		NC	 NORMALLY CLOSED

Figure 6-15. Circuit protection and control devices have standard symbols that represent the devices or equipment and how they connect to sources of power.

ELECTRONIC DEVICE AND LOAD SYMBOLS			
Description	**Device**	**Abbrev**	**Symbol**
AC Motors	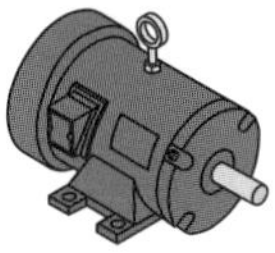	1ϕ	 T1 T2 T = TERMINAL **SINGLE-PHASE, SINGLE- OR DUAL-VOLTAGE**
Lights	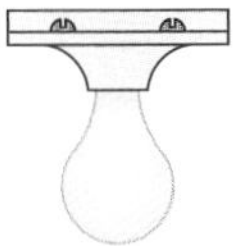	L	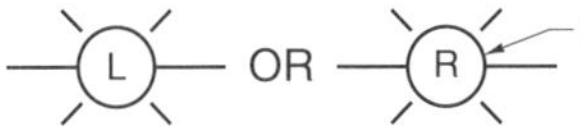 L OR R LETTER INDICATES COLOR A = AMBER G = GREEN R = RED B = BLUE **PILOT LIGHT**
Alarms		AL or BELL	
		AL or HORN	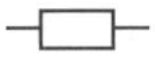
		BUZZ	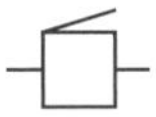
Diode		D	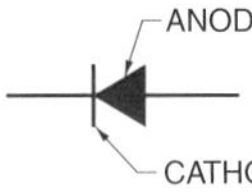 ANODE CATHODE
Light-Emitting Diode		LED	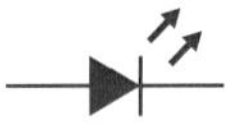
Photodiode		D	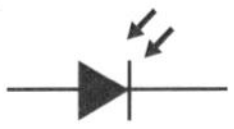

Figure 6-16. Electronic devices and loads have standard symbols that are used in electrical diagrams to convey information to individuals.

A *manually controlled circuit* is any circuit that requires a person to initiate an action for the circuit to operate. For example, line diagrams can be used to illustrate manually controlled circuits that have a single-pole switch (S1) controlling a light in a fixture. **See Figure 6-17.**

Flipping a single-pole switch (S1) to the ON position allows current to pass from the black (hot) conductor (L1) through the closed contacts of the wall switch (S1), through the light (L), and on through the white (neutral) conductor (L2), forming a complete closed circuit for current to flow allowing the light to illuminate. Flipping the switch to the OFF position opens the contacts of the switch, stopping the flow of current to the lamp and turning the light off.

An *automatically controlled circuit* is any circuit that uses a control device to initiate an action for a circuit to operate. For example, an electric motor on a sump pump can be turned on and off automatically by adding a control device such as a normally open (NO) float switch. **See Figure 6-18.** This control circuit is used in basements to control a sump pump to prevent flooding. When water reaches a predetermined level, the float switch closes by sensing the change in water level and automatically starts the pump to remove water.

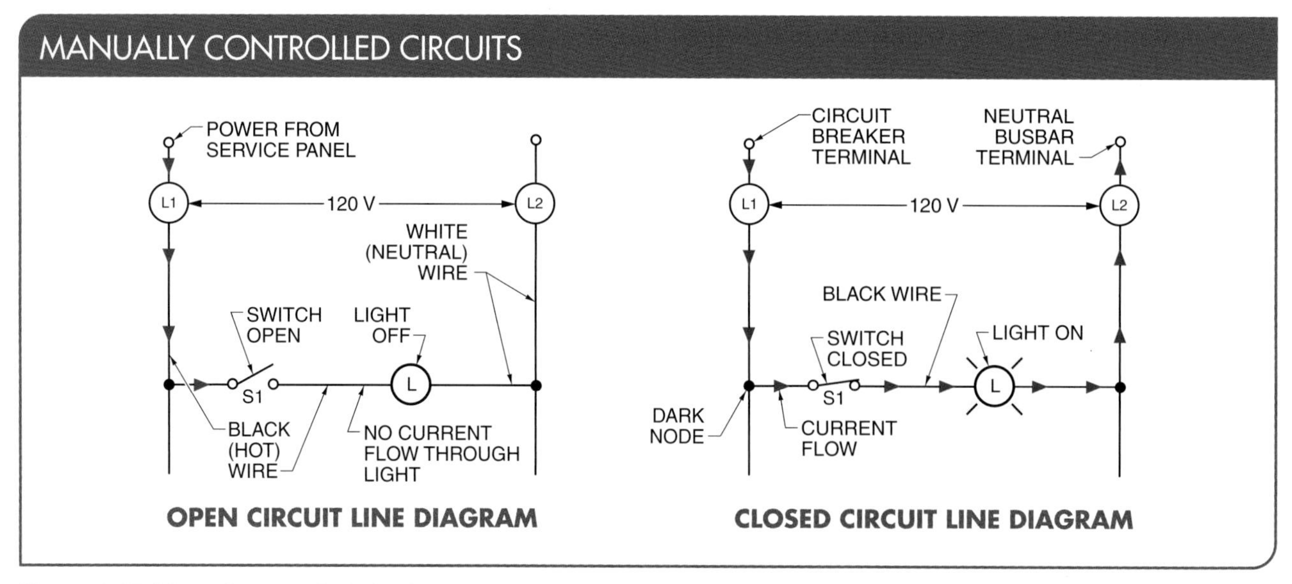

Figure 6-17. Manually controlled circuits can be shown on line diagrams to illustrate how a single-pole switch (S1) controls a light in a fixture.

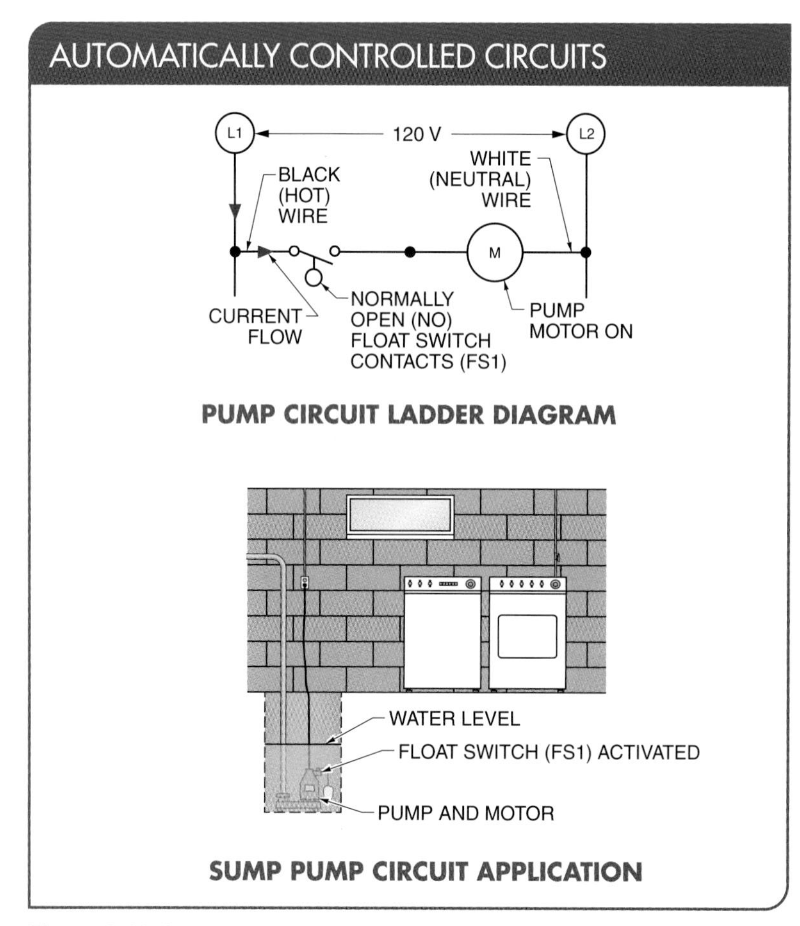

Figure 6-18. Automatically controlled circuits can be shown on line diagrams to illustrate how an electric motor on a sump pump can be turned on and off automatically using a normally open (NO) float switch.

In this circuit, the NO float switch contacts (FS1) determine when current passes through the circuit to start the pump motor. When water reaches a level above the float switch, FS1 contacts close and current passes through the pump motor. The pump removes water until the water level drops enough to open the FS1 contacts, which shuts off the pump motor. A power failure or the manual opening of the contacts prevents the pump from automatically pumping water even after the water reaches the predetermined level.

Wiring Diagrams

A *wiring diagram* is a diagram that shows the wired connections of all components in an electrical device. Wiring diagrams show, as closely as possible, the actual location of each component and the point of connection in an electrical circuit. Wiring diagrams often include details concerning the type of wire and the kind of hardware by which the wires are fastened to terminals.

For example, the wiring diagram for an interior lighting circuit includes the five components of a typical electrical circuit. These components include an AC source from the service panel, an interior light, a single-pole switch for control, a circuit breaker for circuit protection, and a cable for connecting all devices. **See Figure 6-19.** The location or layout of the components is accurately represented by the diagram. All connecting wires are shown connected from one component to another. Wiring diagrams are widely used by individuals when installing, maintaining, and troubleshooting electrical equipment and circuits.

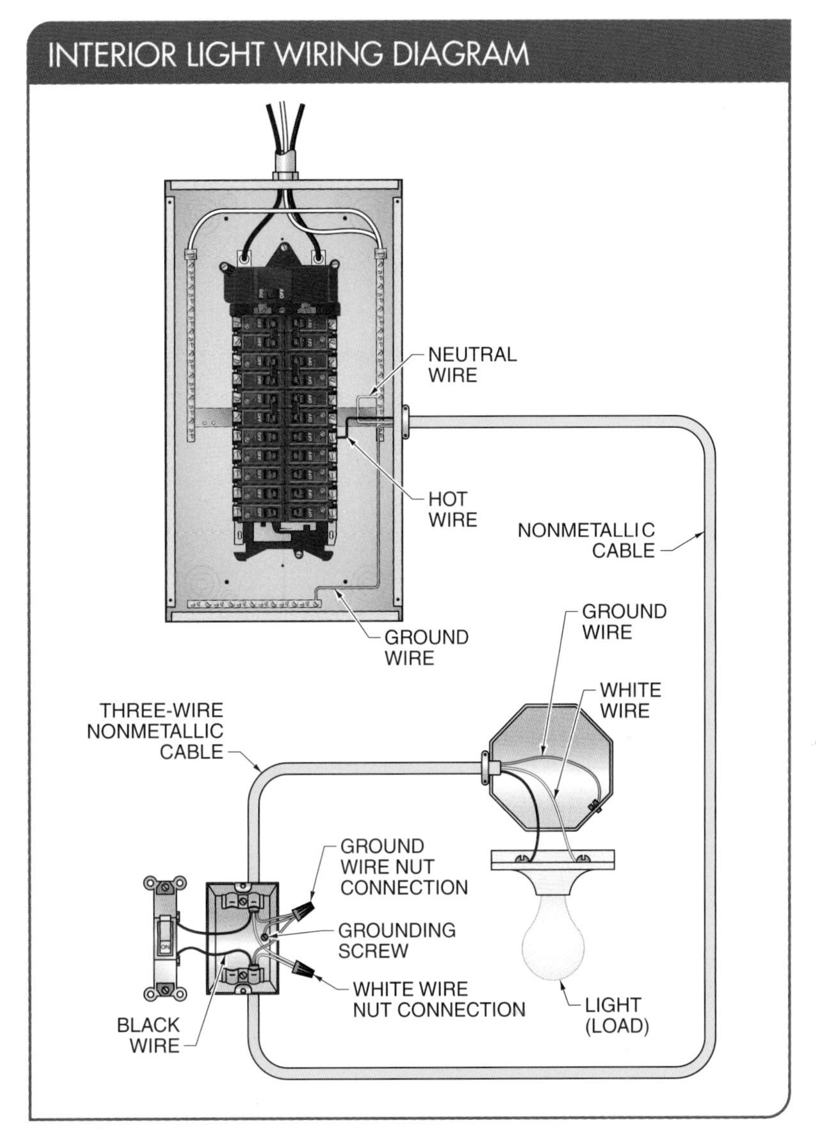

Figure 6-19. Wiring diagrams show the location of each component and the point of connection in an electrical circuit.

> CODE CONNECT
>
> *Per NEC® Section 210.52, lighting and convenience outlets are typically on 15 A circuits, and refrigerators, microwaves, and dishwashers should be on separate 20 A circuits.*

Schematic Diagrams

A *schematic diagram* is a drawing that indicates the electrical connections and functions of a specific circuit arrangement using graphic symbols. Typically, schematic diagrams are associated with electronic circuits found in electrical devices and equipment. They are used to make repairs, troubleshoot problems, or inspect circuits.

Schematic diagrams are intended to show the wiring for the operation of a device in a circuit. However, schematic diagrams are not intended to show the physical size or appearance of any component or device. Schematic diagrams are essential for troubleshooting because they enable an individual to trace a circuit and its functions without regard to the actual size, shape, or location of the component, device, or part. Schematic diagrams can be used for many different applications, such as depicting the connections of devices in a bedroom or representing a simple light circuit from the service panel. **See Figure 6-20.**

An electrical circuit is complete (closed) when current flows from the service panel to the load (lights) and back to the power source. A circuit is incomplete (open) when current does not flow. A broken wire, a loose connection, or a switch in the OFF position stops current from flowing in an electrical circuit.

One-Line Diagrams

A *one-line diagram* is a diagram that uses single lines and graphic symbols to show the current path, voltage values, circuit disconnect, OCPDs, transformers, and service panels for a residential electrical system. **See Figure 6-21.** One-line diagrams are typically used as an overview of power distribution throughout a residence. One-line diagrams show voltage values that enter through the service entrance from the utility transformer, to the service panel, and on to each branch circuit and connected devices.

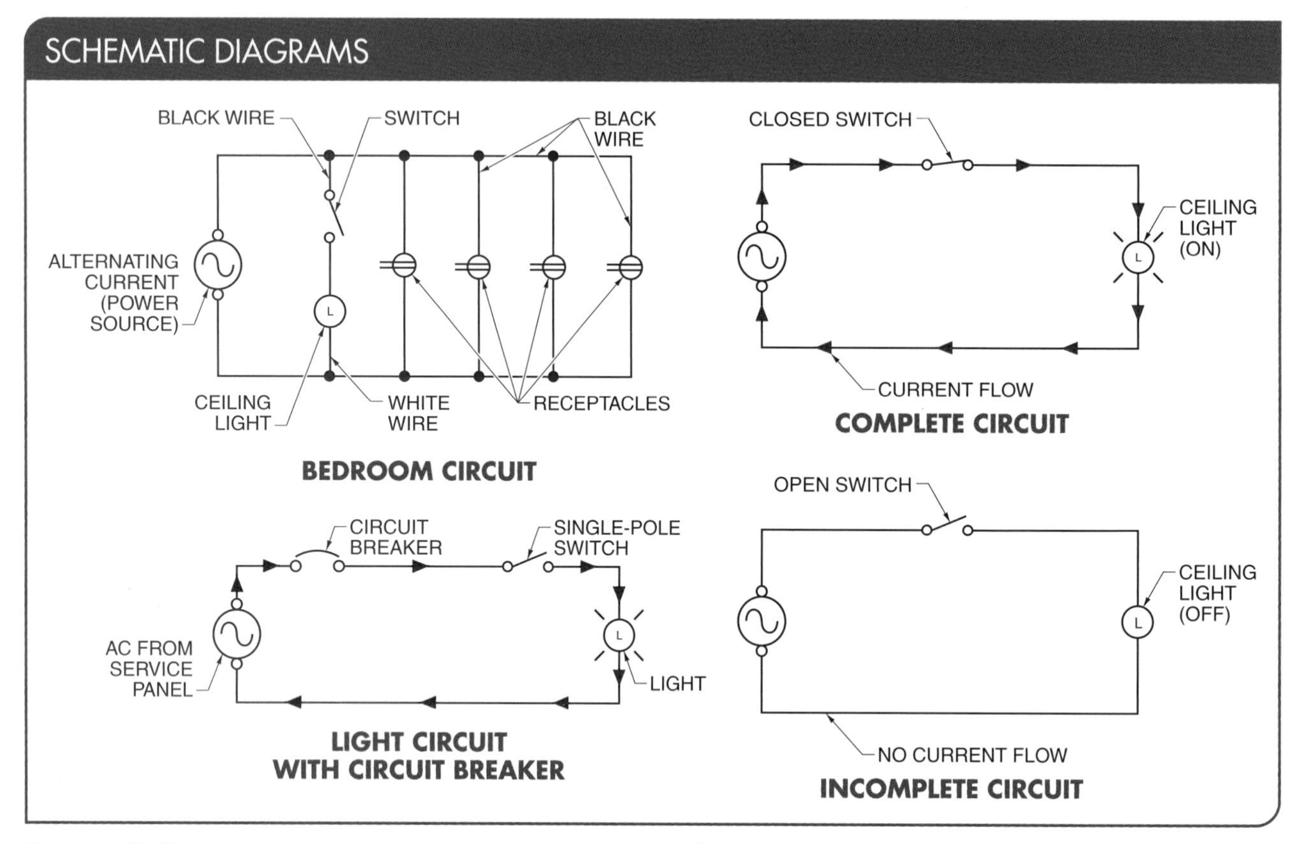

Figure 6-20. Schematic diagrams are intended to show the wiring for the operation of a device, but they are not intended to show the physical size or appearance of the device.

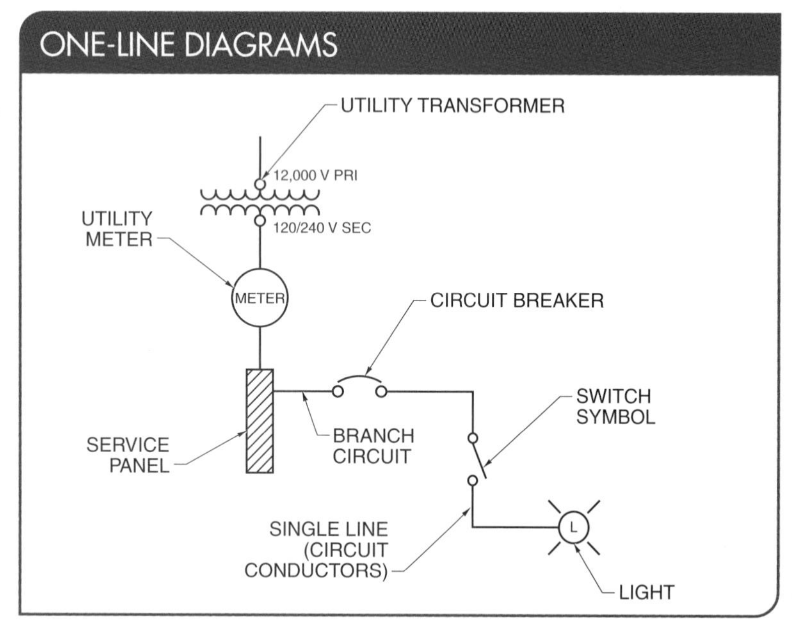

Figure 6-21. One-line diagrams use single lines and graphic symbols to show the current path, voltage values, circuit disconnect, OCPDs, transformers, and service panels for a residential electrical system.

SECTION 6.2 CHECKPOINT

1. What are the five main parts of an electrical circuit?
2. What is a pictorial drawing and what does it indicate?
3. What kind of service is typical for a residential system?
4. What is the difference between a wiring diagram and a schematic diagram?
5. What is a one-line diagram used for and what does it show?

6 Electrical Prints and Diagrams

Chapter Activities

Name ______________________ Date ______________

Electrical Floor Plans

Identify the three different types of plans as shown.

_______________ **1.** Component plan

_______________ **2.** Electrical floor plan

_______________ **3.** Wiring plan

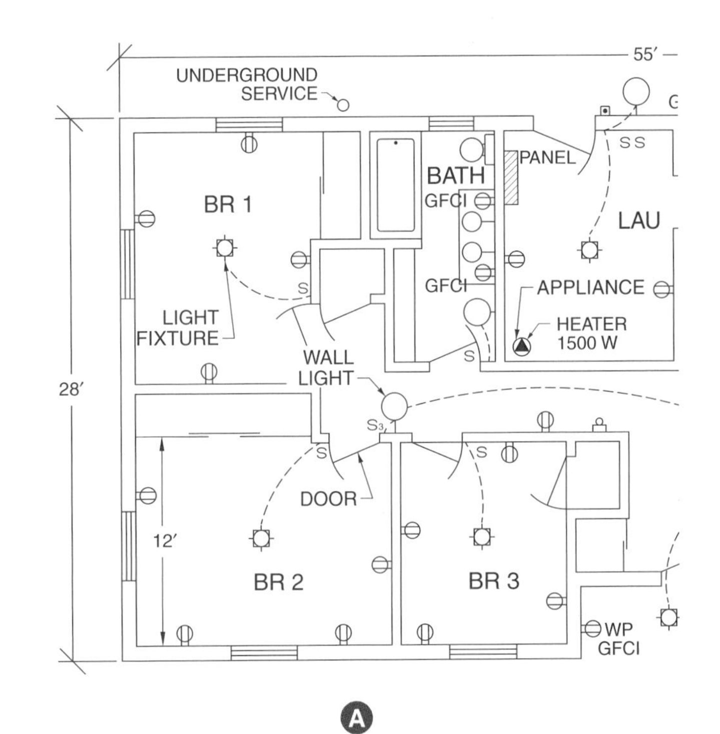

A

Area	Convenience Receptacles	Special-Purpose Outlets
Bedroom(s) Schedule	No space along a wall should be more than 6′ from a receptacle outlet. Any wall space 2′ or larger should have a minimum of 1 receptacle. NEC® 210.52(A).	TV outlet, intercom, speakers (music), and telephone jack
Living Room Schedule	No space along a wall should be more than 6′ from a receptacle outlet. Any wall space 2′ or larger should have a minimum of 1 receptacle. NEC® 210.52(A).	TV outlet, intercom, speakers (music)

B

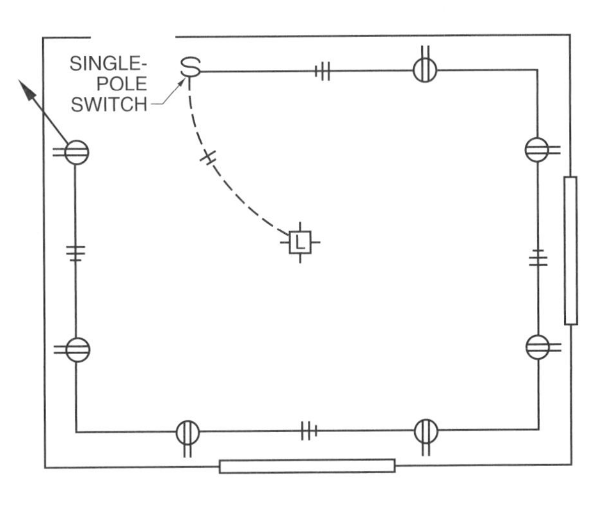

C

Electrical Symbols and Abbreviations

Identify the symbols and abbreviations shown.

_______________ 1. Ground-fault circuit interrupter

_______________ 2. Junction box

_______________ 3. Three-way switch

_______________ 4. Fan outlet

_______________ 5. Duplex receptacle

_______________ 6. Light with lamp holder

_______________ 7. Double-pole switch

_______________ 8. Fluorescent light

_______________ 9. Split-wired duplex receptacle

_______________ 10. Single-pole switch

	Device	Abbrev	Symbol
A		SPST	$S - S_1$
B		DPST	S_2
C		SPDT	S_3
D		FLUOR	Extend rectangle for length of installation
E		L	Ⓛ
F		F	Ⓕ Box listed as acceptable for fan support
G		J	Ⓙ
H		DX RCPT	
I		SPW RCPT	
J		GFCI	GFCI

Electrical Circuit Part Identification

Identify the five main parts of the circuit shown.

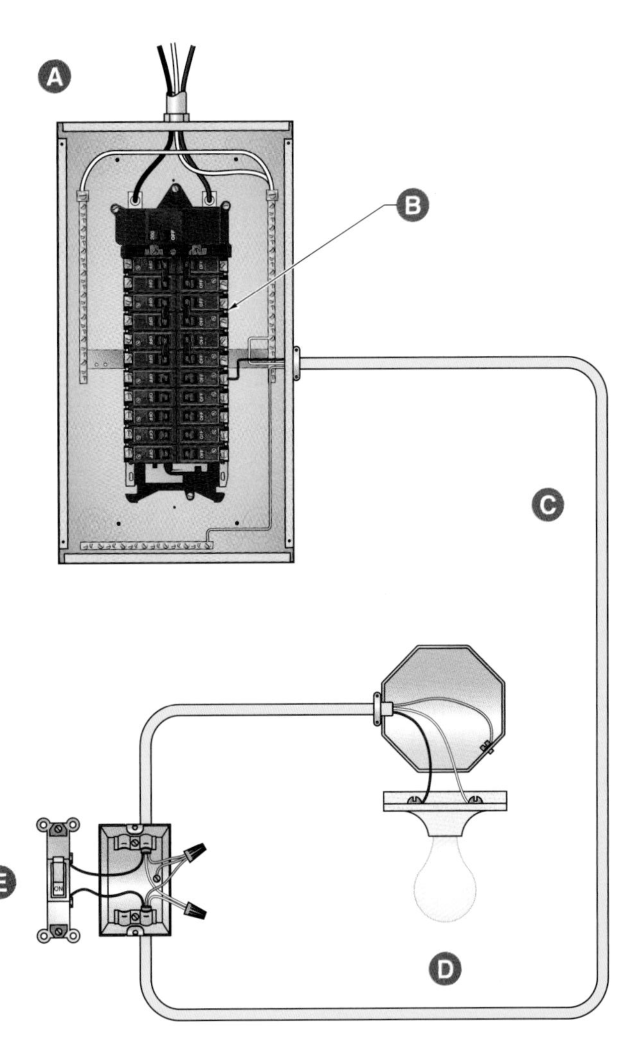

______________________ **1.** Conductors

______________________ **2.** Switch

______________________ **3.** Load

______________________ **4.** Source

______________________ **5.** Circuit protection device

Electrical Diagram Identification

Identify each type of diagram shown.

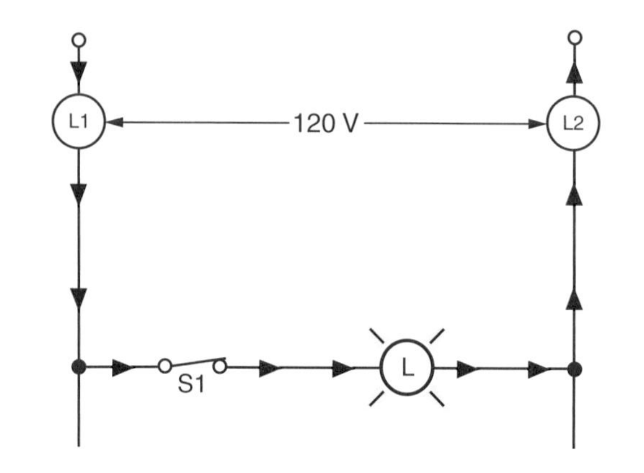

______________________ **1.** ___ diagram

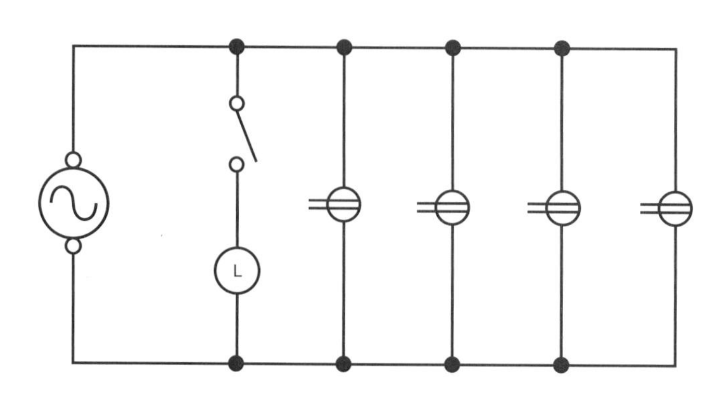

______________________ **2.** ___ diagram

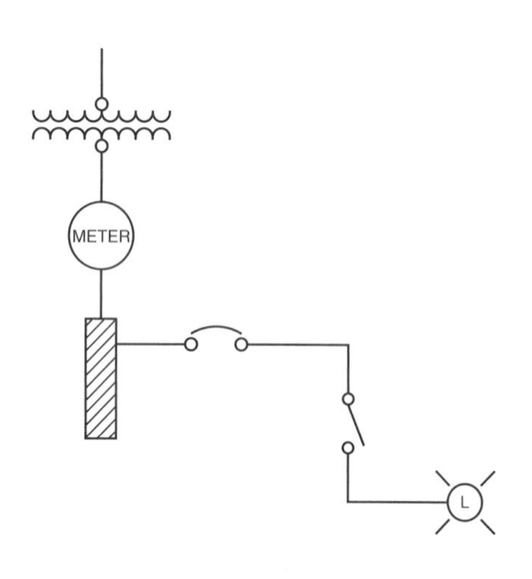

______________________ **3.** ___ diagram

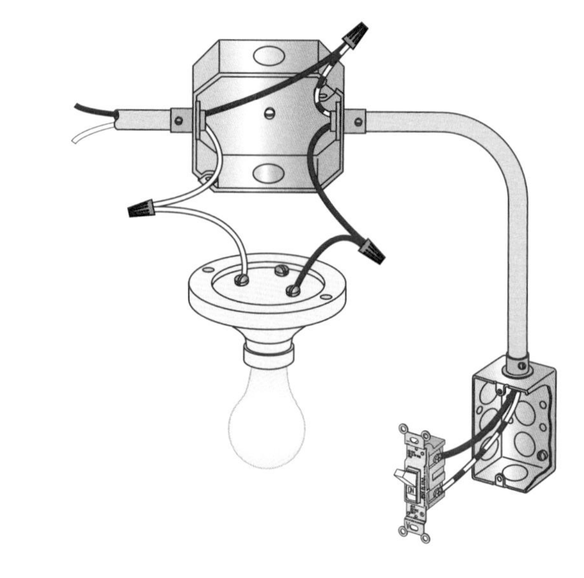

______________________ **4.** ___ diagram

7 Electrical Connections

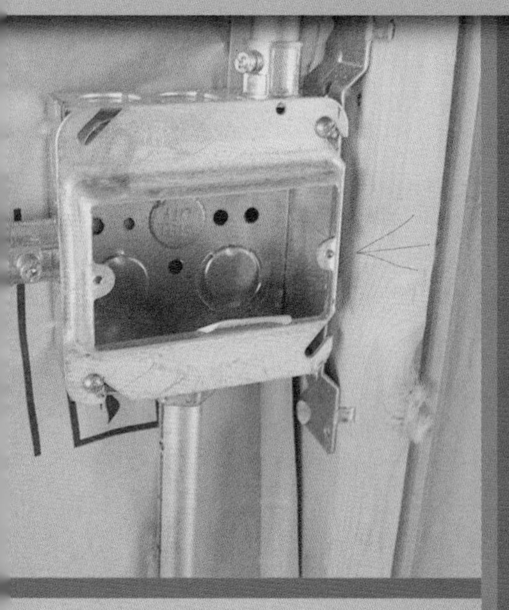

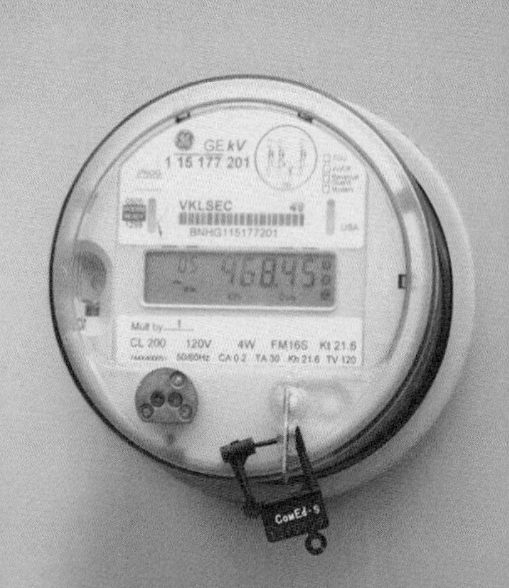

SECTION 7.1—ELECTRICAL SUPPLIES

- List the electrical supplies used for wiring applications in a residence.
- Explain the use of electrical supplies used in wire applications.

SECTION 7.2—SPLICED CONNECTIONS

- Explain the types of electrical splices found in a residence.
- Describe how to remove insulation from larger wires.
- Describe how to tape splices.

SECTION 7.3—SOLDERED CONNECTIONS

- Explain the use of soldering irons.
- Describe how to care and clean soldering iron tips.
- Describe how to solder splices, solder on printed circuit (PC) boards, and prevent bad solder joints.
- Describe how to safely solder.

SECTION 7.4—CONDUCTOR CONNECTORS

- List the types of conductor connections.
- Describe how to install wire connectors.
- Explain the use of back-wired, crimp-type, and split-bolt connectors.

SECTION 7.5—REPLACING AND TESTING PLUGS

- Explain how to replace grounded, ungrounded, heavy-duty, appliance, and large appliance plugs.
- Describe how to test connections using a digital multimeter (DMM).

Learner Resources

ATPeResources.com/QuickLinks
Access Code: 791712

In residential electrical systems, making electrical connections that are mechanically and electrically secure is important to the integrity of the system. Specific electrical supplies, connecting operations, and conductor-connecting devices are needed for such work tasks. Occasions can arise where conductors, cables, cords, or plugs need to be installed or replaced and then connected through splicing or soldering operations. Once these operations are complete, tests can be performed to ensure proper electrical connections.

SECTION 7.1 — ELECTRICAL SUPPLIES

Electrical installations or maintenance tasks involve different types of electrical supplies that are required to make conductor (wire) connections. The typical electrical supplies required to make wire connections in a residence include electrical tape, heat-shrink tubing, wire connectors, and wire markers.

Electrical Tape

Many types of tape can be used to insulate exposed wire in electrical wiring connections. The most common type of electrical tape used in residential wiring connections is vinyl (PVC) electrical tape. **See Figure 7-1.** *Vinyl electrical tape* is a type of plastic tape that has good insulation properties and can insulate up to 600 V per wrap. Vinyl electrical tape can withstand moisture, abrasion, and corrosive environments.

ELECTRICAL TAPE

Figure 7-1. Vinyl (PVC) electrical tape is the most common type of tape used in residential wiring applications.

Heat-Shrink Tubing

Heat-shrink tubing is plastic (polymer) tubing that is designed to shrink (contract) when heated. **See Figure 7-2.** As heat is applied, the plastic tube begins to shrink and becomes a covering that insulates and protects a wire, cable, splice, or electrical connection. Some advantages of heat-shrink tubing include sealing out moisture, combining wires together, creating barriers between corrosive environments, and color-coding of specific wires. Heat-shrink tubing is available in a wide range of colors, sizes, lengths, thicknesses, and temperature ratings. The most common types of heat sources used for heat-shrink tubing are heat guns, blow-dryers, or microtorches.

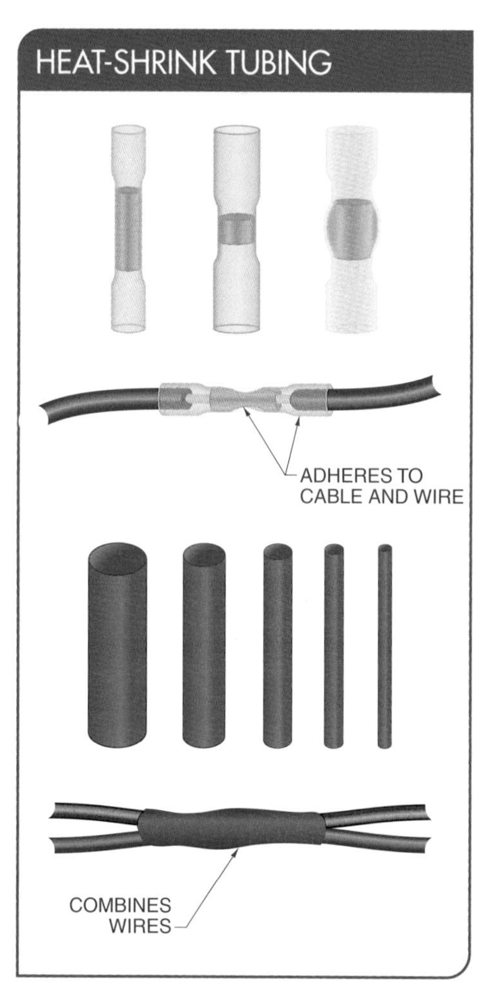

Figure 7-2. Heat-shrink tubing insulates and protects wires, cables, splices, and other electrical connections.

Wire Connectors

A *wire connector* is a plastic device designed to be twisted onto several electrical wires to firmly hold them together and to provide an insulating cover for the connection. Wire connectors are available in several sizes. **See Figure 7-3.** The size of a wire connector is determined by the number and size of the wires (conductors) to be connected. Wire connector manufacturers use color coding to indicate the maximum number of conductors allowed per connection, although the color coding scheme can vary by manufacturer.

Color	Wire Combinations
Gray	300 V Rating (#22 AWG to #16 AWG) MIN: Two #22 wires MAX: Two #16 wires
Blue	300 V Rating (#22 AWG to #14 AWG) MIN: Two #22 wires MAX: Three #16 wires
Orange	600 V Rating (#22 AWG to #14 AWG) MIN: One #18 wire with one #20 wire MAX: Four #16 wires with one #20 wire
Yellow	600 V Rating (#18 AWG to #12 AWG) MIN: Two #18 wires MAX: Four #14 wires with one #18 wire
Red	600 V Rating (#18 AWG to #10 AWG) MIN: Two #14 wires MAX: Two #14 wires with two #12 wires

Figure 7-3. Wire connectors are available in several sizes and can firmly hold several electrical wires together.

Wire Markers

A *wire marker* is a preprinted peel-off sticker designed to adhere to insulation when wrapped around a wire. Wire markers resist moisture, dirt, and oil and are used to identify wires that have the same color but different uses. For example, the two hot black wires (L1 and L2) of a home can each be marked with a different lettered or numbered wire marker. **See Figure 7-4.** Wire markers can also be used with different-colored conductors to further clarify the uses of all wires.

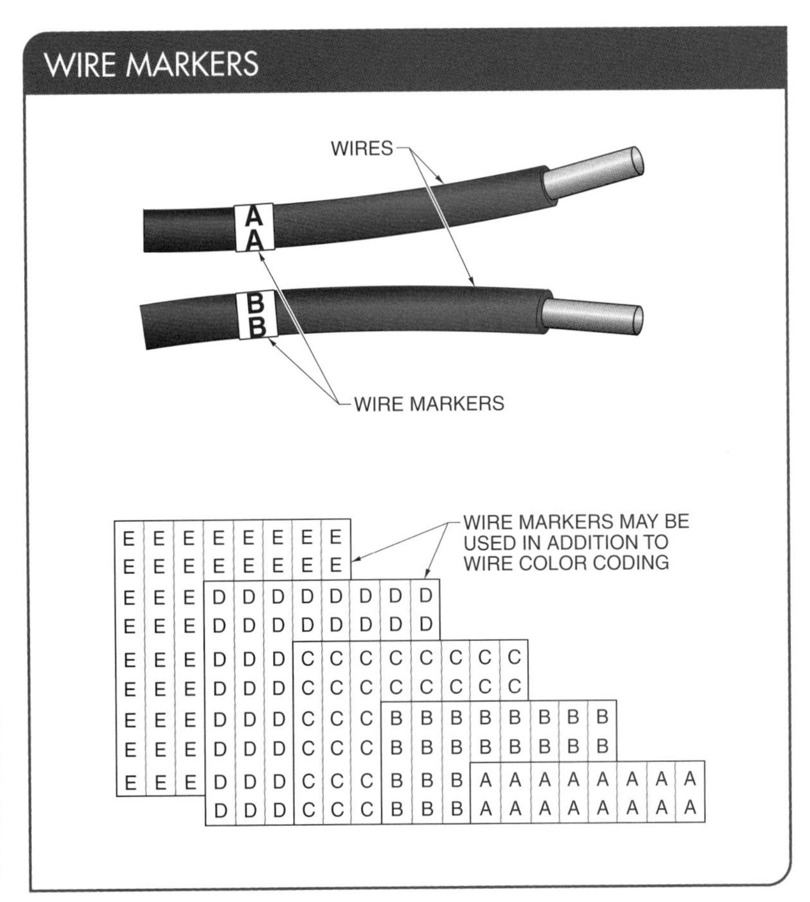

Figure 7-4. Wire markers are used to identify wires that have the same color but different uses.

SECTION 7.1 CHECKPOINT

1. How many volts can vinyl electrical tape insulate with one wrap?
2. What are some advantages of heat-shrink tubing?
3. How is the size of a wire connector determined?
4. What do wire markers resist?

SECTION 7.2—SPLICED CONNECTIONS

A *splice* is the joining of two or more electrical wires by mechanically twisting the wires together or by using a special splicing device. Individuals must be careful when making splices because splices can cause electrical problems. Splices must be able to withstand any reasonable mechanical strain that might be placed on the connection. Splices must also allow electricity to pass through the connection as if the connection were one wire. Many types of wire splices can be found in a residence. Typical residential wire splices include pigtail, Western Union, T-tap, portable cord, and cable splices.

Pigtail Splices

A *pigtail splice* is a type of splice that is used to connect the ends of conductors (wires). Pigtail splices are the simplest electrical splice to make, which makes them commonly used in residential wiring. **See Figure 7-5.** When two wires are joined by a pigtail splice, both wires must be twisted together securely. Twisting the wires together ensures that all the wires are properly fastened. There are two ways to end pigtail splices: bent over and cut off. When a pigtail splice is insulated with electrical tape, the ends must be bent over so that the sharp wire points do not penetrate the tape. When wire connectors are used instead of electrical tape, the ends of the wires are cut off.

CODE CONNECT

Per NEC® Subsection 110.14(B), all splices shall be covered with an insulation equivalent to that of the conductors or with an identified insulating device.

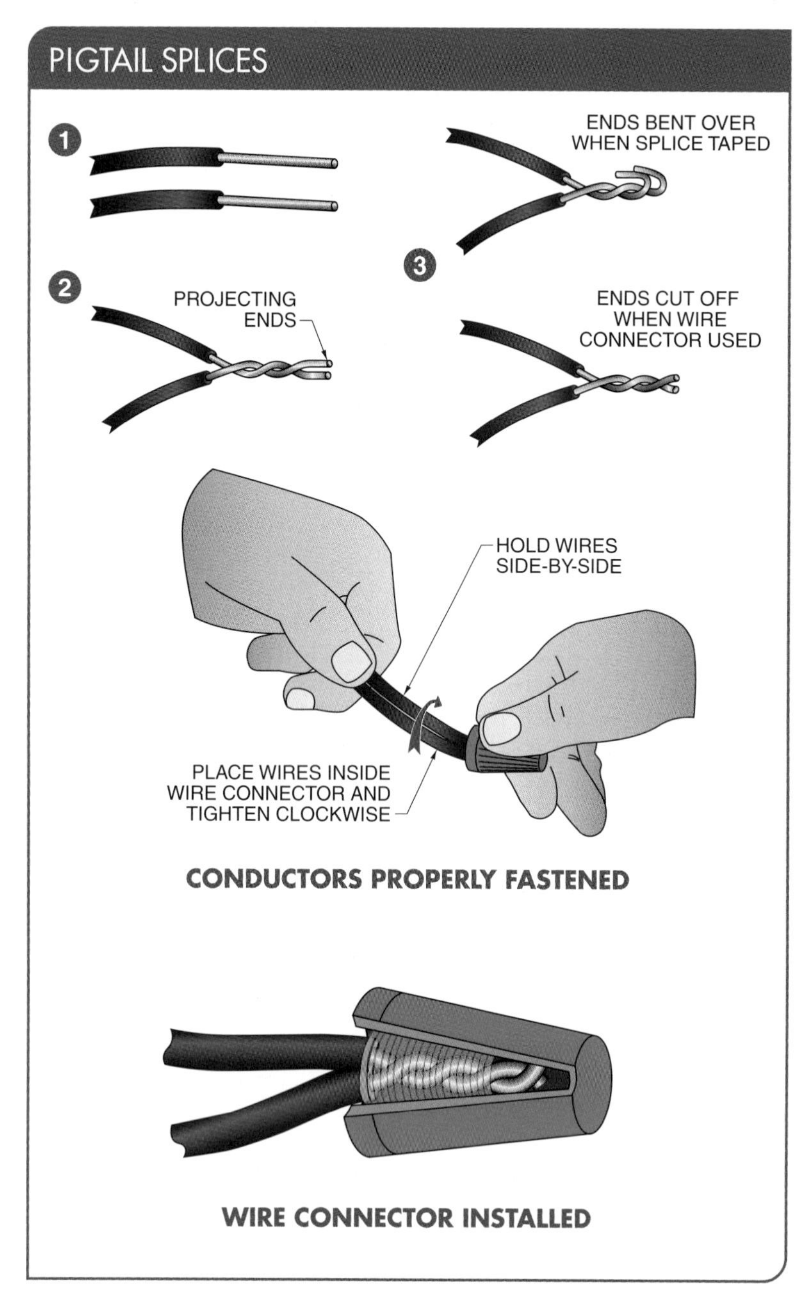

Figure 7-5. Pigtail splices are secured together by twisting the wires together and using either tape or a wire connector to end the connection.

Western Union Splices

A *Western Union splice* is a type of splice that is used when the connection must be strong enough to support long lengths of heavy wire. **See Figure 7-6.** A Western Union splice is made by positioning the wires and then wrapping them around each other. The wire ends are then tightly wrapped on each end of the splice. Before a Western Union splice is insulated with electrical tape, care must be taken to cut any sharp edges from the wire ends. Cutting the sharp edges prevents them from penetrating the tape and exposing the wires.

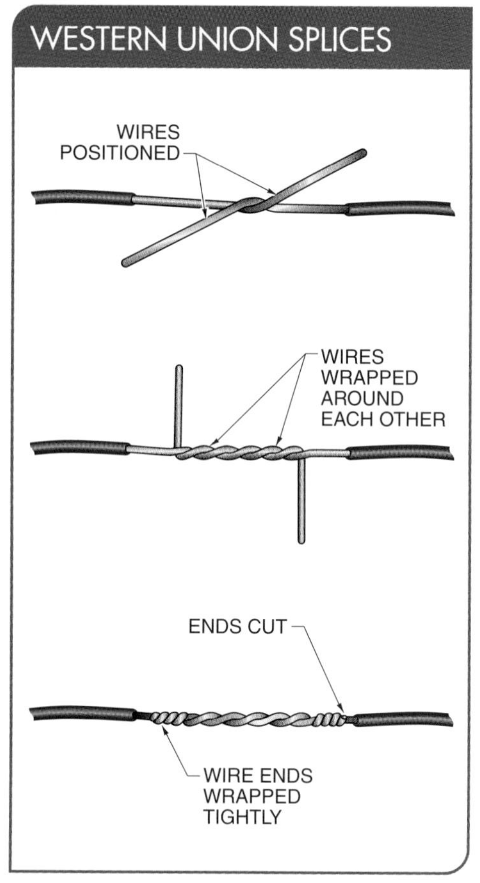

Figure 7-6. Western Union splices are used when a strong connection must be made with thicker wire.

T-Tap Splices

A *T-tap splice* is a type of splice that allows a connection to be made without cutting the main wire. **See Figure 7-7.** For a T-tap splice, a cross section of insulation is stripped from the main wire. The splicing wire making the T-tap is positioned and then wrapped several times around the main wire. Once the splicing wire is wrapped securely, its end is cut off. A T-tap splice is one of the most difficult splices to perform correctly. A good technique and practice is required to ensure proper T-tap splicing.

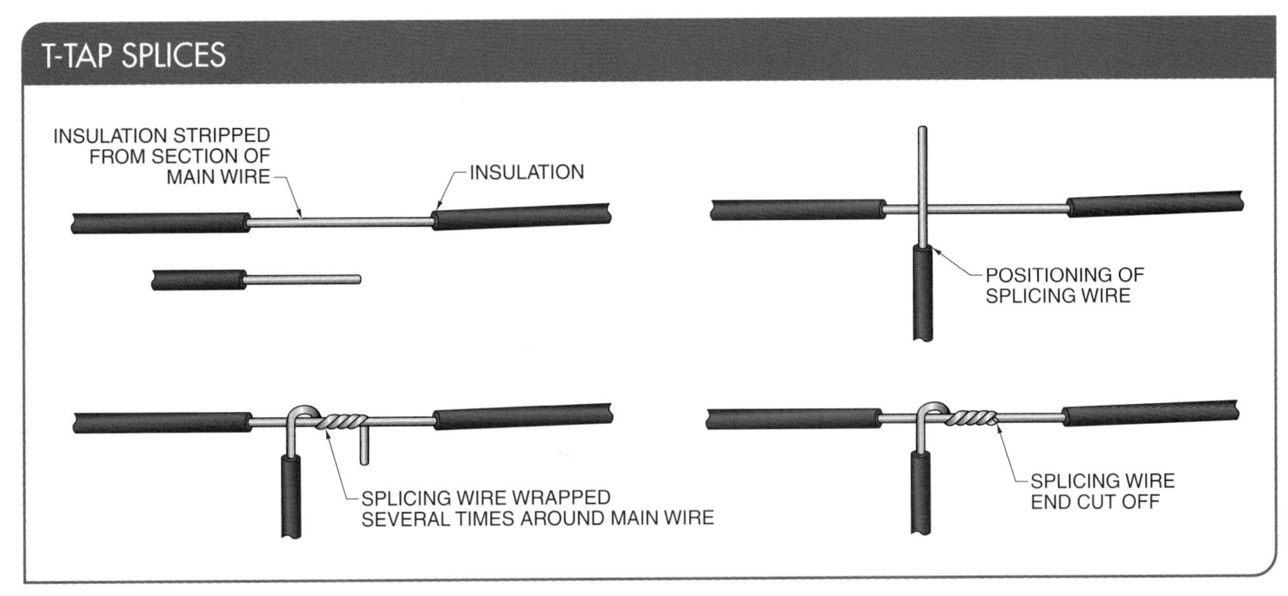

Figure 7-7. T-tap splices are used to make connections without cutting the main wire.

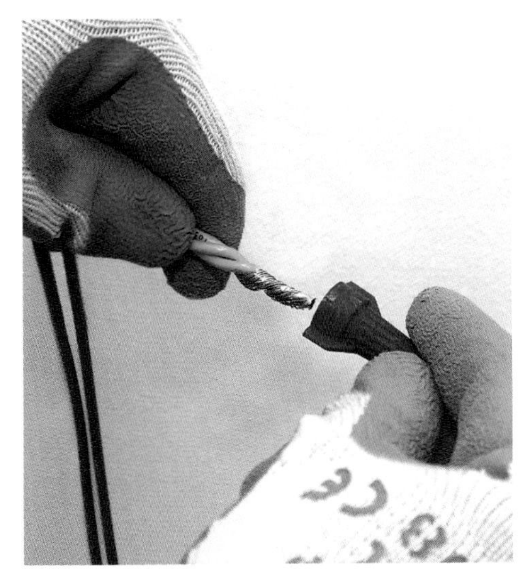

Three grounding conductors are secured together with a pigtail splice, and a red wire connector is used to terminate the ground connection.

Portable Cord Splices

Portable cord splices are a weak type of splice because there is no connector to hold the wires together. **See Figure 7-8.** Portable cords with stranded wires or solid wires can be spliced if the wires are No. 14 AWG or larger. Pigtail or Western Union splices are used to make portable cord splices with two or three conductors. Electrical tape or heat-shrink tubing can be used to cover and insulate the portable cord wires.

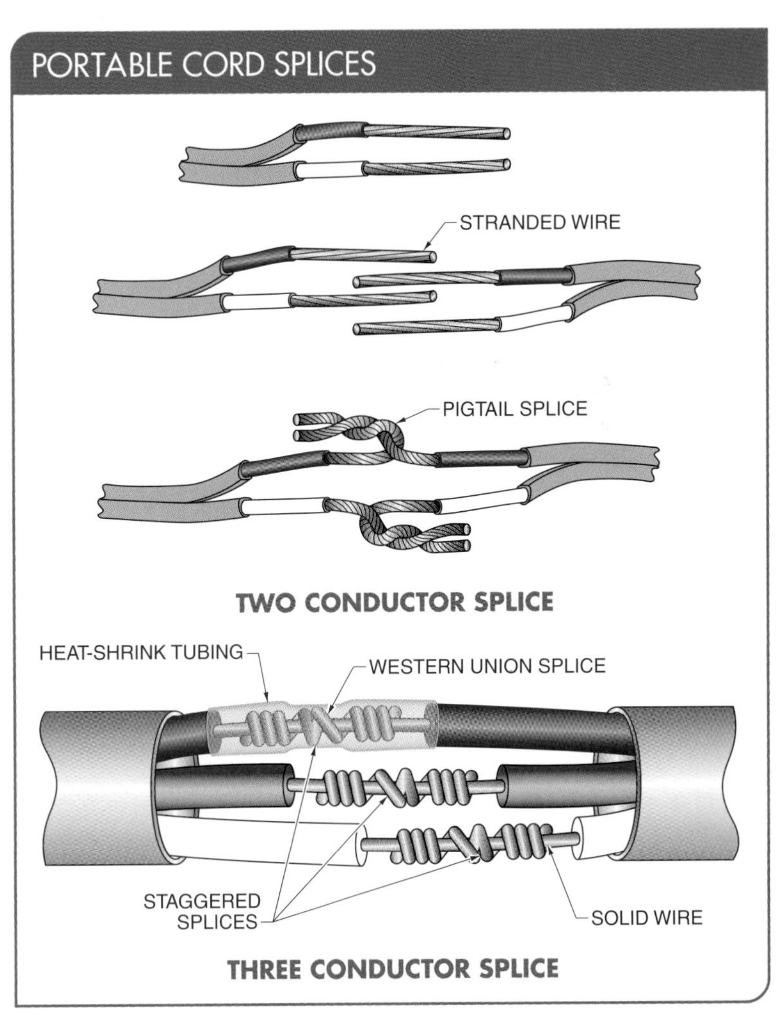

Figure 7-8. Portable cords can be spliced if the wires are 14 AWG or larger.

Cable Splices

Larger stranded cables are not often used in residential wiring. However, cable splices are used in other applications, such as for battery jumper cables and welding cables. When jumper cables or welding cables are broken, the cables can be temporarily repaired through cable splicing. **See Figure 7-9.** To perform a cable splice, the cable wires are separated and then both sides are meshed by alternating each wire. After the sides are meshed together, each individual wire is wrapped around the opposing cable.

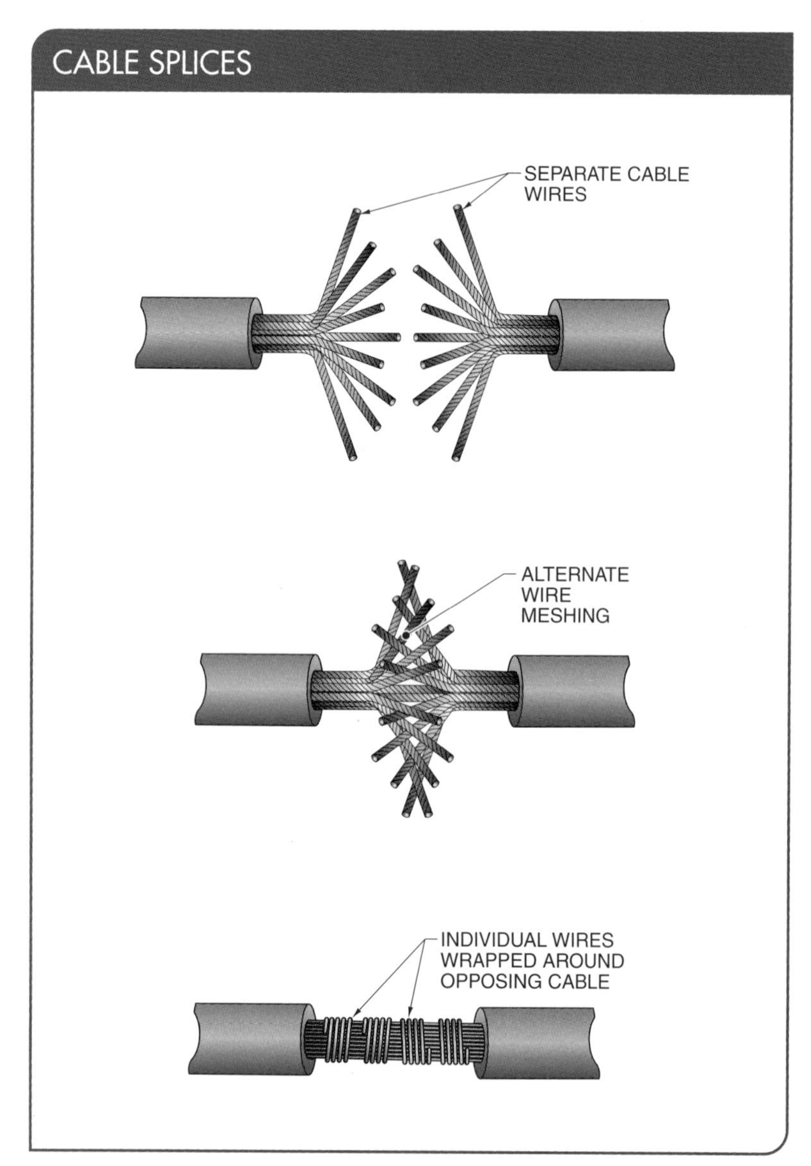

Figure 7-9. Large stranded cables must be carefully spliced and firmly secured using cable splicing procedures.

Removing Insulation from Larger Wires

A wire stripper is the preferred method of removing insulation from the end of most wire. However, when insulation must be removed from a large wire or when tapping into a section of a wire, a wire stripper cannot be used. In such cases, an electrician's knife is used to remove the insulation. **See Figure 7-10.** To remove insulation using an electrician's knife, the following procedure is applied:

1. Cut the insulation at an angle to prevent cutting into the wire with the knife. A nick in the wire weakens it and may cause the wire to break when it is bent. Never make a circling cut at right angles to the wire. Keep fingers away from the blade and always cut away from the body.
2. Cut both sides, leaving a tapered cut on the insulation.

Taping Splices

Splices must be taped in order to protect the wire from oxidation (corrosion) and to insulate the wire to protect individuals from electrical shock. Taping must provide at least as much insulation for the splice as the original insulation. Although one wrap of electrical tape provides insulation protection up to 600 V, several wraps are required to provide a strong insulated connection. **See Figure 7-11.** When electrical tape is used, the tape is stretched as it is applied. Stretching secures the tape more firmly around the electrical connection.

SECTION 7.2 CHECKPOINT

1. What is a splice?
2. Which splice is the most difficult to perform correctly?
3. Portable cords can only be spliced with what wire gauge and larger?
4. What is used to remove insulation from large wires?

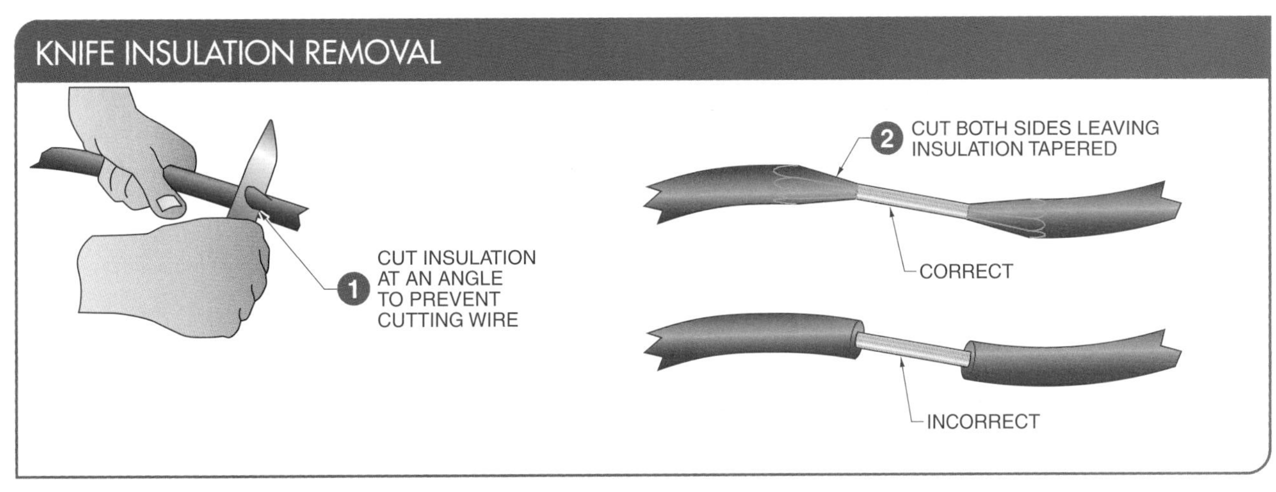

Figure 7-10. An electrician's knife is used to remove insulation from a large wire or tapping into a section of wire.

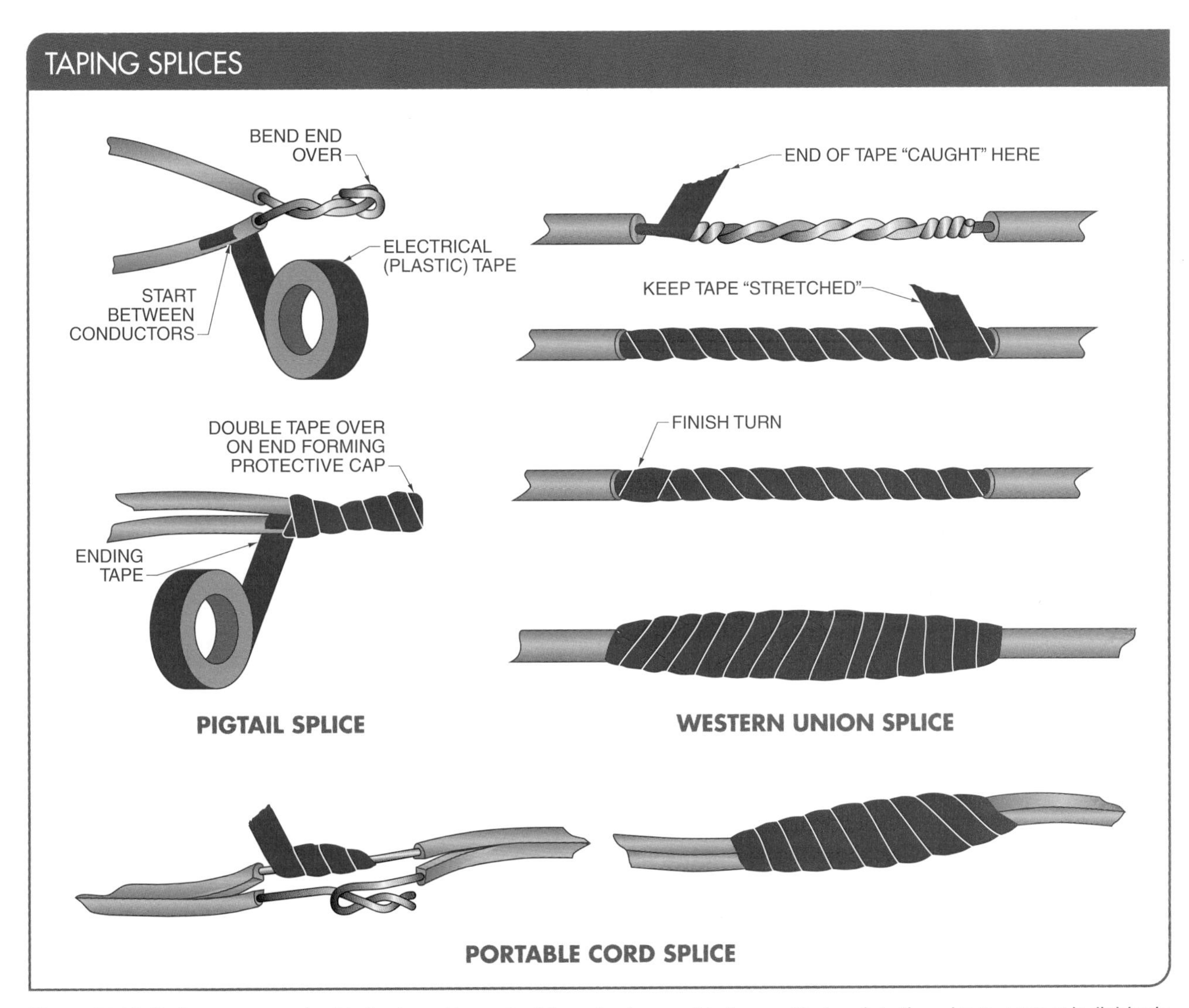

Figure 7-11. Splices are required to be taped to protect the wire from oxidation and to insulate the wire to protect individuals from electrical shock.

SECTION 7.3— SOLDERED CONNECTIONS

Soldering is a process of joining a base metal with a filler metal that has a melting point below that of the base metal. *Solder* is an alloy consisting of specific percentages of two or more metals. **See Figure 7-12.** In electrical work, solder usually consists of tin (Sn) and lead (Pb) or a tin-based, lead-free (Pb-free) alloy. A good solder joint provides a strong electrical and mechanical connection between the two metals.

SOLDER

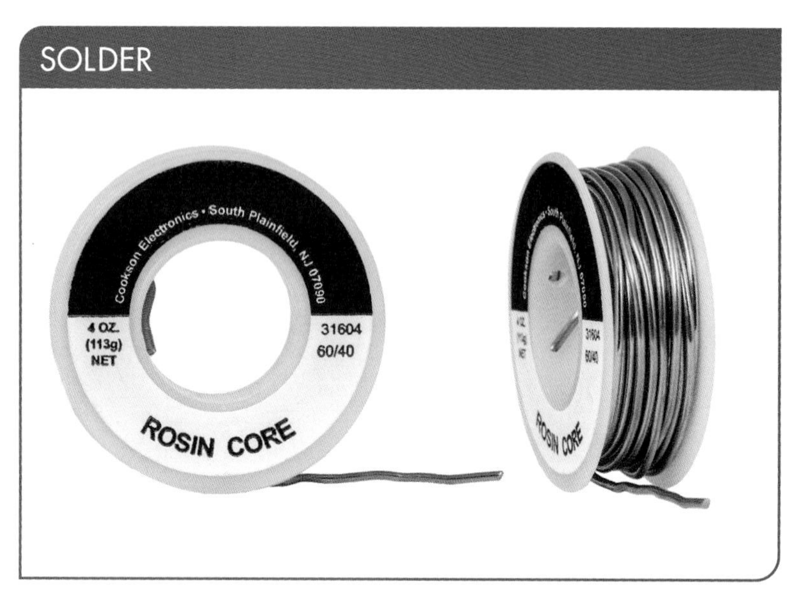

Figure 7-12. Solder is an alloy that consists of specific percentages of metals, usually tin (Sn) and lead (Pb).

The most common type of solder used for electrical connections is lead-based solder wire. Two types of lead-based solder wire are 63/37 solder wire and 60/40 solder wire. These solder wires consist of 63% tin to 37% lead and 60% tin to 40% lead, respectively. A 63/37 solder wire is a eutectic alloy. A *eutectic alloy* is an alloy that has one specific melting temperature with no intermediate stage. In other words, 63/37 solder wire has no plastic or elastic stage between its solid and liquid state. These lead-based solder wires have low melting temperatures. This prevents excessive temperature from being applied to the electrical connection and minimizes unreliable or cold solder joints. *Note:* Melting points of solder alloys are always much lower than the base metals that are being joined.

For a good solder joint, metal surfaces must be clean of surface dirt, oil, and oxides. Oxide is formed on metal surfaces by oxygen and moisture in the air. For example, when copper is exposed to air long enough, the oxide appears as a green tarnish. However, oxide can be removed by using flux during the soldering process. *Flux* is a chemical substance that cleans the soldering surface and promotes the melting of the solder wire. Flux removes oxide by making it soluble, which allows the oxide to evaporate due to the high heat. Flux remains on the surface of the metal during the soldering process to keep oxide from reforming.

CODE CONNECT

Per NEC® Subsection 110.14(B), soldered splices shall be mechanically and electrically secured before being soldered.

Soldering Irons

A soldering iron transfers thermal energy from a solder station (heater) through its tip to the solder joint. The soldering iron tip must heat the solder wire to its melting point before actual soldering can take place. **See Figure 7-13.** Once the solder wire is heated to its melting point, it first liquefies and then solidifies on the joint. The conductivity of a soldering iron tip determines how fast the thermal energy can be transferred from the solder station to the solder joint (connection). The shape and size of the soldering iron tip also affects the transfer of thermal energy.

The correct temperature is required for melting solder wire. However, a major factor in determining the correct temperature is the length and size of the soldering iron tip. The actual shape of the soldering iron tip establishes how well heat is transferred from the tip to the solder joint.

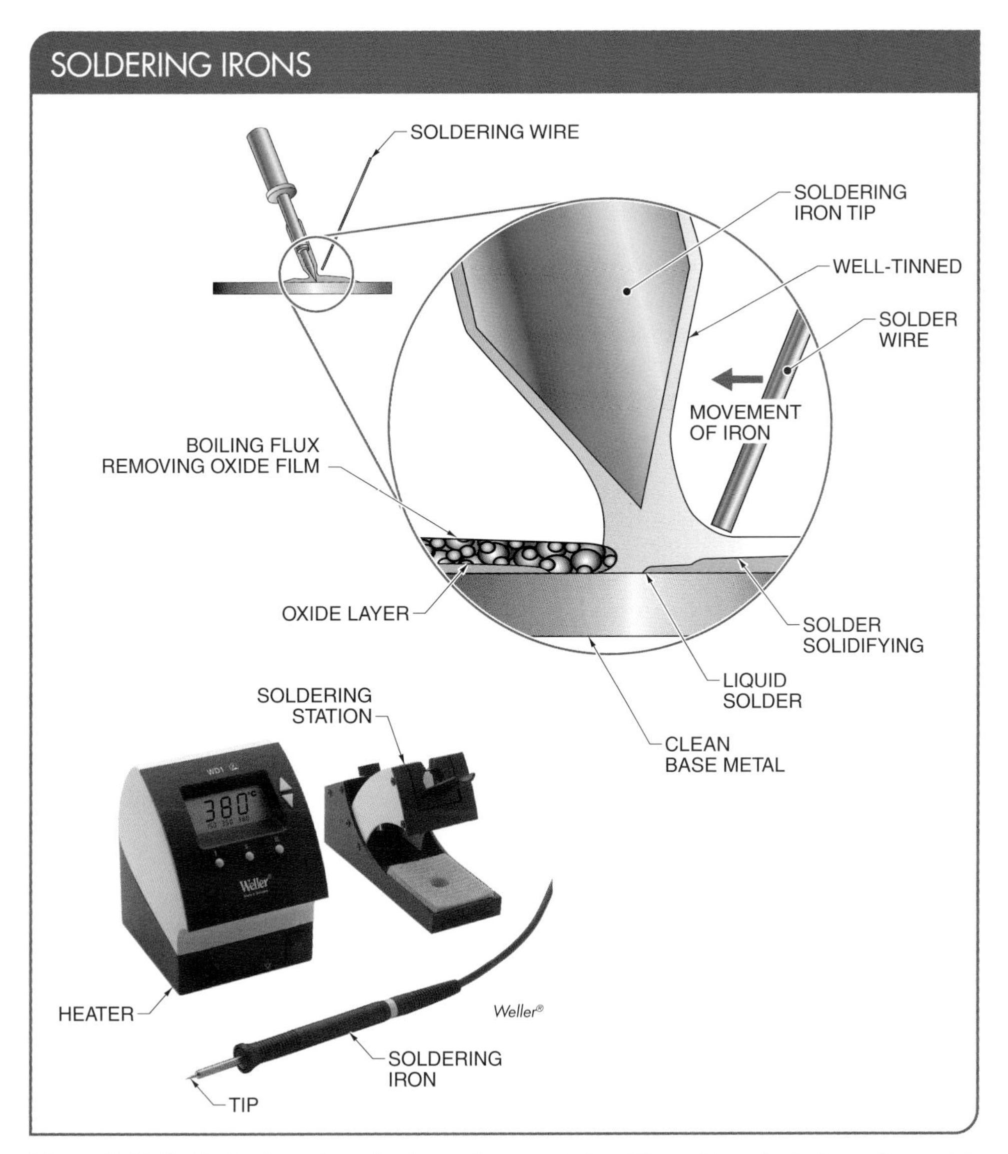

Figure 7-13. Soldering irons transfer thermal energy to heat the solder wire to its melting point.

The correct soldering iron tip used for an application should have a similar shape and dimension to the joint being soldered. For example, when a joint has a long contact area, the soldering iron tip used should have a flat, longer tip to match. Flat soldering iron tips have a large contact area that produces better joint connections than conical soldering iron tips. Flat soldering iron tips also tend to transfer heat more efficiently. A wide range of tip sizes and shapes, such as angled, chiseled, and conical, are available for use in a variety of soldering applications. **See Figure 7-14.**

U.S. Navy

Most soldering applications consist of soldering semiconductor components on circuit boards.

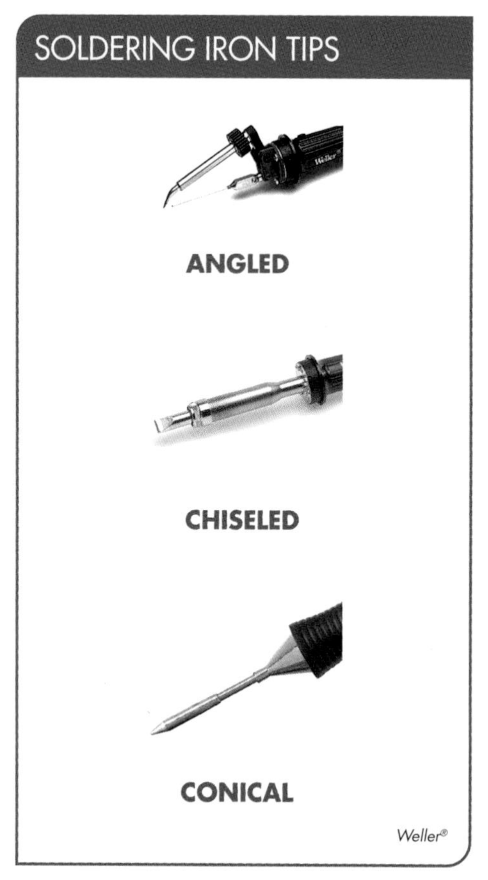

Figure 7-14. A wide range of soldering tip sizes and shapes, such as angled, chiseled, and conical, are available.

solder melts and covers the tip to act as a medium for heat transfer, allowing the solder to leave a continuous, permanent film on the joint. When the soldering iron tip does not provide sufficient heat, the solder will not be able to efficiently transfer to the joint and will only adhere to a tiny area of the joint.

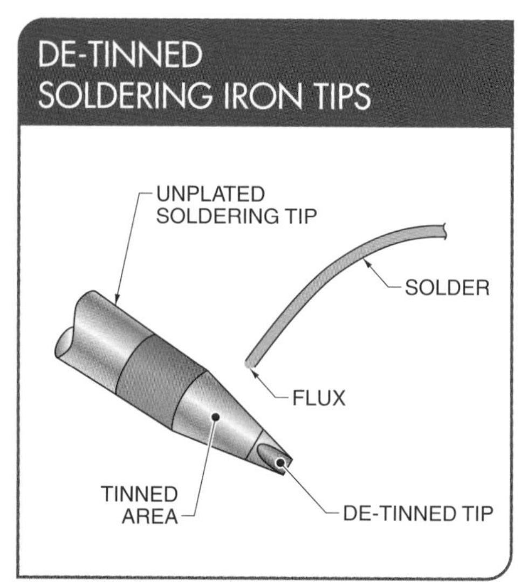

Figure 7-15. A de-tinned soldering iron tip minimizes the ability to accept solder and allow heat transfer.

Soldering Iron Tip Care

One of the most common causes of soldering iron tip failure is the loss of a protective layer of solder on the tip, which causes the soldering iron tip surface to become oxidized. This is commonly referred to as a de-tinned tip. A de-tinned tip minimizes the ability of the tip to accept solder and allow heat transfer. Without a properly tinned tip, the heat transfer is insufficient and the solder wire cannot be joined to the electrical connection. **See Figure 7-15.**

A de-tinned tip will also lose its ability to supply heat to the joint being soldered, which does not allow proper wetting. *Wetting* is the property of solder when it is molten. Wetting occurs when solder reaches its melting point. Wetting can only be achieved on a clean, nonoxidized tip surface. As the soldering iron tip heats,

Soldering Iron Tip Cleaning

All dirt and oil must be removed from the metal surface of a soldering iron tip. Using light abrasives and/or fluxes to remove these contaminants can produce highly durable solder joints and good wetting on soldering iron tip surfaces. Soldering iron tips can be cleaned with a sponge, tip tinner, or polishing bar. **See Figure 7-16.**

Sponges. Sponges can provide a sufficient surface area to allow oxidized contaminants to fall to the bottom of a sponge tray. The best method of cleaning a soldering iron tip is to dampen a sponge and rotate the soldering iron tip on the surface of the sponge. A couple of passes on the sponge surface should remove the contaminants. Synthetic sponges are not recommended for soldering iron tips because they may contain contaminants that could reduce the life of the tip.

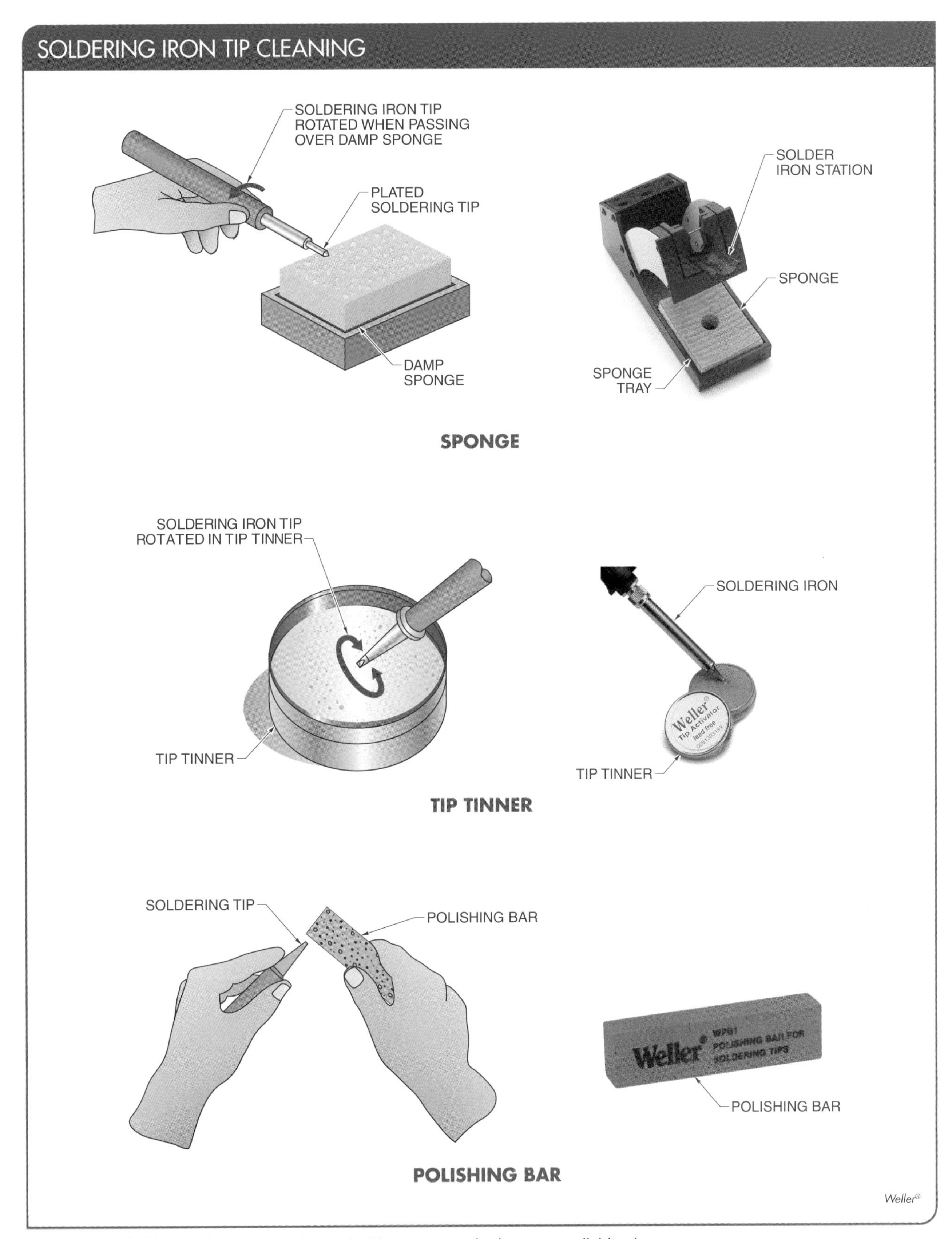

Figure 7-16. Soldering tips may be cleaned with a sponge, tip tinner, or polishing bar.

Tip Tinners. A tip tinner is used to remove light oxidation from soldering iron tips. Cleaning with a tip tinner is necessary when a sponge does not work to remove oxidation. A tip is rotated in the tip tinner until a bright tinning appears on the surface of the tip. Solder should be applied immediately to re-tin the surface of the tip. Tip tinners should not be overused because they will reduce the service life of a soldering iron tip.

Polishing Bars. When heavy oxidation cannot be removed by using a tip tinner, cleaning the tip may require the use of an abrasive polishing bar. The polishing bar is applied to the soldering iron when cool. Care must be taken not to remove the iron plating. Once cleaned, the tip should be re-tinned immediately.

SOLDERING DEVICES

Weller®

SOLDERING IRON

RIDGID®

SOLDERING GUN

PROPANE TORCH

Figure 7-17. Common soldering devices include soldering irons, soldering guns, and torches.

Soldering Splices

An electrical splice should be soldered as soon as the decision to solder has been made and the insulation has been stripped off the wire. The longer a metal wire is exposed to dirt and air, the greater the amount of oxidation will form on the wire and the less chance of achieving a properly soldered connection. Common soldering devices include soldering irons, soldering guns, and torches. **See Figure 7-17.**

Soldering irons and soldering guns are used when electricity is available. A propane torch or minitorch is used to solder larger wires or when there is no electricity at the job site. Whichever heating method is used, solder must be applied to the splice on the side opposite of where the heat is being applied. **See Figure 7-18.** Molten solder flows toward the source of heat. Thus, if the top of the wire is hot enough to melt the solder, the bottom of the wire closest to the heat source will draw the solder down through all the wires. The splice must be allowed to cool naturally without movement. Once cooled, the splice is cleaned of any excess flux with a damp rag and then taped.

Omni Training

Organized soldering workstations can be used to improve work efficiency and save time when soldering.

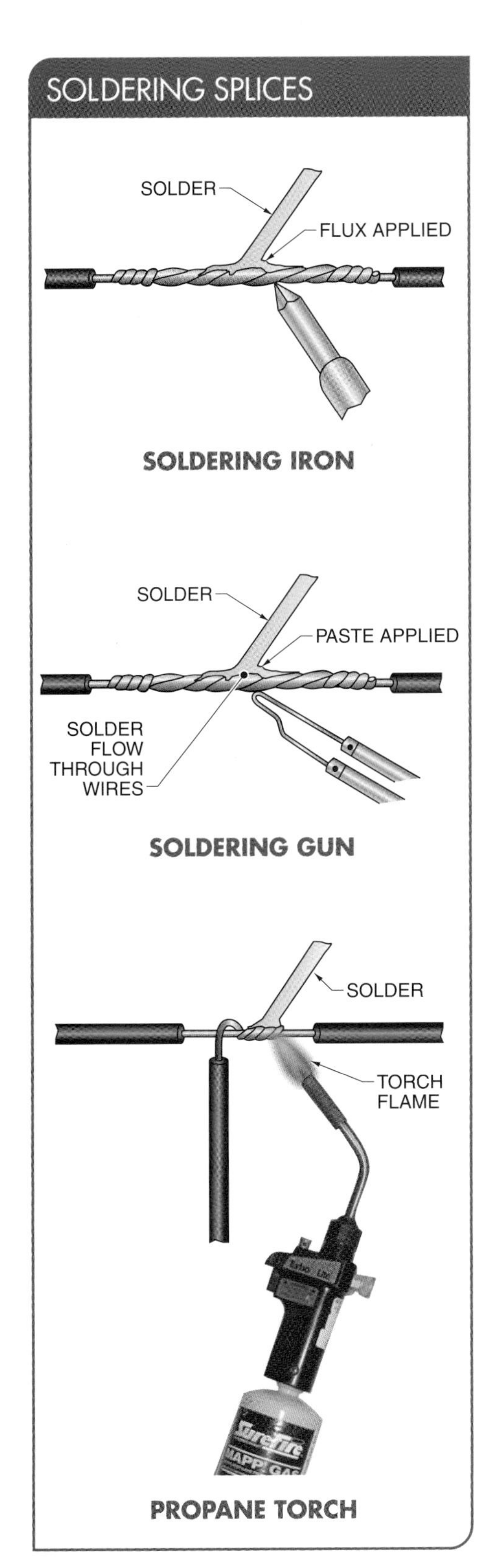

Figure 7-18. Solder must be applied to the opposite of where the heat is applied.

Soldering on Printed Circuit Boards

Advanced soldering skills are essential for repairing printed circuit (PC) boards. When PC boards are repaired, the soldering iron tip should be held against one side of the joint connection on a component lead while the solder wire is applied to the other side. **See Figure 7-19.** Once the solder flows onto the joint, first the solder is removed and finally the soldering iron tip is removed. The joint should have a smooth and shiny finish.

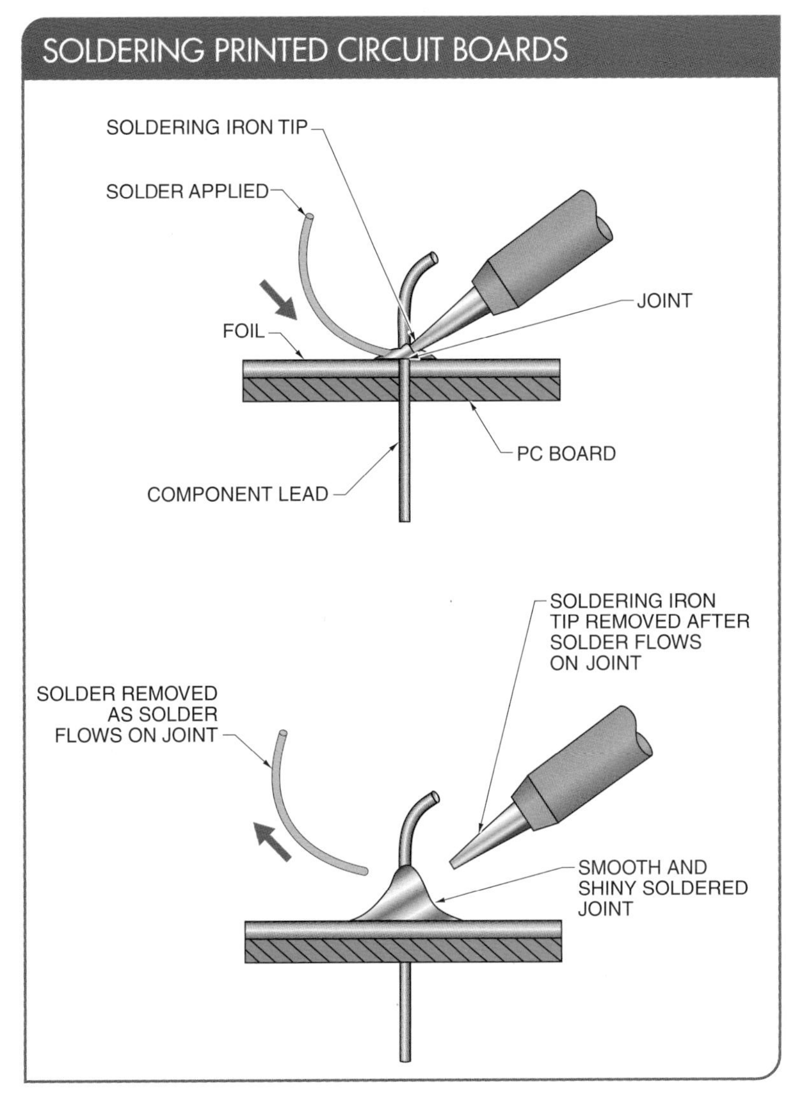

Figure 7-19. When PC boards are repaired, the soldering iron tip should be held against one side of the joint connection while the solder wire is applied to the other side.

Care must be taken to prevent damaging the surrounding parts by overheating them. Semiconductor components such as transistors, integrated circuits (ICs), and resistors are very sensitive to heat. A heat sink, such as an alligator clip, is used to help prevent heat damage. **See Figure 7-20.** The heat sink is placed between the joint being soldered and the component that requires protection. The heat sink absorbs most of the heat produced during soldering.

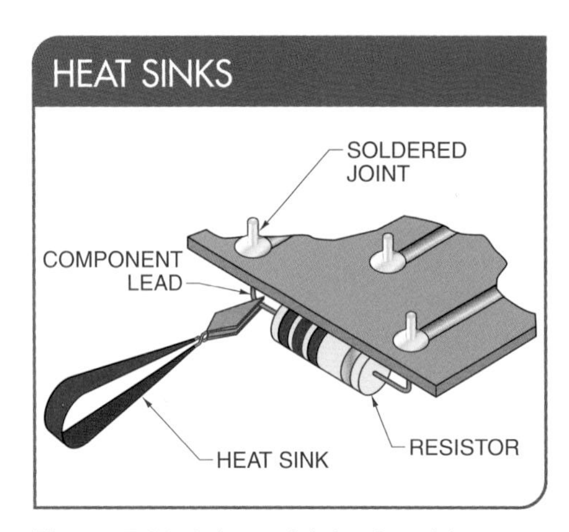

Figure 7-20. A heat sink is placed between the soldered joint and the component that requires protection.

Preventing Bad Solder Joints

Bad solder joints result in improper electrical connections of components or wires and can lead to damaged components or residential circuits. Any joint that exhibits a high resistance will adversely affect the operation of the circuit. Therefore, every effort should be made to prevent bad solder joints. To prevent bad solder joints, the following conditions should be avoided:

- adding too little solder on a joint, which produces a weak joint and causes a component to work loose over time
- adding too much solder on a joint, which could bridge component leads and cause unwanted conducting paths
- providing too little heat while soldering, which produces a cold solder joint and results in a poor joint
- providing too much heat while soldering, which may damage a component
- providing no flux or too little flux while soldering, which does not allow all contaminants to be removed and can result in a poor electrical joint

Solder joints that are overheated generally have a dull, bumpy exterior. A good solder joint is smooth and shiny. **See Figure 7-21.** Overheated joints are often mistakenly called "cold solder joints." A *cold solder joint* is a defective solder joint that results when the parts being joined do not exceed the liquid temperature of the solder wire. A cold solder joint can be identified by jagged shapes on the surface of the joint, which indicate that the solder did not flow properly within the joint connection. Overheated and cold solder joints indicate a lack of proper wetting techniques when solder was applied.

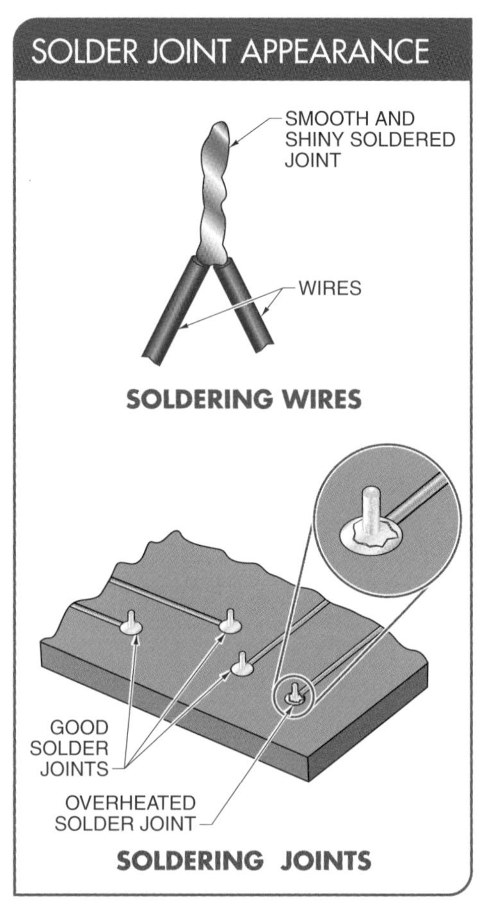

Figure 7-21. A good solder joint is smooth and shiny, while overheated joints have a dull, bumpy exterior.

Soldering Safety

It is necessary to follow safe working practices when soldering. For safe soldering, the following rules must be applied:

- Always return a soldering iron to its stand when not in use and never set it down on a workbench.
- Solder in a well-ventilated area. The smoke formed while soldering is mostly created from the flux and can be toxic or irritating. Keep the head to the side of the soldering work to prevent inhalation.
- Always wash hands after handling solder wire. Solder can contain lead, which is a toxic metal.
- Always wear safety glasses to protect the eyes from splatter.

SECTION 7.3 CHECKPOINT

1. What is the most common type of solder used for electrical connections?
2. What are the three abrasives used to clean soldering iron tips?
3. What soldering device is used to solder larger wires when electricity is not available?
4. What device is used to help prevent heat damage to semiconductor components?

SECTION 7.4—CONDUCTOR CONNECTORS

A *solderless connector* is a device used to firmly join wires without the help of solder. Because solderless connectors are convenient and save time, several types have been developed. Some of the most commonly used solderless connectors approved by UL® (formerly Underwriters Laboratories) are wire connectors, screw terminals, backwired (quick) connectors, crimp-type connectors, and split-bolt connectors.

Wire Connector Installations

Wire connectors have almost eliminated the need for soldering and taping electrical connections. Wire connectors are manufactured in a variety of sizes and shapes to fit most wire applications. For the most effective use of wire connectors, the wires are twisted together clockwise before the wire connector is twisted on and taped. **See Figure 7-22.** When solid and stranded wires are joined together, the stranded wire is twisted around the solid wire and the solid wire is bent back over the stranded wire. Finally, a wire connector is installed.

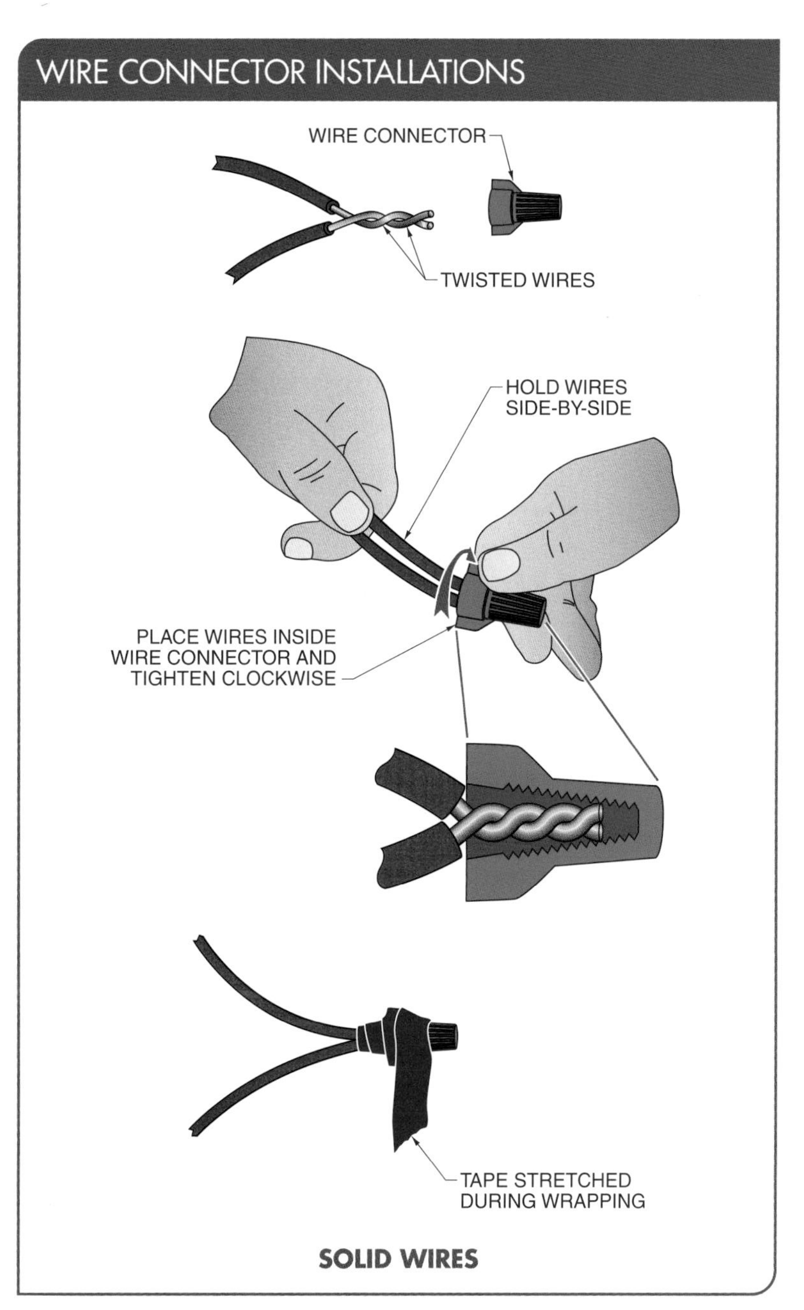

Figure 7-22. Wires are twisted together clockwise before the wire connector or tape is applied.

Wire connector sizes are determined by the size and number of wires to be connected. Wire connectors are commonly used to connect AWG sizes No. 22 through No. 8. The shape of a wire connector is an individual preference and is determined by the manufacturer. To ensure safe connections, every wire connector is rated for minimum and maximum wire capacity. Wire connectors are used to connect both conducting wires and grounding wires. Green wire connectors are used only for grounding wires.

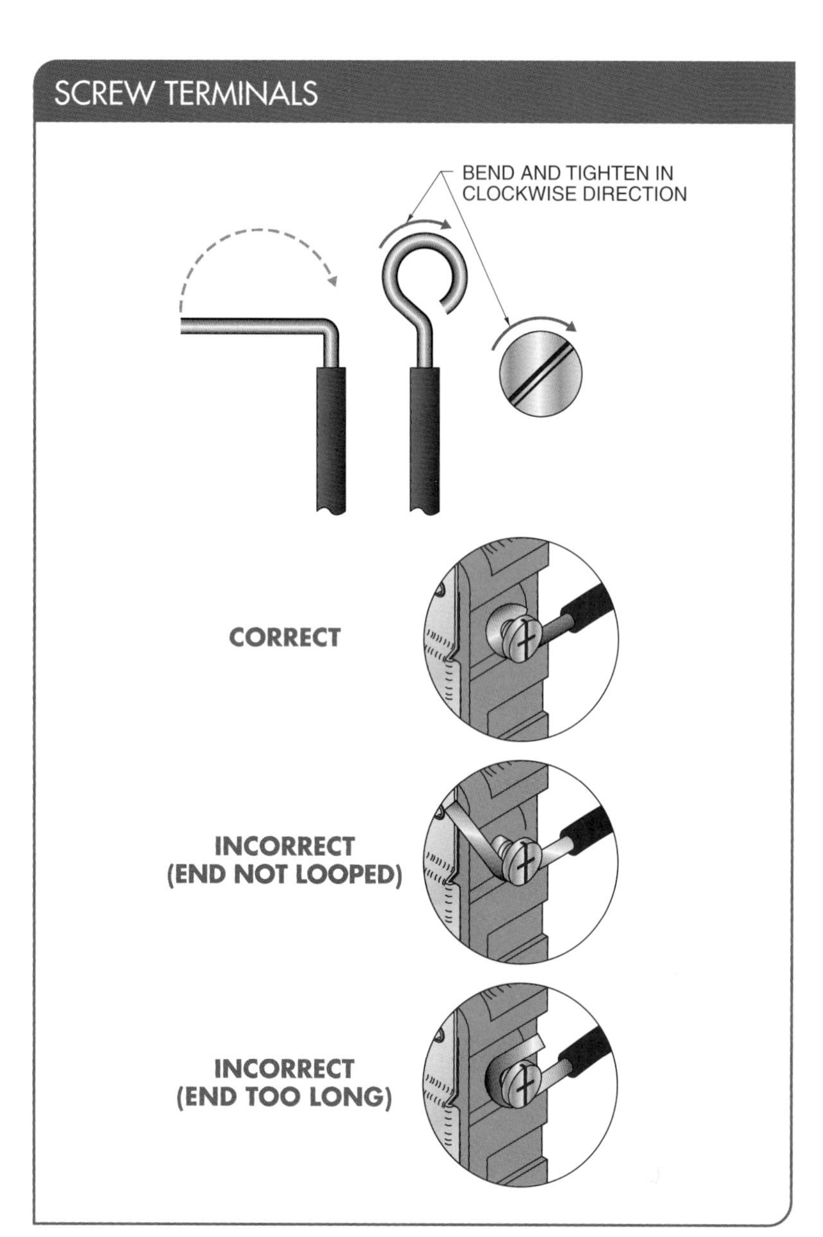

Figure 7-23. Wire should be bent around screws in a clockwise direction.

Screw Terminals

Screw terminals provide a mechanically and electrically secure connection. Since wiring is always attached to electrical equipment with right-hand screws, the wire should be bent around a screw in a clockwise direction. **See Figure 7-23.** When screw terminals are used, the screw draws the wire tight without pushing the wire away from the terminal.

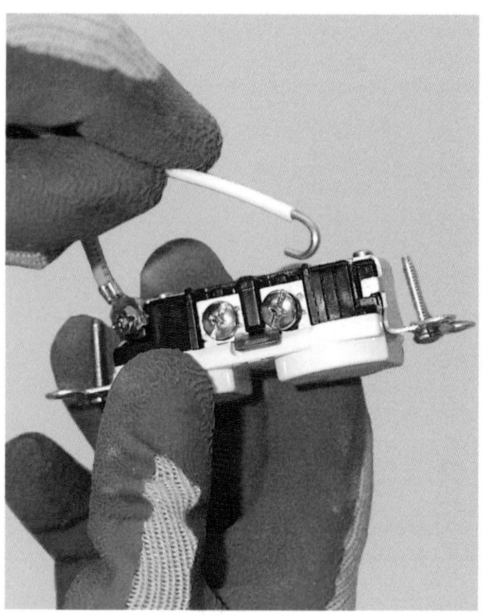

A neutral wire is bent in a clockwise direction to fit the neutral-side screw terminal of a standard receptacle.

Back-Wired (Quick) Connectors

A *back-wired connector,* also known as a quick connector, is a mechanical connection method used to secure wires to the backs of switches and receptacles. The wires are held in place by either spring tension or screw tension. Screw tension is the most secure. **See Figure 7-24.** To remove a wire from a spring-type back-wired connector, a screwdriver or a stiff piece of wire is inserted into the spring opening next to the connection. Pressing down on the spring through the opening releases the wire. Screw-type back-wired connectors release the wire as the screw is loosened.

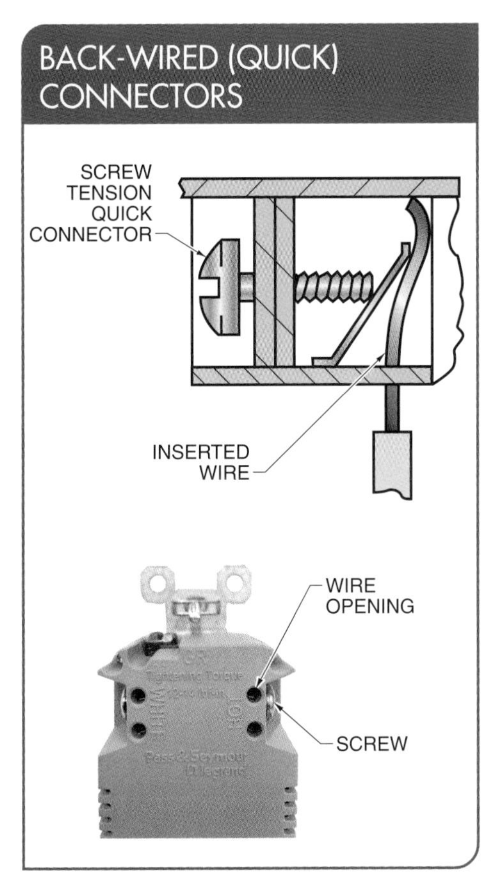

Figure 7-24. Back-wired (quick) connectors secure wires to the backs of switches and receptacles.

CODE CONNECT

Per NEC® Subsection 110.14(A), connections of conductors to terminal parts shall be made by means of pressure (crimp-type) connectors.

Crimp-Type Connectors

A *crimp-type connector* is an electrical device that is used to join wires together or to serve as terminal ends for screw connections. **See Figure 7-25.** Crimp-type connectors are manufactured as insulated or noninsulated. Noninsulated crimp-type connectors are less expensive and are used where there is no danger of shorting the connector to a metal surface. Identification can be a problem when working with crimp-type connectors. To avoid confusion, a wire marker is used to identify each wire.

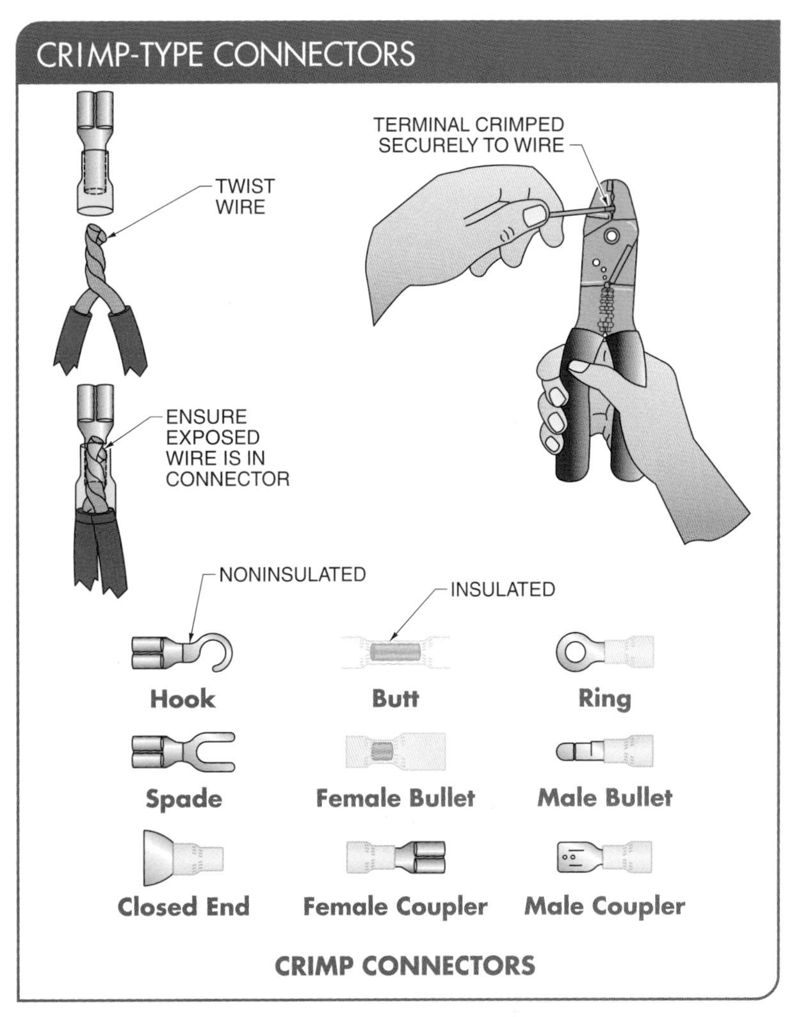

Figure 7-25. Crimp-type connectors are manufactured as insulated or noninsulated and are used where there is no danger of shorting to a metal surface.

Crimp-type connectors are secured into place with a crimp tool or a multipurpose wire stripper. **See Figure 7-26.** In order to install a crimp connector, the following procedure is applied:

1. Insert the correct wire into the correct wire gauge hole size.
2. Squeeze the wire stripper's handles and rotate it around the wire until the insulation is cut.
3. Strip the insulation off of the end of the wire.
4. Twist both wires together.
5. Ensure that the exposed wire is in the crimp connector.
6. Use the wire stripper to crimp the connector securely to the twisted wires.

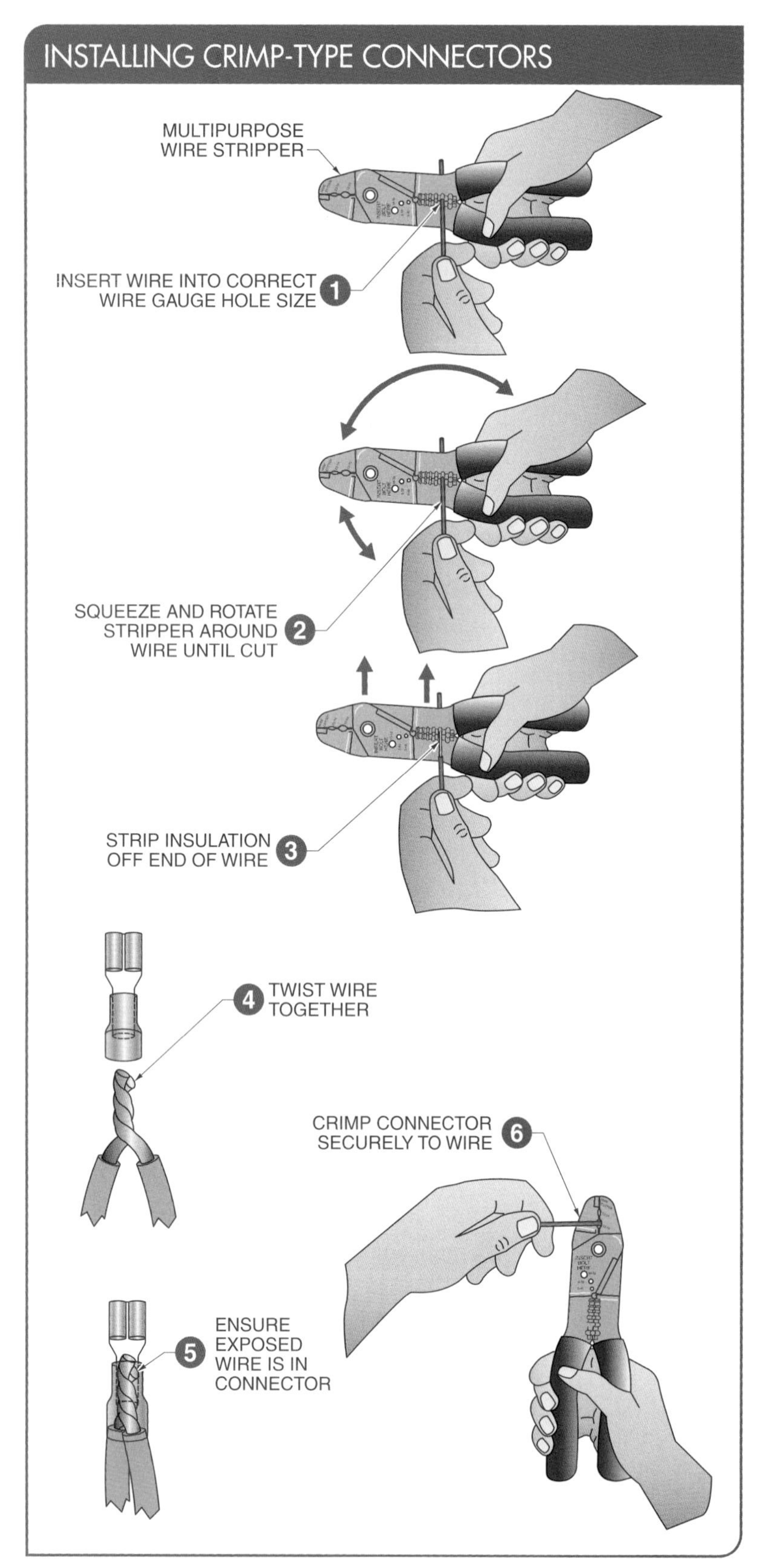

Figure 7-26. Crimp-type connectors are secured into place with a crimp tool or a multipurpose wire stripper.

TECH TIP

Warming a conductor will allow for easier insulation removal.

Split-Bolt Connectors

A *split-bolt connector* is a solderless mechanical connector used for joining large cables. For example, split bolt connectors are used for the large cables in service entrances. A split bolt is slipped over the large wires to be connected so that a tightening nut can be attached. Split-bolt connectors can be used for a tap splice or to splice two large wires. Split-bolt connectors must be made of the same material as the wires to prevent corrosion. **See Figure 7-27.**

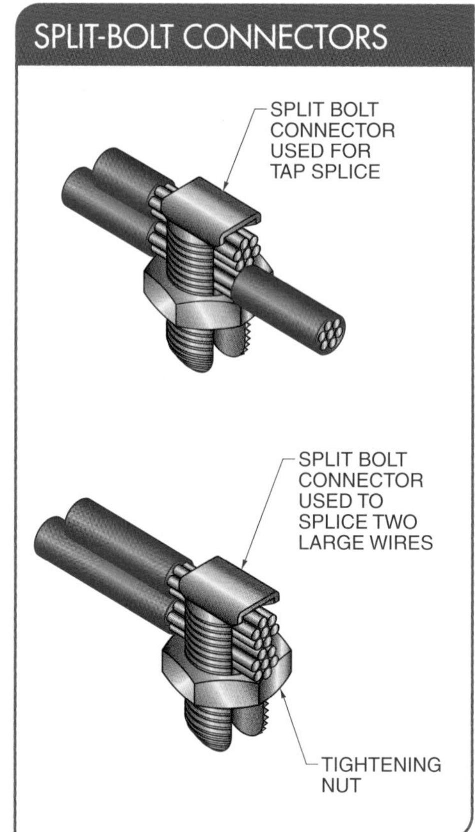

Figure 7-27. Split-bolt connectors join large cables, such as those in service entrances.

SECTION 7.4 CHECKPOINT

1. What is a solderless connector?
2. How should a wire be connected to a screw terminal?
3. Crimp-type connectors are manufactured in what two ways?
4. What are split-bolt connectors used for?

SECTION 7.5—REPLACING AND TESTING PLUGS

A *plug* is a device at the end of a cord that connects equipment to an electrical power supply by means of a receptacle. Plugs sometimes crack, break, or lose their electrical connection when the device is disconnected from the receptacle by pulling on the cord instead of the plug. A plug should always be replaced, never repaired, if it becomes damaged. If the cord is also broken or damaged, both the cord and plug should be replaced. Plugs that are replaced due to damage typically include grounded plugs, ungrounded plugs, heavy-duty plugs, and large appliance plugs.

Grounded Plugs

Grounded plugs should always be used if included with the original equipment. If a grounded plug is not used or the ground wire is not connected, the load will work, but the circuit could be a potential hazard. A damaged grounded plug should always be replaced with a new grounded plug. **See Figure 7-28.** To replace a grounded plug, first the old plug is removed, and then the following procedure is applied:

1. Insert the cord into the plug cover and pull about 8″ of cord through the plug.
2. Cut about 2″ of the plastic sheathing from the cord. Peel back and cut off any paper wrapping. Be careful not to cut into the insulation around the individual wires.
3. Strip about ¾″ insulation from the wire.
4. Insert wires into the plug's wire pockets. The hot terminal is identified by a bronze screw, the neutral by a silver screw, and the ground terminal by a green screw.
5. Tighten the terminal screws. Care must be taken to ensure that no strands are exposed because most cords use stranded wire. Solder the stranded ends if required.
6. Insert the plug into the cover, and then tighten the assembly screws.

Proper PPE must be worn when replacing or testing any electrical equipment such as receptacles, switches, or corded plugs, even when the electrical circuit is not energized.

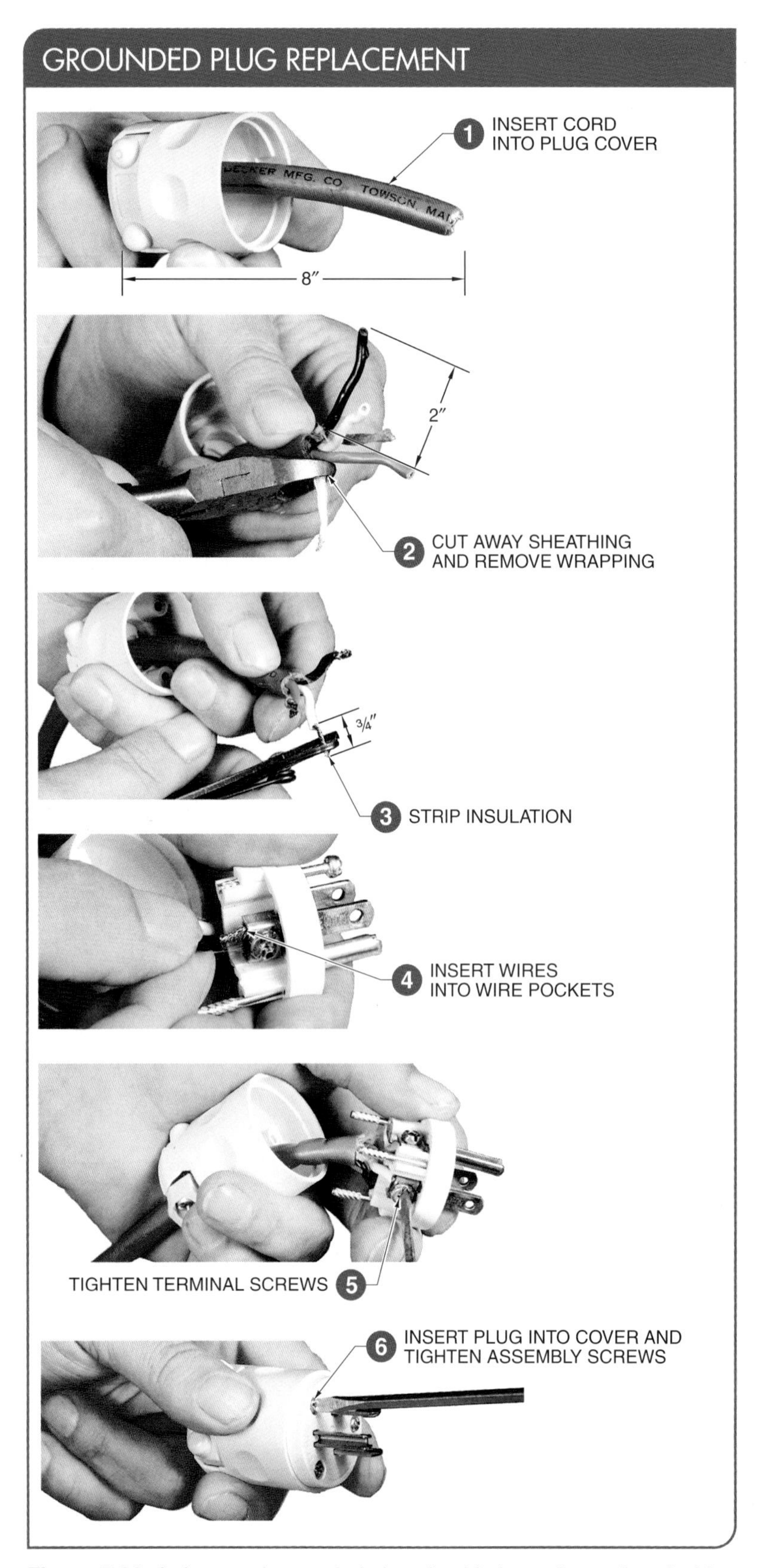

Figure 7-28. A damaged grounded plug should always be replaced with a new grounded plug.

CODE CONNECT

Per NEC® Subsection 300.3(B), all conductors of the same circuit shall be contained within the same cord.

Ungrounded Plugs

Not all electrical equipment uses grounded plugs. However, ungrounded equipment still has the potential to cause electrical shock or electrical fires. If available, ungrounded equipment should always be plugged into receptacles with grounding means or a GFCI receptacle for protection. **See Figure 7-29.** To replace an ungrounded plug, first the old plug is removed, and then the following procedure is applied:

1. Insert the wire into the plug. Pull about 6″ of wire through the plug.
2. Strip about ¾″ insulation from the wire.
3. Tie an underwriter's knot in the two wires. An underwriter's knot is used to relieve the strain on the terminals when the plug is pulled out by the cord.
4. Pull the underwriter's knot tight. The knot should be large enough so that it cannot be pulled back through the plug. If the plug is too large and the knot can be pulled through the plug, a double knot may be required.
5. Wrap the wires around the terminal screws. Ensure the hot (black) wire is connected to the smaller blade if the plug is a polarized plug. A *polarized plug* is a plug in which one blade is wider than the other blade. The wider blade is the neutral terminal and is designed to fit only into the neutral side of a polarized receptacle. The hot terminal is normally identified by a bronze-colored screw and the neutral by a silver-colored screw.
6. Tighten the terminal screws. Care must be taken to ensure no strands of wire are exposed on cords that use stranded wire. Solder the stranded ends if required.
7. Attach the insulated cover that comes with the plug. Most plugs include a screw-on cover.

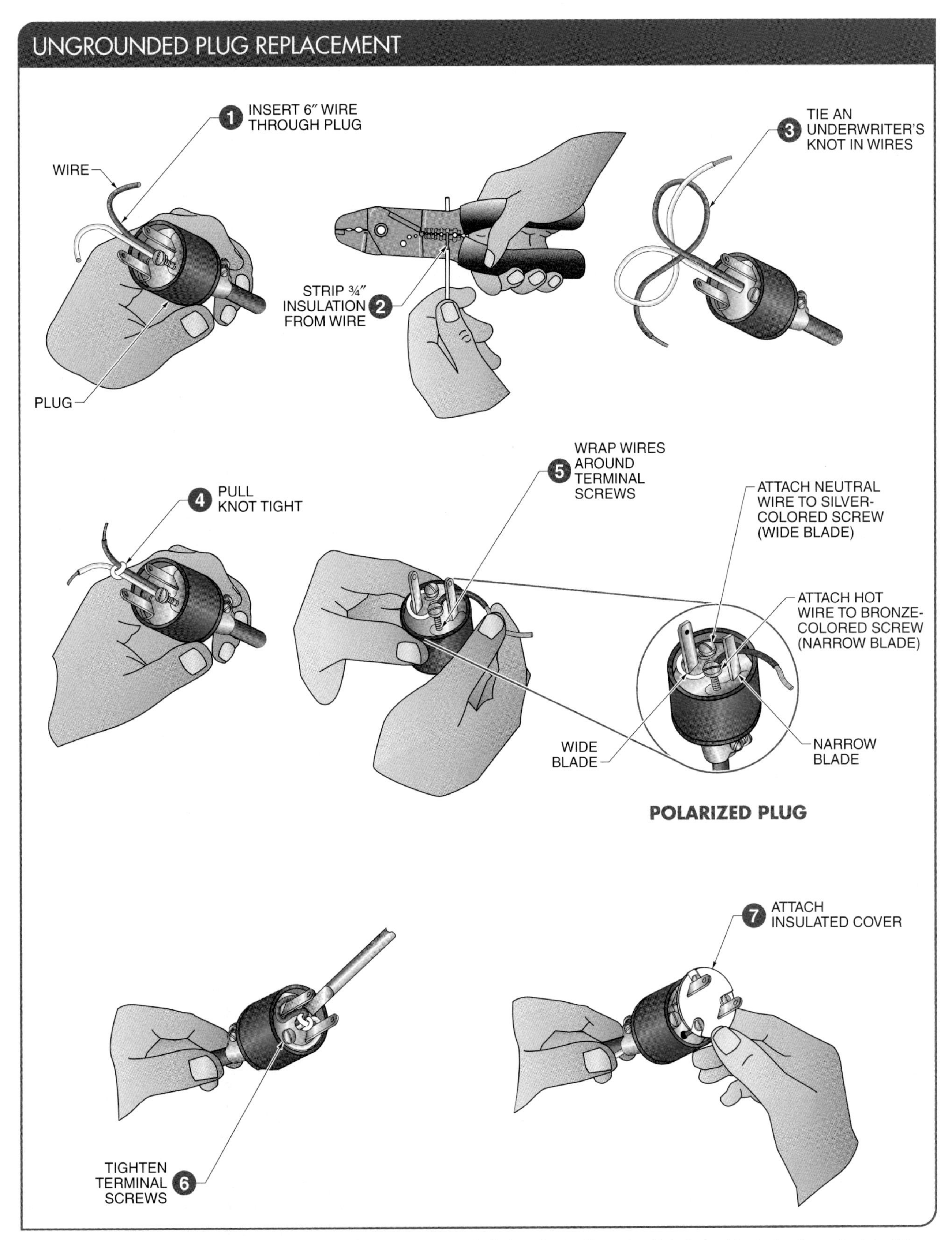

Figure 7-29. Care must be taken when replacing an ungrounded plug due to the potential of electrical shock or electrical fires.

Heavy-Duty Plugs

A *heavy-duty plug* is a plug that is used on high-wattage appliances and equipment that operate on 230 V or 460 V. These plugs may be used on 1ϕ or 3ϕ power equipment. The exact size and number of wires required depends on the individual load that the cord is used to power. **See Figure 7-30.** To replace a heavy-duty plug, first the old plug is removed, and then the following procedure is applied:

1. Insert the cord into the plug and strip about ¾″ insulation from the wires. If stranded, twist each individual wire.
2. Insert the wires into the correct terminal pockets and tighten the terminal screws. The correct terminal screw varies considerably depending on the power supply used.
3. Attach the insulated cover onto the plug and tighten the assembly screws. High-quality plugs are manufactured with screw-on covers.
4. Tighten the strain relief screws.

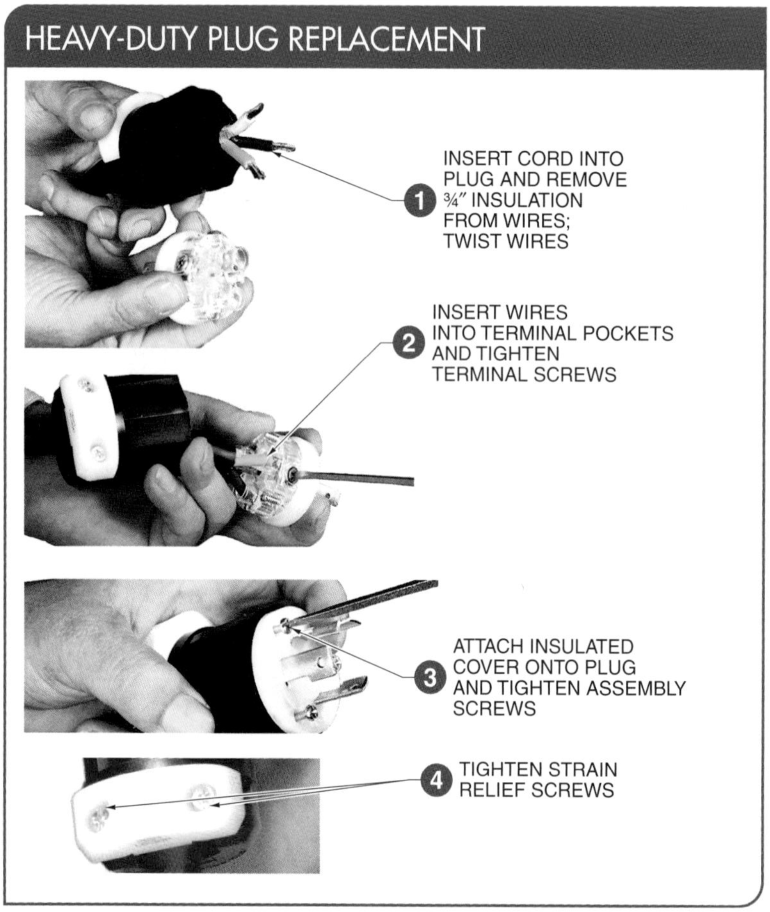

HEAVY-DUTY PLUG REPLACEMENT

1 INSERT CORD INTO PLUG AND REMOVE ¾″ INSULATION FROM WIRES; TWIST WIRES

2 INSERT WIRES INTO TERMINAL POCKETS AND TIGHTEN TERMINAL SCREWS

3 ATTACH INSULATED COVER ONTO PLUG AND TIGHTEN ASSEMBLY SCREWS

4 TIGHTEN STRAIN RELIEF SCREWS

Figure 7-30. Heavy-duty plugs are used with high-wattage appliances and operate on 230 V or 460 V.

Appliance (Heater) Plugs

An *appliance plug,* also known as a heater plug, is a plug used to power an appliance that produces heat, such as an electric grill, roaster, broiler, waffle iron, or large coffeemaker. Appliance plugs fail when the cord is pulled or overheated. Appliance plugs and their cords should be replaced with new ones if they are damaged. **See Figure 7-31.** To replace an appliance plug, first the old plug is removed, and then the following procedure is applied:

1. Strip about ¾″ insulation from the wire and twist the stranded ends into a loop. Do not solder the wires because the plug may get hot enough to melt the solder while the appliance is in use.
2. Wrap the wire around the terminal screws and tighten the terminal screws.
3. Screw the plug halves together.

Large Appliance Plugs

Heavy electrical loads, such as those found on washers, dryers, and ovens, require large 120/220–240 V appliance plugs and receptacles to adequately secure the large wires used for such equipment. A 120/220–240 V appliance plug is typically fed by two colored wires (black and red) for the 220 V connections and a neutral wire (white) for the 110 V connection. This is because the heating element of such equipment may require 220–240 V power, while the lights and controls may require 110 V power.

For 120/220–240 V loads, an appropriate receptacle with a configuration that matches the plug is also required. Typically, there is no grounding slot. Grounding occurs from a grounding wire in the cable to the outlet box and to the service panel. Receptacles can be box mount or surface mount. **See Figure 7-32.**

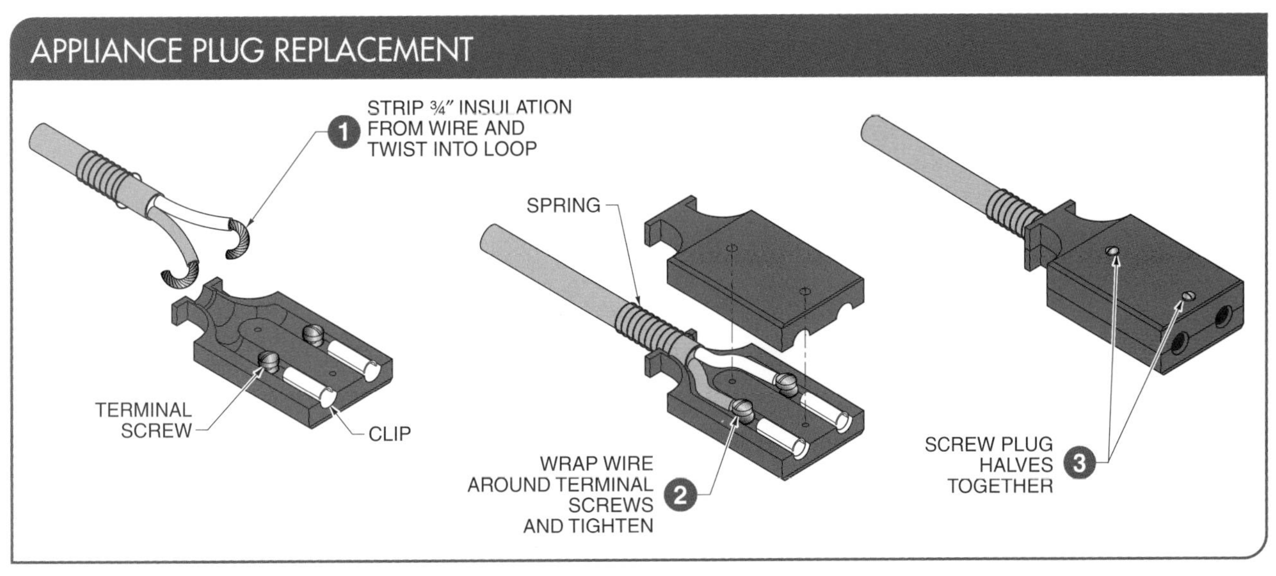

Figure 7-31. Appliance (heater) plugs should be replaced when the cord is damaged due to loose connections or overheating.

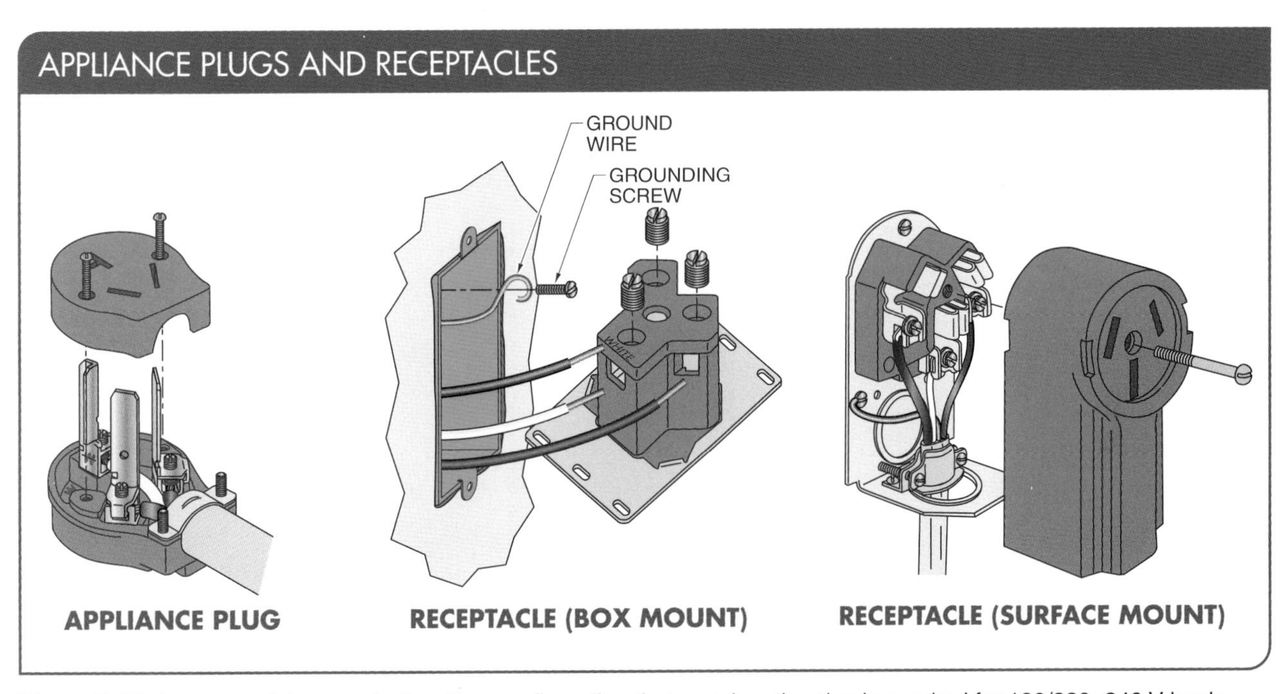

Figure 7-32. An appropriate receptacle with a configuration that matches the plug is required for 120/220–240 V loads.

Testing Connections

Testing connections using a DMM can be useful in isolating problems. Set to the resistance (Ω) function, the DMM can test for loose connections, corrosion, or short circuits. With all power off, the test leads of a DMM are placed across a connection, splice, load, or circuit. The resistance of the connection or circuit being tested is displayed. The resistance measurement should display a low resistance to indicate a good connection. A 0.0 Ω measurement indicates no connection. When a high fixed-resistance value (150.0 Ω or greater) is recorded, the measurement indicates a bad connection. **See Figure 7-33.**

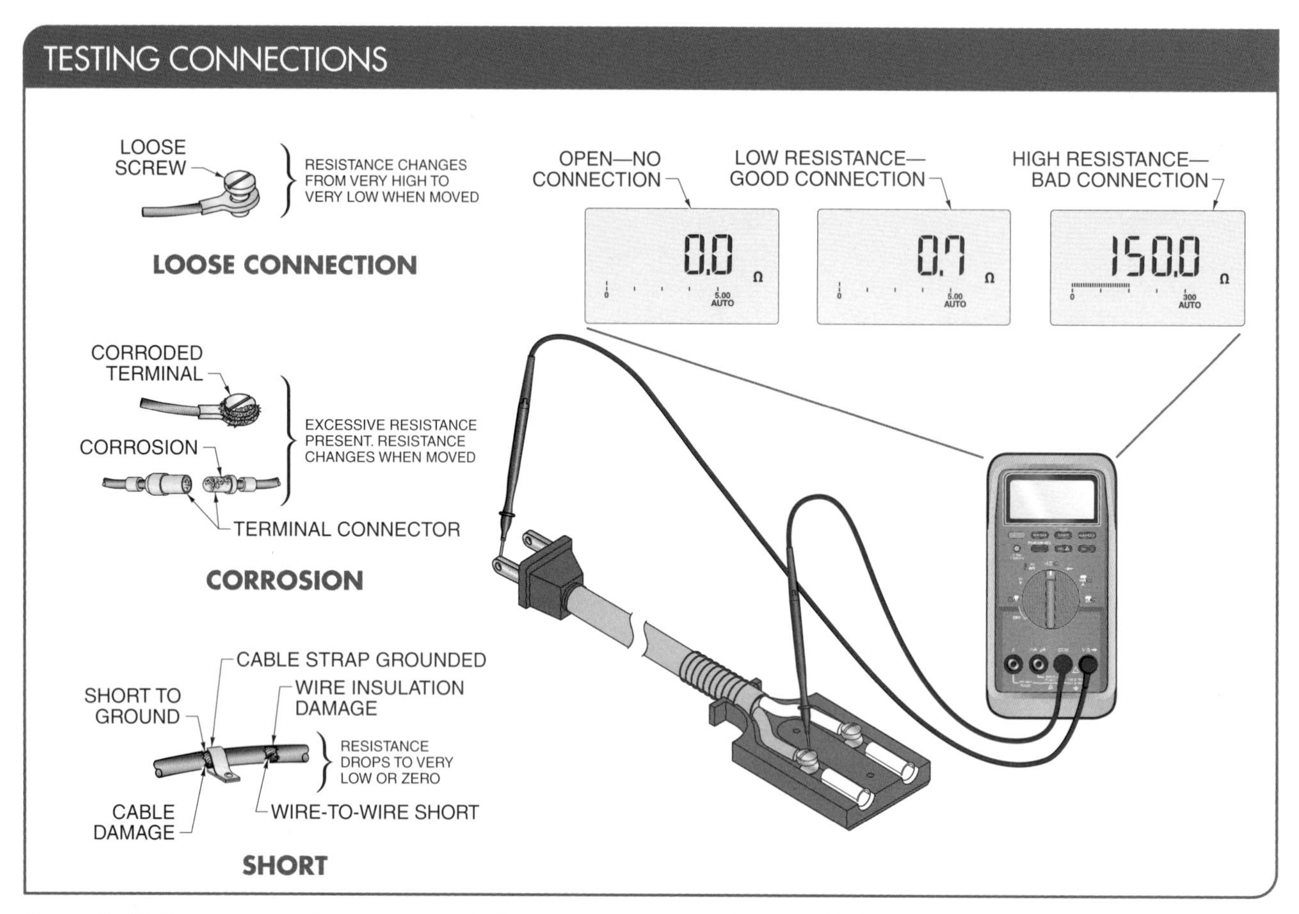

Figure 7-33. Testing connections using a DMM set to resistance can be useful in identifying loose connections, corrosion, and short circuits.

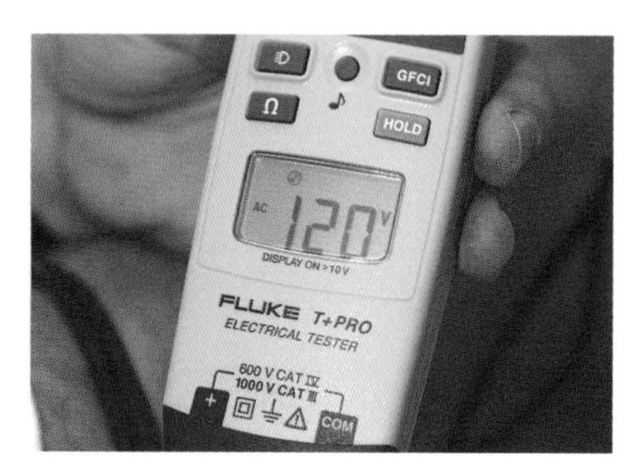

Test instruments with digital displays provide for fast readings of electrical quantities.

SECTION 7.5 CHECKPOINT

1. What is a plug?
2. What amount of voltage do heavy-duty plugs operate on?
3. What kind of plug do washers, dryers, and ovens require?
4. What would a DMM display show when resistance is tested on a proper connection?

Chapter Activities

7 Electrical Connections

Name ______________________________ Date ______________

Spliced Connections

Identify each splice shown.

______________ 1. ___ splice

______________ 2. ___ splice

______________ 3. ___ splice

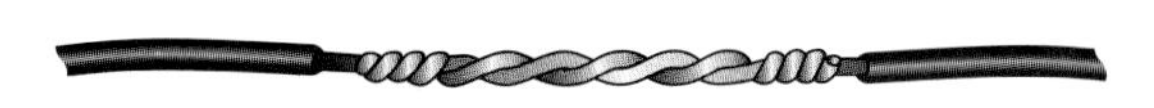

______________ 4. ___ splice

Testing Connections

Determine which DMM reading indicates a good, bad, or no connection based on the resistance displayed.

_______________ **1.** Good connection

_______________ **2.** Bad connection

_______________ **3.** No connection

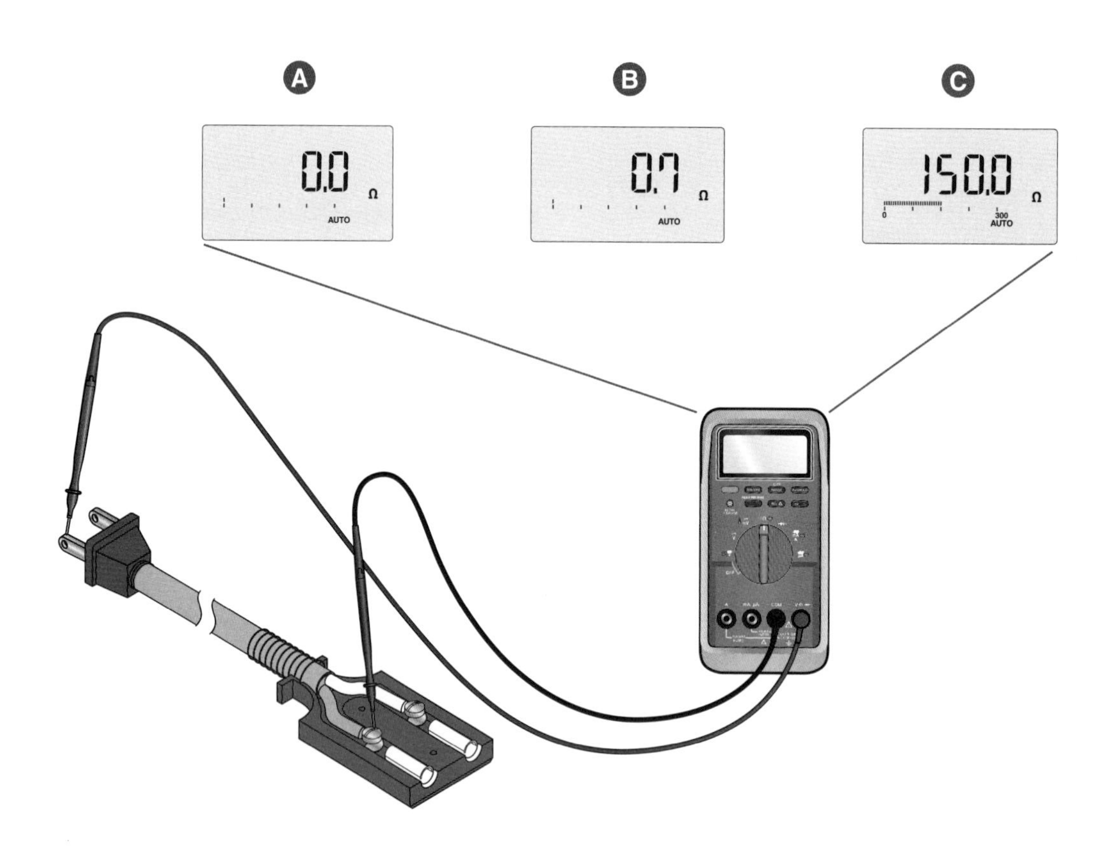

8 Nonmetallic-Sheathed Cable

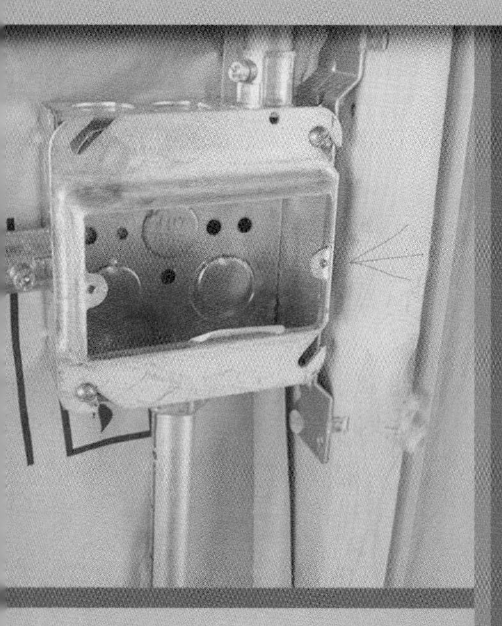

SECTION 8.1—NONMETALLIC-SHEATHED CABLE

- Describe nonmetallic-sheathed (NM) cable.
- List the different types of conductor materials.
- Explain the difference between stranded and solid wire.
- Describe American Wire Gauge (AWG) and the applications of different wire sizes.
- Explain what information is printed on the plastic jacket of NM cable.
- Explain how electrical conductors are rated by the National Electrical Code® (NEC®).
- Explain what the different cable jacket colors signify.

SECTION 8.2—NM CABLE INSTALLATION

- Explain the specifications that must be known when purchasing NM cable.
- Explain how NM cable is prepared for installation.
- Explain how to remove the plastic jacket of an NM cable.
- Explain how to remove the cable wire insulation.
- Explain what the phrase "roughing-in" means.
- Describe the tools used to drill holes in studs and floor joists to route NM cable.
- Explain how cable is routed through studs, through joists, around corners, and through masonry walls.
- Describe how NM cable is secured.

Learner Resources

ATPeResources.com/QuickLinks
Access Code: 791712

Nonmetallic-sheathed (NM) cable is used in residential buildings for different wiring applications per the National Electrical Code® (NEC®). Specific construction materials are used for different applications, which are printed on the cable jacket and the packaging. The appropriate type of NM cable is selected depending on the job requirements, and specific tools are used to prepare and route the cable during installation.

SECTION 8.1 — NONMETALLIC-SHEATHED CABLE

Nonmetallic-sheathed (NM) cable is electrical cable that has a set of insulated electrical conductors held together and protected by a strong plastic jacket. NM cable typically contains two or three insulated wires (conductors) and may include a separate bare ground wire. Information about the cable is printed on the outer jacket. **See Figure 8-1.**

NM cable is popular in residential wiring because it is relatively inexpensive and easy to install. Romex® is a common brand name that is used when referring to NM cable produced by a variety of manufacturers. NM cable is designed for use in corrosive, dry, moist, and damp locations.

Conductors

Electrical circuits and electrical devices are connected using conductors. A *conductor* is a material that has very little resistance and permits electrons to move easily through it. Conductors used in residential electrical circuits include insulated wires bundled together that make up cables.

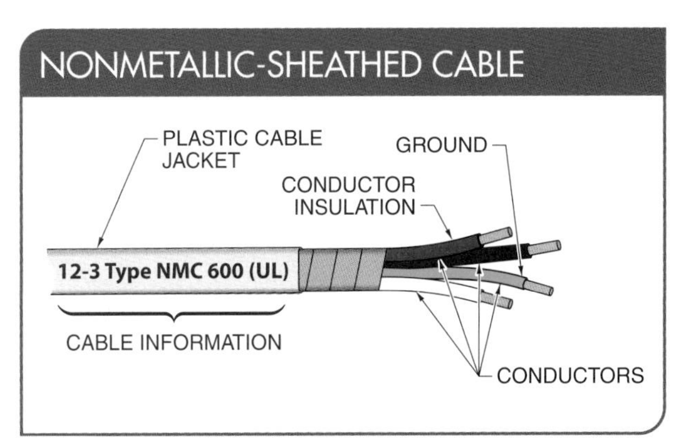

Figure 8-1. The wire size, number of conductors, and ground information is printed on the outer jacket of NM cable.

A *cable* is two or more insulated wires grouped together within a common protective cover. Cable is used to connect individual electrical devices. NM cable wires are enclosed in an insulated cover to protect the wire, increase safety, and meet code requirements. Some individual wires, such as the ground wire, may be bare. **See Figure 8-2.**

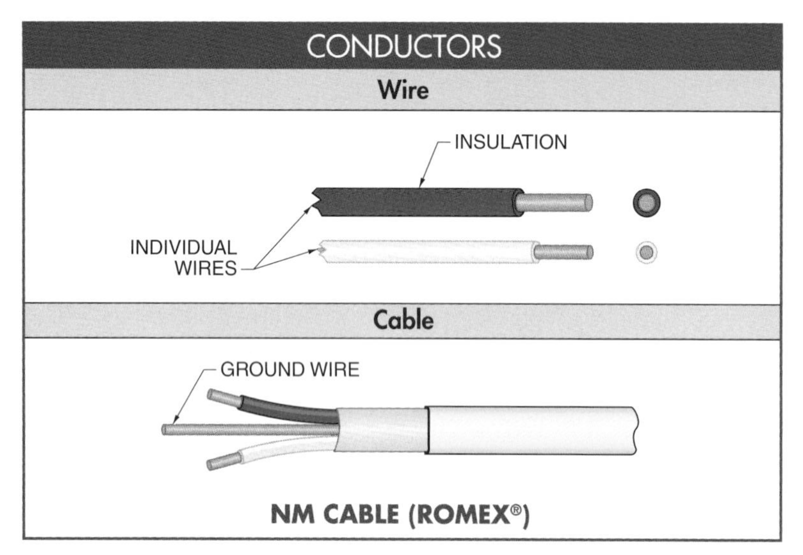

Figure 8-2. Conductors may be individual wires or bundled together to make up cables.

Conductor Material. Conductor materials include copper, aluminum, and copper-clad aluminum. Copper (Cu) and aluminum (Al) are the most commonly used conductor materials for wires. Copper is preferred because it has a lower resistance, takes up less space, and does not oxidize or corrode at connections like aluminum. *Copper-clad aluminum* is a conductor material composed of copper bonded to the outside of an aluminum wire to counter the problem of oxidation.

The current flowing through a conductor must be kept below the American Wire Gauge (AWG) rated limit to maintain a safe system. The larger the cross-sectional area of the conductor, the greater its current-carrying capacity.

Conductor Types. Conductors are available in stranded and solid wire types. A *stranded wire* is an insulated conductor composed of several smaller wires twisted together. A *solid wire* is an insulated conductor composed of only one wire. Stranded wire is used in applications with large wire sizes, such as service entrances and subpanels. When these applications are routed, they require wire with more flexibility than solid wire can provide. Stranded wires are also more durable against vibration or wire movement.

Solid wire is used where it is easy to bend and route wire. Solid wire is less expensive than an equivalent size of stranded wire. Besides the strands of wire, the only noticeable difference between stranded and solid wires is the overall length of their diameters. Stranded wire is slightly longer in overall diameter than solid wire due to the small gaps between strands and the longer-diameter insulation needed to cover the wire.

Conductor Sizes. AWG is a US standard for copper and aluminum wire diameters. The larger the AWG number assigned to a wire, the smaller the physical size of the wire or conductor. A stranded wire occupies more space than a solid wire because the wire gauge is measured by summing the cross-sectional area of the strands. **See Figure 8-3.**

Wires that are smaller than No. 8 AWG may be either solid or stranded. Wires that are No. 8 AWG or larger are stranded. No. 14 and No. 12 AWG wires are used for wiring most lighting circuits and supplying power to standard receptacles. Electrical appliances are wired using No. 10, No. 8, No. 6, and No. 4 AWG wires. Large conductors, such as No. 3 and No. 2 AWG wires, are used to supply power to main service panels.

NM Cable Information

Important information is printed on the plastic jacket of NM cable. The printed information typically includes numbers and letters. The numbers provide information about the size of the wire and the number of conductors in the cable. For example, a mark of "14-2" indicates the cable is No. 14 AWG with two current-carrying conductors, not counting the ground wire.

CONDUCTOR SIZES AND AMPACITIES							
Size AWG or kcmil	Copper Conductor Ampacities			Aluminum Conductor Ampacities			Size AWG or kcmil
	Temperature Rating of Conductor			Temperature Rating of Conductor			
	60°C	75°C	90°C	60°C	75°C	90°C	
	Types TW UF	Types RHW THHW THW XHHW THWN USE	Types RHH THHW RHW-2 THWN-2 XHHW THW-2 XHHW-2 THHN XHH USE-2	Types TW UF	Types RHW THHW THW XHHW THWN USE	Types RHH THHW RHW-2 THWN-2 XHHW THW-2 XHHW-2 THHN XHH USE-2	
14	20	20	25	—	—	—	—
12	25	25	30	20	20	25	12
10	30	35	40	25	30	35	10
8	40	50	55	30	40	45	8
6	55	65	75	40	50	60	6
4	70	85	95	55	65	75	4
3	85	100	110	65	75	85	3
2	95	115	130	75	90	100	2
1	110	130	150	85	100	115	1
1/0	125	150	170	100	120	135	1/0
2/0	145	175	195	115	135	150	2/0
3/0	165	200	225	130	155	175	3/0
4/0	195	230	260	150	180	205	4/0

Figure 8-3. AWG tables contain conductor sizes, temperature ratings, and types of materials.

Letters. The letters printed on the NM cable jacket indicate the designated location. For example, the letters "NM" indicate cable typically used for dry interior wiring. The letters "NMC" indicate cable used for damp locations. The letters "UF" indicate cable used for damp underground locations. Letters such as "TW" and "THW" indicate common insulation types. The letters "W/GRN" mean "with ground." Some older homes may have NM cable without a ground wire installed, but all contemporary homes using NMC should include a ground wire for safety.

Ratings. Insulated wires are rated based on their current-carrying capacity. The current rating depends on the type of wire (Cu or Al), the wire size, the type of insulation used, and the conductor's temperature rating.

The NEC® should be referred to for wire size and ampacity. Some of this information may be printed on the cable jacket. For example, a nonmetallic-sheathed Type-B (NM-B) cable may be UL® listed, with No. 10 AWG copper (Cu) wire and 600 V-rated insulation, made of thermoplastic insulation (T), heat resistant (H), used in wet locations (W), and have a 75°C (167°F) temperature rating. This information may appear on Romex® as "(UL) Type NM-B 10-3 W/GRN 600 V CU THW 75°C."

Colors. All NM cable was originally produced with a white covering. Manufacturers now produce different cables using standardized colors for identification. The commonly used No. 14 AWG (15 A) wire is white, the No. 12 AWG (20 A) wire is yellow, and the No. 10 AWG (30 A) wire is orange. **See Figure 8-4.**

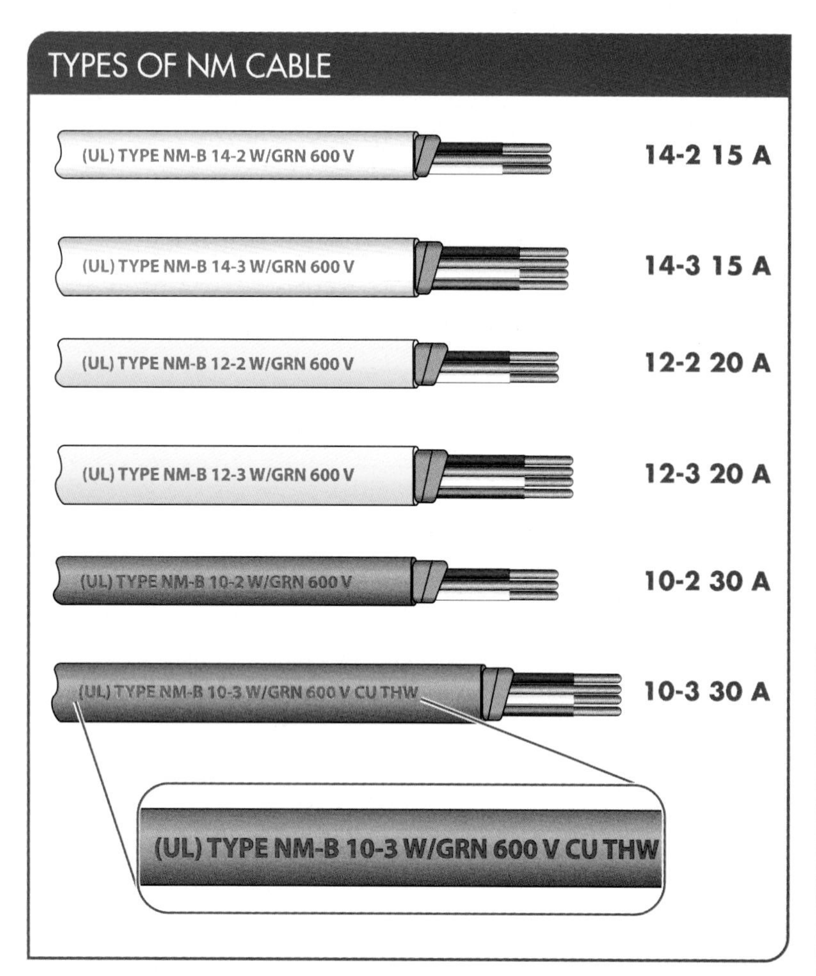

Figure 8-4. Different types of NM cables can be easily identified by color and the printed information on the outer jacket.

SECTION 8.1 CHECKPOINT

1. How many conductors are in a typical NM cable?
2. What is Romex®?
3. What is a conductor?
4. What is a cable?
5. What are the two most common conductor materials used?
6. What is a stranded wire?
7. What is a solid wire?
8. What two AWG conductor sizes are used to wire residential lights and receptacles?
9. What are the three commonly used colors, wire sizes, and ampacity ratings of NM cable?

SECTION 8.2—NM CABLE INSTALLATION

NM cable is found at electrical supply houses, hardware stores, and in some large retail stores with home repair centers. NM cable is typically sold in boxed rolls of 50′ and 250′ and in reels of 1000′. **See Figure 8-5.** When purchasing NM cable, an individual must specify the length and type of cable required. For example, when an application requires 250′ of cable that has two conductors of No. 14 AWG wire with a separate ground wire, an individual should purchase cable that is specified as 250′ of Type NM or NMC 14-2 W/GRN.

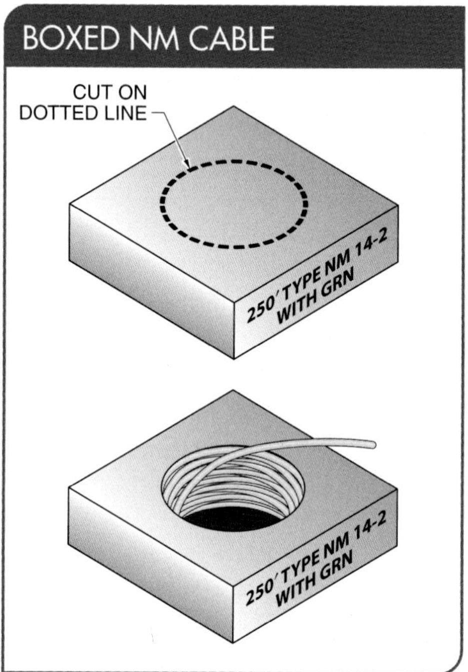

Figure 8-5. NM cable packaging information describes the cable length, type of cable, number of conductors, and if the cable has a ground conductor.

Preparing NM Cable

NM cable must be prepared for installation. Typically, the cable must be cut to length. Then the outer plastic jacket is removed to 8″ from the end. Once the outer jacket is removed, 1″ of insulation is removed from the individual wires.

Plastic Jacket Removal. NM cable is easy to install and requires few tools to perform a quality job. Side-cutting pliers are capable of cutting NM cable quickly to length. After the cable has been cut to length, the ends of its outer plastic jacket must be removed before the cable is inserted into electrical (receptacle or junction) boxes. To open the plastic jacket of NM cable, an electrician's knife is used to cut through the center of the jacket where the ground wire is located. **See Figure 8-6.** The bare ground wire guides the path of the knife blade. At least 8″ of the jacket should be removed from each end of the cable.

When great lengths of cable must be prepared, cable rippers are used to save time. Cable rippers have a small, sharp blade designed to penetrate a short distance into the cable to cut only the outer plastic jacket.

> CODE CONNECT
>
> *Per NEC® Subsection 334.116(A), the overall sheath covering should be flame retardant and moisture resistant.*

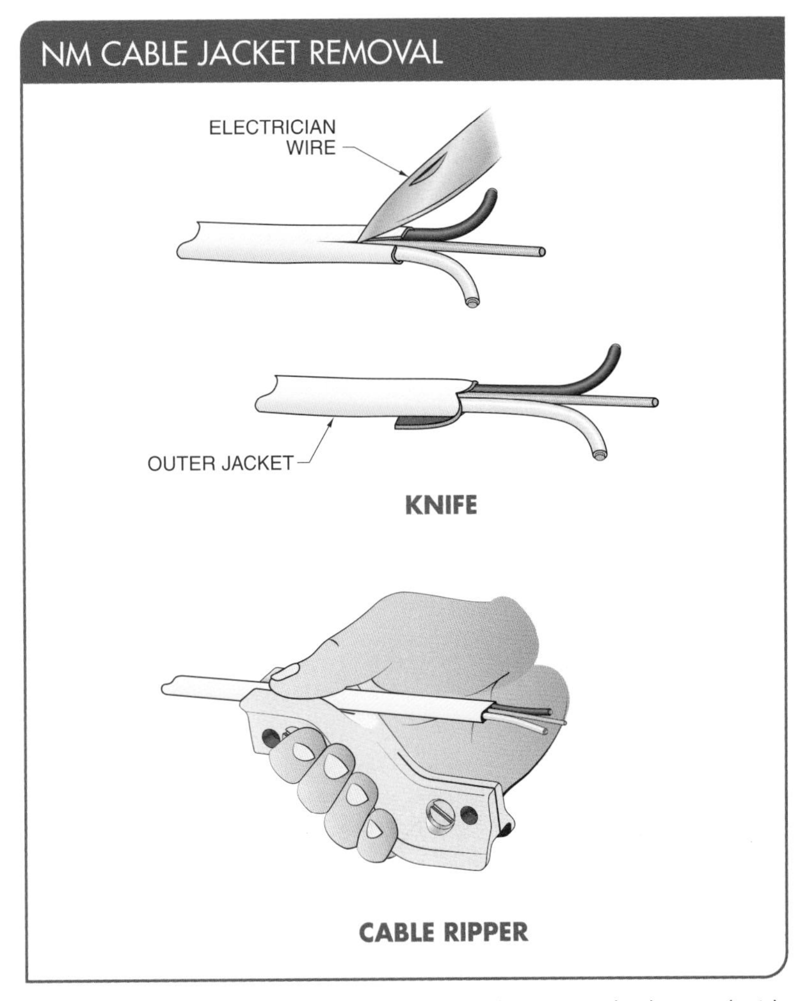

Figure 8-6. The plastic jacket of NM cable can be removed using an electrician's knife or a cable ripper. The cut should always be made away from the hand and not towards the body.

Cable Wire Insulation Removal. Once the outer plastic jacket of NM cable is cut away, insulation from the individual conductors (wires) can be removed. Various multipurpose wire-stripping tools are used to remove insulation from individual wires. **See Figure 8-7.**

Roughing-in NM Cable

Roughing-in is a phrase that refers to the placement of electrical boxes and wires before wall coverings and ceilings are installed. Roughing-in must be performed in such a way that the entire electrical system (wires and boxes) can be easily traced before walls and ceilings are in place. All rough-in work must be carefully checked by the authority having jurisdiction (AHJ). This is normally done by the city, town, or county electrical inspector. Errors in the roughing-in process are difficult to find once the wall coverings are in place.

Installing Cable Runs

A *cable run* is a length of installed cable connecting two electrical devices that are not in immediate proximity to one another. The cable is pulled through holes that were drilled in the wood studs and joists. All holes should be drilled as straight as possible to aid in pulling wires through multiple studs. When all holes have been properly drilled, NM cable can be installed quickly and easily. To obtain the proper length of cable required, the cable should be pulled into position before it is cut. Positioning the cable ensures that enough cable is pulled without creating waste. **See Figure 8-8.**

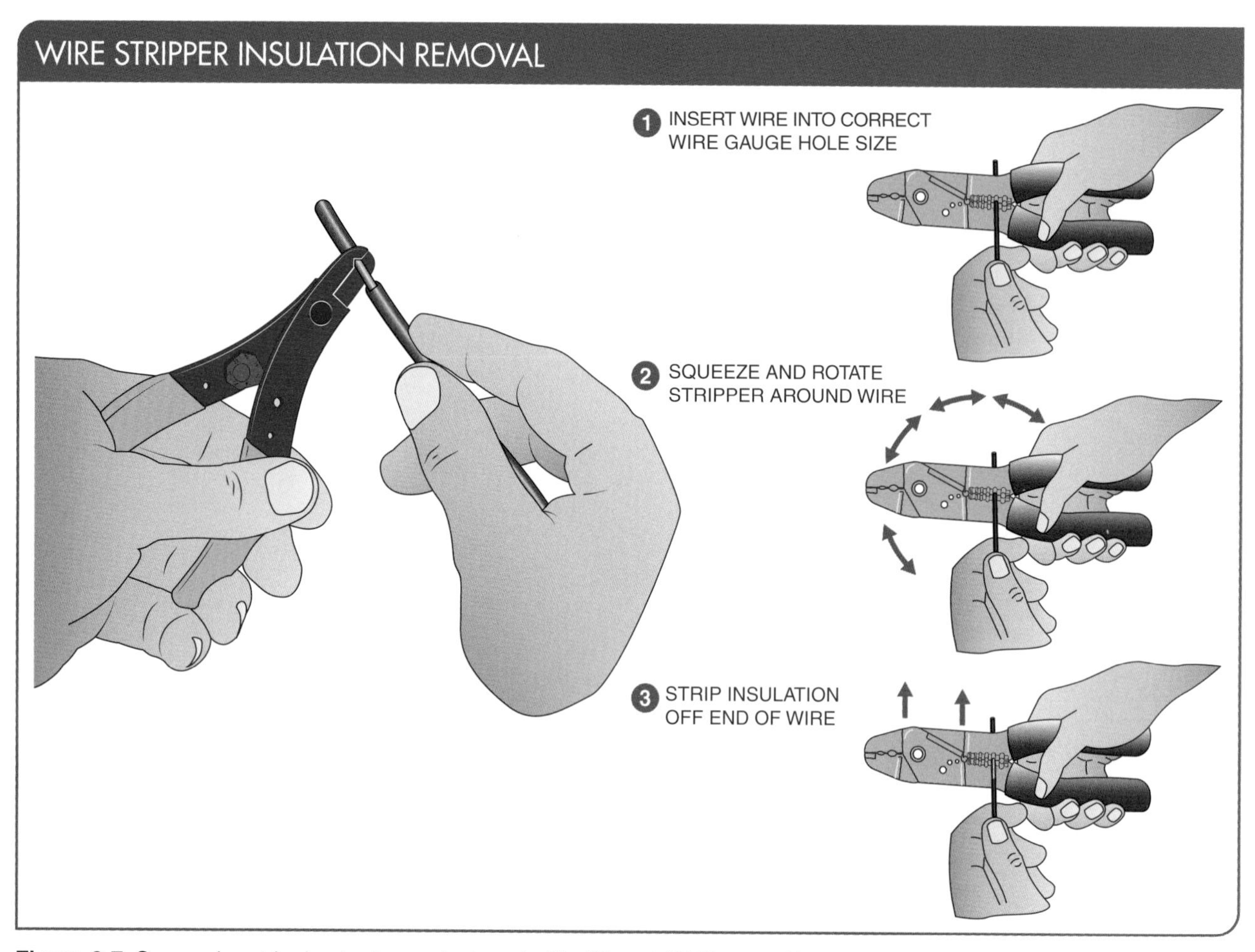

Figure 8-7. Some wire-stripping tools are designed with different AWG capacities to control insulation removal without cutting into the metal conductor.

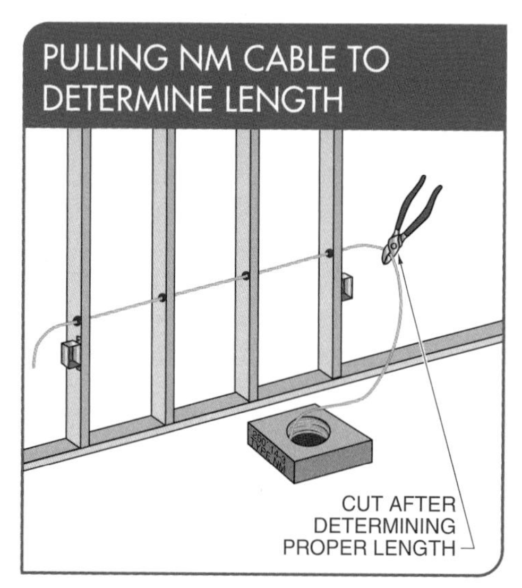

Figure 8-8. Cable should be pulled to the proper position first and then cut to length to prevent waste.

Cable and wires should be stripped before being installed in a box to save time later when outlets and switches are being wired. **See Figure 8-9.** When the cable is stripped, at least 8″ of wire should be left for rough-in purposes. It is easier to cut off excess wire later than to try to add wire.

Drilling Holes. It is best to mount boxes first according to the electrical print and then drill the holes. Mounting boxes first reduces the number of unnecessary holes and saves time.

Holes in studs or floor joists can be made most effectively with an offset drill. Individuals can use offset drills that easily fit between studs and joists. When an offset drill is not available, a heavy-duty electric drill with a spade power bit can be substituted. **See Figure 8-10.**

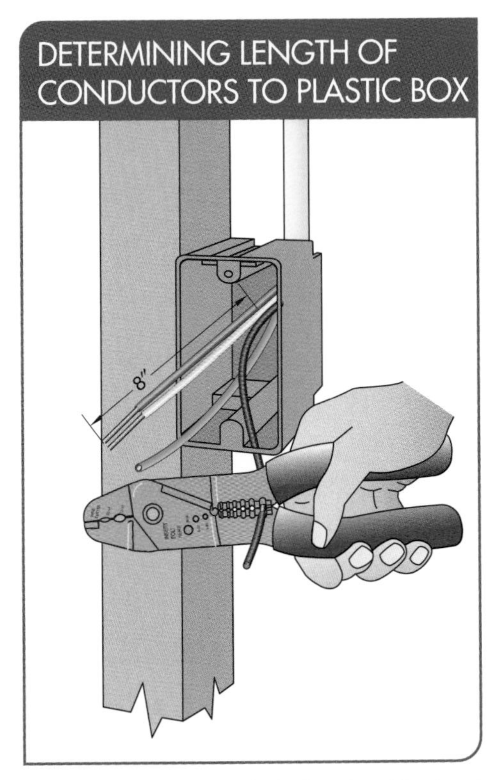

Figure 8-9. At least 8″ of wire should extend beyond the bottom of a box to allow for connections to receptacles and switches.

OFFSET DRILLS

Milwaukee Tool Corporation

Figure 8-10. Offset or right-angle drills are ideal for drilling in tight spots between studs.

CODE CONNECT

Per NEC® Section 334.24, bends in NM cable must be made without damaging the cable. The radius of the curve of the inner edge of any bend during or after installation should not be less than five times the diameter of the cable.

Using Extended Drill Bits. Extended drill bits are extra-long and are used when cables or conduit need to run from one floor to another. An extended drill bit can be used to penetrate the top plate of a wall, obstacles inside a wall, subflooring, and the bottom plate of a wall. **See Figure 8-11.** A fish tape can be used to pull a cable to connect a receptacle to a junction box from the basement to the first floor or from the attic to the floor below.

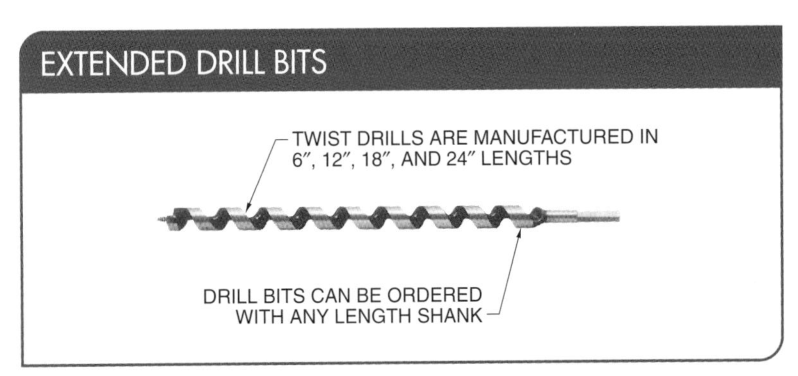

Figure 8-11. Extended drill bits are used to drill holes between walls and floors for routing NM cable.

Routing Cables

Fish tapes are essential when routing cables through ceilings and walls. The proper use of a fish tape allows individuals to run wires that otherwise would be impossible to route. Fish tapes sometimes need to be hooked together. This requires a five-part sequence to properly pull a wire or cable. **See Figure 8-12.** Cable or wire must always be firmly secured to the ends of a fish tape. This ensures that the wire will not release halfway through a pull.

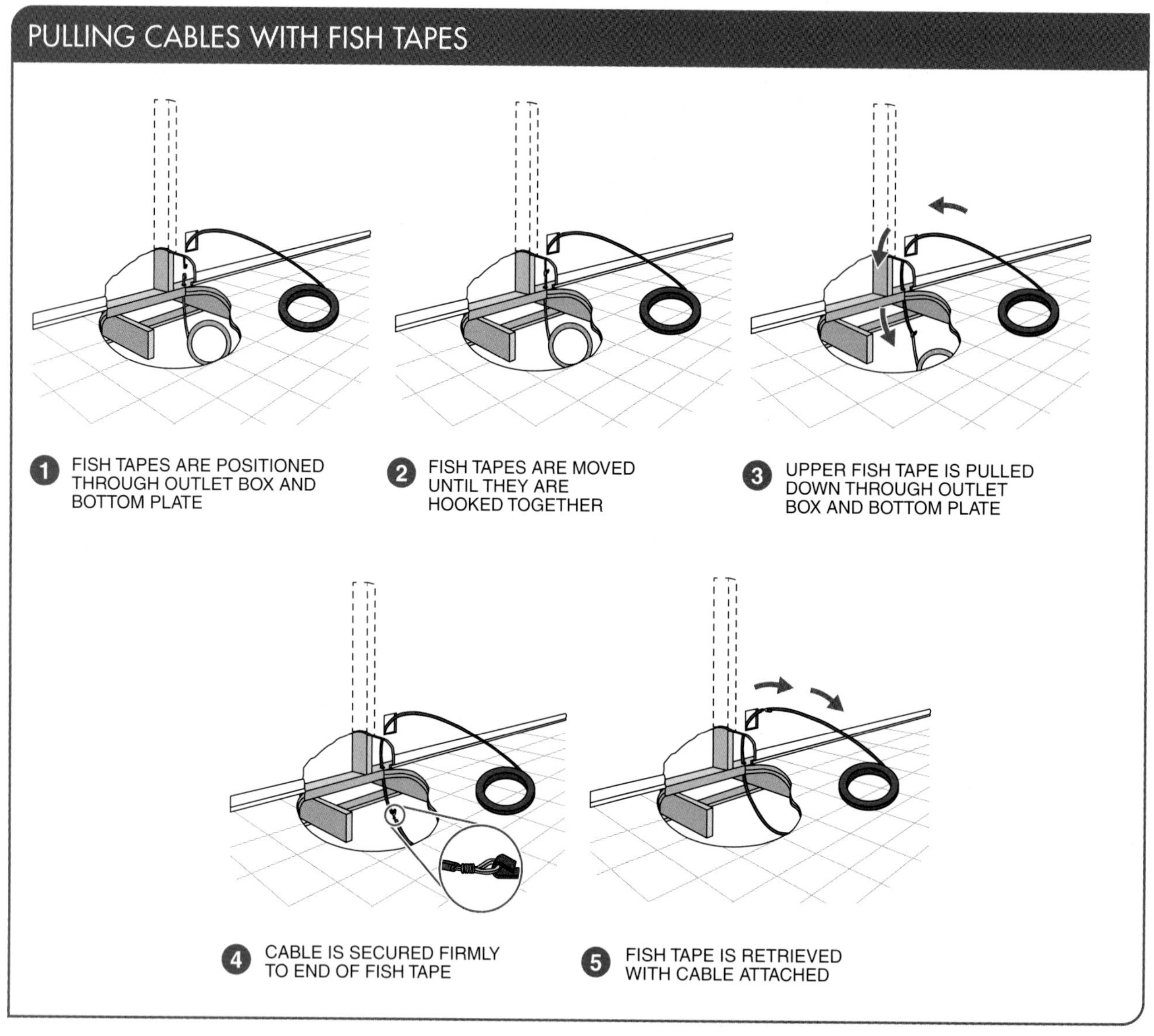

Figure 8-12. A five-step sequence is used when fish tapes need to be hooked together to pull wire or cable.

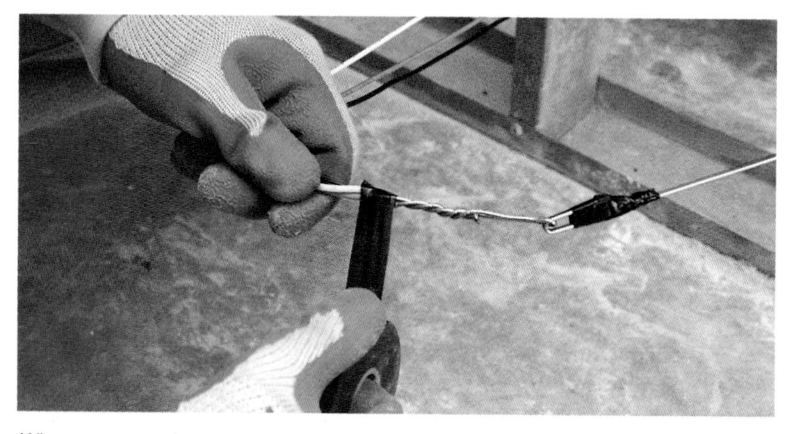

Wires are secured to the end of fish tape so they do not release when pulled through walls.

Fish tapes are used to pull cable along a wall from receptacle box to receptacle box. Back-to-back receptacles can require two fish tapes to get the pull started for a cable. **See Figure 8-13.** A fish tape can also be used to pull cable along the baseboard of a wall from Point A to Point B.

Routing through Metal Studs. Metal studs typically have precut holes for routing conduit and cable. Before any cable or conduit is routed through the holes, plastic bushings must be inserted into each hole to prevent damage to NM cable. **See Figure 8-14.**

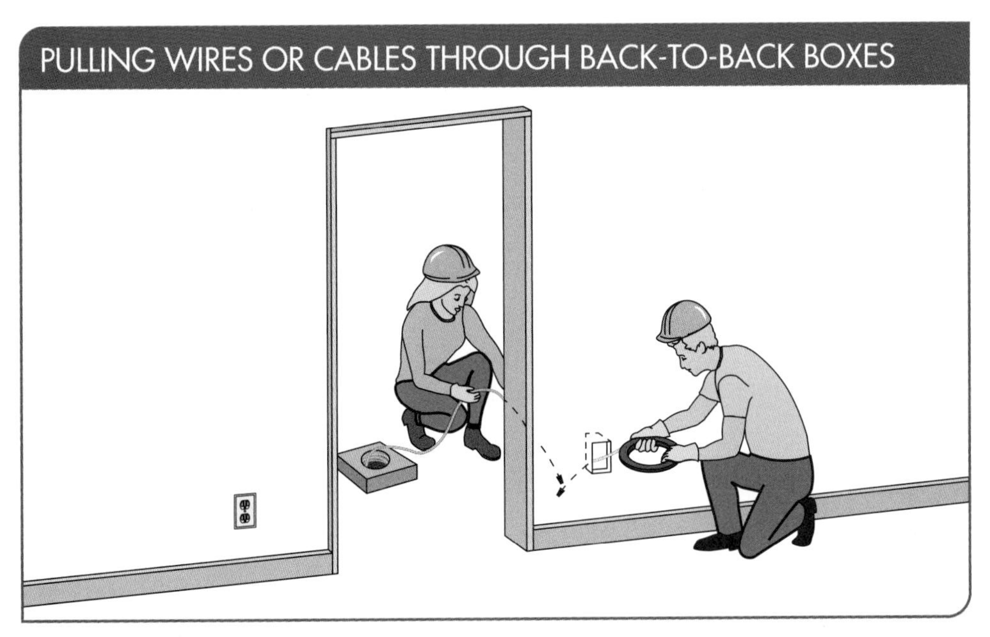

Figure 8-13. Fish tapes are used to pull wire through back-to-back boxes, from box to box, and to pull cable along a baseboard.

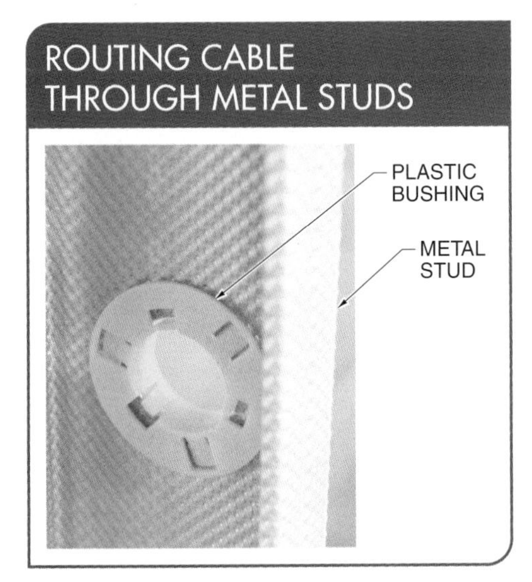

Figure 8-14. Plastic bushings must be inserted into each hole in metal studs to prevent damage to NM cable.

Routing through Wood Studs and Joists. Running NM cable through wall studs and floor joists is the most common way of routing cable. Holes are drilled through the center of the studs and joists to reduce the possibility of a nail penetrating deep enough to puncture the cable. **See Figure 8-15.** When cable is run through drilled holes, no additional support is required.

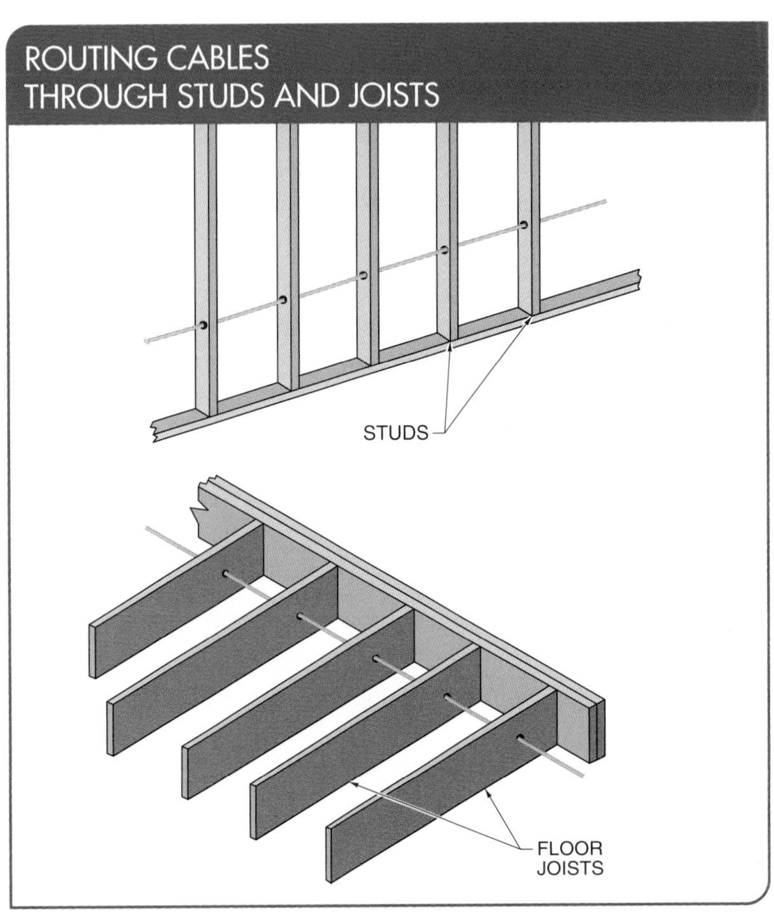

Figure 8-15. NM cable is protected from nail penetration by being pulled through the center of drilled out studs and joists.

In some instances, studs may have to be notched instead of drilled. **See Figure 8-16.** When notches are substituted for holes, the cable run must be protected from nails by a steel plate at least 1⁄16″ thick. A steel plate is a common, inexpensive, and easy-to-install item that is available at most electrical supply distributors.

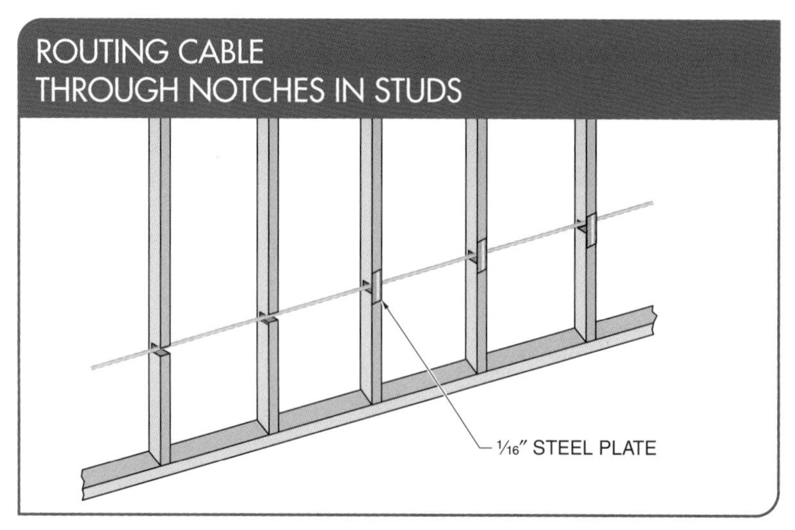

Figure 8-16. NM cable run through notched studs must be protected by steel plates at least 1⁄16″ thick.

Routing around Corners. Routing cables through solid wood corners can sometimes be difficult. Various methods can be used to overcome the problem of solid wood corners. One method to accommodate a cable run is to drill holes at an angle into each side of the solid wood corner. **See Figure 8-17.** Drilling holes into each side will save wire but may be time consuming if the holes do not line up properly. Another method is to notch the corner studs, using steel plates for protection where allowed.

Routing over or under a corner is another method of routing cable. The routing-over method can be accomplished rapidly by one person who does not need to leave the room. The routing-under method requires extra time for a person to go into the basement or crawl space to send the wire back up through the floor.

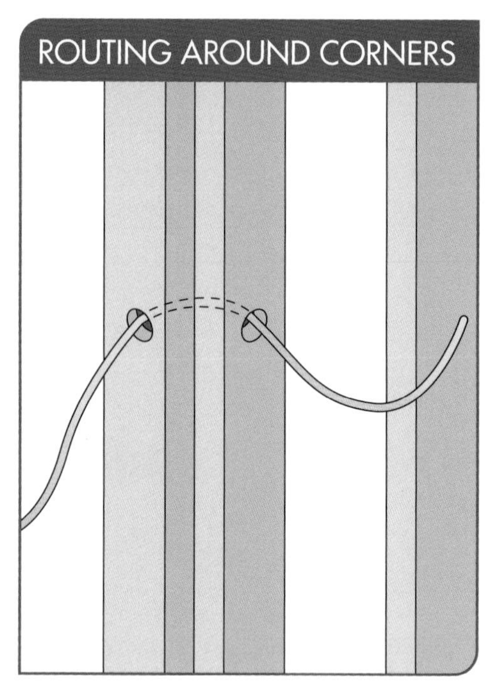

Figure 8-17. To route NM cable around solid wood corners, holes are drilled at an angle into each side.

Routing through Masonry Walls. NM cable may be routed through masonry walls if no moisture is present and the wall is above grade. **See Figure 8-18.** When moisture is present, NMC cable must be used. Neither NM nor NMC cable can be embedded in poured cement, concrete, or aggregate.

CODE CONNECT

Per NEC® Subsection 334.12(B), NM cable shall not be embedded in masonry, concrete, or plaster.

Securing Cable

To properly secure cable, individuals must staple cable along runs, secure ground wires, and use components to secure cable in electrical boxes. NM and NMC cable must be supported or secured (stapled) every 4′-6″ of cable run and within 12″ of a box. **See Figure 8-19.**

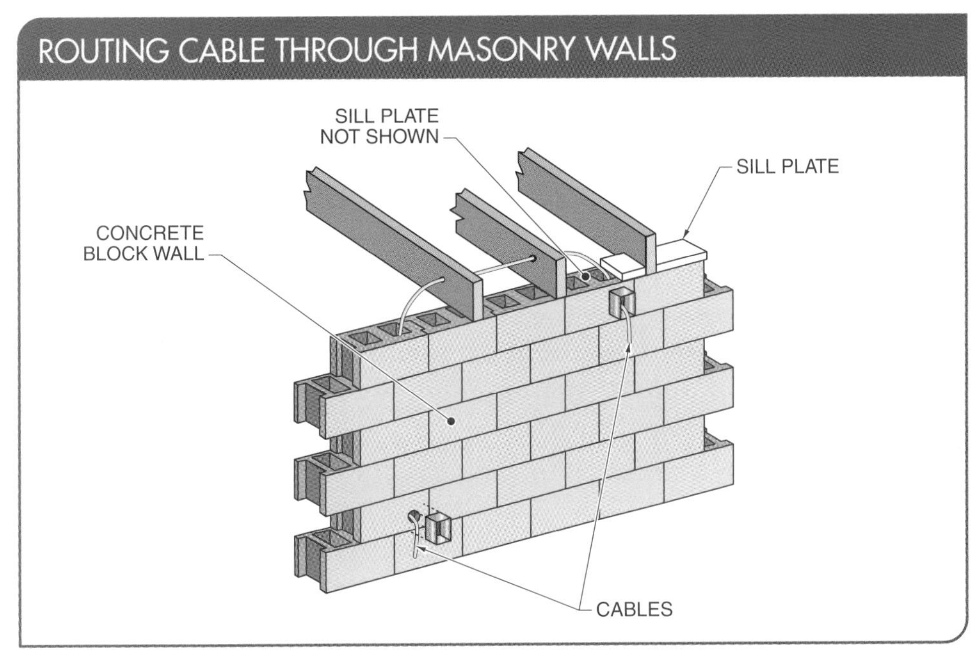

Figure 8-18. NM cable can be routed through dry masonry walls above grade. NMC cable must be used where moisture is present.

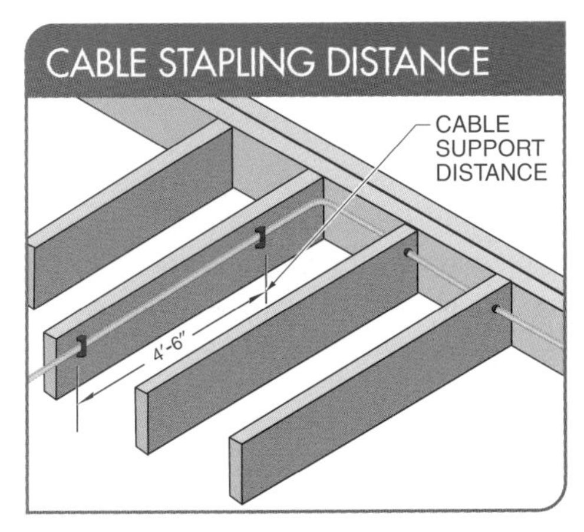

Figure 8-19. All NM type cable runs must be supported or stapled every 4′-6″.

Cable Run Stapling. When cable has been pulled into position, the cable must be securely fastened with special plastic staples. The staples for NM cable typically have a plastic strap that reduces the possibility of damage to the cable. **See Figure 8-20.** The NEC® requires that all cables be secured near an electrical box.

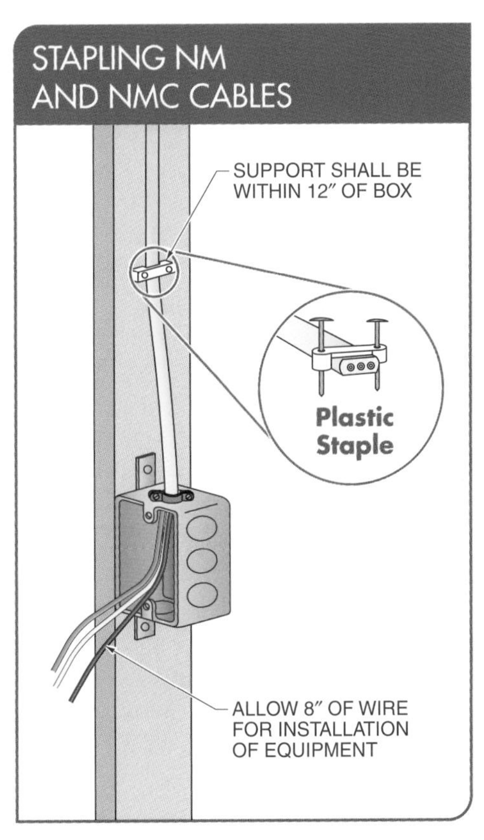

Figure 8-20. All NM type cable must be stapled within 12″ of an electrical box.

Securing Ground Wires. When a cable is secured, a good mechanical and electrical ground must be established to provide electrical protection. The three widely accepted methods of proper residential grounding are component, pigtail, and clip grounding. **See Figure 8-21.**

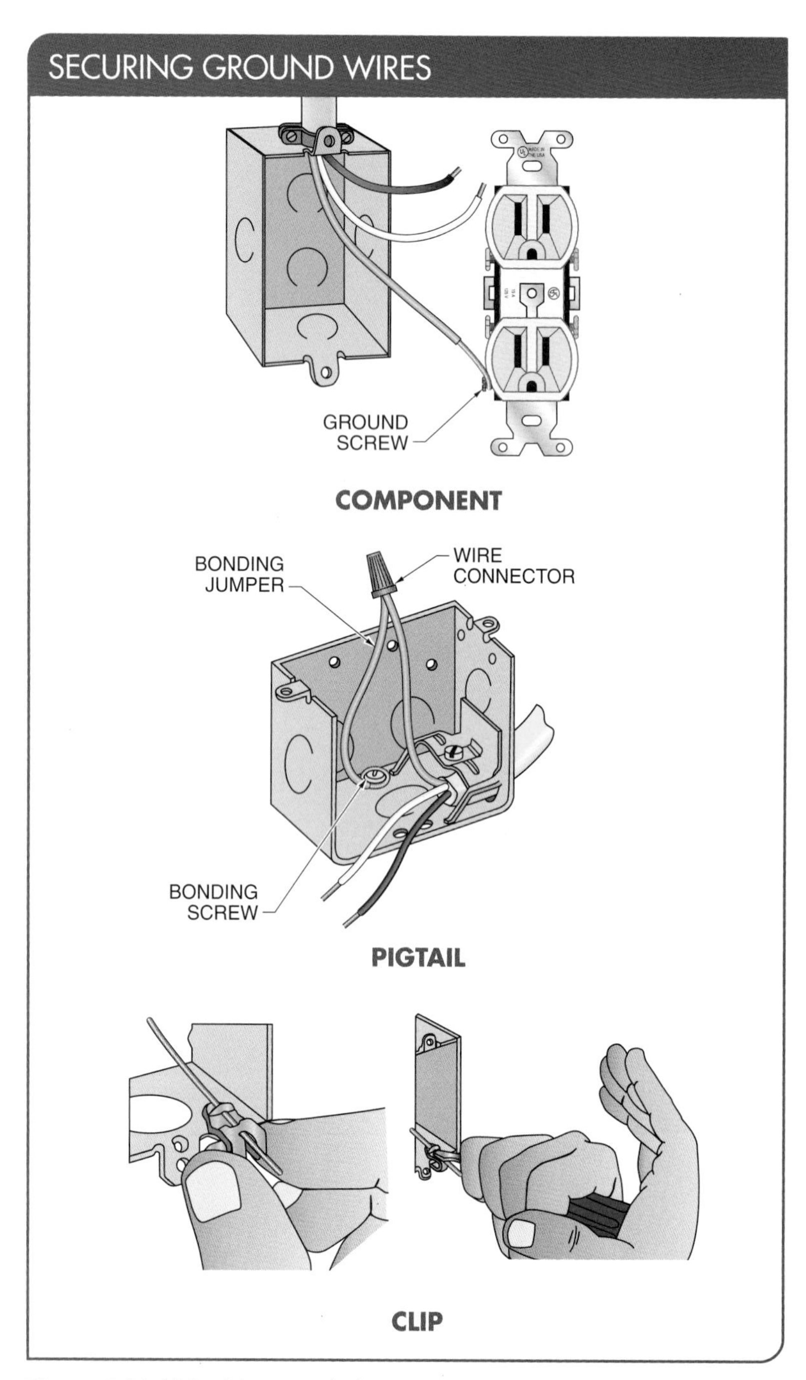

Figure 8-21. NM cable ground wires are secured at electrical boxes to provide electrical protection.

Component grounding is a grounding method where the ground wire is attached directly to an electrical component, such as a receptacle. Component grounding requires that the grounding wire be attached before the electrical component is permanently mounted.

Pigtail grounding is a grounding method where two grounding wires are used to connect an electrical device to a grounding screw in the box and then to system ground. The box ground wire is secured by a screw to a threaded hole in the bottom of the box. Once secured, the cable ground wire is pigtailed to the box ground wire.

Clip grounding is a grounding method where a grounding clip is slipped over the grounding wire from the electrical device. The grounding wire and grounding clip are secured to the box with pressure using a screwdriver. Clip grounding is the method most often used to reduce installer fatigue when there are many ground connections to be made. Clip grounding, like pigtail grounding, can be secured as the cable is put in place.

Securing Cables in Electrical Boxes. Cables may be secured to electrical boxes by clamps. The three types of clamps typically available for NM cables are straight, saddle, and cable clamps. Straight clamps and saddle clamps are part of an electrical box when it is manufactured. Cable clamps are installed individually as needed. **See Figure 8-22.** Locknuts are installed so that the points of the nut point inward to dig firmly into the metal box.

When cable is inserted through clamps, care must be taken not to damage insulation by scraping the cable on the rough surfaces and edges. Too much pressure when cable clamps are tightened can cause the clamp to penetrate the insulation, which may cause a short circuit when the cable is energized. After NM cable and ground wires have been secured, electrical devices are typically connected to the cable and mounted to the electrical box.

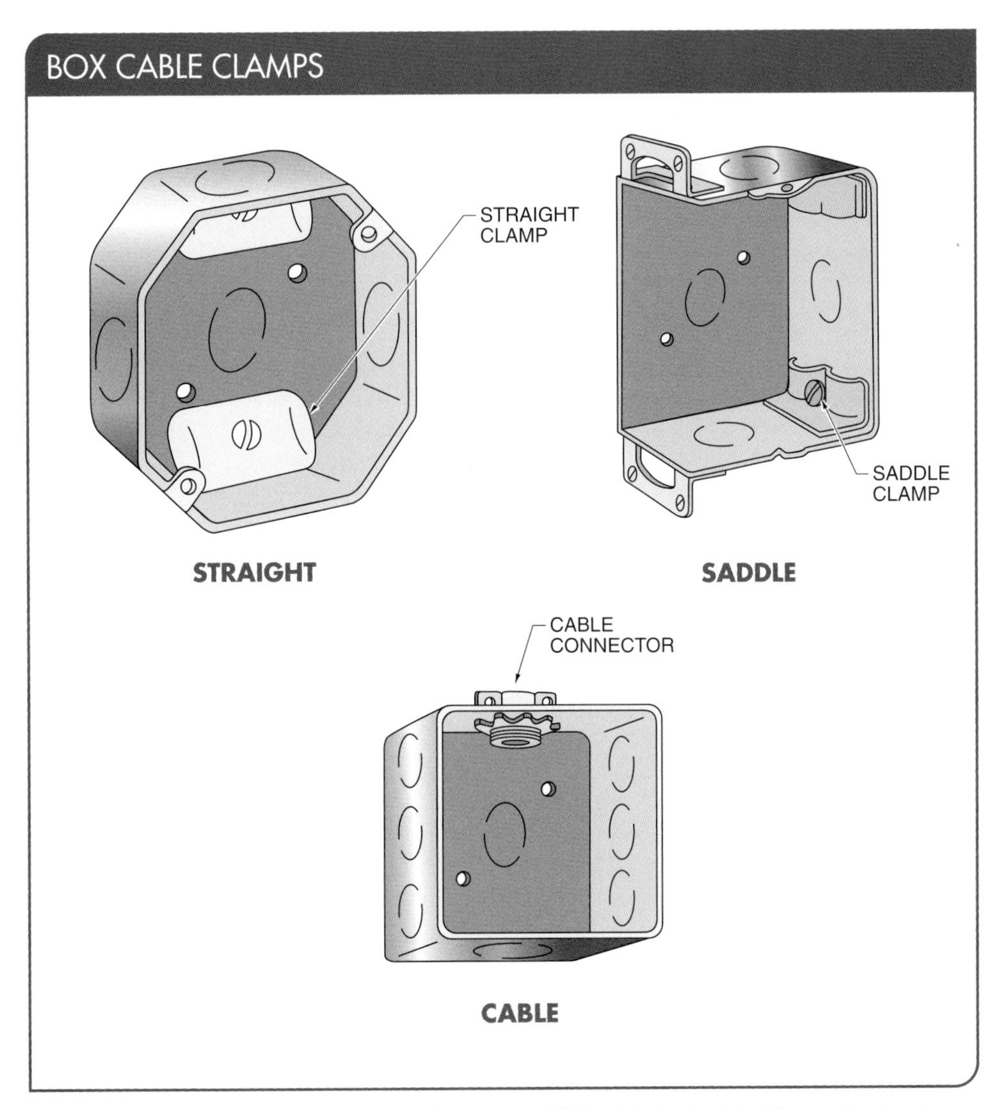

Figure 8-22. Box cable clamps are used to secure NM cable to electrical boxes. Locknuts are installed so that the points of the nut point inward to dig firmly into the metal box.

Connectivity and Folding Back Wires into the Box

Stripping 8″ of wire off the cable end allows devices to be wired to the circuit using properly sized wire connectors to make a good connection. After the wires are connected, the last roughing-in task is to neatly fold and tuck the wires inside the electrical box. Folding and tucking wires inside the electrical box avoids damaging exposed wires as wall coverings are placed and other construction is being performed. **See Figure 8-23.**

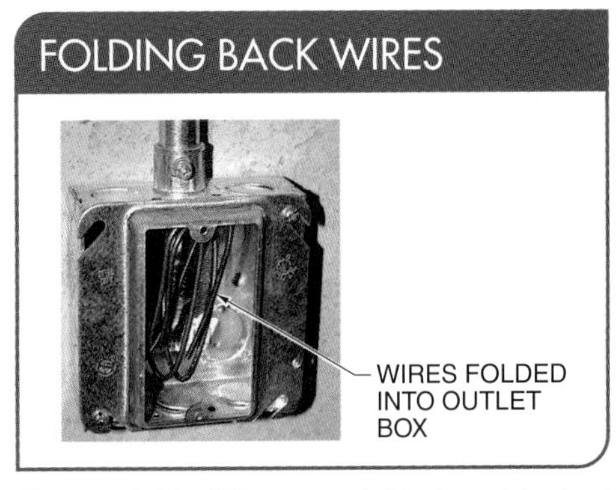

Figure 8-23. Wires are folded and tucked inside electrical boxes before wall or ceiling coverings are installed.

Electrical boxes must be sized appropriately to accommodate wires and wire connectors. The wires should be neatly folded into the box. Bare ground wires must not make contact with hot terminals.

SECTION 8.2 CHECKPOINT

1. How is NM cable prepared for termination?
2. What tools are used to remove the outer plastic jacket of NM cable and the conductor insulation?
3. What does the phrase "roughing-in" mean?
4. What tools are used to drill holes in studs and joists?
5. When are extended drill bits used?
6. What are fish tapes used for?
7. What are the three methods used to secure ground wires?
8. What is the last roughing-in task?

Chapter Activities

8 Nonmetallic-Sheathed Cable

Name ______________________ Date ______________

Nonmetallic-Sheathed Cable Information

Identify the nonmetallic-sheathed cable components shown.

______________ **1.** Cable information

______________ **2.** Conductors

______________ **3.** Conductor insulation

______________ **4.** Ground

______________ **5.** Plastic cable jacket

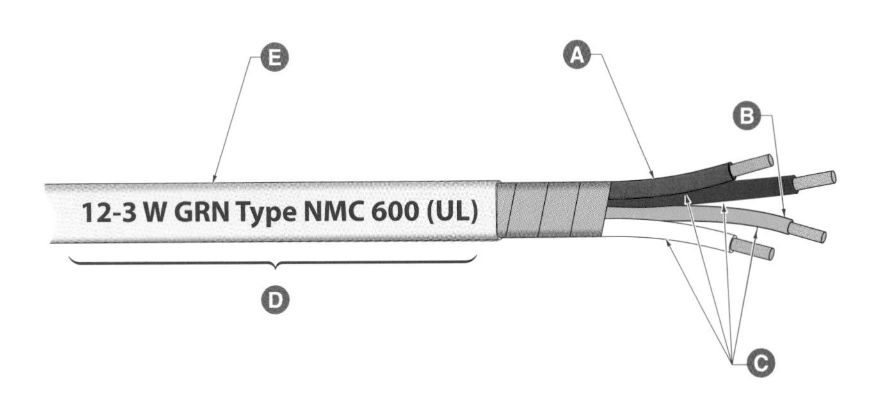

Wiring Receptacles

1. Draw arrows to indicate where the ground, neutral, and hot wires are terminated on the receptacle.

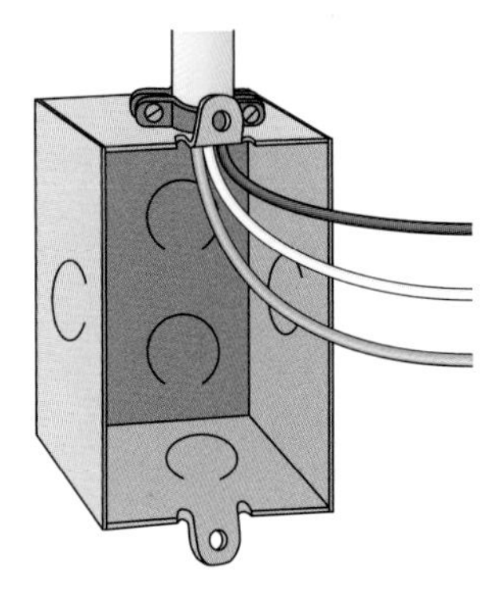

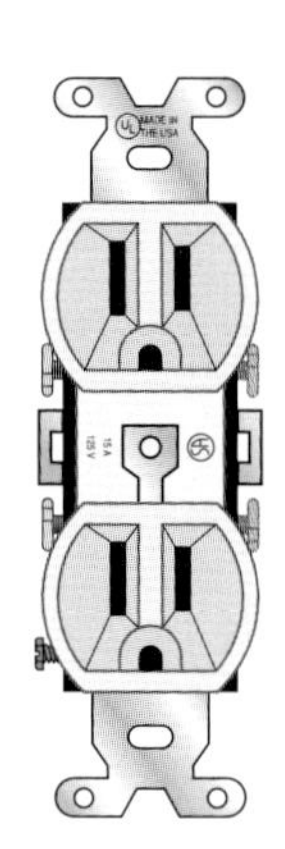

Box Cable Clamps

Identify the box cable clamps shown.

______________________ **1.** Cable

______________________ **2.** Saddle

______________________ **3.** Straight

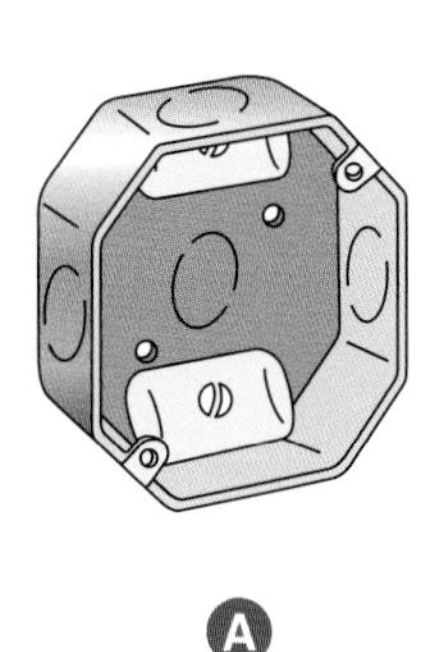

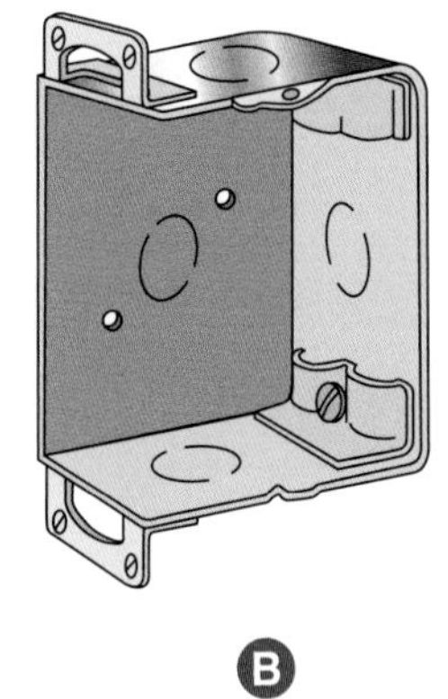

9 Metallic-Sheathed Cable

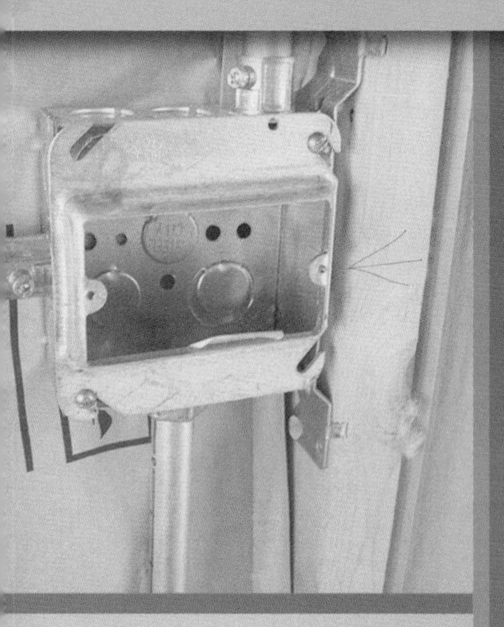

SECTION 9.1—METALLIC-SHEATHED CABLE

- Describe armored cable.
- Differentiate between Type AC, Type ACT, and Type ACL armored cable.
- Describe metal-clad cable.
- Describe Type MC^{AP} cable.

SECTION 9.2 – METALLIC-SHEATHED CABLE INSTALLATION

- Describe how cable is packaged for sale.
- Describe how armored cable is roughed-in.
- Describe how armored cable is pulled.
- Describe how armored cable is cut.
- Define anti-short bushing.
- Describe how armored cable is secured.
- Describe how armored cable is grounded.
- Explain how Type MC and MC^{AP} cables are grounded.

Learner Resources

ATPeResources.com/QuickLinks
Access Code: 791712

In residential construction, nonmetallic-sheathed cable may not be suitable for some wiring installations. Instead, metallic-sheathed cable is used to wire electrical devices in certain locations. Most types of metallic-sheathed cable are constructed with a corrugated metallic sheath to provide physical protection to insulated wires from environmental conditions. For increased protection, some cables can be made fire retardant and waterproof with special coatings.

SECTION 9.1 — METALLIC-SHEATHED CABLE

Metallic-sheathed cable is an alternative type of wiring installed in residential homes. This type of cable protects wiring from fire, vibration, wet conditions such as moisture-stricken areas, and physical harm such as from gnawing pests. The two types of metallic-sheathed cable used for installations are armored cable and metal-clad cable.

Armored cable and metal-clad cable are very similar in design and construction but differ in available sizes and grounding properties. Armored cable is also referred to as the trade name "BX," which is a generic term used by many manufacturers to denote armored cable. However, BX is not a defined term in the NEC®. Both types of metallic-sheathed cable offer thermoplastic insulation, which has efficient thermal characteristics and insulating properties.

CODE CONNECT

NEC® Article 320 (Type AC) and Article 330 (Type MC) identify the use, installation, and construction of metallic-sheathed cable used for residential purposes.

Armored Cable

Armored cable (Type AC) is a cable that consists of two, three, or four individually insulated wires and a bonding strip protected by a flexible metal outer jacket. The outer jacket is made of galvanized steel or aluminum alloy wrapped around the wires in an interlocked, overlapping configuration. Type AC has thermoset insulation covering the wires. Copper Type AC is available in No. 14 to No. 1 AWG, while aluminum Type AC is available in No. 12 to No. 1 AWG. **See Figure 9-1.**

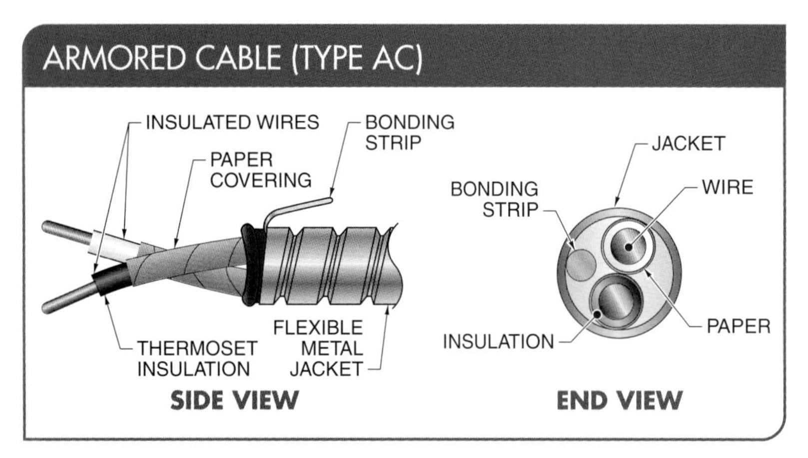

Figure 9-1. Armored cable (Type AC) consists of two, three, or four individually insulated wires and a bonding strip protected by a flexible metal outer jacket.

Metallic-sheathed cable is typically installed along studs and secured to electrical boxes with a cable connector.

In addition to insulation, each current-carrying wire is wrapped in light brown paper. The paper provides additional protection from heat and moisture within the sheath. The paper is removed from the part of the wires that are exposed for connection when the cable is being installed.

A *bonding strip* is an uninsulated conductor inside armored cable that is used for grounding. The bonding strip is in contact with the metallic sheath to ensure a proper conducting (ground) path along the entire length of the cable. This allows the metallic sheath to also be used as a grounding conductor. Bonding strips are typically made of aluminum.

The three types of armored cable used for residential work include Type AC, ACT, and ACL. Article 320 of the NEC® contains detailed information on armored cables.

Type ACT Cable. Type ACT cable is used in dry locations for either exposed or concealed work. Unlike Type AC, Type ACT has thermoplastic insulation (represented by the suffix "T") wrapped around the wires. Although the thermoset insulation of Type AC can withstand higher temperatures than thermoplastic insulation, it cannot be recycled like thermoplastic insulation. Type ACT can be fished through the air voids of masonry walls when the walls are not exposed to excessive moisture. It can also be used for under-plaster electrical extensions.

Type ACL Cable. Type ACL cable is used in damp locations for either exposed or concealed work. These types of work can include underground installations that are embedded in concrete or masonry and areas that are exposed to gasoline or oil. The suffix "L" indicates that the cable has a lead covering wrapped around the wire assembly. The lead covering provides additional protection for masonry and underground applications. **See Figure 9-2.**

Metal-Clad Cable

Metal-clad (Type MC) cable is a cable that consists of two or more individually insulated wires and a separate grounding wire inside a flexible metal outer jacket. Instead of wrapped paper around each insulated wire, Type MC cable has a polyester tape/film around the wire assembly. Type MC cables are made with either solid or stranded copper or aluminum wires and are available in No. 18 AWG to 2000 kcmil copper and No. 12 AWG to 2000 kcmil aluminum. *Note:* There is no limit to the amount of wires found in Type MC cable.

Unlike Type AC, Type MC cable contains a separate insulated grounding wire that serves as an equipment ground for outlet boxes, fixtures, or electrical panels. This does not allow the metal jacket to supplement a ground connection. However, some types of MC cable, such as Type MC^{AP} cable, specify that the metal jacket can be used as a ground connection. Some Type MC cables come with corrugated copper armor or with an outer supplementary corrosion-resistant material. **See Figure 9-3.**

Type MC^{AP} Cable

Type MC^{AP} cable is metal-clad cable manufactured with copper THHN insulated wires inside an interlocked flexible metal outer jacket. The letters in the abbreviation "THHN" denote a particular meaning: "T" to indicate thermoplastic insulation, "HH" to indicate high-heat resistance, and "N" to indicate nylon coating. Nylon coating is resistant to oil and gasoline.

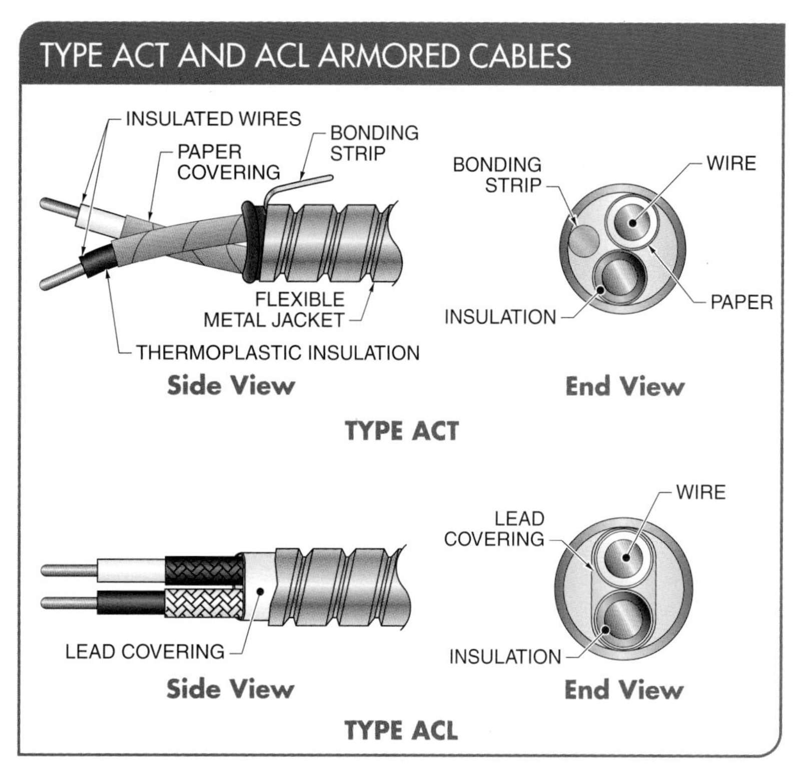

Figure 9-2. Type ACT has thermoplastic insulation, and Type ACL cable has lead-covered insulated wires.

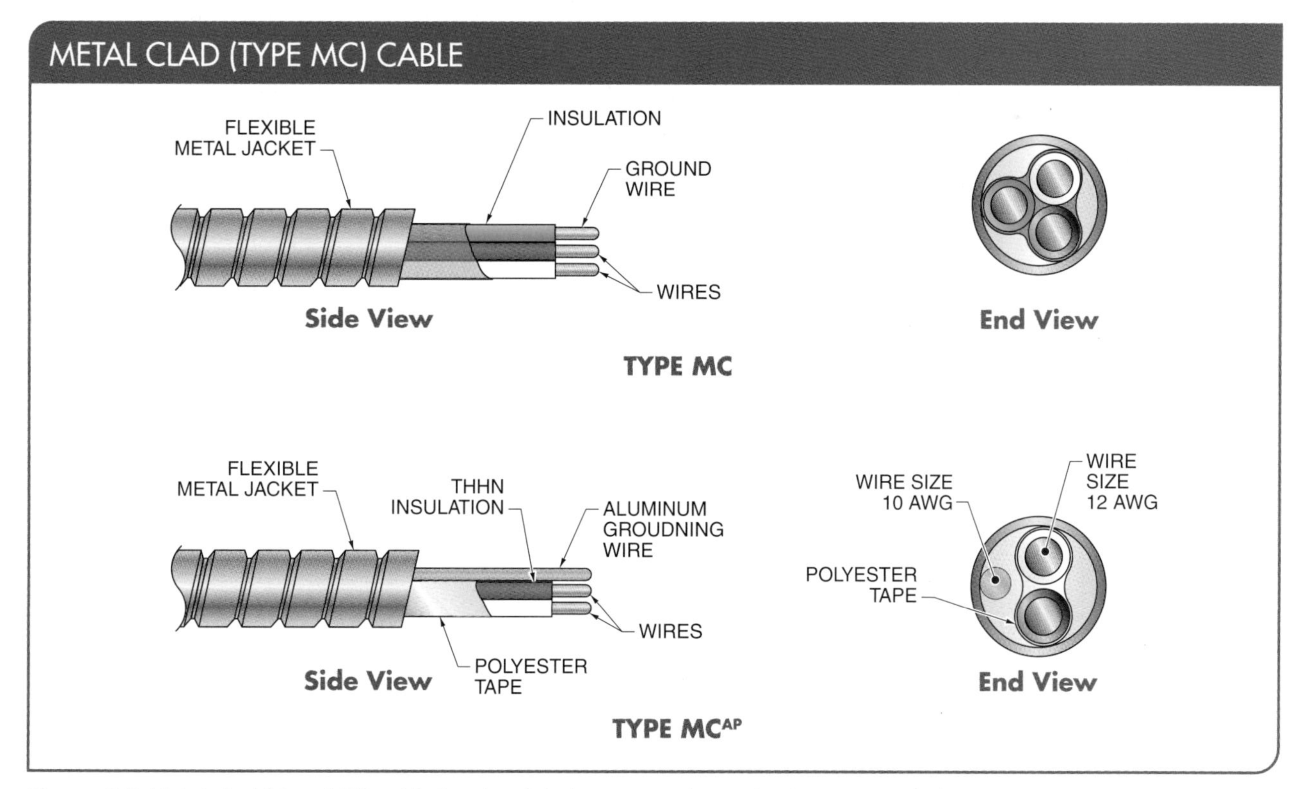

Figure 9-3. Metal-clad (Type MC) cable has insulated copper or bare aluminum ground wires.

Type MC^{AP} cables have a bare aluminum ground wire located outside the polyester tape covering that is in direct contact with the interlocked armor throughout the entire cable length. This aluminum ground wire serves as the grounding connection like the green insulated copper wire used in Type MC cable. The aluminum ground wire is typically two AWG sizes smaller than the insulated copper wires.

SECTION 9.1 CHECKPOINT

1. What is armored cable?
2. What is BX?
3. What is a bonding strip?
4. What are the three types of armored cable used for residential work?
5. What is the difference in ground wires between Type MC and MC^{AP} cable?

SECTION 9.2—METALLIC-SHEATHED CABLE INSTALLATION

Metallic-sheathed cables provide an efficient and convenient way to install wire in new construction and home renovations. Each type of metallic-sheathed cable is prewired, provides mechanical protection, and is flexible to fit in most areas of a home. Depending on the installation requirements, metallic-sheathed cable comes in a variety of lengths. Typically, Type AC, MC, or MC^{AP} cable can be purchased in coils of 25′, 50′, 100′, 125′, 200′, 250′ or reels of 500′ and 1000′. Each coil is shrink-wrapped and color-coded for a specific AWG size and number of current-carrying conductors, making it easier to distinguish each cable type. **See Figure 9-4.**

Installing Type AC, MC, or MC^{AP} cables requires many of the same techniques and equipment used when working with nonmetallic-sheathed cable. However, metallic-sheathed cable requires more careful attention when cutting and splicing due to exposed sharp metal edges. Typically, Type AC is installed throughout a home in limited-space applications where wires would normally be subject to vibration or inadvertent contact. For example, Type AC can be installed under kitchen sinks and in laundry areas, garages, workshops, attics, and any other locations that could have exposed conductors. **See Figure 9-5.**

Exposed runs of Type MC cable are permitted to be installed on the underside of joists as long as the cable is supported and not subject to physical damage.

Roughing-in Type AC Cables

Roughing-in is typically done when walls are open and drywall is not installed in a construction or renovation project. Roughing-in Type AC involves pulling, cutting, securing, and properly grounding the cable.

If Type AC must be bent, the minimum bend radius of the inner curve should be no less than five times the diameter of the cable per Section 320.24 in the NEC®. This measure is to avoid damaging the metallic sheath surrounding the wires. Installed Type AC must also be secured within 12″ of an electrical box. **See Figure 9-6.**

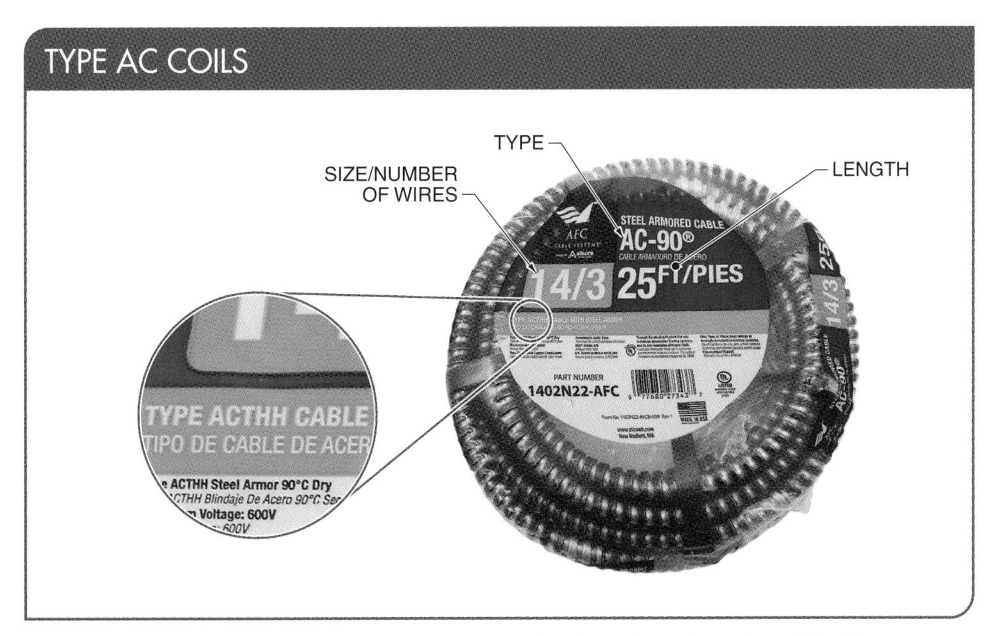

Figure 9-4. Type AC is typically sold in coils of 25′, 50′, 100′, 125′, 200′, and 250′.

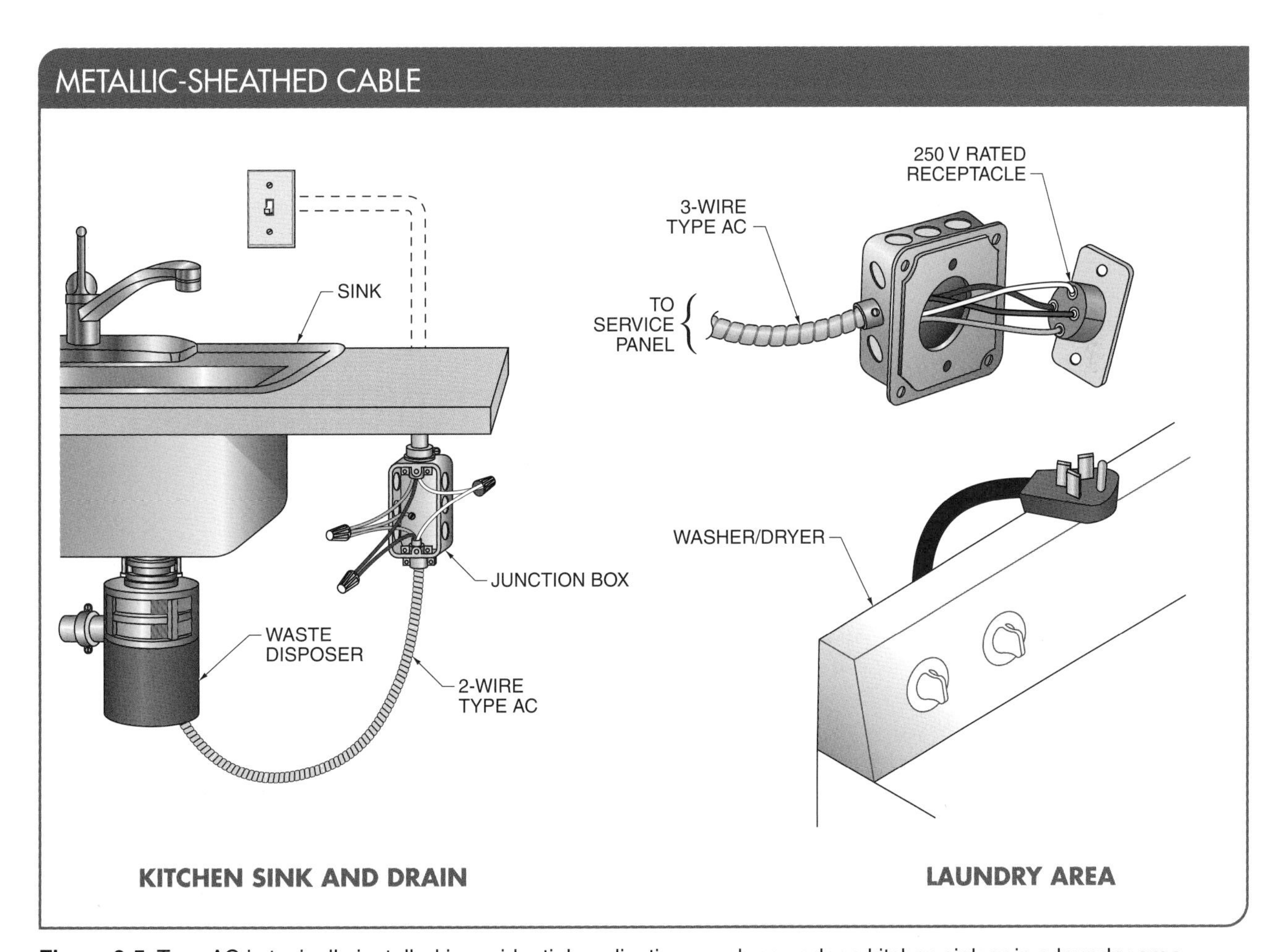

Figure 9-5. Type AC is typically installed in residential applications such as under a kitchen sink or in a laundry area.

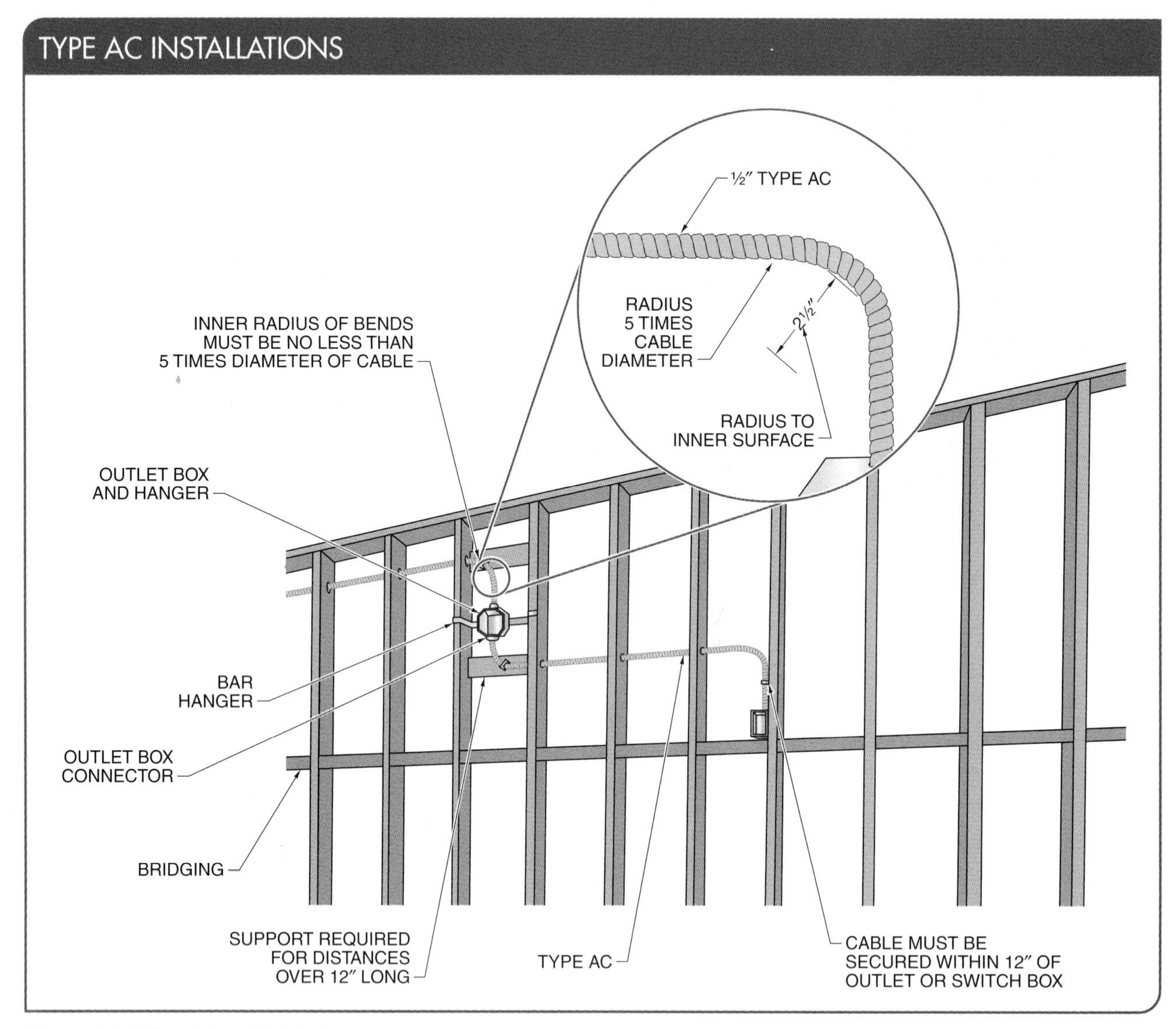

Figure 9-6. When Type AC is being roughed-in, the cable must be pulled, cut, secured, and grounded properly.

Pulling Type AC. To prevent waste when Type AC is roughed-in, the cable should be "run wild" into position before it is cut. A cable that is run wild is measured to the appropriate length between each electrical box to which it terminates. Unwinding a coil of Type AC from the center can keep the cable from twisting and kinking and maintain smoothness when the cable is run wild. A center cut is used to separate the armored cable, and side cuts prepare each end for mounting the cable to electrical boxes. **See Figure 9-7.**

Cutting Type AC. Type AC is cut once the proper measurements have been made for its installation. The two types of tools commonly used to cut Type AC are armored cable cutters and hacksaws. An armored cable cutter is the most effective tool to cut an abundance of Type AC for large renovations. Before any cutting operations are performed, protective gloves must be donned to protect the hands from cuts and scratches from the cutting tools.

If an armored cable cutter is not available, a hacksaw can be used to cut through the metal jacket of Type AC. The first step when using a hacksaw is to cut through the outer armor at a 45° angle about 6″ to 8″ from an end. To avoid damaging wire insulation, care must be taken not to cut too deeply into the jacket of the cable. When the armored cable is cut, the cable is separated by bending and twisting the cut section from the rest of the cable. Type AC can be carefully flexed until the outer armor breaks. Twisting the cable can open the armor enough for side-cutting pliers to be used. After the outer armor is cut through, any metal bending outward must be trimmed. **See Figure 9-8.**

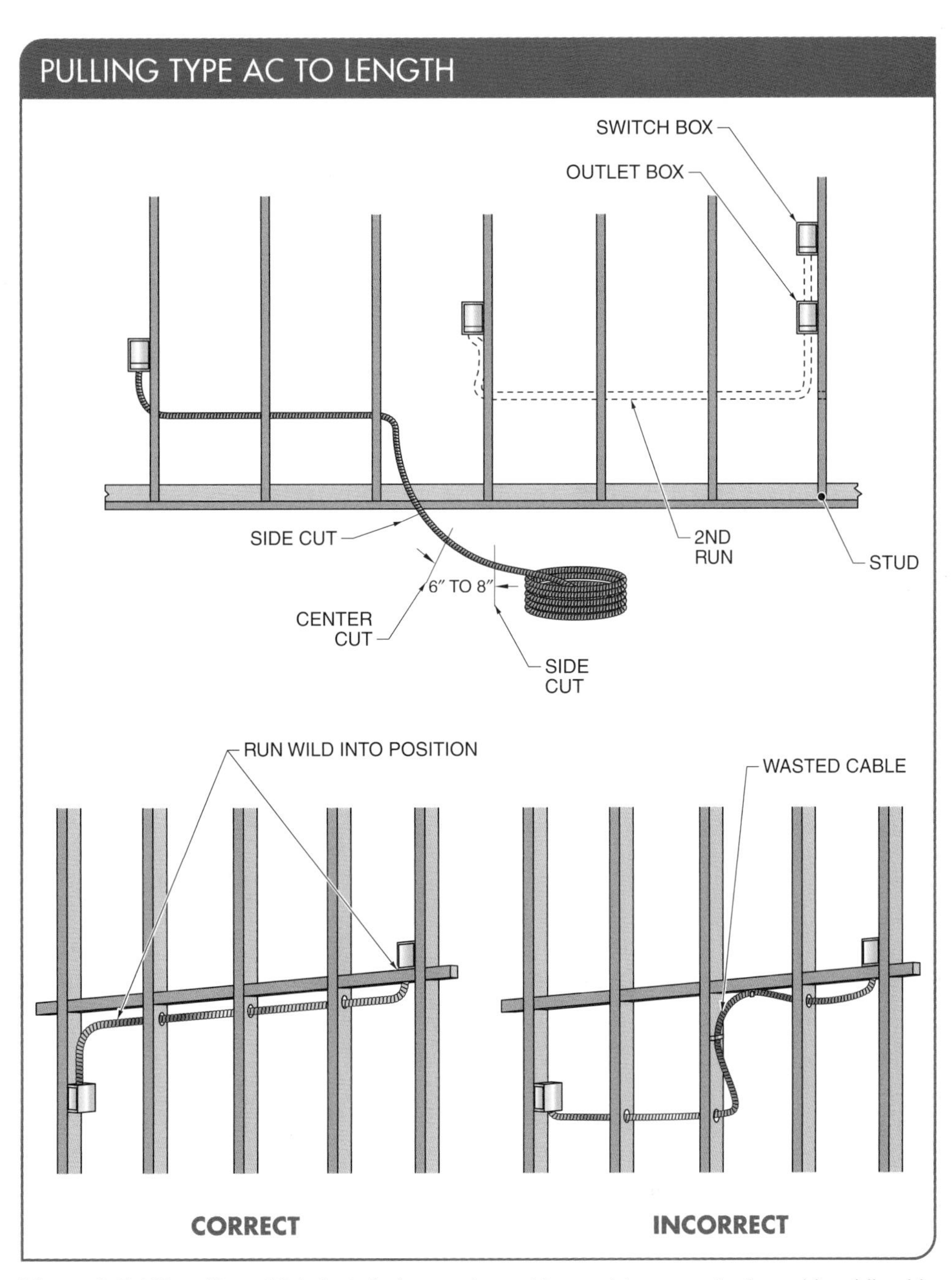

Figure 9-7. When Type AC is installed, a center cut is used to separate the cable while side cuts prepare each end for the cable to be mounted to electrical boxes.

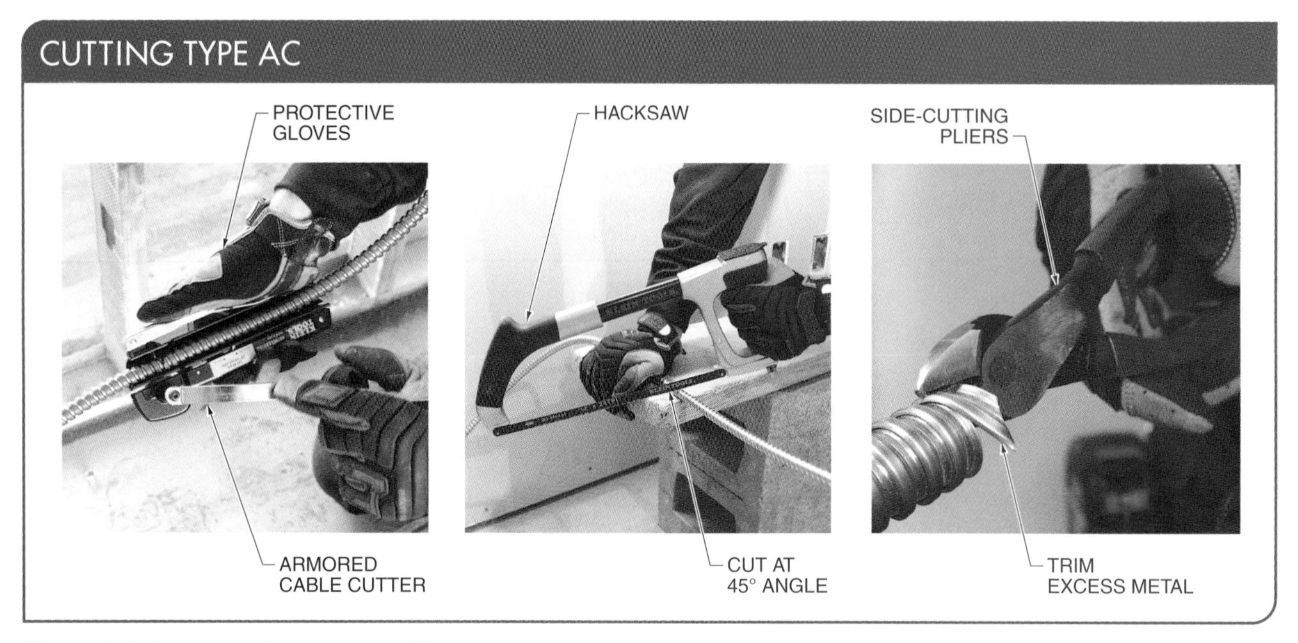

Figure 9-8. Armored cable cutters and hacksaws are two commonly used tools for cutting Type AC.

Installing Anti-Short Bushings. After Type AC is cut, a sharp edge remains on the ends of the metal jacket. To avoid any damage to the insulation of the wires, anti-short bushings must be installed. An *anti-short bushing* is a thermoplastic insulating device used to protect the wires of metallic-sheathed cable. The bushing covers the wires and is inserted into the jacket opening to cover the sharp edges at the ends of the jacket. This will reduce the possibility of damage to wire insulation. **See Figure 9-9.**

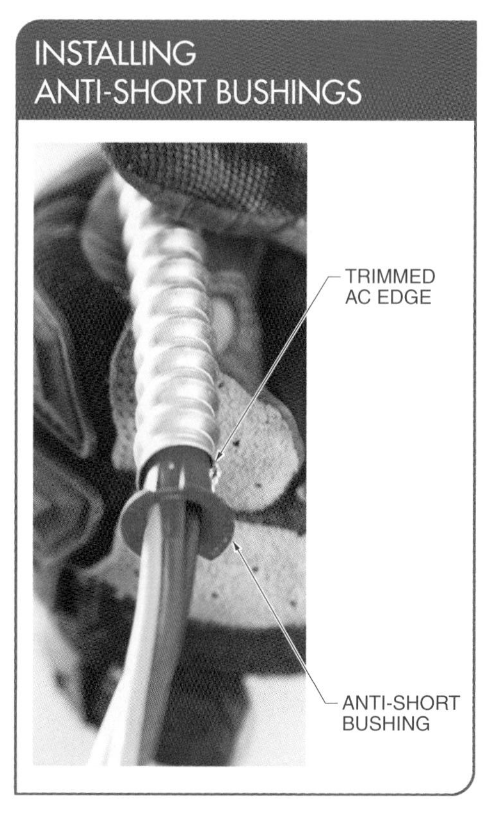

Figure 9-9. Anti-short bushings cover the sharp edges at the ends of the jacket to reduce the possibility of damage to wire insulation.

Securing Type AC. Armored cable can be secured in place with cable clamps or cable connectors. Cable clamps are typically part of a box, but cable connectors are installed before a box is wired. Cable connectors come in several different configurations. The most common configurations are straight-on box entry and 90° box entry connectors. **See Figure 9-10.** A 90° box entry connector is used where Type AC cannot be bent because the minimum permitted bend radius is less than five times the diameter of the cable.

> CODE CONNECT
>
> *Per NEC® Section 320.40, at every point where Type AC terminates at an electrical box, a connector/fitting shall be used with an anti-short bushing to protect wires from abrasion by contact with the metal jacket. The connectors used should allow the anti-short bushings to be visible during an inspection.*

Grounding Type AC. Only metallic electrical boxes can be installed with Type AC to maintain a proper grounding system. Grounding Type AC requires that the metal jacket and bonding wire be firmly secured to the electrical box. The bonding wire is bent back as the anti-short bushing is placed. A typical method used to ground Type AC is to secure the bonding wire between the connector and box and to wrap the bonding strip around the jacket before inserting it into the armored connector. A cable connector is then connected over the end of the cable. The cable, with connector, is then secured to the electrical box with a locknut. **See Figure 9-11.**

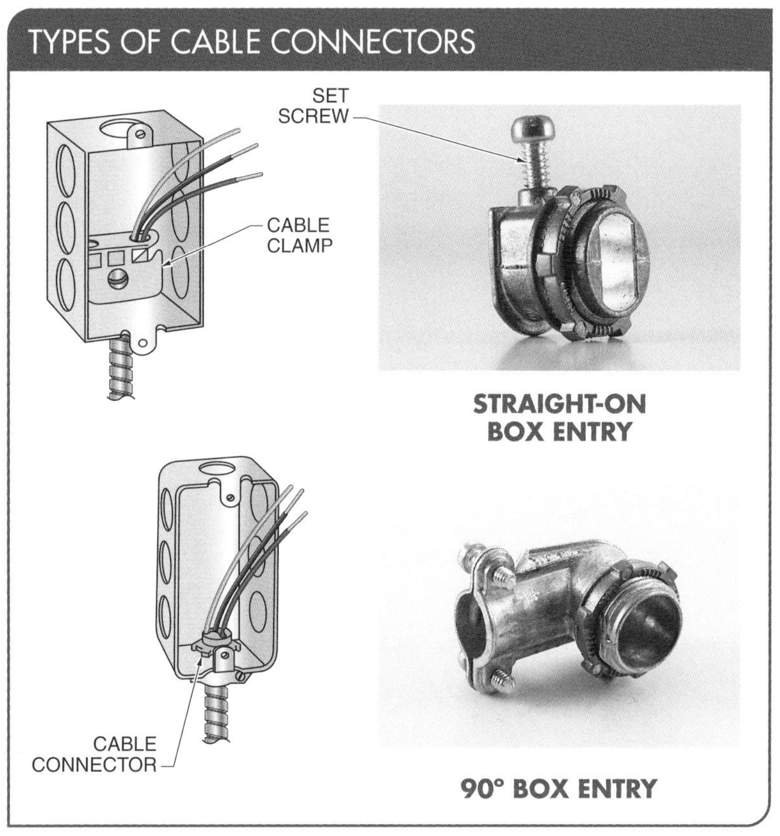

Figure 9-10. Type AC is secured to electrical boxes using cable clamps or cable connectors.

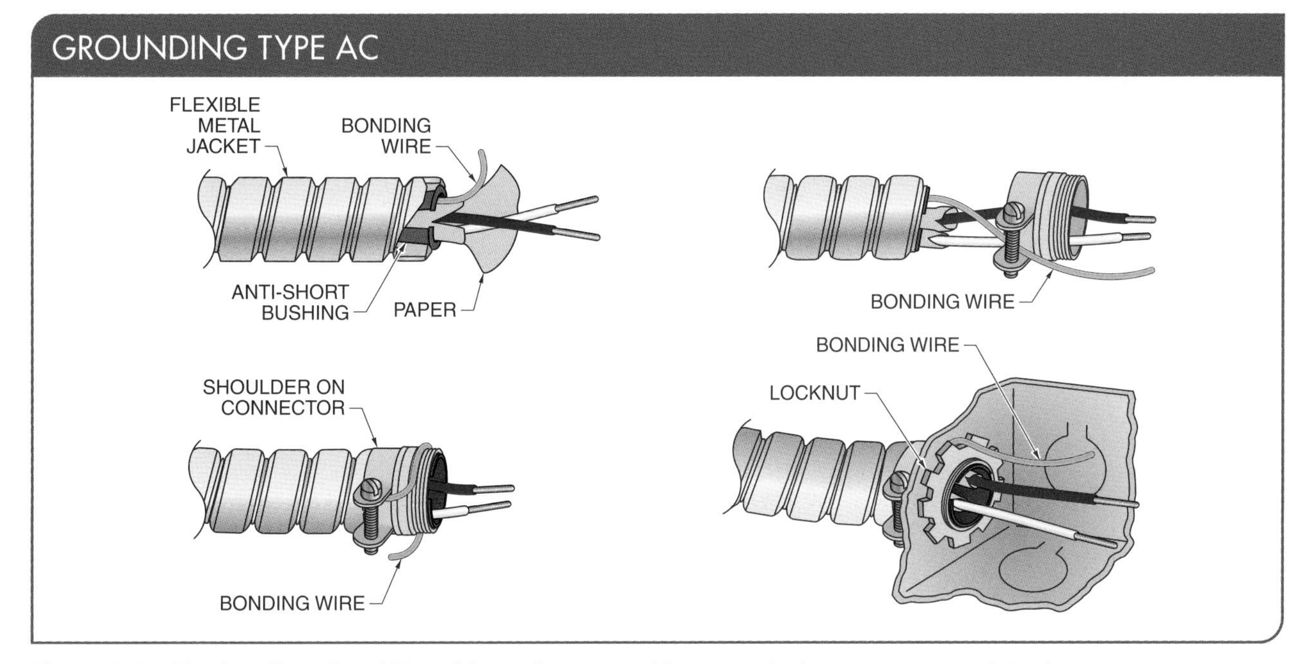

Figure 9-11. The bonding wire of Type AC can be secured between the box connector and the box to secure a proper ground path.

Grounding Type MC and MC^{AP} Cables

Type MC cable has a separate green insulated wire for grounding equipment. The insulated ground wire should be pigtailed with equipment grounds and connected to the electrical box with a grounding screw. Type MC^{AP} cable has a bare aluminum ground wire in direct contact with the interlocked flexible metal outer jacket throughout the entire cable length. This provides a unique built-in ground that eliminates the need to make up equipment grounding wires in every electrical box like those that are necessary for Type MC cable. This saves a significant time during installation. **See Figure 9-12.**

SECTION 9.2 CHECKPOINT

1. In what lengths are Type AC, MC, and MC^{AP} cable typically sold?
2. Within how many inches should Type AC be secured of an electrical box?
3. What two tools are commonly used to cut Type AC?
4. What is an anti-short bushing?
5. What are the most common connectors used to secure Type AC?
6. What is the minimum bend radius of Type AC?

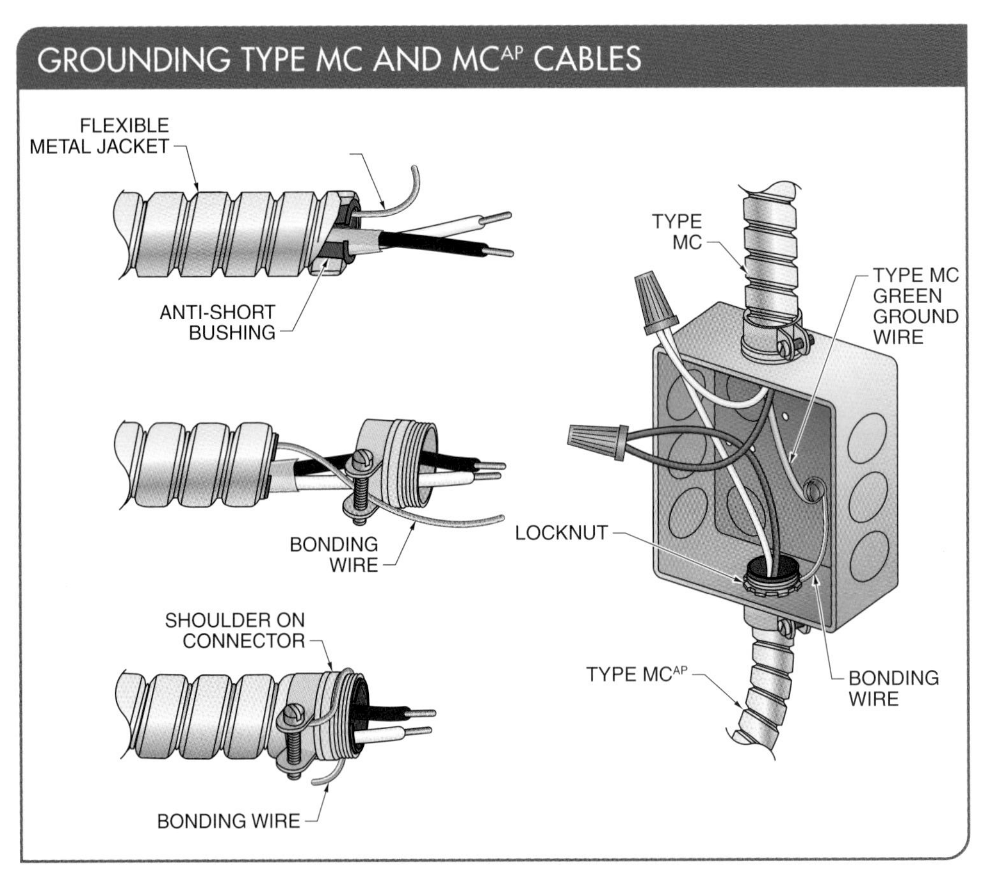

Figure 9-12. The insulated ground wire or bare aluminum wire of Type MC cable should be pigtailed with incoming and outgoing ground wires and then connected to the box with a grounding screw.

Chapter Activities

9 Metallic-Sheathed Cable

Name ______________________ Date ______________

Metallic-Sheathed Cable

Identify each type of the metallic-sheathed cable shown.

______ **1.** Type MC

______ **2.** Type ACL

______ **3.** Type ACT

______ **4.** Type MC^{AP}

End View

A

End View

B

End View

C

End View

D

Installing Type AC

Determine whether each run of Type AC is correct or incorrect.

__________________ **1.** Run A is ______.

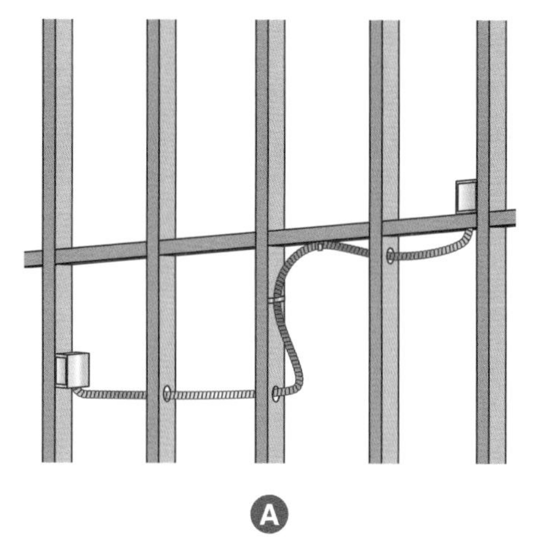

__________________ **2.** Run B is ______.

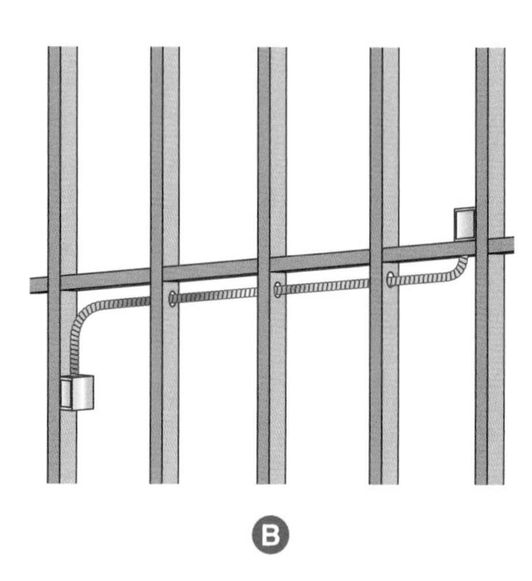

10 Conduit

SECTION 10.1—CONDUIT

- Define "conduit."
- Describe electrical metallic tubing (EMT).
- Describe rigid metal conduit (RMC).
- Describe flexible metal conduit (FMC).
- Describe polyvinyl chloride (PVC) conduit.

SECTION 10.2—CONDUIT INSTALLATION

- Describe how EMT is installed.
- List and describe conduit bends.
- Explain how to secure EMT.
- Describe how RMC is installed.
- Describe how FMC is installed.
- Describe how PVC is installed.
- Explain how wires are pulled through conduit.

Learner Resources
ATPeResources.com/QuickLinks
Access Code: 791712

In residential electrical systems, raceway wiring methods are used to distribute wire in order to power branch circuits and electrical equipment. A typical residential raceway is constructed using conduit. Conduit is specifically designed to hold, protect, and distribute wires throughout a home. Conduit wiring systems must be careful planned. The installation of a sound raceway design requires the knowledge of specialized tools and calculations. Any conduit design layout must be in accordance with the National Electrical Code® (NEC®) and local building codes.

SECTION 10.1—CONDUIT

A *conduit* is a metallic tube that protects and routes electrical wiring. The four most common types of conduit used in residential electrical systems are electrical metallic tubing (EMT), rigid metal conduit (RMC), flexible metal conduit (FMC), and polyvinyl chloride (PVC) conduit. **See Figure 10-1.**

Electrical Metallic Tubing (EMT)

Electrical metallic tubing (EMT) is light-gauge metallic tube used to route wires. EMT is often referred to as "thin-wall conduit" due to it having a wall thickness that is about 40% the thickness of rigid metal conduit. Typically, EMT is used for a complete residential house layout because it is lighter and easy to bend. EMT comes unthreaded, but it can be connected using set-screw, indentation, or compression-type couplings and connectors. NEC® Article 358 explains the use, installation, and construction specifications for EMT.

Rigid Metal Conduit (RMC)

Rigid metal conduit (RMC) is heavy-duty galvanized metallic tube used to route service entrance conductors. For residential electrical systems, RMC is limited mainly to risers for service entrances. Risers allow service entrance cable to be located high above the ground to minimize contact with individuals around the home. Many electrical suppliers carry RMC sizes in various lengths that are prethreaded and ready for use. Standard trade sizes are available in ½″ to 6″ diameters at a standard length of 10′. NEC® Article 344 explains the use, installation, and construction specifications for RMC.

CODE CONNECT

Per NEC® Subsection 358.20(A), Minimum, EMT smaller than metric designator 16 (trade size ½) shall not be used; and per (B), Maximum, the maximum size of EMT shall be metric designator 103 (trade size 4).

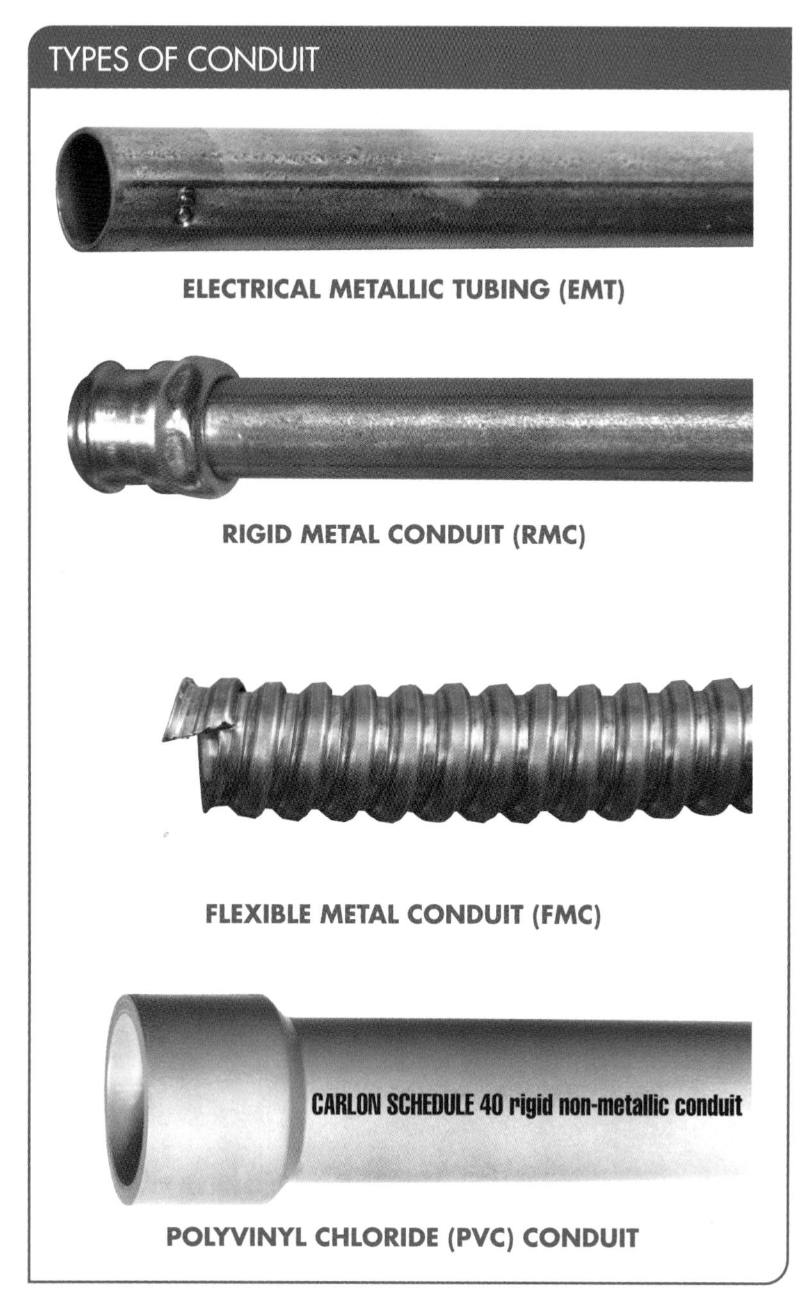

Figure 10-1. Common types of conduit used in residential electrical systems are EMT, RMC, FMC, and PVC conduit.

Flexible Metal Conduit (FMC)

Flexible metal conduit (FMC) is metallic tube of interlocked metal strips that is bendable by hand. Unlike armored or metal-clad cable, wires are routed through FMC after the raceway is installed. FMC is designed to connect to appliances and other equipment. FMC is often used where some type of movement or vibration is present or where other conduit might be difficult to bend. For example, FMC is suited for wiring to an electric motor. FMC is often referred to as the trade name "Greenfield" and is available in trade sizes from ½″ to 4″ diameters. The ⅜″ trade size can be used for lengths not more than 6′. NEC® Article 348 explains the use, installation, and construction specifications for FMC.

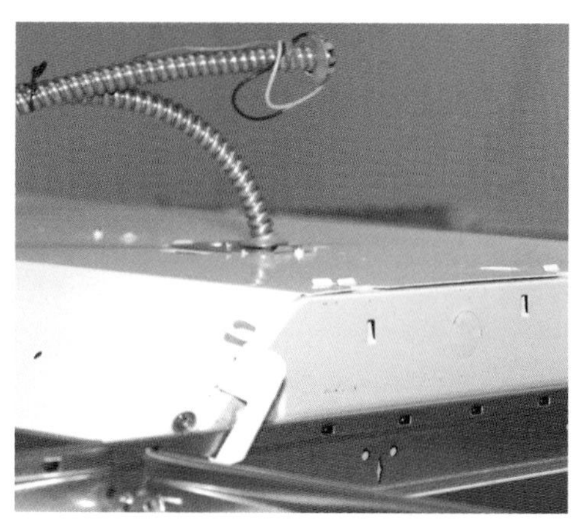

FMC is often used to wire light fixtures in suspended ceilings.

Polyvinyl Chloride (PVC) Conduit

Polyvinyl chloride (PVC) conduit is strong lightweight plastic pipe used for indoor, outdoor, and underground installations. PVC provides good insulation, has high impact resistance and tensile strength, and does not rust from exposure to moisture. PVC conduit is sometimes referred to as "rigid nonmetallic conduit" and is available in trade sizes from ½″ to 6″ diameters. NEC® Article 352 explains the use, installation, and construction specifications for PVC.

The two types of PVC are Schedule 40 (thin wall) and Schedule 80 (heavy wall). Schedule 40 PVC does not have the strength of similarly sized metal conduit and cannot withstand severe physical damage. Schedule 80 PVC is durable and has strength properties similar to RMC. **See Figure 10-2.**

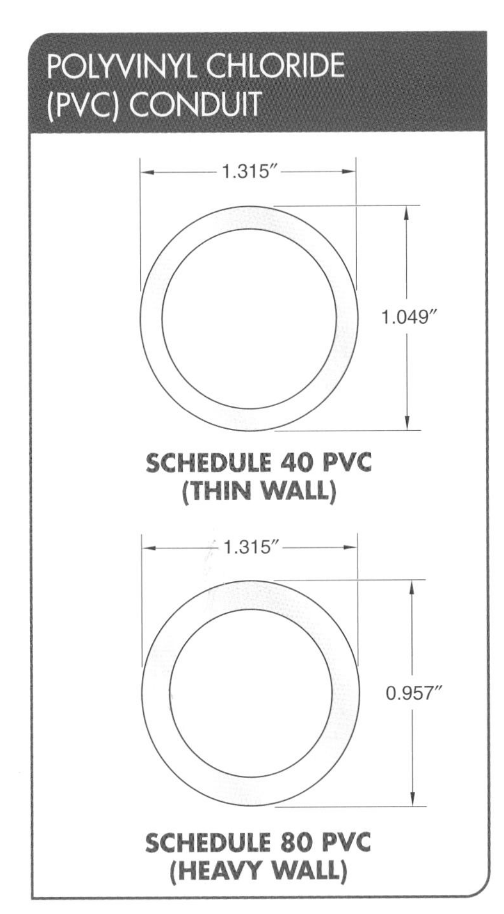

Figure 10-2. PVC conduit is rated as Schedule 40 (thin wall) and Schedule 80 (heavy wall).

SECTION 10.1 CHECKPOINT

1. Define "conduit."
2. What is EMT?
3. What is RMC?
4. What is FMC?
5. What is Greenfield?
6. What is PVC conduit?

SECTION 10.2—CONDUIT INSTALLATION

Each type of conduit system requires that specific installation guidelines be followed. These installation guidelines include gathering the proper materials and tools, understanding the local building requirements, and having knowledge of the NEC®.

Installing EMT

The process of installing EMT involves cutting, bending, and securing conduit sections. EMT is typically purchased in bundles of 100′. Each bundle contains ten lengths of conduit, each 10′ long. Electrical supply houses often have individual 10′ lengths of conduit that can be purchased. Trade sizes for conduit range from ½″ to 4″ in diameter. All sizes of conduit have limitations to the size and number of conductors that are permitted to be installed in the conduit. **See Figure 10-3.**

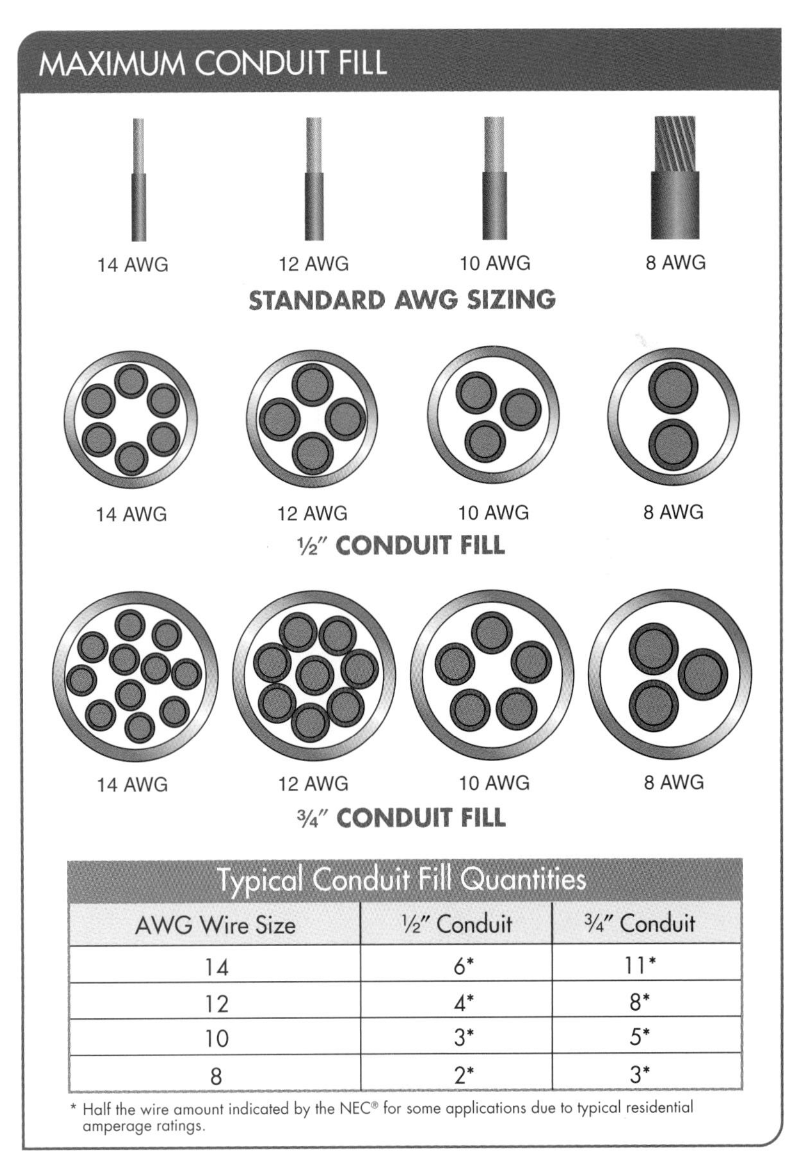

Typical Conduit Fill Quantities

AWG Wire Size	½″ Conduit	¾″ Conduit
14	6*	11*
12	4*	8*
10	3*	5*
8	2*	3*

* Half the wire amount indicated by the NEC® for some applications due to typical residential amperage ratings.

Figure 10-3. All sizes of conduit have limitations to the size and number of conductors that are allowed in the conduit.

Further information on the allowable number of wires per size of conduit can be obtained from Annex C of the NEC®. Conduit is also installed to add to or extend existing systems. When conduit cannot be installed in or through walls, wires must be run externally to the wall or ceiling through surface raceways. **See Figure 10-4.** After EMT is installed, equipment is grounded through the conduit system. Additional information on the installation of EMT can be found in NEC® Article 358.

Cutting EMT. Conduit such as EMT is typically cut to length using a hacksaw. However, all rough edges from cutting must be removed before wire can be pulled through conduit. Removing the rough edges is called "deburring" and is accomplished with a conduit reaming/deburring tool designed for this purpose. **See Figure 10-5.** Conduit can also be deburred with a file or power conduit reamer, which is attached to a drill.

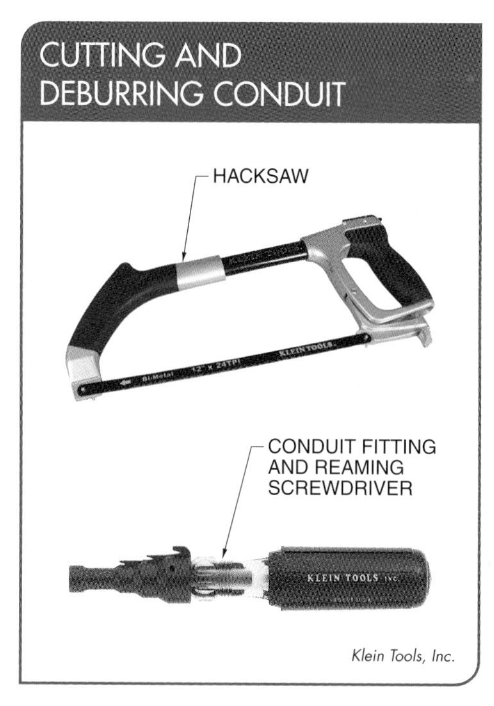

Figure 10-5. Conduit is often cut using a hack saw and deburred using a reaming/deburring tool.

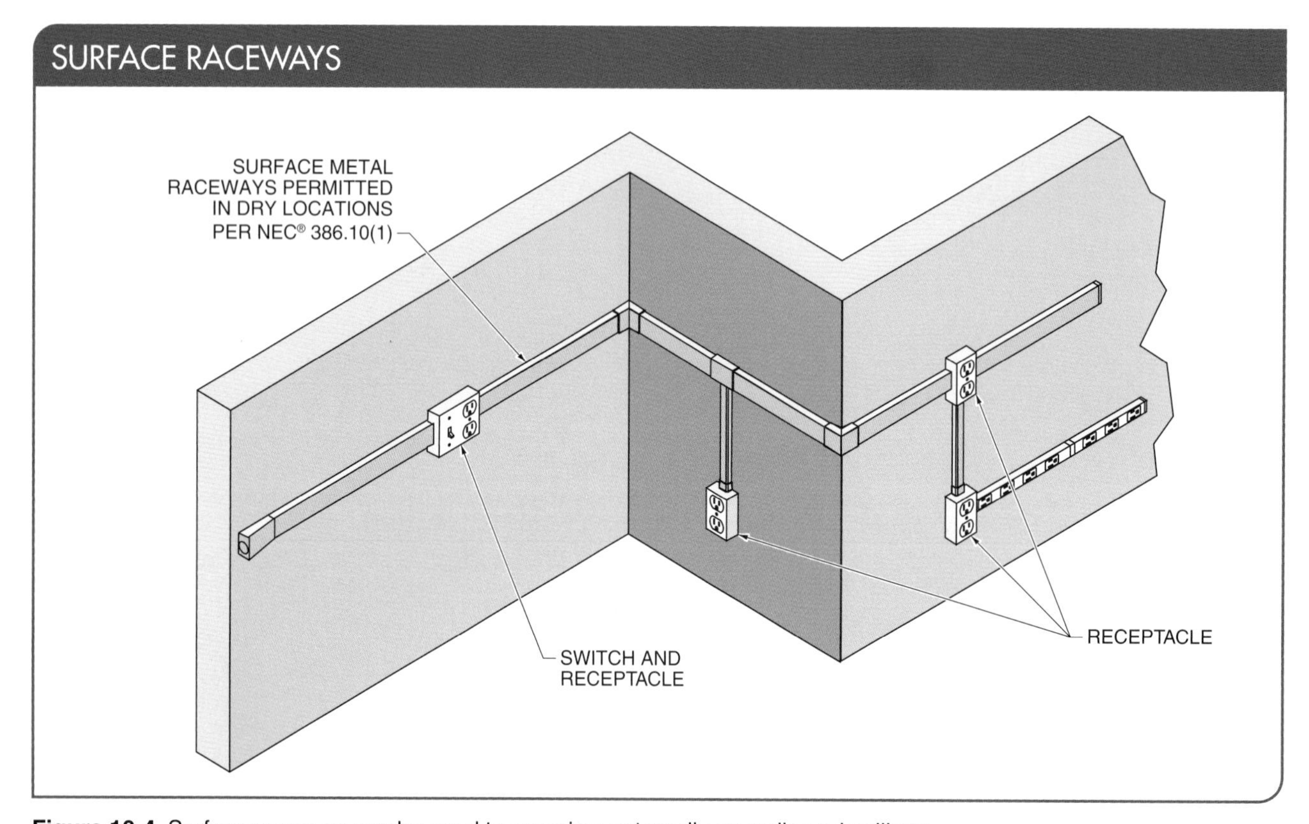

Figure 10-4. Surface raceways can be used to run wires externally on walls and ceilings.

Bending EMT. EMT bends are made using a handheld conduit bender and the proper technique. The parts of a conduit bender include the shoe, handle, hook, and foot pedal. **See Figure 10-6.**

The bender hook holds the conduit in place while foot pressure is applied to the foot pedal. All conduit benders have benchmarks on the shoe to indicate common bend angles. The arrow benchmark is the most used benchmark and is used to make 90° bends. The star benchmark indicates the back of a 90° bend. Many bender shoes have rim notches to indicate the center of a 45° bend. Some bender shoes have benchmarks on the outside. When conduit is parallel to such a benchmark, that specific angle has been achieved. NEC® Section 358.24 states that bends in tubing should be made so tubing will not be damaged and the internal diameter of the tubing will not be effectively reduced.

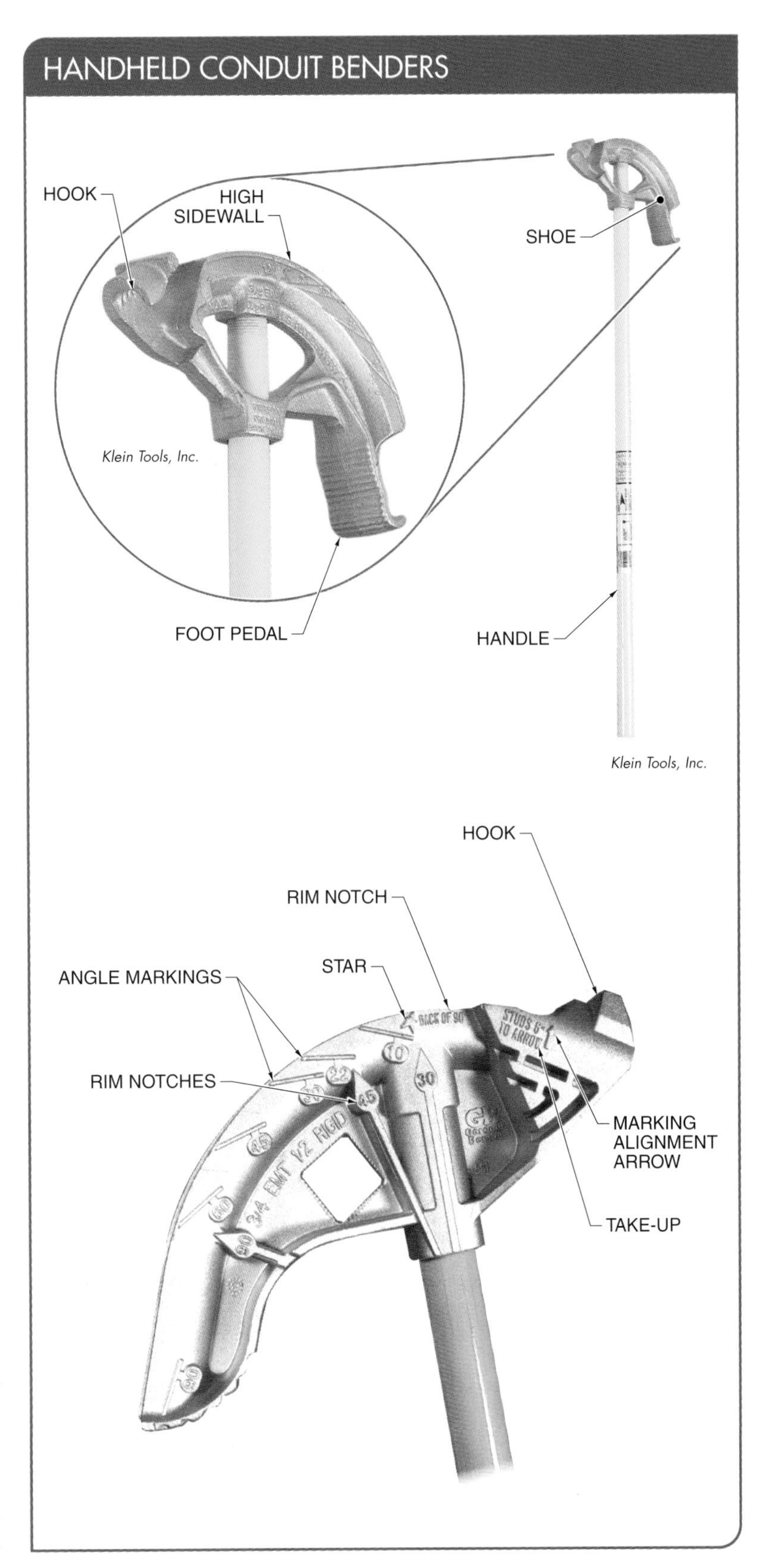

Figure 10-6. Handheld conduit benders are used with accurate measurement, careful alignment, and the proper technique to make bends.

CODE CONNECT

Per NEC® Section 358.24, Bends – How Made, the radius of the curve of any field bend to the centerline of the tubing shall not be less than what is shown in Table 2 in Chapter 9 of the NEC® for one-shot and full-shoe benders.

Types of Conduit Bends

The five main types of conduit bends are 45°, 90°, back-to-back, offset, and saddle bends. NEC® Section 358.26 states that a run between pull points (between outlet and outlet, between fitting and fitting, or between outlet and fitting) shall not contain more than the equivalent of four 90° bends (360° total), including bends located immediately at the outlet box or fitting. Table 2 in Chapter 9 of the NEC® provides the minimum acceptable radius for various sizes of conduit.

Forty-Five Degree Bends. To create a 45° bend, a conduit bender is placed on EMT conduit. The handle is pulled until the angle required (45°) is completed. **See Figure 10-7.** The bender is then removed, releasing the EMT conduit with a smooth-flowing 45° bend. Conduit benders also have a long arc that permits making 90° bends in a single sweep without moving the bender to a new position.

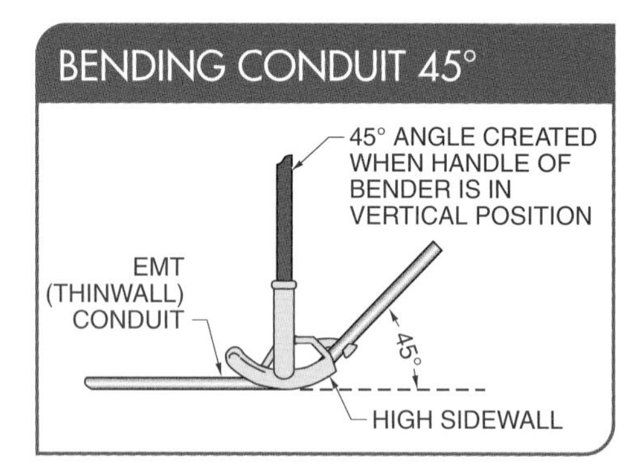

Figure 10-7. Rim notches on the bender shoe indicate the center of a 45° bend, which is achieved when the handle is pulled to the vertical position.

Ninety-Degree Bends. Ninety-degree bends are used to turn corners or to form a predetermined length for use as a stub through a floor that will be added onto later. To place the bending mark in the correct position on the conduit, the take-up value must first be subtracted from the desired stub length. *Take-up* is an adjustment made to a measurement when bending conduit. Since conduit is bent on a radius, the length of conduit in the shoe of the conduit bender must be accounted for. The take-up value can be found in the conduit bender manual or stamped on the bender. For example, ½″ conduit benders have a take-up of 5″.

To make a 90° bend for a stub, the following procedure is applied:

1. Measure the length of the required stub from back edge of conduit to where the conduit should end. For example, a stub may need to be a length of 10″. **See Figure 10-8.**
2. Subtract the take-up from the measured stub length and then mark the conduit at the calculated distance with a pencil. In this example, the 10″ stub will have a mark penciled at 5″ since a ½″ conduit bender has a take-up of 5″ (10 – 5 = 5).
3. Place the bender on the conduit and align the arrow benchmark on the bender with the pencil mark on the conduit. The bender hook should face toward the end of the conduit where the measurement was taken.
4. Place the conduit and bender on a hard surface. Place a foot on bender foot pedal and grasp the handle. Finally, use heavy foot pressure to make the bend.
5. Use hands to guide handle.
6. Check plumb with level.

EMT 90° bends can be run to a light fixture and through joists.

Back-to-Back Bends. Back-to-back bends consist of two 90° bends in the same length of conduit. These bends may be required in the opposite direction or in the same direction, 90° to the left and right. Back-to-back 90° bends can be used to connect two outlet or panel boxes that are a known distance apart or to fit conduit in a confined space between obstructions.

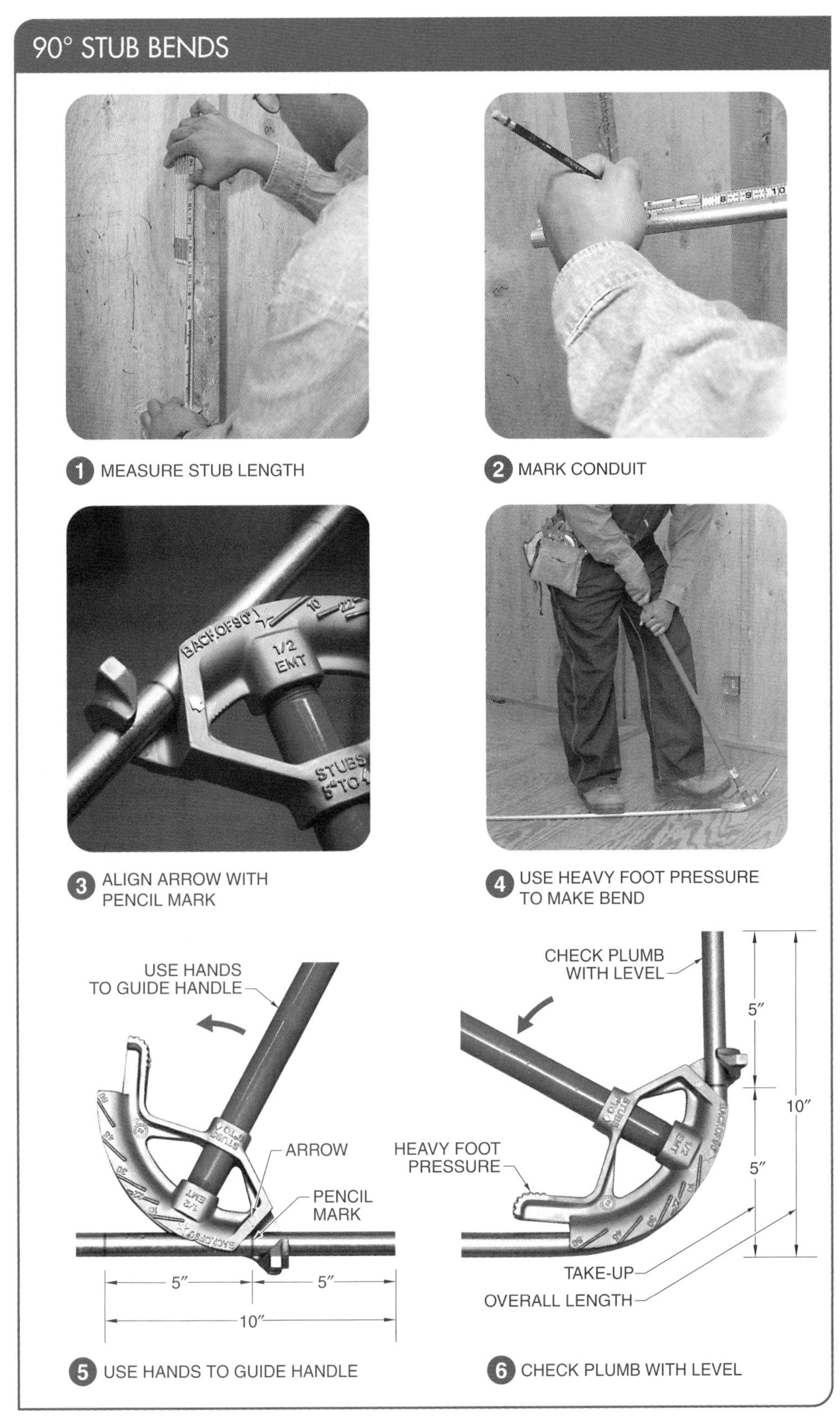

Figure 10-8. A pencil is used to mark the correct position for a 90° bend on the conduit after subtracting the take-up value from the desired stub length.

Three measurements are needed to make a back-to-back 90° bend: the required lengths of the two side stubs and the back-to-back distance. The back-to-back distance is measured from the back of one bend to the back of the other. Back-to-back distance can also be measured between both knockouts on outlet or panel boxes. **See Figure 10-9.** To make a back-to-back 90° bend in the same direction, the following procedure is applied:

1. Measure the required lengths of the two side stubs and the back-to-back distance.
2. Use the first measured stub length and the conduit bender's take-up value to make the first bend. (Use the same procedure for making a 90° bend.)
3. Then subtract the take-up value from the back-to-back distance to calculate the distance for making the pencil mark for the second bend. Place a straightedge ruler along the first stub, extending to the back of the first 90° bend. Measure from the straightedge away from the first bend and place a pencil mark at the calculated distance.
4. Place the bender on the conduit and align the arrow with the pencil mark. It is very important to keep the two bends in the same plane. Carefully sight along the bender and conduit to make sure the second bend is aligned with the first. Make certain the bender hook is pointing toward the first bend and then make the second bend.
5. Check the bend with a level. Measure the length of the second stub and place a pencil mark on the conduit. The stub length can be measured from the ground or from a straightedge placed on the back of the second bend.
6. Once the 90° bends are complete, conduit can be cut to length, reamed, and installed.

CODE CONNECT

Per NEC® Subsection 358.28(A), Reaming, all cut ends of EMT shall be reamed or otherwise finished to remove rough edges; and per (B), Threading, EMT shall not be threaded.

Offset Bends. An *offset* is a double conduit bend with two equal angles bent in opposite directions in the same plane in a conduit run. Offsets are often used to bypass obstructions or enter a knockout in a box or enclosure. The location of the bend is calculated based on the distance to the obstacle and the angle chosen for the bend.

The angle for the bend of an offset is a compromise based on the space available and the ease of pulling wire. Common offset angles are 5°, 10°, 15°, 22½°, 30°, and 45°. Each offset angle has an associated shrink constant and distance multiplier that are figured into the calculations for making offset bends. **See Figure 10-10.** A *shrink constant* is the reduction in distance that a conduit can run per inch of offset elevation. Every bend angle has a shrink constant value.

The first step in making an offset bend is to measure the offset rise. Measurements should be from the bottom of the conduit at the floor or wall to the bottom of the conduit where it will cross the obstacle. After the offset rise is measured the bend angle can be chosen. Although smaller bend angles take up more room to rise up to the obstacle (offset rise) than larger bend angles, it is easier to pull wires through smaller bend angles than larger bend angles.

Once the bend angle is chosen, the shrink can be calculated as *shrink = shrink constant × offset rise*. The resulting value is used to place the first pencil mark on the conduit. For example, a 30° bend is chosen to go over a 4½″ obstacle (offset rise) that is at a distance of 10″. The shrink constant for a 30° bend is 0.27. The shrink is then calculated as *shrink* = 0.27 × 4.5 = 1.215, which is rounded to 1⅛″. Since the distance to the obstacle is 10″, the first mark is placed at 11⅛″ (10 + 1⅛ = 11⅛)

Next the distance between bends can be calculated as *distance = multiplier × offset rise*. Since a 30° bend has been chosen for the offset rise of 4½″, the multiplier is 2.00. The distance between bends is then calculated as *distance* = 2 × 4.5 = 9. Therefore, the second pencil mark on the conduit is placed 9″ from the first mark.

Figure 10-9. The required lengths of the two side stubs and the back-to-back distance must be known to make a back-to-back 90° bend.

The first bend for an offset is made using a two-step procedure. **See Figure 10-11.** To make the first bend, the following procedure is applied:

1. Place the bender on the conduit with the hook facing away from the second mark and line up the arrow on the shoe with the first pencil mark.
2. With conduit squarely on the floor, use heavy foot pressure on the foot pedal of the conduit bender and grasp the handle until the bend is completed at the marks on the bender.

> CODE CONNECT
>
> *Per NEC® Subsection 358.30(A), Securely Fastened, EMT shall be securely fastened in place at intervals not to exceed 3 m (10′).*

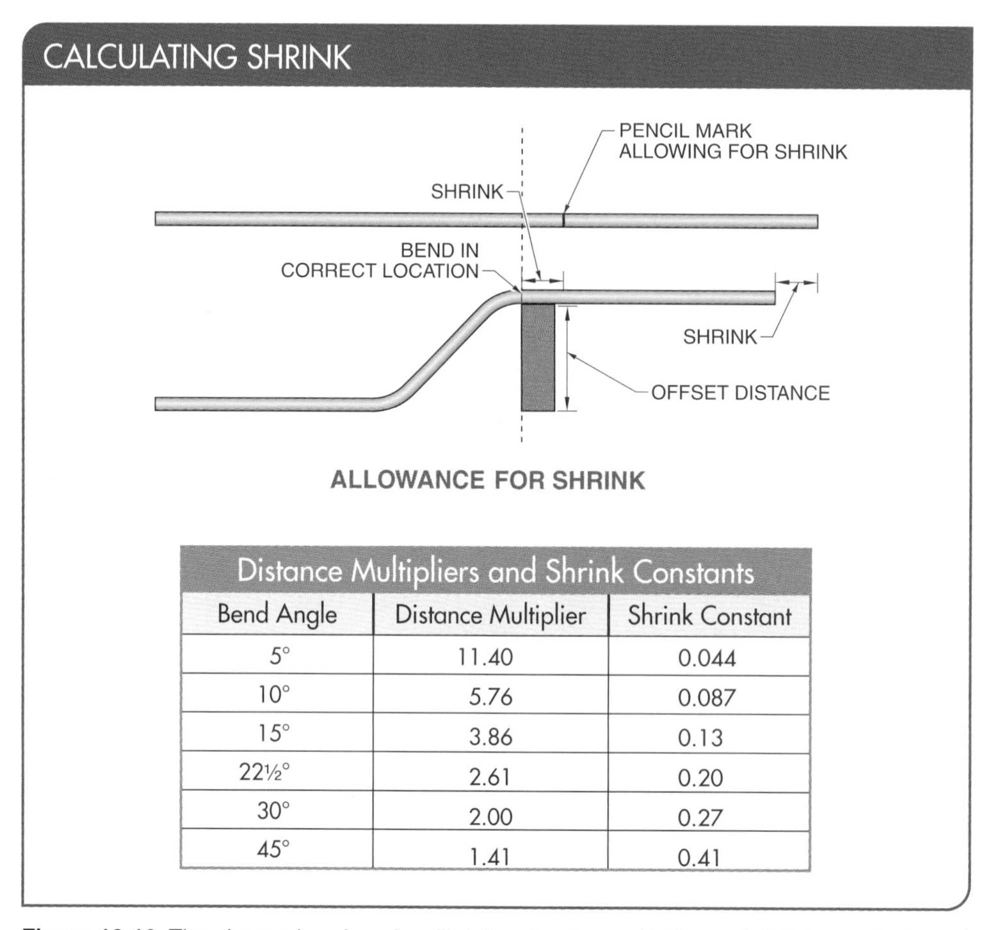

Distance Multipliers and Shrink Constants

Bend Angle	Distance Multiplier	Shrink Constant
5°	11.40	0.044
10°	5.76	0.087
15°	3.86	0.13
22½°	2.61	0.20
30°	2.00	0.27
45°	1.41	0.41

Figure 10-10. The chosen bend angle will determine the multiplier and shrink constant used to calculate shrink and the distance between bends.

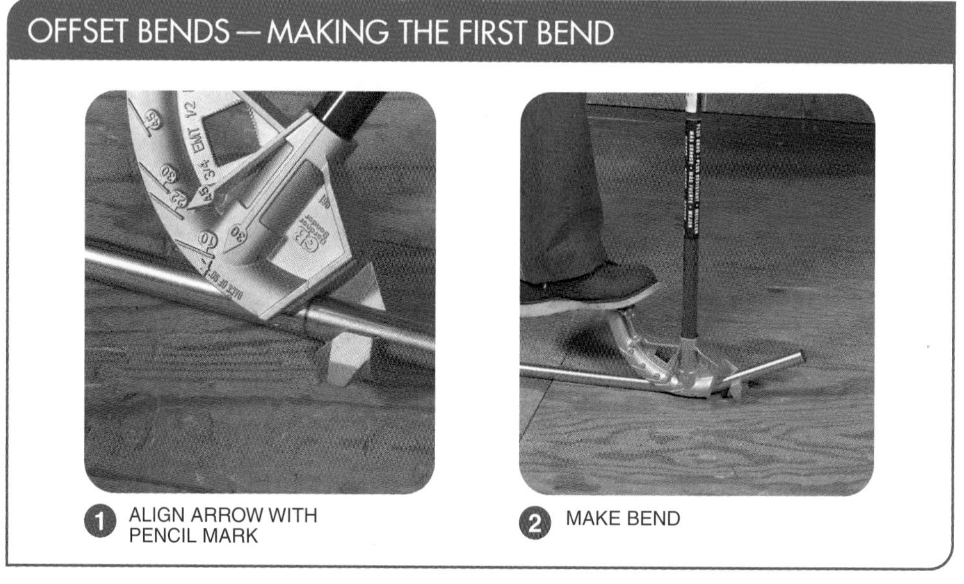

Figure 10-11. The first bend for an offset is made with the shoe of the bender on the floor and the conduit squarely on the ground.

Next, a six-step procedure is used to make the second bend for an offset. **See Figure 10-12.** To make the second bend, the following procedure is applied:

1. Turn the conduit bender upside down so that it rests on the handle. Move the conduit in the bender until the second mark is aligned with the arrow on the bender.
2. Rotate the conduit in the bender so that it is now turned 180° from its original position. Sight down the conduit and align the first bend relative to the bender handle. The second bend must be on the same plane as the first bend.
3. Start the second bend by applying hand pressure as close to the bender shoe as possible.
4. Make a small bend of 5° to 10° and check the alignment of the two bends.
5. Turn the conduit bender over and complete second bend on the floor using the foot pedal, making sure the conduit does not move or rotate in the bender.
6. Measure the accuracy of the bend with a level and rule (ruler).

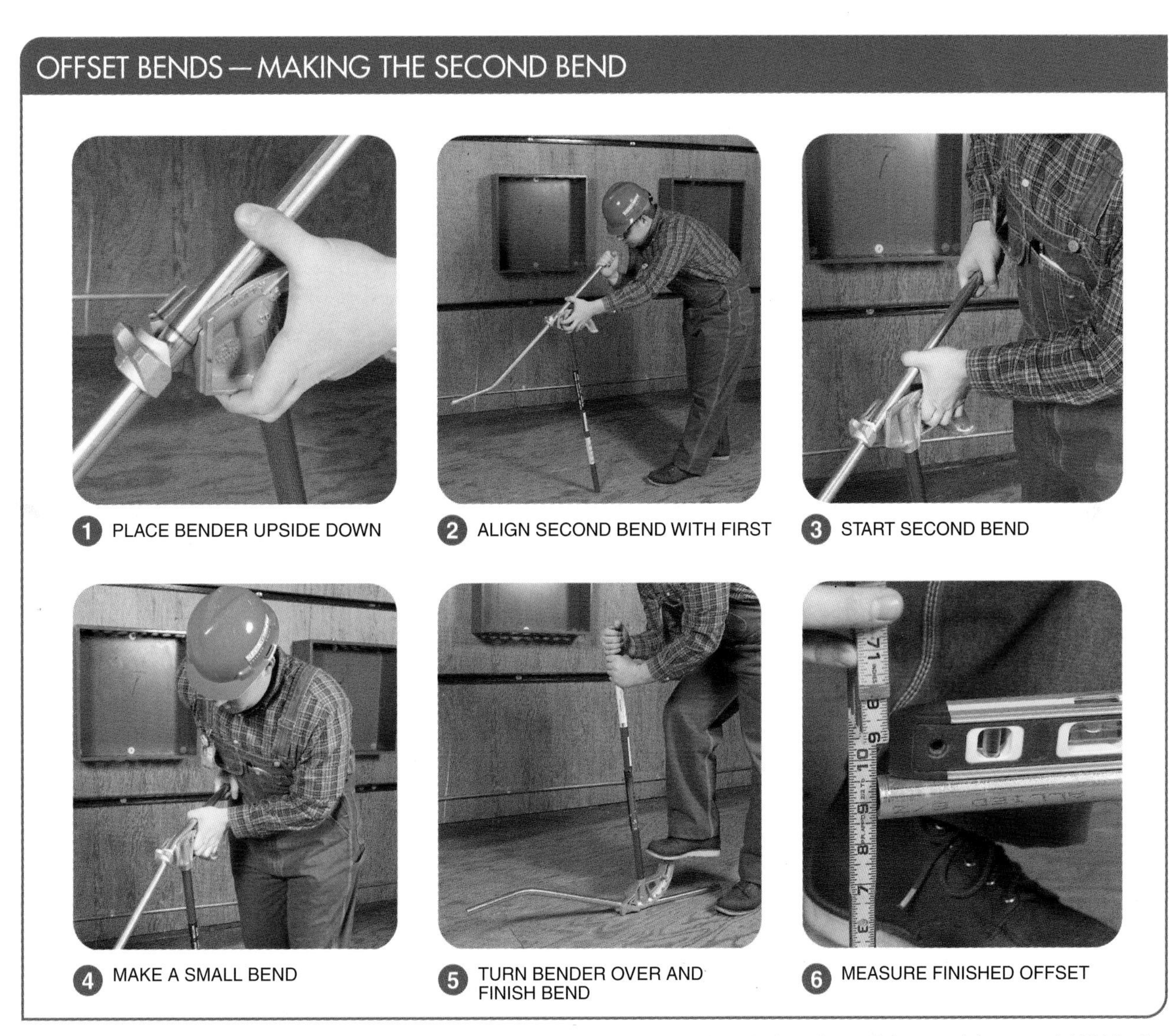

Figure 10-12. To make the second bend for an offset, the bender is rested on its handle and the conduit rotated 180° in the bender shoe and aligned with the first bend.

Saddle Bends. A saddle bend consists of three or four bends that are used to bypass obstructions and return the conduit to its original level. The number of bends used for the saddle bend depends on the size and shape of the obstruction. A three-bend saddle consists of a center bend that has double the angle of the side bends. A four-bend saddle consists of two offset bends with a length of straight conduit between the offsets.

A three-bend saddle is typically used for cylindrical obstructions that are smaller than the bend radius of the center bend. For example, the bend radius of a ½″ conduit bender is 4″. This means that a three-bend saddle would fit tight around an 8″ diameter pipe.

There are four variables that are taken into consideration when making saddle bends: the rise of the obstruction, the width of the obstruction, the distance to the obstruction, and the choice of bend angles. After these variables are determined, the shrink can be calculated.

For three-bend saddles, the distance to the obstruction is measured to the center of the obstruction. If it is not convenient to measure to the center of the obstruction, the distance to the obstruction's edge and the width of the obstruction can be measured. The distance to the center of the obstruction is half the width plus the distance to the edge of the obstruction.

For example, the most common choice for a center bend is 45° and 22½° for both side bends. A three-bend saddle with a 45° center bend, 22½° side bends, and an offset rise of 4″ has a shrink constant of 0.20. Therefore, shrink can be calculated as *shrink* = 0.20 × 4 = 0.8, which is rounded to ¾″. This means that if the end of the conduit to the center of the obstruction measures 32″, then ¾″ should be added and a mark should be penciled at 32¾″ for the calculated center bend.

For the 22½° side bends, the distance multiplier is 2.5. **See Figure 10-13.** Therefore, the distance between the bends can be calculated as *distance* = 2.5 × 4″ = 10″. Then, pencil marks for the side bends are placed 10″ from each side of the pencil mark for the calculated center bend.

THREE-BEND SADDLE MULTIPLIERS	
Side Bend Angle	**Distance Multiplier**
15°	3.8
22½°	2.5

Figure 10-13. A distance multiplier based on the chosen angle of the side bends is used to calculate the distance between the bends.

After the three pencil marks are made, the center bend can be made. **See Figure 10-14.** To make the center bend for a saddle bend, the following procedure is applied:

1. Place the conduit in the bender, align the pencil mark for the center bend with the 45° benchmark, and make the bend using foot pressure on the foot pedal.
2. The two 22½° side bends are made in the air. The bender is placed on the conduit with the bender hook facing the center bend and the pencil mark aligned with the arrow benchmark. The conduit is rotated 180° and the first side bend is made.
3. For the second side bend, the bender is removed from the conduit and reversed, and the arrow aligned with the pencil marks for the other side. The bender hook is pointed at the center bend for both side bends. Each new bend must be in the same plane as the center bend and the bender handle.

A four-bend saddle is made when conduit needs to change elevations to clear a large obstruction in a run. Four-bend saddles are two offset bends placed in the same plane in the same run. The four variables for making a four-bend saddle are the same as for making a three-bend saddle: the rise of the obstruction, the width of the obstruction, the distance to the obstruction, and the choice of bend angles. The shrink must then be calculated and added to the distance to the center of the obstruction. Also, locating the center of an obstruction for a four-bend saddle is very similar to locating the center of an obstruction for a three-bend saddle.

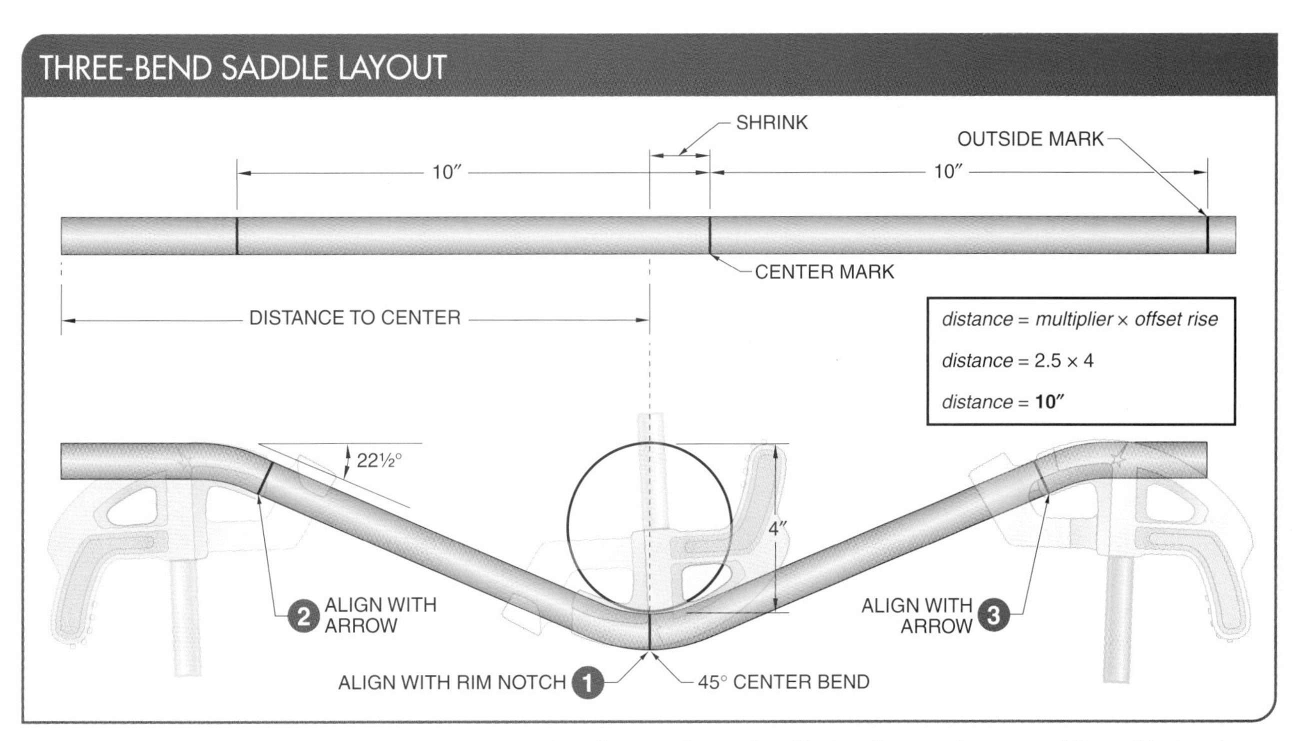

Figure 10-14. For a three-bend saddle, the center bend is usually made with the shoe on the ground. Two side bends are made with the shoe in the air and the bender handle on the ground.

Three-bend saddles can be run over PVC pipe parallel to each other.

A common method for placing the pencil marks for a four-bend saddle is called the edge of obstruction method. In this method, all bends are made relative to the location of the edge of the obstruction. The distance multiplier and shrink constant are the same as for making offset bends.

> CODE CONNECT
>
> *Per NEC® Section 358.120, EMT shall be clearly and durably marked at least every 3 m (10′) as required in the first sentence of 110.21(A).*

The most common angle for the offset bends of a four-bend saddle is 30°, which has a distance multiplier of 2 and a shrink constant of 0.27. Therefore, for an obstruction with a 7″ offset rise, the distance between bends is 14″ ($2 \times 7 = 14$) and the shrink is 1¾″ ($0.27 \times 7 = 1.89$, rounded to 1¾″). If the width of the obstruction is 12″, then the first mark is placed at the distance from the beginning of the conduit to the nearest side of the obstruction plus the shrink. **See Figure 10-15.** The second mark is placed 14″ from the first mark toward the beginning of the conduit. The third mark is placed 12″ away from the first mark, toward the far end of the conduit. Finally, the fourth mark is placed 14″ away from the third mark, toward the far end of the conduit.

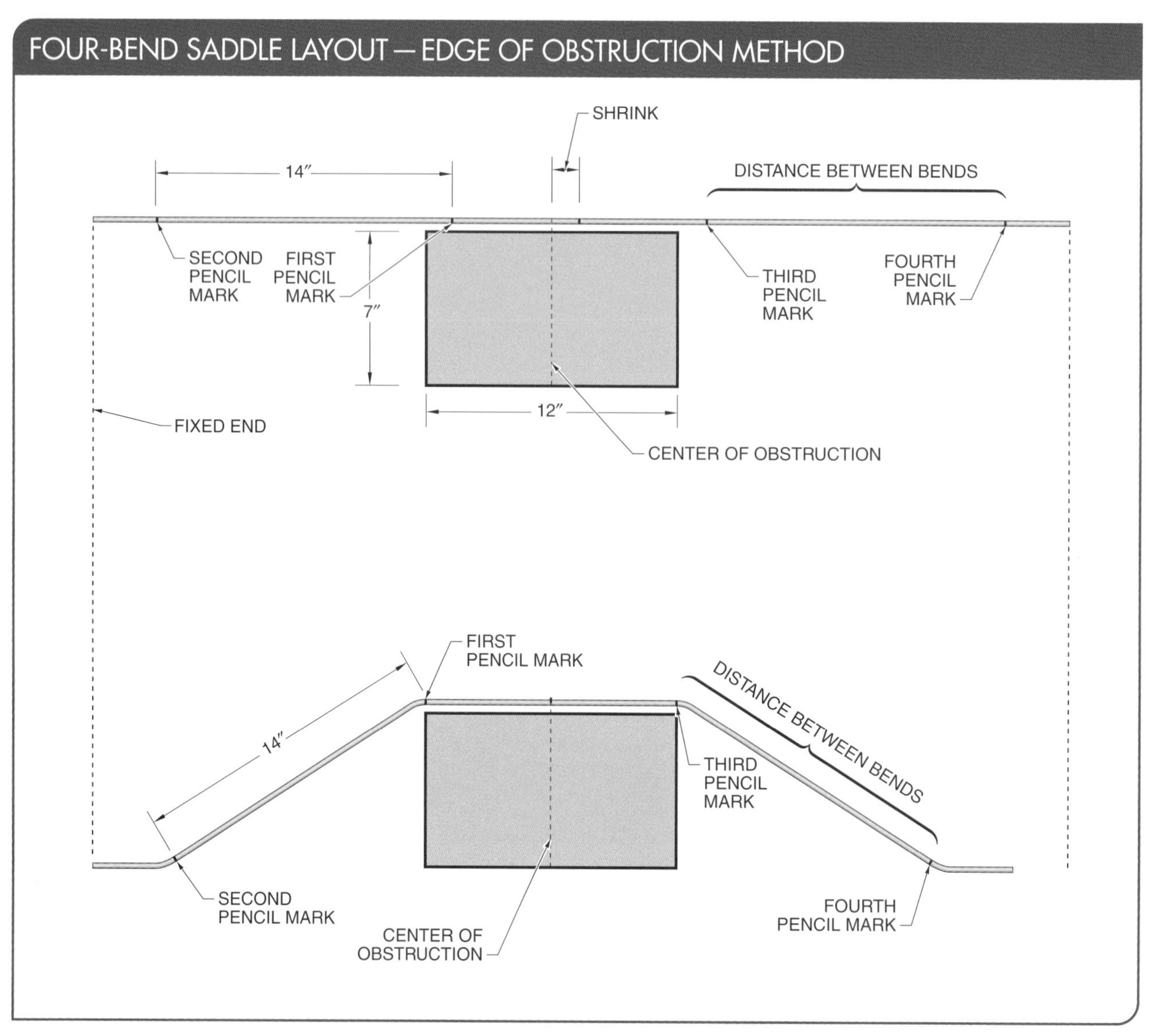

Figure 10-15. In the edge of obstruction method, the measurements and layout of the pencil marks for the four bends are made relative to the leading edge of the obstruction.

> **TECH TIP**
>
> *When pencil marking conduit, run the mark 360° around the conduit so it is easy to see, making sure the measurement is correct.*

After the pencil marks are placed on the conduit, the bends can be made. The offset bends for a four-bend saddle are normally made with the bender hook pointing toward the center of the obstruction. **See Figure 10-16.**

To make the offset bends, the following procedure is applied:

1. Place the bender on the conduit with the bender shoe on the ground facing the center of the saddle. Align the arrow benchmark with the first mark on the conduit. Make the first bend of the first offset using the foot pedal.
2. Turn the conduit bender upside down and rotate the conduit 180°. Align the arrow benchmark with the second mark. Carefully align the conduit and

start the second bend of the first offset in the air. Complete the bend on the ground using the foot pedal.

3. Remove the bender from the conduit and reverse the bender. Place the bender back on the conduit at the third mark, with the bender shoe facing the center of the saddle. Carefully align the bends with each other. Make the first bend of the second offset using the foot pedal.
4. Turn the bender upside down and rotate the conduit 180°. Align the arrow benchmark with the fourth mark. Carefully align the conduit with the previous bends and make the second bend of the second offset.

FOUR-BEND SADDLES

1 ALIGN PENCIL MARK WITH ARROW BENCHMARK

2 TURN BENDER UPSIDE DOWN AND MAKE SECOND BEND

3 MAKE FIRST BEND OF SECOND OFFSET

4 ROTATE CONDUIT IN BENDER AND MAKE LAST BEND

Figure 10-16. The offset bends for a four-bend saddle are normally made at 30° and with the bender hook pointing toward the center of the obstruction.

Securing EMT

The NEC® requires that EMT be installed as a complete system and securely fastened in place at least every 10′ and within 3′ of each outlet box, junction box, cabinet, or fitting. **See Figure 10-17.** Compression couplings ensure conduit is installed into smooth-flowing pathways through which wires can be easily pulled. Straps, connectors, and couplings are used to secure conduit in place and attached it to electrical boxes and other conduit.

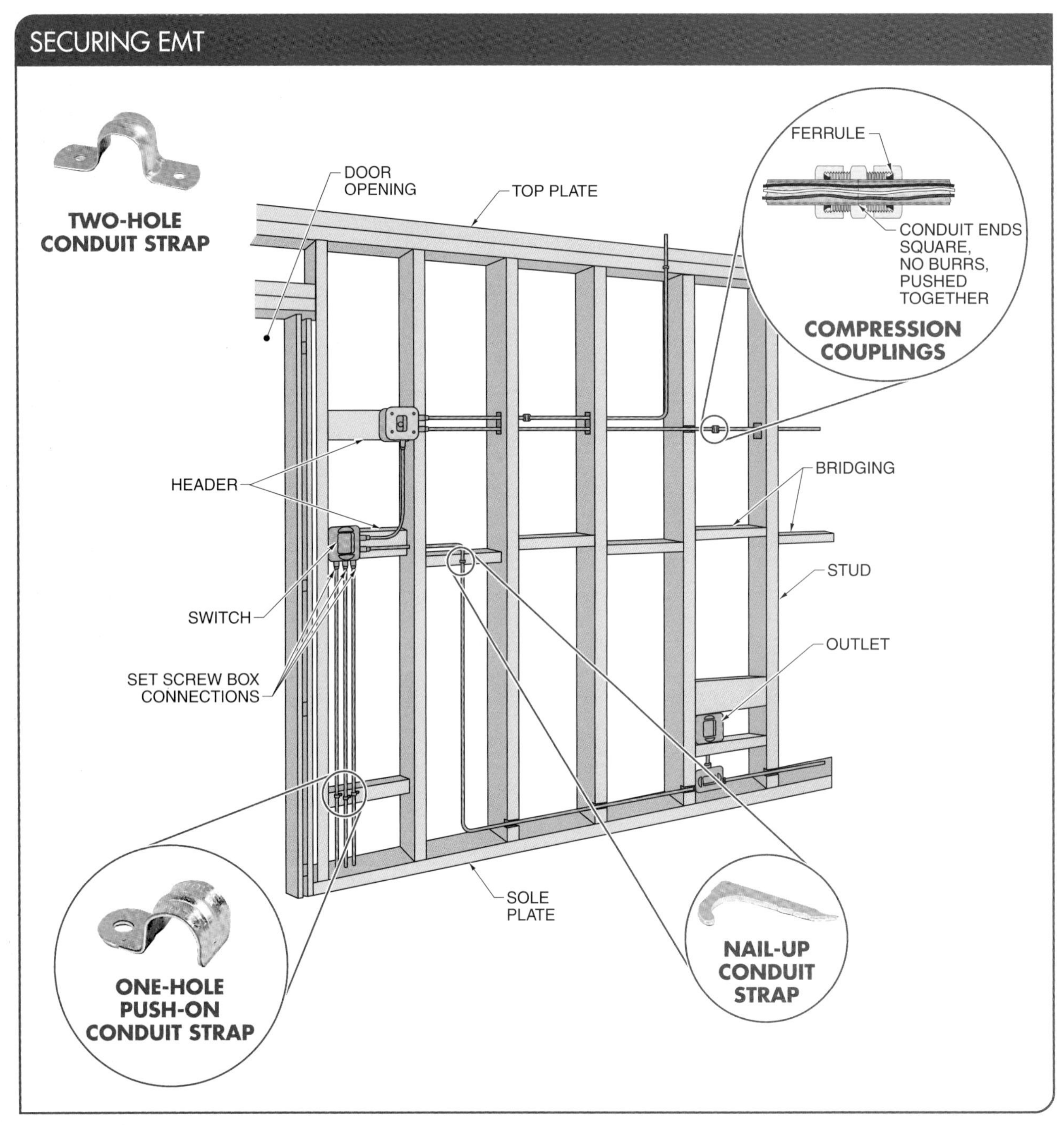

Figure 10-17. The NEC® requires that EMT be secured every 10′ and within 3′ of each outlet box, junction box, cabinet, or fitting. Compression couplings connect two lengths of conduit for smooth-flowing pathways through which wire is pulled.

EMT conduit is firmly secured to electrical boxes using compression, indenter, and set screw connectors. Each type of connector has a distinctly different technique for securely holding the conduit. Various types of connectors are accepted by the NEC® for residential use. Applications that require conduit to be run where water is present require the use of special liquidtight conduit and watertight conduit connectors. **See Figure 10-18.**

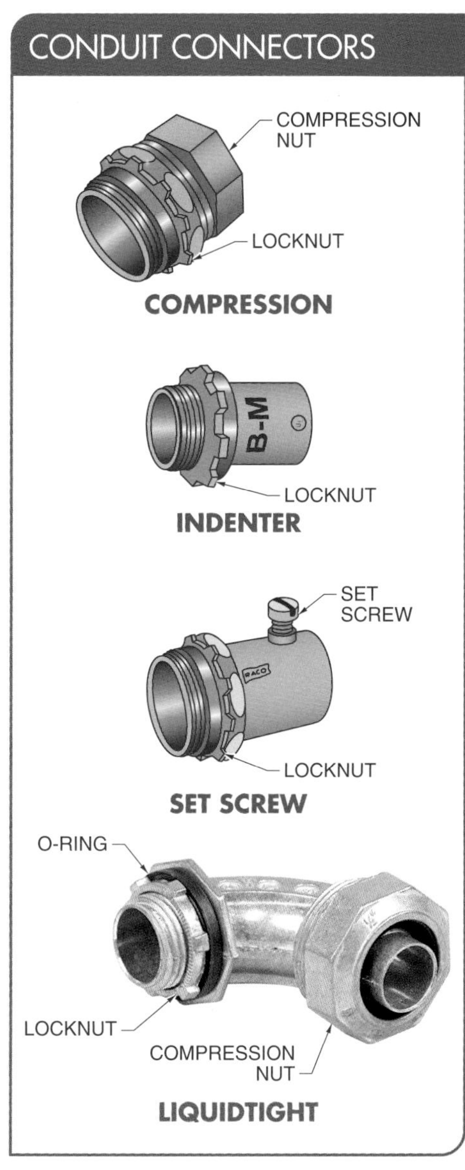

Figure 10-18. Compression, indenter, set screw, and liquidtight connectors secure conduit to electrical boxes.

A *compression connector* is a type of box fitting that firmly secures conduit to a box by utilizing a nut that compresses a tapered metal ring (ferrule) into the conduit. As the compression nut is tightened, the nut forces the ferrule into the conduit, locking the conduit in position. Compression connectors can be loosened to remove the conduit and can be reused multiple times to attach the same piece of conduit.

An *indenter connector* is a type of box fitting that secures conduit to a box with the use of a special indenting tool. **See Figure 10-19.** Indenter coupling connectors are not reusable and must be cut off with a hacksaw to be removed.

Figure 10-19. Indenter connectors are crimped with a special tool.

A *set screw connector* is a type of box fitting that relies on the pressure of a screw against conduit to hold the conduit in place. Set screw connectors are also reusable.

Unlike metallic and nonmetallic cable, conduit must be joined together using conduit couplings when a run exceeds 10′ in length. A *conduit coupling* is a type of fitting used to join one length of conduit to another length of conduit and still maintain a smooth inner surface. **See Figure 10-20.** Conduit couplings are similar in design to back-to-back conduit connectors and also operate on the principles of compression, indentation, and pressure.

Installing RMC

Installing RMC requires many of the same skills used to install EMT. The primary difference is that RMC is heavier and more difficult to bend than EMT. Often, hydraulic machinery is used to bend RMC instead of an RMC conduit bender called a hickey. Due to the difficulty in bending, RMC is typically threaded. The threads allow threaded fittings to be used instead of bending the conduit. **See Figure 10-21.**

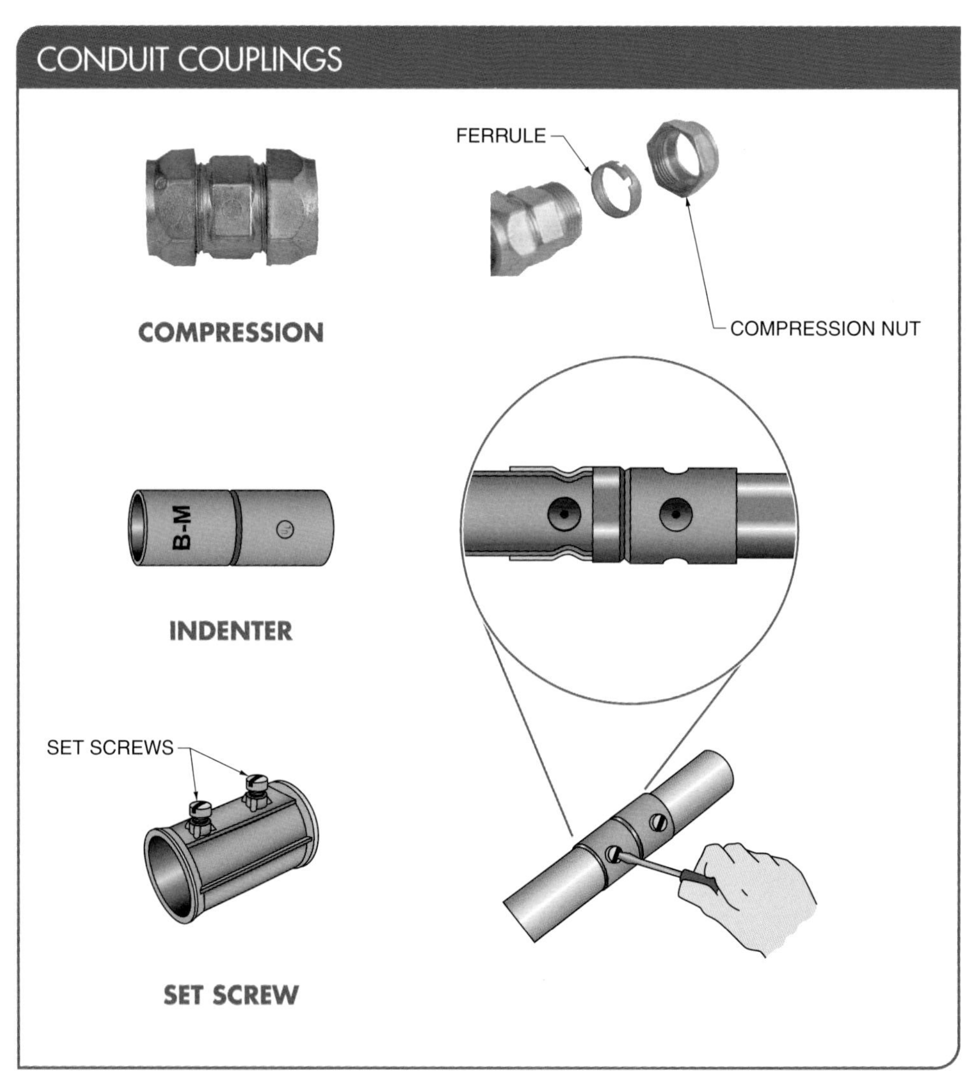

Figure 10-20. Conduit couplings join one length of conduit to another for a smooth inner surface through which to pull wire.

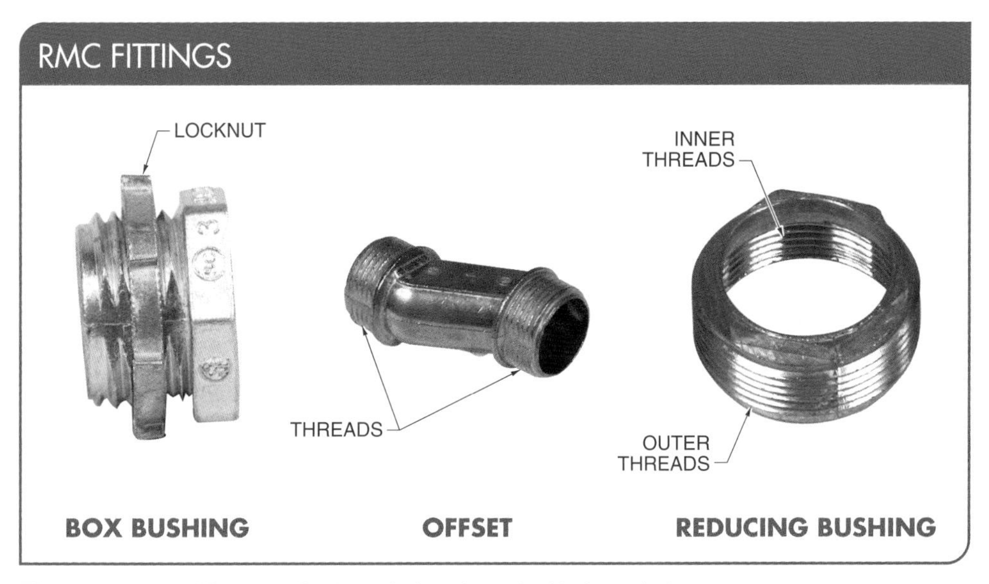

Figure 10-21. RMC is usually threaded and used with threaded connectors.

For service installations, RMC can be purchased precut and prethreaded. Electrical supply houses or hardware stores often stock a variety of lengths and sizes of RMC ready for use. Electrical supply houses or hardware stores also carry stock fittings, such as 45° and 90° bends. When special lengths are required, RMC can be cut and threaded by a supply house for an additional fee. After RMC is installed, equipment is grounded through the conduit system. Additional information on the installation of RMC can be found in NEC® Article 344.

Installing FMC

The installation of FMC involves using similar techniques as installing metallic-sheathed cable. FMC is a metal conduit without wires that can be bent by hand. FMC differs from metallic-sheathed cable in that wires must be pulled through FMC after installation. **See Figure 10-22.** After FMC is installed, equipment is grounded through the conduit system to a separate ground wire. Additional information on the installation of FMC can be found in NEC® Article 348.

Installing PVC

Schedule 40 PVC conduit has a larger inside diameter than Schedule 80 PVC conduit, so it is easier to pull wires through it. Although Schedule 80 has the same outside diameter as Schedule 40, it has a thicker wall and therefore a smaller inside diameter. Schedule 80 PVC conduit is often installed in high-traffic areas where conduit might be exposed.

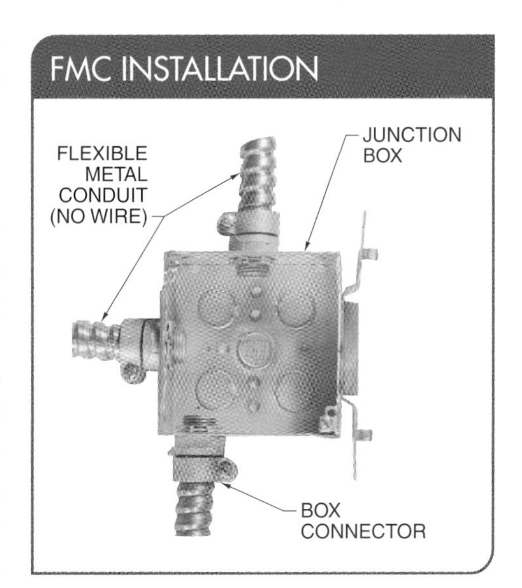

Figure 10-22. FMC can be bent by hand and wires must be pulled through it after it is installed.

Installing PVC conduit involves cutting, deburring, dry-fitting, and solvent-cementing the conduit together. Like EMT conduit, there should be no more than a total of 360° of bends per run. After PVC is installed, equipment is grounded through the conduit system to a separate ground wire. Additional information on the installation of PVC conduit can be found in NEC® Article 352.

> CODE CONNECT
>
> *Per NEC® Subsection 352.20(A), Minimum, PVC conduit smaller than metric designator 16 (trade size ½) shall not be used; and per (B), Maximum, PVC conduit larger than metric designator 155 (trade size 6) shall not be used.*

Solvent-Cementing PVC. Solvent-cementing is the process of fusing plastic conduit and fittings together by softening the adjoining surfaces through a chemical reaction. A chemical solvent is applied to soften both surfaces to be joined together. The surfaces are then forced together to form a solid joint that is as strong as the conduit walls. Three considerations must be taken into account to achieve a watertight and airtight seal: the application of appropriate PVC primer, verification that the sections to be joined together have a good interference fit, and the application of the proper installation techniques.

The appropriate primer and solvent cement must be used for PVC. A *primer* is a chemical agent that cleans and softens a surface and allows solvent cement to penetrate more effectively into the surface. *Solvent cement* is a chemical agent that penetrates and softens the surface of plastic pipe and fittings and fuses them together. Precautions for the safe handling of solvent cements and primers include the following:

- Avoid breathing solvent cement vapors and keep the work area properly ventilated.
- Keep cleaners, primers, and solvent cements away from ignition sources, sparks, heat, and open flames.
- Keep storage containers tightly closed when the product is not in use.
- Dispose of all cloths and rags used for cleaning excess material.
- Wear proper PPE such as goggles and neoprene gloves.
- Use proper application tools.

The proper preparation and installation techniques must be used to ensure watertight and airtight joints. **See Figure 10-23.** To prepare and solvent-cement PVC conduit and joint fittings, the following procedure is applied:

1. Cut PVC squarely with a chop saw, universal saw, or PVC pipe cutter.
2. Smooth the pipe ends with a deburring tool, hand file, or utility knife.
3. Ensure that the pipe ends are clean and dry-fit the pipe and joint fitting together to ensure a proper interference fit. The pipe should be able to be inserted only about halfway into the socket of the joint fitting.
4. Apply primer first to the inside diameter (ID) of the joint fitting socket and then to the outside diameter (OD) of the pipe to the depth that it will be seated in the fitting. Apply a second coat of primer to the socket fitting ID. *Note:* The application of too much primer results in a puddle, which can cause flow restrictions.
5. Wait 10 sec to 15 sec, then apply solvent cement with a brush or roller in the same manner as the primer application. Apply a second coat of solvent cement to the pipe OD. *Note:* The application of too much solvent results in a puddle, which can cause flow restrictions.
6. Fit and position the pipe and fitting together before the solvent cement evaporates, ensuring that pipe is properly seated in the fitting socket. Once properly seated, turn the pipe ¼ turn and hold it and the fitting in place for 15 sec to 30 sec to allow the solvent cement to bond the two surfaces together.
7. Apply a bead of solvent cement around the entire diameter of the pipe and fitting.
8. Remove any excess solvent cement with a clean, dry cloth.

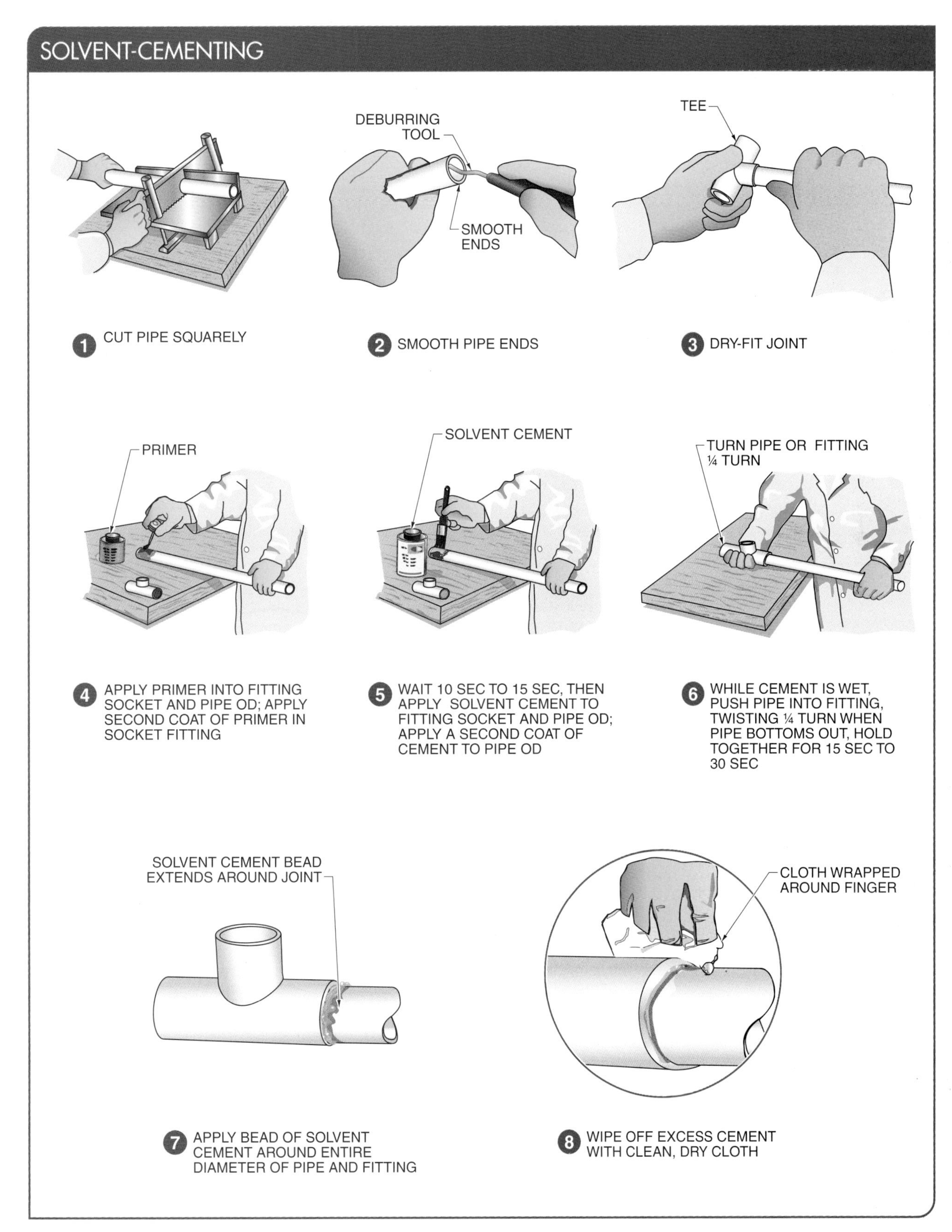

Figure 10-23. Proper installation techniques must be used to ensure watertight and airtight PVC conduit joints.

Pulling Wires

There are two types of fish tapes commonly used to pull wires through conduit: rigid fish tape and polyethylene fish tape. A rigid fish tape is used for pulling wires through conduit, walls, and ceilings. Polyethylene fish tape is used to pull wires within conduit systems. **See Figure 10-24.**

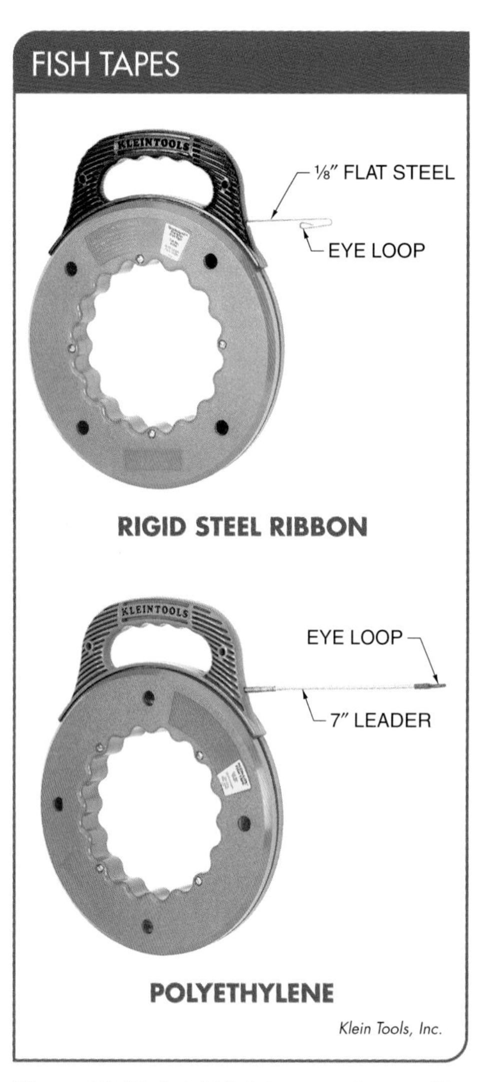

Figure 10-24. A rigid fish tape can be used for pulling wires through conduit, walls, or ceilings, and a polyethylene fish tape can be used for pulling wires through conduit.

Pulling grips and lubricants can also be used to help pull wires through conduit. A *pulling grip* is a device that is attached to a fish tape to allow more leverage. A *lubricant* is a wet or dry compound that is applied to the exterior of wires to allow them to slide easier.

SECTION 10.2 CHECKPOINT

1. What are the five main types of EMT conduit bends?
2. What are the three types of conduit connectors used for EMT?
3. What is a conduit coupling?
4. How is RMC usually connected and why?
5. What is the process of fusing plastic conduit and fittings together?
6. What is primer?
7. What is solvent cement?
8. What are two types of commonly used fish tapes?

Chapter Activities

10 Conduit

Name ______________________ Date ______________

Conduit Types

Identify each type of conduit shown.

______________ **1.** EMT

______________ **2.** PVC

______________ **3.** FMC

______________ **4.** RMC

A

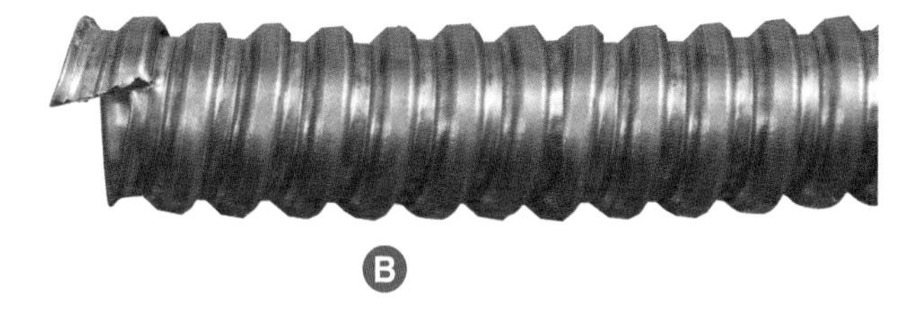

B

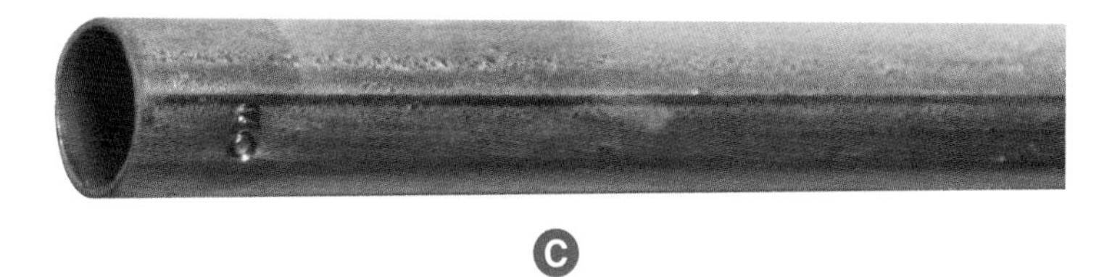

C

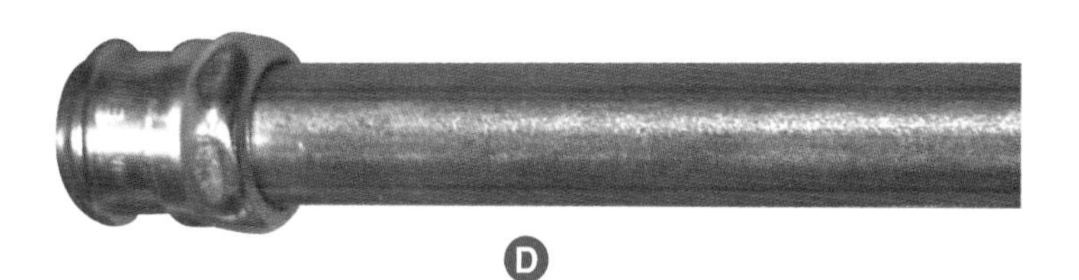

D

Conduit Connectors

Identify each type of conduit connector shown.

________________ **1.** Set screw

________________ **2.** Indenter

________________ **3.** Liquidtight

________________ **4.** Compression

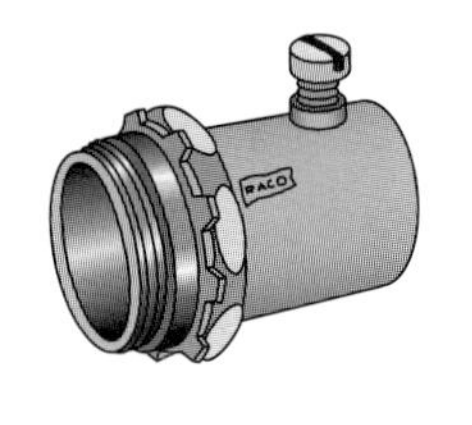

A

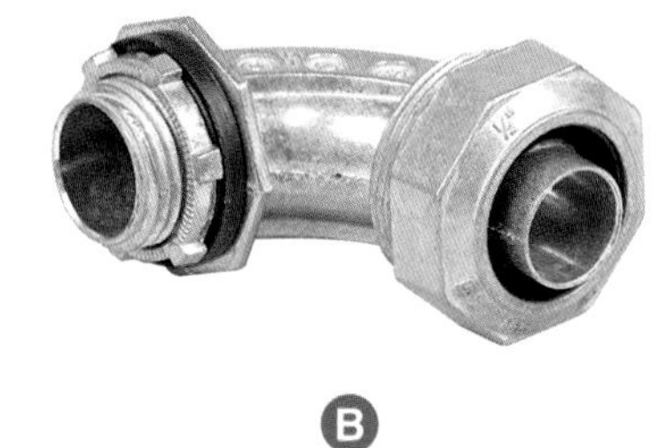

B

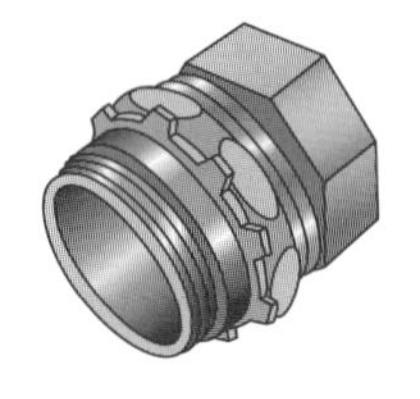

C

D

Conduit Couplings

Identify each type of conduit coupling shown.

________________ **1.** Indenter

________________ **2.** Set screw

________________ **3.** Compression

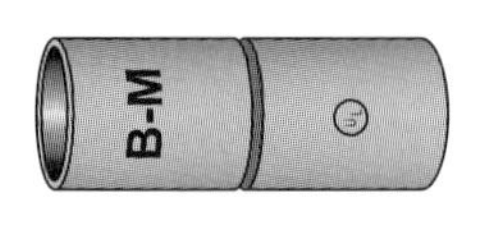

A

B

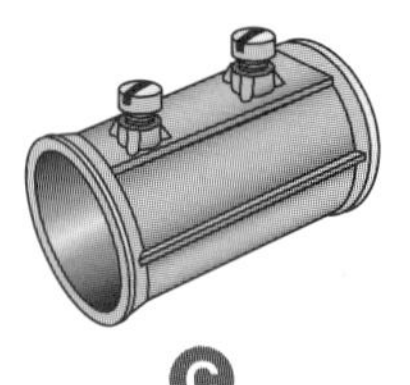

C

11 Receptacles

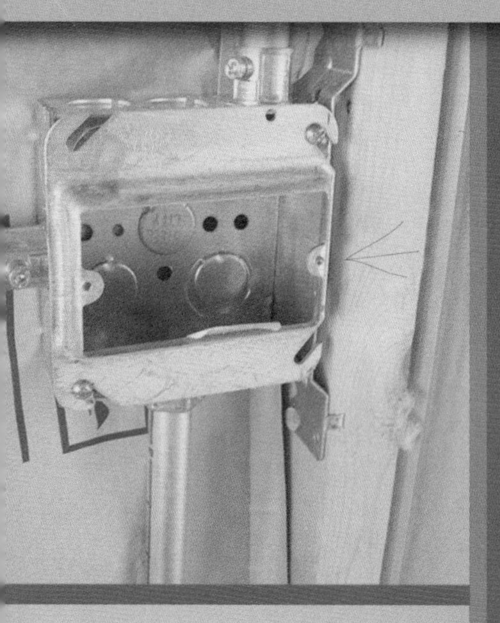

SECTION 11.1—RECEPTACLES

- Identify the four basic types of receptacles used in a residence.
- Identify standard receptacle ratings, terminal color designations, polarization, and grounding connections.
- Describe how to test standard receptacles.
- Explain how to wire split-wired receptacles.
- Describe ground-fault circuit interrupter (GFCI) receptacles.
- Describe arc-fault circuit interrupter (AFCI) receptacles.
- Describe universal serial bus (USB) receptacles.
- Explain the uses of receptacle faceplate shapes and positions.

SECTION 11.2—RECEPTACLE BOXES

- Identify receptacle box shapes.
- Explain how to size receptacle boxes.
- Describe how to install a receptacle box.
- Identify receptacle box extenders and covers.

Learner Resources

ATPeResources.com/QuickLinks
Access Code: 791712

Receptacles are the main device used to provide power to appliances. For this reason, they are found in almost every location both inside and outside a home. Certain types of receptacles are installed for appliances with standard or higher power ratings. The standard receptacle is mainly used and wired in branch circuits. Some receptacles can be constructed to be switch operated, whereas other types are mandatory in specific locations. Advances in technology have allowed for new capabilities of power and control in receptacles used for certain applications, such as USB and smart receptacles.

SECTION 11.1—RECEPTACLES

A *receptacle,* or outlet, is an electrical contact device used to connect equipment with a cord and plug to an electrical system. The four basic types of receptacles that may be found in a residence are standard, isolated-ground, ground-fault circuit interrupter (GFCI), and arc-fault circuit interrupter (AFCI) receptacles. Additional types of receptacles that may be found in a residence include tamper-resistant, universal serial bus (USB), and smart receptacles. **See Figure 11-1.**

Standard Receptacles

Standard receptacles are constructed with a long (neutral) slot, a short (hot) slot, and a U-shaped (ground) slot. The most common standard receptacle in a residence is the duplex receptacle. **See Figure 11-2.** A *duplex receptacle* is a standard receptacle that has two outlets for connecting two different plugs. A two-wire cable attaches to receptacles at screw terminals or push-in fittings (quick connectors). A connecting tab between the two hot screw terminals provides electrical power to both outlet spaces for plugs. The connecting tab allows for both outlets to be powered when a hot wire is connected to one hot terminal screw. A solid copper or green insulated wire is attached to the green (grounding) screw terminal.

Standard Receptacle Ratings. Standard receptacles are marked with standard ratings for maximum voltage, amperes, and frequency. Manufacturers also stamp the receptacle with markings of standards organizations, such as UL® and CSA, to indicate compliance with national codes and standards.

Standard receptacles are marked with ratings, such as 125 V, 15/20 A, and 60 Hz. Receptacles marked 20 A, 125 V are typically required for appliance circuits. Lighting and other branch circuits typically have 15 A, 125 V receptacles installed. Receptacles are also marked with the appropriate type of wire to be used, such as CU for copper wire only, CU-CLAD for copper-coated aluminum wire, and CO/ALR for copper-aluminum wire. The type of wire needed is marked as either solid or stranded.

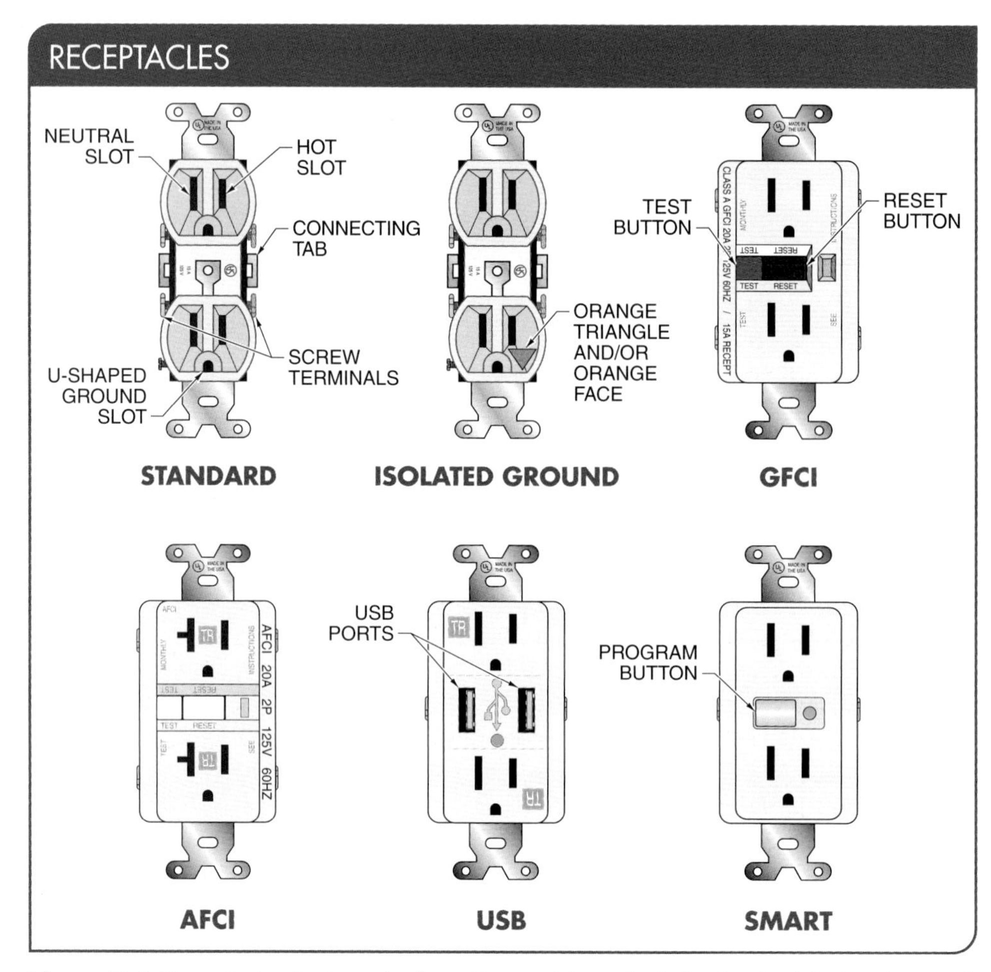

Figure 11-1. Receptacles that may be found in a residence include standard, isolated-ground, GFCI, AFCI, USB, and smart receptacles.

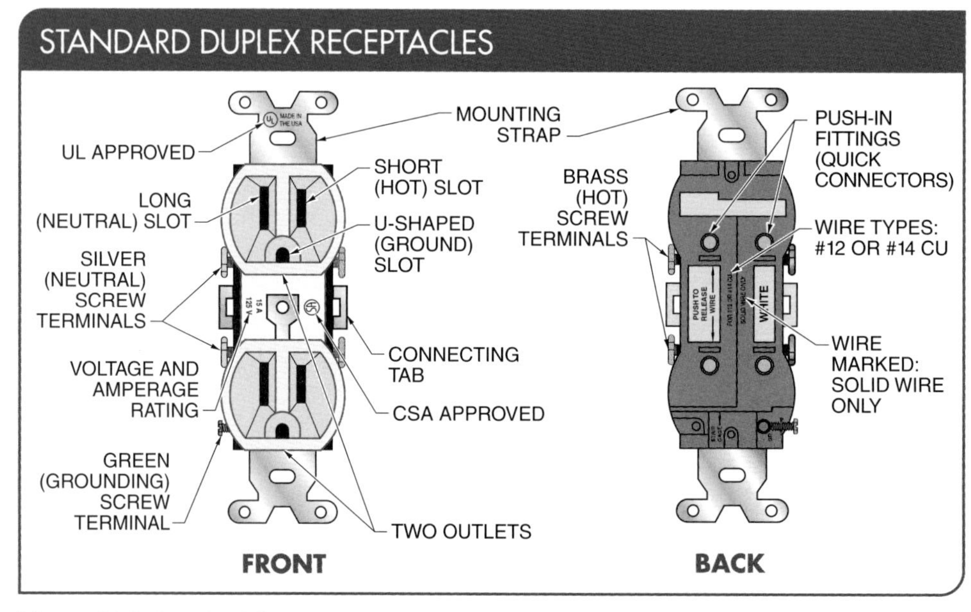

Figure 11-2. Standard duplex receptacles have two outlets for connecting two different plugs and are constructed with a long (neutral) slot, a short (hot) slot, and a U-shaped (ground) slot.

CODE CONNECT

Per NEC® Table 210.21(B)(3), the receptacle rating should not be less than the branch circuit rating.

Standard Receptacle Terminal Color Designations. Brass, silver, and green colors are typically used to designate the terminal screws as either hot, neutral, or ground. A brass-colored screw indicates the hot terminal, a silver-colored screw indicates the neutral terminal, and a green-colored screw indicates the ground terminal.

When a 125 V-rated duplex receptacle is being wired, the black (hot) wire is connected to the brass terminal screw, the white (neutral) wire is connected to the silver terminal screw, and the green or bare (ground) wire is connected to the green terminal screw. **See Figure 11-3.** When a duplex receptacle is properly wired, the hot wire powers the short slot and the neutral wire provides the path back to the power source from the long slot. This allows for proper polarization.

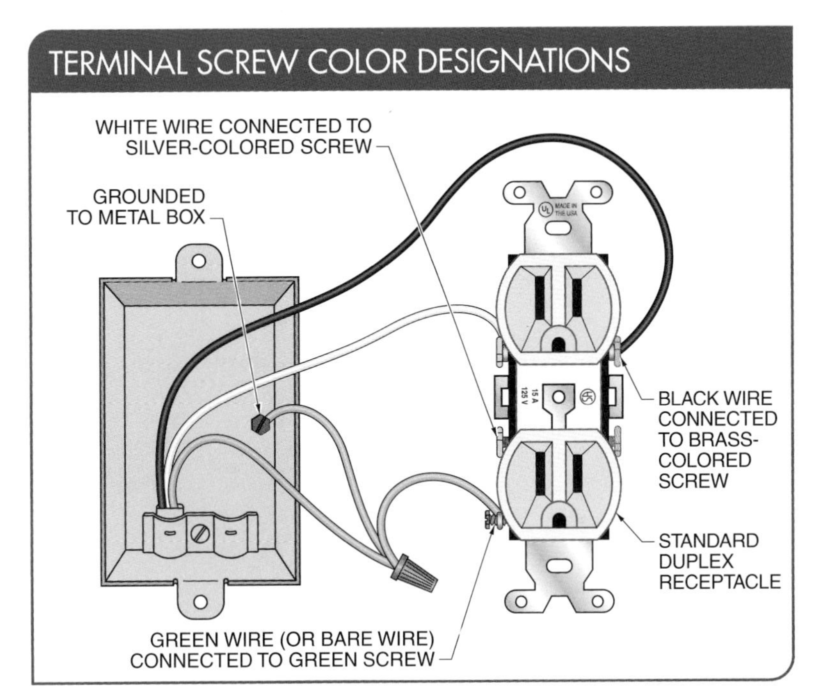

Figure 11-3. When a 125 V-rated duplex receptacle is being wired, a brass-colored screw indicates the hot terminal, a silver-colored screw indicates the neutral terminal, and a green-colored screw indicates the ground terminal.

Standard Receptacle Polarization. Most receptacles that are installed in a residence are polarized. A *polarized receptacle* is a standard receptacle that consists of one short (hot) slot and one long (neutral) slot. Polarized receptacles can only power polarized plugs. Polarized plugs have one blade that is wider than the other, which allows them to be inserted in only one way. The wider blade is the neutral connection and the thinner blade is the hot connection.

If a receptacle is wired correctly, the short slot provides power to the hot blade and the long slot connects to the neutral blade in a plug. This keeps the external surfaces of devices, such as toasters, microwaves, and entertainment equipment, de-energized on the neutral/ground side of the circuit. If a receptacle is not polarized or is wired incorrectly, a switch, such as one that controls a light, can allow the neutral wire or the device's surface to be energized, thus posing an electrical hazard.

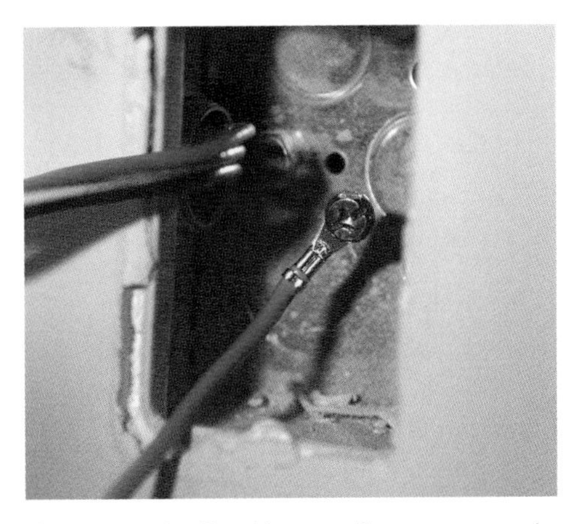

Grounding pigtails with grounding screws can be purchased in different wire sizes for terminating to electrical boxes.

Standard Receptacle Grounding. A residential grounding system typically includes all metal wiring, metal boxes, conduit, water pipes, and non-current-carrying metal parts of most electrical equipment. To ground a standard receptacle, the green screw terminal on the receptacle is connected to the common grounding screw in the metal outlet box. This is done with a ground bonding jumper connection. **See Figure 11-4.**

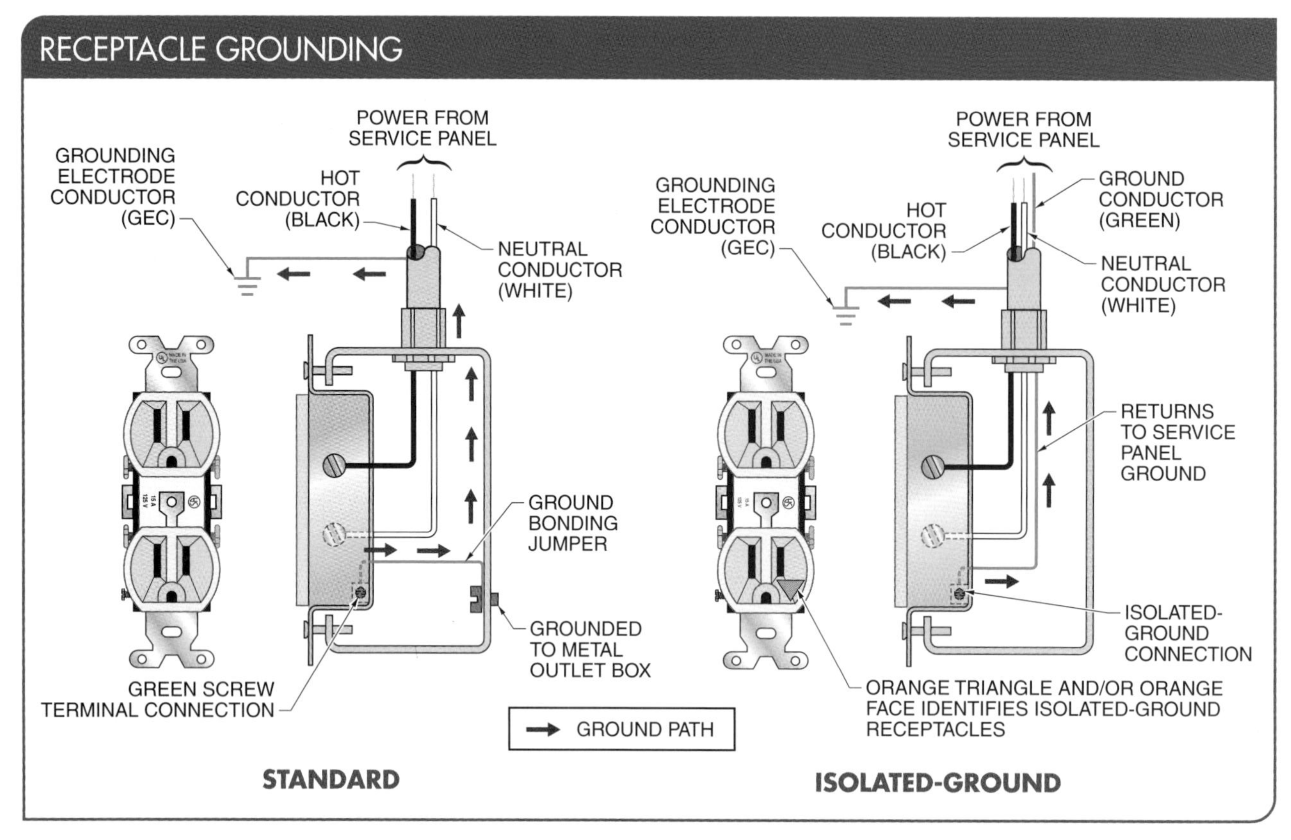

Figure 11-4. To ground a standard receptacle, the green screw terminal on the receptacle is connected to the common grounding screw in the metal outlet box. This is done with a ground bonding jumper connection.

The receptacle ground becomes part of the residential grounding system when a piece of electrical equipment with a grounded plug is inserted into the receptacle. During a ground fault, the fault current travels through the ground bonding jumper to the metal outlet box and back to the grounding electrode conductor (GEC) at the service panel and to the grounding electrode. This diminishes the fault current into the earth.

A residential grounding system can also act as a large antenna and conduct electrical noise away from electrical equipment. *Electrical noise* is unwanted signals that are present on a power line. Electrical noise typically enters a residence through power supply lines or from noise-generating equipment, such as motors. Electrical noise can cause interference in computers, electronic devices, and home-entertainment systems.

An isolated-ground receptacle is used to minimize problems in applications or areas of high electrical noise. An *isolated-ground receptacle* is a special standard receptacle that minimizes electrical noise by providing its own grounding path. Isolated-ground receptacles are identified by an orange-colored faceplate and/or an orange triangle on the face of the receptacle. A separate ground conductor is run with the circuit conductors in an isolated-ground circuit. During a ground fault, the isolated-ground connection takes the fault current back to the GEC at the service panel and ends in the grounding electrode.

Standard Receptacle Testing. Receptacles can be tested for proper operation with either a test light or receptacle tester. A *receptacle tester* is an electrical test instrument that is plugged into a standard receptacle to determine whether the receptacle is properly wired and energized. **See Figure 11-5.** Receptacle testers are constructed with indicator lights and corresponding light codes to show certain wired connections or opens. A test light can be used for testing a receptacle for proper voltage levels. A receptacle that is operating correctly has a nominal voltage level of 120 V between the hot slot to the neutral slot or the hot slot to the ground slot.

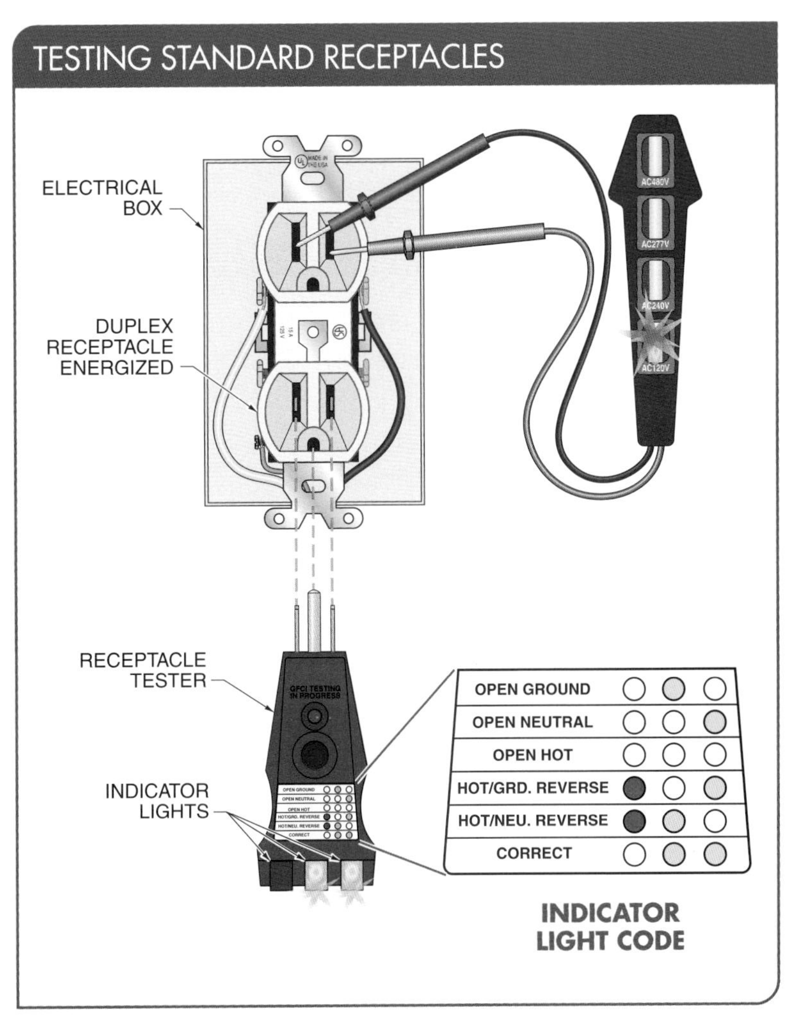

Figure 11-5. Receptacles can be tested for proper operation with either a test light or a receptacle tester. A receptacle that is operating correctly has a nominal voltage level of 120 V between the hot slot to the neutral slot or the hot slot to the ground slot.

CODE CONNECT

Per NEC® Subsection 406.3(D), isolated-ground receptacles used to reduce electromagnetic interference should be identified by an orange triangle on the face of the receptacle.

Split-Wired Receptacles

A duplex receptacle can be converted into a split-wired receptacle by removing the connecting tab between the hot screw terminals. A *split-wired receptacle* is a standard duplex receptacle that has had the tab between the two brass-colored (hot) terminals removed, while the tab between the two silver-colored (neutral) terminals remains in place.

Switch-Controlled Outlets. Split-wired receptacles can be used in lighting applications that allow for one outlet to be powered and the other outlet to be controlled by a switch. **See Figure 11-6.** A typical application of a switch-controlled outlet is a lamp that is controlled by a switch through the outlet.

Double-Pole Circuit Breaker Connections. Locations such as kitchen countertops may have two high-wattage appliances that are connected to the same circuit receptacle. This can cause an overload at the receptacle if the same branch circuit is used to power both outlets. To eliminate this problem, a double-pole circuit breaker can be used to power one split-wired receptacle. In this circuit, both ungrounded (hot) wires are de-energized when an overload occurs.

Four conducting wires are used when wiring from the double-pole circuit breaker to the split-wired receptacle. Two hot wires, one red and one black, are run from the circuit breaker and connected to the two brass-colored screw terminals on the side from which the connecting tab has been removed. One neutral wire is connected to one of the silver-colored screws and run back to the neutral busbar. A green (or bare copper) wire is connected to the green terminal screw and tied together with a wire connector to the grounded metal outlet box terminal screw and the green wire from the main ground busbar. **See Figure 11-7.**

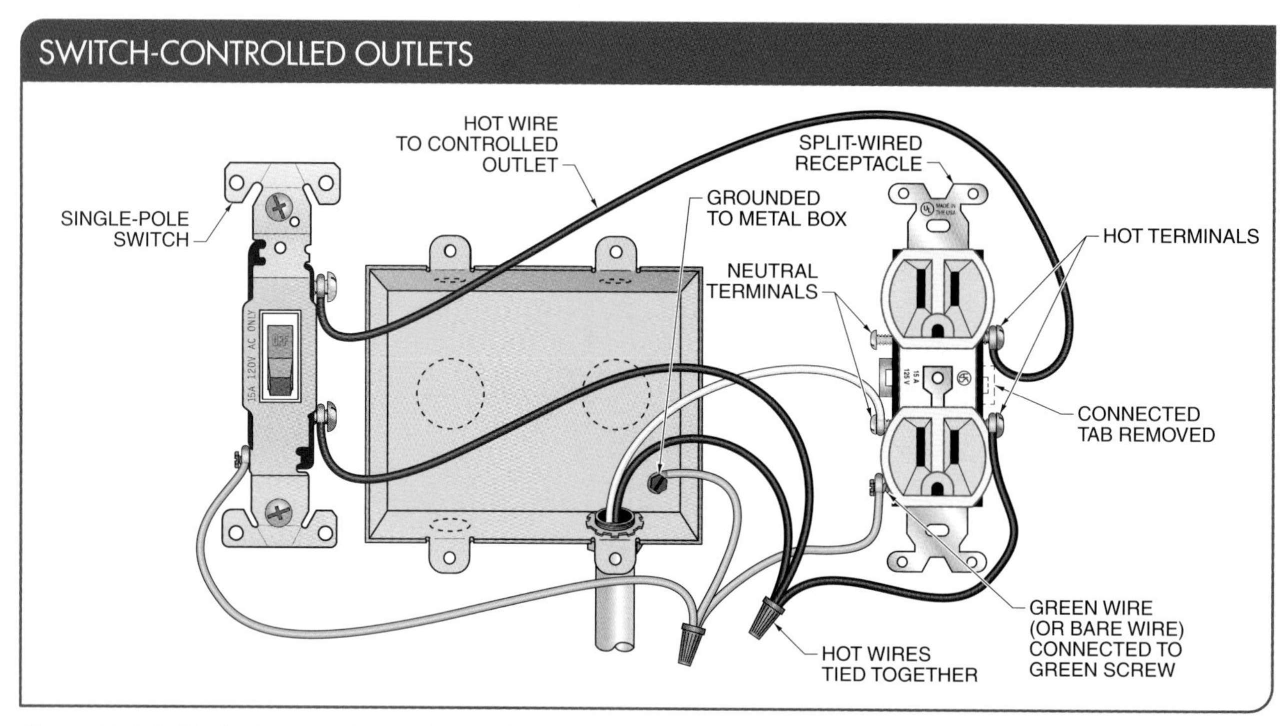

Figure 11-6. Split-wired receptacles can be used in lighting applications that allow for one outlet to be powered and the other outlet to be controlled by a switch.

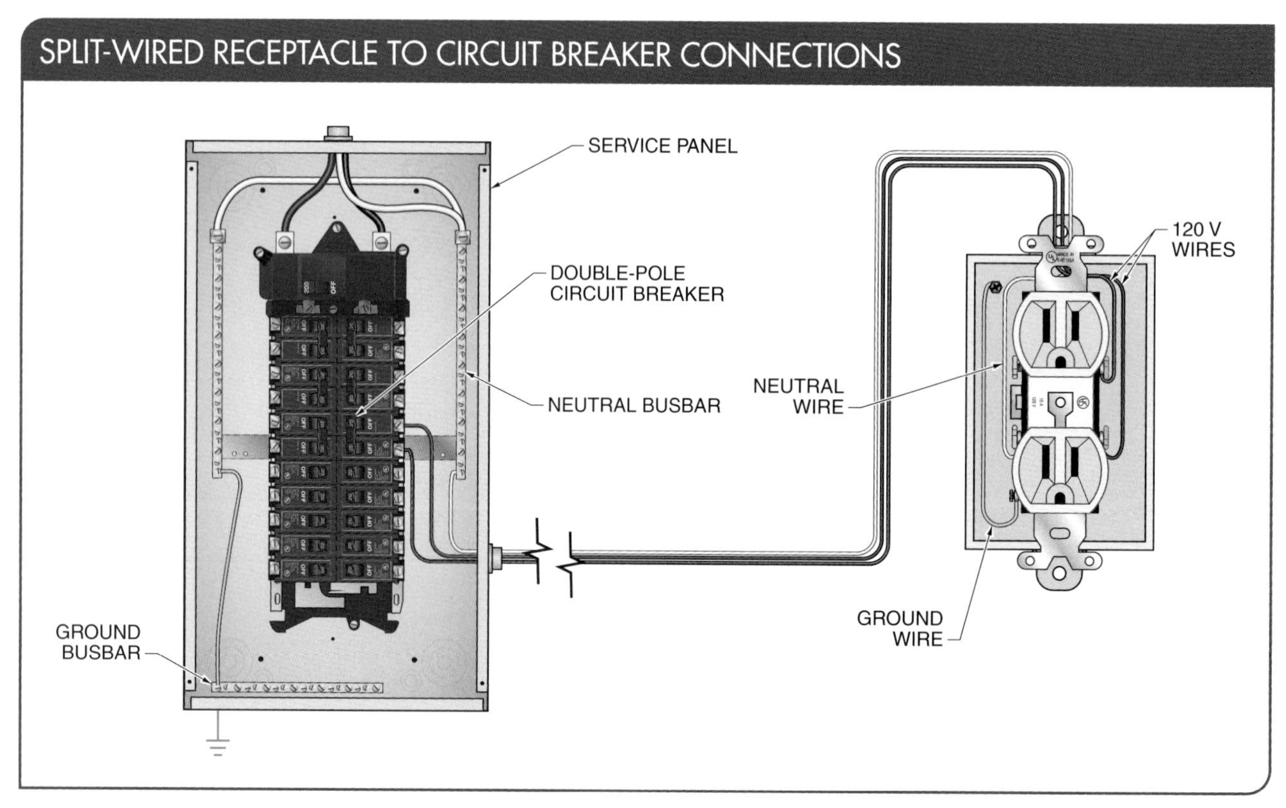

Figure 11-7. Double-pole circuit breakers can be used to power high-wattage appliances that are connected to the same split-wired receptacle.

Ground-Fault Circuit Interrupter (GFCI) Receptacles

A *ground-fault circuit interrupter (GFCI) receptacle* is an electrical receptacle that offers protection by detecting ground faults and then quickly disconnecting power from the circuit. **See Figure 11-8.** A *ground fault* is any amount of current above the level that may deliver an electrical shock. Ground faults occur when an energized conductor comes into contact with the frame or enclosure of electrical equipment or ground connection. GFCIs detect a low level of current unbalance between a hot/neutral conductor and ground wire, instantaneously opening the circuit in response to the ground fault. GFCIs provide greater protection than standard or isolated-ground receptacles.

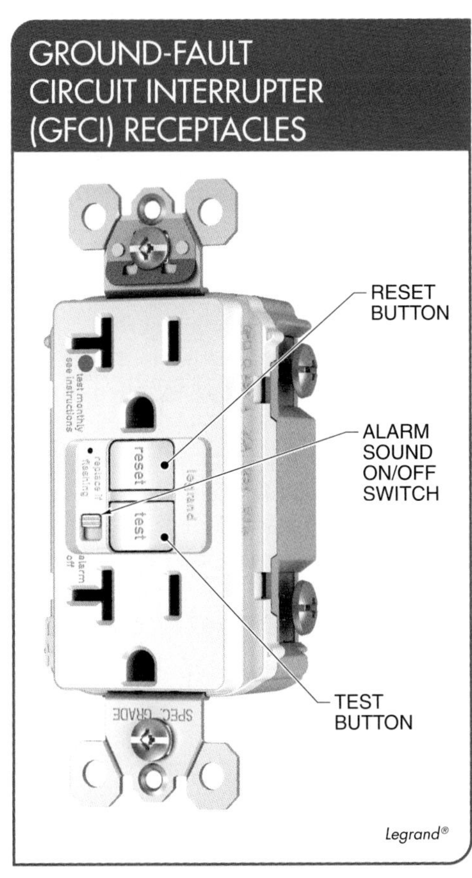

Figure 11-8. GFCIs detect a low level of current unbalance between a hot/neutral conductor and ground wire, instantaneously opening the circuit in response to the ground fault.

GFCI receptacles can also be installed in place of existing ungrounded receptacles. A GFCI receptacle installed in a former two-wire (no grounding outlet) circuit will provide electrical shock protection. GFCIs sense an imbalance of current between the hot wire and the neutral wire instead of relying on a grounding wire. The GFCI trips if there is an imbalance of at least 6 milliamps (mA). When a GFCI is installed on a two-wire circuit, it must be permanently marked with "No Equipment Ground."

If a GFCI receptacle is not available for replacing an existing ungrounded receptacle, a GFCI circuit breaker can be installed in the service panel for circuit protection. The GFCI circuit breaker replaces the existing circuit breaker in the service panel, providing the same circuit protection as a GFCI receptacle. All protected receptacle outlets on the circuit must be marked with "No Equipment Ground."

GFCI Receptacle Terminal Designations. GFCIs include line and load terminal designations. Line terminals are designated for the incoming hot and neutral connections from the power source. The load terminals are designated for the outgoing hot and neutral connections to the next receptacle in the circuit. The next receptacle in the circuit can be a standard single or duplex receptacle that will be protected by the GFCI when connected to it. The standard receptacle must be marked with a label indicating that it is GFCI protected. **See Figure 11-9.**

GFCI Receptacle Testing. GFCI receptacles include a test button and reset button for verifying proper operation and to restore power when tripped. When the test button on an operational GFCI is pressed, it initiates the GFCI to trip and open the circuit, removing power from the outlets and receptacles downstream. When the reset button is pressed, the power is restored to the outlets and circuit receptacles. Testing can also be done with a GFCI receptacle tester. A *GFCI receptacle tester* is a test instrument that is plugged into a GFCI receptacle to determine whether the receptacle is properly wired and energized. **See Figure 11-10.**

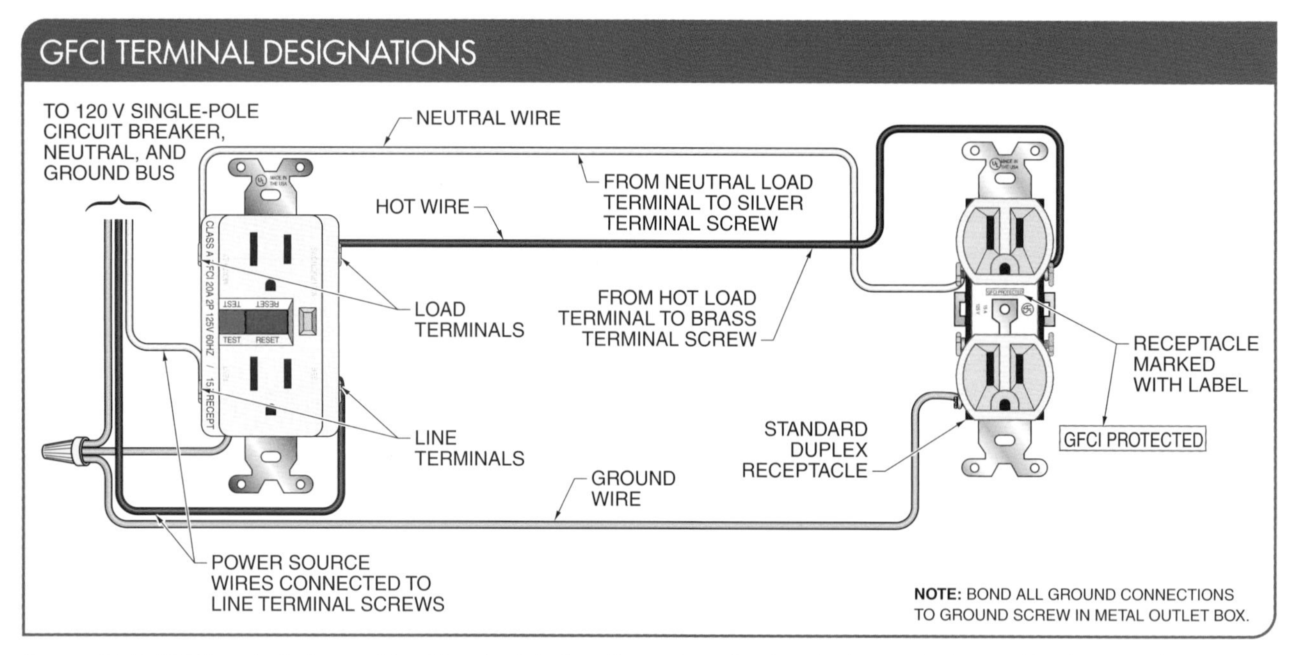

Figure 11-9. GFCIs include line and load terminal designations where line terminals are designated for the incoming hot and neutral connections from the power source and load terminals are designated for the outgoing hot and neutral connections to the next receptacle in the circuit.

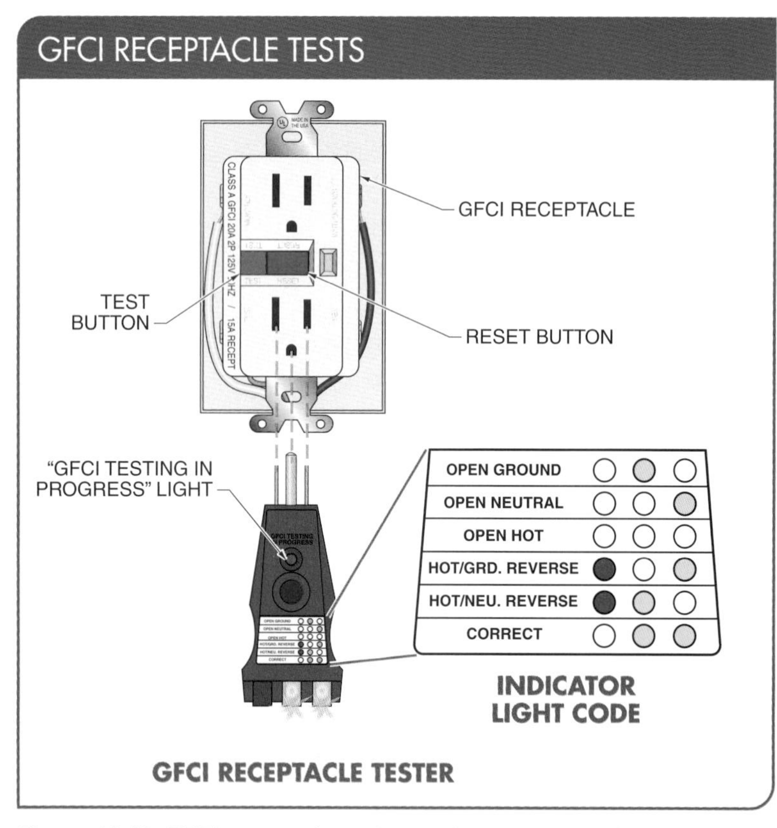

Figure 11-10. GFCI receptacle testing can be done with a GFCI receptacle tester that is plugged into a GFCI receptacle to determine if the receptacle is properly wired and energized.

Arc-Fault Circuit Interrupter (AFCI) Receptacles

An *arc-fault circuit interrupter (AFCI) receptacle* is a fast-acting electrical device that detects and opens a circuit in response to an electrical arc. **See Figure 11-11.** AFCIs are installed to provide a significant level of protection against arc faults and prevent electrical fires. AFCIs are installed in bedrooms and are the first device installed in a branch circuit to protect all other devices downstream. Similar to a GFCI, an AFCI has both a test button to check its operation and a reset button to restore operation. Some AFCIs provide additional circuitry that self-tests the receptacle every 3 sec.

> CODE CONNECT
>
> *Per NEC® Subsection 210.12(A), the maximum length between an arc-fault circuit interrupter (AFCI) and the first outlet should not exceed 50′ for No. 14 AWG wire or 70′ for No. 12 AWG wire.*

ARC-FAULT CIRCUIT INTERRUPTER (AFCI) RECEPTACLES

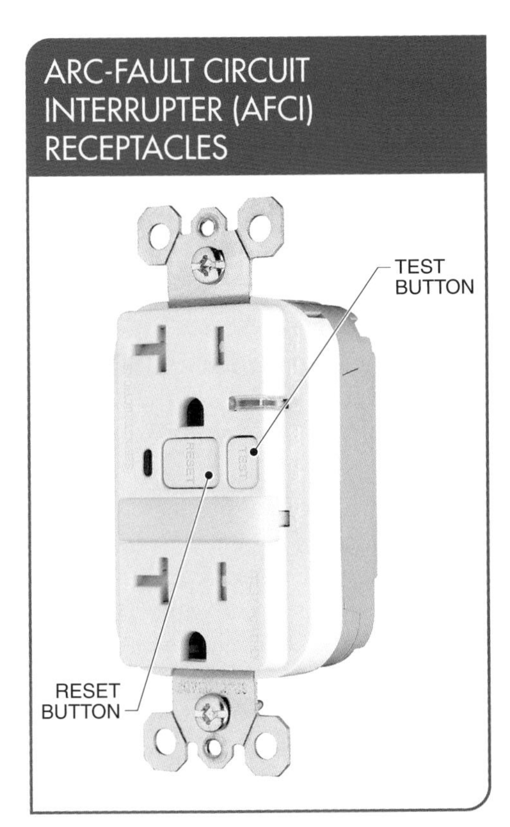

Figure 11-11. AFCIs are installed to provide a significant level of protection against arc faults to prevent electrical fires and are the first receptacles installed in a branch circuit to protect all other devices downstream.

TAMPER-RESISTANT RECEPTACLES

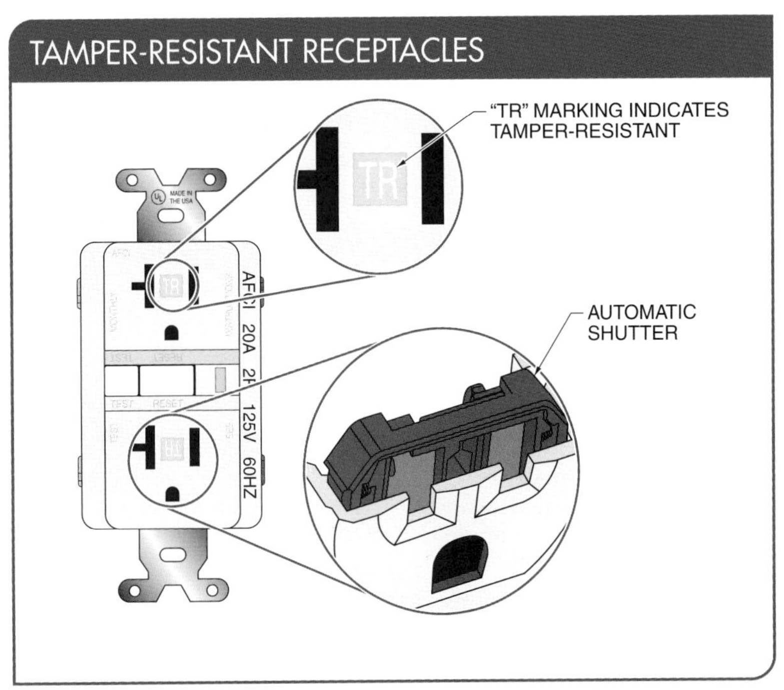

Figure 11-12. In new dwelling unit construction, all 125 V, 15/20 A receptacles shall be tamper-resistant and have a "TR" marking that is clearly visible on the faceplate of the receptacle when installed.

Tamper-Resistant Receptacles

The NEC® requires tamper-resistant receptacles in all new construction of dwelling units. This includes single dwelling units providing complete and independent living facilities for one or more persons. All 125 V, 15/20 A receptacles shall be tamper resistant in these dwellings. UL®-listed devices must have a "TR" marking, meaning tamper resistant, clearly visible on the faceplate of the receptacle when installed. **See Figure 11-12.**

Tamper-resistant receptacles have an internal mechanism that limits access to electrically live components within the receptacle. Tamper-resistant receptacles resemble standard receptacles, but they include automatic shutters that admit plugs but block the insertion of foreign objects, such as hairpins, keys, and paper clips.

Dual-Voltage Receptacles

In most residences, dual-voltage receptacles are used to supply high-power loads. Typically, a residence will contain 125/250 VAC and 250 VAC dual-voltage receptacles. Depending on the application, a 125/250 VAC-rated receptacle is used for supplying 120 V and 240 V power to appliances, while a 250 VAC-rated receptacle only supplies 240 V power.

250 VAC Receptacles. A 250 VAC-rated receptacle is typically not used to supply electricity to standard appliances. Based on updates to the NEC®, 250 VAC receptacles can pose an electrical shock hazard because they do not have a neutral wire. However, 250 VAC outlets are still used in older dwellings and can be mainly reserved for motor-driven power tools. When wired, a 250 VAC receptacle requires two hot wires and one ground wire that are connected to a 240 V double-pole circuit breaker fed from the service panel. A 250 VAC receptacle can have two configurations designed at a 15 A or 20 A current rating. **See Figure 11-13.**

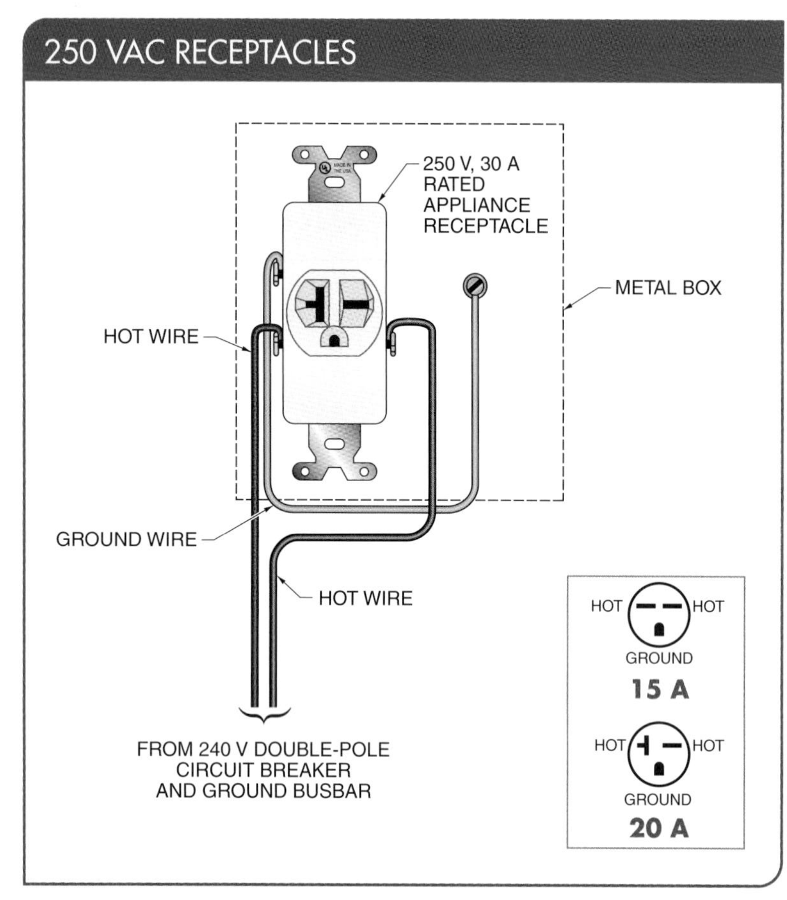

Figure 11-13. When wired, a 250 VAC receptacle requires two hot wires and one ground wire that are connected to a 240 V double-pole circuit breaker fed from the service panel.

125/250 VAC Receptacles. Most high-power appliances in a residence require a 125/250 VAC-rated receptacle. These appliances can include ovens, ranges, cooktops, clothes dryers, washers, water heaters, hot tubs, furnaces, air conditioners, and electric vehicle chargers. For example, hot tubs require two voltages to enable the heating element to operate on 240 V and the circulating pump motor to operate on 120 V. A 125/250 VAC receptacle requires two hot wires, one neutral wire, and one ground (green or bare) wire fed from a 240 V double-pole circuit breaker at the service panel. A 125/250 VAC receptacle can have four configurations designed at 15 A, 20 A, 30 A, or 50 A current ratings. **See Figure 11-14.**

CODE CONNECT

Per NEC® Subsection 406.9(B), receptacles of 15/20 A, 125/250 V installed in wet locations must have waterproof enclosures. An outlet box hood installed for weatherproof protection should be marked "extra duty."

Universal Serial Bus (USB) Receptacles

A *universal serial bus (USB)* is an industry-standard device composed of a connector, cable, and connector port for the communication of data and/or power supply between electronic devices. A *USB receptacle* is a receptacle that is constructed with USB connector ports for charging electronic devices.

USB receptacles are installed to eliminate the need for plug-in USB AC-to-DC adapters that charge electronic devices around a home. USB receptacles are available in two types. One type is a combination 125 V receptacle with one or more USB DC charging ports. The other type has only DC charging ports. **See Figure 11-15.** USB receptacles differ from standard receptacles by size and use of USB DC charging ports.

USB Receptacle Sizes. A USB receptacle is larger than a standard receptacle. USB receptacles typically range from 1⅜″ to 1⅞″ in depth, whereas a standard receptacle is typically 1$\frac{1}{10}$″ in depth. **See Figure 11-16.** *Note:* Electrical manufacturers typically have their own set of standard sizes for assembled receptacles.

A receptacle box already installed in a home may be too small to install a USB receptacle. Before a USB receptacle is purchased, it is necessary to determine the location where the receptacle will be replaced and to see if the receptacle box is large enough to accommodate the new USB receptacle. The depth and width of the receptacle box should be measured. These dimensions are then used to purchase an appropriately sized USB receptacle or a new receptacle box.

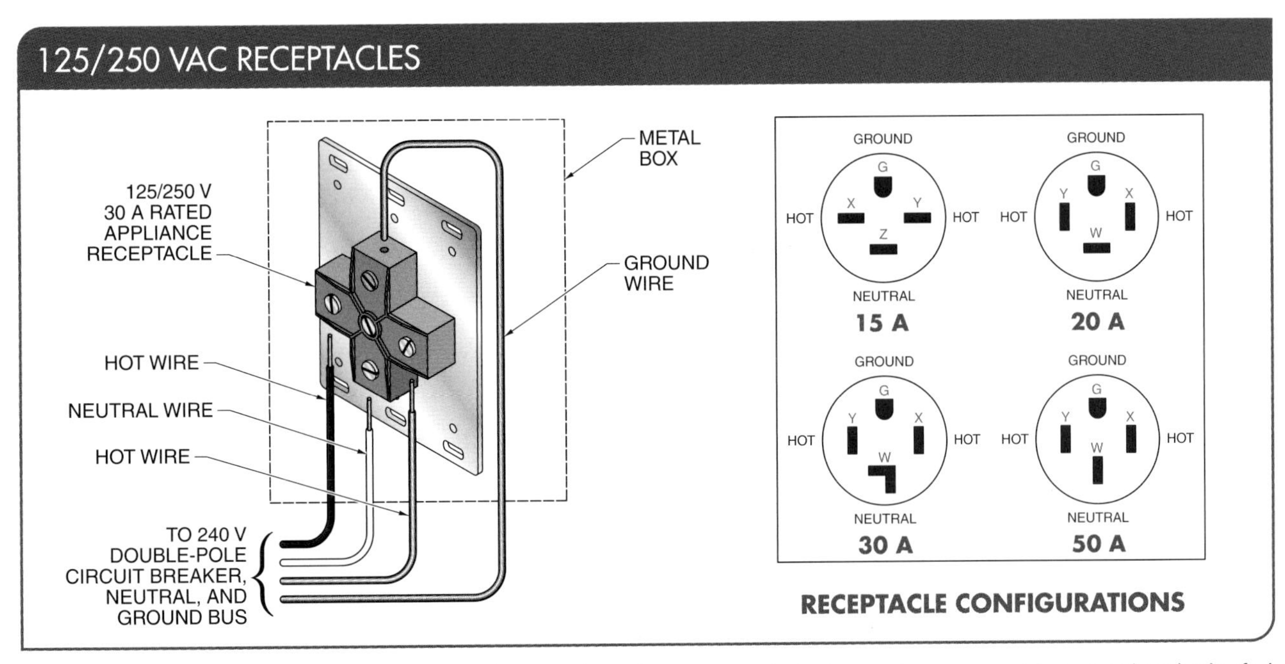

Figure 11-14. A 125/250 VAC receptacle requires two hot wires, one neutral wire, and one ground (green or bare) wire fed from a 240 V double-pole circuit breaker at the service panel. A 125/250 VAC receptacle can have four configurations designed at 15 A, 20 A, 30 A, or 50 A current ratings.

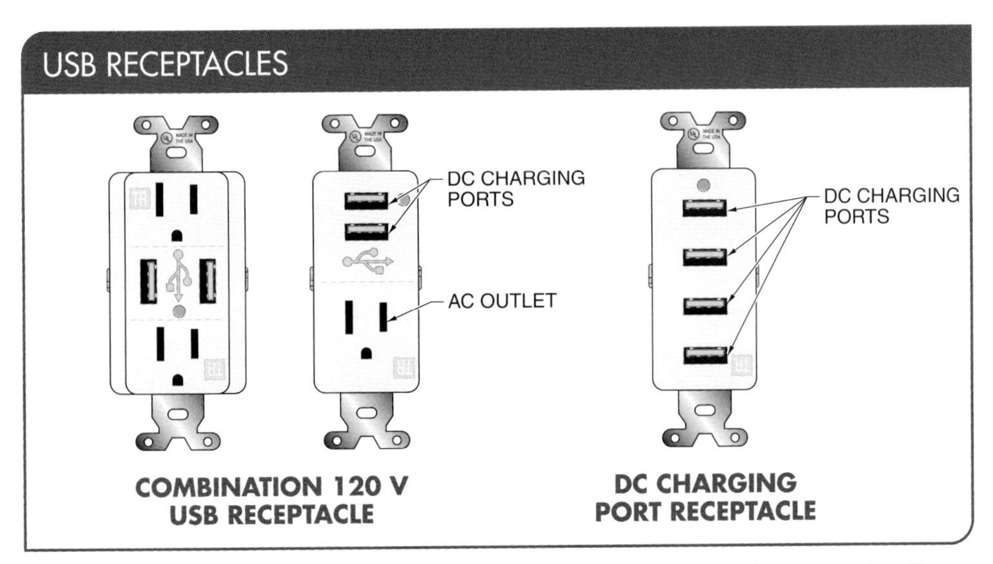

Figure 11-15. Two types of USB receptacles include a combination 125 V receptacle with one or more USB DC charging ports and another type with only DC charging ports.

A USB receptacle is installed in the same manner as a 15 A or 20 A standard receptacle. One feature that an individual must check on a USB receptacle is the location of the terminal screws on the side. The terminal screws should not interfere with the width of the receptacle box when wires are connected during installation. Also, an individual must check any back-mount connections or pigtail connections that require space for wire connectors. It is also necessary to check that the appropriate AWG size is used to support a 15 A or 20 A USB receptacle and that all the NEC® code and local code requirements are met before and after installation.

USB RECEPTACLE SIZES

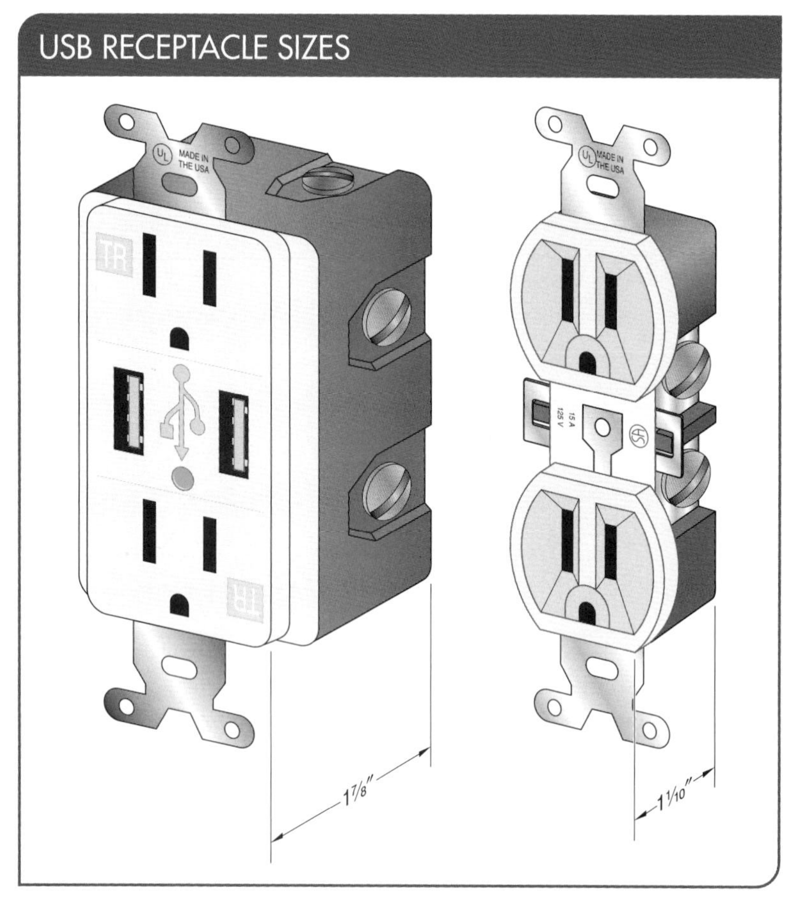

Figure 11-16. USB receptacles typically range from 1⅜″ to 1⅞″ in depth, whereas standard receptacles are typically 1⅒″ in depth.

DC Charging Port Rates. It is important to check the DC charging port amperage rates when selecting either type of USB receptacle. DC charging ports can have a maximum amperage rate of 5 A or less. Some USB receptacle DC ports split their maximum amperage capacity. For example, if the maximum amperage capacity is 3 A between two USB DC ports, each DC port can only charge 1.5 A when two electronic devices are plugged in. The larger the load the device places on the outlet, the slower the device will charge. Some outlets have a high-amperage port and a low-amperage port to charge large and small devices simultaneously. *Note:* Connecting a device such as a cell phone to a higher amperage DC port will not harm the device, but it will charge the device faster.

CODE CONNECT

Per NEC® Subsection 406.2(F), 125 V, 15/20 A receptacles with a USB charger must have Class 2 circuitry integral with the receptacle.

Smart Receptacles

A *smart receptacle* is a receptacle controlled remotely by a control device using specific wireless signals. **See Figure 11-17.** Control devices for a smart receptacle can include cell phones, tablets, or computer-based systems. Smart receptacles are compatible with lamps, appliances, and electrical loads that are turned on and off periodically. An advantage of smart receptacles is that switching devices on and off can be done remotely through a wireless connection.

SMART RECEPTACLES

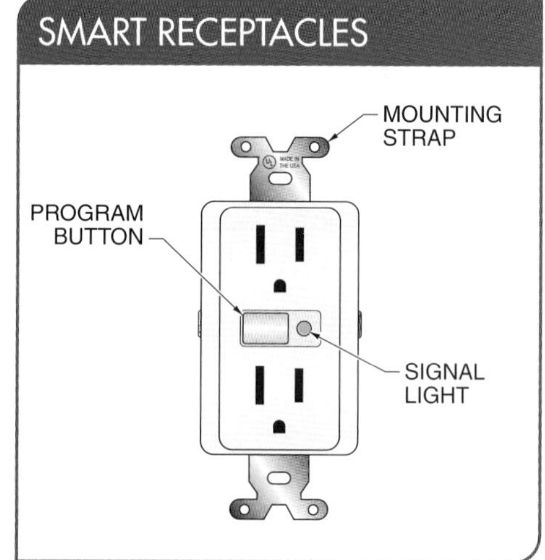

Figure 11-17. Smart receptacles are compatible with lamps, appliances, and electrical loads that are turned on and off periodically by control devices such as cell phones, tablets, or computer-based systems.

Smart receptacles are also available as portable plug-in outlets. **See Figure 11-18.** Portable smart plug-in outlets, or smart plugs, can be inserted into existing standard receptacles. The portable smart plug-in outlet can immediately connect to a software application installed on a control device.

Receptacle Faceplate Shapes and Positions

Receptacle faceplates can convey important information through the shape and position of the outlet on the receptacle. The shape and position of an outlet's connection slots are used to differentiate between hot, neutral, and ground connections, as well as voltage and amperage ratings.

For example, a combination high/low-voltage receptacle has a shape and position for a 125 V, 15 A outlet and a 250 V, 20 A outlet. **See Figure 11-19.** The 125 V, 15 A polarized outlet has connection slots that are vertical and at different lengths. The 250 V, 20 A outlet has connection slots that are horizontal and the same length. *Note:* Other 20 A outlets may have T-shaped neutral slots.

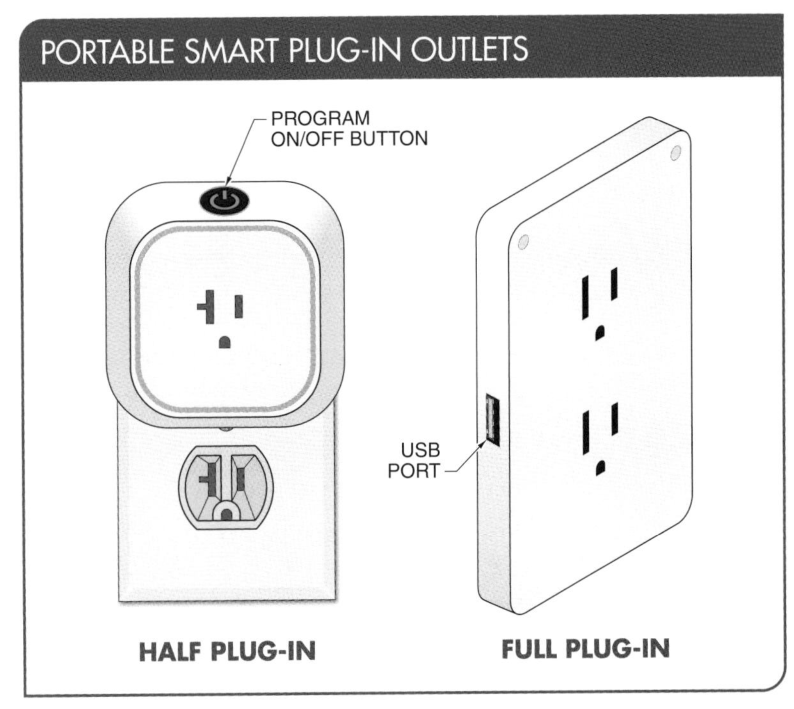

Figure 11-18. Smart plug-in type outlets can be inserted into existing standard receptacles and can immediately connect to a software application installed on a control device.

Outdoor receptacles require weatherproof covers instead of faceplates.

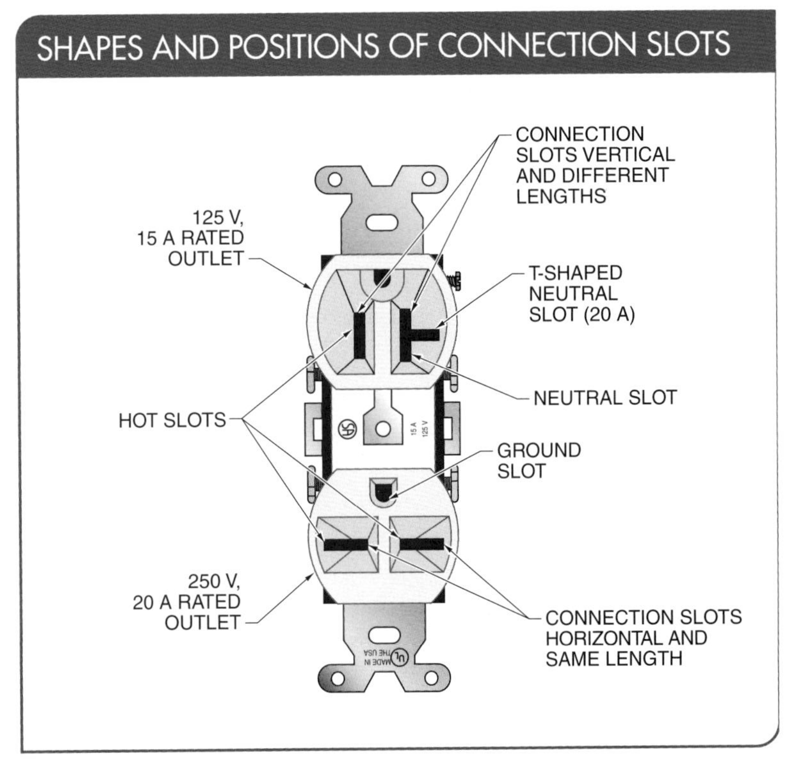

Figure 11-19. A combination high/low-voltage receptacle has a 125 V, 15 A polarized outlet with connection slots that are vertical at different lengths and a 250 V, 20 A outlet with connection slots that are horizontal and the same size.

Voltage and amperage ratings can be differentiated by the configuration of connection slots positioned on a receptacle. **See Figure 11-20.** There are a large number of configurations and wiring diagrams for receptacles used in residential construction based on the type of application. Most receptacles found in a residence include 15 A and 20 A, double-pole, three-wire grounding types. Other heavy-duty types of receptacles can include 20 A, three-pole, three-wire or four-wire grounding types.

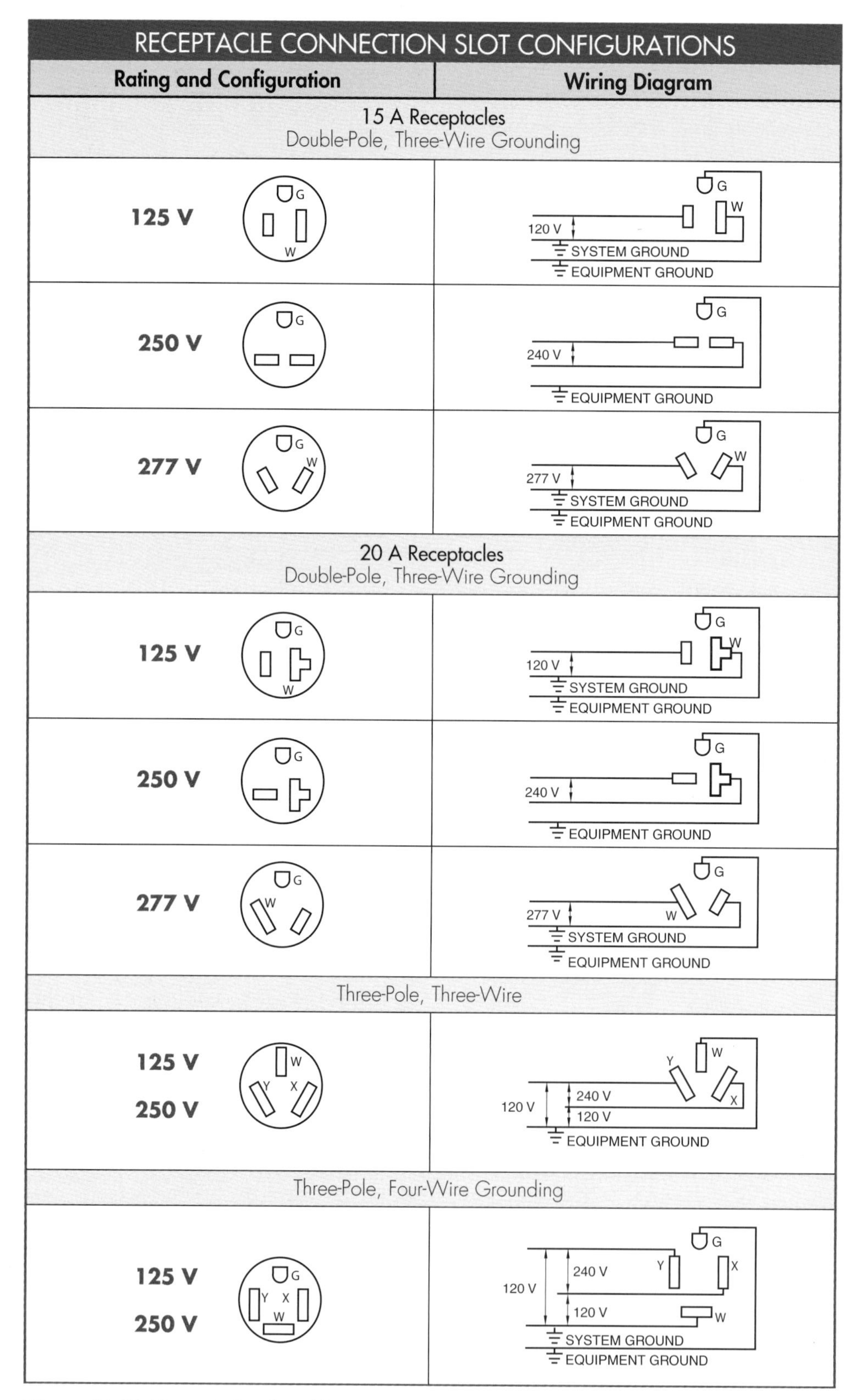

Figure 11-20. Receptacles found in a residence include 15 A and 20 A rated, double-pole, three-wire grounding types. Other heavy-duty types of receptacles can include 20 A, three-pole, three-wire types and 20 A, three-pole, four-wire grounding types.

Receptacle shapes are designed for connecting straight blade or locking blade plugs. However, locking receptacles are more common in commercial and industrial applications than in residential applications. Typical residential straight and locking receptacles include three-prong and four-prong straight and locking receptacles. **See Figure 11-21.**

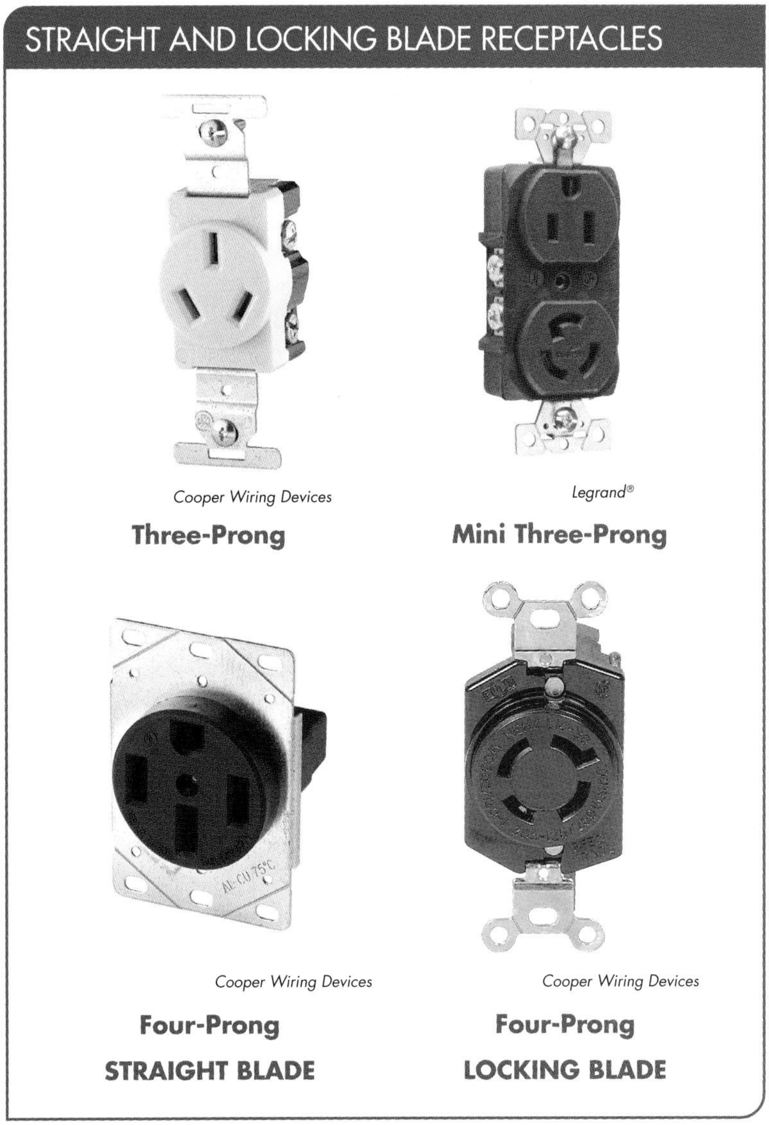

STRAIGHT AND LOCKING BLADE RECEPTACLES

Cooper Wiring Devices

Three-Prong

Legrand®

Mini Three-Prong

Cooper Wiring Devices

Four-Prong

STRAIGHT BLADE

Cooper Wiring Devices

Four-Prong

LOCKING BLADE

Figure 11-21. Typical residential straight and locking receptacles include three-prong and four-prong straight and locking receptacles.

SECTION 11.1 CHECKPOINT

1. What types of receptacles may be found in a residence?
2. What is the most common receptacle used in a residence?
3. What are the color designations for the receptacle terminal screws?
4. Define "ground fault" and describe a ground fault situation.
5. What is a split-wired receptacle?
6. What type of terminal designations are used on GFCIs?
7. What are dual-voltage receptacles used for?
8. What kind of appliances are used with 125/250 VAC-rated receptacles?
9. Define "universal serial bus" and "USB receptacle."
10. What kind of devices are compatible with smart receptacles?

SECTION 11.2—RECEPTACLE BOXES

A *receptacle box* is an enclosure designed to house and protect switches, receptacles, and wiring connections. **See Figure 11-22.** Depending on the specific application, number of wires, and size of wires, receptacle boxes can have various shapes, sizes, and standard accessories. Receptacle boxes can either be manufactured as metallic or nonmetallic types.

Receptacle boxes and junction boxes must be installed at every point in the electrical system where wires are spliced or terminated on a device. A *junction box* is an enclosure designed to house a termination of electrical wires. Wires in junction boxes can either be spliced together or combined using a wire connector. Receptacle boxes are installed so that every outlet in a system is accessible for tests, repairs, or renovations. Receptacle boxes should never be completely covered to the point that they are inaccessible.

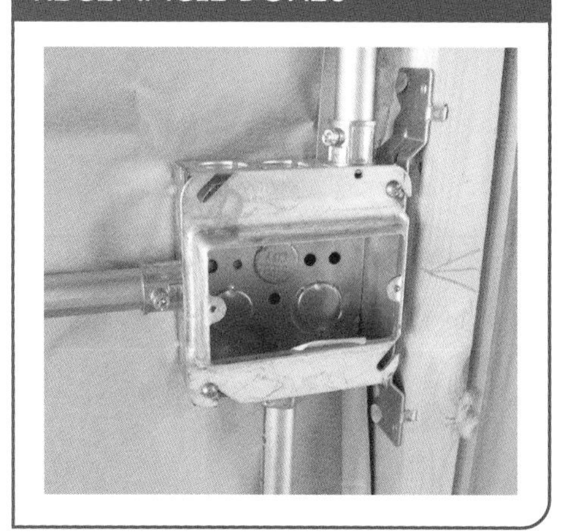

Figure 11-22. Receptacle boxes are designed to house and protect switches, receptacles, and wiring connections.

Receptacle Box Shapes

Typically, square, octagonal, and rectangular receptacle boxes are used in residential wiring. **See Figure 11-23.** Although there is no standard regarding the proper location of each shape of receptacle box, octagonal boxes are typically installed in ceilings for light fixtures or fans. Square and rectangular boxes are typically installed for switches and receptacles on walls or in floors. Any of the three receptacle box shapes can be used as junction boxes where spliced connections are made.

Square, octagonal, and rectangular boxes come in various widths and depths and have various knockout arrangements. A *knockout (KO)* is a round indentation stamped on a receptacle box that allows for quick installation of fittings to secure cables, conduit, or metallic cable. KOs are located on the top, bottom, back, and sides of receptacle boxes to provide access from multiple directions. **See Figure 11-24.** To remove a KO from an outlet box, the following procedure is applied:

1. Gather a flathead screwdriver, hammer, and needle-nose pliers.
2. Use the hammer to lightly tap the holding end of the flathead screwdriver to lodge the flathead into the KO slot.

   WARNING: Do not use a hammering force on the flathead screwdriver.
3. Apply torque to the flathead screwdriver handle to bend the KO open.
4. Use the needle-nose pliers to pry the bent KO back and forth to remove.

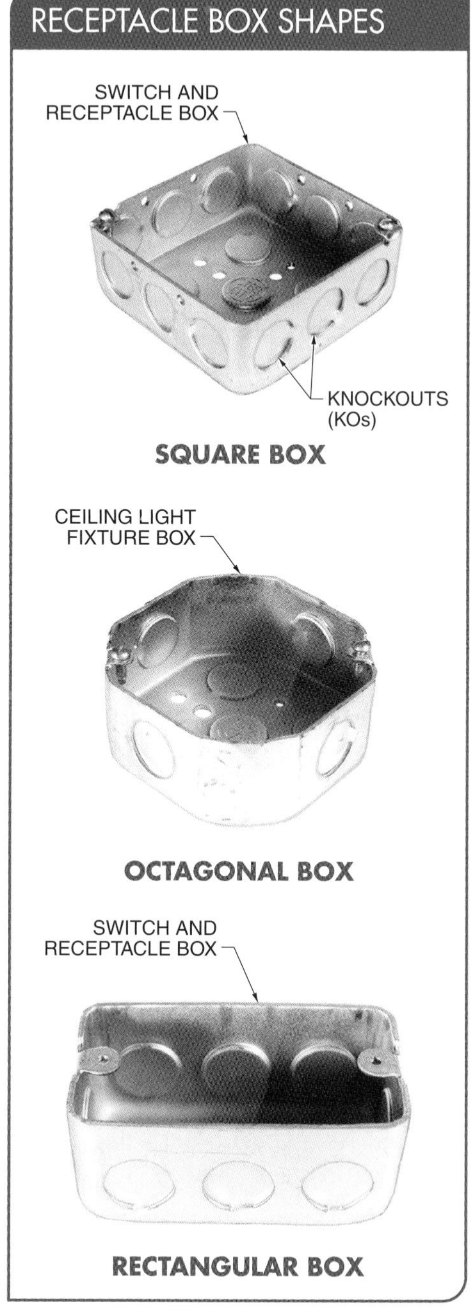

Figure 11-23. Receptacle boxes used in residential wiring applications can be square, octagonal, or rectangular.

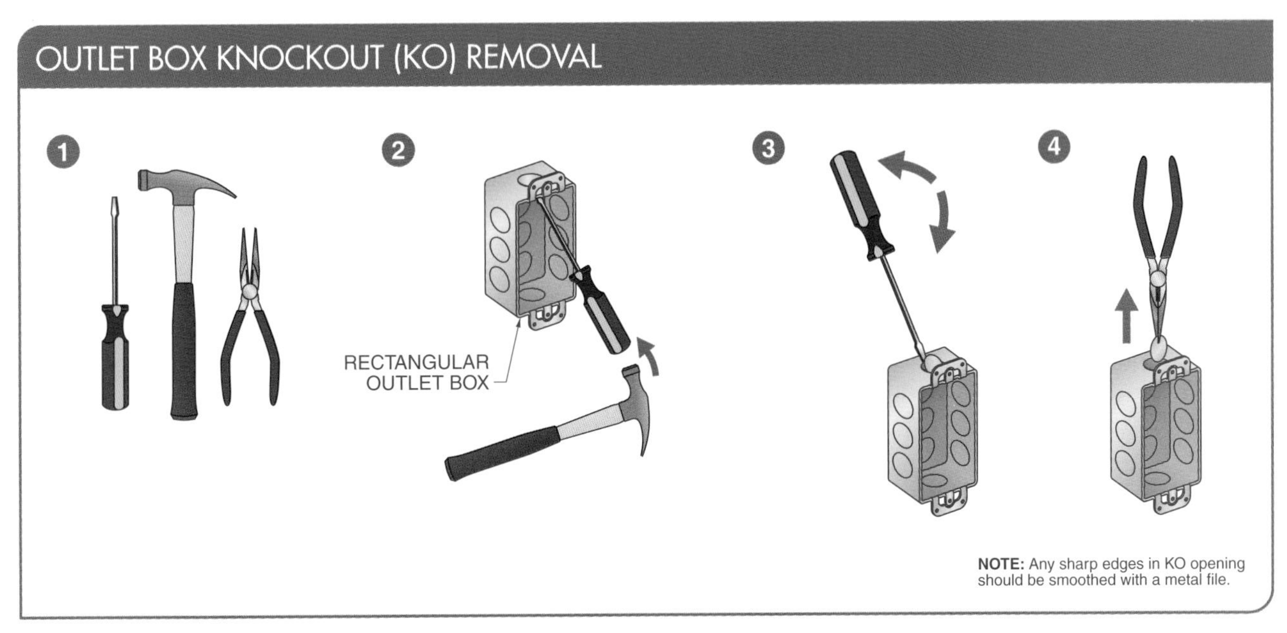

Figure 11-24. Knockouts (KOs) are located on the top, bottom, back, and sides of receptacle boxes to provide access from multiple directions.

KOs on nonmetallic boxes can be removed before or after a box is mounted. Any sharp edges that may penetrate the insulation on wires or the sheathing that surrounds nonmetallic cable should be removed. Burrs or sharp edges can typically be removed by rotating a screwdriver around the KO hole. **See Figure 11-25.**

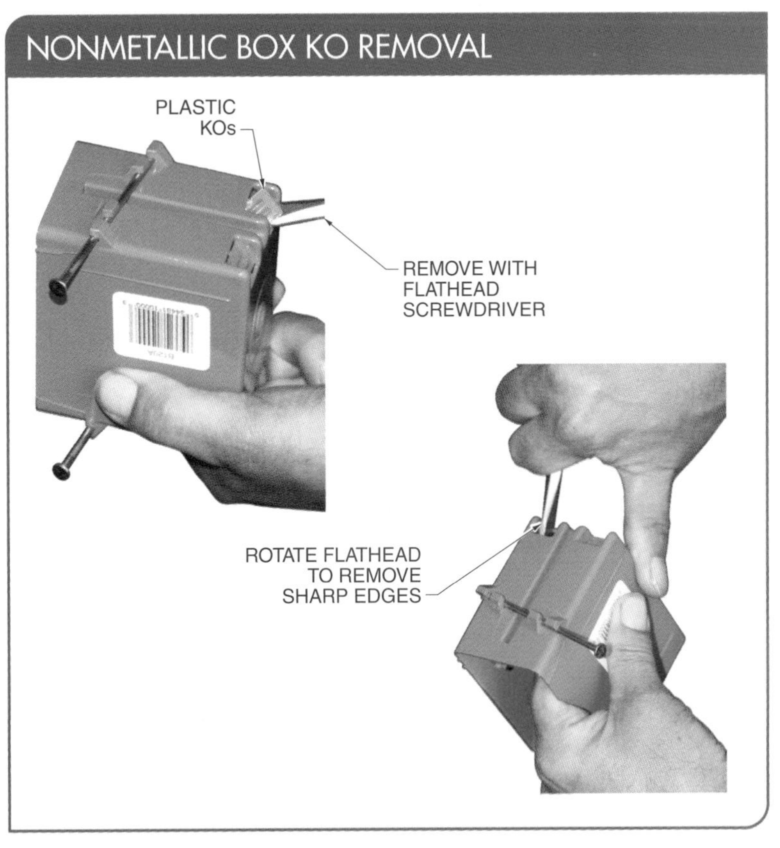

Figure 11-25. KOs on nonmetallic boxes can be removed before or after a box is mounted. Burrs or sharp edges can typically be removed by rotating a screwdriver around the KO hole.

Receptacle Box Sizes

Choosing the appropriate size of outlet box is based on the number of wires and the AWG size of the wires entering the box. A greater number of wires and larger AWG size requires a larger or deeper box. Typically, outlet boxes are marked with their volumes on the outside, but manufacturer specification sheets should be consulted when ordering boxes to find the appropriate box volume per cubic inch capacity (cu in.). Additional information can be obtainWed from Article 314 in the NEC® for metal box volumes and volume allowances per conductor. Larger boxes may be necessary where dimmers, combination switch and receptacle outlets, or USB receptacles are going to be installed.

Receptacle Box Installation

To install a metal or nonmetallic outlet box in a wall, the proper location must first be determined. Then a hole must be cut in the plaster or drywall. The outlet box is then mounted to either a stud or the bracing attached to studs. Cable or wire must then be routed to the appropriate box and location for connection.

Box Location. Before an outlet box is installed in a room, a hole must be cut in the existing wall or ceiling. To avoid unnecessary holes and the associated repairs, individuals must carefully determine the proper location for a box before cutting.

In residential construction, wall studs and ceiling joists are laid out with 16″ on-center (OC) spacing, which means the distance between the center point of each stud is 16″. Advanced framing methods that allow 24″ OC stud spacing may be used as well. A *stud* is an upright wood member that extends from the bottom to the top plates of a framed wall. A *joist* is a horizontal wood member placed on edge to support a floor or ceiling.

> CODE CONNECT
>
> *NEC® Subsection 406.5(G), Receptacle Orientation, states that receptacles must not be installed in a face-up position unless they are listed for countertop or worksurface applications.*

Two methods can be used to locate a stud. One method uses an electronic stud finder to locate the stud. The presence of the stud causes the display of the stud finder to indicate the location of each stud edge. Another method uses a small finishing nail driven into the wall at 2″ intervals until a stud is located. The disadvantage of driving finishing nails every 2″ is that several small holes will require patching after a stud is located. **See Figure 11-26.**

Stud finders can detect the location of wood and metal studs.

Box Hole Cutting. Once an area is found for proper box installation, an individual will typically trace the pattern of the outlet box on the wall. A pattern ensures that cutting does not go outside the traced outline. Drilling each corner of the outline establishes a place to start and end cutting. A drywall saw is typically used to cut into plaster walls. Wallboard or drywall is typically cut with a utility knife or a drywall saw. A more-efficient method for cutting an opening is to use a zip router. **See Figure 11-27.**

Box Mounting. Receptacle boxes are typically mounted onto studs or joists and secured by nails or screws. Nails or screws can be part of a box or driven through mounting accessories on the box. Mounting accessories include frontal straps, side straps, side brackets, or bar hangers. When a box must be mounted between studs or joists, bar hangers are used. Almost all ceiling outlet boxes are mounted with bar hangers. **See Figure 11-28.**

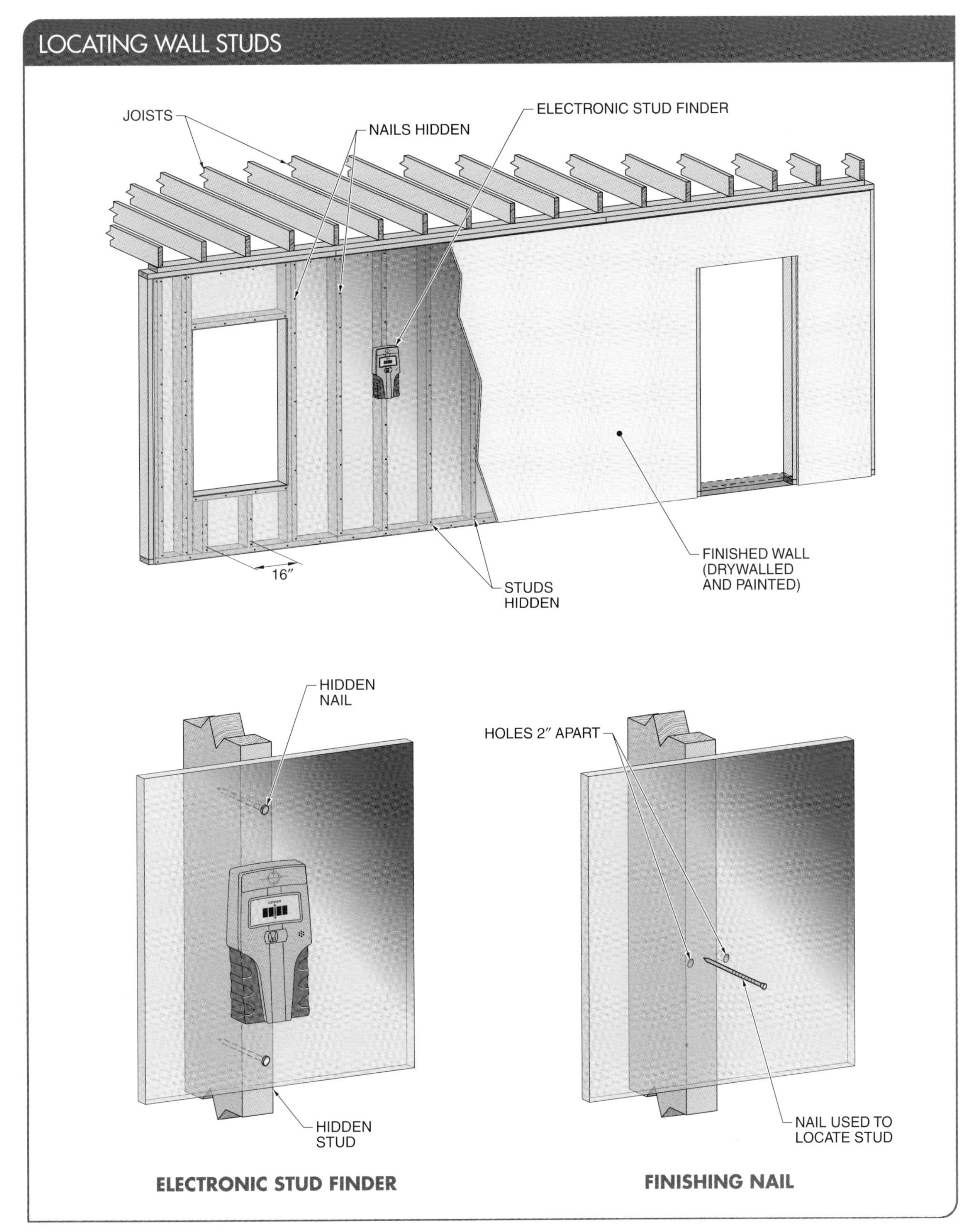

Figure 11-26. To find wall studs, electronic stud finders can be used or nails can be driven into drywall at 2″ intervals until a stud is located.

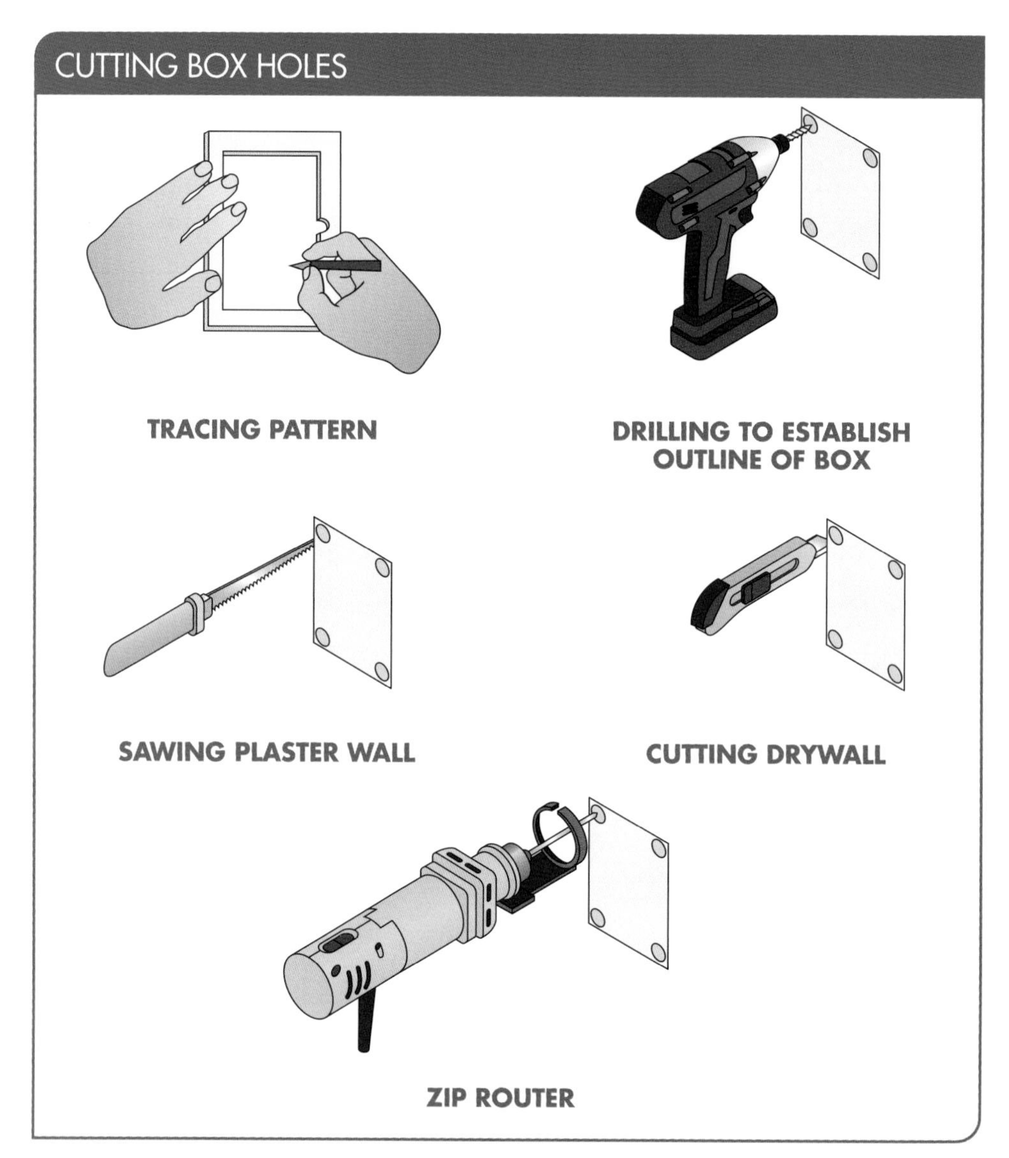

Figure 11-27. Wallboard or drywall is typically cut with a utility knife or drywall saw. Drywall saws are typically used when cutting into plaster walls, although a more-efficient method for cutting an opening is to use a zip router.

When nonmetallic boxes are mounted, the box should be positioned against a stud so that the front of the box will be flush with the finished surface of the wall. For example, if the finished wallboard is ½″ drywall, the box should be positioned ½″ past the face of the stud. If the wallboard will be covered by tile, the thickness of the tile must also be added to the thickness of the drywall in order to maintain proper flush surface. **See Figure 11-29.**

When a nonmetallic box is mounted between studs, either a bar hanger can be used or the outlet box can be attached to a fire block. A *fire block* is a horizontal member between studs that slows down the passage of flames in case the structure catches fire. Fire blocks are installed with the top edge 46″ above the floor. When the outlet box is attached to a fire block, the face of the box is flush with the finished wall. The box is mounted to the wooden fire block with nails or screws.

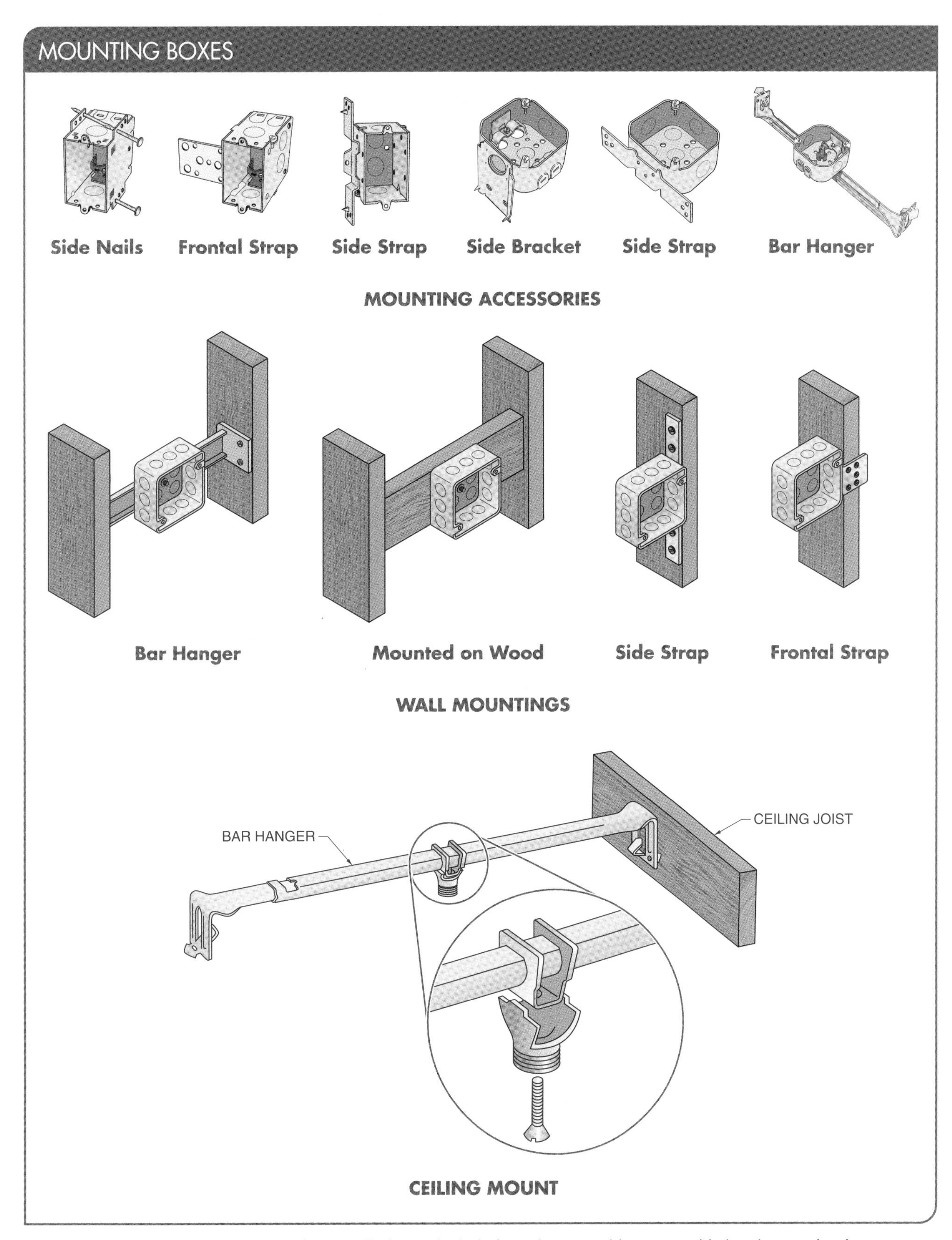

Figure 11-28. Mounting accessories for metallic boxes include frontal straps, side straps, side brackets, or bar hangers.

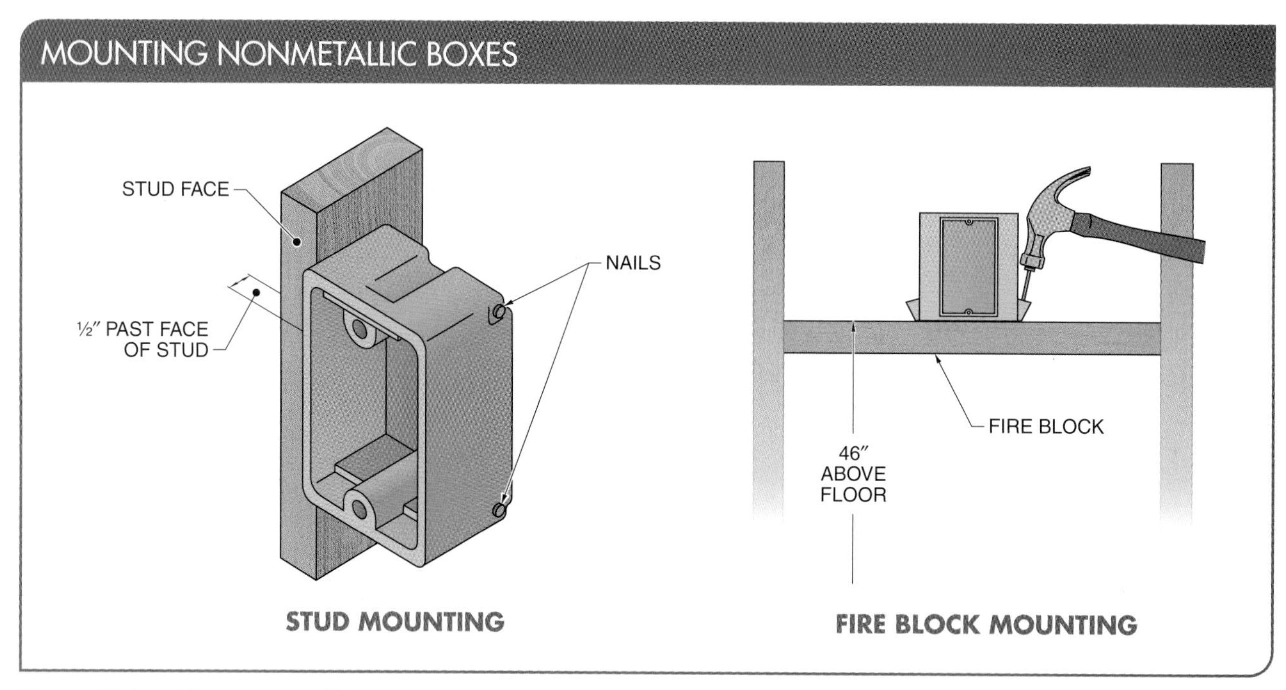

Figure 11-29. When nonmetallic boxes are being mounted, the box should be positioned against a stud so that the front of the box will be flush with the finished surface of the wall.

To aid in renovations, manufacturers produce receptacle boxes with mounting accessories, such as tabs and L clamps, to secure the boxes to existing structures. **See Figure 11-30.** One common feature that manufacturers include on rectangular boxes are folding tabs on two of its sides that allow the box to be firmly secured from the inside of the wall opening.

L clamps on the front of the rectangular box are used to keep it from being pulled through the drywall hole as the tabs are tightened with threaded screws. **See Figure 11-31.** To install a rectangular box in an existing wall, the following procedure is applied:

1. Adjust the L clamps with a flathead screwdriver to be flush with wall.
2. Fold the side tabs and then insert the outlet box into the wall opening.
3. Open the folding tabs by pushing on the screw-tab attachment.
4. Tighten the screws with a flathead screwdriver to secure the box to the wall.
5. Feed wires from the power source to the installed outlet box.

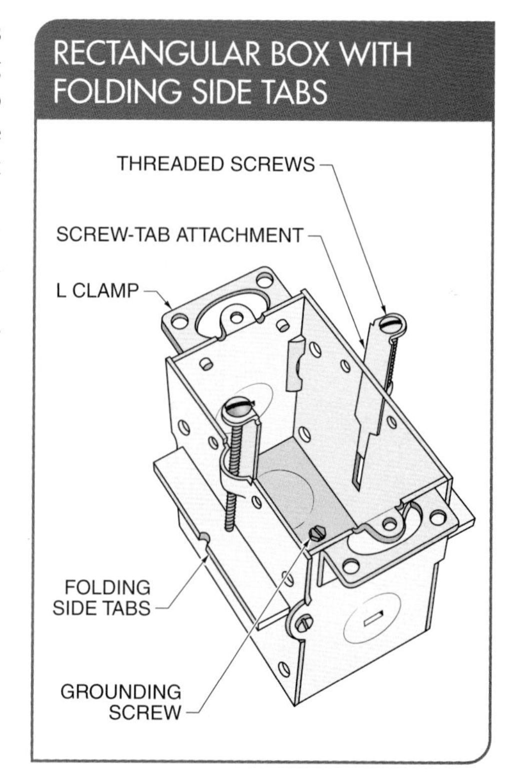

Figure 11-30. Manufacturers produce receptacle boxes with mounting accessories, such as tabs and L clamps, to secure boxes to existing structures.

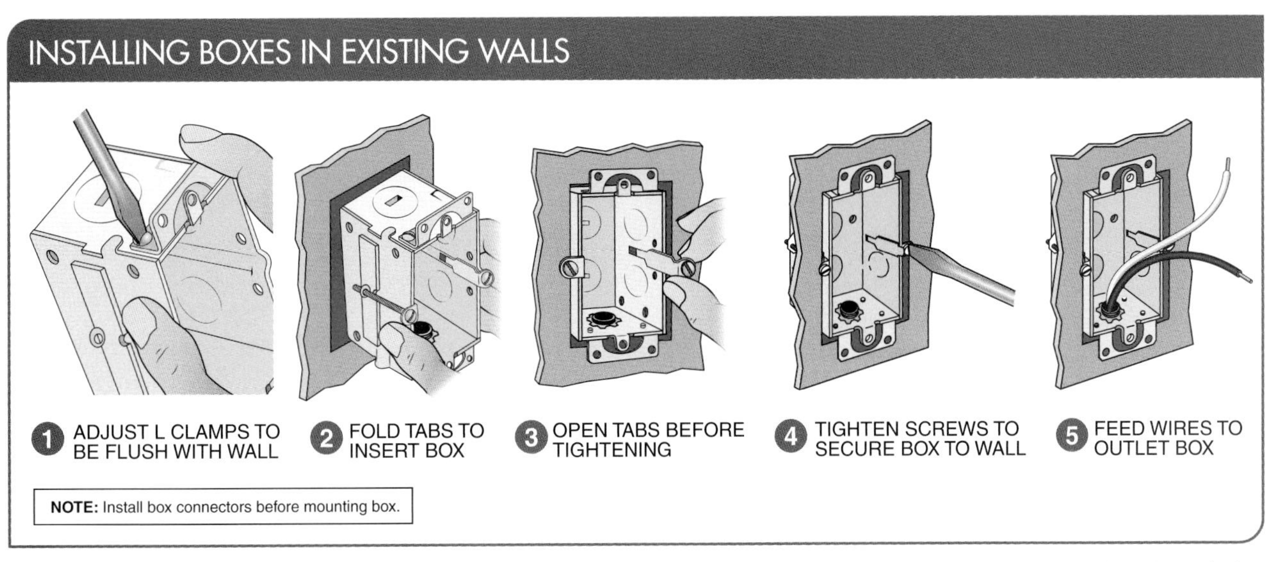

Figure 11-31. L clamps on the front of a rectangular receptacle box are used to keep the box from being pulled through the drywall hole as the tabs are tightened with threaded screws.

Another method of securing rectangular boxes is the use of metal stampings. **See Figure 11-32.** Individual metal stampings are positioned in the wall and the tabs are bent over the box. The L clamps keep the box from being pulled through the opening. Rectangular boxes are useful for adding switches and receptacles to finished walls.

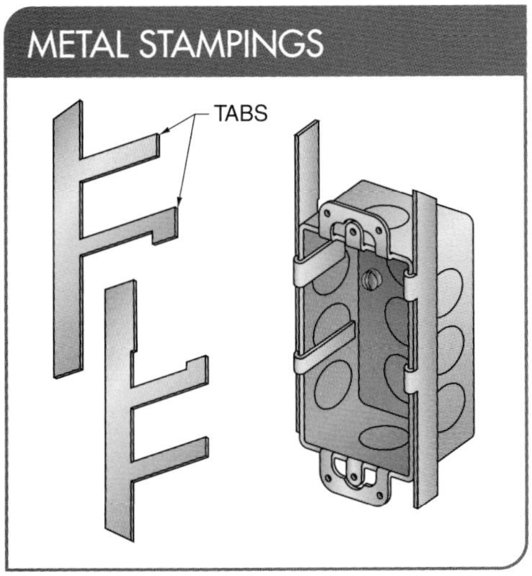

METAL STAMPINGS

TABS

Figure 11-32. Receptacle boxes can also be secured by positioning individual metal stampings in the wall and bending the tabs over the box. L clamps are used to keep the box from being pulled through the opening.

Receptacle Box Extenders and Covers

Outdated receptacles can be replaced with new receptacles when old homes are renovated. New receptacles, such as USB receptacles, GFCIs, and AFCIs, typically require box extenders. **See Figure 11-33.** A box extender provides additional room in a box for all connections without cramming the outlet box. If a box extender is not used, the pressure on the wires can cause connections to come apart or overheat during operation.

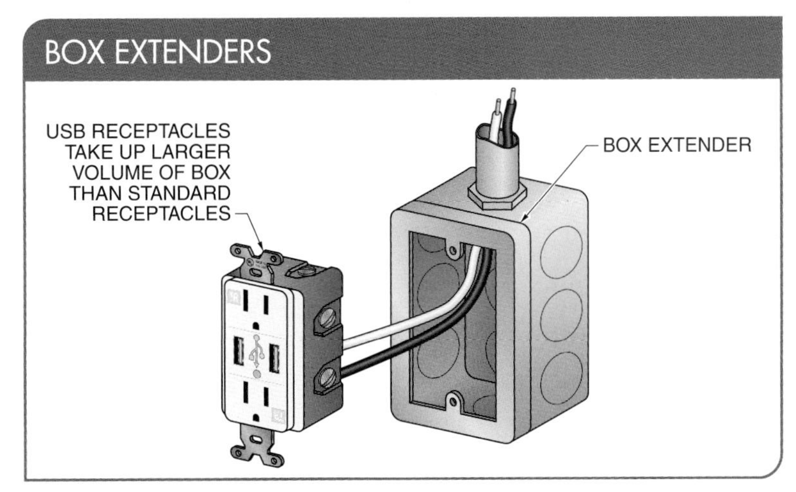

Figure 11-33. A box extender provides additional room in a box to allow USB receptacles, GFCIs, or AFCIs to fit in standard outlet boxes.

Once a receptacle is installed with or without a box extender, the outlet must be properly covered. **See Figure 11-34.** Receptacle covers include indoor and outdoor types. Covers located indoors include duplex, single, combination switch and receptacle, and GFCI/AFCI types. Receptacles located outdoors and exposed to weather require weatherproof covers, which provides coverage of the faceplate. Weatherproof covers can be metallic or hard plastic, and can protect single, duplex, or multiple receptacle installations.

SECTION 11.2 CHECKPOINT

1. Where must receptacle boxes and junction boxes be installed in a residential electrical system?
2. What are the three shapes of receptacle boxes?
3. What article in the NEC® has information regarding metal box volumes and volume allowances per conductor?
4. Describe how wall studs and ceiling joists are laid out.
5. What can be used to mount nonmetallic boxes between studs?

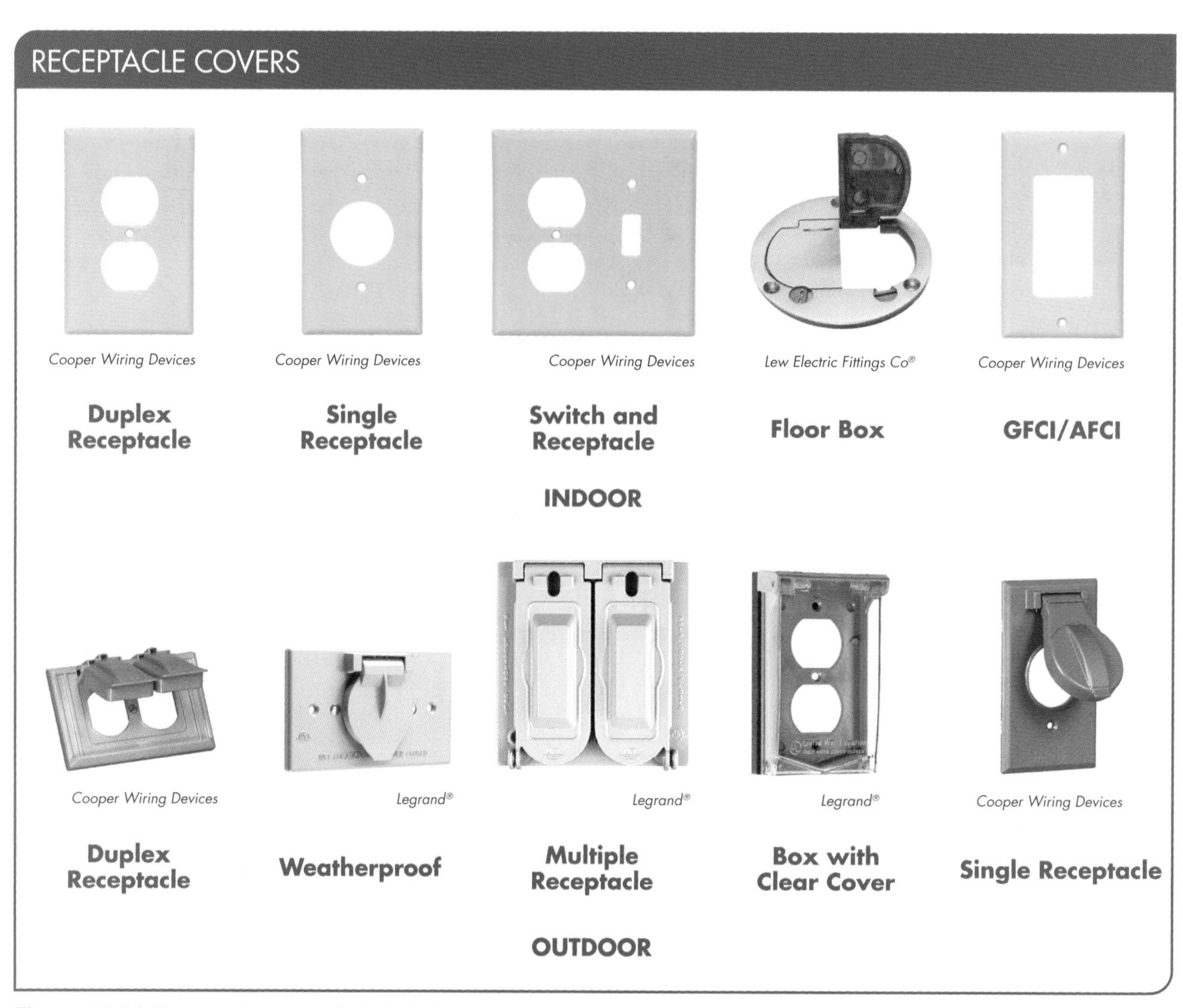

Figure 11-34. Receptacle covers include indoor and outdoor types for duplex, single, combination switch and receptacle, and GFCI/AFCI receptacles. Receptacles that are located outside and exposed to weather require weatherproof covers, which provides coverage of the faceplate.

Chapter Activities

11 Receptacles

Name ____________________ Date ____________

Receptacle Types

Identify the different receptacles shown.

_______________ **1.** AFCI

_______________ **2.** Isolated ground

_______________ **3.** GFCI

_______________ **4.** Smart

_______________ **5.** Standard

_______________ **6.** USB

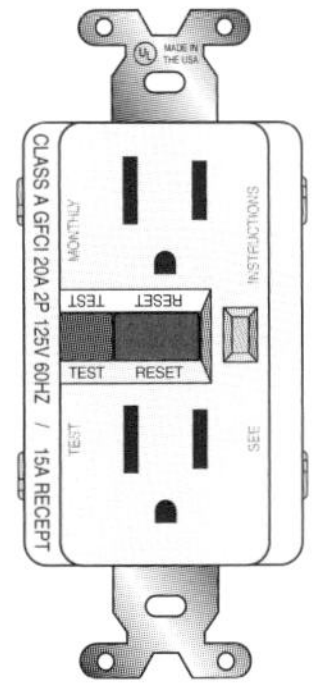

A 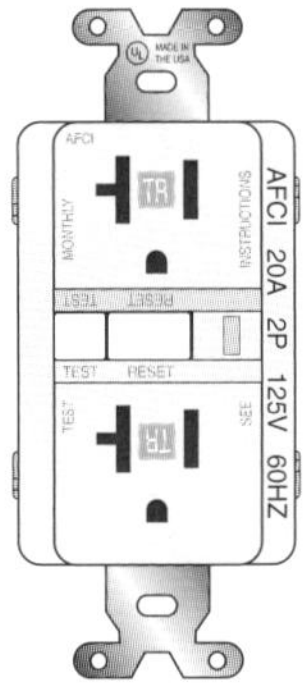 B 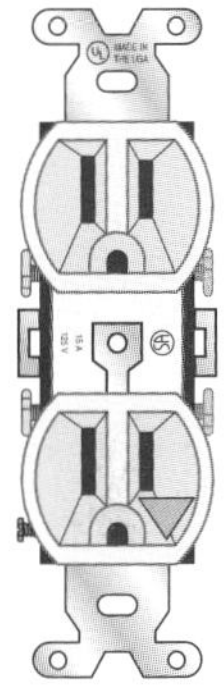 C

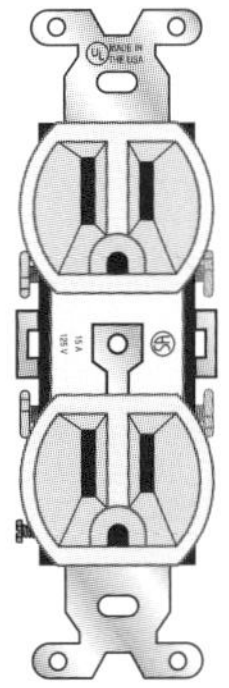

D  E 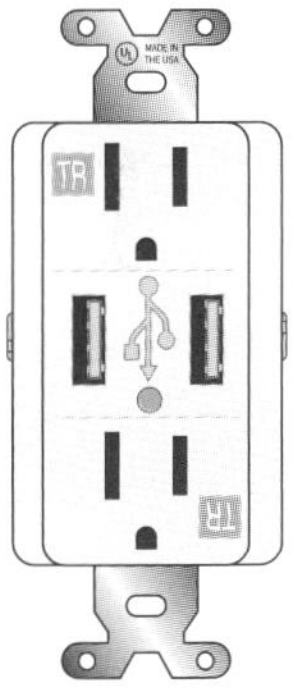 F

Testing Standard Receptacles

1. Draw lines from the test leads to the proper receptacle slots to test for voltage.

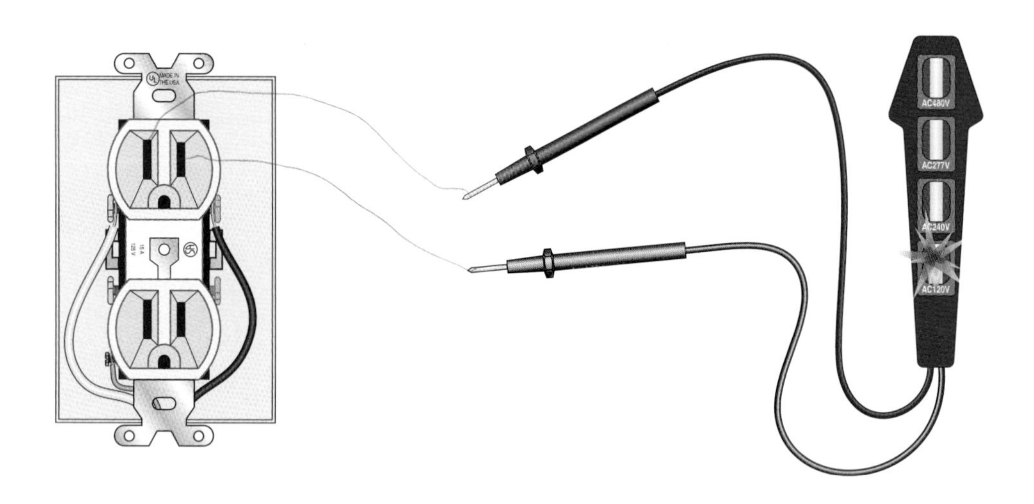

Split-Wired Receptacles

1. Draw lines from the neutral, hot, and ground wires to the terminal screws to demonstrate how to wire a switch-controlled split-wired receptacle.

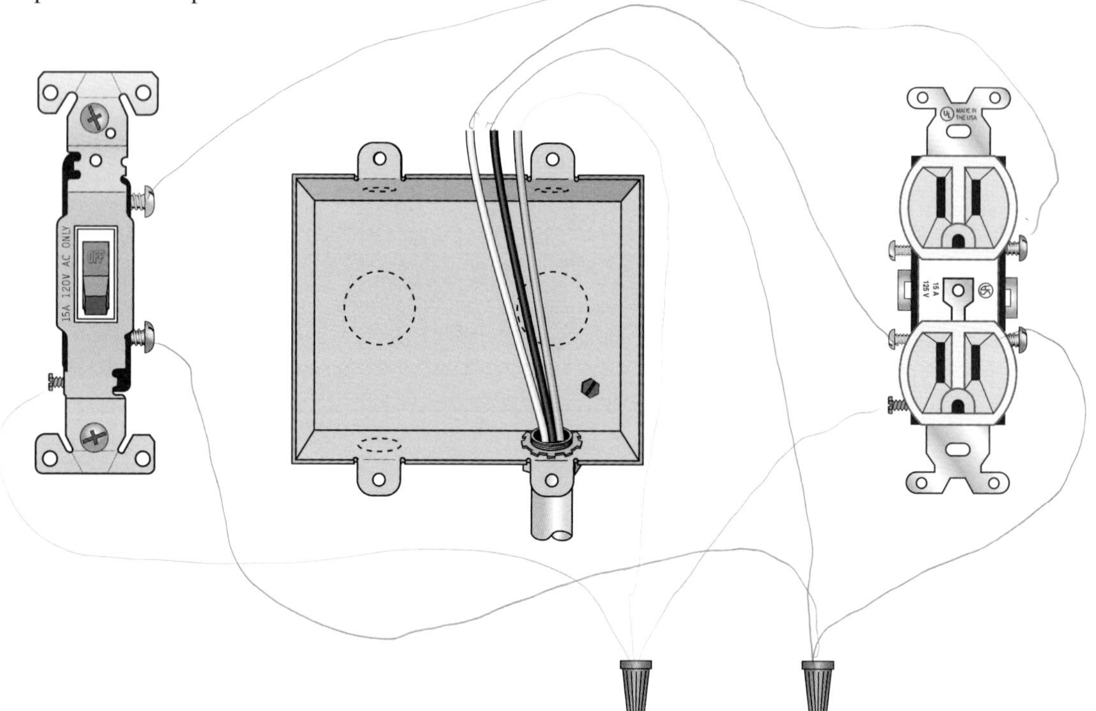

12 Switches and Branch Circuit Installation

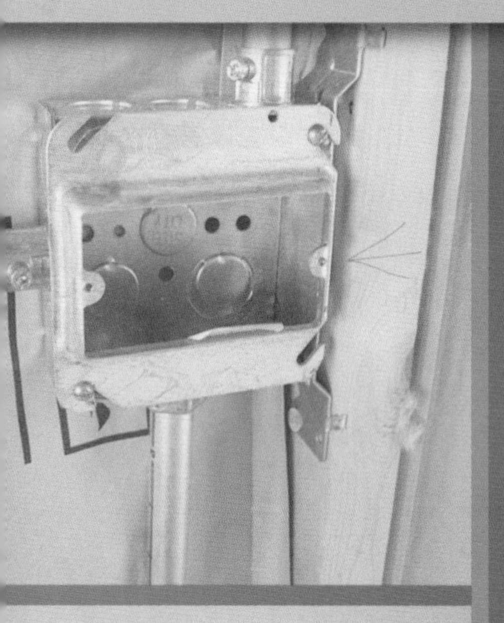

SECTION 12.1—SWITCHES

- Identify the types of technical information found on switches.
- Describe the color designations for switch terminal screws.
- Identify the types of manually operated switches.
- Explain how a timer switch and motion sensor switch operate.
- Describe how to install switch light circuits containing recessed lights.

SECTION 12.2—BRANCH CIRCUIT INSTALLATION

- Define "branch circuit."
- Explain how branch circuits are rated.
- Calculate the amount of receptacles that can be installed on branch circuits.
- Describe bedroom/closet, bathroom, and hallway branch circuits and garage branch circuits.
- List the guidelines that should be followed for kitchen branch circuits.

Learner Resources

ATPeResources.com/QuickLinks
Access Code: 791712

In a residential electrical system, switches are the control for lighting circuits and other appliances. Many different types of switches can be found throughout a dwelling and can be wired into different branch circuits for particular operations. Single-pole, three-way, and four-way switch circuits can be used to control electrical devices, such as receptacles and lighting, throughout a dwelling. Specific branch circuits require GFCIs to protect electrical devices in different rooms of a dwelling. Understanding how to install these circuits with proper information from electrical plans helps to create sound branch circuits for a sound electrical system.

SECTION 12.1—SWITCHES

A *switch* is an electrical device used to control loads in an electrical circuit. **See Figure 12-1.** In a dwelling, many types of switches are used to direct the flow of current to control lighting and receptacles (outlets). Typically, manually operated switches are used. A manually operated switch is toggled to the ON position (closed circuit) to direct current to a load and to the OFF position (open circuit) to disconnect the current to the load. However, other control switches include dimmers, timers, and motion sensors.

Switch Technical Information

Technical information can be found on the markings on a switch. This information includes standard organization labels, voltage ratings, current ratings, wire compatibility, and wire gauge ratings. To ensure safe operation and compliance with the NEC®, the technical information of a switch must be known and understood before the switch is installed.

Typically, switch markings are stamped on the front of a switch's metallic mounting strap. However, some marked information, such as wire compatibility ratings and wire gauge ratings, can be found on the back of the switch. Switches may also include an engraved strip gauge marking to show the length of wire insulation that may need to be removed for back wiring. **See Figure 12-2.**

Standard Organization Labels. The UL® label and the CSA label are the two standard organization labels found on the metallic mounting straps of switches. The *UL® label* is a stamped icon indicating that an electrical device has been approved for consumer use by Underwriters Laboratories Inc. (UL®). When an electrical device passes an extensive testing program, it is listed in a UL® category as having met the minimum safety requirements. The *CSA label* is a stamped icon indicating that extensive tests have been conducted on an electrical device by the Canadian Standards Association (CSA). Both labels are found on many electrical devices, indicating compliance with national codes and standards in both Canada and the United States.

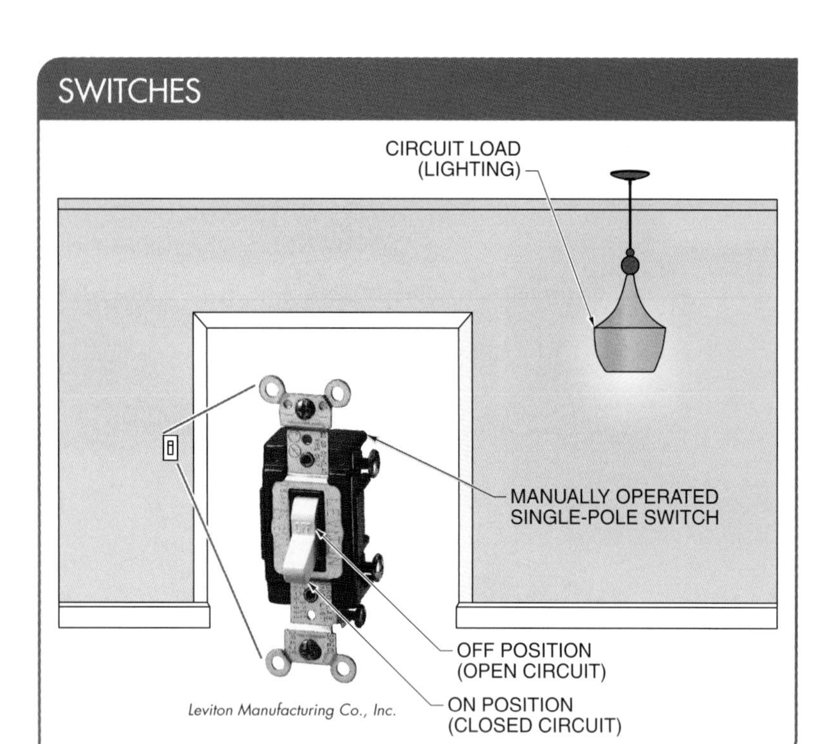

Figure 12-1. Switches direct the flow of current to control loads such as lights, receptacles (outlets), and appliances.

Voltage and Current Ratings. Voltage and current ratings are always marked on a switch. Depending on the type of application, switches may have different ratings. Typical single-pole switches, such as the switches that control lighting circuits, have a voltage rating of 125 V and a current rating of 15 A or 20 A. *Note:* Some switches are only specified for alternating current (AC) use only.

Wire Compatibility Ratings. Wire compatibility ratings are stamped on the front of a switch's metallic mounting strap or on the back of the switch. Wire compatibility ratings are represented as abbreviations of copper (CU), copper-clad (CU-CLAD), or aluminum (AL). Typically, switches stamped with CU and CU-CLAD can only be connected to copper and copper-clad wires. Switches marked with AL can only be connected to aluminum wire. However, when a switch is specified with the abbreviation CU/AL, the switch can be connected to either copper or aluminum wires.

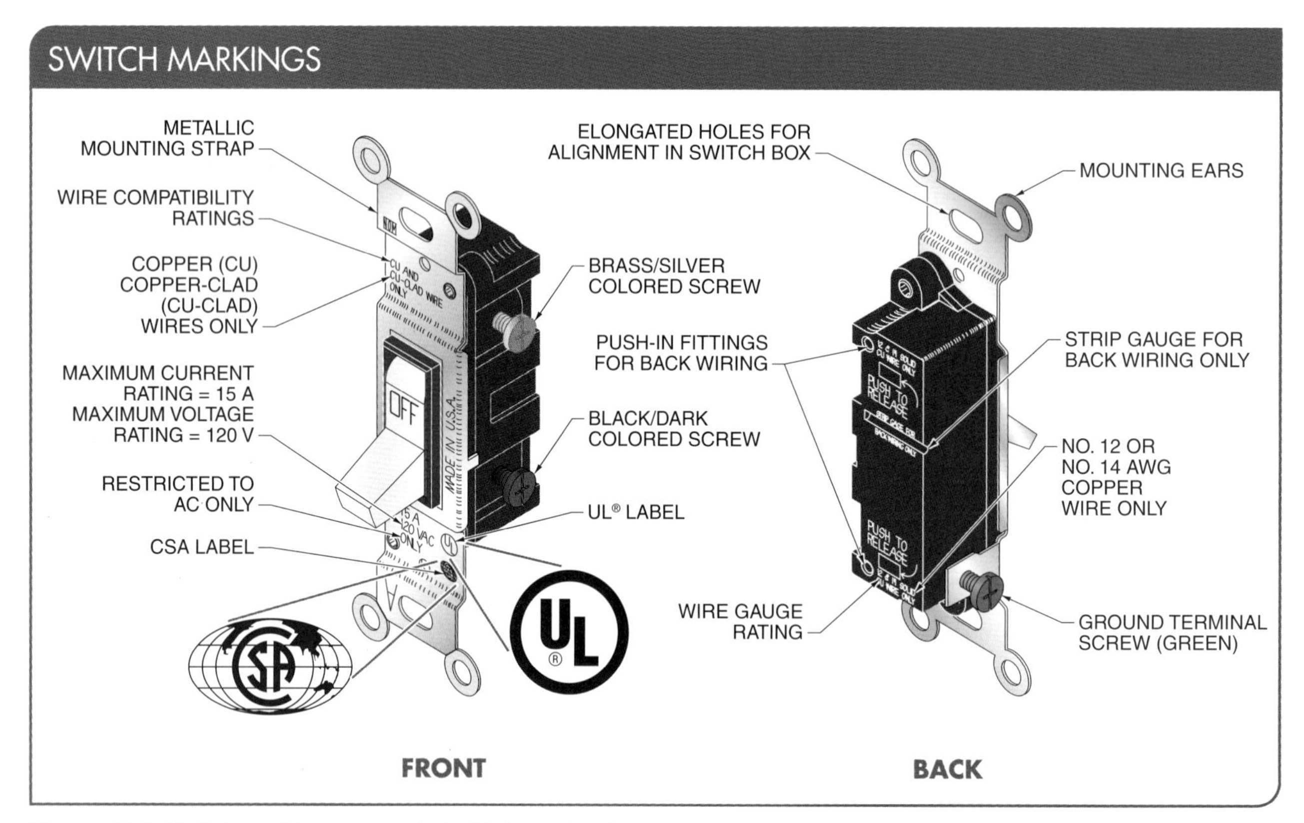

Figure 12-2. Switch markings are technical information found on the front and back of switches and include organization labels, voltage, current, wire compatibility, and wire gauge ratings.

Wire Gauge Ratings. Wire gauge ratings are stamped on the back of switches near the terminal screws or push-in fittings. Manually operated switches used in dwellings have wire gauge ratings stamped in American Wire Gauge (AWG) sizes. Wire sizes No. 12 or No. 14 AWG are typically used to connect switches rated at 120 V and 15 A. Switches can also be marked to indicate the use of solid wire only.

Some switches have a strip gauge located on the back portion near the wire gauge ratings. A *strip gauge* is a short groove engraved in a switch to indicate the length of insulation that must be removed from a wire to fit it into a push-in fitting. Using the strip gauge can ensure the appropriate amount of insulation has been removed from a wire so that a sound connection is made without exposing extra bare wire.

Switch Terminal Screw Color Designations

Switch terminal screws have color designations to identify the proper connections for wire. Typical household single-pole, three-way, and four-way switches use black, brass, silver, and green colors to identify certain terminal screws. **See Figure 12-3.**

Black (or the darker colored screw) indicates the common terminal screw. The common terminal screw is where the wire with incoming current from the power source is attached. Brass- or silver-colored screws indicate the traveler terminal screws. Unlike the common terminal screw, the traveler terminal screw is where the wire with outgoing current going to power a load or continue along the circuit is attached. A green-colored screw is used to indicate the ground terminal screw. The ground terminal screw is attached to the mounting strap of the switch.

Manually Operated Switches

The most common types of manually operated switches used in dwelling applications are the single-pole (two-way), three-way, four-way, double-pole, and dimmer switches. The type of switch used depends on the number of switch locations required to control lighting, receptacles, or appliances. Manual switches are connected into a circuit in the positive or hot side (leg).

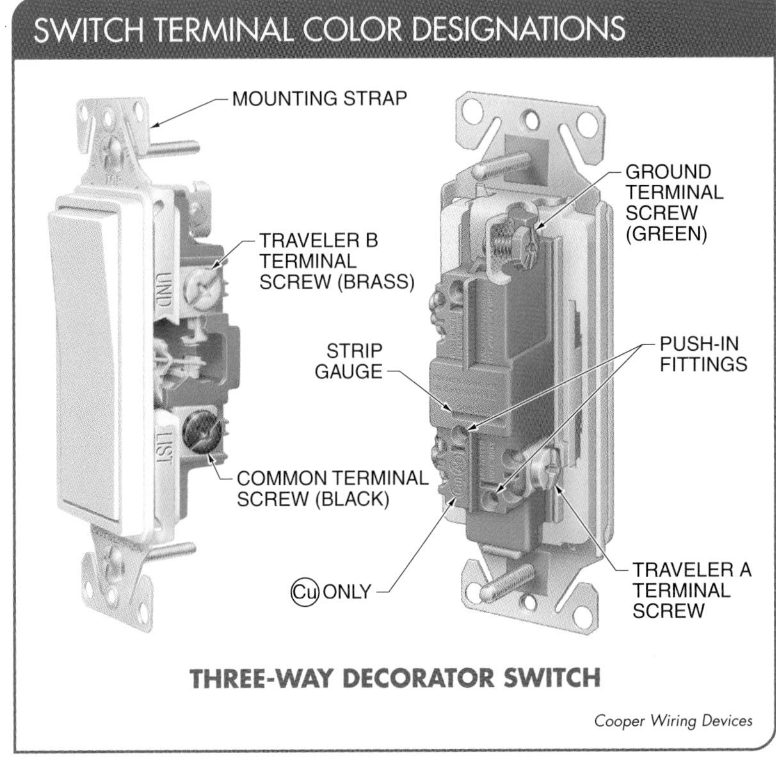

Figure 12-3. Switch terminal screws use color designations such as black, brass, silver, and green to identify proper connections for wire.

Single-pole, three-way, and four-way switches are also specified according to their number of poles and throws. The function of each switch can be represented using control symbols. **See Figure 12-4.**

A *pole* is the number of completely separate circuits into which a switch can feed current. A single-pole (SP) switch can only carry current through one circuit at a time, whereas a double-pole (DP) switch can carry current through two circuits at a time.

A *throw* is the number of closed switch positions per pole. A single-throw (ST) switch can control only one circuit, whereas a double-throw (DT) switch can control two circuits. With a DT switch, the two circuits are mechanically connected to open or close simultaneously, but they are electrically insulated from each other.

MANUAL SWITCH CONTROL SYMBOLS				
Description	**Device**	**Abbrev**	**Symbol**	**Terms**
Single-Pole Switch	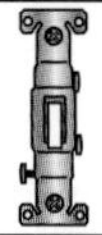	SPST	S/S_1	SINGLE-POLE, SINGLE-THROW
Double-Pole Switch	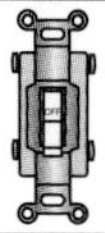	DPST	S_2	DOUBLE-POLE, SINGLE-THROW
Three-Way Switch	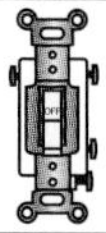	SPDT	S_3	SINGLE-POLE, DOUBLE-THROW
Four-Way Switch	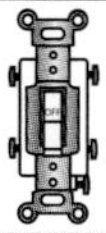	DPDT	S_4	DOUBLE-POLE, DOUBLE-THROW

Figure 12-4. The functions of single-pole, three-way, and four-way switches are specified according to their number of poles and throws, which can be represented using control symbols.

Single-Pole Switches. A *single-pole switch* is a single-pole, single-throw (SPST) electrical control device. Single-pole switches have two distinct positions. The switch allows current to flow in the ON (closed) position and does not allow current to flow in the OFF (open) position. The ON and OFF position markings are marked on the toggle component of the switch. **See Figure 12-5.**

SINGLE-POLE SWITCHES

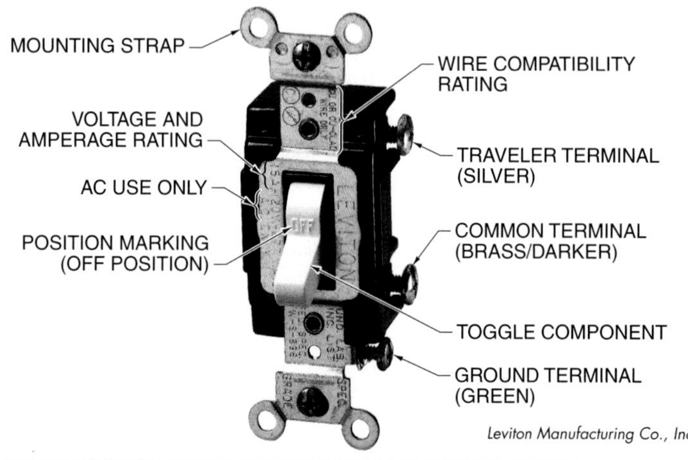

Figure 12-5. Single-pole switches have two distinct positions: the ON (closed) position that allows current to flow and the OFF (open) position that does not allow current to flow.

In a dwelling, single-pole switches are mainly used to control light circuits. Because single-pole switches have one source of power and one load to control, replacing these switches can be easier than replacing three- or four-way switches. **See Figure 12-6.** *Note:* Proper PPE must be worn before replacing any switch, even when the circuit is not energized. To replace a single-pole switch, the following procedure is applied:

1. At the service panel, turn off the circuit breaker feeding current to the switch.
2. Remove the wall plate and existing single-pole switch.
3. Straighten the wires from the receptacle box.
4. Trim the wires and strip insulation as required. Bend loops at the end of wires and connect them to the terminal screws on the switch.
5. Connect the ground wire from the green terminal screw on the switch to the receptacle box bonding screw.
6. Fasten the new single-pole switch and wall plate to the receptacle box.
7. Return to the service panel and turn on the circuit breaker.

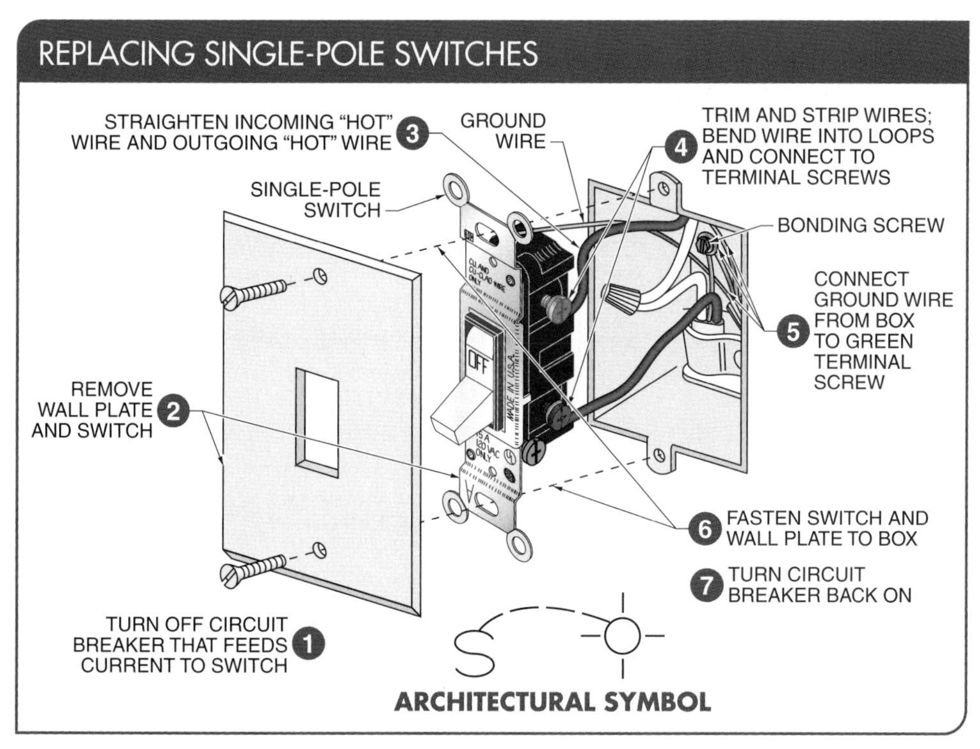

Figure 12-6. Single-pole switches are used to control light circuits and are easier to replace than three- or four-way switches. The proper PPE must be worn when replacing switches.

A common lighting circuit in a home is called a switch loop circuit. A switch loop circuit allows a light to be controlled by a single-pole switch from a remote location. When a switch loop circuit is wired, two wires are used to connect the power source to the switch and light fixture. **See Figure 12-7.** To wire a switch loop circuit, the following procedure is applied:

1. At the light fixture box, connect the black (hot) wire from the power source to the white (neutral) wire from the switch box with a wire connector. *Note:* Mark the white wire with black electrical tape to indicate that it carries current to the switch.
2. Connect the white wire from the power source directly to the light fixture neutral terminal.
3. Connect the black wire from the switch box directly to the light fixture brass terminal.
4. At the switch box, connect the marked white wire to the switch common terminal.
5. Connect the black wire in the switch box to the traveler terminal.
6. Connect the green wire from the switch ground terminal to the cable green wire, and connect the cable green wires at the light fixture box.

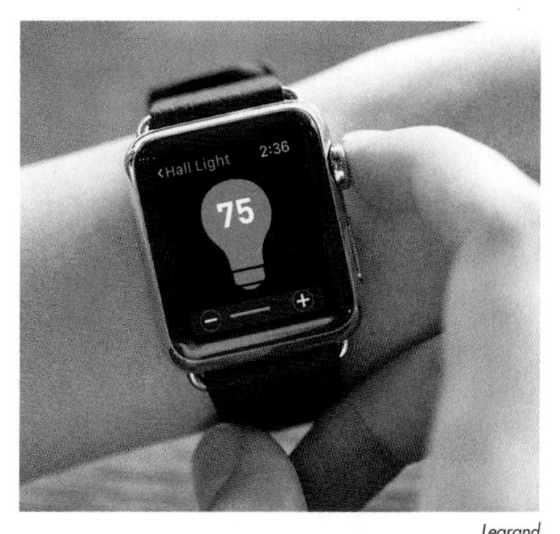

Legrand

Many original equipment manufacturers (OEMs) now provide smart home technology that allows a homeowner to control light switching functions through the use of wearable smart devices.

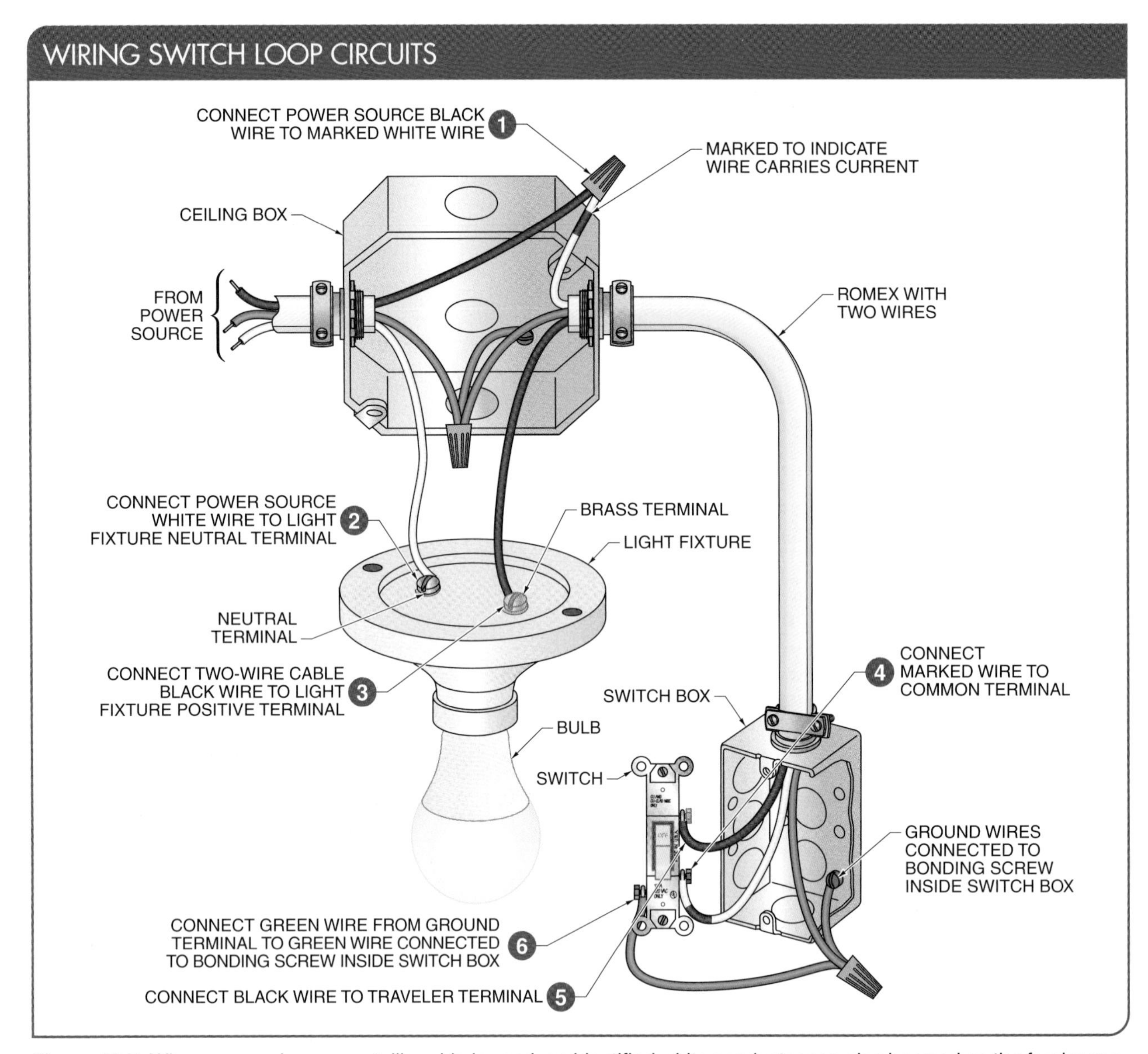

Figure 12-7. When armored or nonmetallic cable is used, an identified white conductor can also be used as the feeder conductor to a switch. When switch loop circuits are wired, two-wire cable is typically used for the connection.

Three-Way Switches. A *three-way switch* is a single-pole, double-throw (SPDT) electrical control device. Three-way switches are commonly used to control lights from two switch locations. Four terminals can be identified on a three-way switch: the common, traveler A, traveler B, and ground terminals. **See Figure 12-8.**

The common terminal is the single terminal screw at bottom end of the switch. The common terminal is easily identified because it is black or darker in color than the traveler terminals. *Note:* The power source (hot) wire of a circuit is always connected to the common terminal. The traveler terminals are typically silver or brass and are the connecting points for wires to connect another three-way switch. The ground terminal is green and is attached to the mounting strap. Three-way switches do not have a designated ON or OFF position marked on the toggle component because the common terminal always feeds current to one of the traveler terminals.

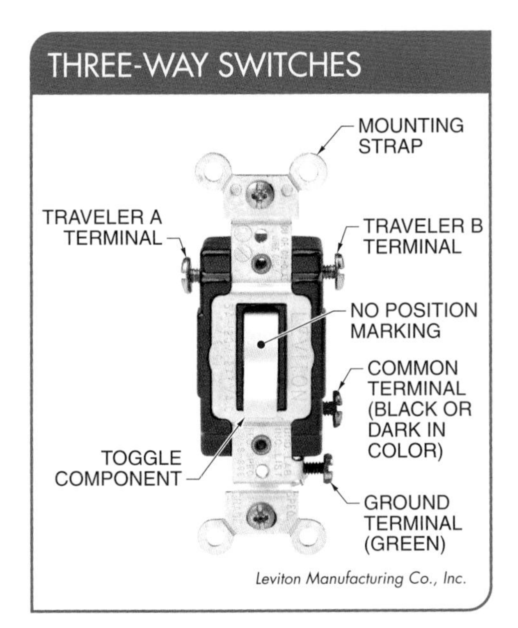

Figure 12-8. Three-way switches use four terminals identified as the common, traveler A, traveler B, and ground terminal to control lights from two switch locations.

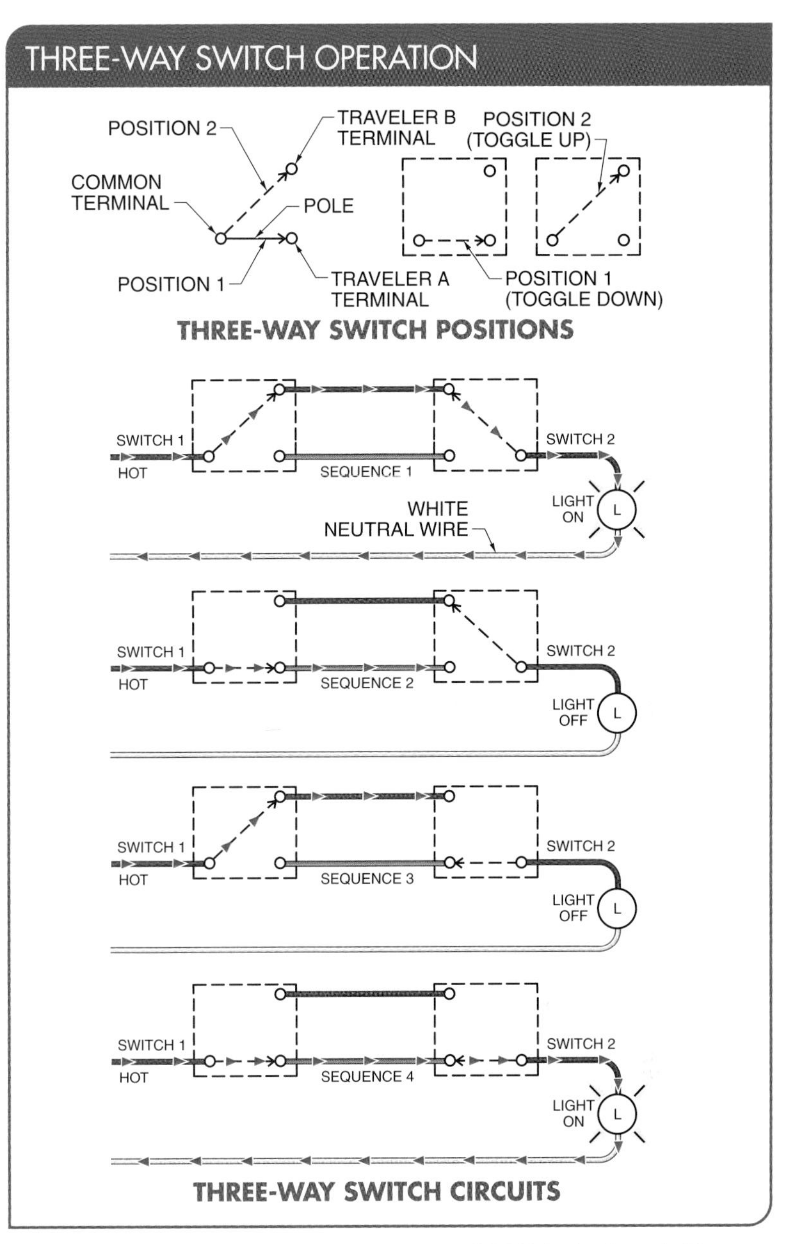

Figure 12-9. A three-way switch operates as a SPDT device by switching one pole to feed current between two positions. The position of both three-way switches determines whether the circuit is complete.

A three-way switch operates as a SPDT device by switching one pole to feed current between two positions (two circuits). When the toggle is down, contact is made with terminal traveler A (position 1). When the toggle is up, contact is made with terminal traveler B (position 2). The pole position of both connected three-way switches determines whether or not the circuit is complete (the light is on). **See Figure 12-9.**

When both three-way switches are in position 1 or position 2, the conducting path is complete and the light turns on. When one switch is in position 1 and the other is in position 2, the conducting path is open and the light turns off. Since either three-way switch can open or close the conducting path, the load can be turned off and on from two locations.

Typically, three-way switch circuits are used for hallways, stairways, and rooms with two entrances/exits. A lamp that is used to light a staircase (six steps or more) must have a three-way switch at the top and bottom of the staircase. Rooms that have two entrances/exits require two three-way switches to control the light(s) from each entrance/exit.

To wire a three-way switch circuit, the power source wires can be run from the service panel to the location of the light fixture at the ceiling box. In this case, the power source black wire feeds current to the location of switch 1. Switch 1 then feeds current to the location of switch 2. Three wires (two black and a red) are fed from the ceiling box to both switch locations. **See Figure 12-10.**

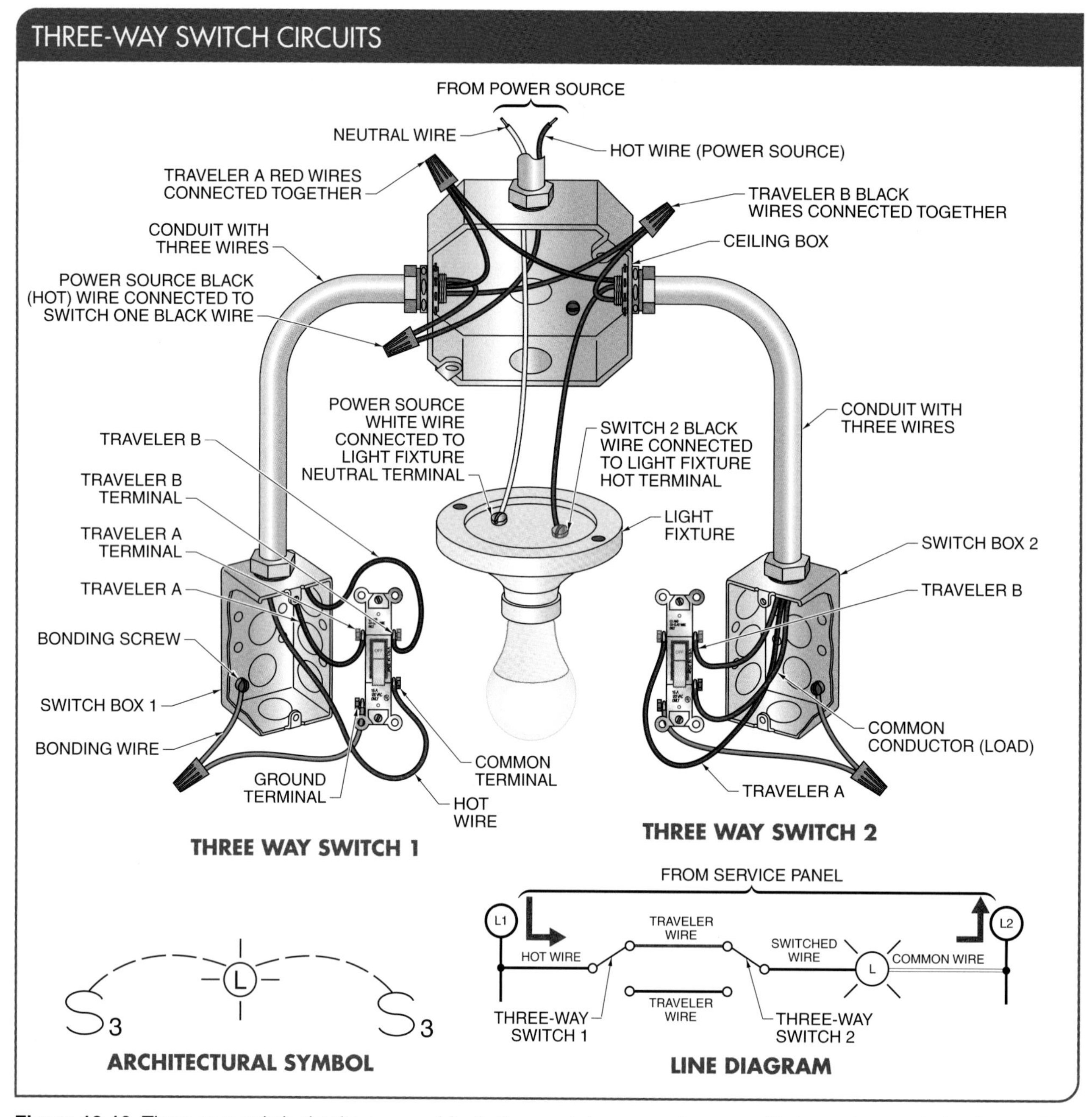

Figure 12-10. Three-way switch circuits are used for hallways, stairways, and rooms with two entrances/exits. Proper PPE must be worn before attempting to wire a three-way switch circuit.

At the ceiling box, the power source black (hot) wire is connected to the black wire being fed to switch 1, and the black wire fed from switch 2 is connected to the light fixture hot terminal. The power source connected white (neutral) wire is connected directly to the light fixture neutral terminal. The two traveler A red wires are connected together, and the two traveler B black wires are connected together. The green wires at switch 1 and switch 2 are connected to a bonding screw inside both receptacle boxes.

At both switch 1 and switch 2 locations, the black (hot) wire is connected to the common terminal. The red wire is connected to the traveler A terminal, and the other black wire is connected to the traveler B terminal. The green wire is connected to the switch ground terminal and bonded to the switch box.

Four-Way Switches. A *four-way switch* is a double-pole, double-throw (DPDT) electrical control device. Four-way switches are used between two three-way switches to control a light from three switch locations. Two or more four-way switches can be connected between two three-way switches to provide control of a load from more than three locations.

The two types of four-way switches are the through-wired type and the cross-wired type. **See Figure 12-11.** Both switch types have a total of five terminals: traveler 1, traveler 2, traveler 3, traveler 4, and a ground terminal. The difference between the through-wired type and the cross-wired type is the position of the poles when the toggle is switched.

The positions of the switch contacts in relation to each other determine whether a load is on or off. Four-way switch circuits can provide three-location control and four-location control. A circuit with two three-way switches and one four-way switch has four possible combinations to power a light. A circuit with two three-way switches and two four-way switches has six possible combinations to power a light. **See Figure 12-12.**

For example, a room that contains three entrance/exits would require a four-way switch circuit to turn a light on or off from any of the three entrance/exits. A four-way switch is connected between two three-way switches in the circuit to control the light from all three entrance/exits. Two wires (black and white) are fed in the circuit to connect the service panel to the light fixture, but two black wires are fed from the light fixture box to the first three-way switch. Three wires are fed to connect the first three-way switch to the four-way switch as well as the four-way switch to the last three-way switch.

To wire a four-way switch circuit, certain connections must be made at each receptacle box. Starting at the light fixture box, the power source white wire terminates at the light fixture neutral terminal. The power source black wire is connected to the switch 1 black wire. The black wire connected at the switch 1 common terminal, from switch 1, is terminated at the light fixture positive terminal.

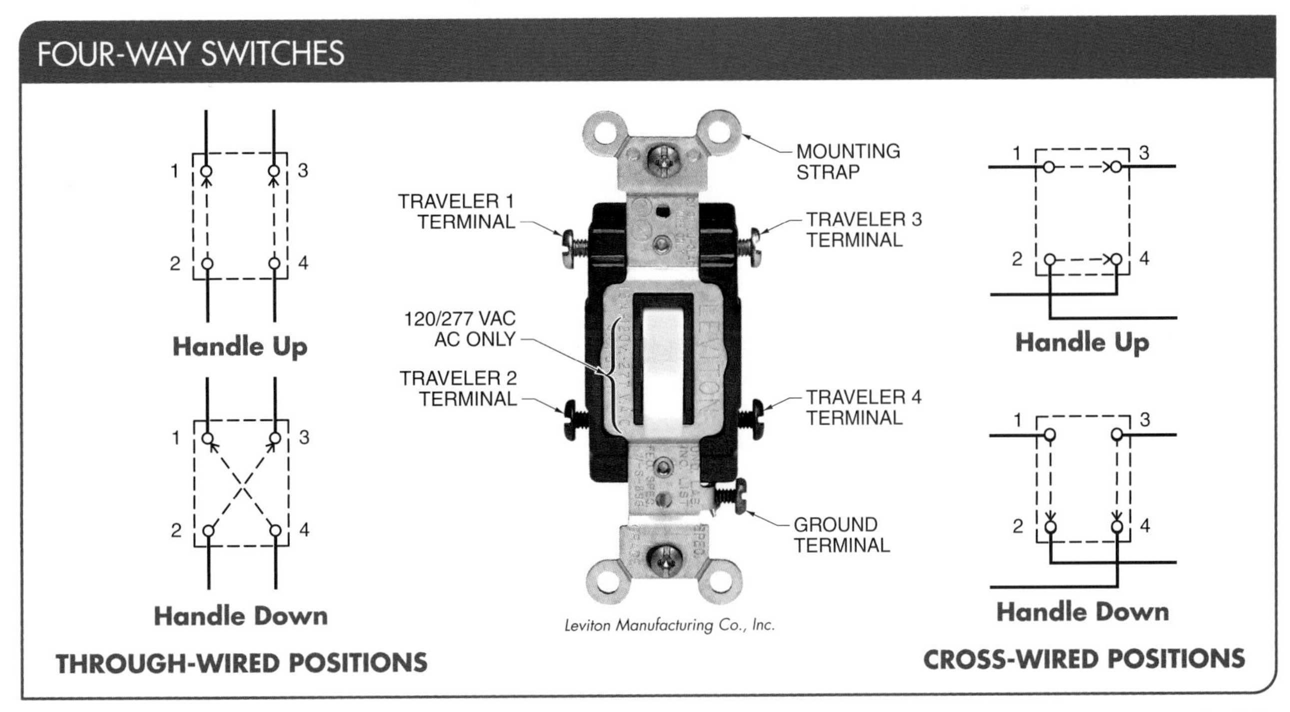

Figure 12-11. Through-wired and cross-wired four-way switches are used between two three-way switches to control a light from three switch locations.

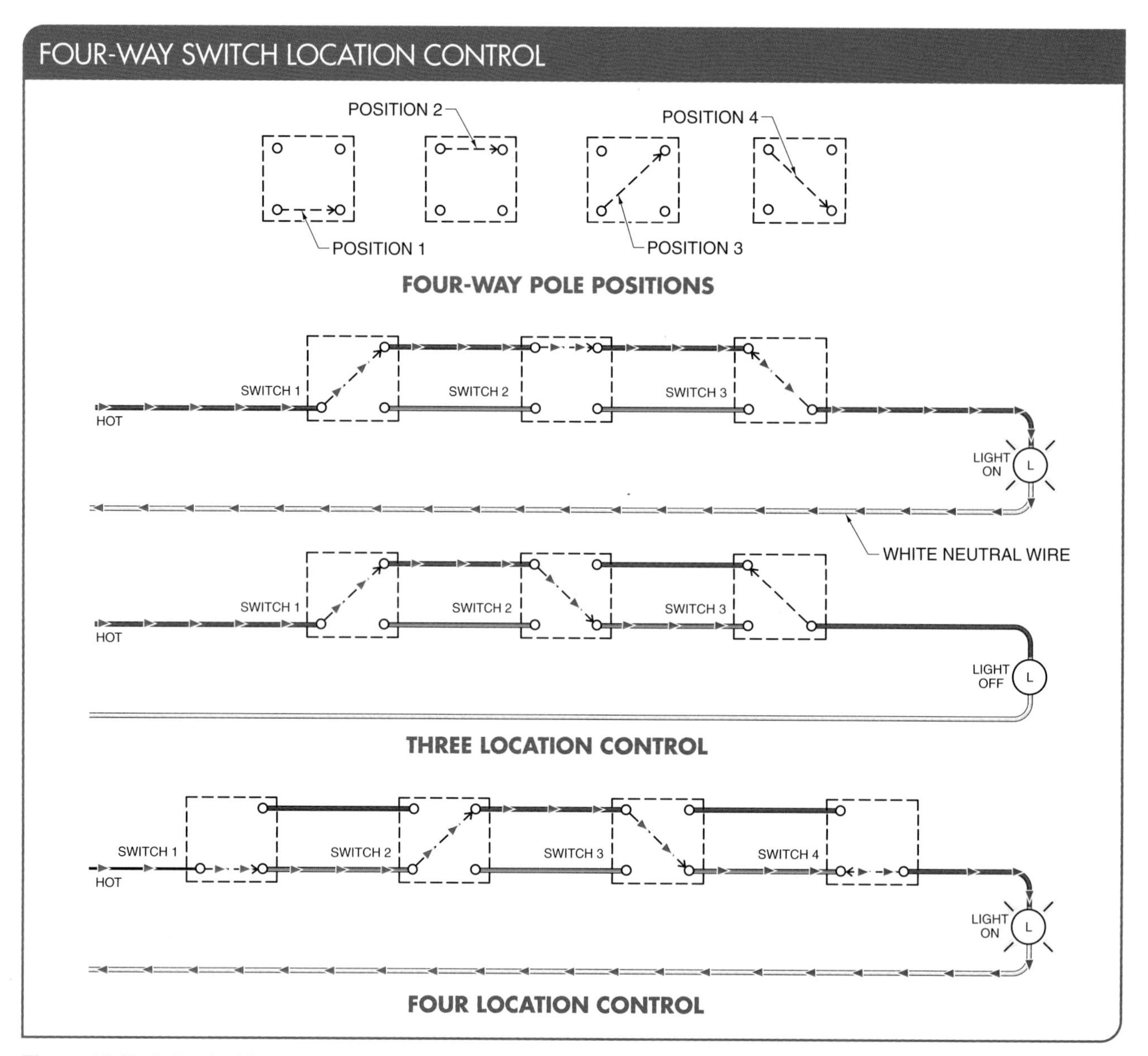

Figure 12-12. A circuit with two three-way switches and one four-way switch has four possible combinations to power the light, whereas a circuit with two three-way switches and two four-way switches has six possible combinations to power the light.

Next, at the switch 1 receptacle box, the two wires fed from the light fixture are connected to the first three-way switch and the three wires fed to the four-way switch. The black wire is connected to the black wire fed to the four-way switch. This connection feeds current to the next switch location. The black wire from the source is connected to the three-way switch (switch 1) common terminal. The red wire is connected to the traveler A terminal, and the remaining black wire is connected to the traveler B terminal. The green wires are connected together and bonded to the receptacle box.

At the four-way switch receptacle box, the three wires fed from three-way switch 1 are connected to the four-way switch and the three wires fed to three-way switch 2. The black wire from switch 1 is connected to the corresponding black wire from switch 2. The red wire from switch 1 is connected to the four-way switch at traveler terminal 1, and the other black wire from switch 1 is connected to traveler terminal 3. The red wire from switch 2 is connected to traveler terminal 2, and the other black wire from switch 2 is connected to traveler terminal 4. The green bonding wires are connected to the switch ground terminal and bonded to the receptacle box. **See Figure 12-13.**

At the last receptacle box, the three wires are connected to three-way switch 2. The black wire is connected to the common terminal, the other black wire is connected to traveler B, and the red wire is connected to the traveler A.

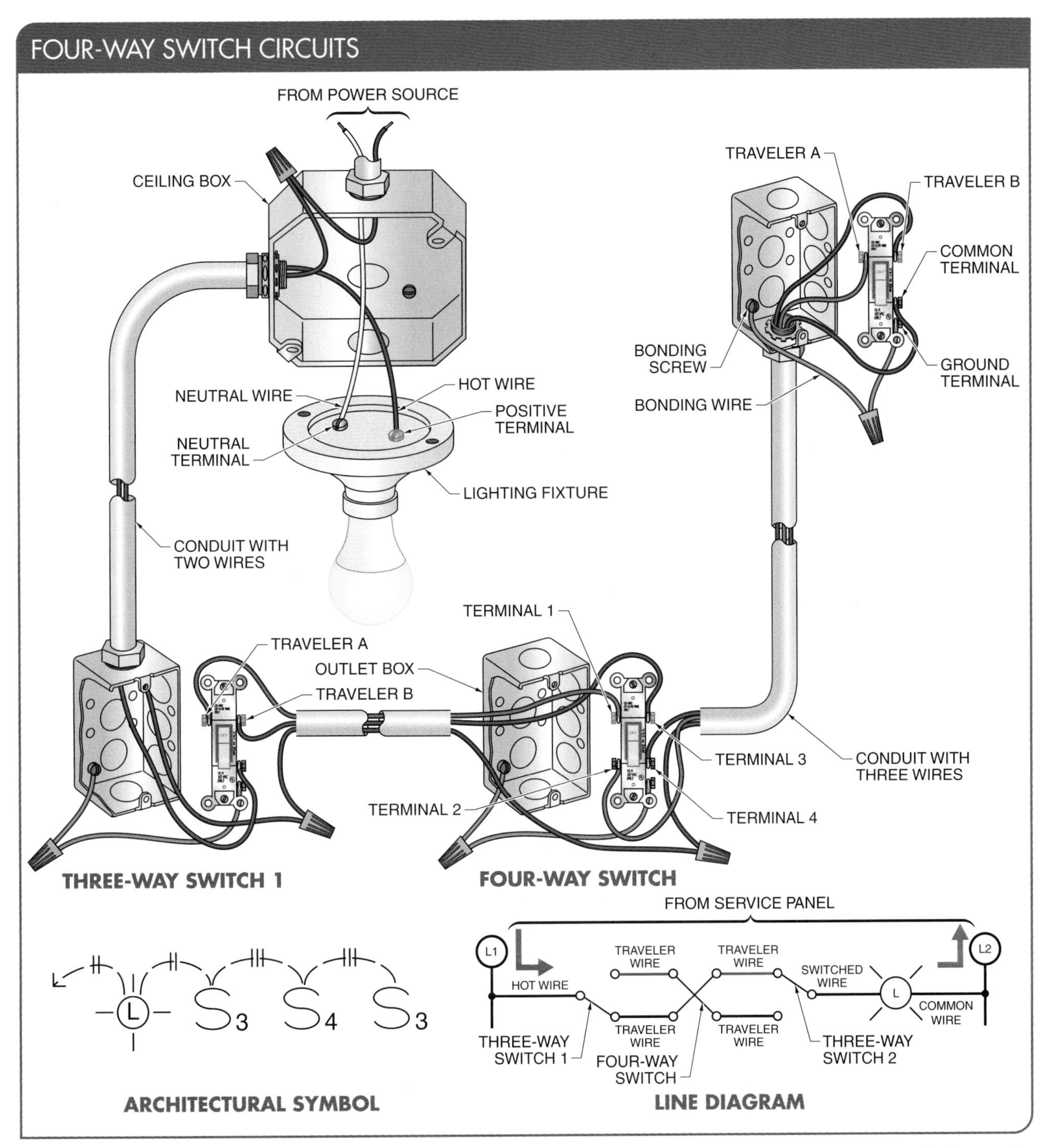

Figure 12-13. To wire a four-way switch circuit, certain connections must be made at each outlet box. Proper PPE must be worn before attempting to wire a four-way switch circuit.

Double-Pole Switches. A *double-pole switch* is a double-pole, single-throw (DPST) electrical control device. Double-pole switches are designed to control two separate 120 V loads. Since both inputs have 120 V, they also have the capability to control a 240 V load. The construction of a double-pole switch is essentially two single-pole switches that mechanically operate together as one switch. Double-pole switches are similar to four-way switches, except they have ON and OFF position markings on the toggle component. **See Figure 12-14.**

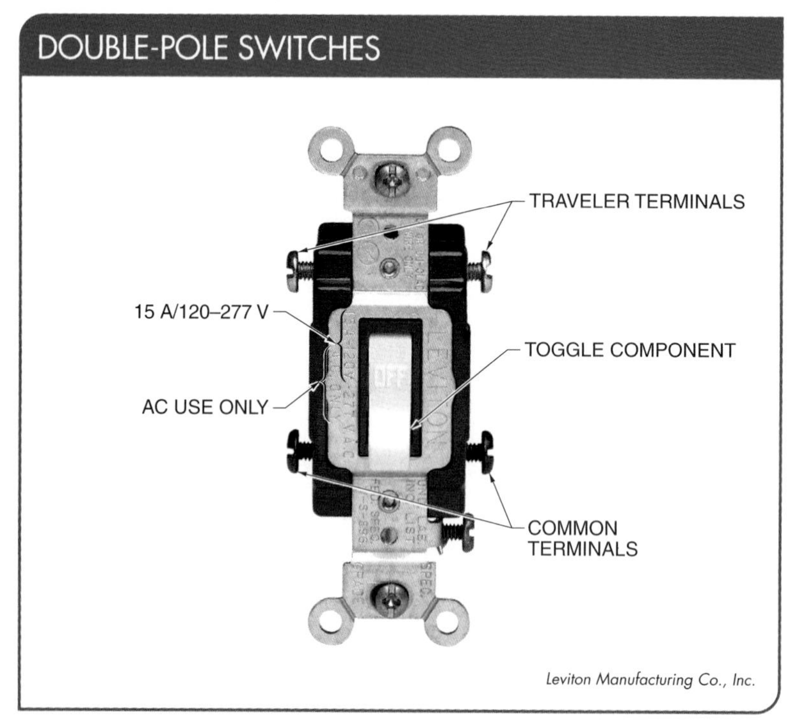

Figure 12-14. The construction of a double-pole switch is essentially two single-pole switches that mechanically operate together as one switch and designed to control two separate 120 V loads or one 240 V load.

A double-pole switch can control a single 250 VAC, 30 A receptacle. This installation uses two wires and a ground wire cable from the double-pole switch to the outlet location. At the double-pole switch outlet box, the power source cable red and black wires are connected to the traveler terminals. The red and black wires are used to feed power to the double-pole switch common terminals. A green wire is connected from the switch ground terminal and bonded to the outlet box bonding screw.

At the receptacle location, the red wire is attached to the one hot terminal and the black wire is attached to the other hot terminal. A green wire is connected from the receptacle ground terminal and bonded to the box bonding screw. **See Figure 12-15.**

Dimmer Switches. A *dimmer switch* is an electrical control device that is used to control light brightness by adjusting the voltage level applied to the light. Dimmer switches can be used in living rooms, dining rooms, and other areas in which variable light level is desired. Dimmer switches can be constructed in different styles. Some conventional styles include a knob, while other more contemporary styles include a lever and wireless control. Each type of dimmer switch is used to turn a light on or off and adjust the full range of light brightness. **See Figure 12-16.**

When a dimmer switch is installed on a fixture with multiple bulbs (such as a chandelier), the dimmer wattage rating must be high enough to handle the lighting load. In order to verify that the dimmer can handle the load, the wattage of each bulb is multiplied by the number of lights in the fixture. This determines the total wattage of the fixture with multiple bulbs.

For example, a chandelier may have six 60 W bulbs for a total of 360 W. Most dimmer switches are rated at around 600 W and should handle the lighting load for this 360 W chandelier. If the wattage is higher than the rating of the dimmer switch, a higher-rated dimmer switch is needed for the installation. *Note:* Do not use a dimmer switch to control a fan, as it can degrade the operation of the fan motor.

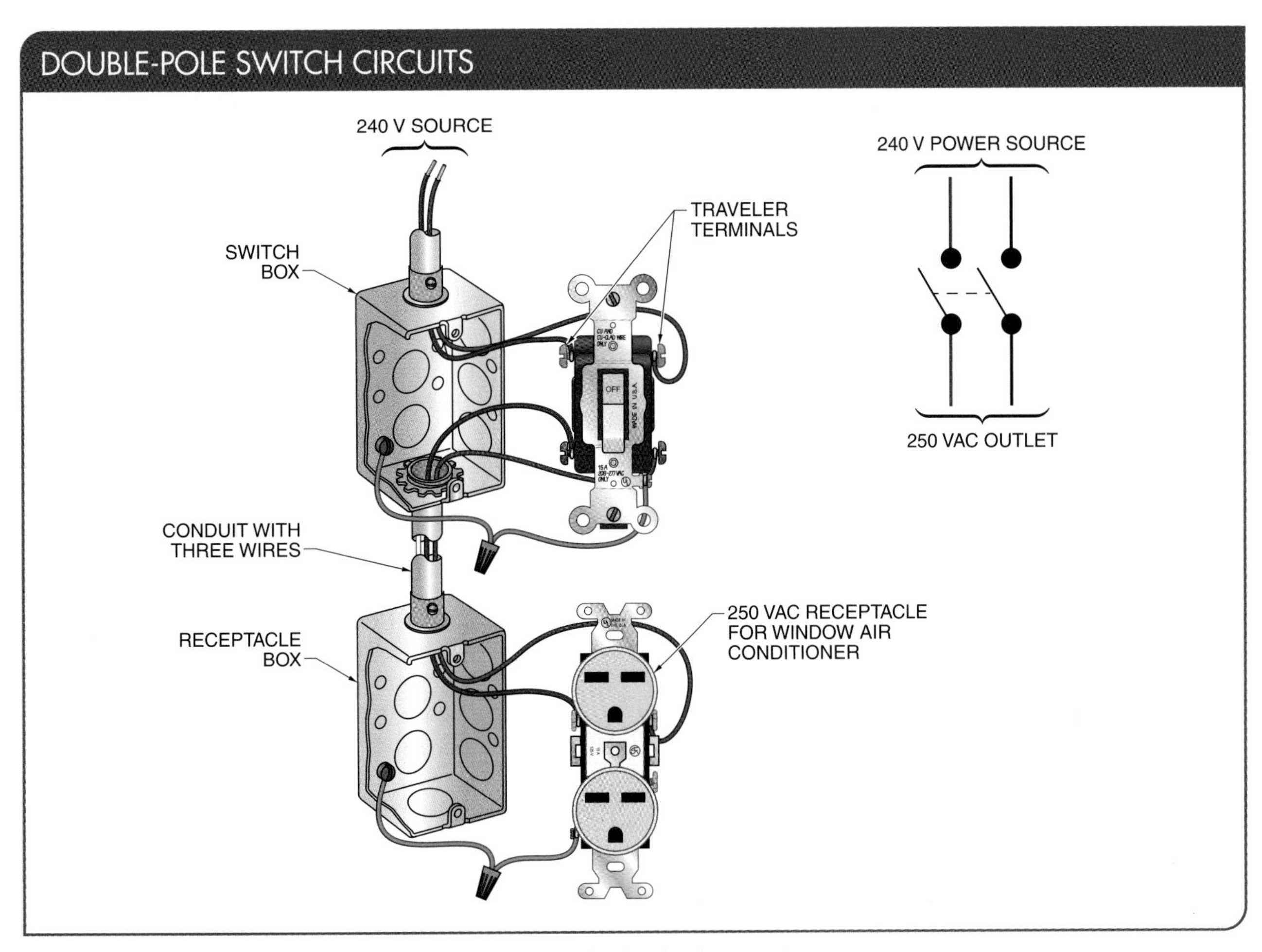

Figure 12-15. A double-pole switch opens two lines or circuits simultaneously.

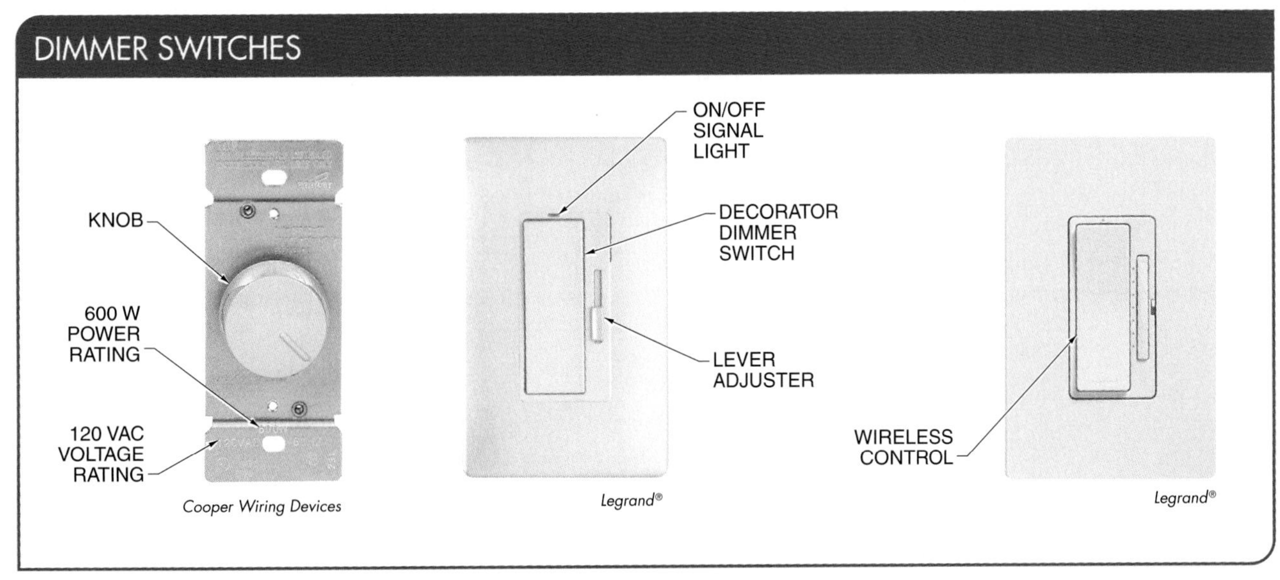

Figure 12-16. Conventional knob-style or contemporary lever-style dimmer switches can be used in living rooms, dining rooms, and other areas in which variable light level is desired.

A dimmer switch circuit is connected in the same manner as a standard single-pole switch circuit. However, dimmer switches are larger than single-pole switches and use prewired extensions rather than terminal screw connections for connection purposes. **See Figure 12-17.** To install a dimmer switch to control a light, the following procedure is applied:

1. Shut off the power feeding the light circuit.
2. Test the circuit with a voltage indicator to verify that there is no power present.
3. Remove the wall plate and existing switch.
4. Trim and strip wires on the dimmer switch.
5. Connect the white neutral wires together with a wire connector.
6. Connect the black wire from the power cable to the black wire on the dimmer switch.
7. Connect the ground wire to the outlet box bonding screw.
8. Connect the dimmer red wire to the black wire going to the light.
9. Fold the wires into the outlet box and fasten the dimmer switch.

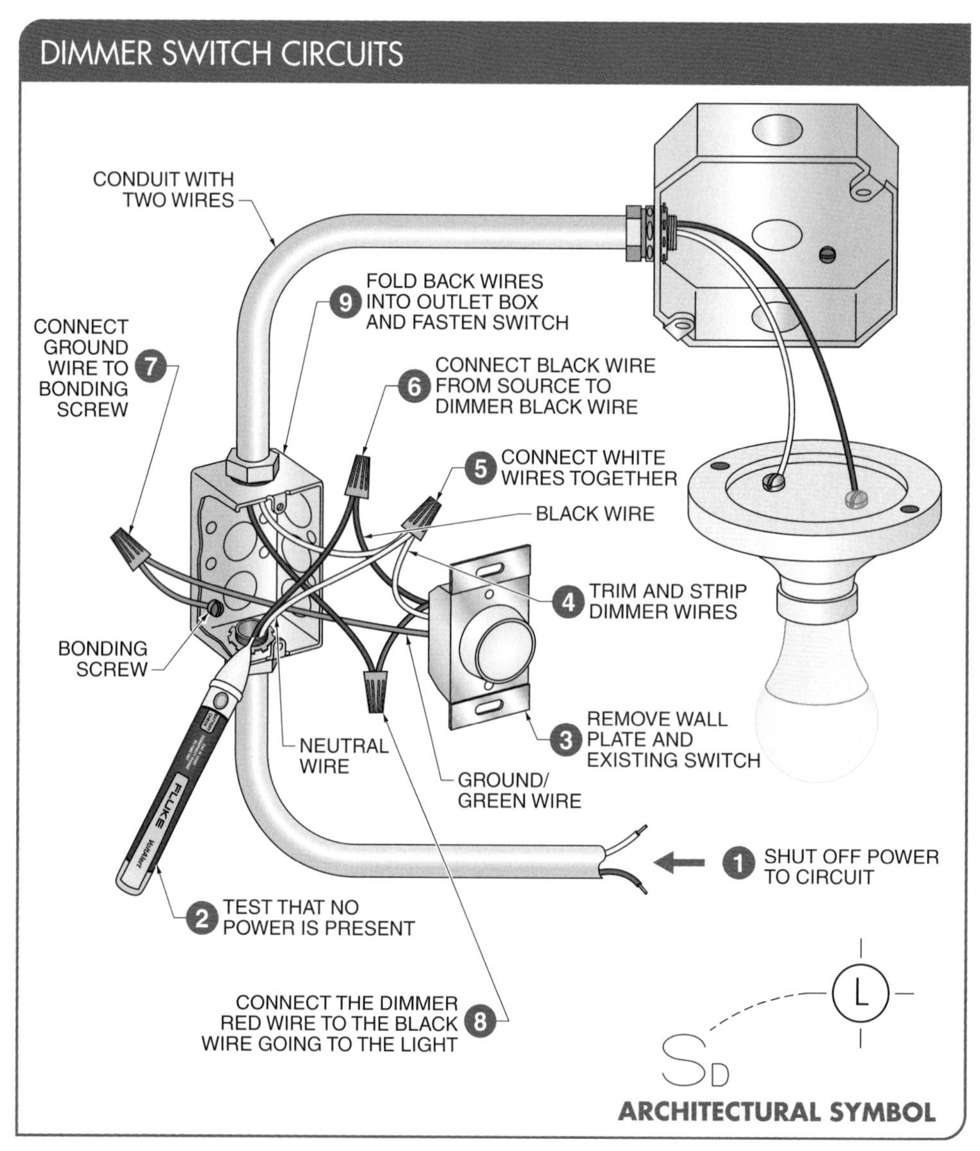

Figure 12-17. A dimmer switch circuit is connected in the same manner as a standard single-pole switch circuit, but it uses wires rather than terminal screw connections for connection purposes.

Timer Switches

A *timer switch* is a programmable electrical control device that automatically turns off after an elapsed amount of time. Typically, a timer switch has preset buttons that are each configured with a timing limit. Timer switches range in configured timing limits based on the manufacturer's design. A typical timer switch can have six configured timing buttons and a regular ON/OFF pushbutton. Like dimmer switches, timer switches have prewired extensions for connection purposes. **See Figure 12-18.**

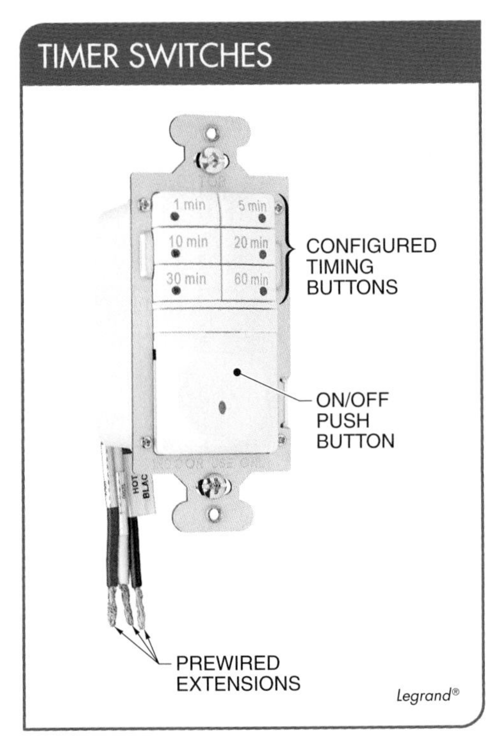

Figure 12-18. Timer switches use wires for connection purposes and have preset buttons that are each configured with a timing limit.

Motion Sensor Switches

A *motion sensor switch* is an electrical control device that automatically turns on or off based on the occupancy or vacancy of a room space. **See Figure 12-19.** Two types of motion sensors that can be used in dwellings are the occupancy sensor and the vacancy sensor. An occupancy sensor operates by turning a light on or off automatically based on presence in the space. Whereas, a vacancy sensor can automatically turn a light off, but it must be manually turned on. Both motion sensor switches can save energy by automatically switching lights off when no one is detected in the area.

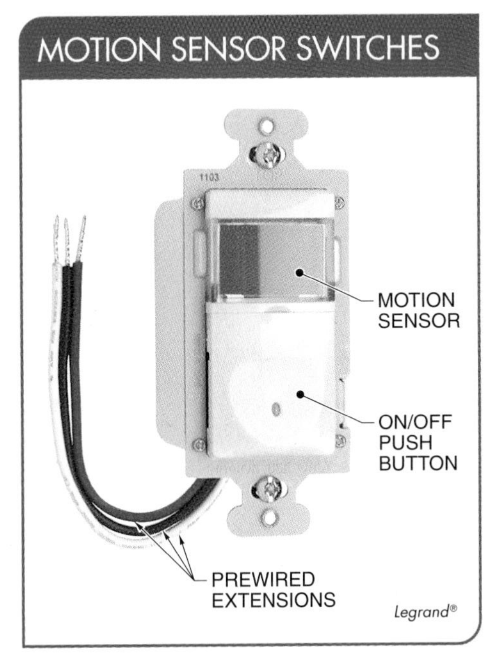

Figure 12-19. The two types of motion sensors are the occupancy sensor and the vacancy sensor; each can save energy by automatically switching lights off when no one is detected in the area.

Installing Switch Light Circuits

Switches are the main control device for lights throughout a dwelling. An example of one of the main applications for switches is the control of recessed lights in ceilings. Recessed lights are installed along ceilings between floor joists during dwelling construction or installed in appropriate ceiling areas during renovations.

During home construction, cable should be run or wire should be pulled through conduit from the switch to the recessed light fixture location. To pull wire through conduit, the hooked end of a fish tape is pushed through the switch electrical box opening all the way to the light fixture junction box. At the junction box, another fish tape is hooked to the first fish tape. When the two hooked ends are joined, the first fish tape is reeled to pull the conjoining fish tape to the switch box. Once the fish tape from the ceiling light fixture is inside the wall and ceiling conduit, a continuous length of wire is attached to the hooked end and pulled to the light fixture junction box. **See Figure 12-20.**

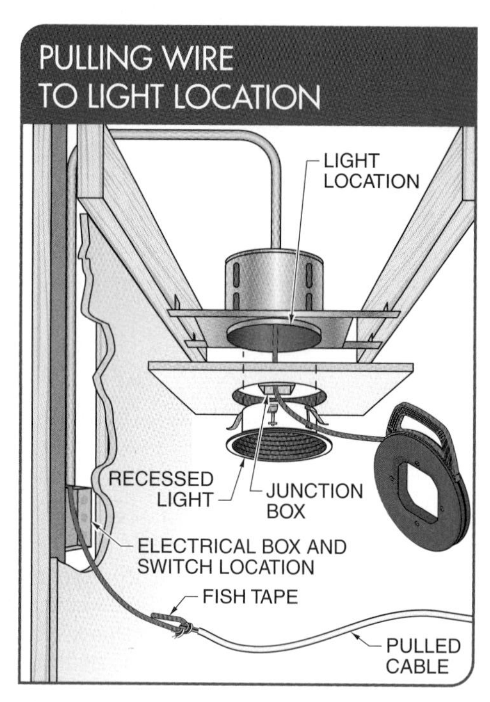

Figure 12-20. Cable should be run or wire should be pulled through conduit from the switch to the recessed light fixture location during home construction.

Before a recessed light can be installed for renovations, it must be verified that there are no obstructions where the light fixture is to be placed, there is enough room to accommodate the entire light fixture assembly, and there is cable in the area for wiring. Once an area meets these objectives, an opening can be cut using a drywall saw or a drill with a hole saw attachment. **See Figure 12-21.** Hole saws are ideal for making holes in drywall when creating several openings for multiple lights. The opening created by a hole saw must provide a snug fit for the canister of the light assembly.

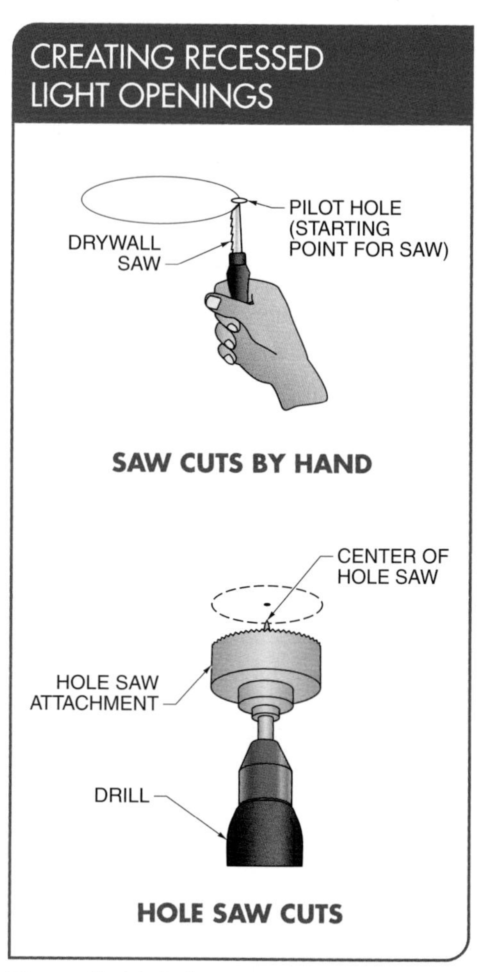

Figure 12-21. Before a recessed light can be installed for renovations, it must be verified that there are no obstructions where the light fixture is to be placed, there is enough room to accommodate the entire light fixture assembly, and there is cable in the area for wiring.

Wiring Switch Light Circuits. Recessed lights are manufactured with prewired junction boxes attached to the housing assembly. Once the cover plate of the junction box is removed, the cable can be secured and connected to the box. The cable must be connected so that the cable cannot be pulled out of the junction box. Standard wiring techniques are used to strip and connect the wires of the cable—white to white, red to red or black, and ground to ground. In some cases, the wire connector should be taped for additional security to ensure the connections do not come undone when folded back into the junction box. **See Figure 12-22.**

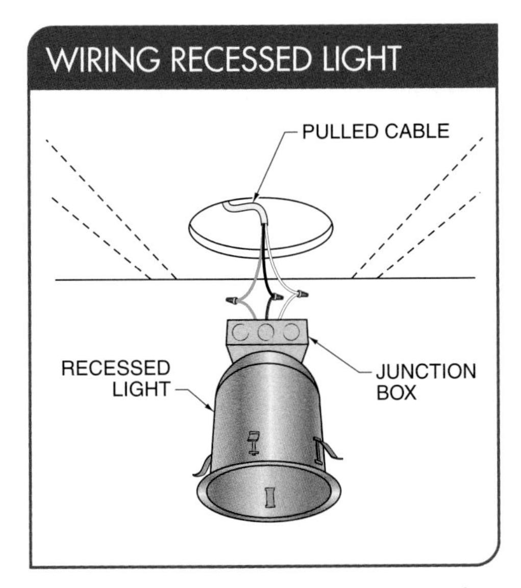

Figure 12-22. Recessed lights are manufactured with prewired junction boxes attached to the housing assembly. Standard wiring techniques are used to strip and connect the wires of the cable—white to white, red to red or black, and ground to ground.

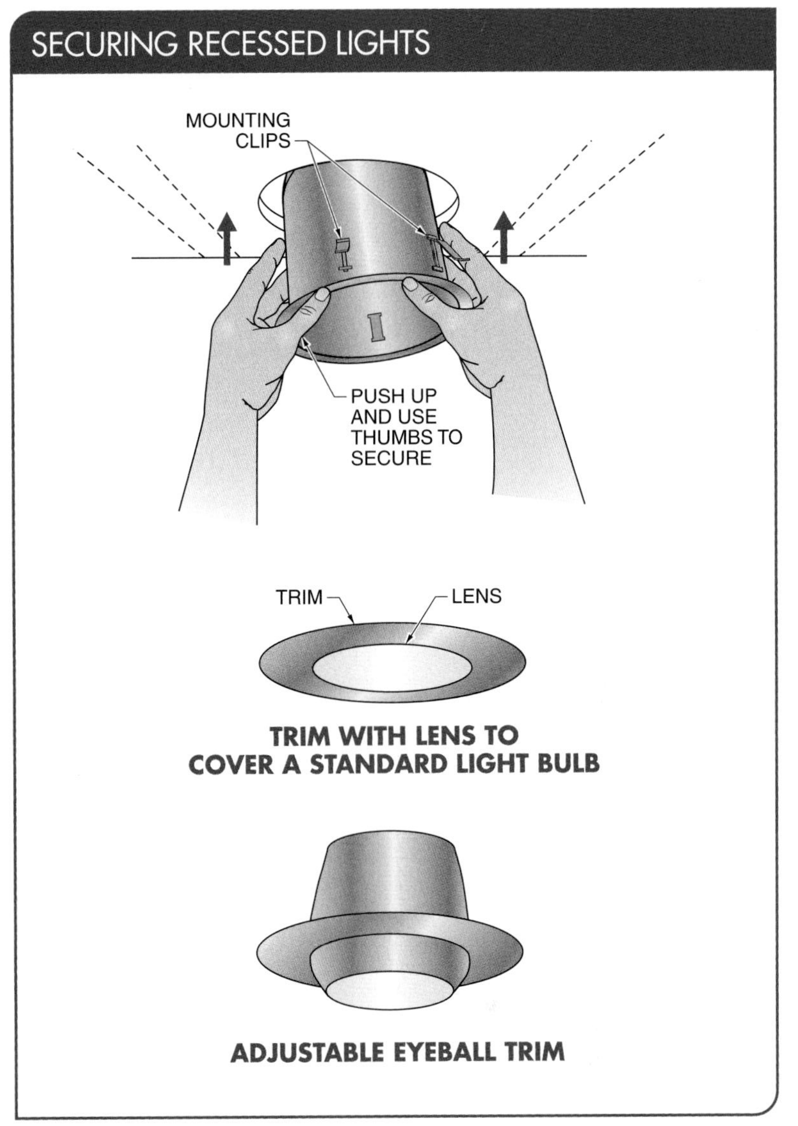

Figure 12-23. Recessed lights designed for renovations are equipped with mounting clips that clamp or secure the fixture to the ceiling.

Securing Recessed Lights. Recessed lights designed for renovations are equipped with mounting clips that clamp or secure the fixture to the ceiling. The mounting clips push down against the drywall to secure the fixture. Once the recessed light fixture is pushed up into the hole, the clips are secured. The mounting clips make an audible sound (a short click) when pushed into clamping position. Pressing a thumb against the clip on the inside of the fixture can help the process of securing a light fixture. Some fixtures require the aid of a screwdriver. **See Figure 12-23.**

Some recessed light fixtures are manufactured with built-in pieces of trim. Typically, the fixtures and trim are separated and reattached after the light fixture is secured in the ceiling. Before the trim is reattached, it must be ensured that the trim is the same size and make as the recessed light fixture. Types of trim can include trim with lenses that are used for fixtures with standard light bulbs and eyeball trim that swivels so the light emitted can be pointed in any direction to highlight a part of a room.

Insulating Recessed Lights. For safety reasons, the correct type of recessed light should be chosen to help avoid damage to the light and eliminate potential fire risks. Recessed light fixtures and housings are rated either as noninsulated ceiling (NIC) or insulated ceiling (IC) fixtures. For fire safety, NIC-rated fixtures must have at least 3″ of clearance around the fixture (including the wiring box) from thermal insulation. An IC-rated ceiling fixture can make direct contact with thermal insulation without creating a fire hazard. **See Figure 12-24.**

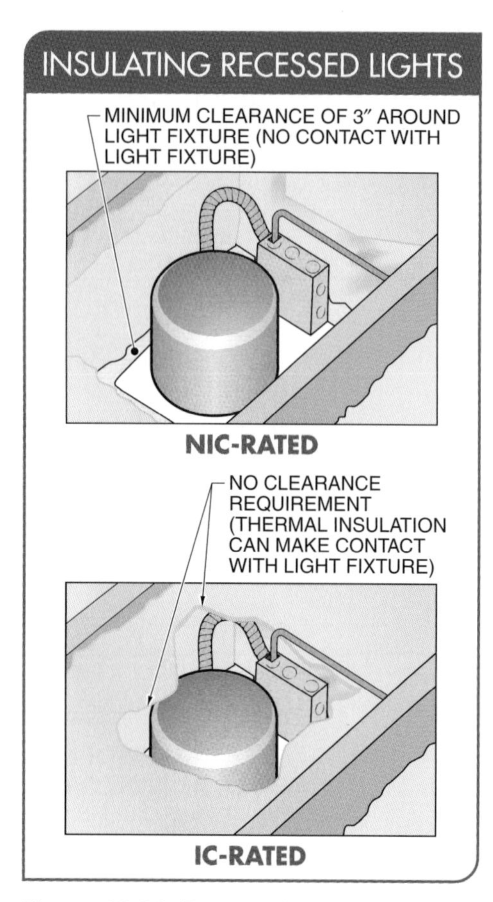

Figure 12-24. Recessed light fixtures and housings are rated either as noninsulated ceiling (NIC) or insulated ceiling (IC) fixtures.

SECTION 12.1 CHECKPOINT

1. Define "switch."
2. What are the two standard organization labels found on the mounting straps of switches?
3. What size wire is used to connect switches rated at 120 V and 15 A?
4. What are the color designations for the terminal screws on switches?
5. What is a single-pole switch?
6. Where are three-way switches typically used?
7. What are two types of four-way switches?
8. What is a timer switch?
9. What are the two types of motion sensor switches used in dwellings?
10. Where are recessed lights typically installed?

SECTION 12.2—BRANCH CIRCUIT INSTALLATION

Electrical devices are installed in locations shown on electrical floor plans. The objective of learning to interpret and follow electrical floor plans is the installation of safe, reliable, and NEC®-compliant branch circuits that are ready for use in residential electrical systems. Parts of an electrical floor plan may consist of construction layouts, device layouts, and wiring plans.

Construction Layouts

A *construction layout* is an isometric drawing of a branch circuit in a dwelling. Construction layouts indicate the placement of electrical devices, electrical boxes, and cable or conduit runs. Proper planning using a construction layout may result in substantial savings through the elimination of long runs of wire and unnecessary equipment.

Construction layouts, as well as device layouts and wiring plans, can help to clarify the approximate installation location, provide guidance, and show the necessary equipment needed to wire switches, receptacles, lights, and appliances in any room of a dwelling. The placement of devices can be directly correlated to the construction layout, which includes receptacle and ceiling boxes between stud walls, floor joists, and conduit or nonmetallic-sheathed cable for wiring. Typical construction layouts of branch circuits can be used to help install and wire residential electrical systems for bedrooms, bathrooms, hallways, kitchens, and garages.

Individuals can use construction layouts to determine where to place electrical boxes and how many boxes are required. Construction layouts may also help in determining a bill of materials (BOM) to approximate costs of branch circuit installations. Most construction layouts for residential applications use wires in conduit, metallic-sheathed cable, or nonmetallic-sheathed cable for device and equipment connection.

Device Layouts

A *device layout* is an exploded view of wires and electrical device connections. Device layouts provide detailed representations of the connections for switches and receptacles in an electrical circuit. When the same type of device is installed at multiple receptacle boxes throughout a branch circuit, the type of device can be given a letter designation on a wiring plan to indicate each installation is wired exactly the same way.

Wiring Plans

A *wiring plan* is an electrical floor plan that indicates the type, placement, and connection of all electrical devices required to wire an electrical circuit. Wiring plans provide enough information to successfully wire each room. All residential electrical systems vary, but most systems have same types of devices. The design of the branch circuit for one room might require only minor changes to be used for another room. Wire runs are laid out on a wiring plan in a very linear and definitive sequence to make the plan easier to interpret. Starting from the home run, a line is shown connected from each device symbol to another device symbol.

Residential Branch Circuits

A *branch circuit* is an electrical circuit that spans from the service panel and throughout a dwelling to power electrical devices. For residential branch circuits, the size of the circuit breaker (an OCPD) determines the branch circuit rating. Typically, circuit breakers for residential use are rated at 15 A or 20 A.

The NEC® does not provide a limit to the number of receptacles that can be installed on a branch circuit. However, per the NEC®, each receptacle in a branch circuit must be calculated at 180 VA (volt-amperes). The number of receptacles permitted in a branch circuit can be determined by dividing the rating of the circuit breaker (15 A or 20 A) by 1.5 A (180 VA ÷ 120 V [nominal voltage level of receptacles] = 1.5 A). For example, a 15 A branch circuit can supply power to 10 receptacles (15 A ÷ 1.5 A = 10), and a 20 A branch circuit can supply power to 13 receptacles (20 A ÷ 1.5 A = 13). **See Figure 12-25.**

Typical branch circuits in dwellings supply either combination loads or single loads. For example, lighting and multiple devices connected to receptacles are combination loads. Single loads include HVAC units, food waste disposers, trash compactors, sump pumps, and other similar appliances or equipment.

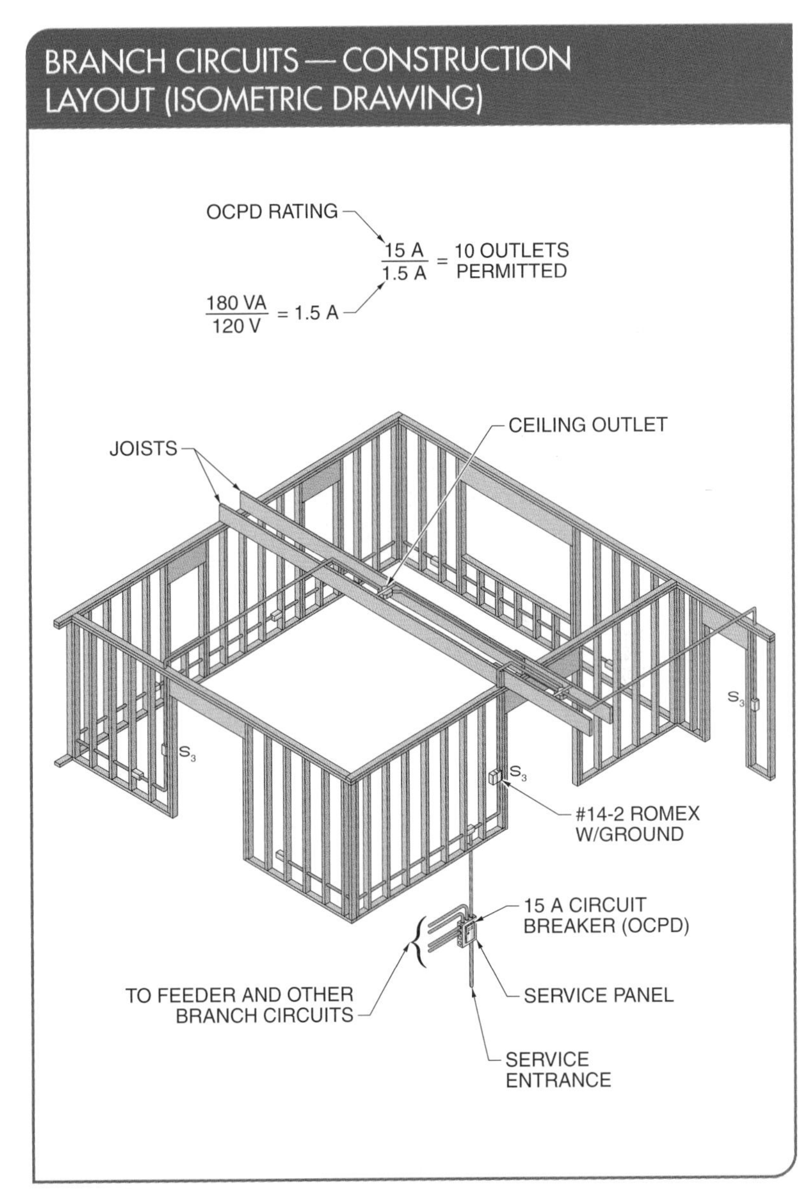

Figure 12-25. A branch circuit is an electrical circuit that spans from the service panel and throughout a dwelling to power electrical devices. The number of receptacles permitted in a branch circuit can be determined by dividing the rating of the circuit breaker (15 A or 20 A) by 1.5 A.

Bedroom/Closet Branch Circuits

In a typical bedroom, lights are controlled by single-pole switches and provided power from receptacles. Lights are installed in the ceiling and switches are installed by the entrance/exit. **See Figures 12-26.**

Often, a bedroom includes a closet. A bedroom with a closet may have two single-pole switch locations. One switch is inside the bedroom next to the entrance/exit and the other switch is outside the closet door. **See Figure 12-27.**

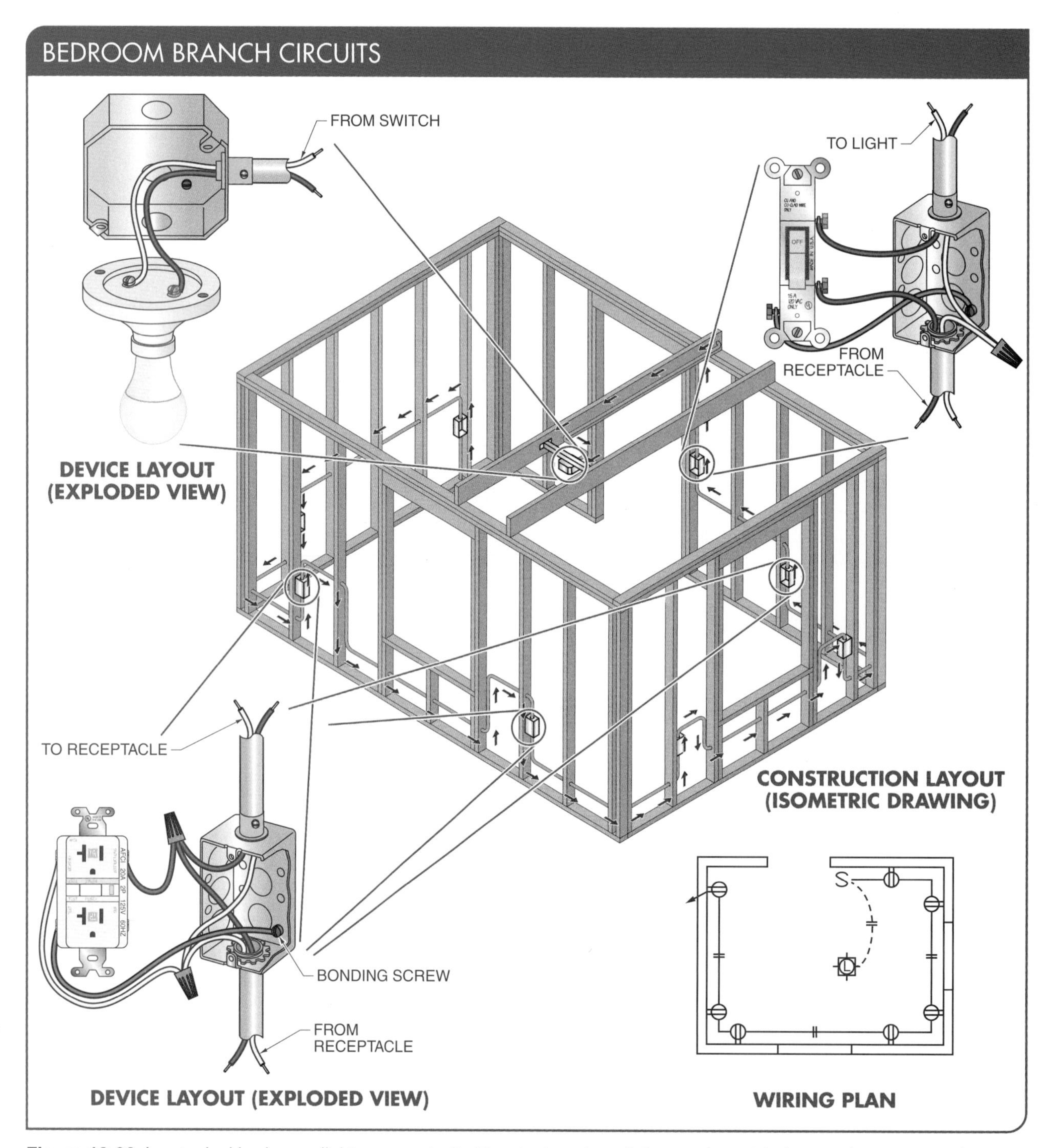

Figure 12-26. In a typical bedroom, lights are controlled by single-pole switches and provided power from receptacles.

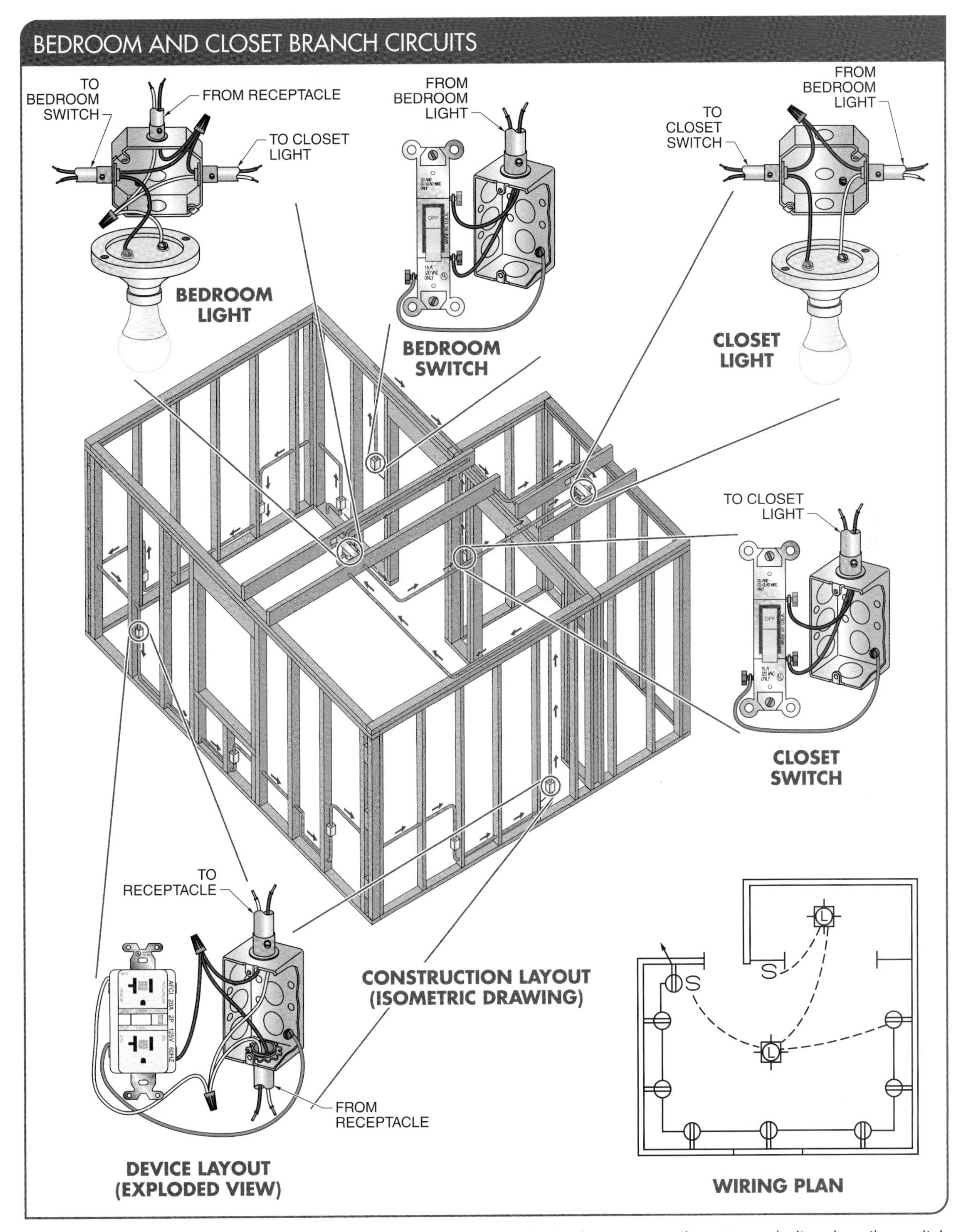

Figure 12-27. A bedroom with a closet may have one switch inside the bedroom next to the entrance/exit and another switch outside the closet door.

Bathroom Branch Circuits

A bathroom in a dwelling may consist of a few different electrical devices in one branch circuit. The types of electrical devices used in bathroom branch circuits include switches, receptacles, GFCIs, lights, and exhaust fans. In a bathroom branch circuit, a GFCI has to be the first device installed to protect all other devices during a ground fault. This is required by the NEC®. Depending on the size and arrangement of a bathroom, multiple branch circuits may need to be wired. However, one branch circuit can provide all the switches, receptacles, lights, and fans that are needed for a relatively small bathroom. **See Figure 12-28.**

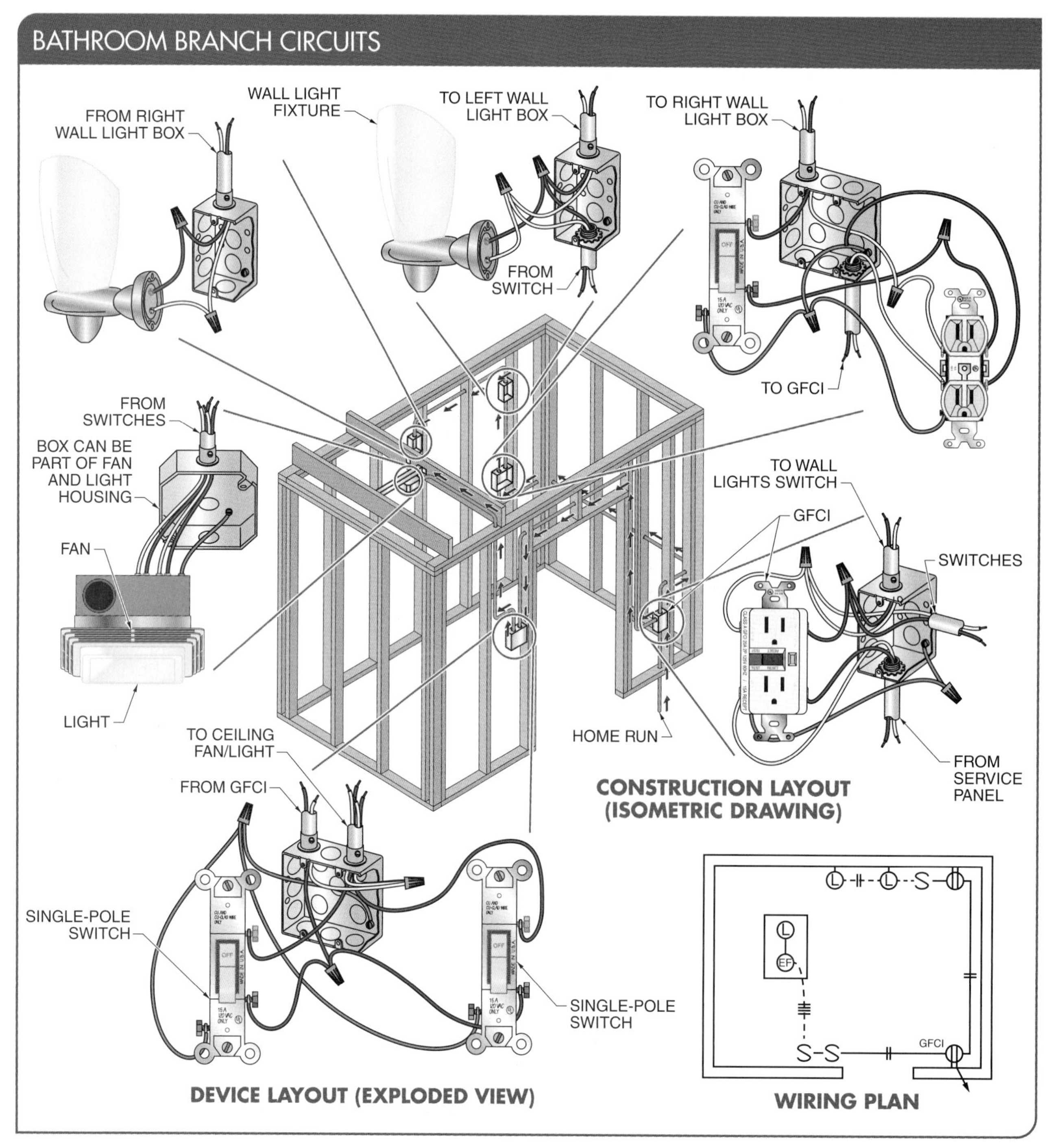

Figure 12-28. The types of electrical devices used in bathroom branch circuits include switches, receptacles, GFCIs (which have to be the first device installed to protect all other devices during a ground fault), lights, and exhaust fans.

Hallway Branch Circuits

Hallway branch circuits may contain a variety of switches to control one or more lights. Typically, there are combination-load circuits that consist of four-way switch circuits, three-way switch circuits, and three-way switch, single-light control circuits. All of these circuits control a single light or a combination of lights from different points in a hallway.

Three-Way Switch, Single-Light Control. In a hallway with a single light, a three-way switch circuit can be installed to power the light from both entrance/exit locations. Power for the circuit starts at the light fixture and is run to the first switch. Then, a red wire and a black wire are run to the next three-way switch. **See Figure 12-29.**

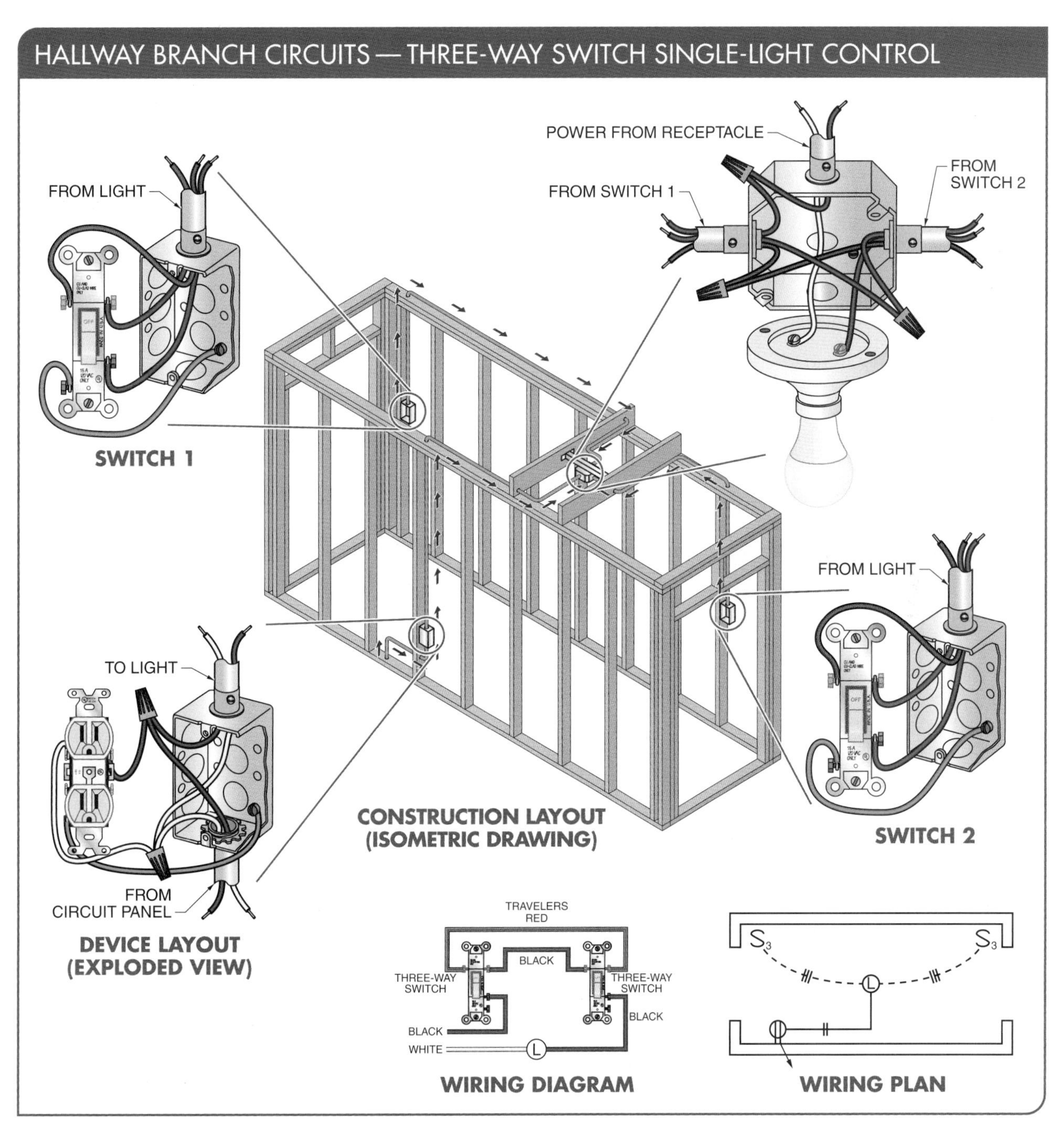

Figure 12-29. In a hallway with a single light, a three-way switch circuit can be installed to power the light from both entrance/exit locations.

Three-Way Switch Control Installation. In a hallway with two entrances/exits, two hallway lights are switched together and controlled by a three-way switch circuit. Power is supplied to one three-way switch first and ran to the next three-way switch, which provides the power for both lights. **See Figure 12-30.**

Four-Way Switch Control Installation. In a hallway with three entrances/exits, two hallway lights can be controlled by switches from each door using a four-way switch circuit. Two three-way switches and one four-way switch are wired, with the four-way switch connected between both three-way switches. **See Figure 12-31.**

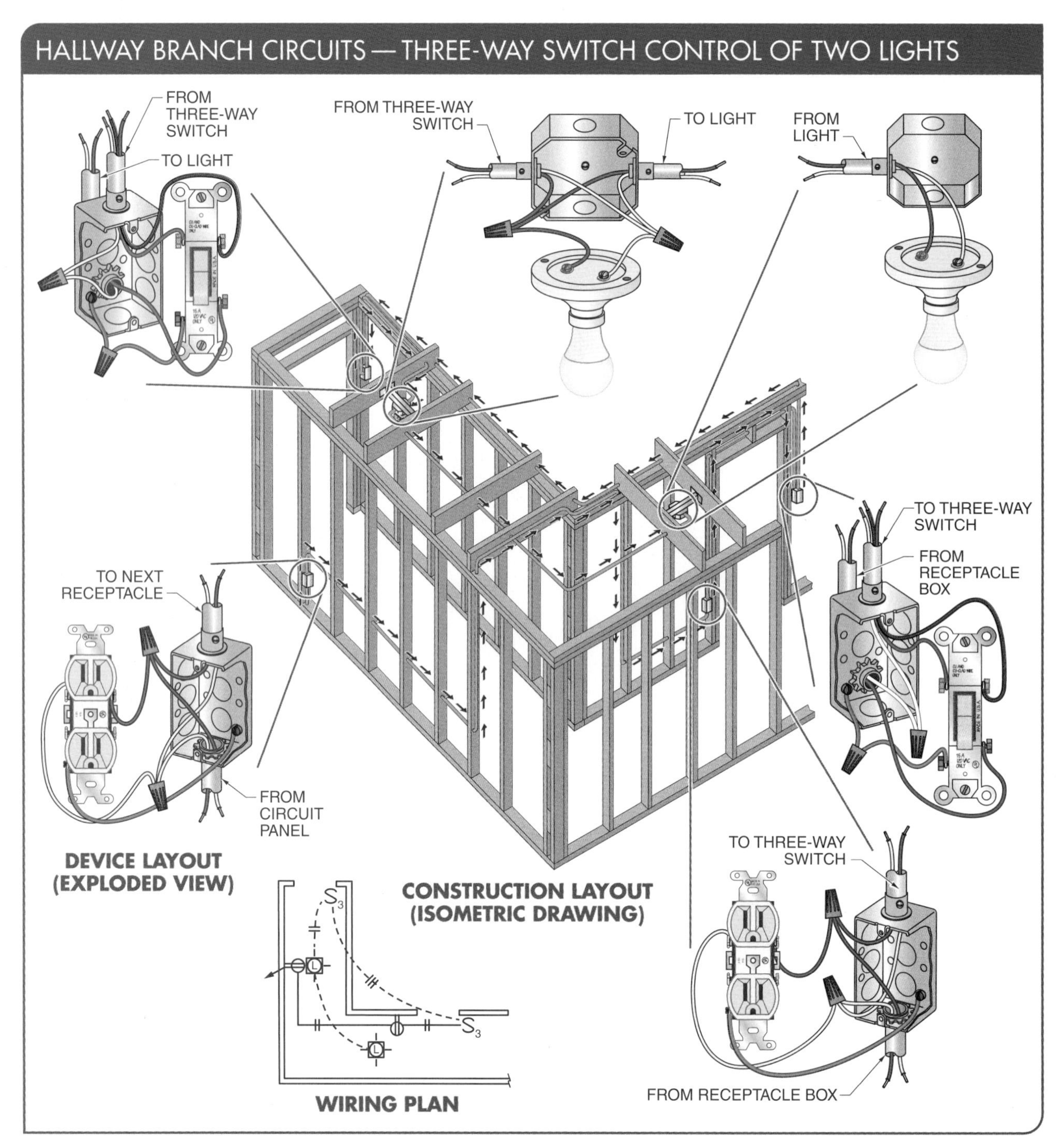

Figure 12-30. In a hallway with two entrances/exits, two hallway lights are switched together and controlled by a three-way switch circuit.

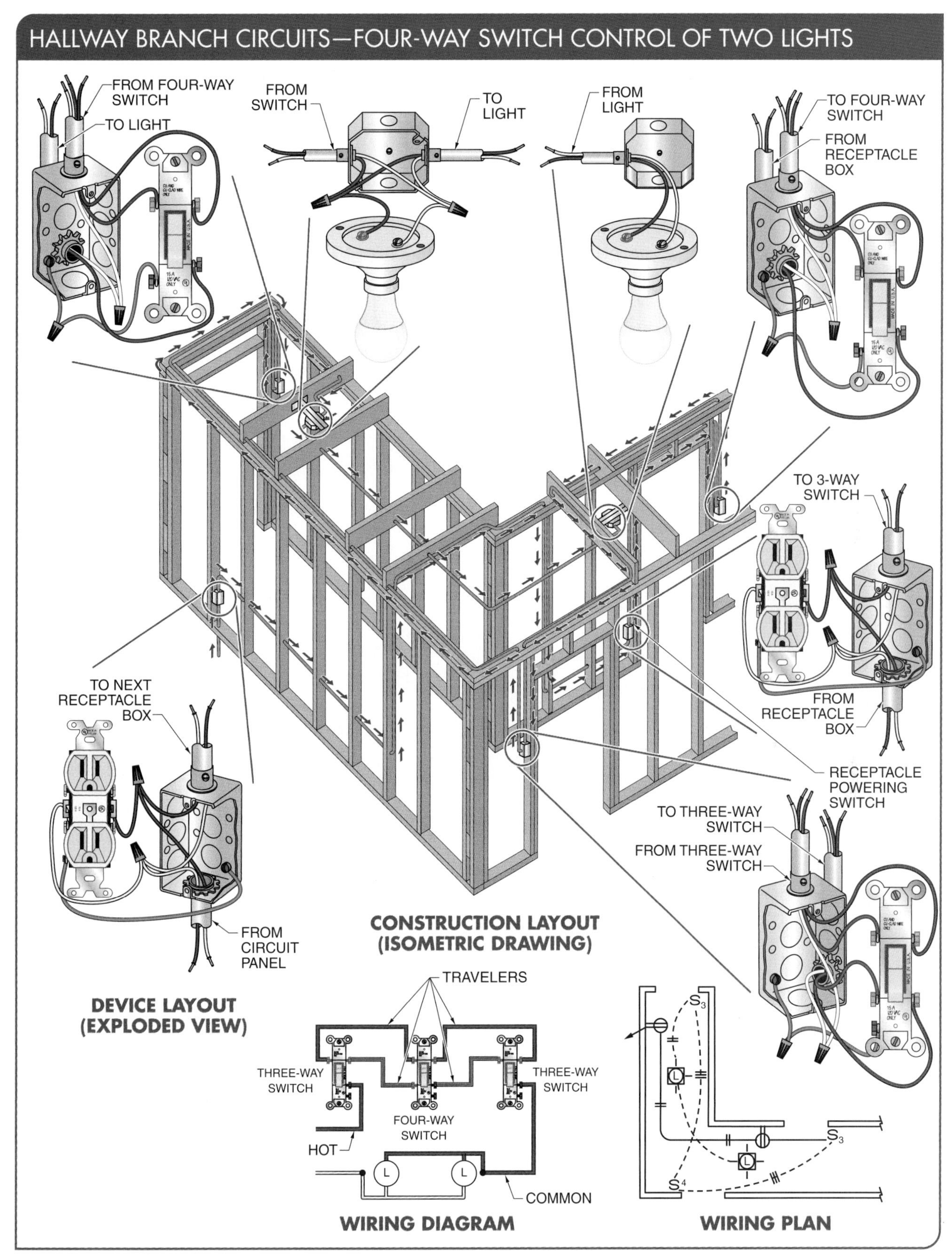

Figure 12-31. In a hallway with three entrances/exits, two hallway lights can be controlled by switches from each door using a four-way switch circuit.

Kitchen Branch Circuits

In a dwelling, a kitchen may have combination-load circuits for receptacles and lighting and single-load circuits for major appliances. The types of appliances include food waste disposers, ovens, refrigerators, and dishwashers. **See Figure 12-32.** For any kitchen branch circuit installation, the following list of guidelines that should be followed include:

- GFCI receptacles are required every 4′ of continuous counter space that is 12″ wide or greater.
- Receptacle height above a countertop cannot exceed 18″.
- Receptacles used for cooktops, ovens, compactors, or refrigerators cannot be more than 4′ from appliances.
- Refrigerator receptacles typically must be 40″ above the floor and connected to a 15 A or 20 A dedicated circuit.
- Food waste disposer receptacles or junction boxes cannot be more than 3′ from the disposer.
- Dishwasher receptacles or junction boxes cannot be more than 4′ from the dishwasher.
- Any wall space that is 2′ in length, but less than 6′, requires a receptacle.

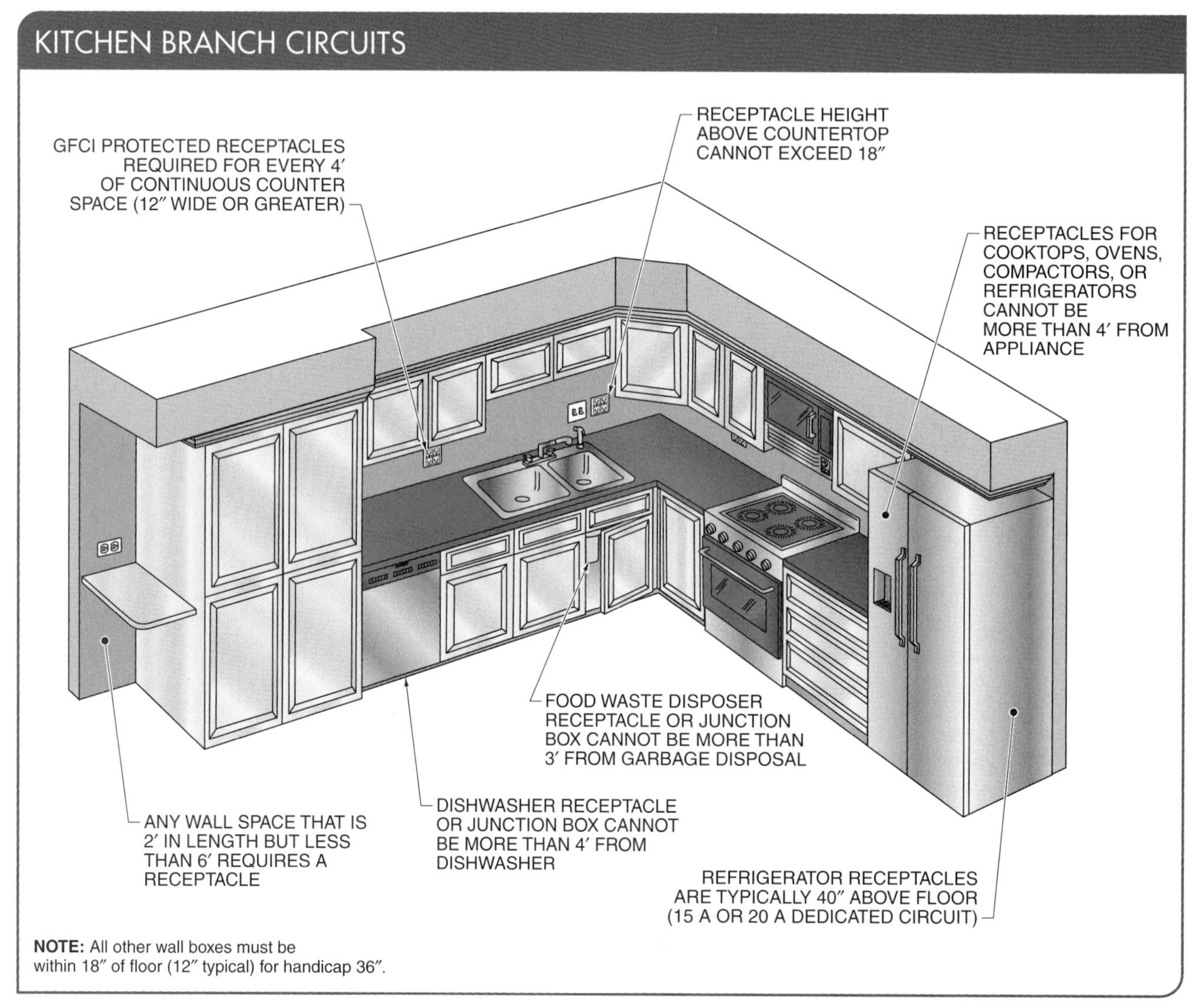

Figure 12-32. In a dwelling, a kitchen may have combination-load circuits for receptacles and lighting and single-load circuits for major appliances, such as food waste disposers, ovens, microwaves, refrigerators, dishwashers, and compactors.

250 VAC Receptacle Installation. To power a 250 VAC receptacle in a kitchen, a single-load circuit must be installed. A 250 VAC receptacle powers an electric range with a four-prong cord. Three wires (black, red, and white) are ran from a subpanel to the location of the receptacle box and connect to the 250 VAC receptacle. A ground bonding wire connects from the receptacle ground slot to the bonding screw inside the receptacle box. **See Figure 12-33.**

GFCI Receptacle Installation. In kitchen branch circuits, GFCI receptacles must be installed every 4′ of countertop space if the countertop is 12″ or greater. A combination-load circuit can be run from a subpanel to feed a GFCI that protects the receptacles connected to the load side of the GFCI. **See Figure 12-34.**

CODE CONNECT

Per NEC® Subsection 210.4(A), all conductors of a branch circuit shall originate from the same service panel.

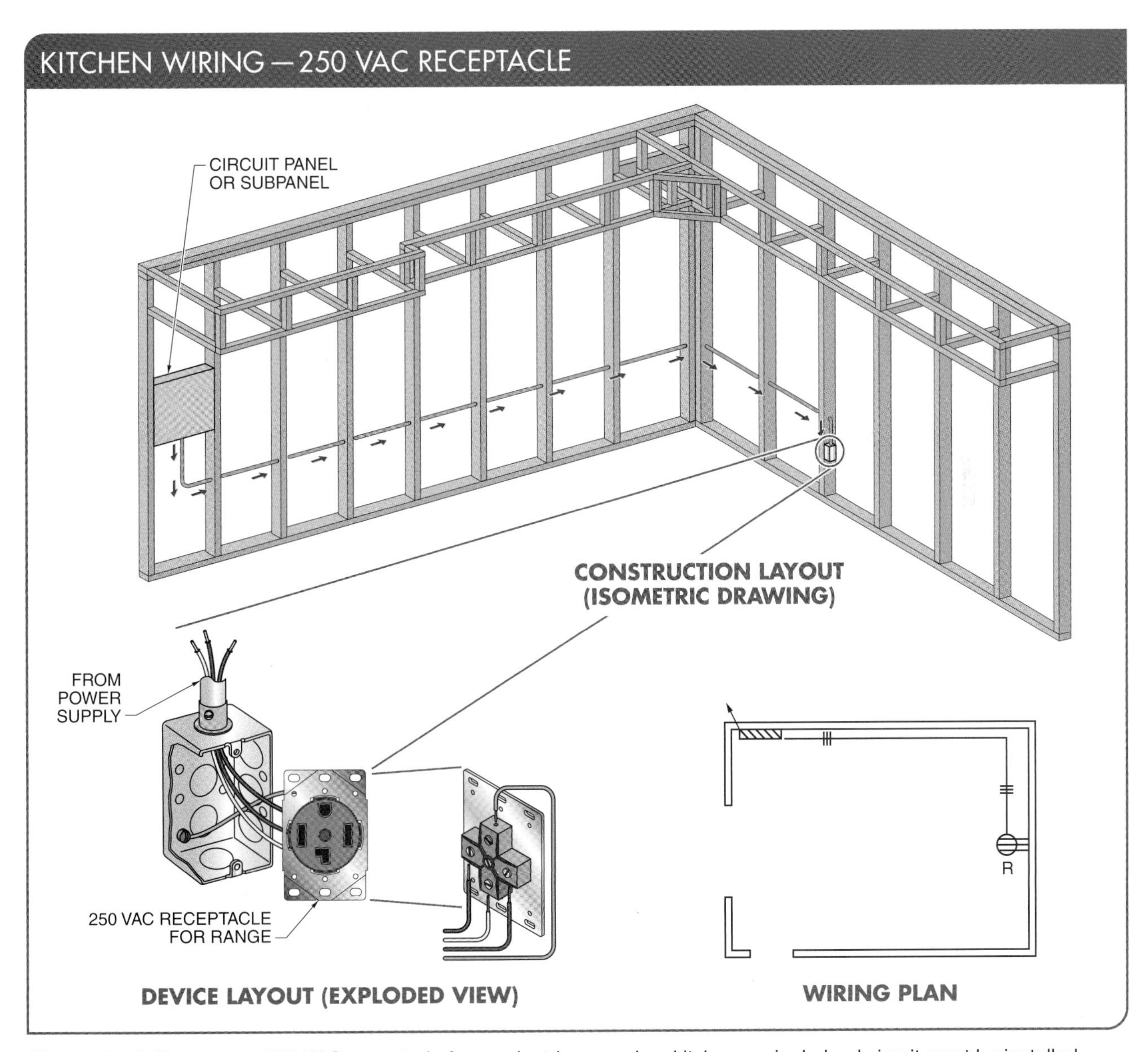

Figure 12-33. To power a 250 VAC receptacle for an electric range in a kitchen, a single-load circuit must be installed.

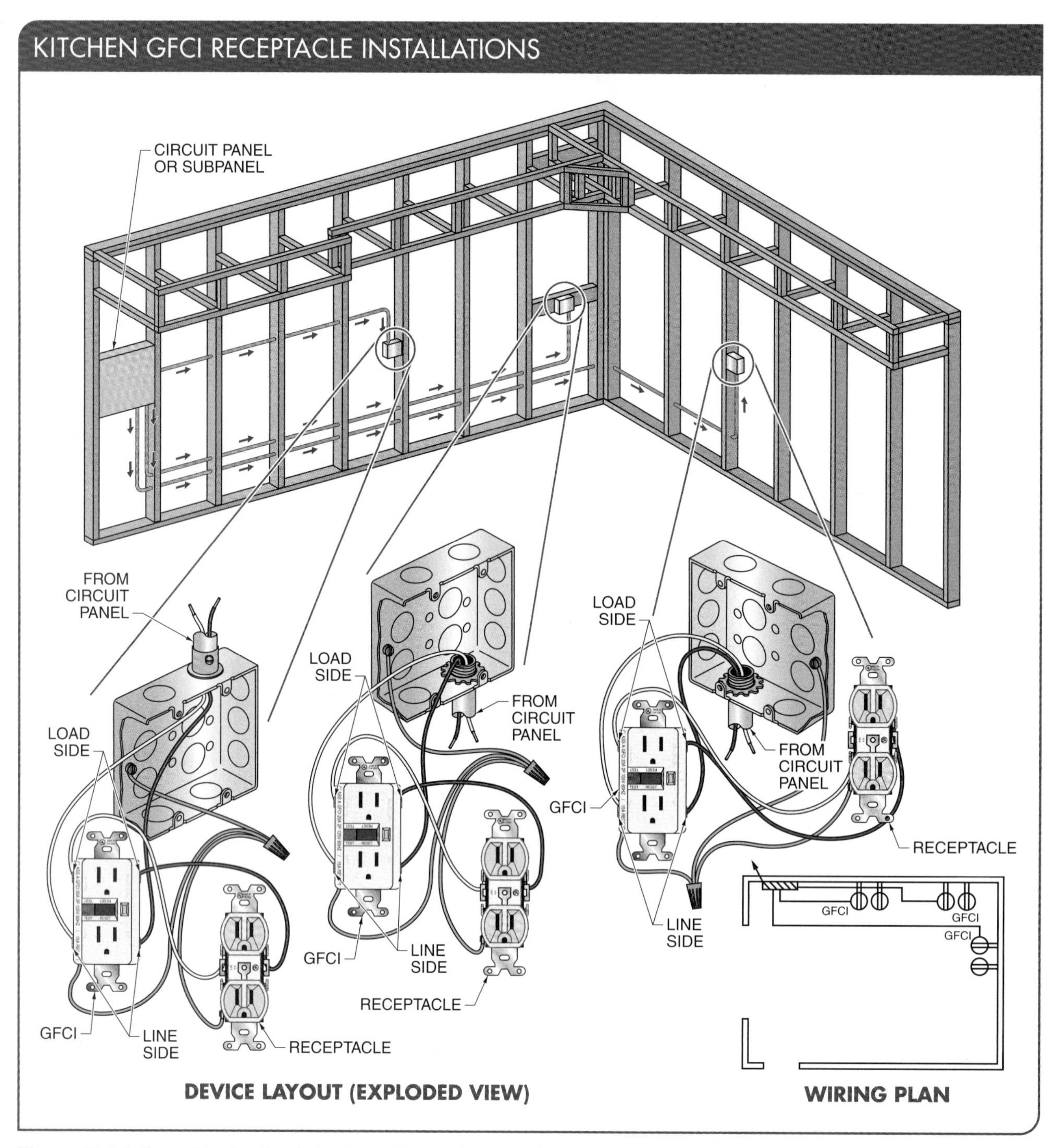

Figure 12-34. A combination-load circuit can be run from a subpanel to feed a GFCI that protects countertop receptacles connected to the load side of the GFCI.

An example installation of a single-load kitchen branch circuit includes a GFCI to supply power to a dishwasher. An example installation of a combination-load kitchen branch circuit includes a GFCI-protected switch that controls a food waste disposer, a line of receptacles for countertop use, and a receptacle that powers a refrigerator. The food waste disposer is the only device controlled by a switch. **See Figure 12-35.**

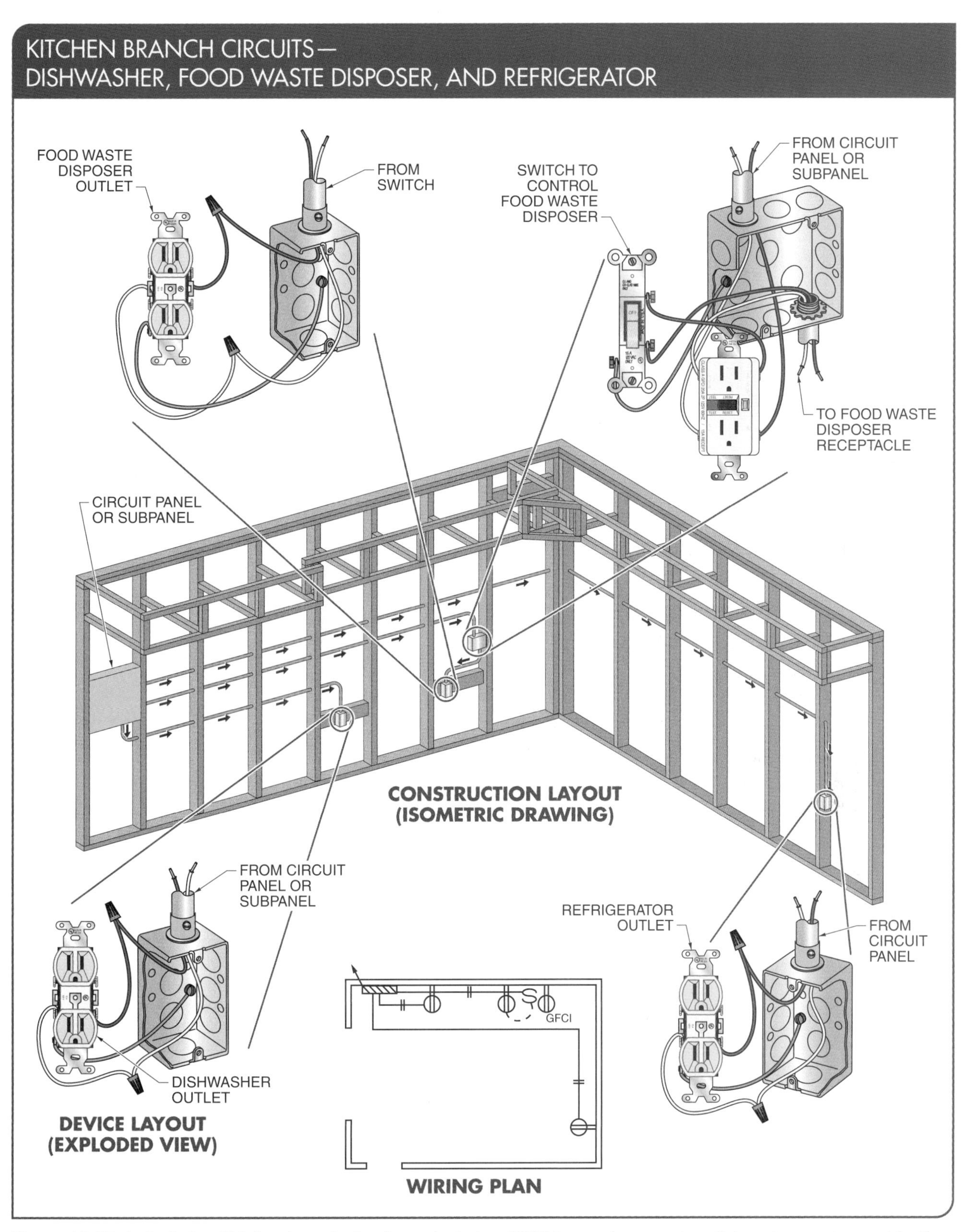

Figure 12-35. A single-load kitchen branch circuit may include a GFCI to supply power to a dishwasher, and a combination-load kitchen branch circuit may include a GFCI-protected switch that controls a food waste disposer, a line of receptacles for countertop use, and a receptacle that powers a refrigerator.

Lighting and Exhaust Fan Installation. In a kitchen, two separate runs from a subpanel can be used to power lights and an exhaust fan. A combination-load circuit provides power for the lights and a single-load circuit provides power for the exhaust fan. A GFCI protects the lighting circuit, and another GFCI protects the exhaust fan. *Note:* The supply power must connect to the line side of the GFCI and devices connect to the load side to protect devices and circuits down line. **See Figure 12-36.**

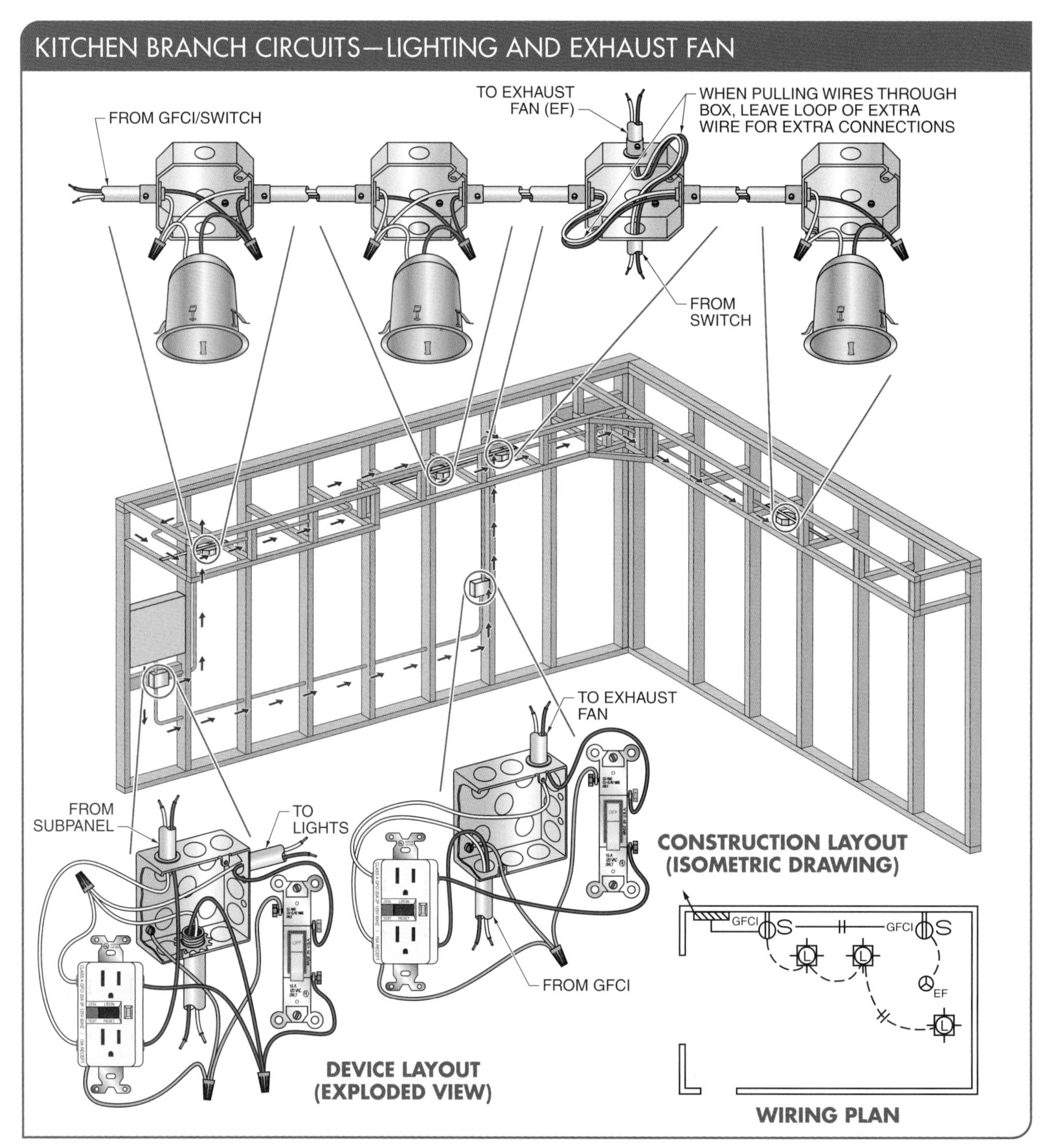

Figure 12-36. In a kitchen, a combination-load circuit can provide power for the lights and a single-load circuit can provide power for the exhaust fan.

Garage Branch Circuits

In garage branch circuits, the power may be supplied to a single-pole switch and a three-way switch. The single-pole switch controls an outdoor side entrance/exit light and the three-way switch circuit controls two interior lights. The single-pole and three-way switch circuits are protected by a GFCI, which is supplied power first. **See Figure 12-37.**

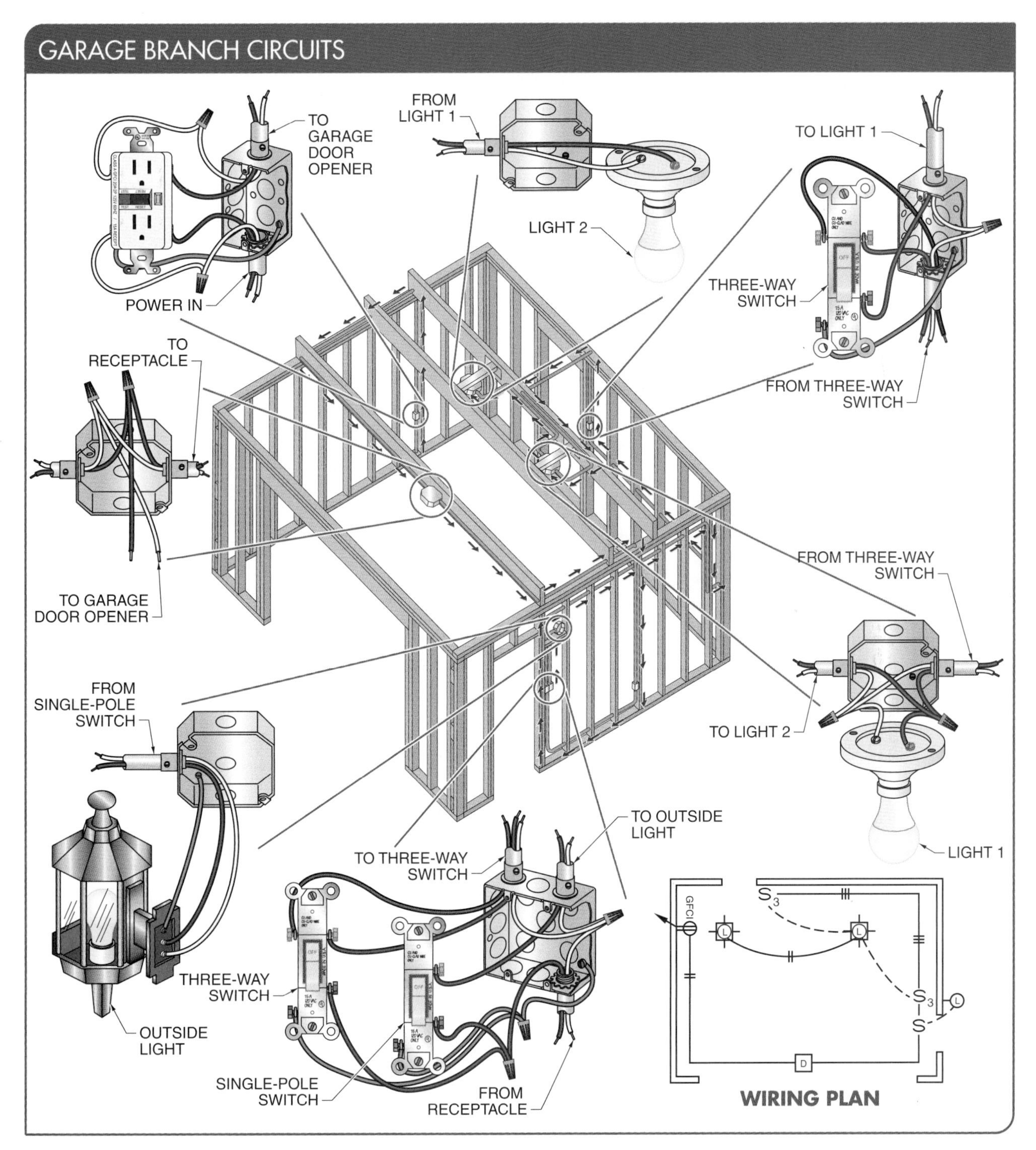

Figure 12-37. In garage branch circuits, a single-pole switch can control an outdoor side entrance/exit light and a three-way switch circuit can control two interior lights.

CODE CONNECT

NEC® Article 410, Part XI, provides information on the proper air-space clearances surrounding a light fixture.

Power may also need to be supplied to additional outdoor lighting, such as a porch light and two lights at the front of the garage. The porch light can be controlled by a single-pole switch to control the lights at the front of the garage. **See Figure 12-38.**

SECTION 12.2 **CHECKPOINT**

1. Define "branch circuit."
2. What electrical device determines the circuit rating for branch circuits?
3. What is the number of receptacles permitted on a 15 A branch circuit and on a 20 A branch circuit?
4. What is the first electrical device installed in a bathroom branch circuit?
5. List the guidelines that should be followed for kitchen branch circuits.

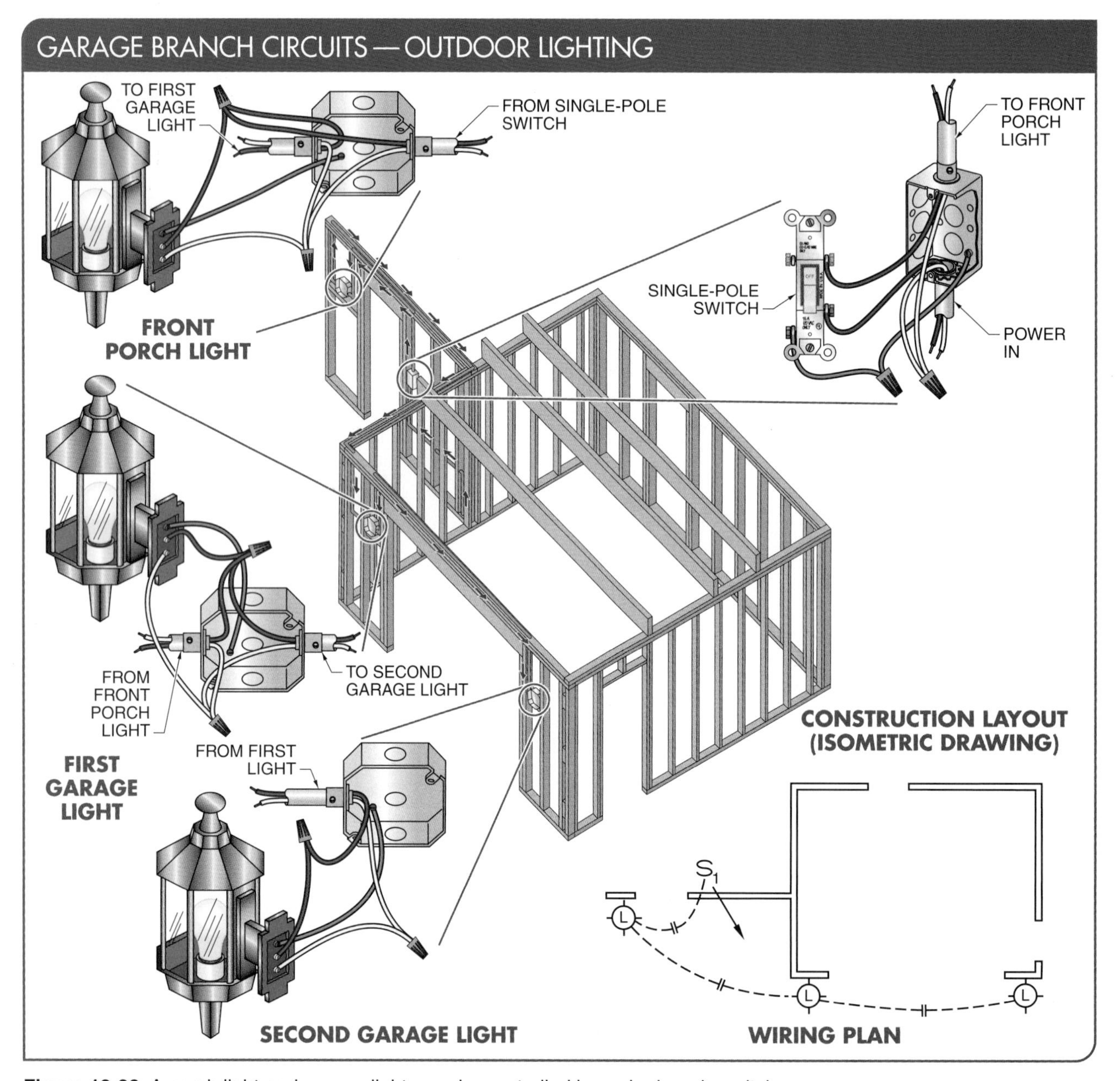

Figure 12-38. A porch light and garage lights can be controlled by a single-pole switch.

Chapter Activities

12 Switches and Branch Circuit Installation

Name ______________________ Date ______________

Switches

Identify the components of the single-pole switch shown.

_____________ 1. Mounting strap

_____________ 2. Wire compatibility ratings

_____________ 3. Maximum current and voltage ratings

_____________ 4. CSA label

_____________ 5. UL® label

_____________ 6. Common screw

_____________ 7. Traveler screw

_____________ 8. Wire gauge rating

_____________ 9. Mounting ear

_____________ 10. Green terminal screw

_____________ 11. Strip gauge

_____________ 12. Push-in fittings

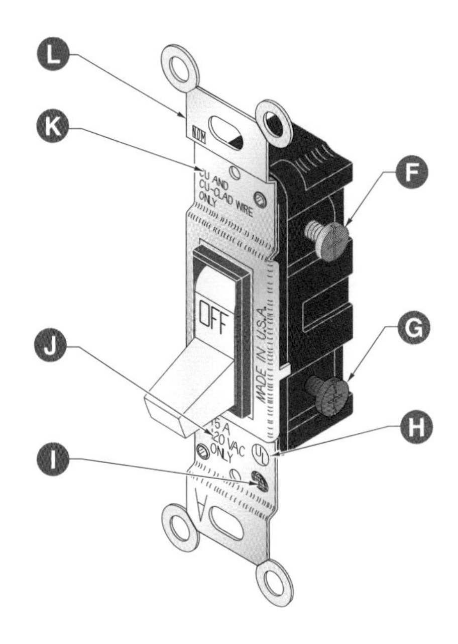

FRONT

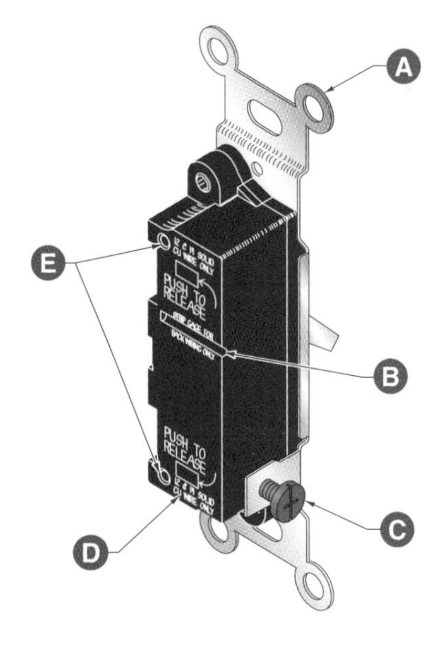

BACK

Switch Loop Circuits

1. Draw lines to wire a switch loop circuit. Indicate the color of each wire.

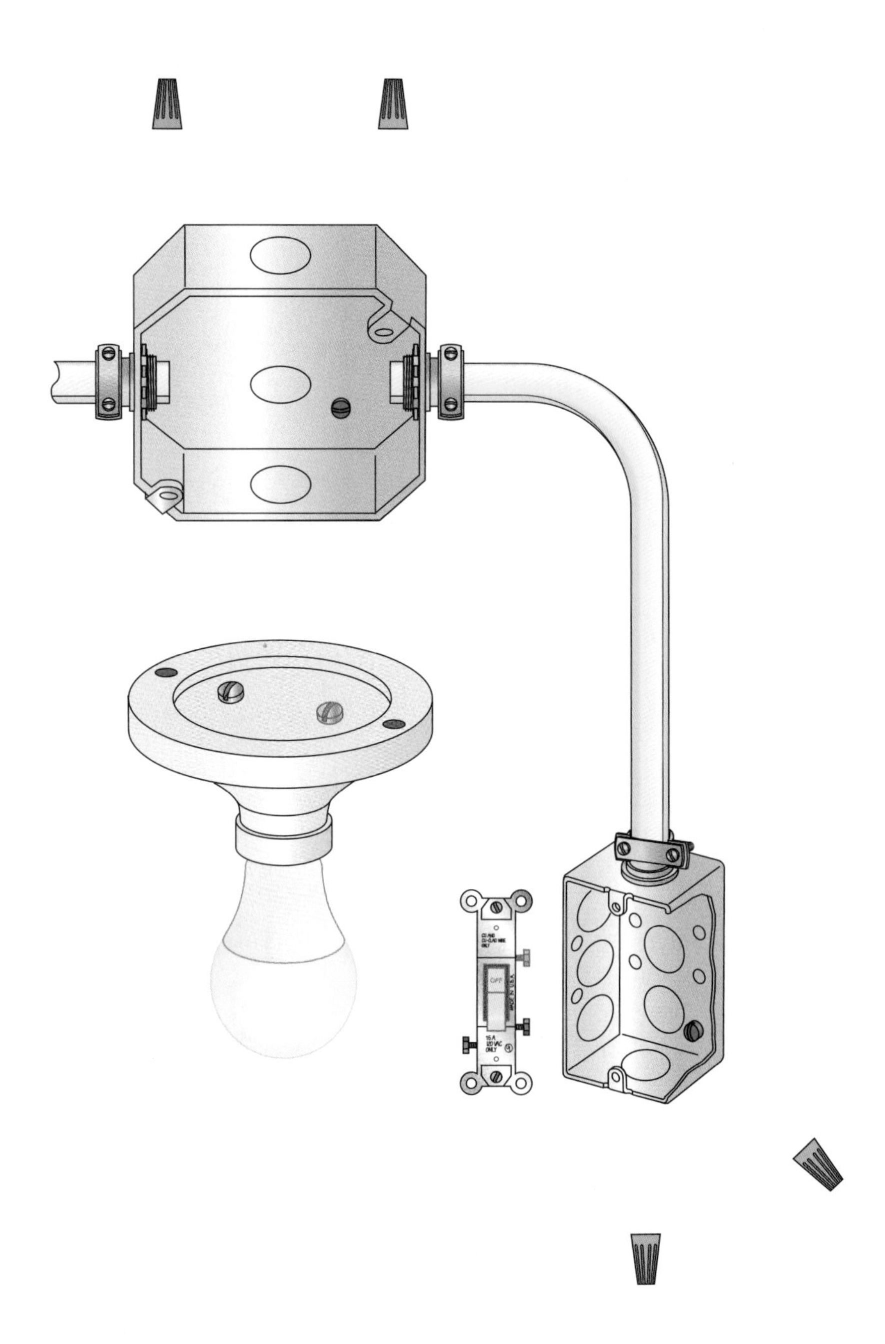

Three-Way Circuits

1. Draw lines to wire a three-way circuit. Indicate the color of each wire.

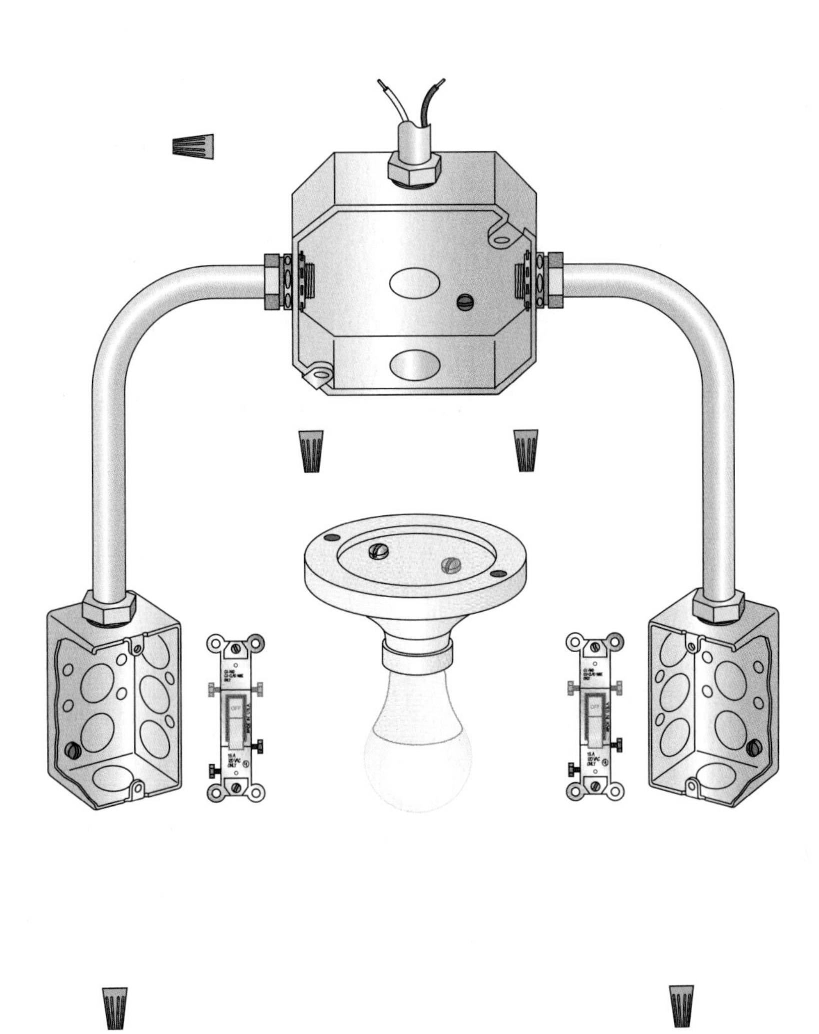

Four-Way Circuits

1. Draw lines to wire a four-way circuit. Indicate the color of each wire.

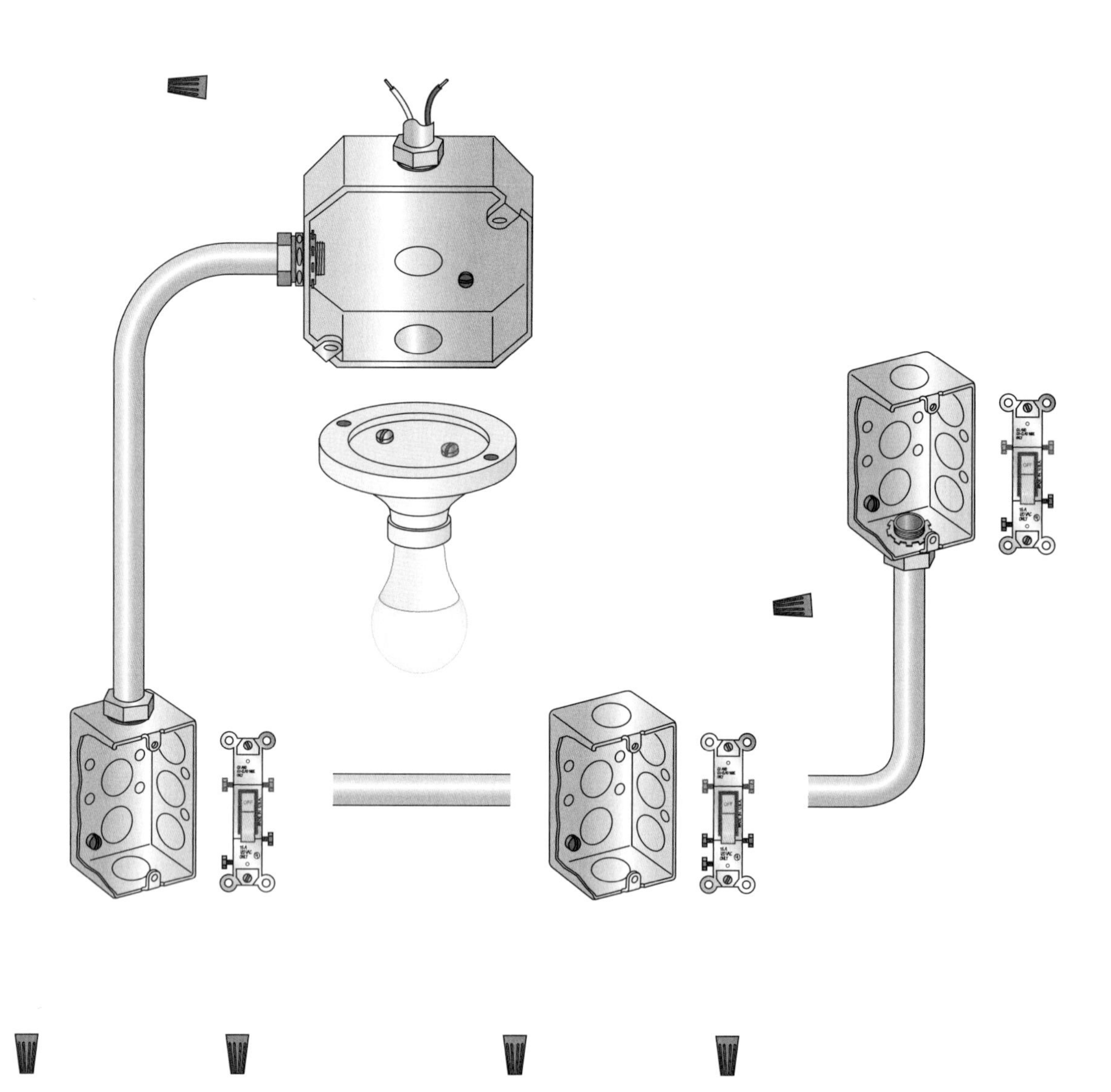

13 Smart Home Infrastructure

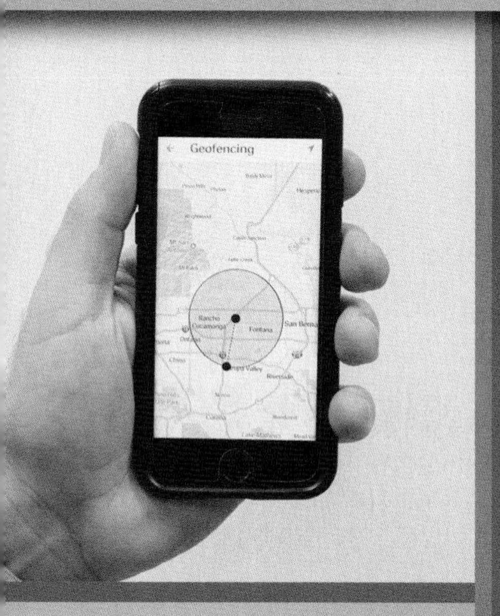

SECTION 13.1—SMART HOME INFRASTRUCTURE

- Explain how the industrial, scientific, and medical (ISM) frequency bands and protocols are used by wireless-controlled devices.
- Explain the purpose of the Institute of Electrical and Electronics Engineers (IEEE) and alliances.
- Describe the types of transmitters, receivers, and transceivers.

SECTION 13.2—SMART HOME KEY COMPONENTS

- Describe the key components of a smart home.
- Describe how internet service providers (ISPs) connect a variety of devices.
- Describe the types of wireless Wi-Fi routers.

SECTION 13.3—WIRELESS SIGNALS AND PROTOCOLS

- Describe wireless signal range.
- List the types of signal interference.
- Describe Bluetooth technology.
- Explain wireless signal data rates.

SECTION 13.4—SMARTPHONE WI-FI SWITCHING

- Explain smartphone wireless switching.
- Describe common applications of the early uses of wireless Wi-Fi switching.

SECTION 13.5—CREATING WIRELESS SMART HOME SYSTEMS

- Explain how to create a wireless smart home system.
- Describe safety upgrades for legacy homes.
- Explain switch upgrades for safety and convenience.

SECTION 13.6—SMART HOME HUBS

- Define "smart home hub" and describe how one operates.
- List the key features considered when selecting a home automation system.
- Explain voice-activated virtual assistants and how to perform their setup via smartphones.

SECTION 13.7—MESH TECHNOLOGY

- Explain the need for mesh technology and its application in smart homes.
- Describe a mesh technology system.
- Describe how mesh routers operate.

Learner Resources

ATPeResources.com/QuickLinks
Access Code: 791712

In a conventional dwelling, branch circuits are physically wired from electrical device to electrical device. The infrastructure of a legacy wiring system in a dwelling can be reconfigured to incorporate wireless-controlled devices. In a smart home, the infrastructure of an electrical system can be designed to accommodate the use of wireless-controlled electrical devices. Installing wireless-controlled devices can be a better alternative in a dwelling where it is costly and not practical to run new wiring. However, newer homes or homes under construction can provide an opportunity to design a wireless-controlled electrical system in part with the conventional electrical system.

SECTION 13.1—SMART HOME INFRASTRUCTURE

Smart home infrastructure includes electrical devices, equipment, and software designed to provide security, energy management, and a convenient lifestyle for a homeowner using smart devices. The terms "connected home" and "smart home" are often interchanged in the context of home automation. The installation of a wireless-controlled system is developed around the fundamentals of wireless controls and basic automation principles. *Note:* Not all electrical receptacles and switches need to be automated. It is important not to automate certain circuits and devices, such as those protected by GFCIs or AFCIs.

All installations of wireless-controlled devices must be in accordance with the NEC® and local building regulations. When wireless-controlled devices are used, careful consideration must also be given to the unintended consequences created by automatically powering on and off unattended appliances and equipment. For example, a portable heater should not be plugged into a wireless-controlled receptacle. Wireless-controlled switches should not be used to control the ON/OFF functions of any medical or life-support systems.

Wireless vs. Wired Systems

Wireless-controlled devices offer economic benefits due to savings in installation, labor, and maintenance costs compared to wired electrical systems. Wireless systems eliminate the risks of improperly wired connections and terminations, which can be potential sources of electrical problems and fire hazards. Wireless controls can be used to link devices found in a variety of systems, such as HVAC, lighting, fire alarm, and security systems. Other advantages of wireless systems include portability, freedom of movement, and elimination of exposed cables. Disadvantages of wireless systems include concerns about hacking and radio interference due to weather or other frequencies. **See Figure 13-1.**

WIRED VS. WIRELESS SYSTEMS	
Wired Advantages	**Wireless Advantages**
Hardwired systems are very reliable.	Wireless device prices may decrease as they are accepted and used in volume.
Wired systems are not hackable.	Wireless systems offer convenience and remote access with Internet connection.
Sensors tend to be more generic and may connect to a wide range of wired systems.	Wireless devices take less time to install because there is no drilling or wires to pull, reducing cost of installation.
Wires can carry information signals over long distances with little interference in larger homes.	Wireless sensors are easy to replace.
An entire wired system can be installed with a battery back-up system.	Wireless devices can be taken out if the owner moves to another location.
Wired devices are relatively inexpensive compared to wireless devices because they have been around a long time.	Some sensors may be powered by energy-harvesting technology, reducing the need for batteries.
Wired signals cannot be reflected, absorbed, or even blocked by common furnishings like a fish tank or file cabinet.	Additional sensors can be added and upgraded easily after new construction is complete.
	Wires cannot be cut and critical connections/terminations cannot become eroded.

Figure 13-1. Wired and wireless systems should be carefully compared before purchases are made.

Development of Wireless Communication

The use of mechanical switches in a dwelling is economical and practical because the length of the wires is relatively short and the wires carry the actual current to operate devices. However, this type of control proved to be impractical for utilities as they began developing longer transmission lines. Since the utilities required a different solution, they decided to use power-line low-frequency communication to control remote equipment in the early development of their distribution networks.

Power-line low-frequency communication has been used by utility companies for years. Beginning in 1922, control signals were sent over transmission lines to turn on and off equipment at remote locations. The frequency range utilities used to control equipment remotely was between 100 Hz to 5000 Hz.

Residential Power-Line Low-Frequency Communication. In 1975, power-line low-frequency communication technology was applied to residential wiring systems and was referred to as "X10 technology." X10 technology uses a frequency of 120 Hz to send information over an existing dwelling wiring system using a low-frequency carrier. Other systems, such as HomePlug™ and Universal Powerline Bus, also use residential power-line low-frequency communication as their base technology. Residential power-line low-frequency communication is still in use and can provide solutions for certain applications, such as wireless control of lighting circuits or improving a wireless internet signal. **See Figure 13-2.**

Insteon is a company that has developed technology to help bridge the gap between power-line low-frequency communication and the higher RF communication control systems that are dominating the smart home market. Insteon technology allows X10 users to transition to RF wireless technology, while still using some X10 products.

> TECH TIP
>
> *X10 devices utilize a home's existing power-line wiring for communication. X10 products can be used to control lamps and appliances, replace wall switches, and monitor doors and windows.*

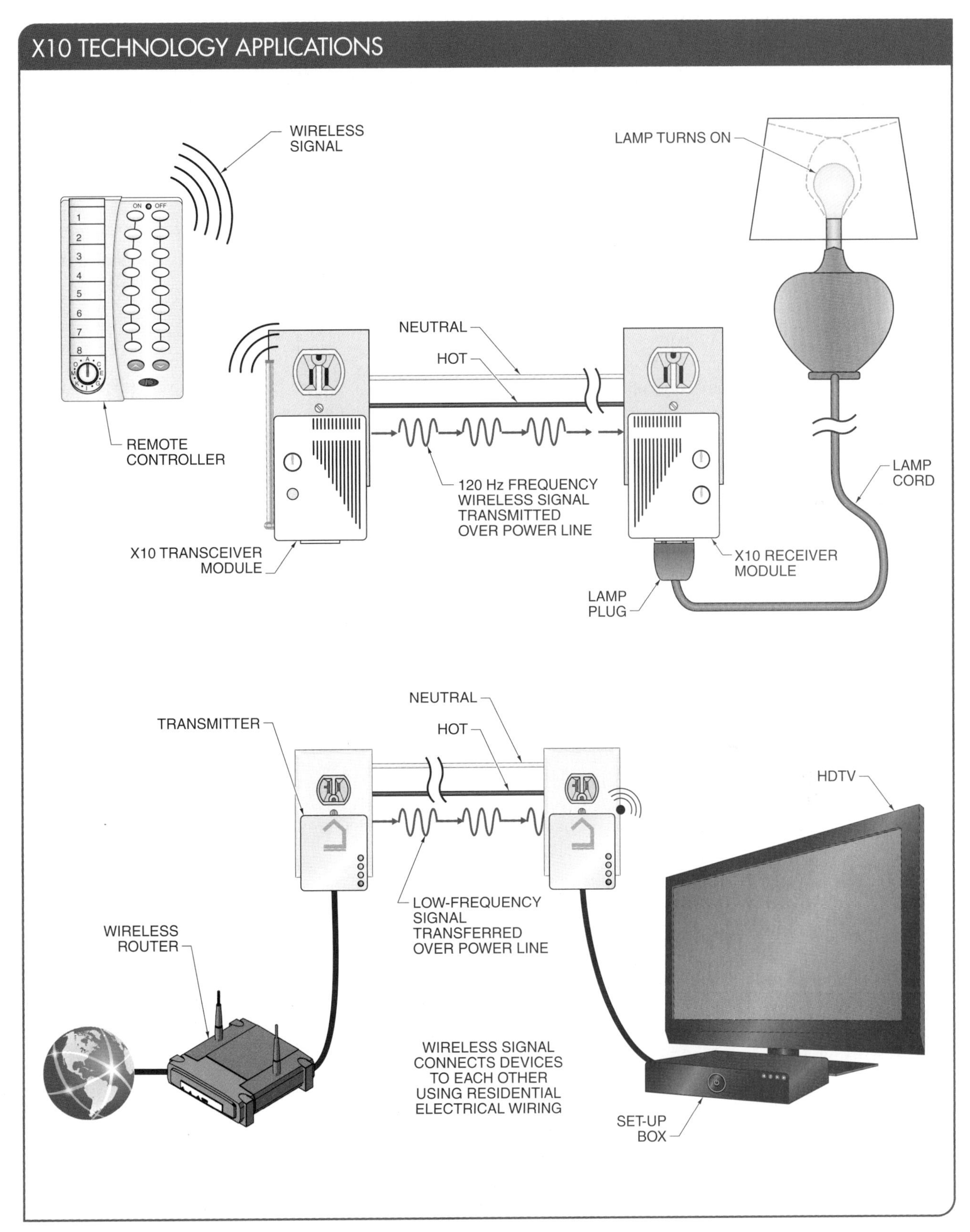

Figure 13-2. X10 technology used a frequency of 120 Hz to send information over an existing dwelling wiring system using a low-frequency carrier.

Wireless RF Controlled Devices

Wireless RF controlled devices use higher RFs to send and receive control information between electrical devices. RF devices require no wiring between them to send and receive information (data). RF is measured in hertz (Hz), which represents the number of cycles per second a radio wave is transmitted. One hertz equals one cycle per second. Radio waves range from kilohertz (kHz), which are thousands of cycles per second, to megahertz (MHz), which are millions of cycles per second, to gigahertz (GHz), which are billions of cycles per second. Microwaves are a type of radio wave with an even higher frequency. RFs are not visible to the human eye and cannot be heard. The distance over which RF signals can travel depends on their wavelength, transmitter power, and the type of antenna used.

Industrial, Scientific, and Medical (ISM) Frequency Bands

The RF bands used in the design of home automation devices are part of the industrial, scientific, and medical (ISM) frequency bands. The ISM frequency bands were created by the Federal Communications Commission (FCC). The ISM frequency bands are intended to support the communication between industrial, scientific, and medical devices.

> TECH TIP
>
> *Smart home devices communicate using a specific code or language called a protocol or platform. There are different protocols for wired and wireless technology. Examples of different protocols include Ethernet, USB, X10, infrared, Wi-Fi, and Bluetooth.*

Until recently, only the industrial frequency band has been used by most manufacturers of wireless-controlled devices as their transmission frequency. With the development of wearable wireless devices and numerous other mobile devices, manufacturers are now looking at the scientific and medical bands to support communication for some of these applications. The ISM frequency bands are used by manufacturers because it allows them to legally develop and market products without obtaining a broadcasting license from the FCC.

All home automation devices must transmit data within the ISM bands to comply with FCC regulations. The ISM frequency bands used for home automation are 900 MHz, 2.4 GHz, and 5 GHz. The most used ISM frequency band is the 2.4 GHz band for Wi-Fi technology.

There are other types of FCC-licensed wireless communication frequencies, including AM radio (10 MHz), FM radio (100 MHz), television (470 MHz to 800 MHz), cellular phones (850 MHz to 1900 MHz), and satellites (3.5 GHz). These frequencies are used and regulated by the FCC to prevent them from interfering with other vital services that use RF-based communication, such as ambulances, fire trucks, and law enforcement vehicles. **See Figure 13-3.**

Protocols

A *protocol* is an industry-standard machine language that enables the exchange of information between wireless-controlled devices. All wireless-controlled devices are designed with a specific protocol (or language) for that device. It is important to remember that a protocol is a language and a frequency carries the information from device to device. When protocols are the same, devices can communicate with each other. If the protocols are different, the devices will not be able to communicate with each other.

Protocols fall into two categories: proprietary protocols and open protocols. Proprietary protocols are developed and owned by the original equipment manufacturer (OEM) and only that protocol is used with their products. For example, Apple's HomeKit is a proprietary protocol that only works with Apple products or products approved by Apple.

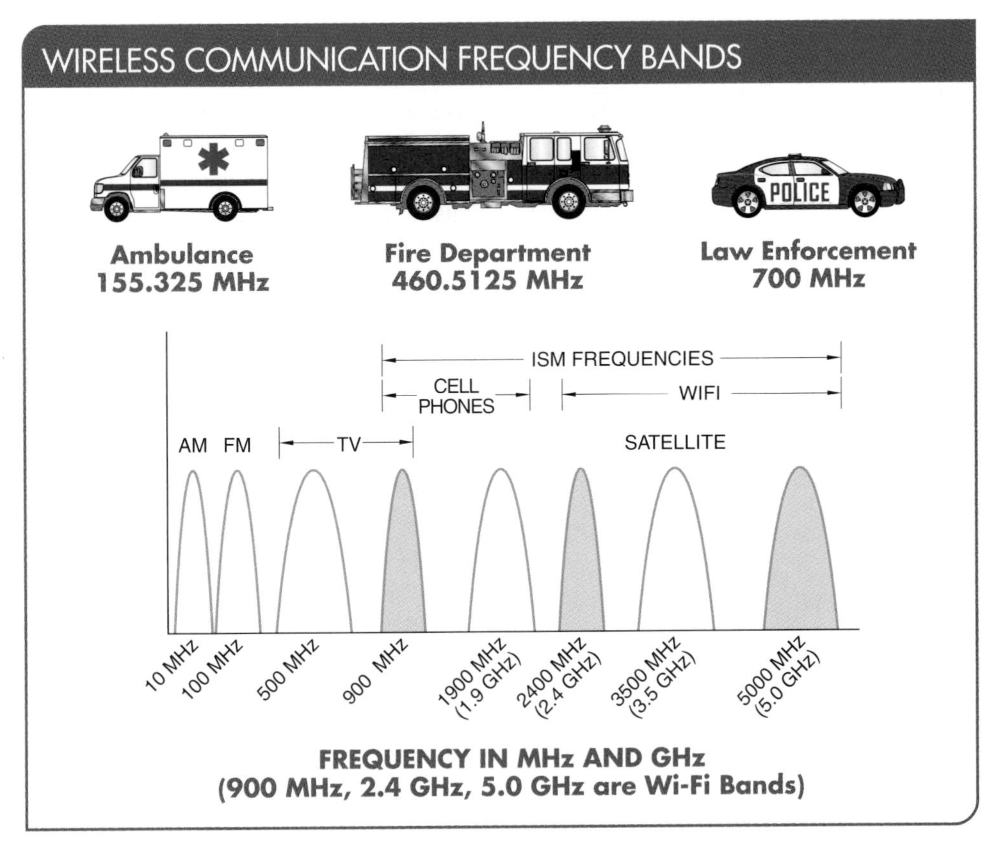

Figure 13-3. All home automation devices must transmit data within the ISM bands to comply with FCC regulations. The ISM frequency bands used for home automation are 900 MHz, 2.4 GHz, and 5 GHz.

Unlike proprietary protocols, which are owned and protected by a company, open protocols are supported with products and services from many different companies and organizations. For example, Wi-Fi is an open protocol that can be used with many different manufacturers' products. This provides homeowners with a much wider choice of devices that can be installed to meet specific applications or needs. The advantages of open protocols include the following:

- supported by multiple manufacturers' software vendors
- widely available third-party software for user interfaces
- easier communication with subsystems and controllers
- active community groups for support and leverage with vendors
- ability to add capabilities in the future

Institute of Electrical and Electronics Engineers (IEEE) Standards

The FCC regulates frequency ranges to protect against interference with critical servers and, in particular, military communication, which is part of US national security. The FCC also has jurisdiction in the commercial radio and television industries. However, when it comes to setting the operating standards for the ISM frequency bands, the FCC has chosen the Institute of Electrical and Electronics Engineers (IEEE) for oversight.

The IEEE is a worldwide organization composed of engineers, technology professionals, medical doctors, physicists, and allied professionals. It is the largest technical professional association in the world.

The IEEE Standards Association develops standards for a wide range of industries, such as power and energy, biomedical and health care, telecommunications, and home automation, through specific committees.

The standards for home automation are maintained by the IEEE 802 LAN/MAN Standards Committee. This committee develops local area network (LAN) and metropolitan area network (MAN) standards that allow smart grids and smart devices to communicate. The IEEE 802 standards are a series of networking standards that evolve over time. For example, versions of the standard for Wi-Fi are tagged with IEEE 802.11 and are often followed by one or more letters to indicate other characteristics of specific versions. The first Wi-Fi standards were 802.11a and an 802.11b. The 802.11a standard was the fastest standard for data transfer, but it was more expensive and was used primarily for business networks. The 802.11b standard was slower but more affordable, so it was used for personal/home networks.

There are many versions of the IEEE 802 standard, and more developments will be incorporated as technology advances. **See Figure 13-4.** These standards are often labeled on the packages of routers and other smart devices, which should be referenced before being purchased. Checking package labels will ensure that a device being purchased has the latest standard with all of the up-to-date features.

Alliances

To regulate specific protocols that are based around the IEEE standards, manufacturers have banned together in alliances to certify that their protocol and equipment meet the standards for their alliance. Alliances are built around protocols and frequency bands. Protocols must operate in one or more of the ISM frequency bands of 900 MHz, 2.4 GHz, and 5 GHz. A protocol is typically named after the alliance, such as Wi-Fi, Bluetooth, ZigBee, Z-Wave, or Thread, among others.

All of these alliances typically have a website that list the member manufacturers that have developed products around their specific protocol. These members also typically have websites to provide additional information regarding the wireless products they sell within the alliance, along with the features and benefits of these products. These websites also direct consumers to where these products may be purchased. Comparing alliances is a good way to evaluate similar products before any items are purchased.

Alliance Certification. Each alliance maintains standards and provides certifications for their products. Alliance certification is done by accredited testing labs to ensure that a product has been tested against standards to comply with regulations of safety and performance. Alliance certification also tests the electrical and mechanical components of a device to verify that they operate

IEEE 802.11 STANDARDS			
Standards	Band Frequency	Maximum Data Transfer	Year Created
802.11a	5 GHz	54 Mbps	1999
802.11b	2.4 GHz	11 Mbps	1999
802.11g	2.4 GHz	54 Mbps	2003
802.11n	2.4 and 5 GHz	600 Mbps	2009
802.11ac	5 GHz	1.3 Gbps	2013
802.11ac Wave 2	5 GHz	3.47 Gbps	2015

Figure 13-4. The IEEE 802 standards are often labeled on the packages of routers and other smart devices and should be referenced before the devices are purchased.

as specified. Certification labels are usually embedded or labeled on a device to verify it has met approved standards.

When devices from different manufacturers are added to a home automation system, research should be done before purchasing each new smart home device. The devices must match the protocol and frequency of the existing system. This compatibility is referred to as "interoperability" in the home automation industry. *Note:* Alliances can be a great source for researching the interoperability of the products they certify.

Interoperability

The National Institute of Standards and Technology (NIST) defines "interoperability" as the capability of two or more networks, systems, devices, applications, or components to exchange and readily use information securely, effectively, and with little or no inconveniences to the user.

Another term used with interoperability is "interchangeability." *Interchangeability* is the ability of an automated system to operate correctly when one device is substituted for another device. For example, interchangeability is when a damaged lighting control device from one vendor can be replaced with a similar lighting control device from another vendor and still communicate with all other original devices in the system as the original control.

Interoperability depends upon a device to use the same protocol (language) and operate at the same frequency. The devices produced by manufacturers for alliances such as Bluetooth and Wi-Fi are considered to be compatible. When devices are purchased outside of these protocols or frequencies and put into the same network, they will not necessarily be compatible. These devices will not communicate with each other because the language is different, or the frequency may also be different. When a larger system is purchased or new devices are to be added to an existing system, interoperability will be one of the most important considerations.

Transmitters, Receivers, and Transceivers

Wireless-controlled devices have no physical hardwired connection between other devices in the system. Wireless-controlled devices depend on RF transmitters, RF receivers, or RF transceivers to operate. A *transmitter* is an electronic device that sends out RF signals. A *receiver* is an electronic device that receives RF signals. A *transceiver* is an electronic device that both transmits and receives RF signals.

Types of transmitters include cell and radio towers, which send audio, video, and digital signals. Types of receivers include radios, car antennas, or television antennas that receive audio and video signals to produce sound or create images on a screen. Types of transceivers include cell phones, tablets, and computers that send and receive data from various sources. **See Figure 13-5.**

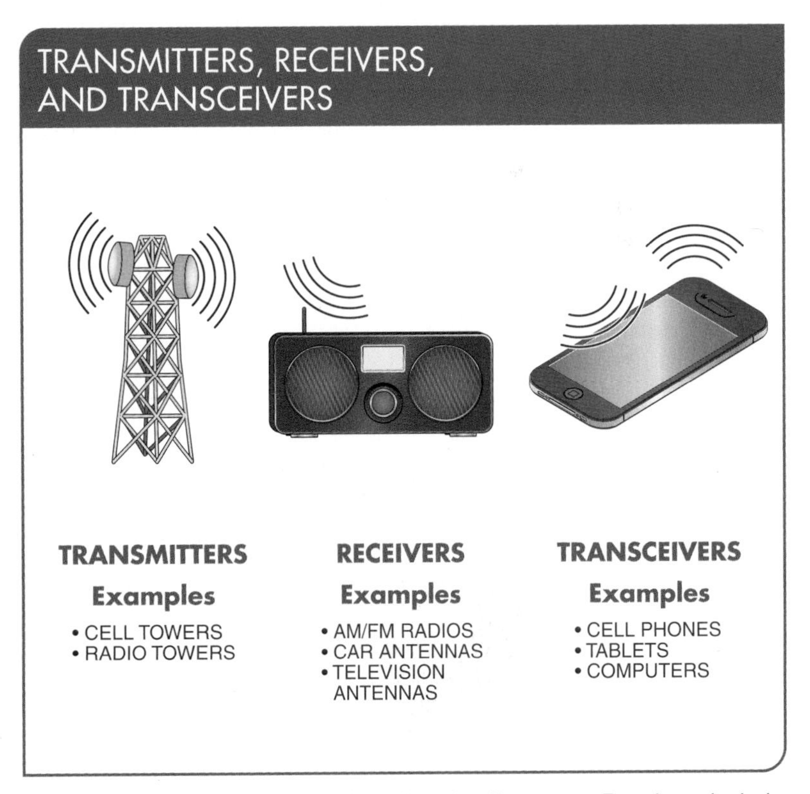

Figure 13-5. Transmitters include cell and radio towers. Receivers include radios, car antennas, and television antennas. Transceivers include cell phones, tablets, and computers.

Transmitting RF Signals

Wireless communication operates by the transmission of RF signals through the air. An RF signal consists of electromagnetic waves that are generated when electrical energy travels through a piece of metallic material such as an antenna. Antennas can both send and receive information at different frequencies.

> TECH TIP
>
> *A transmitter antenna converts electricity into radio waves that travel through the air at the speed of light (about 300,000 km per second). A receiver antenna converts the radio waves to an electric current, re-creating the original signal. Transmitter and receiver antennas are often very similar in design.*

The most common type of antenna used in home automation RF applications is the omnidirectional antenna. An omnidirectional antenna operates by transmitting a signal in all directions. Other antennas include point-to-point and parabolic antennas, which are used in applications where there is a need to transmit concentrated RF signals to a specific location, such as a satellite. **See Figure 13-6.**

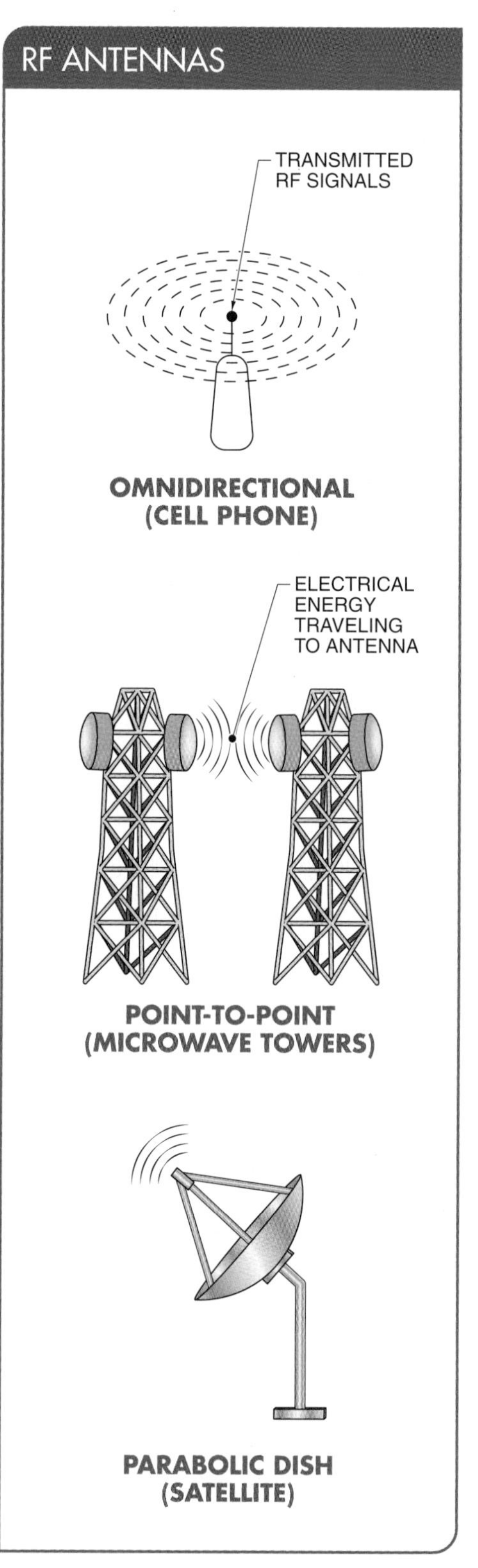

Figure 13-6. The most common type of antenna used in home automation RF applications is the omnidirectional antenna. Other antennas include point-to-point and parabolic antennas.

SECTION 13.1 CHECKPOINT

1. What was the first frequency technology applied to residential wiring systems?
2. What are the ISM frequency bands used for home automation, and which band is most common?
3. What is the name of the committee that maintains the IEEE standards for home automation?
4. What are some of the protocols named after alliances?
5. List examples of transceivers, receivers, and transmitters.

SECTION 13.2—SMART HOME KEY COMPONENTS

The key components of a smart home include access to an internet service provider (ISP), a modem, and a Wi-Fi router. **See Figure 13-7.** These components are combined with smart devices and apps to create a variety of smart applications for controlling multiple devices in a smart home. An *app* is a specialized software program that can be downloaded to a device, such as smartphone or tablet, to perform a specific function.

Internet Service Providers (ISPs)

The internet is an enormous source of information. A modem connecting to an internet service provider (ISP) is used to provide access to this information in the home. An *internet service provider (ISP)* is a company that provides access to the internet. ISPs use a variety of connecting technologies to deliver an internet connection to the end user, including coaxial cable, fiber optic, wireless cellular, and satellite. A satellite connection is used in remote or rural areas where no other type of service is available. **See Figure 13-8.**

Modems

A *modem* is a network device that transforms outgoing digital signals from a computer into analog signals that are transmitted over wires and that transforms incoming analog signals to digital signals for a computer. A modem is the hardware device that allows a homeowner to actually connect with the ISP. The word "modem" is derived from the device's ability to both *mo*dulate and *dem*odulate RF signals. In effect, the modem acts as a translator of incoming and outgoing information to and from the internet. Modems are often combined with routers.

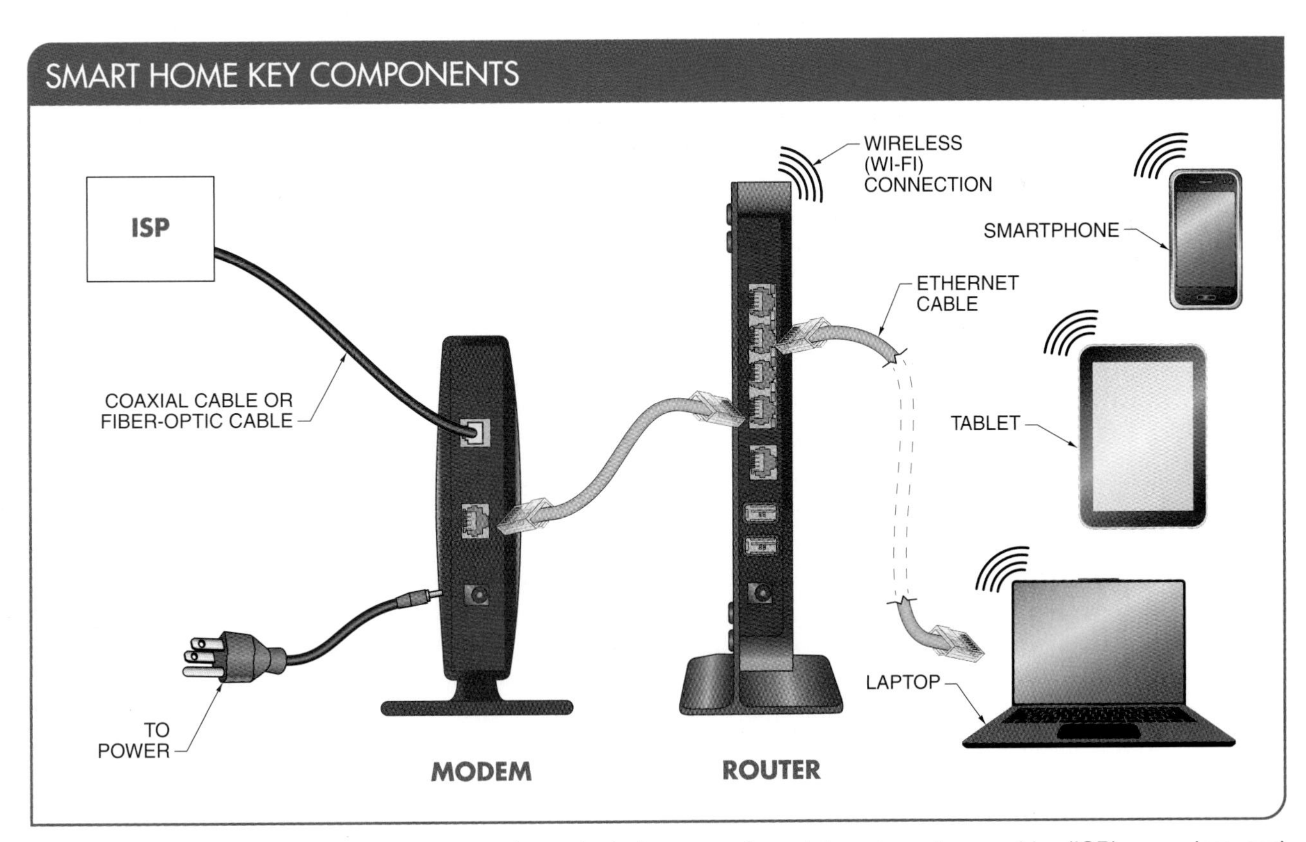

Figure 13-7. The key components of a smart home include access to an internet service provider (ISP), a modem, and a Wi-Fi router.

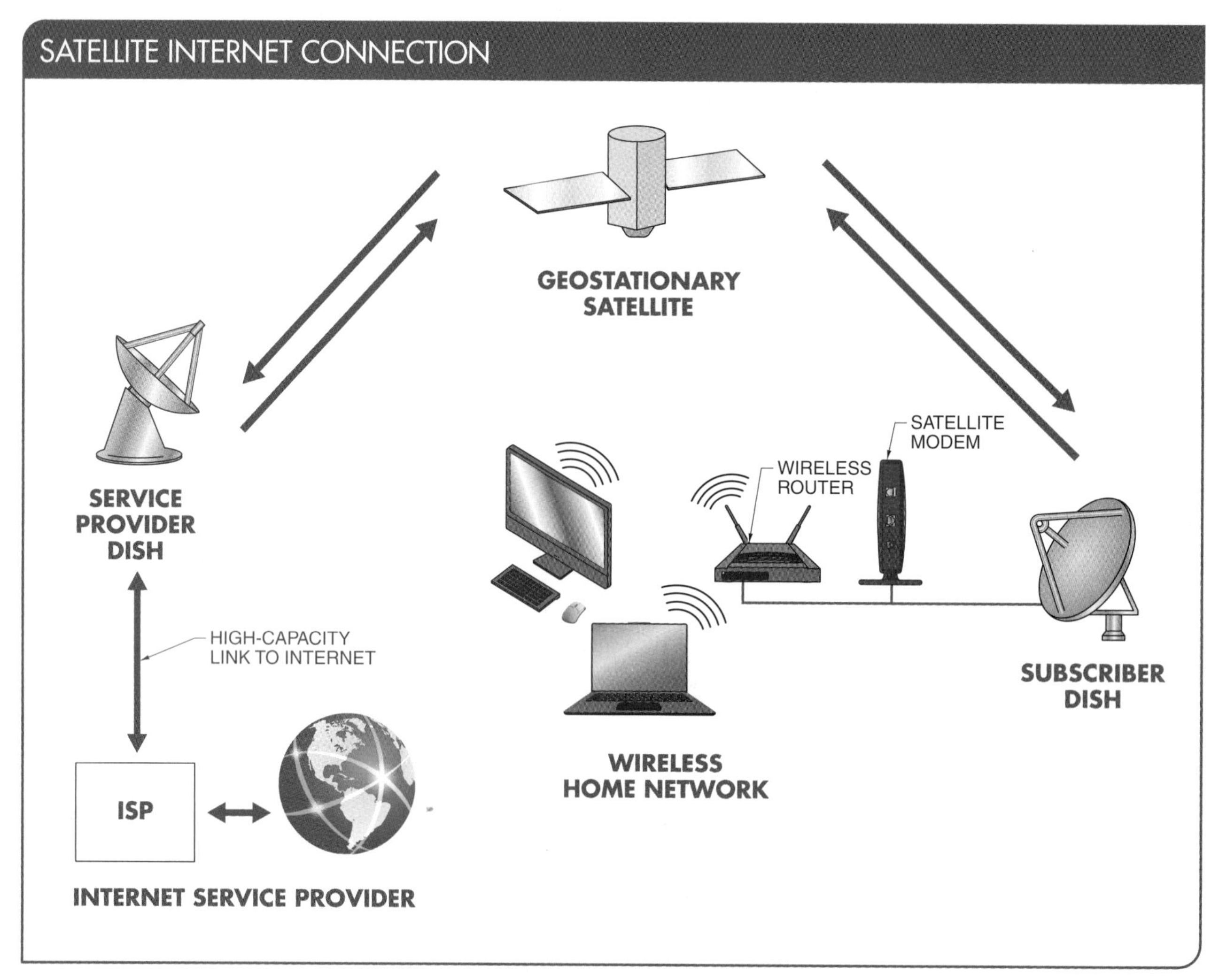

Figure 13-8. ISPs use a variety of connecting technologies, including coaxial cable, fiber optic, wireless cellular, and satellite technology, to deliver an internet connection to the end user.

Wireless Wi-Fi Routers

A *wireless Wi-Fi router* is a network device that routes data between internet-connected devices and the internet. A wireless Wi-Fi router consists of a radio transceiver, antennas, and operational firmware. *Firmware* is a software program permanently embedded into a modem like a computer chip to give permanent instructions. The operating protocol for Wi-Fi is established in the firmware. The firmware will usually come installed as an integrated circuit ready to perform when the router is plugged into the modem.

When the router is being set up, one of the first things it does is obtain an internet protocol (IP) address so the internet knows where to send information. IPs are the language used to communicate. The address refers to the unique number that is linked to all the online activity coming from a specific location. The IP address for a computing device is given to it by the ISP. When a computer is moved from one location to another location, the IP number will change because the ISP will change. This all happens automatically because of the way in which the internet was designed.

The current worldwide system that provides this process is called internet protocol version 6 (IPv6). This was updated from IPv4 because IPv4 was running out of addresses. IPv6 can provide trillions of addresses to meet the demand for the growing number of applications.

The second thing a router does is provide a second address for the local area network (LAN) so that the devices in a LAN can communicate with each other. With millions of new devices being connected to the internet from smart home devices, many of the new routers will be assigning IP addresses to each device. IPv6 and its ability to handle trillions of new addresses will be able to handle providing unique addresses for all of these devices.

Wireless Wi-Fi routers function as wireless access points (WAPs). When a modem is connected to a router, a WAP is created, which allows a Wi-Fi signal to be broadcast throughout a home. The abbreviation "WAP" has been used for a long time to describe the connection. However, another, newer abbreviation of "AP," which is short for "access point," may be used.

Most homes typically need only one wireless Wi-Fi router with an access point to provide Wi-Fi coverage. Once the network cable is plugged into a wireless Wi-Fi router, radio waves transmit information that traditionally travels on only two frequency bands: 2.4 GHz or 5 GHz. *Note:* The Wi-Fi Alliance has added 900 MHz to provide access to this additional band, which will require router upgrades to use this frequency.

Wireless Wi-Fi routers will often have two or more antennas on the side or back of their cases. Some router models house the antennas inside the enclosure. Three types of routers are single-band, dual-band, and tri-band routers. **See Figure 13-9.**

Single-Band Routers. A *single-band router* is a wireless router that only broadcasts one of the Wi-Fi frequencies. The most frequently used single band is 2.4 GHz. This frequency is good for basic internet applications such as surfing the web, sending emails, and downloading files. The advantages of single-band routers include their low purchase price and compatibility with almost all 2.4 GHz devices. The disadvantages of single-band routers include lower speed, significant signal interference that reduces performance in the 2.4 GHz band, and limited features compared to other router types.

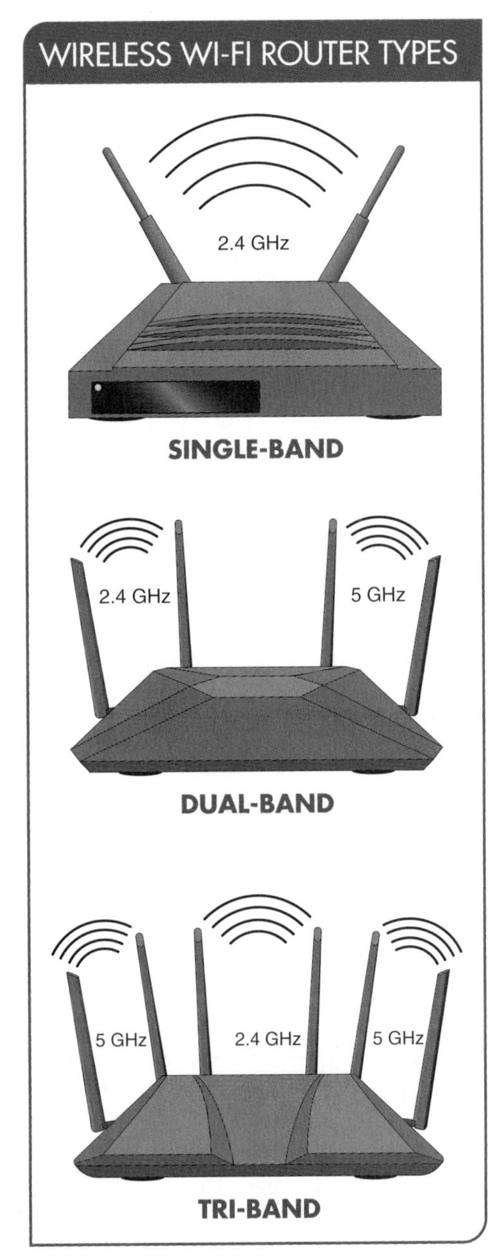

Figure 13-9. Wi-Fi routers will often have two or more antennas on the side or back of their cases. The three types of routers are the single-band, dual-band, and tri-band routers.

Dual-Band Routers. A *dual-band router* is a wireless router that simultaneously broadcasts at 2.4 GHz and 5 GHz. Devices that operate on a 5 GHz Wi-Fi signal will connect to the 5 GHz band, and devices that operate on a 2.4 GHz Wi-Fi signal will connect to the 2.4 GHz band. The advantage of a dual-band router is that it can essentially host two different Wi-Fi networks at once, providing additional bandwidth compared to a single-band router. The 5 GHz signal is better for streaming video and online gaming applications. A disadvantage of the dual-band router is that the 5 GHz signal has a shorter range than the 2.4 GHz signal in homes with many walls, doors, pieces of furniture, and other signal-blocking obstructions.

Tri-Band Routers. A *tri-band router* is a wireless router that simultaneously broadcasts three signals, which include one signal at 2.4 GHz and two separate 5 GHz signals. A tri-band router uses the two 5 GHz signals instead of broadcasting a third different frequency. A tri-band router automatically sorts devices into different networks based on frequency, which offers more speed to share among devices. Tri-band routers are typically used for high-end video streaming and gaming. Advantages of tri-band routers include less interference and more available bands. Disadvantages of tri-band routers include a significantly higher cost than entry-level routers and limited availability of devices compatible with 5 GHz.

SECTION 13.2 CHECKPOINT

1. What are the three key components of a smart home?
2. What devices are used to connect ISPs to end users?
3. Define "wireless Wi-Fi router" and explain what happens when it is set up.

SECTION 13.3—WIRELESS SIGNALS AND PROTOCOLS

For wireless devices to be effective, the signals they send and receive must be strong enough to provide clear and precise information to the device they are controlling. The router should be located as close to the center of the home for best results. However, the kitchen may not be the best location due to microwave appliances and cabinets filled with signal-blocking materials. Also, if a router is placed at one end of a residence, the signal strength can be reduced as the distance from the router increases. **See Figure 13-10.**

TECH TIP

A mobile device that is internet capable can be used to show the Wi-Fi signal strength in different rooms of a dwelling through the settings menu. Another option is to download a Wi-Fi analyzer or Wi-Fi scanner to view signal strength.

Wireless Signal Range

Frequency plays a big part in how far a wireless signal will travel. The 900 MHz frequency provides the longest distance of travel between the three frequencies. The 5 GHz provides the shortest distance of travel. The distance of travel for 2.4 GHz is about 600′. However, these distances are subjective, as there is no definite distance that is totally accurate until the wireless devices are placed in the residence. Location and interference in a particular location determines the actual distance of travel.

Wireless Signal Interference. Wireless signal interference cannot be avoided entirely. However, wireless signal interference can be minimized by relocating wireless networking hardware or using specialized antennas. The most common types of signal interference are reflection, absorption, scattering, refraction, and electrical interference. **See Figure 13-11.**

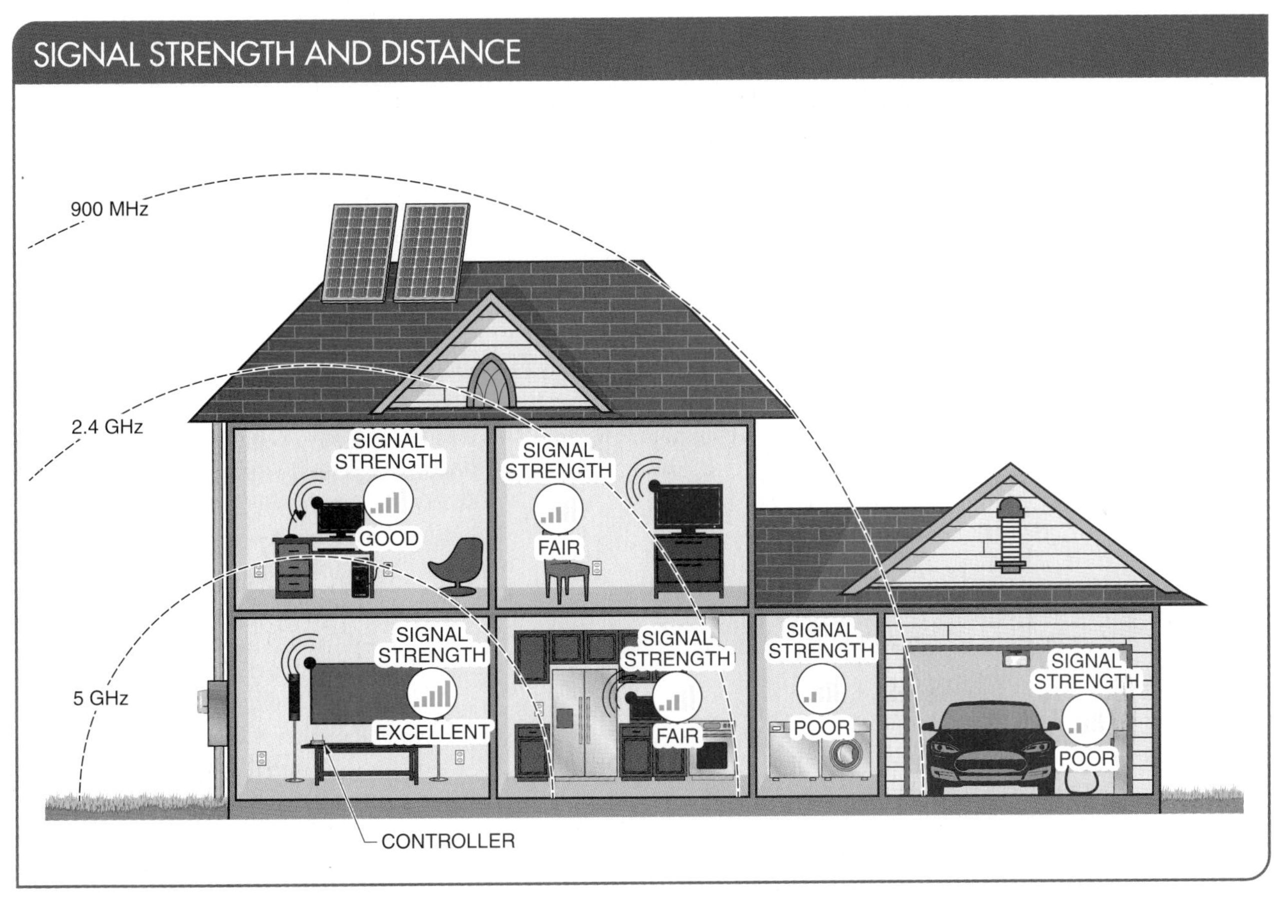

Figure 13-10. To get the best Wi-Fi connection in a dwelling, routers should be placed in a central location and not surrounded by metal objects.

RF SIGNAL INTERFERENCE ISSUES	
Type of Interference	**Definition/Description**
Reflection	A wireless signal is a radio wave. Just like light waves, the signal can bounce off certain surfaces. If a large amount of reflection occurs, signals can be weakened and cause interference at the receiver.
Absorption	Absorption is based on materials converting the signal's energy into heat. Different materials naturally have different absorption rates. Wood and concrete can reduce the signal strength dramatically because of how much these materials absorb. Glass can both absorb and reflect wireless signals. Absorption can result in poor connections and slower than expected speeds.
Scattering	Dust, humidity, uneven surfaces, and other composition of materials can also cause a signal to scatter. These factors can have a significant impact on signal reliability and strength.
Refraction	Refraction is the bending of a wave when it enters a medium. For example, glass or water can refract waves. Because a signal changes direction in traveling from sender to receiver, this can cause lower data rates, and more missing data.
Electrical Interference	Electrical interference comes from devices such as computers, refrigerators, ceiling fans, fluorescent lighting fixtures, and many other motorized devices. The impact that electrical interference has on the signal depends on the proximity of the electrical device to the wireless access point.

Figure 13-11. The most common types of signal interference are reflection, absorption, scattering, refraction, and electrical interference.

Selecting the proper channel for a wireless Wi-Fi router can significantly improve Wi-Fi coverage and performance in a residence. The 2.4 GHz frequency of routers has multiple overlapping channels. However, when channels overlap, it is similar to several people talking at once—not everyone will be heard as well as when they are speaking alone. This also holds true for channel overlap. To minimize interference, it is best to choose one of three non-overlapping channels available in the 2.4 GHz range: channels 1, 6, and 11. **See Figure 13-12.**

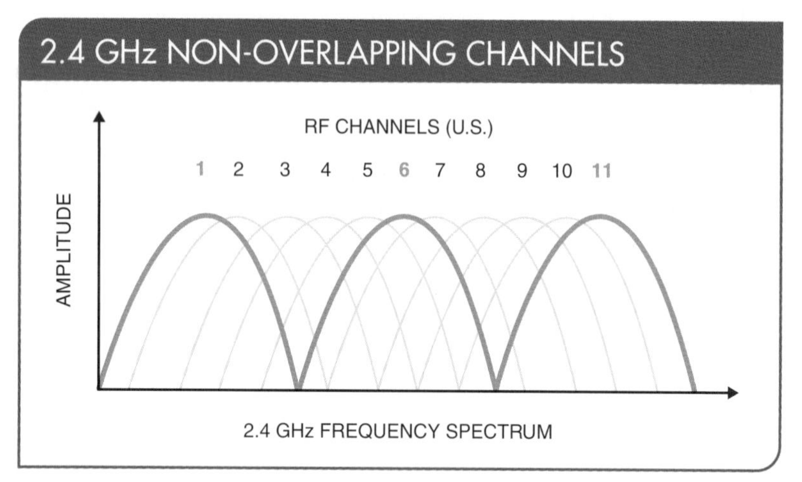

Figure 13-12. Changing the Wi-Fi signal to a non-overlapping channel helps to prevent signal interference.

Many routers automatically select one of the three non-overlapping channels when activated. If there is channel interference due to a neighboring router, the channel can be changed to another channel so it does not compete. Wireless signal interference from outside sources can occur in densely populated areas. Wireless signals from one dwelling can penetrate another dwelling and interfere with that dwelling's network when both routers are set on the same communication channels. If the neighboring dwellings each use a different channel from the non-overlapping channels 1, 6, and 11, cross-network interference can be eliminated.

Bluetooth Technology

Bluetooth technology is used with hands-free phone-calling in automobiles and is often used for music and video streaming. Bluetooth technology has built-in security available, such as data encryption and PIN codes for authentication. Bluetooth technology requires no fixed infrastructure and is simple to install and set up.

When Bluetooth devices are within range of each other, they automatically connect and communicate if their standards are compatible. Bluetooth-enabled electronic devices connect wirelessly through short-range networks, known as piconets or personal area networks (PANs). A *piconet* is a network that allows a device to simultaneously communicate with several other devices. The most common example of a piconet is a hands-free headset used with a cell phone.

There are now two versions of Bluetooth: the standard version of Bluetooth that has been around for years and the newer version called Bluetooth Low Energy (BLE). "Bluetooth Smart" is an industry name often used for the devices technically classified as BLE. BLE technology was designed to reduce power consumption and cost. The major emphasis of Bluetooth is low-power consumption. The big advantage for BLE is that its design permits devices to use both the BLE protocol and the standard Bluetooth protocol.

BLE is ideal for new devices such as wearables or for tiny beacons imbedded in new consumer products. Start-up companies as well as and large, established companies, such as Apple and Google, are developing tiny Bluetooth transmitters that can be embedded in devices used in the healthcare, fitness, and security markets.

Wi-Fi Alliance – 900 MHz

The Wi-Fi Alliance is adding the 900 MHz frequency range to their frequency standards as they focus on innovations for improving wireless device communication. The 900 MHz frequency range has been used in some portable phones, AV equipment, and lighting controls for years. The 900 MHz frequency range has also been the established frequency range for the Z-Wave Alliance and others.

This new Wi-Fi Alliance technology is called "HaLow." HaLow will compete with the low-frequency alliances, such as Bluetooth with their BLE technology. All of the alliances continue to position themselves in the smart home market and the "Internet of Things (IoT)" marketplace. The Wi-Fi Alliance indicates that a compatible router is the only hardware needed to be operational with their new standard and technology. However, this would require that phones and routers be upgraded with new Wi-Fi chips in order to work with HaLow products.

Wireless Signal Data Rates

Higher frequencies allow faster transmission of data. Higher data rates are also known as bandwidth. The higher bandwidth of 5 GHz means that information is able to download and upload data faster for high-bandwidth applications such as games and streaming videos. These programs perform much smoother and faster on 5 GHz compared to 900 MHz or 2.4 GHz. The trade-off is that 5 GHz signals do not travel as far as 2.4 GHz signal or the 900 MHz signal.

Wireless Device Power Consumption

The trade-off in transmission power for a stronger signal is the amount of power consumed by that frequency. Many home wireless devices use a battery as the power source. Some manufacturers are trying to reduce power consumption to lengthen battery life in wireless devices, while other manufacturers are considering "energy harvesting," which requires no batteries. Energy harvesting uses the operation of the device itself as an energy source or obtains (harvests) energy from the surrounding environment. Since Wi-Fi consumes a lot of power, manufacturers of small devices are using the Bluetooth low-power devices. However, the tradeoff is that Bluetooth has a lower travel range.

SECTION 13.3 CHECKPOINT

1. What are the most common types of signal interference?
2. What are the three non-overlapping channels available in the 2.4 GHz frequency range and why are they used?
3. What are the two versions of Bluetooth?
4. What is the trade-off between the 900 MHz and 2.4 GHz signals and the higher data rate of the 5 GHz signal?

SECTION 13.4—SMARTPHONE WI-FI SWITCHING

Smartphone technology has dramatically impacted home automation. The computing power and storage of a smartphone allows it to be a home automation controller using only an existing home router. Switching, such as turning a light or appliance on or off, can now be controlled remotely with this Wi-Fi-enabled device. **See Figure 13-13.** The control information is stored in a smartphone app.

Disadvantages of using a smartphone for a controller include the additional storage required on the smartphone and the additional battery drain when using Wi-Fi through the phone. Wi-Fi has a much higher power consumption than Bluetooth devices. This is a reason why the Wi-Fi Alliance is adding 900 MHz, which has a lower power consumption.

Apps and Smartphones

Many smartphone control apps include a demo version or a free trial that can be experienced before a purchase is made. A smartphone control app must be compatible with the type of smartphone used to interface with the end product. The app interface should be user-friendly and easy to understand.

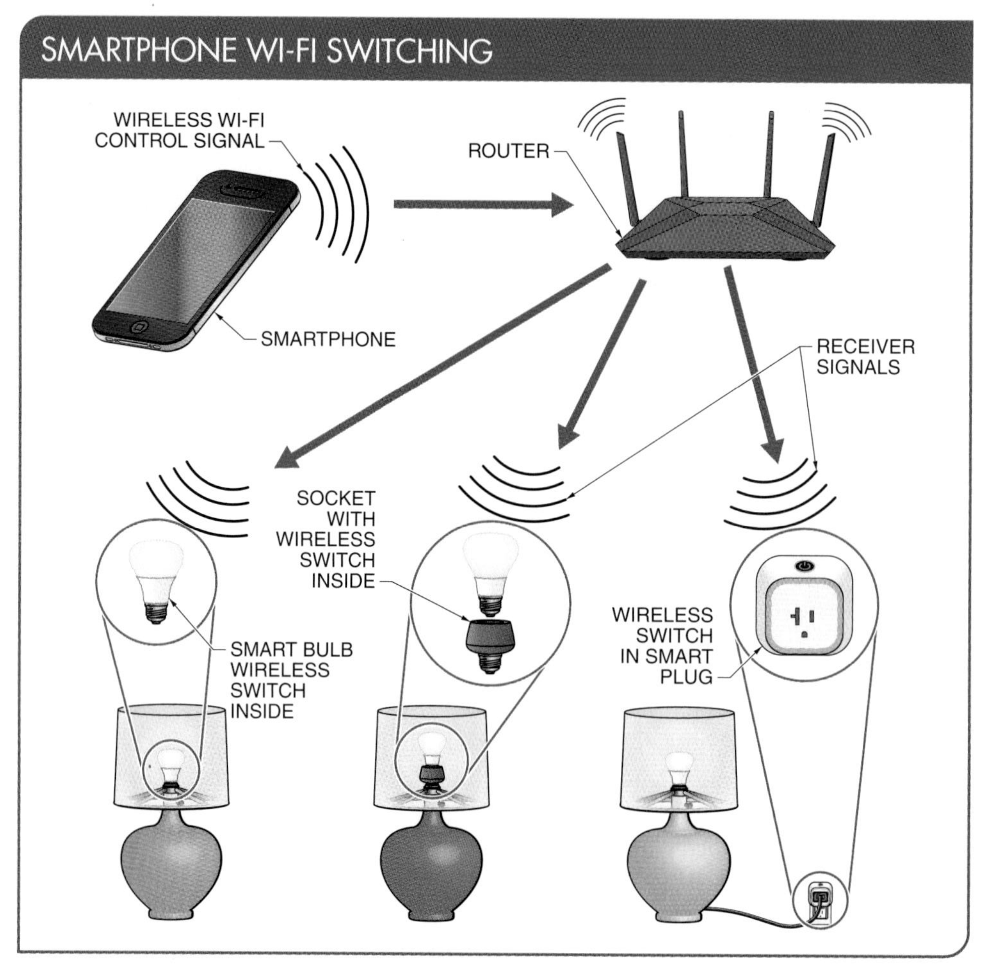

Figure 13-13. The computing power and storage of a smartphone allows it to be a home automation controller using only an existing home router.

The app should also be able to be used on multiple platforms, including tablets and laptops without issues. In a connected home, the smartphone and the app control the smart device. The two largest app providers are Google Play for Android devices and the Apple App Store for iOS devices.

Wi-Fi Wireless Switch Applications – 2.4 GHz

Common applications of the early use of wireless Wi-Fi switching using only a router include the switching of smart bulbs, smart plug wireless switches, wireless ceiling fixture switches, and wireless wall switches. Wireless capability can be as simple as installing a connected smart bulb or a standard bulb into a connected smart light bulb socket. Connected smart plugs can also be installed for control at the receptacle. A *smart bulb* is a light bulb that contains a Wi-Fi receiver and switching device in the actual bulb. A *smart light bulb socket* is a Wi-Fi receiver and switching device installed in a light bulb socket that screws into standard lamps before a light bulb is installed in the smart socket.

In most cases, these devices will contain a wireless receiver and solid-state switch. If the bulb can send back a signal to the homeowner that the bulb has been turned on, then the device has a transceiver and solid-state switch.

Smart Plug Wireless Switches

With a regular lamp and a standard bulb, it is more likely the lamp outlet would be switched using an existing outlet. In that case, a Wi-Fi receiver and switch installed in a smart plug, which is then plugged into the receptacle. **See Figure 13-14.** Smart plugs allow simple transition to wireless switching for a regular receptacle. Once a smart plug is installed and the app is activated, control is possible anywhere a Wi-Fi connection can be established. Connected smart plugs can also control other plugged-in devices, such as small appliances. Connected smart plugs continue to evolve as new intelligence (programming) is added. For example, with additional programming, a smart plug can monitor the energy consumption of the device plugged into it.

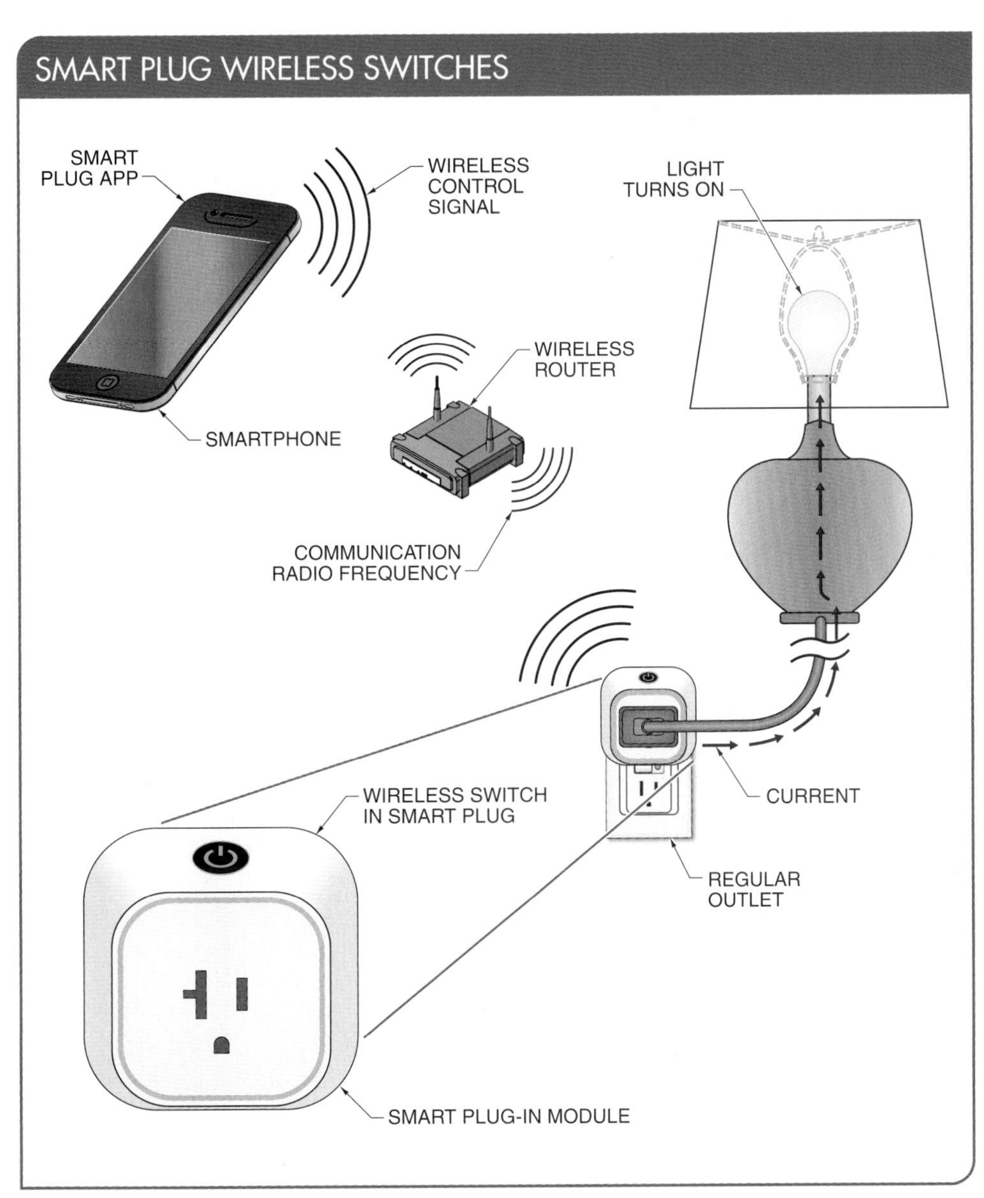

Figure 13-14. Smart plug wireless switches enable a smartphone with a smart plug app to control a regular lamp and a standard bulb.

Wireless Ceiling Fixture Switches

If there is need for a more permanent installation of a wireless switch, such as a ceiling light, kits are available for installation. **See Figure 13-15.** Power must be available at the ceiling fixture when installing wireless switching kits. At the light fixture, the hot wire must be connected to the wireless switch module (receiver) and the white wire from the module to the light fixture. This type of switch is useful to control porcelain pull chain sockets that may be difficult to reach.

Wireless touch-type switches do not need any wires or connections because their ultra-low-power RF transmitters use tiny batteries that have life spans of many years. Companies are also using energy-harvesting technology that does not require any batteries at all. Wireless switches can be manually switched at their location as well as being remotely switched using a smartphone.

> TECH TIP
>
> *Multiple wireless transmitters can command a single receiver to control a light.*

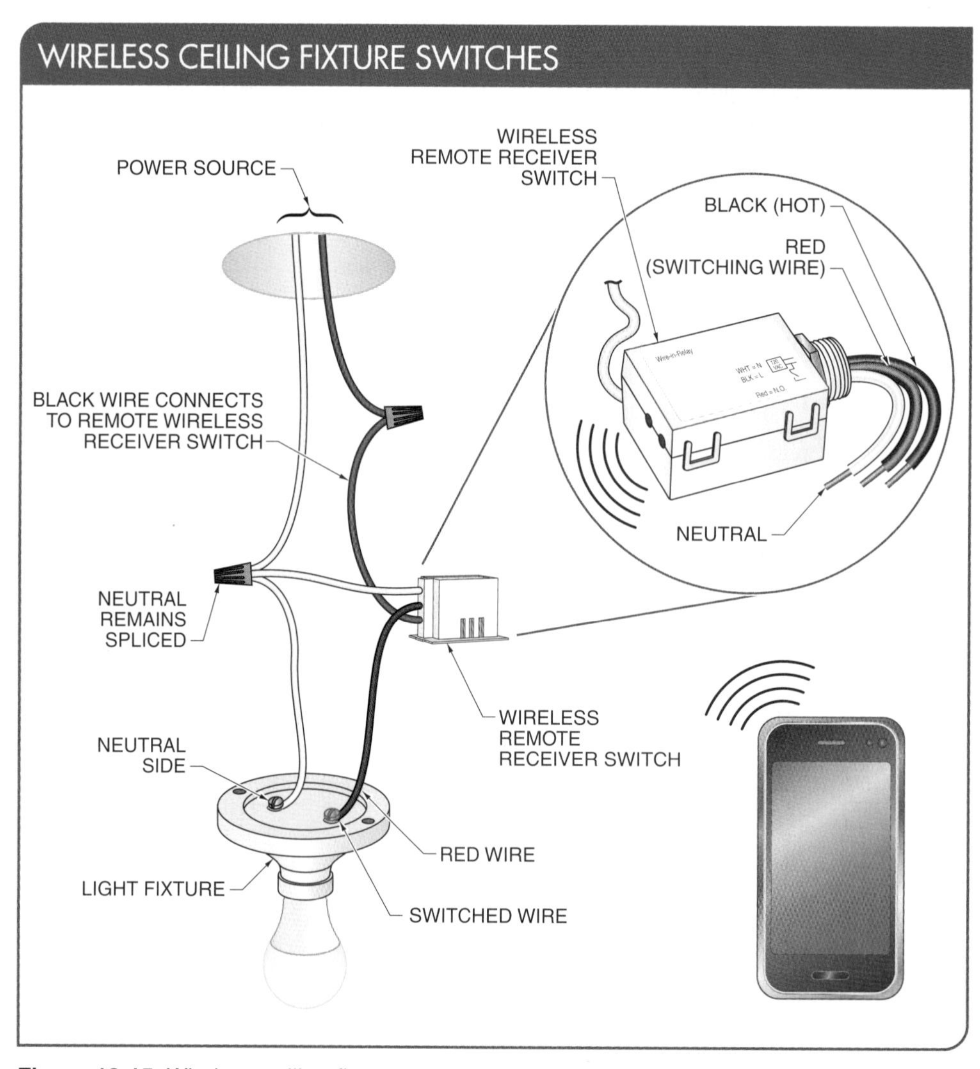

Figure 13-15. Wireless ceiling fixture switches can be used to control a light fixture using a smartphone.

Installing Connected Wireless Wall Switches

A wireless external switch enclosure can be used if a wall switch enclosure is not available or another switch needs to be added. **See Figure 13-16.** To mount a wireless wall switch, an adhesive-backed or screw-in mounting plate is fastened to the wall and the wireless wall switch is attached to the mounting plate.

Relay-Controlled Circuits

A control relay may be needed when large appliances or devices that draw power are installed in a residence. A wireless switch can be used to control an electromechanical relay or a solid-state relay. A wireless relay switch can be used to remotely control a water heater. **See Figure 13-17.**

To wire a relay-controlled circuit, a 10-3 nonmetallic-sheathed cable is wired from a double-pole circuit breaker inside the service panel to the control relay to provide a 240 V, 30 A circuit. A remote-controlled wireless switch is hardwired from the service panel using a 10-2 cable and is connected to the control relay. The wireless switch can be remotely activated to the ON and OFF positions from a smartphone. Finally, all ground wires should be spliced and bonded to the appropriate junction and receptacle boxes.

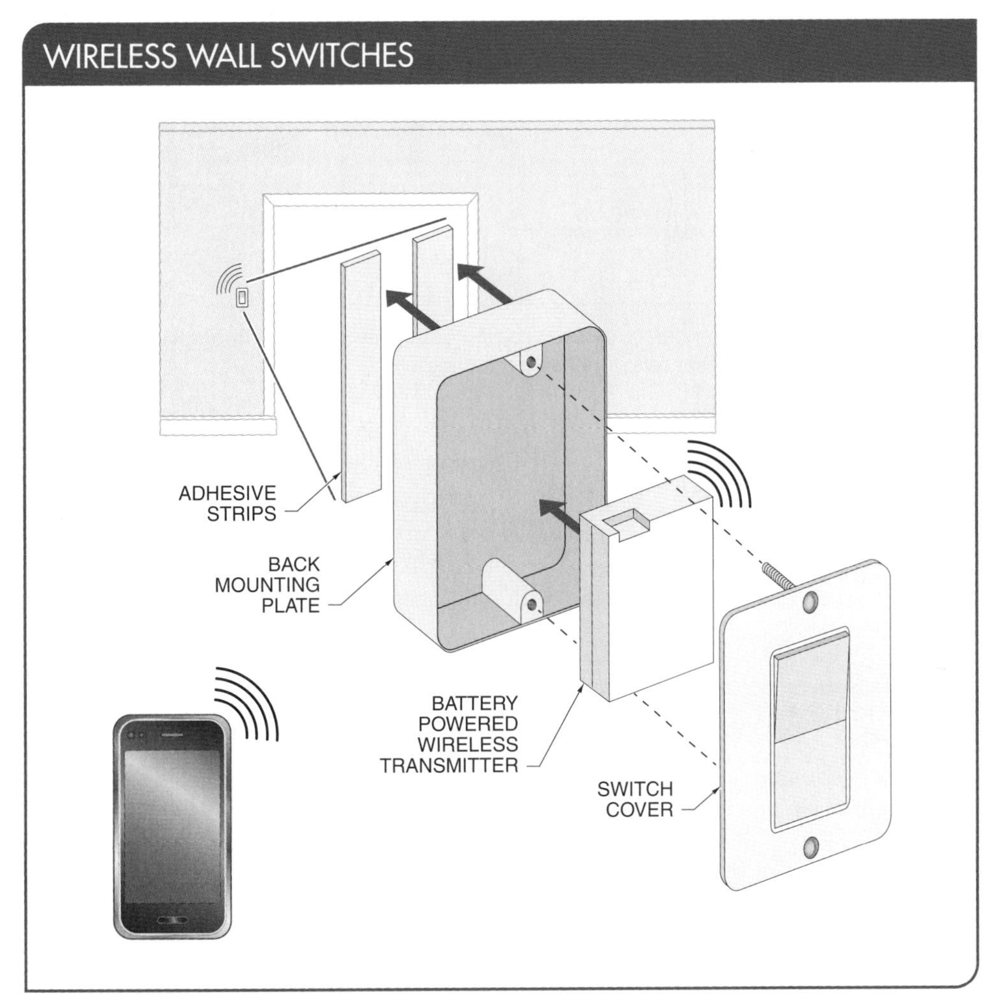

Figure 13-16. A wireless external switch enclosure can be used if a wall switch enclosure is not available or another switch needs to be added.

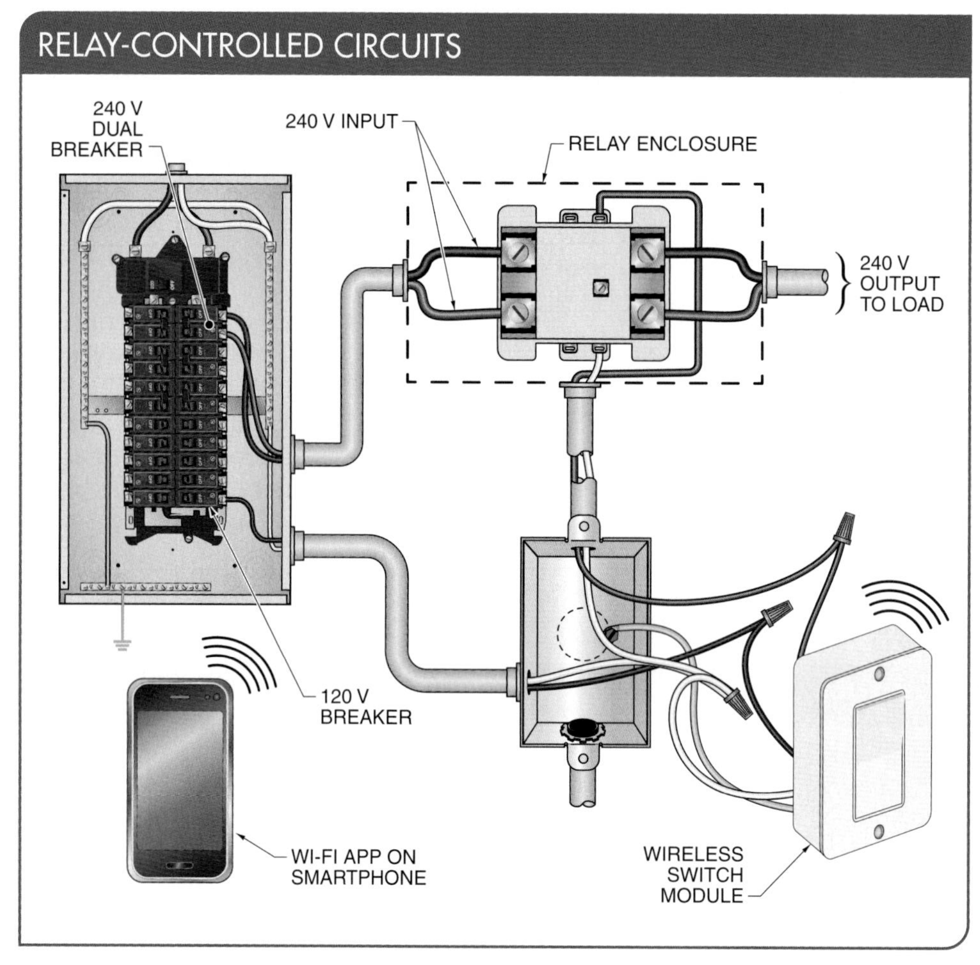

Figure 13-17. Large appliances or devices can be controlled remotely through an electromechanical relay or a solid-state relay and a wireless switch.

SECTION 13.4 CHECKPOINT

1. What are the disadvantages of using smartphones as controllers and what action is the Wi-Fi Alliance taking to address these disadvantages?
2. What is a smart bulb?
3. How does a smart plug control a lamp?
4. How is a remote-controlled wireless switch hardwired?

SECTION 13.5—CREATING WIRELESS SMART HOME SYSTEMS

When a wireless smart home system is created, each room should be looked at individually to determine what is needed for safety, security, and lifestyle. A list of devices that would improve each individual room, such as bedrooms, hallways, and the kitchen, should be created. Then, vendors that provide interoperable devices for the desired applications should be researched. In this chapter, several room layouts will be used to show wireless switches and outlets as direct replacements for legacy devices.

Safety Upgrades

Some of the most important safety upgrades are the addition of GFCIs and AFCIs. In a legacy home, the addition of GFCIs and AFCIs is a direct replacement upgrade. In new construction, these receptacles are required by the NEC®. AFCIs are required in bedrooms, dining rooms, living rooms, and other habitable areas. **See Figure 13-18.**

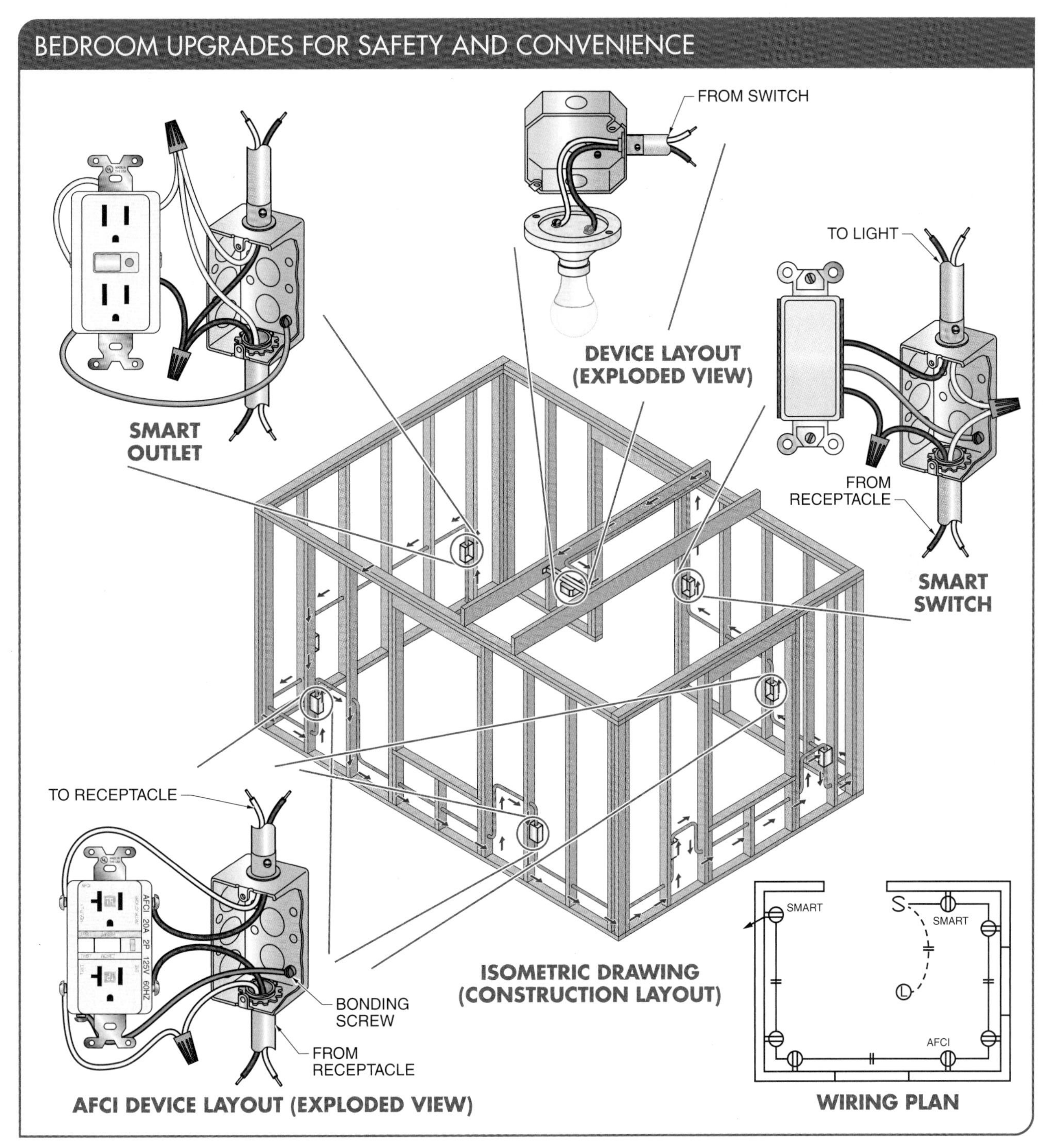

Figure 13-18. Upgrades to the bedrooms of legacy dwellings can include GFCI and AFCI receptacles, which are required for new construction.

Convenience Upgrades

For convenience, a wireless switch, a wireless dimmer switch, or a switch for fan control can directly replace a manual toggle switch. Also, a standard receptacle can be replaced with a receptacle that includes USB charging ports. Even though a bedroom switch can be made wireless, the closet switch may remain as a manual toggle switch. Since the closet may not be used much, especially in a spare bedroom, the cost of replacement compared to convenience may not be worth the installation cost. **See Figure 13-19.**

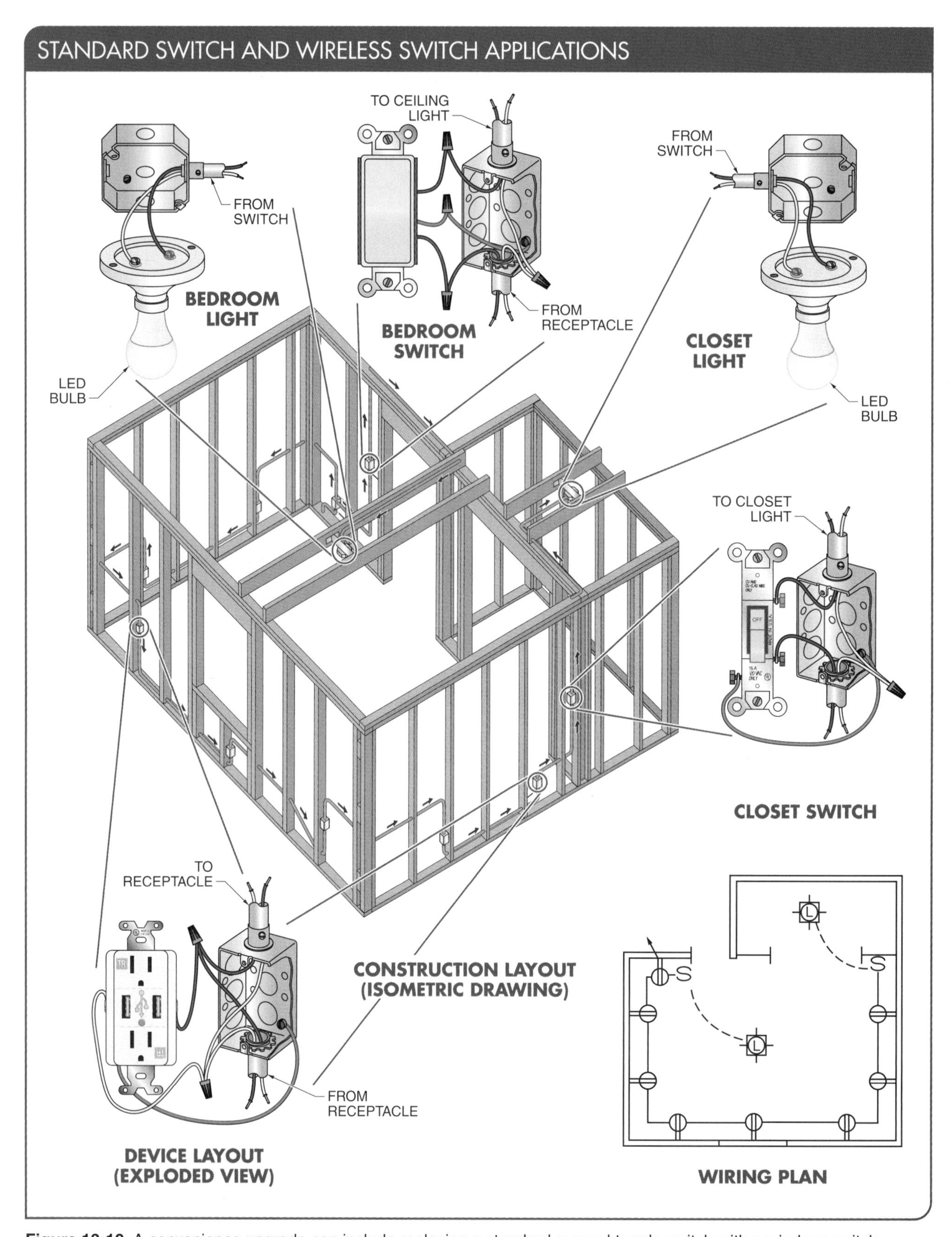

Figure 13-19. A convenience upgrade can include replacing a standard manual toggle switch with a wireless switch.

Convenience upgrades can also be made in the kitchen. For example, in the kitchen, the installation of USB receptacles can provide convenient charging as well as power supplies for some DC applications. **See Figure 13-20.**

> TECH TIP
>
> *USB receptacles have two DC charging ports with 15 A or 20 A outlets. USB receptacles can be conveniently installed around kitchen counters or kitchen islands.*

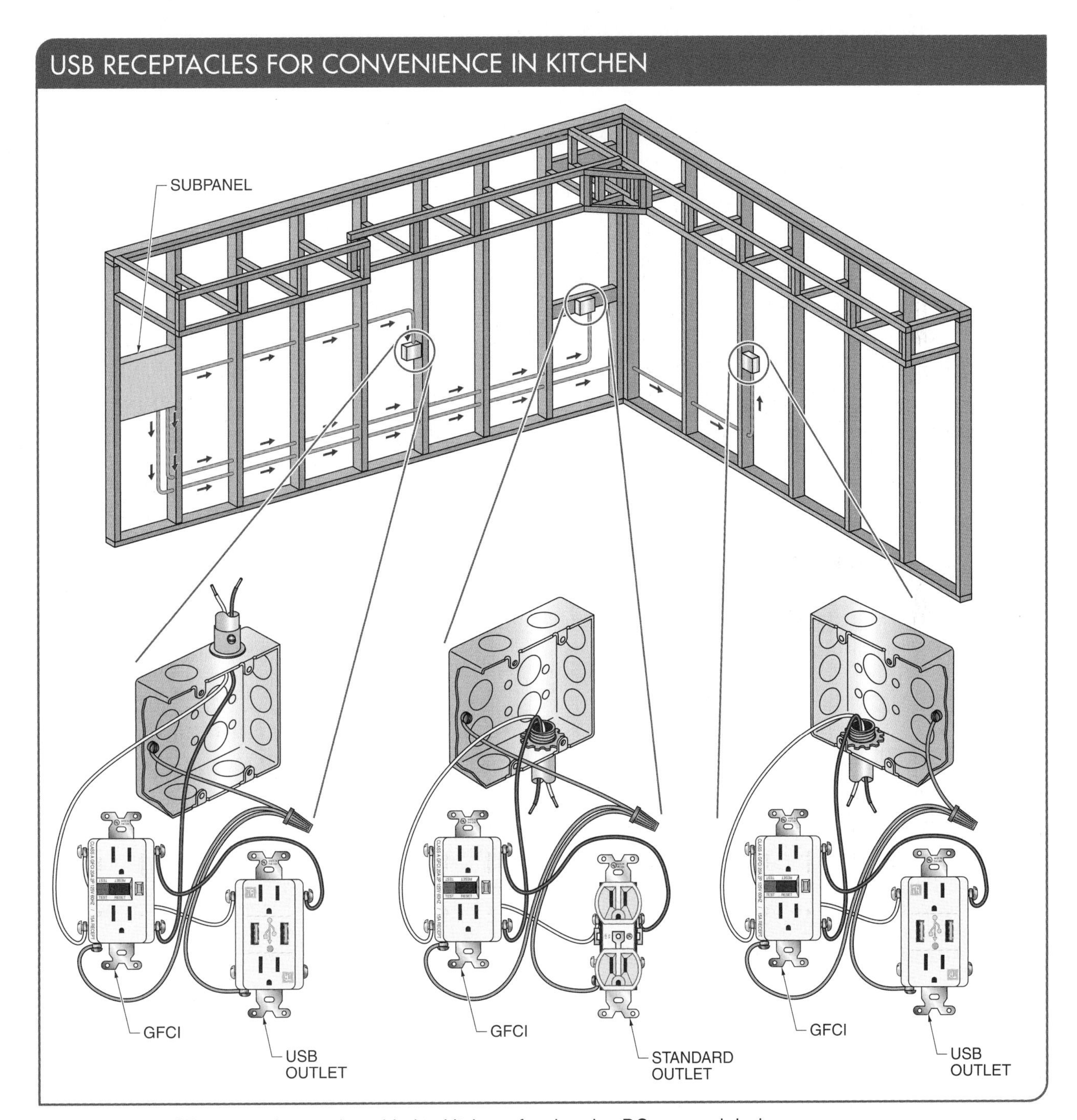

Figure 13-20. USB receptacles can be added to kitchens for charging DC powered devices.

Wireless control of hallway switches can be convenient when an individual is moving throughout a residence at night. This is also true of long hallways or connected hallways. When three-way and four-way switches are replaced with wireless switches, it is important to pay attention to changes in the wiring diagrams. Some wireless switch configurations eliminate a connection to the remote receiver's switch. **See Figures 13-21 and 13-22.**

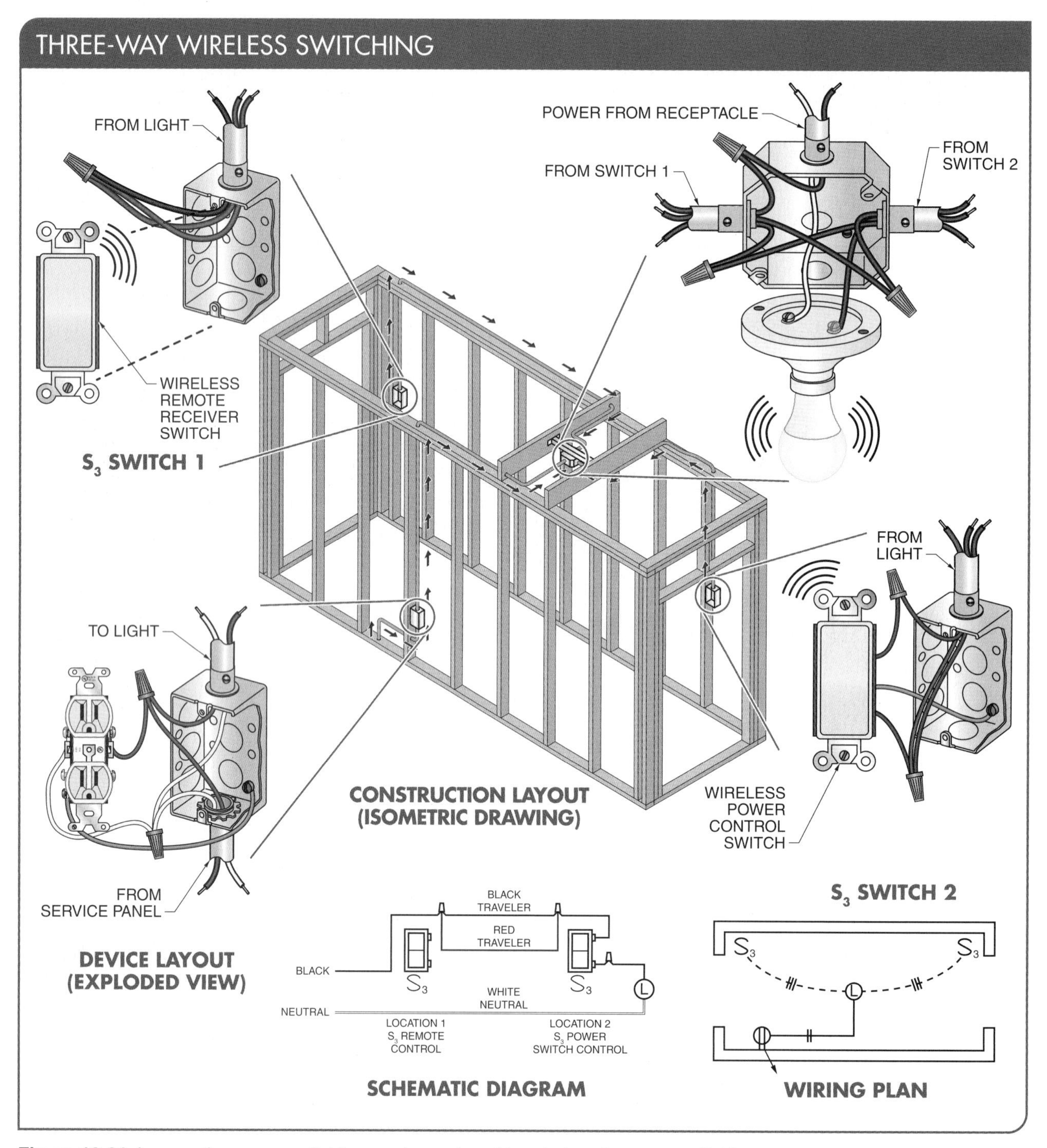

Figure 13-21. Legacy three-way switching can be replaced by wireless four-way switching.

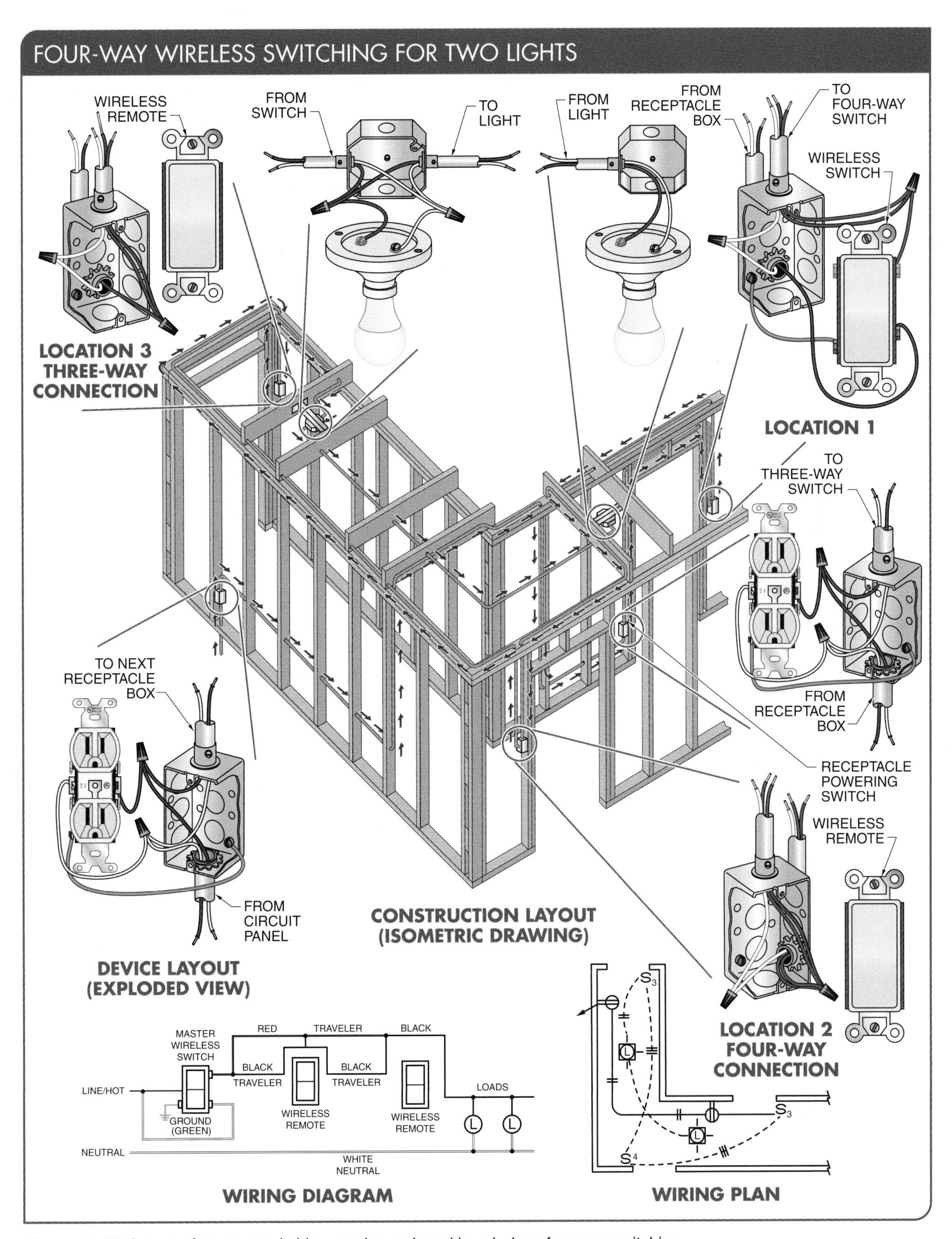

Figure 13-22. Legacy four-way switching can be replaced by wireless four-way switching.

SECTION 13.5 CHECKPOINT

1. When a wireless smart home system is created, what should be considered for each individual room?
2. What are some of the most important safety upgrades?
3. What are some convenience upgrades that can be added to a residence in the bedrooms, kitchen, and hallways?

SECTION 13.6—SMART HOME HUBS

A *smart home hub* is a hardware device that connects multiple devices on a home automation network and controls communication between each device. Although the terms "smart controller" and "hub" are often used interchangeably, the term "hub" is most often used in the smart home industry.

Smart Home Hub Operation

The operation of a smart home hub is relatively simple. A signal or command is sent from a smartphone through a router to the hub and then to the smart device that is being controlled. **See Figure 13-23.** The main reason for adding a smart home hub is to reduce the strain the addition of new devices puts on the wireless network.

Selecting the right smart home hub is important for the best operation of a smart home system. A smart home hub will determine the present capabilities of a smart home system, as well as future expansion of the system. A smart home hub is the interface for most of the wireless devices inside the home. With this interface, hub-connected devices should be capable of being activated remotely through a personal computer, tablet, or smartphone.

A smart home hub uses the apps associated with the appropriate types of input to control an output. Most smart home hub functions are able to be updated by software when new features are developed. Due to its importance, the quality and the ease of use of the software menu is an essential factor to consider when choosing a system.

There are other key features of a home automation system that homeowners should consider when selecting a smart home hub. **See Figure 13-24.** Features such as low power consumption, strong signal strength, data security, interoperability, and smartphone compatibility should be considered.

Homeowners can also make a choice concerning how involved they want to be in the creation and set up of the smart home system. The best smart home hubs combine hardware and software in order to pull everything into one place for interoperability and intuitive control. In many cases, the more sophisticated systems require a professional installation. There are a number of manufacturers that design dedicated smart home systems and some of these hubs can support several smart home technologies. In other words, consumers may be able to mix and match products from different vendors and have all of these features work correctly together.

Smart Home Hub Selection Issues

A major issue surrounding smart home hubs is how manufacturers use hubs to control their devices. For example, some manufacturers of light bulbs require the use of their proprietary hub. As another example, one refrigerator manufacturer now includes a proprietary modem, controller, and hub for control.

In some cases, smart home manufacturers lock homeowners into purchasing only their devices if the hub is proprietary. This can complicate the setup of future home automation if the manufacturer does not make devices for each of the many applications needed in the home. For this reason, a homeowner must thoroughly research the smart home hub or hubs that will be used, especially since new devices are added as the industry grows.

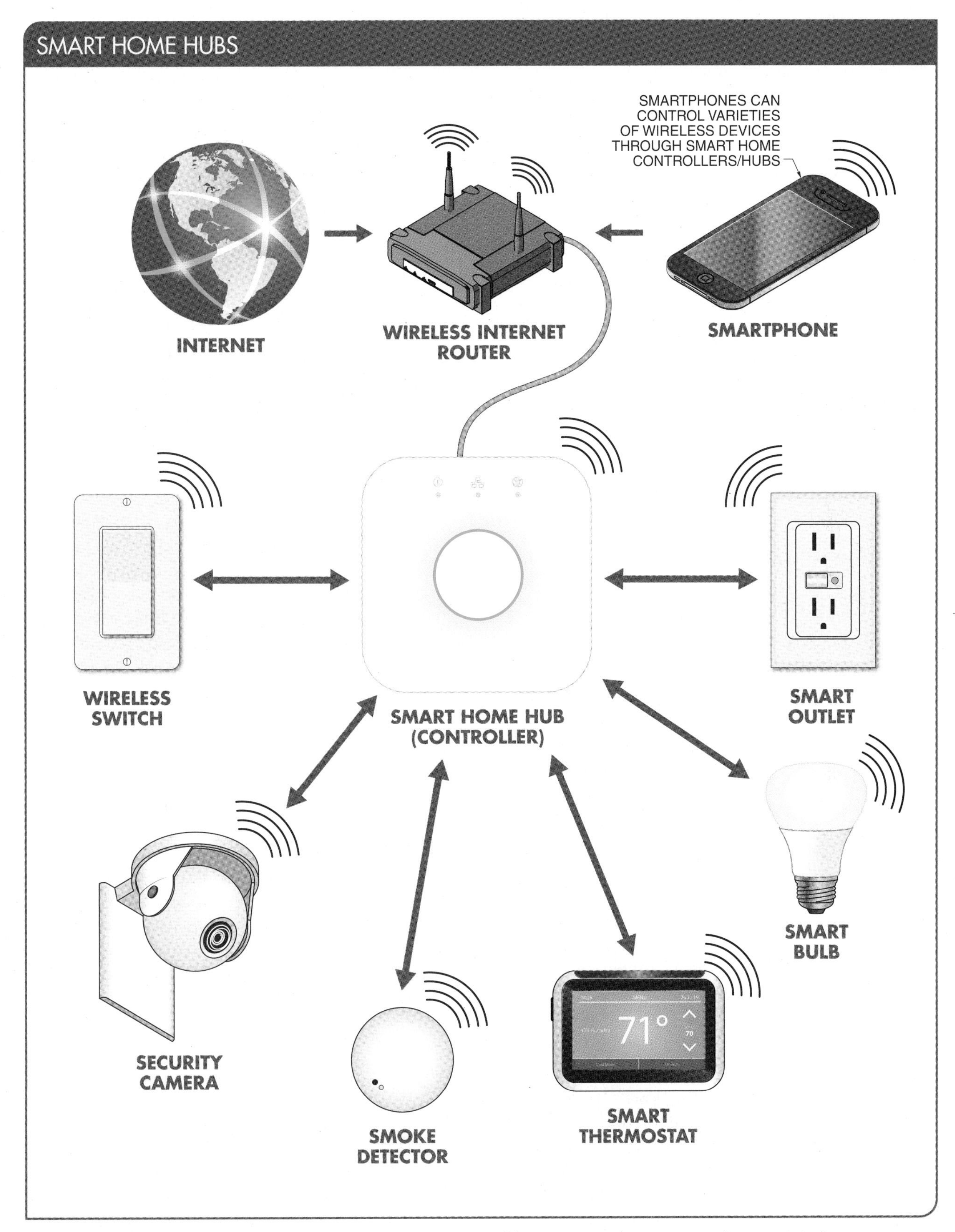

Figure 13-23. A smart home hub can send a signal or command from a smartphone to control a smart device.

KEY FEATURES OF HOME AUTOMATION SYSTEMS	
Low power consumption	Wireless transmitter and receiver should require very little power so that devices could go for months or even years without a battery replacement or recharging. Devices may also be powered by energy harvesting.
Strong signal strength	Signals would pass through walls and floors easily inside and outside a home without interfering with other wireless networks.
Data security	Signals would be encrypted for security reasons and the user would be able to easily add devices to the network.
Interoperability	All the devices on the same network would be able to "talk" to each other with complete interoperability.
Handling capabilities	The control system software (protocol) should be able to handle dozens or even hundreds of devices on a single network. The homeowner should be able to easily add devices to the network.
Central display menu	Display system will provide centralized menu for smartphone or smart home control display.
Cell phone compatibility	Software system should be compatible with Android and IOS smartphones as well as other wireless input devices.

Figure 13-24. Certain key features should be researched before a smart home hub is purchased for a home automation system.

Voice-Activated Virtual Assistants

Advances in voice-activated virtual assistants have blurred the communication lines between connected devices and smart home hubs. Voice-activated virtual assistants can be used to control a single device, such as a smart thermostat or a smart TV. They can also be used to control multiple devices in a smart home system. **See Figure 13-25.**

A voice-activated virtual assistant eliminates the need to carry a smartphone from room-to-room in a home. The strategic placement of a voice-activated virtual assistant throughout the home allows many devices to be controlled through spoken words, such as lights to be turned on, thermostats to be set, and music to be played on demand. The smart capabilities of voice-activated virtual assistants are developing rapidly, as they increasingly are used as solutions for specific situations in smart systems.

Setting Up a Smart Virtual Assistant via Smartphone. Most single smart devices connect only to a Wi-Fi router without the need of a smart hub. This makes the installation process easy for those individuals who are just learning how to use smart devices. This simple process usually involves plugging in the device, downloading and opening the device's companion app on a smartphone or tablet, and then connecting the device to the Wi-Fi network via the app. From there, the device is ready to use. The only major difference when setting up a smart voice-activated virtual assistant is that after the app is downloaded onto the smartphone or tablet, the virtual assistant must be turned on so it can configure or prepare itself for the app download.

Selecting Smart Devices

When a smart device must be selected for purchase, the buyer must be aware of the version of IEEE 802 standard that the device meets. Some manufacturers include their own proprietary protocols in a device. These modifications to open standard protocols are created by individual manufacturers and may not be endorsed by their associated alliance because of their lack of interoperability.

TECH TIP

Security can be added to voice-activated devices during setup configuration and by purchasing a Wi-Fi router to protect against malware. It is important not to use the voice assistant to remember private information such as passwords or credit card numbers.

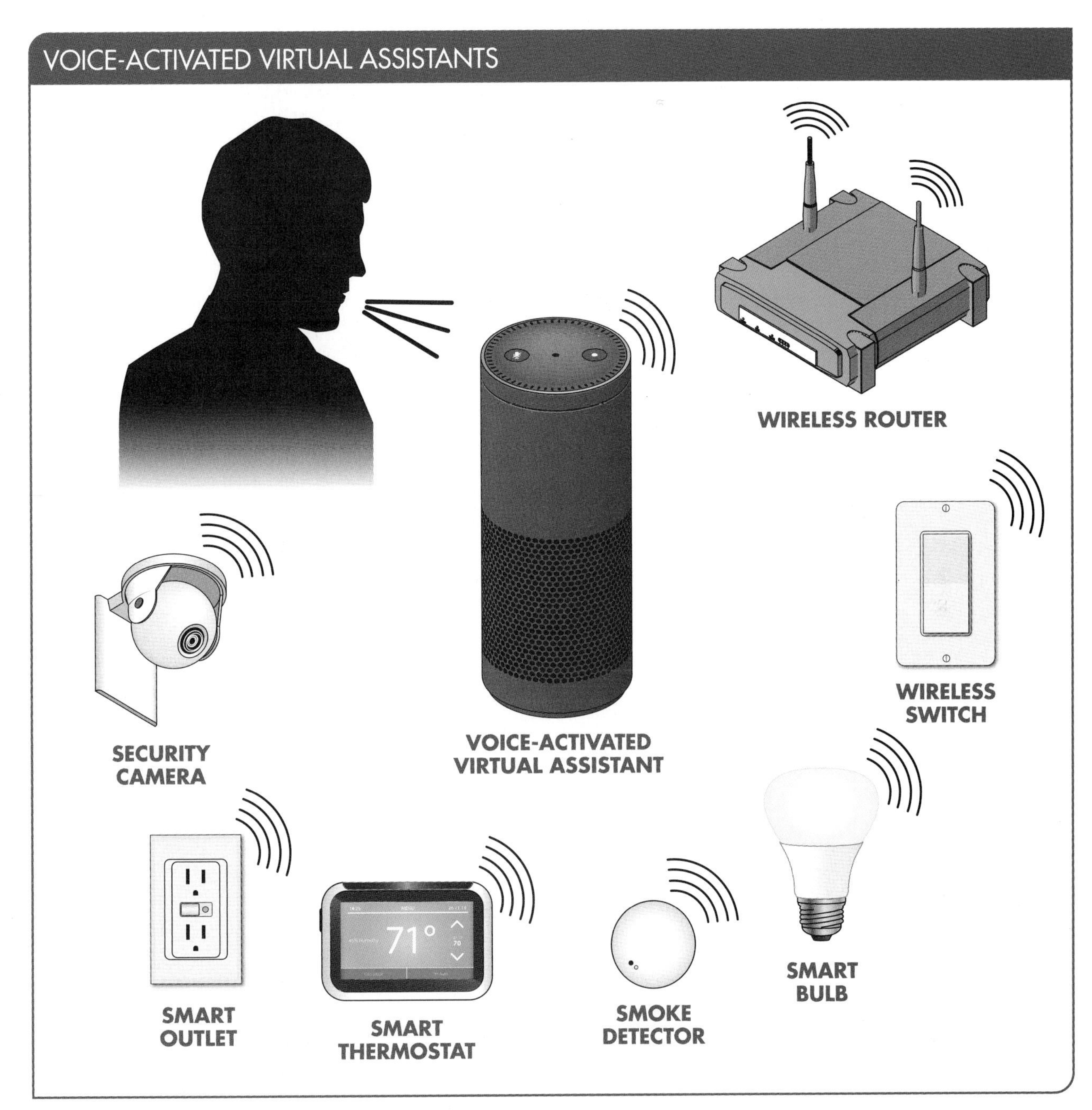

Figure 13-25. Voice-activated virtual assistants can be used to control a single device or to control multiple devices in a smart home system.

For example, although Z-Wave and Zigbee are two of the largest alliances, individual manufacturers may have made minor protocol changes in their own proprietary devices. While any Z-Wave device should be able to communicate with any other Z-Wave device, this may not always be the case if minor protocol changes were implemented by a specific manufacturer of those devices. The following guidelines can help an individual choose compatible devices:

- Look for official manufacturer support lists.
- Visit the alliance network website for the latest updates.
- Look at the product packaging for a label that indicates what the device supports or is compatible with.
- Check the device manufacturer website or perform an internet search about the product/brand for further information if the product box does not provide any information on what it supports.

Smart devices with proprietary systems can be a viable choice. Sometimes a single proprietary system used throughout a smart home can be reliable and easy to manage. A proprietary system can also keep the use of smart hubs to a minimum. Like all systems, proprietary systems have some issues to consider. For example, the choices for adding to a proprietary system are limited. Also, if the manufacturer does not survive in a competitive market, an entirely new system may need to be purchased. Third-party manufacturers may claim compatibility or interoperability that does not work.

SECTION 13.6 CHECKPOINT

1. What is a smart home hub?
2. What can be controlled by voice-activated virtual assistants?
3. What must a buyer know when purchasing a smart device?

SECTION 13.7—MESH TECHNOLOGY

Mesh technology has been introduced at two levels, namely at the device level and more recently at the router level. *Mesh technology* is a system that extends the distance that a signal can travel within the system without increasing the power of each transmitter. Manufacturers introduced mesh technology into the home automation market as a way to improve the quality and distance of a wireless signal. When mesh networks are proprietary, interoperability must be considered when choosing wireless equipment.

TECH TIP

Mesh networks tend to be more static because their nodes do not move much, allowing data to hop from one device to another until it eventually reaches its destination. A smartphone signal from the backyard may travel through a mesh network to control the garage door opener.

Mesh Technology Operation

Each wireless device in a residence has a transmission range. Many of these devices can only transmit a signal for short distances. Mesh networks have moved into residential applications as the need for smart home technology has grown. A mesh network is made up of small RF transceivers called "wireless mesh nodes" that use existing Wi-Fi standards to communicate between each other. Although devices may use the same frequency (carrier), they may use a different protocol (language). This would not allow devices to communicate between mesh networks.

In a wireless network, all transceivers are on the same radio frequency and each can potentially receive each other's transmitted messages and pass them on. The mesh network takes advantage of the fact that many of these transceivers (nodes) are relatively close to each other. This allows the overlapping transmission ranges of the RF transceivers to form a Wi-Fi mesh that connects them all. **See Figure 13-26.** Any transmitting node only needs to be in transmission range to reach the nearest node to receive and transmit the same signal. Each node is connected to multiple other nodes, and messages travel in a variety of paths to reach their destinations.

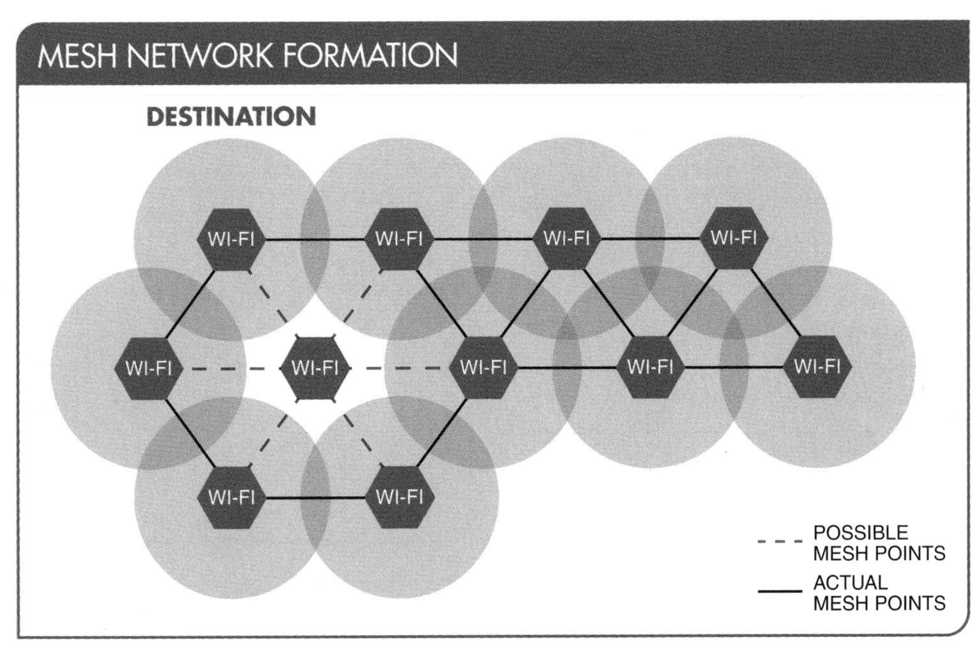

Figure 13-26. In a wireless mesh network, all transceivers are on the same RF grouping and each can potentially receive each other's transmitted Wi-Fi signals and pass them on.

In a dwelling, multiple Wi-Fi nodes can work together as a mesh network to cover the entire residence with a strong wireless signal to send data. The strength of a mesh network is that each node can operate as a simple repeater. If a node receives a message that is not for itself, it passes the message on to its neighbor. Through multiple repeats or "hopping," the message reaches its destination. Technically, the more devices that are in a residence, the more powerful and stronger the network. However, all systems do have a maximum number of "hops" before the signal is too weak to be useful.

An example of a residential use of a mesh network is a smartphone that is being used in the backyard of a residence to send a signal to the garage door. The signal is first sent by the associated app on the smartphone to the mesh network, which then sends the message from node to node until it reaches the garage door opener. If mesh technology were not used, the signal may not be strong enough to reach and activate the opener.

Mesh Routers

A residence with a traditional Wi-Fi router may have "dead zones," where there is no signal or a weak signal. A router based on mesh technology can help eliminate these dead zones. Also, mesh technology devices can prevent or minimize interference by automatically switching channels when communicating with each other.

While a traditional Wi-Fi router transmits a signal from a single location, a mesh router utilizes the multiple nodes in a system to send and receive signals. One node links to the modem and acts as the mesh router. Then, this node, acting as the router, sends its signal to one or more other nodes, which receive the signal and send it out again. This can eliminate the dead zones associated with signals sent from a single location.

Mesh routers can significantly improve the wireless Wi-Fi coverage of large residences with many rooms. Coverage depends on the manufacturer of the mesh node/router, which, in most cases, is approximately 1500 sq ft for one device. Depending on the size of the residence, one, two, or three extra devices may be necessary. Each device also requires AC power. A mesh network system can be started by simply connecting the node that will act as the mesh router to the modem and connecting the additional nodes in rooms or areas where coverage is needed.

Since smart devices, such as smartphones, smart watches, and tablets, are used from room to room, mesh networks will increasingly be used in smart home systems because they can accommodate this movement. Companies, such as Netgear, Linksys, Google, eero, and Luma, are releasing more affordable mesh technology for the home. With the availability of mesh technology for the home increasing, homeowners should thoroughly research the latest options before committing to a purchase.

SECTION 13.7 CHECKPOINT

1. What makes up a mesh network?
2. How does a mesh router system eliminate dead zones?

Chapter Activities

13 Smart Home Infrastructure

Name ______________________________ Date ________________

RF Antennas

Identify the types of RF antennas shown.

__________ **1.** Parabolic dish (satellite)

__________ **2.** Omnidirectional

__________ **3.** Point-to-point

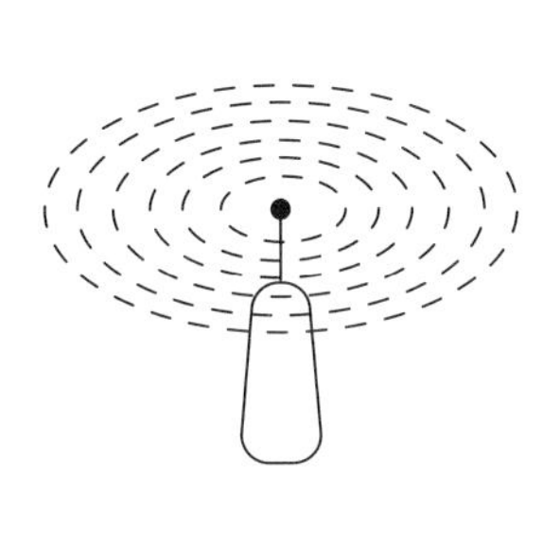

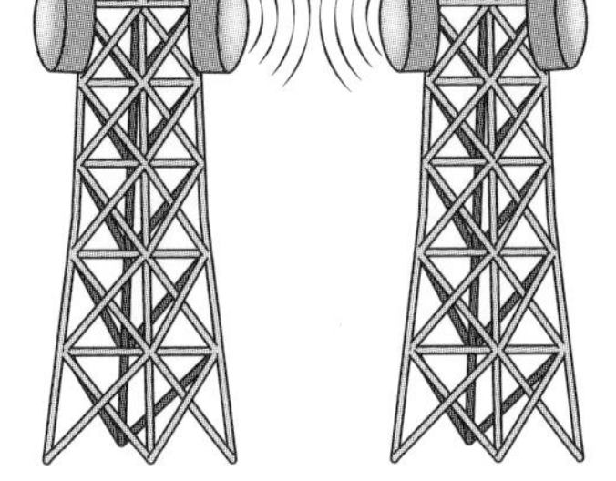

A B C

Wireless Wi-Fi Routers

Identify the types of wireless Wi-Fi routers shown.

__________ **1.** Dual-band

__________ **2.** Single-band

__________ **3.** Tri-band

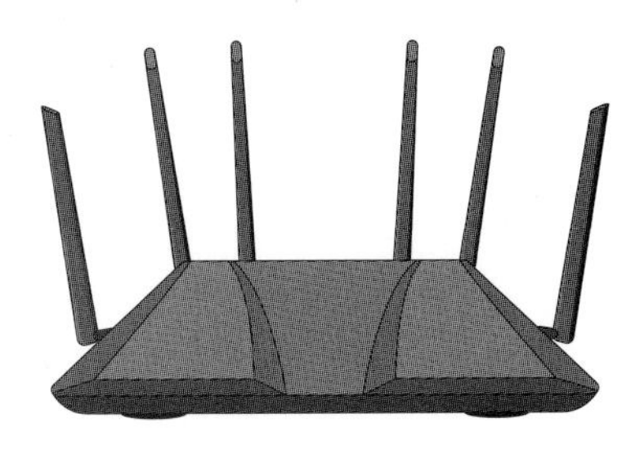

B C

Relay-Controlled Circuits

1. Draw lines to wire a relay-controlled circuit. Indicate the color of each wire.

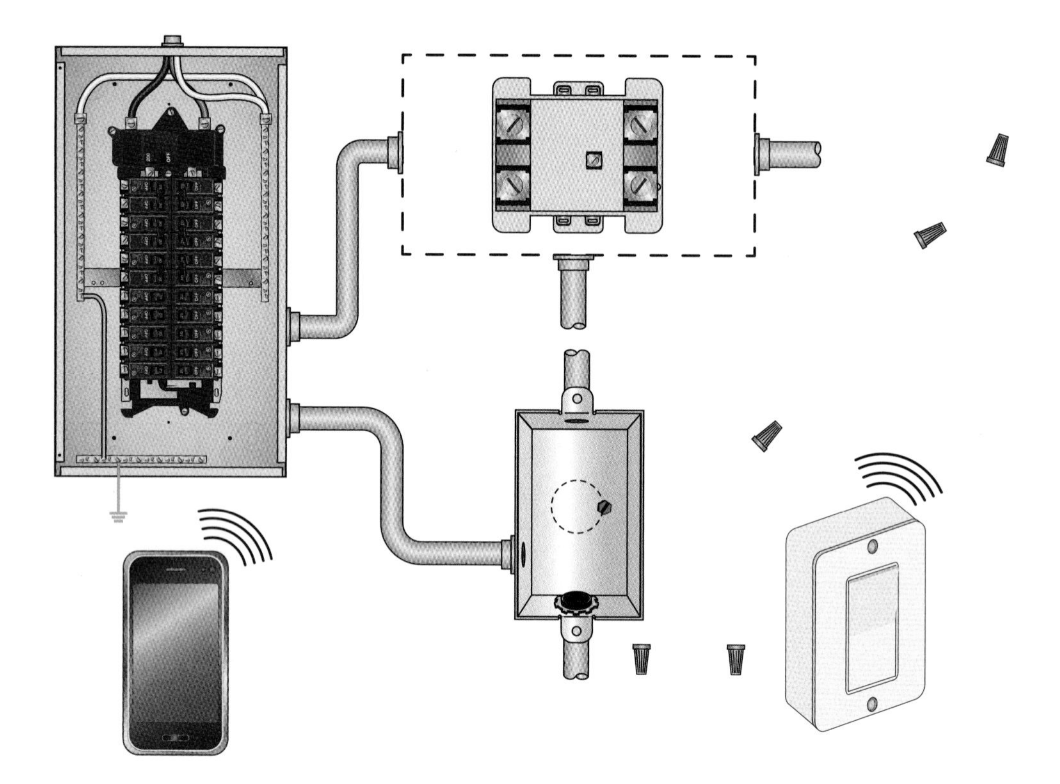

14 Security Systems and Smart Home Applications

SECTION 14.1—LEVELS OF SECURITY

- Identify the three levels of security.
- Describe the four key components of a smart home security system.
- Describe three common display technologies used for user interface devices.

SECTION 14.2—PERIMETER SECURITY

- Describe perimeter security.
- Describe motion detectors.
- Explain outdoor lighting.
- Describe smart doorbells.
- Describe pool safety systems.

SECTION 14.3—INTERIOR SECURITY

- Describe interior security.
- Identify types of interior sensors and detectors.
- Describe door/window sensors.
- Identify types of pressure sensors and garage protection.

SECTION 14.4—ENTRY/EXIT SECURITY

- Describe entry/exit security.
- Describe keypads.
- Describe smart locks.
- Describe keyless systems.
- Define "biometric verification."

SECTION 14.5—VIDEO SURVEILLANCE

- Define "video surveillance."
- Describe security camera types.
- Explain security camera placement.
- Describe outdoor movement filtering.
- Describe cloud storage.

SECTION 14.6—SECURITY SYSTEM INSTALLATION

- Explain the role of a 24/7 security monitoring service.
- Explain how to plan the installation of a security system.
- Explain how sensors can be mounted.

Learner Resources

ATPeResources.com/QuickLinks
Access Code: 791712

A smart home security system provides methods of verification and surveillance and senses activity in and around a dwelling to provide different levels of security. Smart home security systems can be purchased as do-it-yourself (DIY) systems or professionally installed systems. Smart home security planning includes reviewing systems and devices, identifying areas that need protection, and considering accessibility and ease of use for individuals and families.

SECTION 14.1—LEVELS OF SECURITY

Smart home security systems can provide three different levels of security that monitor the outside of a property and the inside of a dwelling. Each level of security provides varying levels of protection and monitors whether the homeowner is on the premises or away. The three levels of security include perimeter security (monitoring property), interior security (monitoring the inside of a dwelling), and entry/exit security (monitoring for authorized/unauthorized entry/exit). **See Figure 14-1.**

A comprehensive smart home security system may incorporate all three levels of security or a combination of them. Additional security devices can be added to secure and monitor access to cabinets for firearms or valuable possessions.

Smart Home Security System Devices

A smart home security system includes four key devices: a control panel, input sensors and detectors, output devices, and interface devices. The control panel is considered the brain of a smart home security system because it manages the system. All input sensors and detectors, output devices, and interface devices are connected to the control panel. **See Figure 14-2.**

Control Panels. A control panel receives information from the input sensors and detectors as well as the interface devices. This information is either stored in the control panel for further processing or used to activate an output device. When the control panel is put into service, it is either set up by the manufacturer or programmed by the installers. The control panel should be installed so it is accessible for service but not readily accessible to other people. Control panels are often located in a basement or in a locked closet away from an entry point.

A hardwired smart home security system is powered by a DC power supply and has a back-up battery system. A wireless smart home security system also uses DC power, but it differs from a hardwired system in that each sensor may require its own battery.

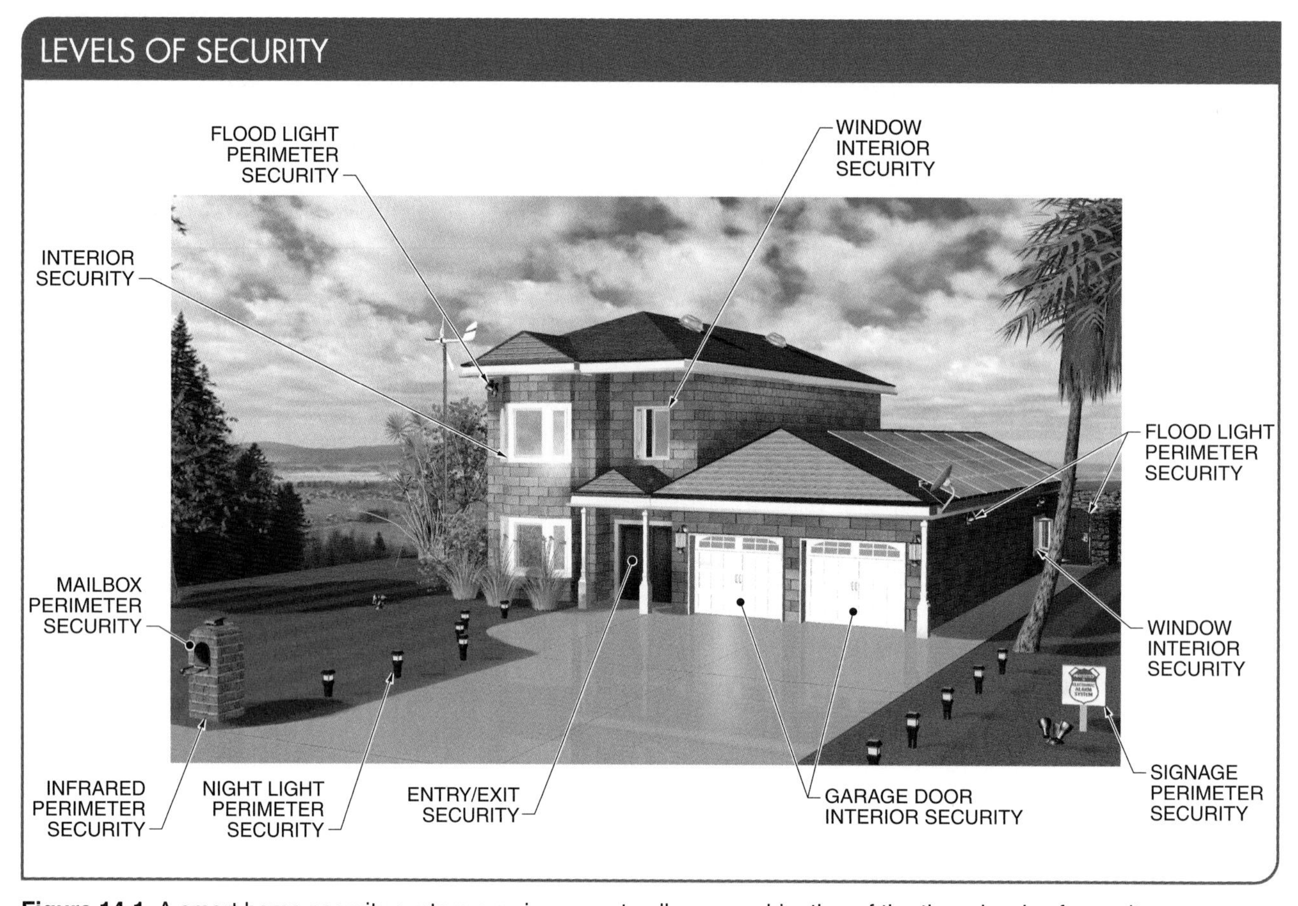

Figure 14-1. A smart home security system may incorporate all or a combination of the three levels of security.

Input Sensors and Detectors. Input sensors and detectors are designed to identify changes in the environment, such as sound, heat, light, temperature, moisture, and motion. The information gathered by a sensor is processed by the electronic circuitry within the detector and sent to the control panel for processing.

Output Devices. Traditionally, the output devices for security systems were standalone devices such as sirens, bells, and flashing lights. Today, smart home security systems can include combined input and output devices (input/output devices). **See Figure 14-3.** For example, an outdoor light may also be a motion detector, camera, speaker, and siren, all in one device. With this combination, the homeowner can see and verbally communicate with a visitor, as well as activate an alarm if needed, from anywhere there is a Wi-Fi connection. Text message notifications can be received via email, smartphones, tablets, computers, and smart TVs.

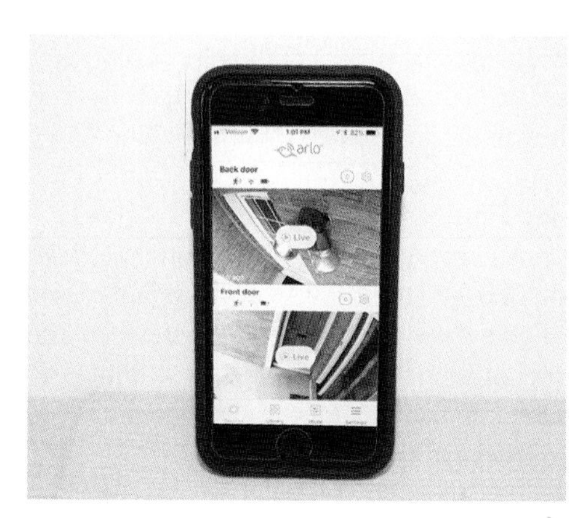

Most home security camera systems stream directly to a smart device, such as a smartphone, for the uninterrupted monitoring of a property.

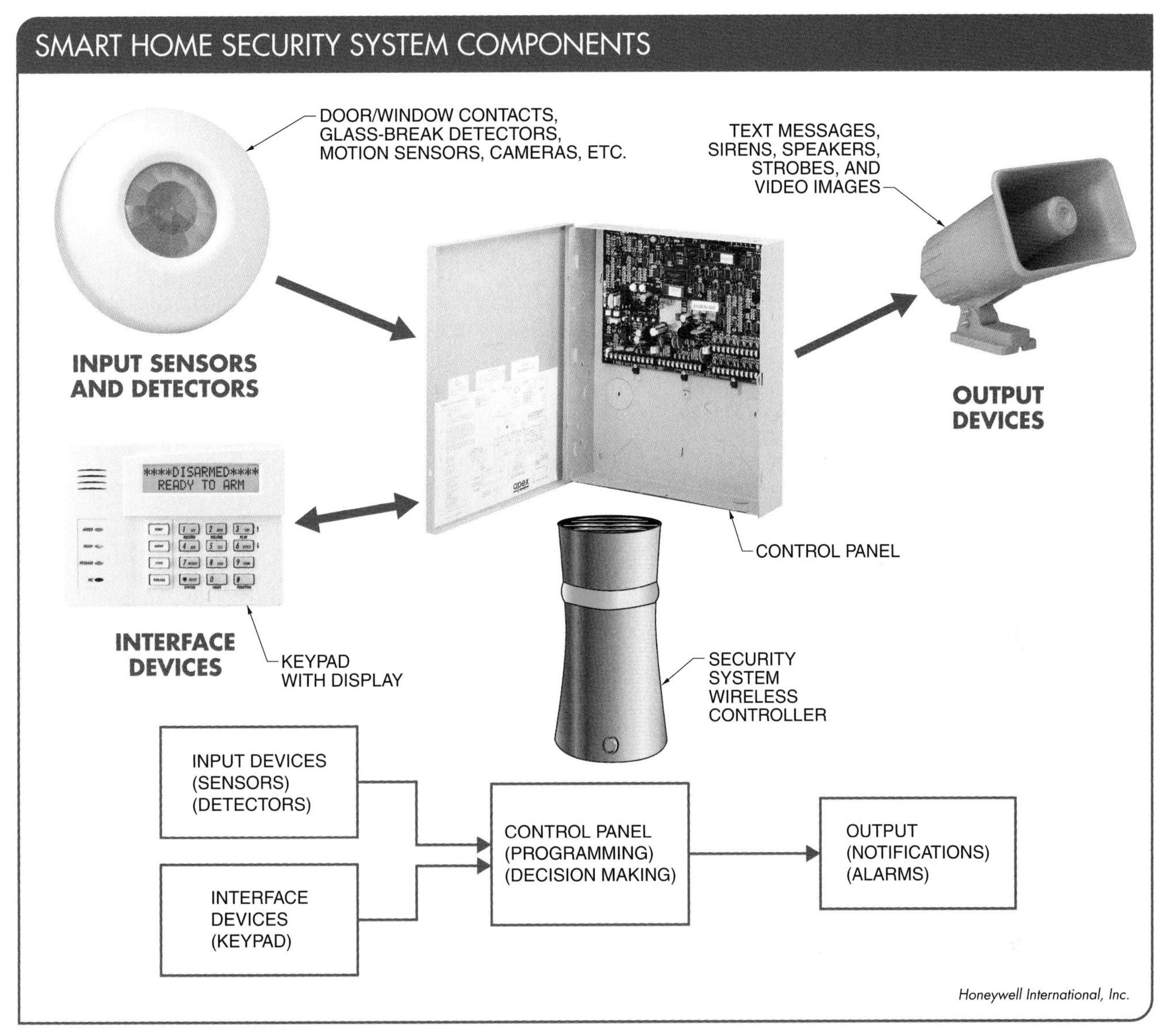

Figure 14-2. All input sensors and detectors, interface devices, and output devices are connected to the control panel.

Interface Devices. A smart home security system user interface device is used to launch one or more applications. The most common user interface devices are keypads, digital panel displays, smartphones, and tablets. Interfaces can be customized to display information about the condition in the residence or to control other smart devices. Some user interface devices require programming by a professional to ensure that the displayed information is easy to understand.

In most security systems, information regarding the status of the system is provided through a digital panel display. A *panel display* is a thin, lightweight screen used to project images. Smart home security systems often include a panel display with a keypad to indicate whether the system is armed or disarmed. A touch-screen panel display allows a user to scroll through different menus. Touch-screen panel displays are often used as the primary user interface and are usually placed in a main part of a house, such as the kitchen. Portable touch-screen panel displays can also be used in bedrooms, family rooms, and home theaters. **See Figure 14-4.**

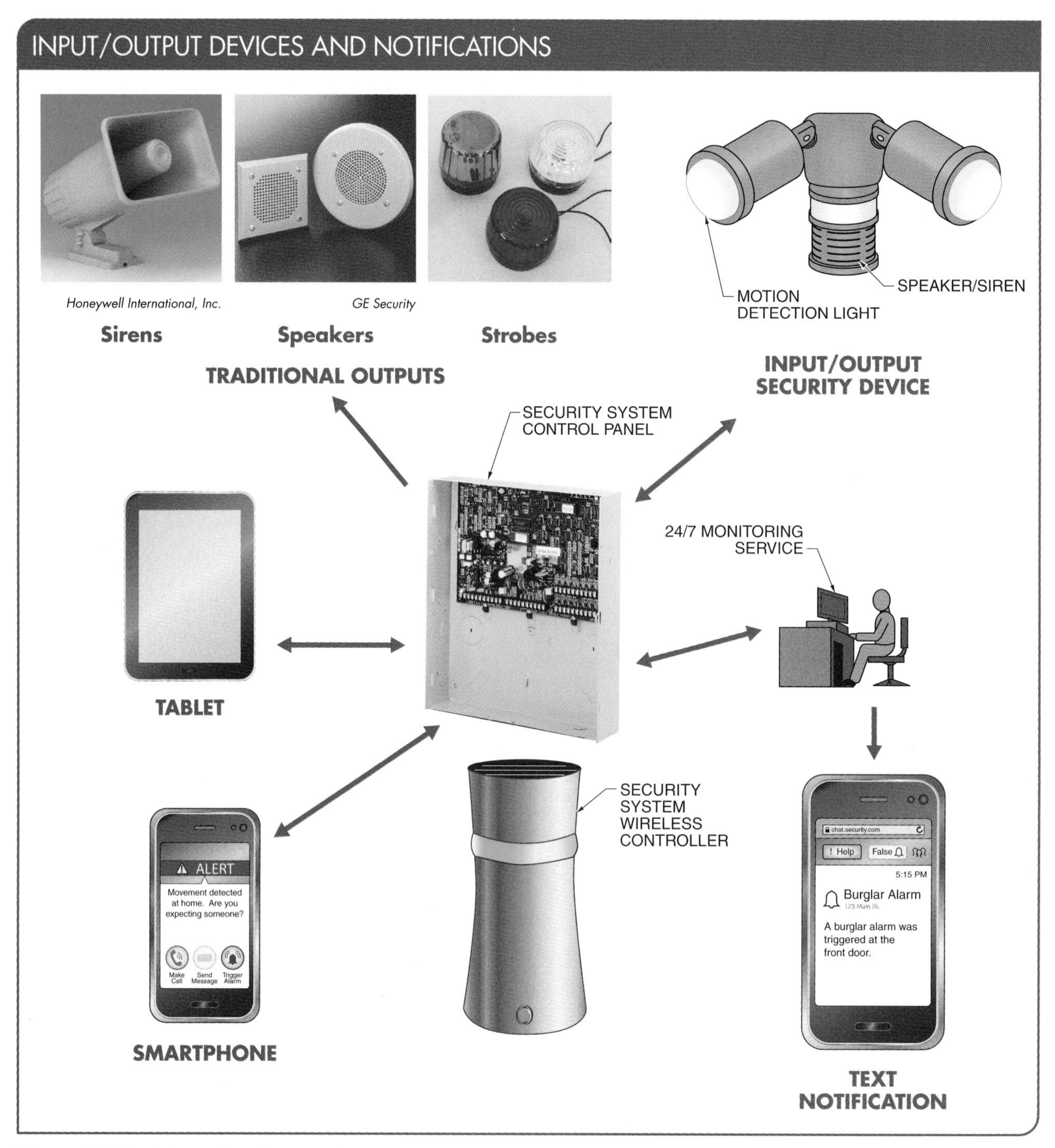

Figure 14-3. Beside traditional output devices, smart home security systems can include combined input and output devices (input/output devices) that can also send text message notifications via email, phone, computer, or smart TV to the homeowner.

> TECH TIP
>
> *OLEDs are organic and provide a natural light source to the display, allowing for OLED screens to be large and lightweight as well as to retain consistent color from wide viewing angles.*

There are several types of display technologies available for user interface devices. The three most common display technologies available today include light-emitting diodes (LEDs), liquid crystal displays (LCDs), and organic light-emitting diodes (OLEDs).

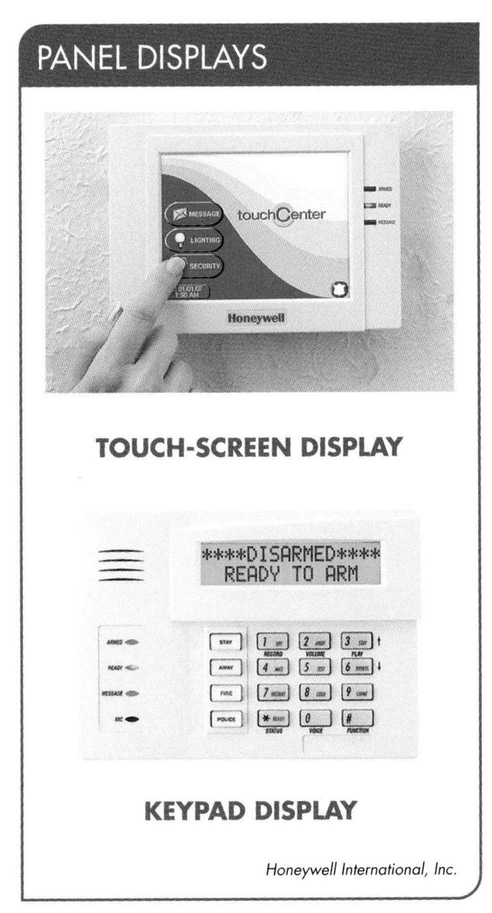

Figure 14-4. Panel displays can be touch-screen displays or keypad displays.

Displays on security panels and devices, such as thermostats, are often presented on a seven-segment display. Each segment of a seven-segment display has several LEDs, but a plastic overlay makes the individual LEDs glow as a unit. For example, the number 3 would be activated on the display by energizing segments a, b, c, d, and g. **See Figure 14-5.**

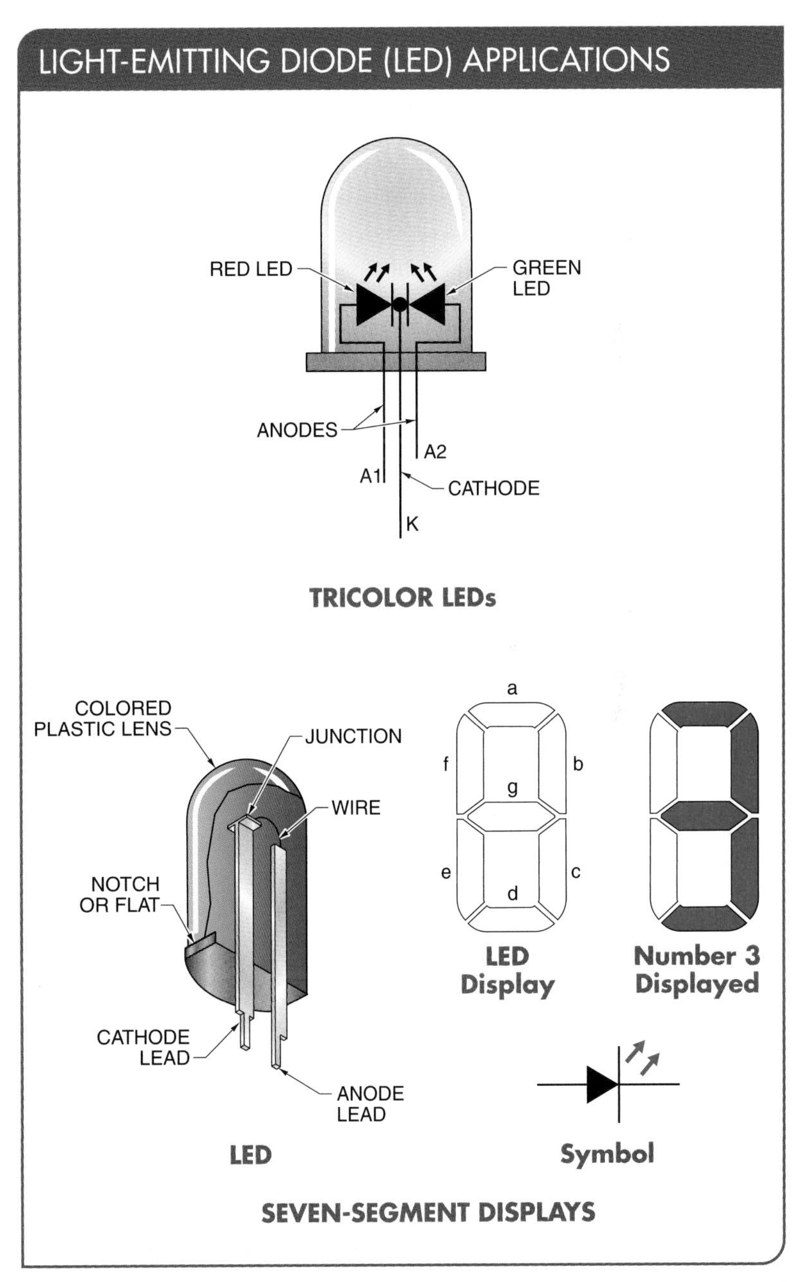

Figure 14-5. A tricolor LED can produce red, green, and yellow light depending on which leads on the LED are connected to a power source. Displays on security panels and devices are often presented on a seven-segment display.

A *light-emitting diode (LED)* is a device that emits a specific color of light when DC voltage is applied across a semiconductor junction. The wavelength of the energy emitted, and therefore the color, depends on the junction materials. The most common LED indicator light colors are red, amber, yellow, and green.

LEDs can also be constructed with multicolor capabilities. LEDs with this capability are called tricolor LEDs because they produce red, green, and yellow light depending on which leads on the LED are connected to the power source. If both the red and green LEDs in a tricolor LED are powered, the combination creates yellow light. LEDs have long life spans of 50,000 hr or more and are resistant to physical shock and vibration.

A *liquid crystal display (LCD)* is a flat, alphanumeric display that uses liquid crystals to display information without directly emitting light. LCDs require less power to operate than LEDs and are less prone than LEDs to wash out under strong ambient light. However, LCDs cannot be read in the dark and must be illuminated by an external light source.

An LCD consists of front and rear pieces of glass separated by a liquid crystal material. The key to LCD operation is the liquid crystal that fills the space between the front and rear glass. The molecules of liquid crystal are normally parallel to the glass. However, when a voltage is applied to a liquid crystal, its molecules rotate 90° to alter the light passing through. **See Figure 14-6.**

TECH TIP

OLED technology was developed by the Eastman Kodak Company in 1987.

An *organic light-emitting diode (OLED)*, also known as a light-emitting polymer (LEP), is a thin film of carbon-based organic molecules that create light with the application of voltage. OLED technology provides clear, bright displays on electronic devices and uses less power than conventional LEDs. OLED displays are available as single-color, multicolor, and full-color displays. OLEDs are thinner than the width of human hair and can be easily flexed and shaped. **See Figure 14-7.**

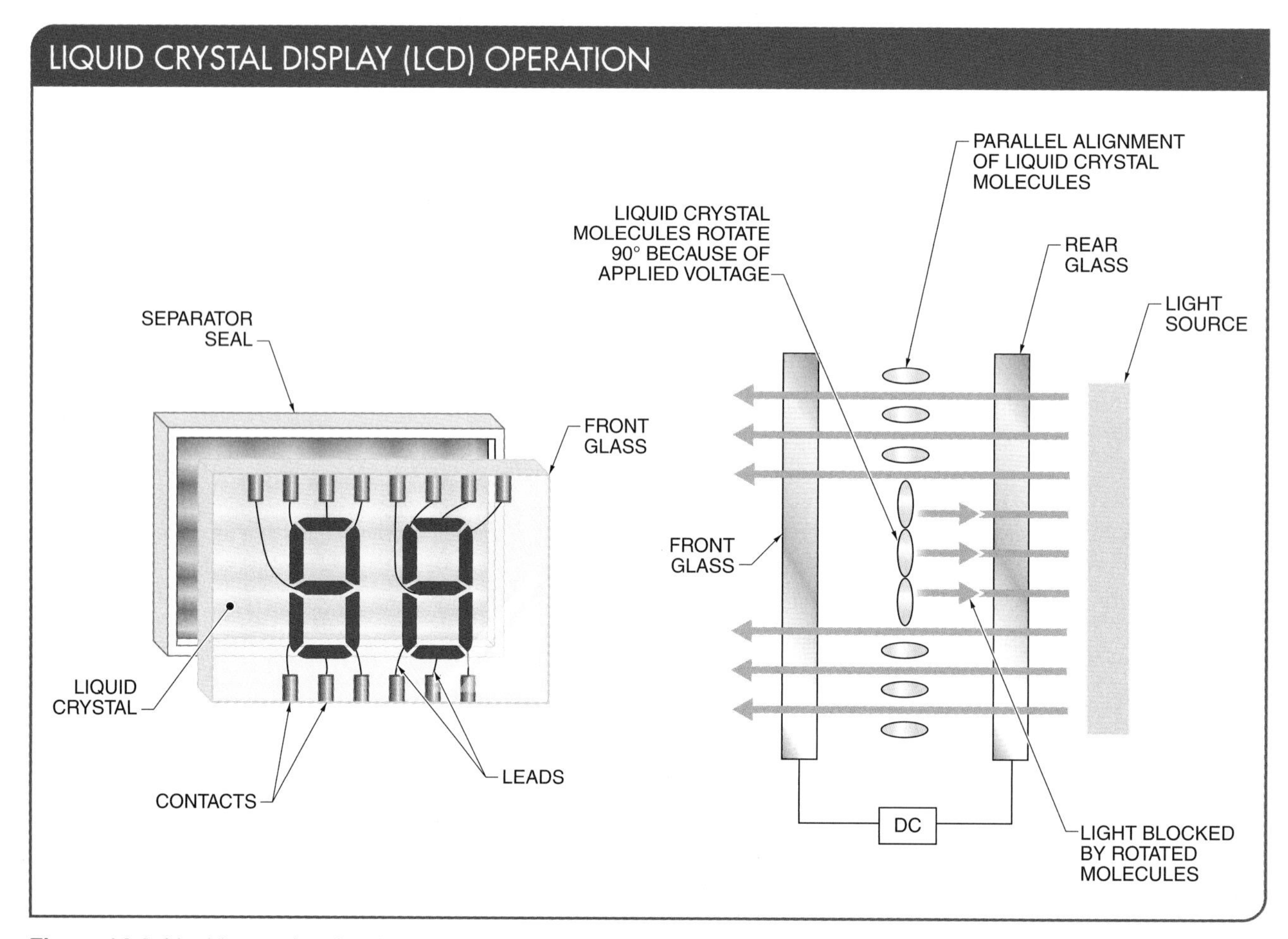

Figure 14-6. Liquid crystal molecules are normally parallel to the glass, but when a voltage is applied to the liquid crystal, the molecules rotate 90° to alter the light passing through.

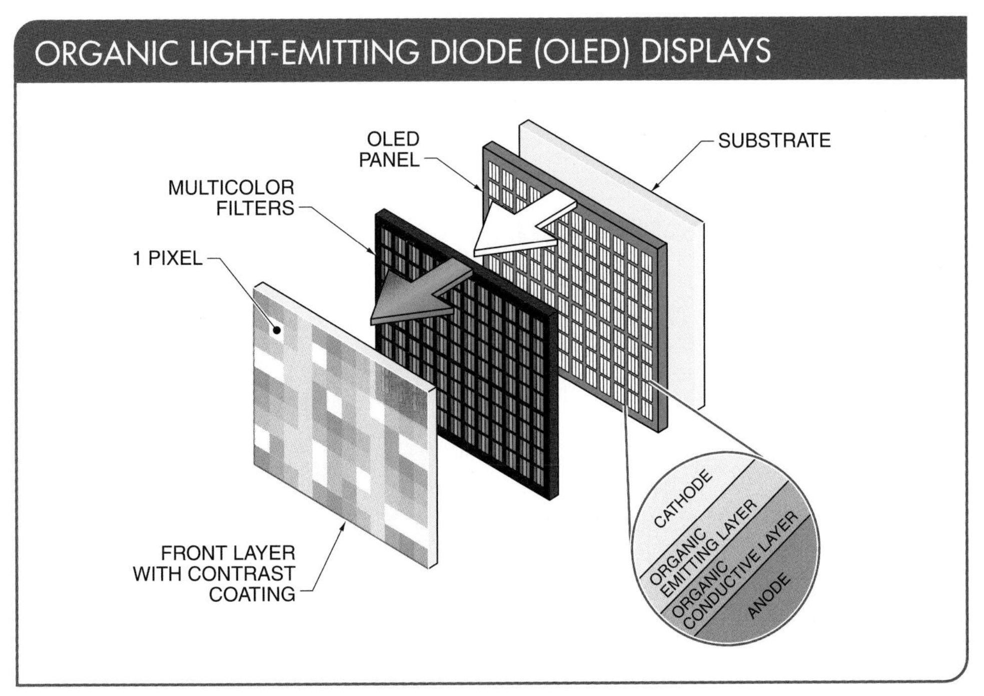

Figure 14-7. OLED technology provides clear, bright displays using less power than conventional LEDs.

OLED technology is being used in many smart TVs. A smart TV is sometimes referred to as a connected TV or hybrid TV because it is Wi-Fi enabled and can connect to the internet. A smart TV is much like a combination of a computer and a flat-screen television set. A smart TV can display information and images from various sensors that are part of a smart home security system installed in a residence.

SECTION 14.1 CHECKPOINT

1. What are the three levels of security?
2. What are the four key components of a smart home security system?
3. What are the most common interface devices?
4. What are the three most common display technologies?

Perimeter security systems can have hardwired camera systems that monitor the major entry access points of a property.

SECTION 14.2—PERIMETER SECURITY

Perimeter security detects people at entry access points in order to provide residents with advanced notice that someone is on their property. The five major entry access points protected by perimeter security include front doors, first floor windows, back doors, basements, and garages. **See Figure 14-8.** Perimeter security devices used to monitor these entry access points include motion sensors, outdoor lighting, enhanced perimeter security devices, indoor light activation devices, smart doorbells, and pool safety devices/systems.

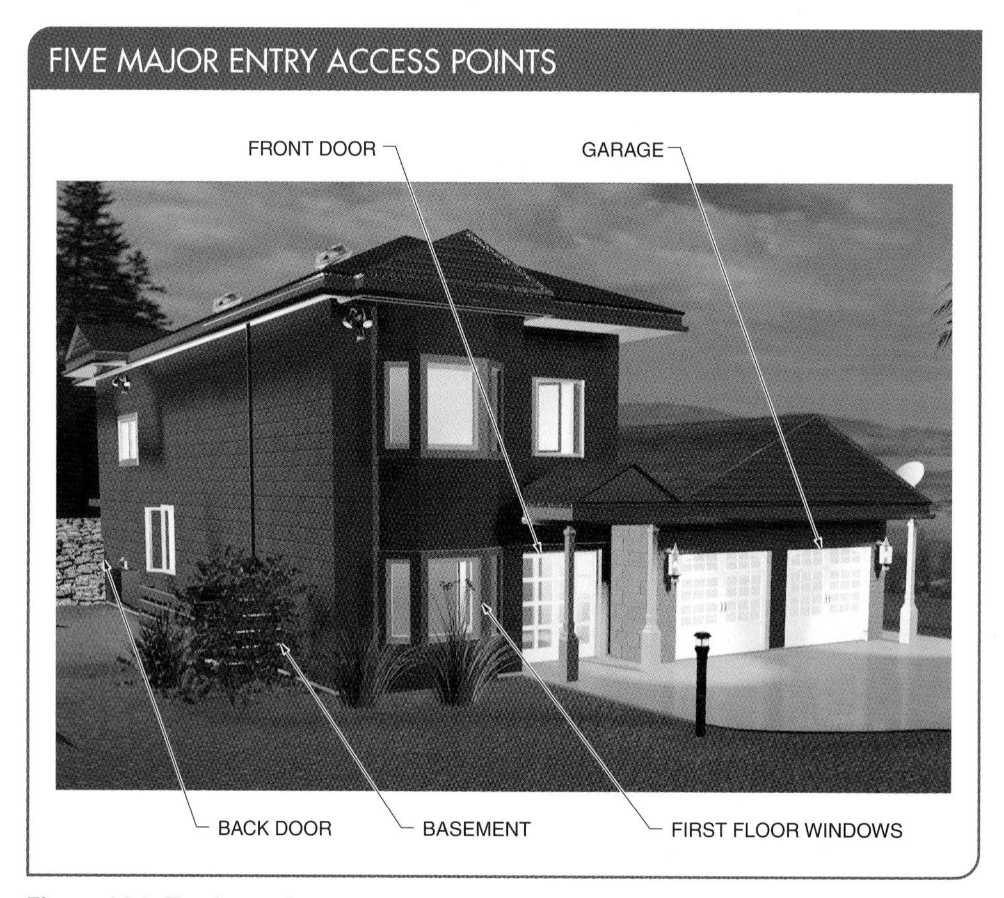

Figure 14-8. The five major entry access points to a residence are the front door, first floor windows, the back door, the basement, and the garage.

Motion Detectors

A *motion detector* is a device that uses a sensor to detect movement and trigger a signal. Motion detectors are the most commonly used outdoor security devices for perimeter security. Because many motion detectors are outdoor devices, local weather conditions should be considered when choosing the type of device to be used.

Motion detectors can be passive detectors (which do not transmit a signal), active detectors (which transmit a signal), or a device that uses both active and passive technology. Active detectors transmit sound waves and radio waves to detect an object. Passive detectors detect abrupt changes in the environment. Four common types of motion detectors are passive infrared (PIR) detectors, ultrasonic detectors, microwave detectors, and dual-technology detectors.

Passive Infrared Detectors. A *passive infrared (PIR) detector* is a device that senses the difference in temperature between humans or animals and the background space. PIR detectors scan the detection area in a fan-shaped pattern.

A pyroelectric sensor is the component in a PIR detector that detects the change in temperature. A *pyroelectric sensor* is a sensor that generates a voltage in proportion to a change in temperature. Electronic components within the PIR detector analyze the signal received. If the signal exceeds a certain threshold, the device communicates with the control panel.

The ability of the PIR detector to detect infrared (IR) radiation, or heat, depends on the distance between the pyroelectric sensor and the object, the speed of the object, and whether the object is within the sensor's field of view. Any obstructions between the object and the sensor's field of view may cause the IR sensor to be less effective. **See Figure 14-9.**

TECH TIP

PIR detectors convert the changes in incoming IR light to electric signals.

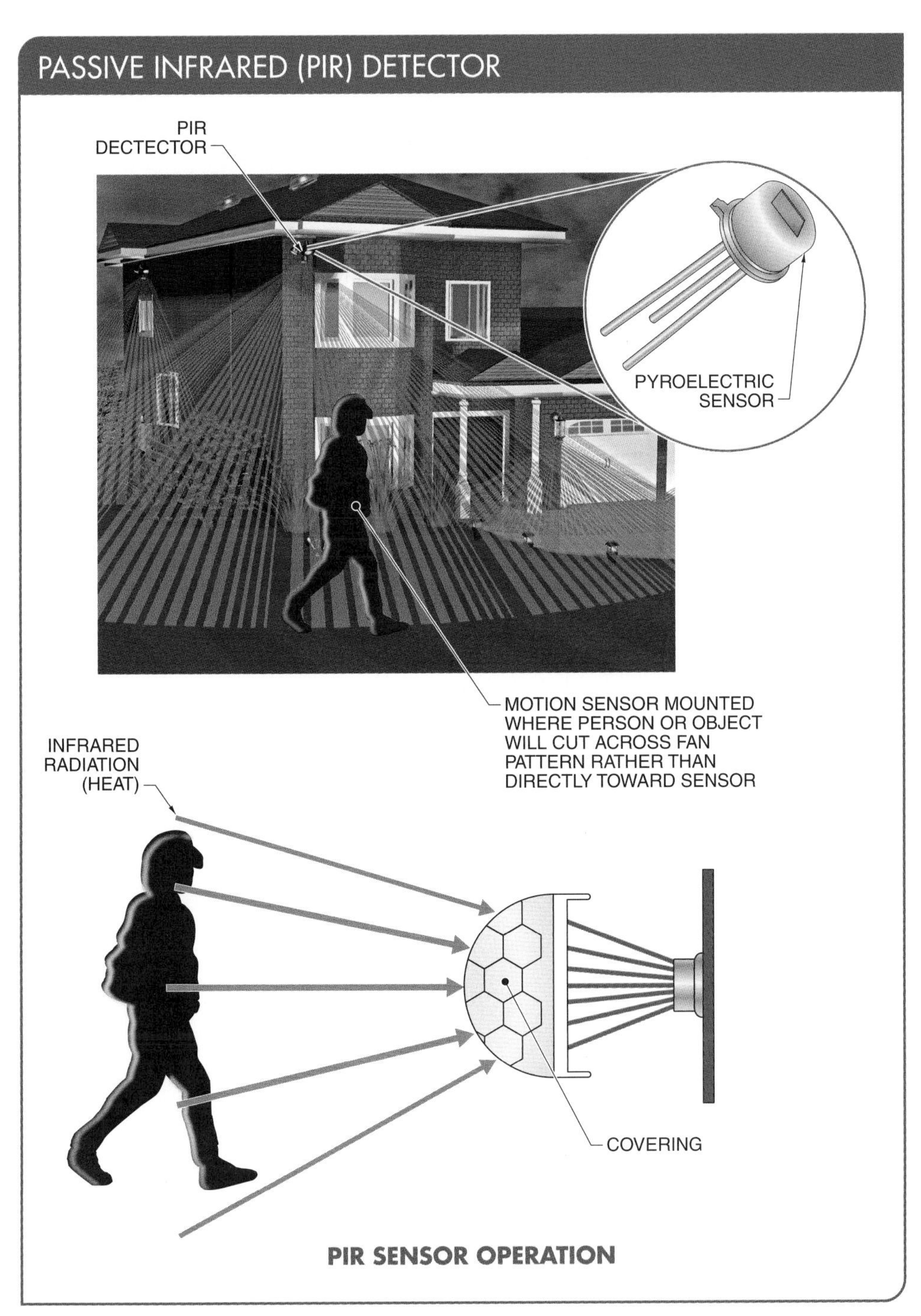

Figure 14-9. PIR detectors contain pyroelectric sensors that react to IR radiation (heat) by generating a voltage. PIR detectors must be mounted where there is an unobstructed view of the area to be detected.

At short distances (10′ or less), a pyroelectric sensor can detect small movements of an object, such as hand motions. As the distance between the object and the sensor increases, larger motions are required to trigger the sensor. When an object, such as a person, is 30′ to 40′ away from the sensor, the sensor only detects large movements, such as when the person is walking. Other factors that affect the effectiveness of a PIR detector are the size of an object, the temperature difference between the object and the environment, and weather conditions such as rain, snow, and fog.

Ultrasonic Detectors. An *ultrasonic detector* is a device that detects a moving object by transmitting sound waves and receiving the reflected waves off of the object. Active ultrasonic detectors emit sound waves at frequency levels above that of human hearing (above 20 kHz). **See Figure 14-10.**

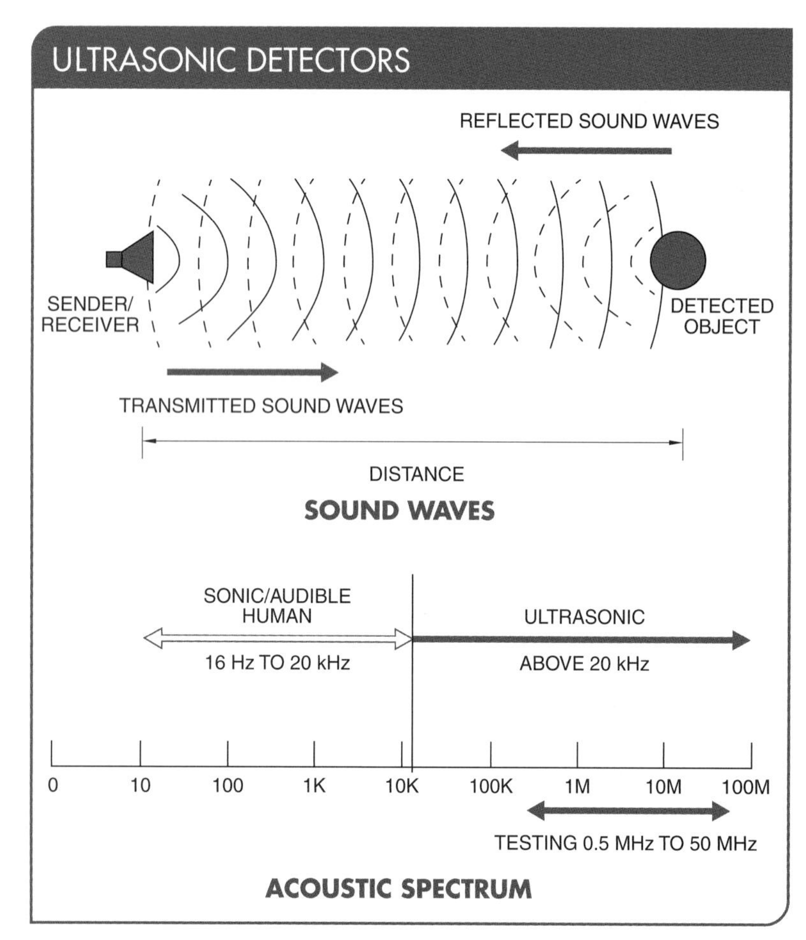

Figure 14-10. An ultrasonic sensor measures the distance to an object using inaudible high-frequency sound waves.

> **TECH TIP**
>
> *A proximity-detection ultrasonic detector detects an object passing within its preset range, while the detection point is independent of target size, material, or reflectivity.*

Microwave Detectors. A *microwave detector* is a device that detects a moving object by transmitting electromagnetic (EM) radio waves and receiving the reflected waves off of the object. **See Figure 14-11.** Microwave detectors operate based on the Doppler effect. This means that any object moving toward or away from the sensor will cause the frequency of the reflected waves to change slightly.

A microwave detector has three main components: a transmitter, a receiver, and an electronic alarm circuit. Some microwave detectors contain circuitry that enables the device to sense whether the object is moving toward the detector or away from the detector.

Dual-Technology Detectors. A dual-technology detector combines both active and passive sensors and only activates when motion is detected by both sensors. Since dual-technology detectors activate only when both sensors detect a moving object, false alarms are greatly reduced. If movement is detected by only one sensor, no signal is sent to the alarm.

Most manufacturers combine PIR sensors and microwave sensors because PIR sensors are more sensitive to lateral motion and microwave sensors are more sensitive to forward motion. This double trigger helps to avoid false alarms. A dual PIR/microwave detector first uses its passive sensor (PIR sensor) because it requires less energy. When the PIR sensor is tripped, the active sensor (microwave sensor) turns on. If the microwave sensor is also tripped, the alarm sounds. The downside to dual-technology detectors is that they may not provide the same level of security as a single sensor since they must be double triggered.

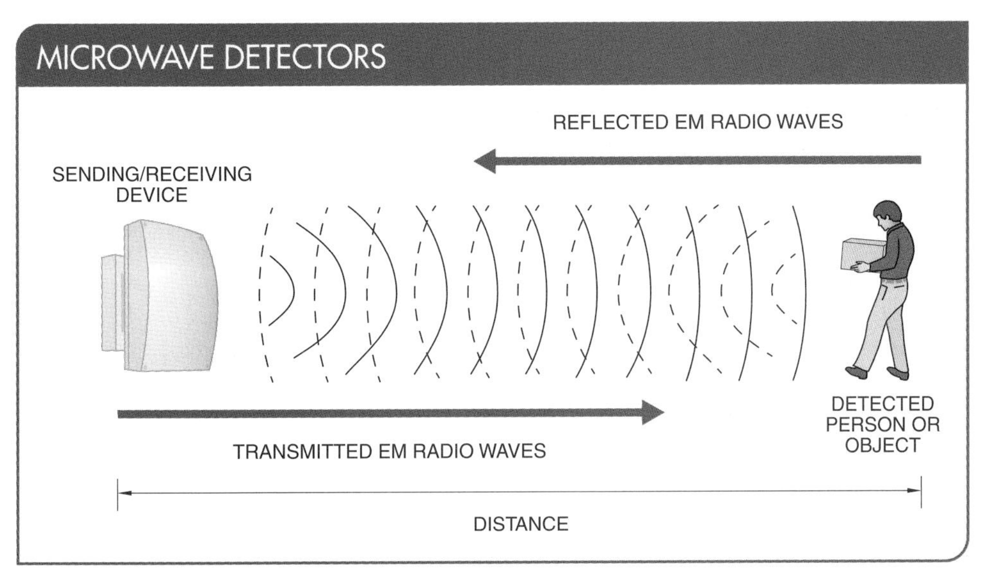

Figure 14-11. Microwave detectors transmit EM radio waves, and a moving object causes a frequency shift in the reflected waves.

Outdoor Lighting

Outdoor lighting is used to illuminate driveways, walkways, entry access points, and landscaping features, as well as to deter trespassers. LED, incandescent, and fluorescent lamps are commonly used for outdoor lighting. Motion detectors and photocells are often used to control solar lights and low-voltage lighting. Motion detectors and photocells can control individual lights or groups of lights. Motion detectors can be positioned to monitor doors, driveways, gates, or any entry access point. **See Figure 14-12.**

Photocells. A *photocell* (photoconductive cell) is a light-activated control switch that varies resistance based on the intensity of the light striking it. A photocell is formed by a thin layer of semiconductor material deposited on an insulator. Photocell leads are attached to the semiconductor material, and the entire assembly is sealed in glass. **See Figure 14-13.**

When light strikes a photocell, electrons are freed and the resistance of the material decreases. When the light is removed, the electrons and holes recombine and the resistance increases. The resistance of a photocell ranges from several million ohms resistance in total darkness to less than 100 Ω at full light intensity. Photocells are used when time response is not critical. Photocells are often used to control street lighting and yard lighting.

Solar Lights. A *solar light* is a light composed of an LED light, solar panel, battery, and a charge controller. A *solar panel,* also known as a PV panel or module, is an assembly of solar cells. A *solar cell,* also known as a PV cell, is an electrical device that converts sunlight (solar radiation) into electricity and stores it in a battery. The charge controller controls the current flow to the battery.

Landscape lighting can be controlled from a separate control panel.

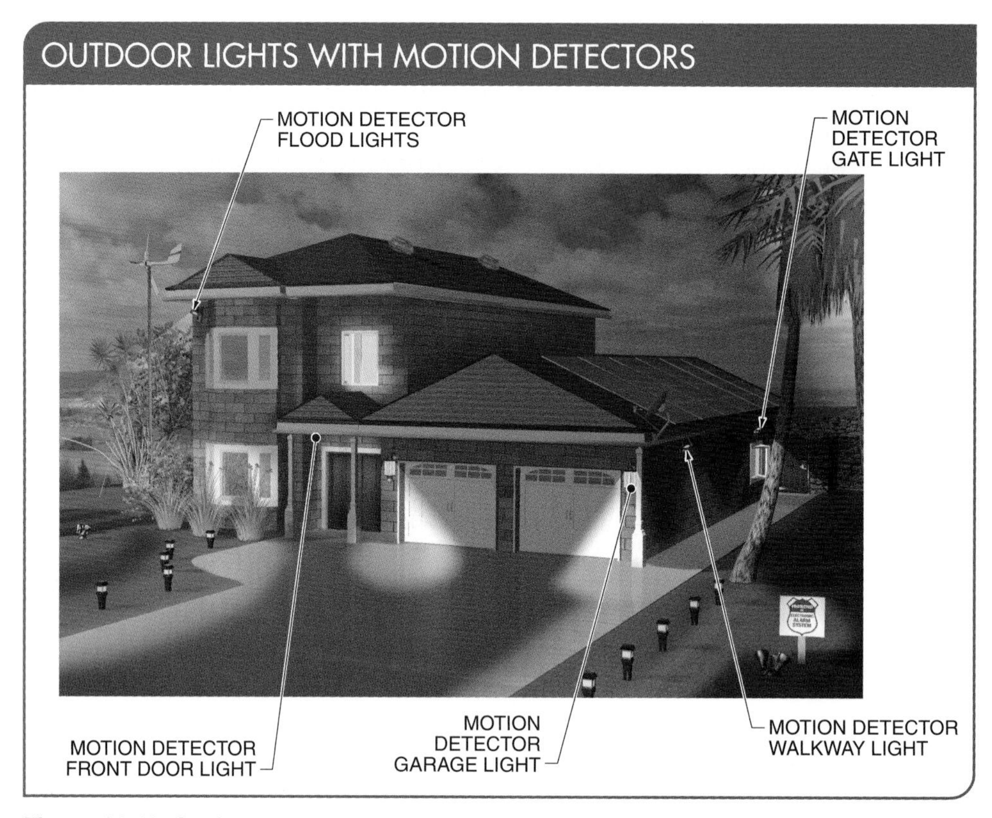

Figure 14-12. Outdoor motion detectors are often positioned to monitor major access points and are combined with outdoor controlled lighting to deter potential trespassers.

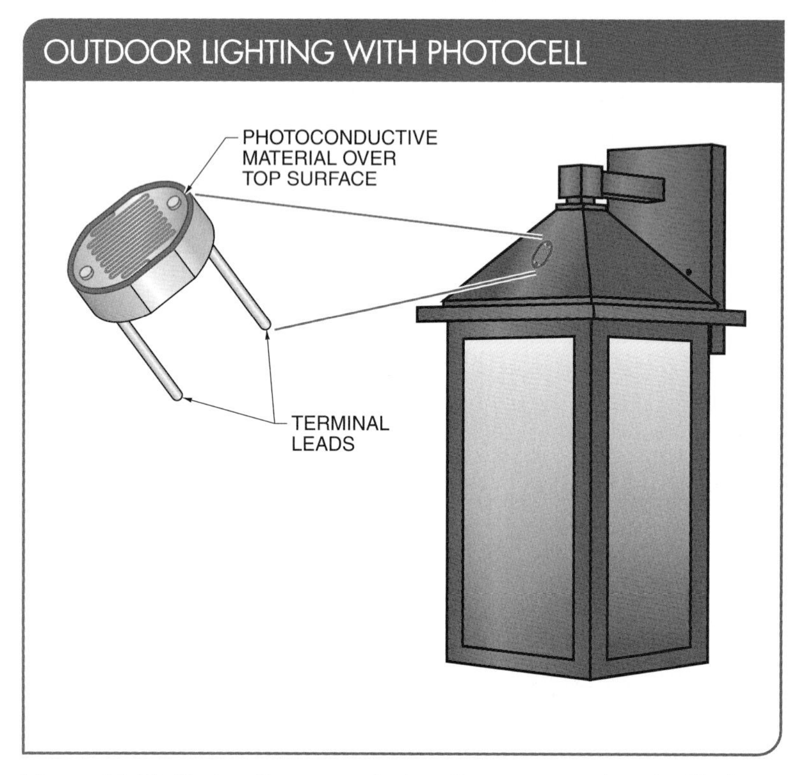

Figure 14-13. Photocells are used when time response is not critical and are often used to control street and yard lighting.

A solar light operates on electricity from a battery switched on by a motion detector or photocell. **See Figure 14-14.** Solar lights are often used in areas that do not require continuous illumination but rather only need to be illuminated a few times at night. For example, a solar light may be placed at the front door since illumination is only needed for a few minutes when the homeowner is at the front door or to deter trespassers.

Low-Voltage Outdoor Lighting. *Low-voltage outdoor lighting* is a system that uses 12 V or 24 V to power low-voltage lamps to illuminate walkways, landscaping, or dark areas on a property. Up-lighting trees with low-voltage floodlights can provide an attractive landscape accent as well as security. Adding downlight fixtures or grazing lights along paths, steps, or stairways can help reduce trip hazards. Illuminating paths to each entrance allows the homeowner to see visitors at night before opening the door. Illuminated path lighting should overlap so that there are no areas left in the dark.

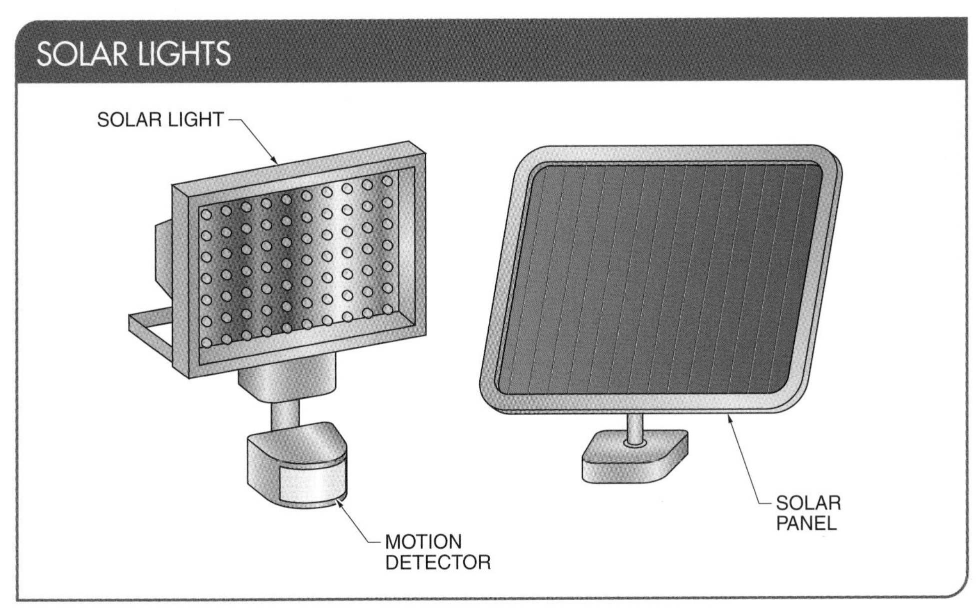

Figure 14-14. Solar lights can be activated by motion detectors or photocells and are often used in areas that do not require continuous illumination.

Powering low-voltage lighting that is installed outdoors usually only requires plugging a prewired transformer into a receptacle. Some transformers can be controlled by a photocell to turn on the lights once a certain level of darkness is reached and to turn them off once daylight arrives. The sensitivity of the photocell should be adjusted so that the lights turn on only when necessary. If the timer is a separate unit from the transformer, additional connections must be made. Low-voltage cables only need to be buried in a shallow trench 4″ to 6″ deep. **See Figure 14-15.**

Enhanced Perimeter Security Devices

Perimeter security can be enhanced by adding special-purpose security devices. These devices include driveway sensors, wireless gate sensors, and wireless mailbox sensors. Enhanced perimeter security devices provide additional information concerning objects on a driveway, within a yard, or placed in a mailbox.

Driveway Sensors. A *driveway sensor* is a device that is designed to detect objects entering a driveway and transmit a signal to a receiver. Driveway sensors can be programmed to activate a doorbell system or to send a message to a cell phone, computer, or television screen. A driveway sensor can also activate outdoor lighting and provide light for a video surveillance camera.

The two most common types of driveway sensors are IR sensors and magnetometers. **See Figure 14-16.** An IR sensor projects an invisible beam that detects vehicles entering the driveway. The IR sensor can be located anywhere along the driveway. A magnetometer detects the metal in vehicles as they enter the driveway. When a vehicle passes over or near the magnetometer, the detector buried in the driveway or next to the driveway sends an alert to a receiver unit inside the residence. Magnetometers are not triggered by people or animals.

Wireless Gate Sensors. A *wireless gate sensor* is an accelerometer device that detects tilt or vibration movements and transmits a wireless signal to a receiver. Wireless gate sensors detect any movement of a sliding or swinging gate. A wireless gate sensor is weather resistant and can be powered from a built-in solar panel and/or a battery, which eliminates the need for cables. An active IR sensor is also used to sense when a person opens and walks through the gate. **See Figure 14-17.**

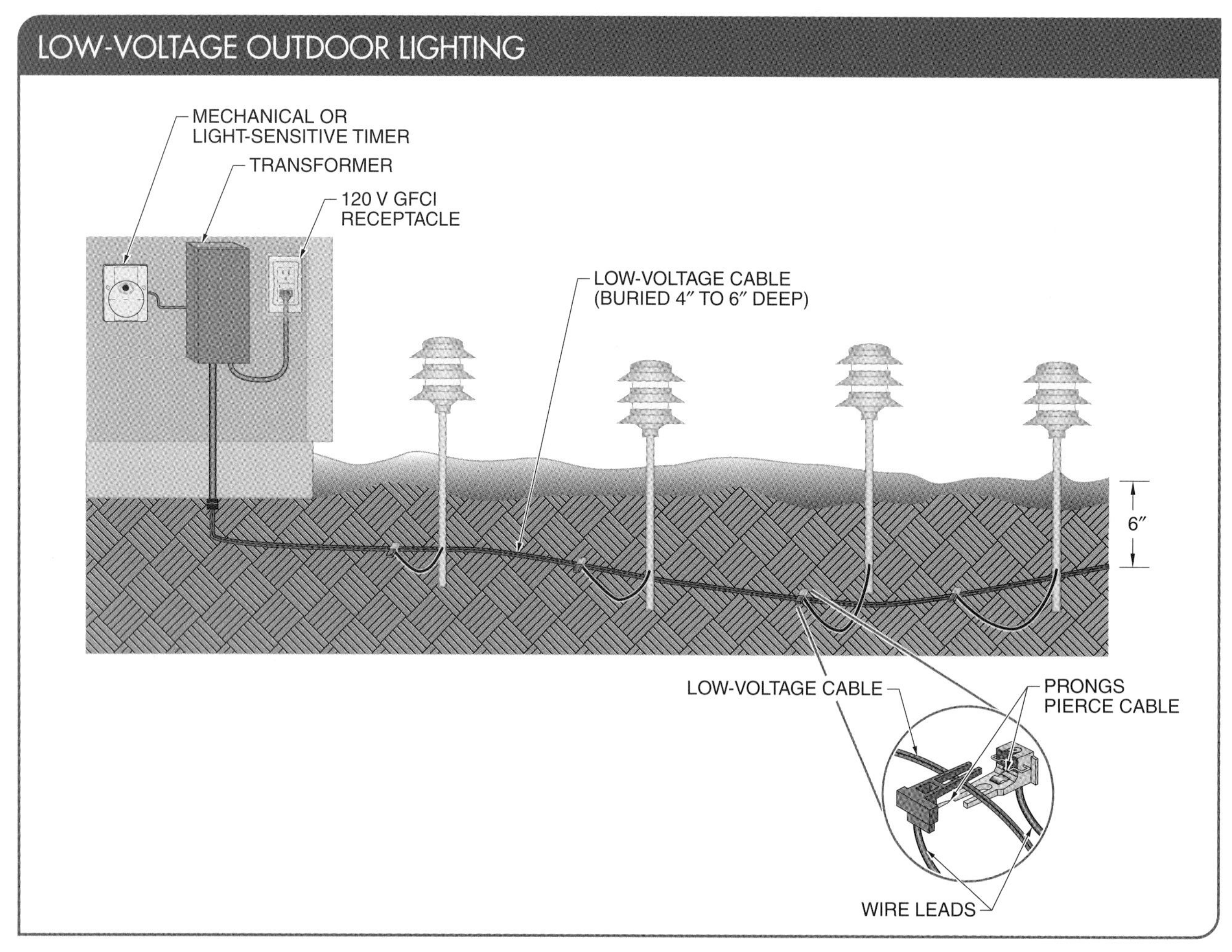

Figure 14-15. Low-voltage cables only need to be buried in a shallow trench 4″ to 6″ deep.

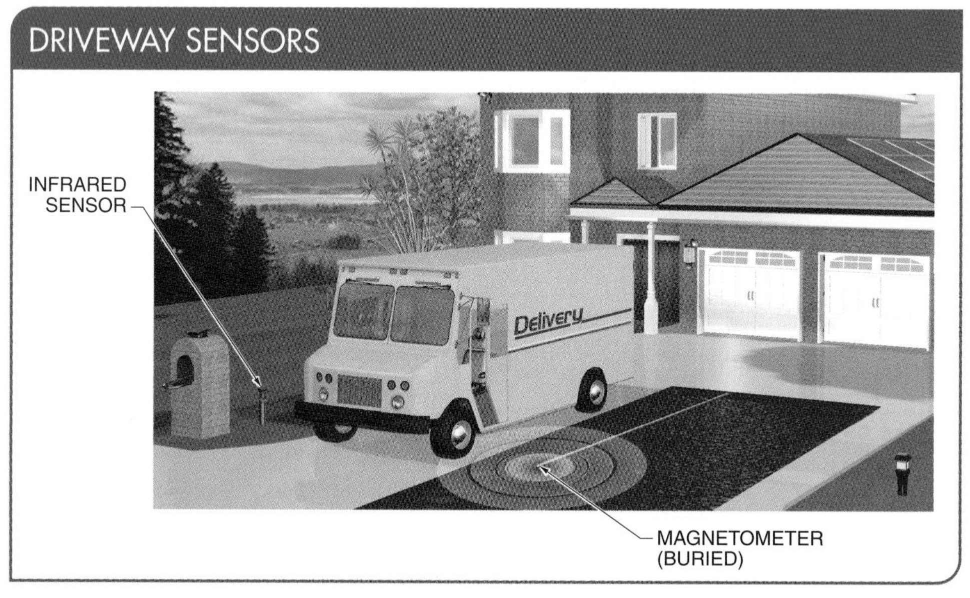

Figure 14-16. IR sensors project an invisible beam to detect objects entering the driveway, while buried magnetometer sensors detect vehicles entering the driveway but not people or animals.

> TECH TIP
>
> *Driveway sensors are typically installed during the construction phase of a driveway.*

Wireless Mailbox Sensors. A *wireless mailbox sensor* is a sensor that wirelessly transmits a signal to a receiver inside a dwelling to provide notification that mail is present in the mailbox. The notification can be a sound, flashing light, or text message to a cell phone. A wireless mailbox sensor can also be triggered by a driveway sensor that notifies a resident returning home of the presence of mail. Wireless mailbox sensors are battery operated, and the receiver is equipped with a 6 VDC adapter inside the dwelling. The transmitters in wireless mailbox sensors typically usually have a range of 200′ to 250′.

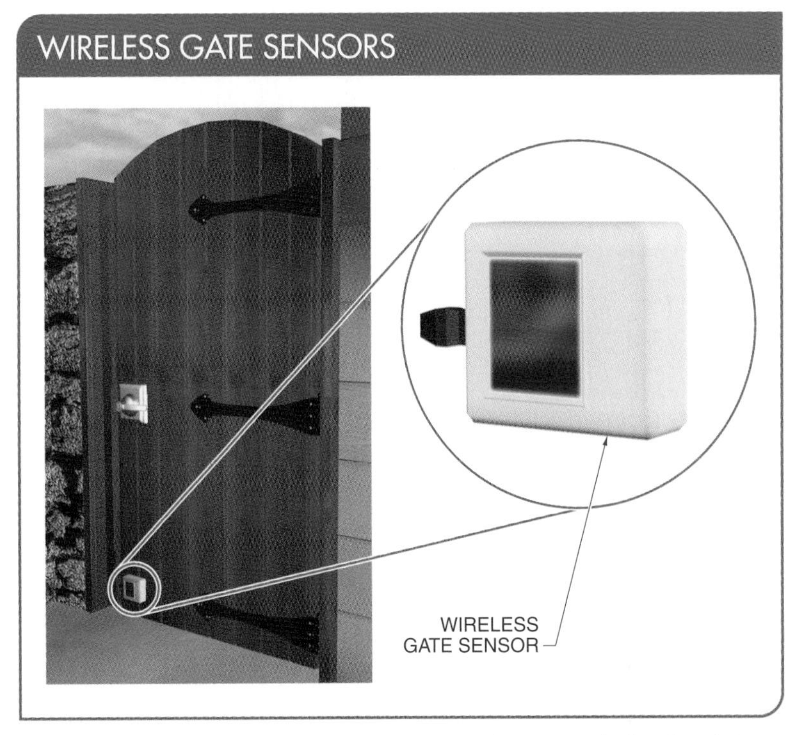

Figure 14-17. A wireless gate sensor is an accelerometer device that detects the movement of a sliding or swinging gate.

Automatic Indoor Light Activation Devices

Although automatic perimeter lighting helps deter trespassers from approaching a residence, additional deterrence can be achieved by connecting wireless motion sensors to interior lighting. This achieves the effect of someone being home. **See Figure 14-18.**

Smart Doorbells

A *smart doorbell* is a Wi-Fi-enabled doorbell that includes video and audio capabilities and can send a signal to the homeowner when activated. A smart doorbell is considered part of perimeter security. Smart doorbells look and operate similar to standard doorbells. **See Figure 14-19.**

A smart doorbell has a camera and a speaker and can alert a homeowner through their mobile device, computer, or smart TV that the front or back door is being approached. When the doorbell is rung, a video image becomes available via Wi-Fi and the person at the door can be verbally greeted. Smart doorbells can be combined with driveway sensors to provide a greater amount of information to the homeowner when someone approaches the dwelling.

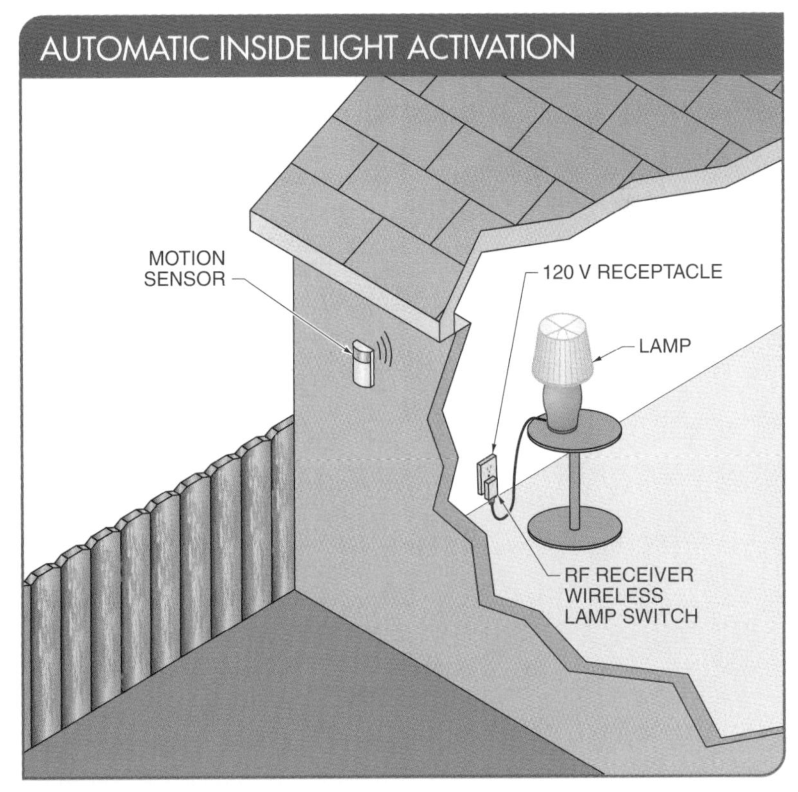

Figure 14-18. Perimeter security motion detectors can be used to turn lights on inside the residence when someone approaches.

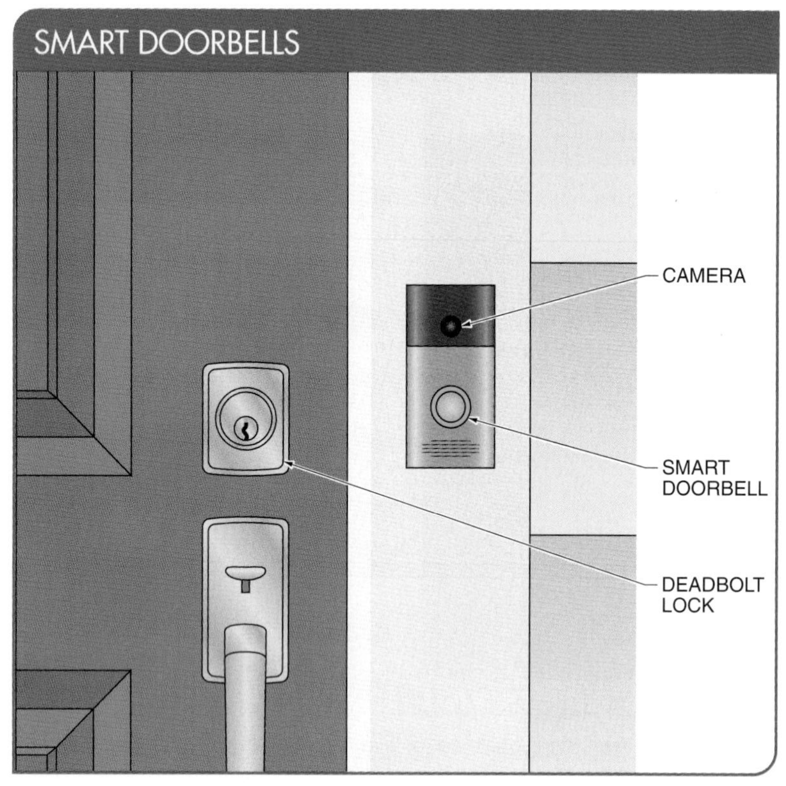

Figure 14-19. Smart doorbells include a button that allows visitors to announce their presence and a camera that allows the homeowner to view a defined area in front of the door using their internet-connected device.

Pool Safety Systems

Pool safety systems can help prevent accidental drownings and damage to pool equipment. Pool safety systems usually require an initial electrical inspection to ensure they meet code requirements. Pool safety can be accomplished using in-pool sensors, pool water level sensors, and pool access door sensors.

In-Pool Sensors. In-pool sensors are mounted on the edge of a pool or in the pool. In-pool sensors are battery powered and detect water displacement when an object weighing more than 18 lb enters the pool. When detection occurs, a signal is sent to a remote receiver. In-pool sensors can be used with pool covers or solar blankets and can be adjusted to increase or reduce their sensitivity.

Pool Water Level Sensors. A low water level in a pool risks damage to the filter system and pump. In contrast, overfilling a pool wastes water and reduces filter and skimmer effectiveness. A water level sensor is used to maintain the correct water level in a pool.

Pool Access Door Sensors. A pool access door sensor may be attached to doors that have direct access to the pool. When the pool access door is opened, an audible alarm sounds or an alert can be sent to a wireless device. These alarms are turned on and off using a key or keypad and are either battery powered or use an AC adapter.

SECTION 14.2 CHECKPOINT

1. What types of devices are used for perimeter security?
2. What are four common types of motion detectors?
3. What is a photocell?
4. What types of sensors can be used to enhance perimeter security?
5. Through which three devices can a smart doorbell alert a homeowner?
6. What are three sensors used for pool safety protection?

SECTION 14.3—INTERIOR SECURITY

Interior security detects when the home envelope is breached and sends alerts to the residents. Interior security devices sense motion and noise from doors and windows, pressure, carbon monoxide (CO), water, temperature, and humidity. When a garage is considered part of the home, additional safety and security requirements must be considered.

Interior security detection devices include door/window sensors, glass-break sensors, motion detectors, pressure sensors, CO sensors, water sensors, security cameras, and temperature/humidity sensors. In addition, if a resident detects a security breach, they can manually send an alert using a panic button or security key fob to signal a monitoring service, sound an audible alarm, or activate a recording system. **See Figure 14-20.** The control panel installed in a residence can also have a device inside of it to sense tampering and automatically send an alarm to a monitoring service.

Door/Window Sensors

The easiest access points into a residence are the doors and windows. A *door/window sensor* is a magnetic switch device that indicates whether a door or window is closed or open. The contacts of a door/window sensor consist of a magnet that is mounted on the door or window and a reed switch that is mounted on the door or window frame. When the door or window is fully closed, the two halves of the device are near each other and the magnet pulls the reed switch contacts closed. When the magnet and reed switch separate as the door or window is opened, the contacts open. **See Figure 14-21.**

Adhesive can be used to keep the sensors in place, although some sensors are designed to be screwed into the door or window frame. Sensors are either wired directly into the alarm system or use battery power to function wirelessly. Wired sensors require more time and labor to install than wireless sensors, but the batteries in wireless sensors may need to be checked frequently to ensure they are providing power.

Glass-Break Sensors. Glass-break sensors are sensors that detect vibrations or the sound waves of breaking glass. Glass-break sensors are used to protect windows that cannot be opened, such as picture windows. Two types of glass-break sensors are shock sensors and acoustic sensors.

Shock sensors detect vibrations when glass breaks or shatters. Shock sensors are mounted directly to the glass or the frame of the window. Acoustic sensors are designed to detect the exact sound waves of breaking glass and are usually mounted on ceilings or walls. Acoustic sensors can monitor a wider area than shock wave sensors, such as a large bay window or a sliding glass door. **See Figure 14-22.**

Figure 14-20. Interior security detectors use sensors to detect changes in the environment and automatically send information to a control panel. An individual can manually send information to the control panel with a panic button or key fob.

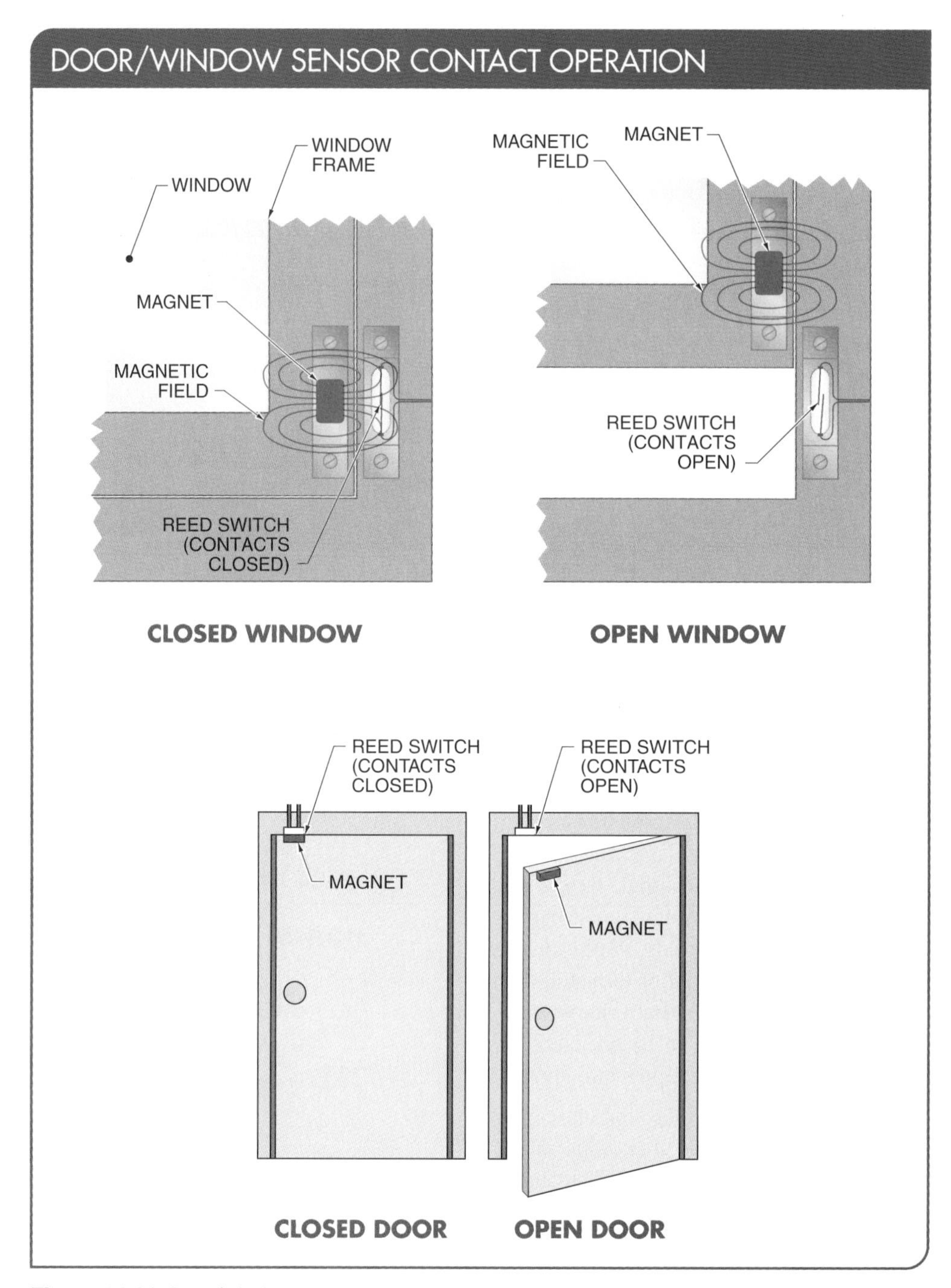

Figure 14-21. Door/window sensor contacts consist of a magnet and a reed switch.

Interior Motion Detectors

Motion detectors are used inside a residence to detect and possibly deter an intruder. Interior motion detectors detect movement even after the initial entry is made. Interior motion detectors used in combination with window/door sensors and outdoor motion detectors provide a comprehensive security system. Interior motion detectors operate basically the same as outdoor motion detectors. The main difference is that outdoor motion detectors are designed to work in harsh weather environments.

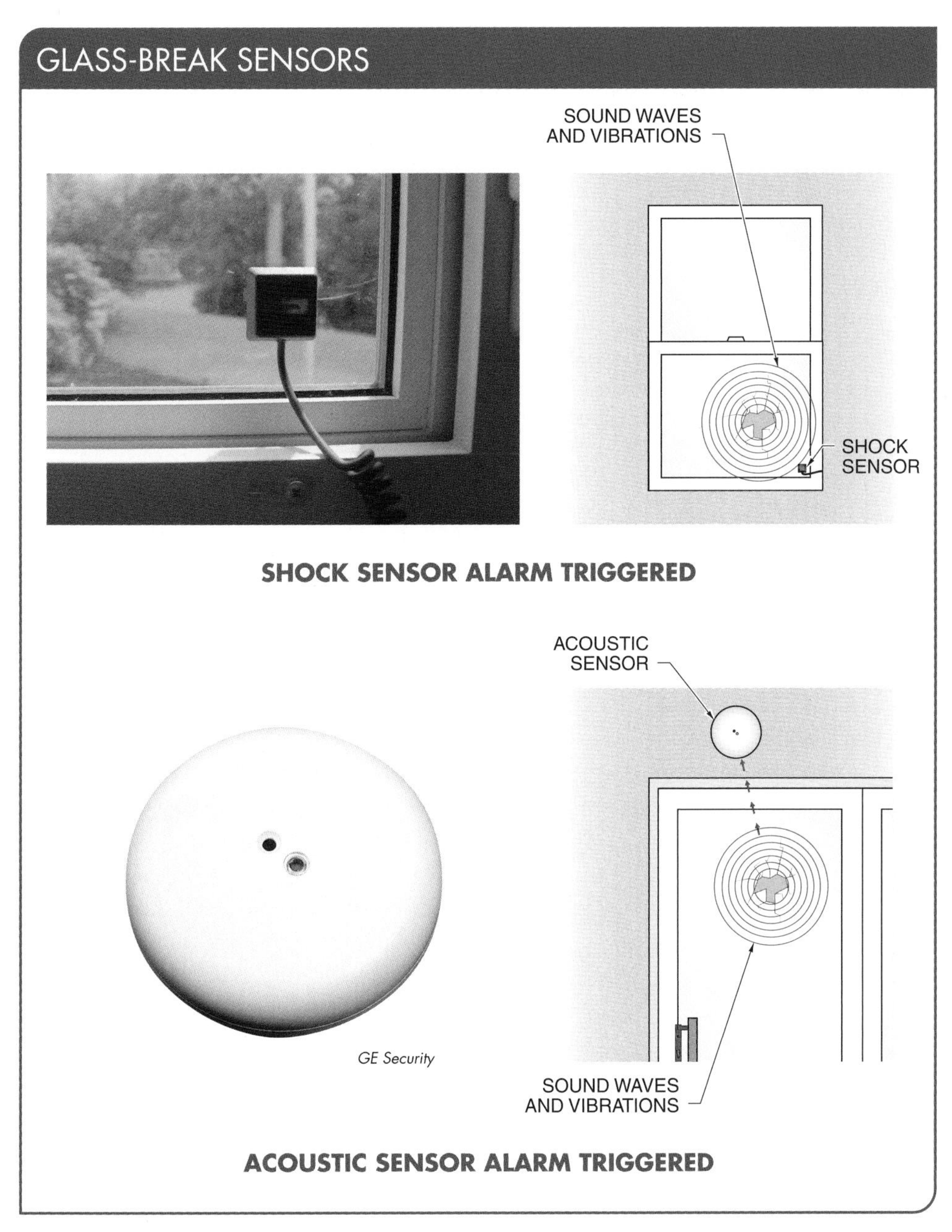

Figure 14-22. Shock sensors detect vibrations and are mounted on windows or window frames. Acoustic sensors detect the exact sound waves of broken glass and are mounted on ceilings or walls.

For example, IR detectors work better indoors where weather conditions will not trigger false alarms. An active IR detector is a combination transmitter-receiver that is recessed into a wall. The transmitter projects an invisible pulsating beam at a specially designed reflector on the opposite wall. An interruption of the beam signal causes the receiver to activate the alarm system. Active IR detectors are not as popular as PIR detectors because they require installing two pieces of equipment. **See Figure 14-23.** A motion detector is selected based on its ability to operate in harsh weather environments, its range and angle of coverage, its sensitivity, and if it has a battery backup.

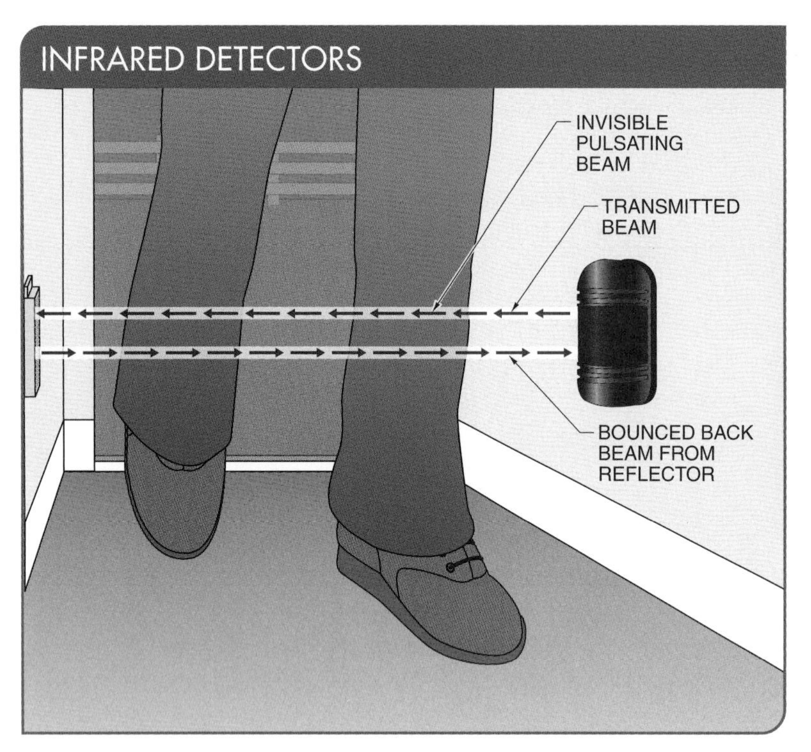

Figure 14-23. IR detectors recessed into the wall transmit an invisible pulsating beam at a reflector to the opposite wall. An interruption of the reflected beam activates the alarm system.

Interior Motion Detector Placement. Interior motion detectors should be placed where an individual must walk through the sensor detection pattern to move through the dwelling. Interior motion detectors are often not placed in every level of a dwelling. Second-floor windows, inaccessible windows, or windows that are too small for someone to fit through are often not included in alarm coverage, unless there is a deck or stairway leading to them. Interior motion detectors should be placed to protect residents and areas containing valuables, such as the master bedroom. **See Figure 14-24.**

Pressure Sensors

A *pressure sensor* is a device that measures pressure and converts it to an electrical signal. Pressure sensors detect the presence of a person in a floor area by detecting the pressure of their feet as they walk. Two common types of pressure sensors are floor mats and stress sensors.

A *floor mat* is a sensing device that detects the weight of a person walking on it. Floor mats are typically installed under carpet in locations where an individual is likely to travel, such as in a doorway or at the bottom or top of a staircase. Floor mats trigger an alarm when armed and the measured weight exceeds a setpoint, such as 60 lb.

A *stress sensor* is a sensing device that measures the weight of a load over a specific area. A stress sensor can be glued with epoxy under a floor joist or floor truss. **See Figure 14-25.** An electronic processing circuit analyzes the measured stress signals and initiates an alarm when the stress exceeds a setpoint. Similar to floor mats, a stress sensor is installed in common or high-traffic areas, but it can also be installed in outdoor areas to protect an area such as a deck or a boat dock.

Garage Protection Sensors

Garage protection sensors are designed to identify changes in pressure, motion, and the air. Garage protection sensors are used to monitor overhead doors, service doors, and windows, as well as the air inside of a garage. The four main types of garage protection sensors are pressure sensors, IR sensors, tilt switch sensors, and CO sensors. **See Figure 14-26.**

Garage Door Pressure Sensors. A garage door opener is built with a pressure (force) sensing circuit. The circuit senses when pressure is applied against the bottom of the garage door as it closes. The garage door automatically reverses when the pressure reaches a set limit. The close force limit can be adjusted on the garage door opener and should be set to the minimum pressure to protect individuals and avoid damage to property.

IR Sensors. IR sensors control garage doors by transmitting a beam of IR light 6″ above the ground or floor from one side of the door to a receiver on the opposite side of the door. If the beam is obstructed by a person or object, the garage door will not operate. If someone steps into the beam when the door is closing, the door stops and reverses direction. If the sensors are malfunctioning, one of the sensor lights will flash. **See Figure 14-27.**

Tilt Switch Sensors. A *tilt switch sensor* is a device that detects inclination or orientation. Tilt switch sensors use a metal ball that rolls and shorts two poles together to complete a circuit. When an overhead garage door moves, a tilt switch sensor alerts the control panel which can send an alert that the garage door is open or closed.

IR sensors are installed at the base of garage doors to protect individuals and prevent property damage.

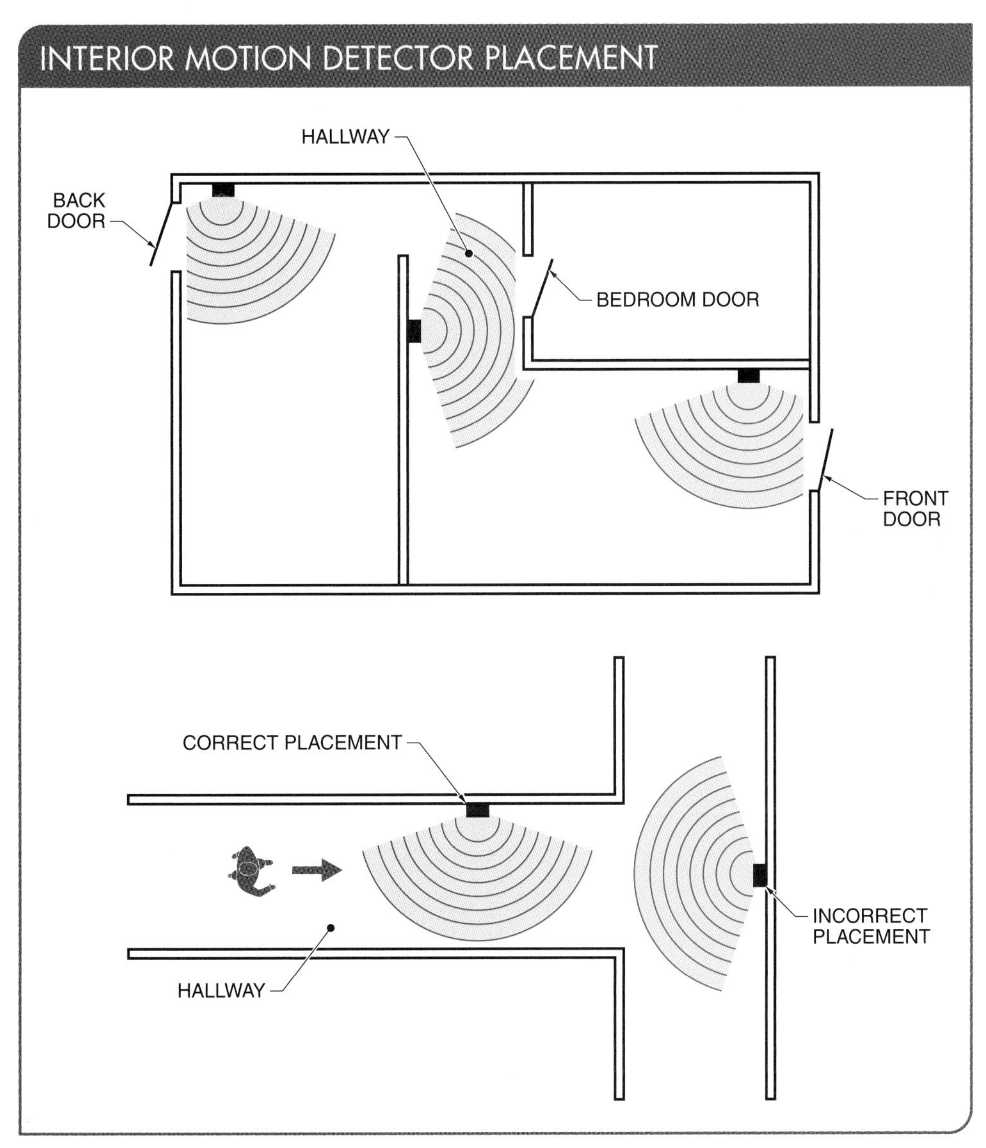

Figure 14-24. Interior motion detectors are placed in areas where an individual must walk through the detection pattern to move through the dwelling.

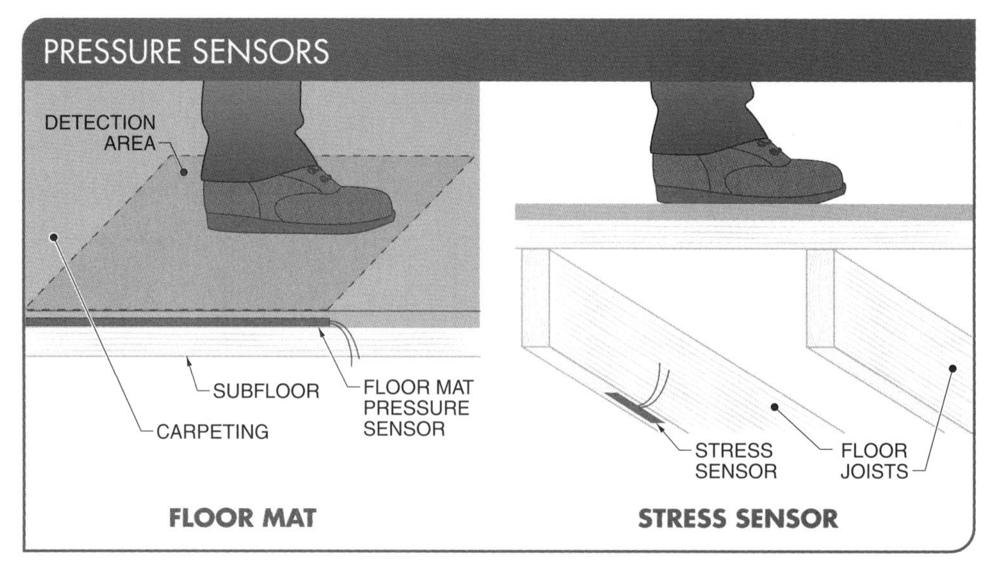

Figure 14-25. Floor mats and stress sensors are two common types of pressure sensors.

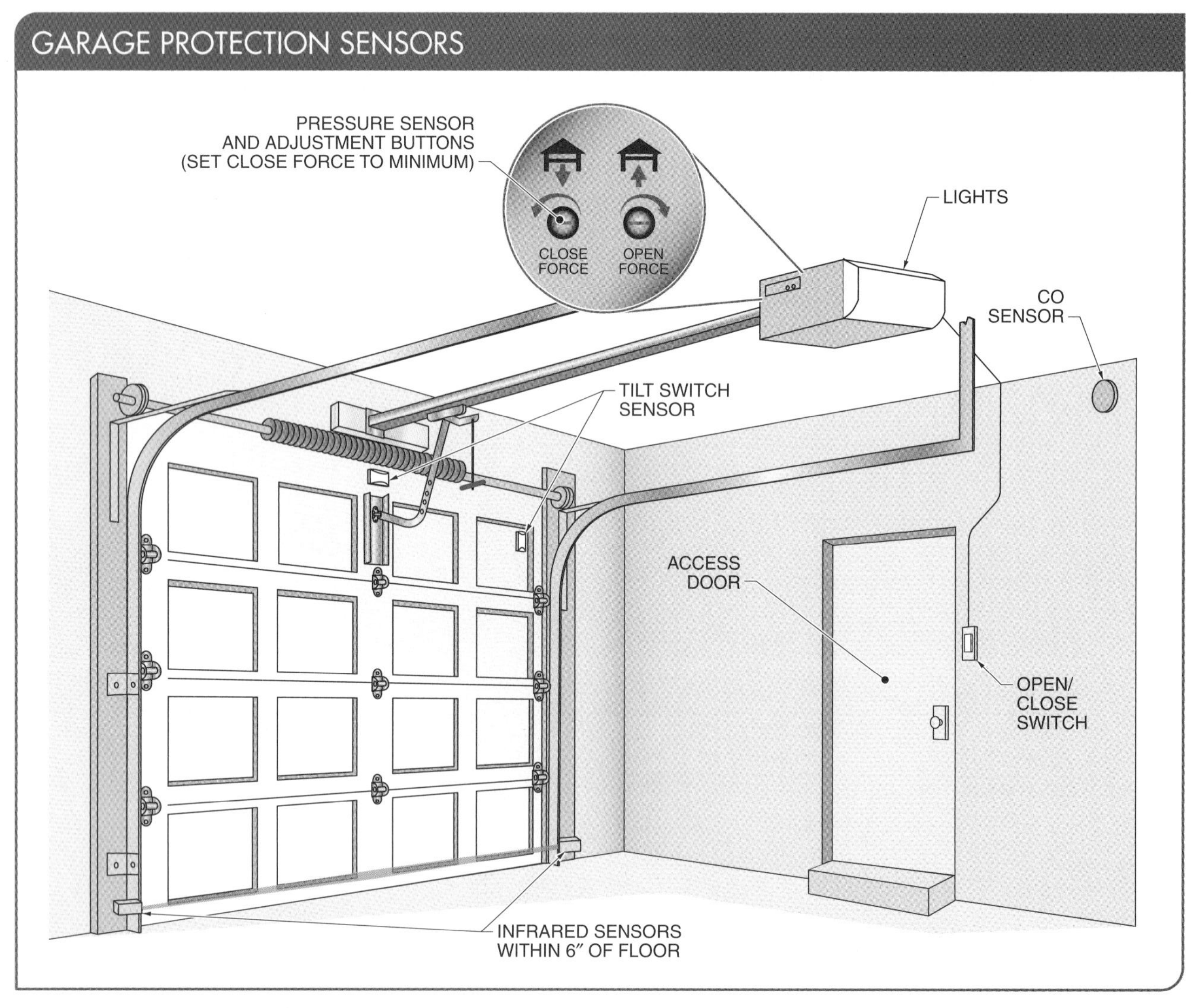

Figure 14-26. Garage protection sensors include pressure sensors, infrared sensors, tilt switch sensors, and carbon monoxide sensors.

CO Sensors. A *carbon monoxide (CO) sensor* is a device that detects CO gas. If the presence of CO in a garage reaches a dangerously high level, the CO sensor will automatically open the garage door to let fresh air in and CO out. This situation usually occurs when the car is being warmed up with the garage door closed.

> TECH TIP
>
> *Since 2006, more than two dozen deaths have occurred from keyless vehicles running in garages.*

Garage Door Openers

A *garage door opener* is a motorized device that opens and closes a garage door. Garage door openers are controlled by a switch on the garage wall or by remote control. A mounted switch can be a pushbutton or a keypad requiring a numeric code. Remote controls can be kept in a car, attached to a key ring, or installed as an app on a smartphone. Garage door openers have traditionally been controlled by direct radio transmitter/receiver systems, but many openers can now be controlled using Wi-Fi as well. **See Figure 14-28.**

Garage Door Safety. A garage is either attached to or detached from a dwelling. In either case, safety considerations include the type of garage door opener, safety issues relating to the garage door opening and closing, the access door to the home, and the presence of CO. The federal government mandates that companies producing garage door openers install safety devices. Manufacturers have added a sensor that allows the resident to remotely check if the garage door is open or closed.

Structural Damage Protection

Interior security should also extend to maintaining the structural integrity of the dwelling envelope (the boundary and materials separating the interior from outdoor elements). Structural damage protection involves using sensors to protect a dwelling against damage from water, humidity, and freezing temperatures. The structure of a dwelling can be protected by using water sensors, humidity sensors, and freeze sensors.

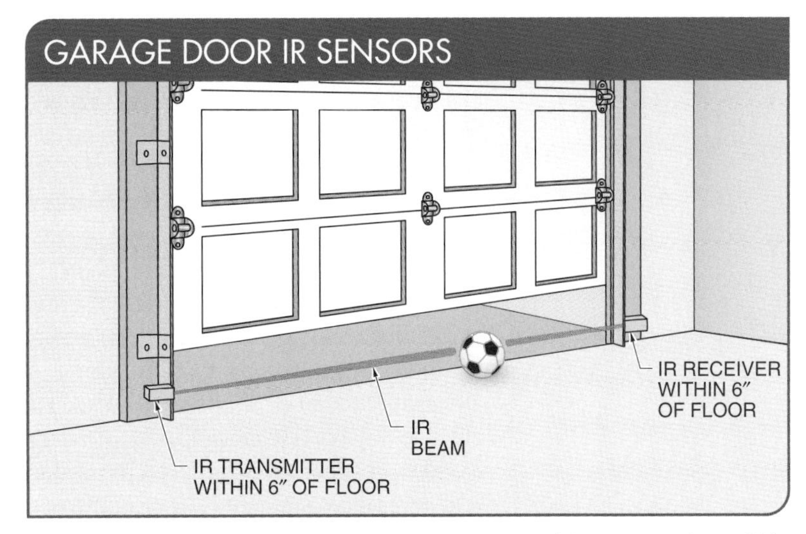

Figure 14-27. IR sensors are placed on either side of the garage door within approximately 6″ of the floor. If the beam is obstructed by an object when the door is closing, the door stops and reverses direction.

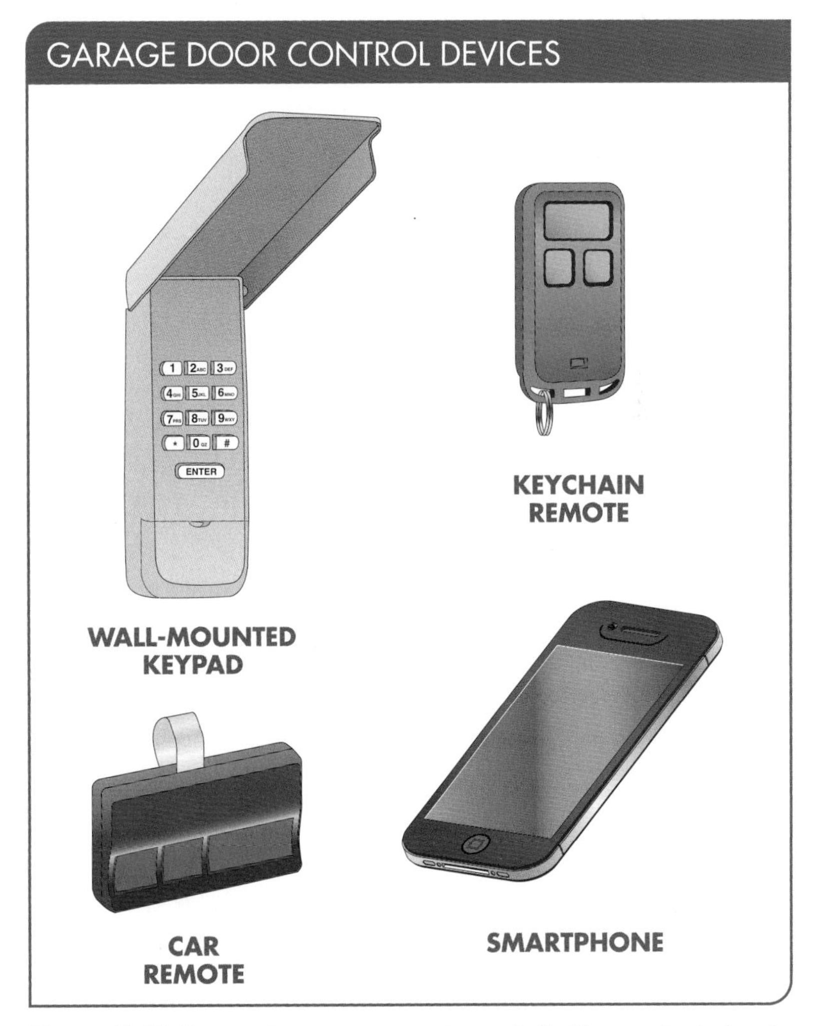

Figure 14-28. Garage door openers can be controlled by remote devices in a car, keychain remotes, apps installed on smartphones, or from keypads or wall-mounted switches.

Water Detection Systems. Water damage is one of the most common home insurance claims and an example of how sensors can be used to protect property against damage. A water sensor placed in an area where leaks may occur will send a wireless signal when moisture is present before damage occurs. A water detection system also automatically shuts off the water supply line if a leak is detected near a toilet, water heater, sump pump, washing machine, or sink. Notification of a leak, as an alert or alarm, can be sent via email to any Wi-Fi-enabled device. Devices with Wi-Fi connections can also be used to remotely adjust the settings and controls on certain systems. **See Figure 14-29.**

Humidity Sensors. A *humidity sensor* is a sensor that detects the amount of moisture that the air can hold at a given temperature. High humidity can lead to mold and mildew, causing allergic reactions and damaging wood and drywall. In most cases, humidity alarms are used in high-humidity locations such as Florida or in other tropical environments. Humidity sensors can provide a variety of warnings and can be monitored remotely.

Freeze Sensors. A *freeze sensor* is a sensor that detects low-temperature conditions and provides a warning before freezing occurs. If the temperature drops below 41°F, the

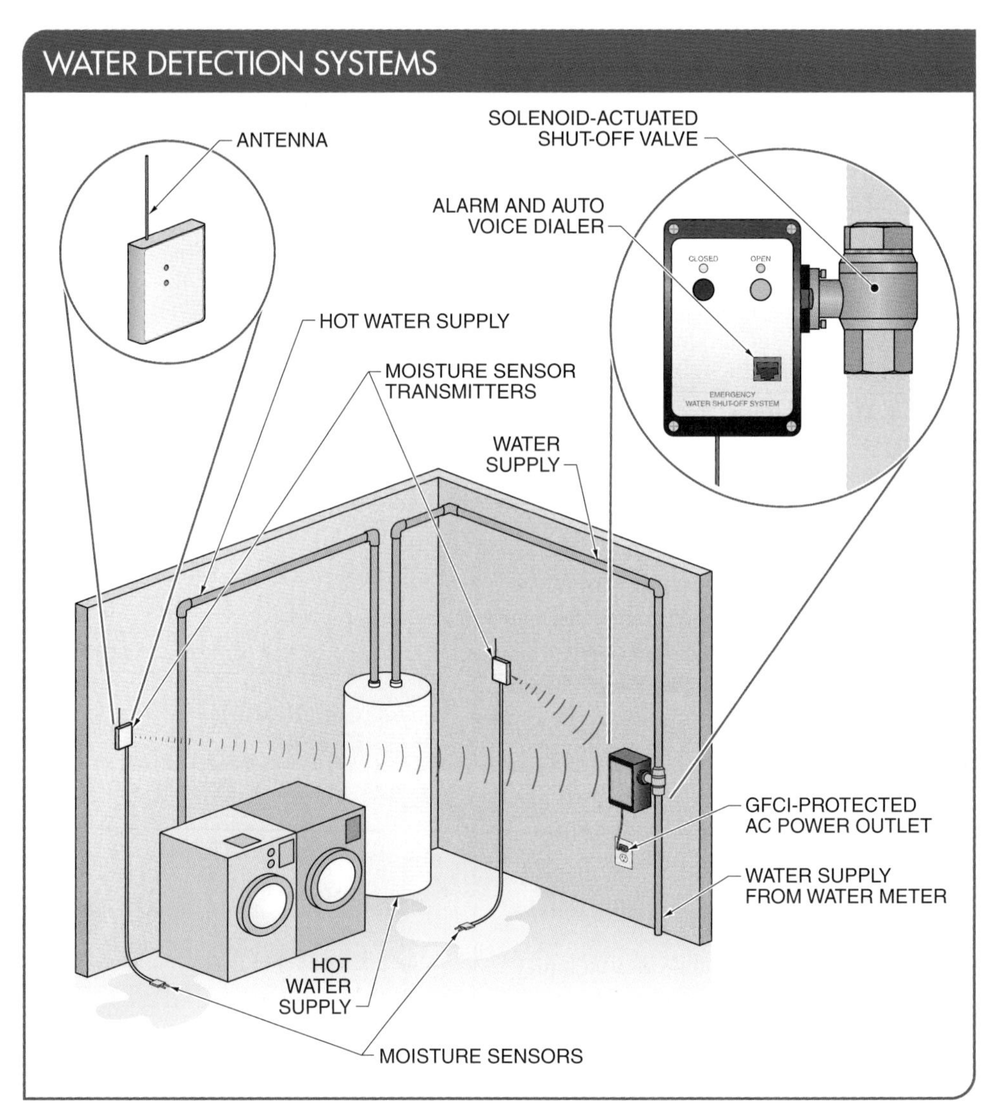

Figure 14-29. Water detection systems detect leaks and automatically shut off the water supply to toilets, water heaters, sump pumps, washing machines, and sinks to prevent water damage.

sensor signals an alarm. Freeze sensors are often located near pipes that may be at risk of freezing.

> ***SECTION 14.3*** CHECKPOINT
>
> 1. What can interior security devices detect?
> 2. What are two types of glass-break sensors?
> 3. What are the four main types of garage protection sensors?
> 4. What types of sensors can be used for structural damage protection?

SECTION 14.4—ENTRY/EXIT SECURITY

Smart home security includes verifying the authorized entry/exit of a dwelling. *Verification* is the process of identifying whether a person is qualified or allowed to enter a residence. Verification involves the collection of specific information from the authorized user. This information is compared to information necessary for entrance into a residence. Once verified by the system, the person can enter the residence.

A failed verification results in a denial of entry. Several attempts at verification may also trigger a security system response to alert the homeowner via a text message or contact the alarm system provider if the residence is remotely monitored. The entrances/exits of a residence can provide security through keypads, mechanical locks, and smart locks. Keypads use a code for verification. Mechanical locks use a key for verification. Smart locks use fingerprints or other physical features of an individual for verification.

Keypads

Keypads may include pushbuttons, LED status indicators, and LCDs. A keypad activates the security system when a code is entered. Generally, keypads are located near entrances and are used to arm and disarm the security system. Illuminated keys provide greater visibility. **See Figure 14-30.**

Figure 14-30. Keypads are used to arm and disarm smart home security systems and often include LED indicators and LCDs.

Stay/Away Arming Modes. Security systems come with two basic arming modes: stay or away. Stay mode turns off all internal motion sensors but keeps door and window sensors active. When the security system is in the stay mode, the residents can move about the house but will still be notified if any entry points are breached. When the security system is in the away mode, the entire system is activated. The away mode is used whenever the house is left and no one else is inside. The authorized access code must be entered on the keypad first before selecting stay or away. Entering the authorized access code only (without selecting stay or away) will disarm the system.

Mechanical Locks

A *mechanical lock* is a lock that includes a tumbler and matching key to permit the lock to be opened. Some standard mechanical locks also incorporate a keypad that uses a number code to provide access. Keypads are available as mechanical pushbutton keypads or digital keypads.

A basic mechanical lock can be upgraded by replacing or retrofitting the lock. A retrofit can be accomplished by placing a motor control kit over the existing lock. The motor control enables the mechanical lock to be controlled directly from a keypad or remotely through wireless controls. **See Figure 14-31.**

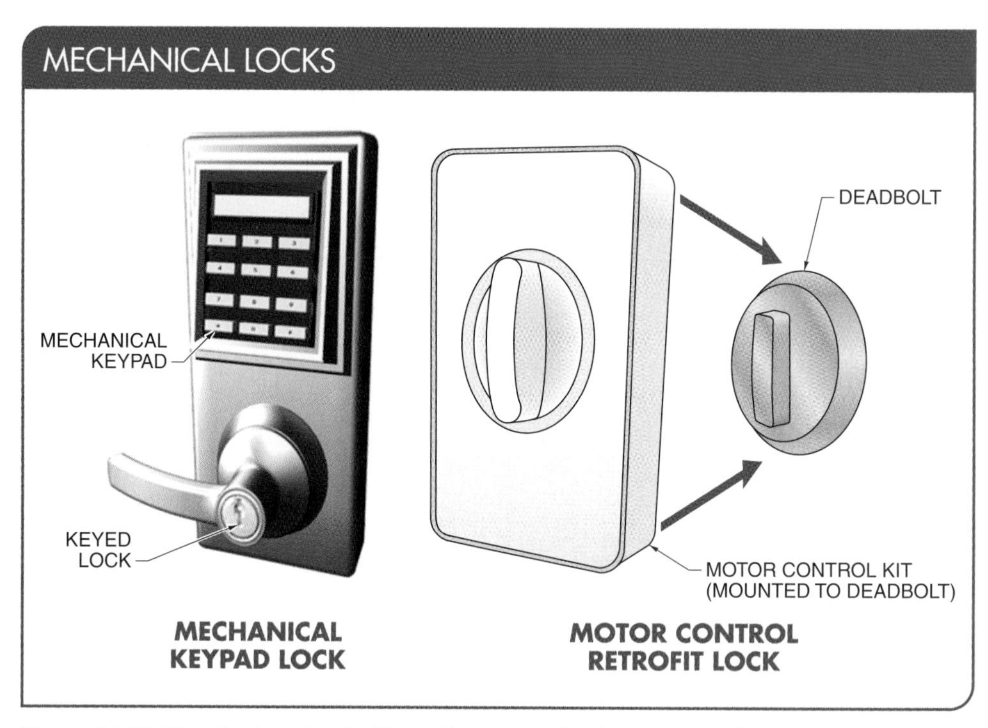

Figure 14-31. Standard mechanical keyed locks can also incorporate a keypad or be retrofitted with a motor control kit that is placed over the existing lock.

Smart Locks

A *smart lock* is an electromechanical device that locks and unlocks a door when it receives instructions from an authorized device using a wireless protocol and a cryptographic key to execute the authorization process. Smart locks are available in a variety of styles and enabled with different technologies. Most smart locks are constructed from hardened steel and tamper-resistant materials for additional security. The internal construction of smart locks includes electronic devices, circuitry, and a motorized drive. When access is approved, the lock/motor assembly operates the locking mechanism to turn the lock bolt.

In addition to its convenience, a smart lock can send a text or email when the lock is opened or closed, as well as which authorized passcode was used. Smart locks can also be installed on interior doors to protect valuables locked in other areas in the house. Smart locks also include capabilities for voice command, touch command, smartphone verification, tamper alarms, shared access, and history tracking.

Smart Lock Controls. Smart lock controls include sensors and a program to process information. Since most access information is stored locally within the lock, the information can be easily updated as needed. Smart locks are battery powered and often come with a low battery warning.

A smartphone can be used as a means of verification to unlock a smart lock. **See Figure 14-32.** Smartphone technology allows a homeowner to send digital keys to an individual who needs temporary access to the residence. A smartphone can also receive notification when an individual arrives or leaves the residence. Smart locks can also trigger smart home products, such as remote lighting control, as the resident moves closer to them.

Smart Lock Access. Smart locks can be opened in different ways. Some smart locks are Bluetooth or Wi-Fi enabled and work through a smartphone. A Bluetooth-enabled lock can be unlocked when an authorized person with a smartphone walks up to it. A Wi-Fi-enabled lock can be unlocked remotely from a smartphone.

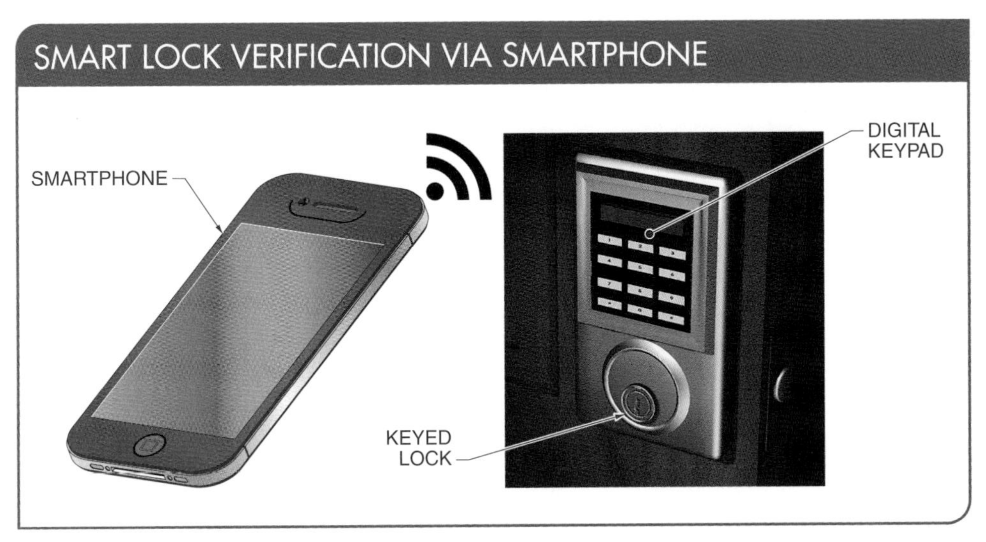

Figure 14-32. A smartphone can be used as a means of verification to unlock a smart lock. The homeowner can also send digital keys to visitors who need temporary access.

Some smart locks use radio frequency identification (RFID), which requires a person to carry a key fob. In this case, the smart lock detects the key fob and automatically unlocks the door. Other smart locks are enabled with biometric technology that can identify the resident and open the door. Biometric technology, such as fingerprint scanning or voice recognition, identifies characteristics that are specific to an individual. Smart locks can also include keypads that use a numeric code as a back-up entry method in case power or the wireless connection is lost. A keypad code that expires after one use can also be given to an individual to enable temporary access.

Smart Lock Security. Cyber security is constantly evolving as both new threats and new security capabilities arise. To reduce security risks, smart lock manufacturers encrypt the smart lock communication system. Several manufacturers encrypt their smart locks using the 128-bit advanced encryption standard (AES) algorithm, which is the same encryption programming used by banks and for classified government data.

Many smart locks come with their own dedicated app that can be used to lock and unlock the door through a smartphone or tablet. Some smart locks can also send text and email notifications to a resident to provide an alert when someone else locks or unlocks the door. Many smart locks can be integrated with home automation systems and devices as well as security cameras to provide additional security measures.

Smart Lock Installation. Smart locks and their apps often come with instructions for homeowners to install and program their own smart locks. Some smart locks have their own proprietary hardware, requiring that the old lock be removed during installation. Other smart locks are designed to be installed over existing deadbolts. Most smart locks are designed to work with the standard lockset openings in typical doors.

Keyless Systems

Keyless systems are based on remote wireless controls that transmit an encrypted numeric code to a lock for entry through a door. The door can be unlocked by being close enough to the keyless lock to wirelessly transmit the code or with the push of a button. Keyless systems use a key fob, which replaces the numeric keypad. A key fob allows a person to arm and disarm the security system before going through the door. Security system providers can usually set up multiple key fobs for individuals or a family. Two types of key fobs are Bluetooth fobs and RFID fobs.

Bluetooth Fobs. Some wireless key fobs are designed to use Bluetooth Low Energy (BLE 4.0) technology since they have a low power requirement. A keyless fob includes a small battery that lasts for many years. Key fob hardware is more expensive than a lock and key. Since fobs must be programmed, they cannot be duplicated like a key.

RFID Fobs. RFID uses EM radio waves to automatically identify and track RFID tags attached to objects. RFID tags contain electronically stored information. Keyless locks using RFID only respond to the one key fob that matches the code for that lock. An RFID key fob can operate a lock from a pocket or purse, freeing an individual's hands. Unauthorized individuals are not able to unlock a door with other key fobs.

Biometric Locks

Biometric locks differ from regular locks because they use biometric verification to unlock the device instead of a key. *Biometric verification* is a means by which a person can be identified by evaluating, or reading, one or more distinguishing biological traits. A *biometric reader* is a specialized device that analyzes a person's physical or behavioral characteristics and compares these characteristics to a database for verification. Basic biometric reader technologies include facial recognition and fingerprint scanning. Some biometric readers can also scan the retina of the eye or identify speech or handwriting.

Biometric Facial Recognition. Biometric facial recognition technology looks at certain characteristics of a person's face as it is scanned by a camera. Iris and retinal scanning technology looks at specific characteristics of the human eye, such as the coloring of the iris or the pattern of the internal blood vessels of the retina, which varies with each person. **See Figure 14-33.** Because power failures can lead to temporary loss of verification data, it is important to maintain a back-up system in place should a power failure occur.

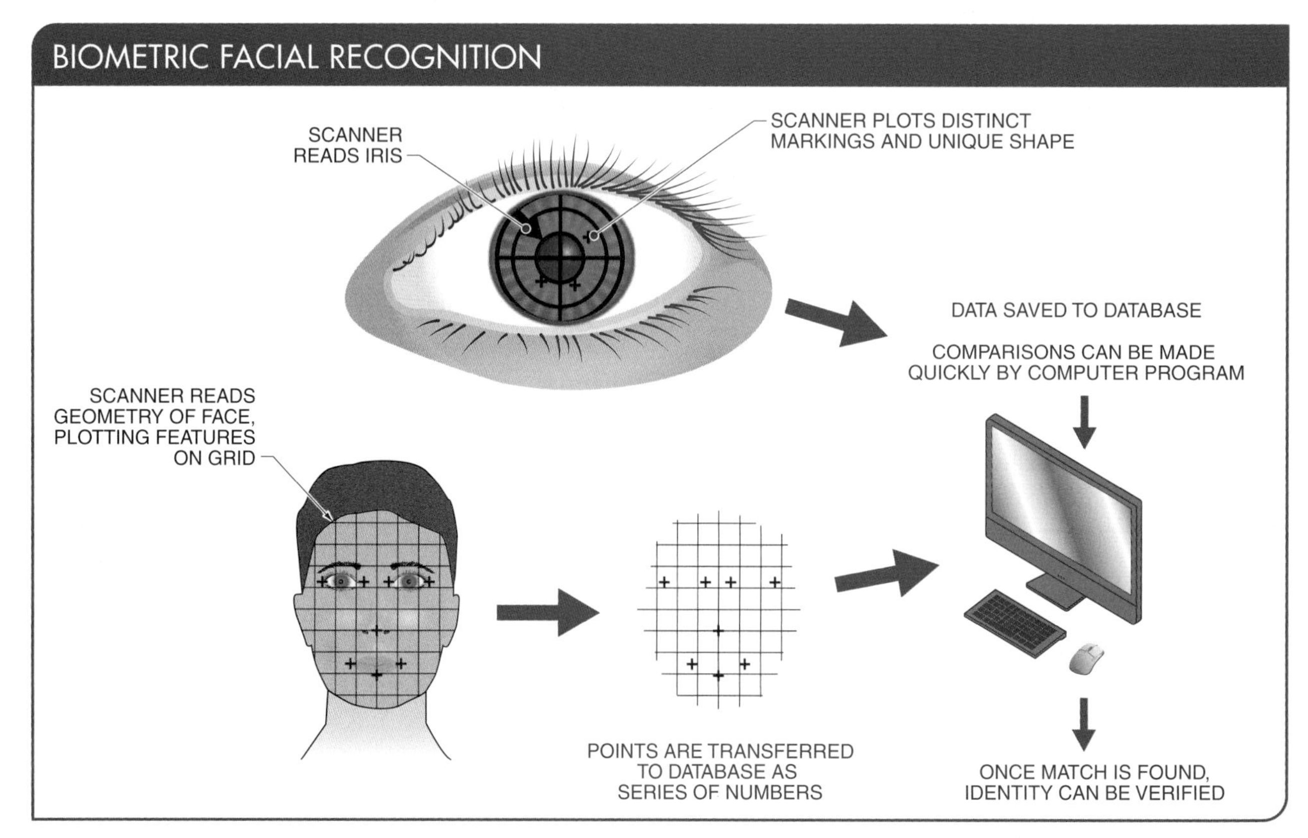

Figure 14-33. Iris and retinal scanning technology looks at specific characteristics of the human eye, and biometric facial recognition technology looks at certain characteristics of a person's face.

Fingerprint Scanning. Fingerprint scanning provides security with a high level of customization. With the swipe of a finger, each authorized individual is automatically identified and presented with a personalized interface and access privileges. An individual must present the same fingerprint for authorization by the scanner before access is approved. **See Figure 14-34.**

Figure 14-34. Fingerprint scanning identifies authorized individuals along with their access privileges.

SECTION 14.4 CHECKPOINT

1. What is verification?
2. How is a keypad used to provide entry/exit security?
3. What is a smart lock?
4. Name two types of keyless fobs.
5. What is biometric verification?

SECTION 14.5—VIDEO SURVEILLANCE

Video surveillance is the monitoring and recording of activity in an area or building. A *security camera* is a video surveillance camera that is used to record activity, which can help detect and prevent crime. Law enforcement and security personnel have been using security cameras for video surveillance for many years.

Video surveillance can be part of a smart home security system that is monitored 24/7 by a paid service or monitored by the homeowner via the internet or cellular phone service. Cameras that are part of a security system are often referred to as "security cams," while internet-connected cameras used to monitor property by a homeowner are often referred to as "webcams." Knowing how video surveillance will be used at a residence can help determine the specifications, features, and type of cameras that will be needed.

Security Camera Specifications

Security cameras are similar to photographic cameras in that light enters a camera through its lens and the image is recorded by a sensor called a charge-coupled device (CCD). A *charge-coupled device (CCD)* is a light-sensitive device that captures light and converts it to digital data. Security cameras with a CCD of a size between ¼″ and ⅓″ typically provide good image quality. A larger CCD does not necessarily result in a higher quality image, but the camera can gather more light in dimly lit locations.

The automatic iris controls the amount of light and the length of time the light can enter the camera. An *automatic iris* is a motorized adjustable aperture that controls the amount of light passing through the lens of a camera to its CCD. Outdoor cameras must have an automatic iris to accommodate changing illumination levels from morning to night and on bright days versus overcast days. The size of the automatic iris opening is called the aperture (f-stop). The lower the f-stop number, the larger the opening in the lens and the more light is allowed through the lens.

The term "lux" is used to describe how well a camera can capture video on low light. One lux (one lumen per square meter) is the amount of light cast by one candle per square meter. A *lux rating* is the rating of the amount of light, or lux, that falls on an object being viewed by a camera. The lower the lux rating of a security camera, the less light is needed for the camera to produce usable pictures (video).

Video cameras record a series of images in rapid succession, which is called the frame rate. *Frame rate* is the number of images that are transmitted each second. The human eye cannot recognize the individual images above a rate of roughly 24 frames per second (fps), which creates the illusion of motion when recorded images are played.

The level of detail that a camera can see is referred to as its resolution. *Resolution* is a measure used to describe the sharpness and clarity of an image or picture. Resolution is specified by the horizontal and vertical pixel size or by the number of equivalent television lines (TVL) in the image. Typical pixel size values for security cameras may be 160 × 120, 320 × 240, or 640 × 480. An ultra-high-resolution camera with facial recognition technology may provide resolution up to 1600 × 1200 pixels. Resolution specified relative to broadcast television uses a TVL number. A standard (non-HD) television image is composed of 480 horizontal lines. A TVL of 480 implies that the image will have the same quality as a non-HD black and white television.

Security Cameras Features

There are several security camera features to consider for residential video surveillance. These features include internet connection, motion detection, recording capability, sound (speakers/microphone), movable versus fixed mount, field of view, resolution, camera size, weather resistance (indoor/outdoor), and night vision capability.

If a security camera has access to the internet, a dwelling can be monitored directly from any web interface. Internet protocol (IP) security cameras can allow the viewing of an area through panning, tilting, or zooming (a capability referred to as "PTZ"). Any device with a web browser such as a computer, tablet, or smartphone can serve as a viewing device for an internet-connected camera. **See Figure 14-35.**

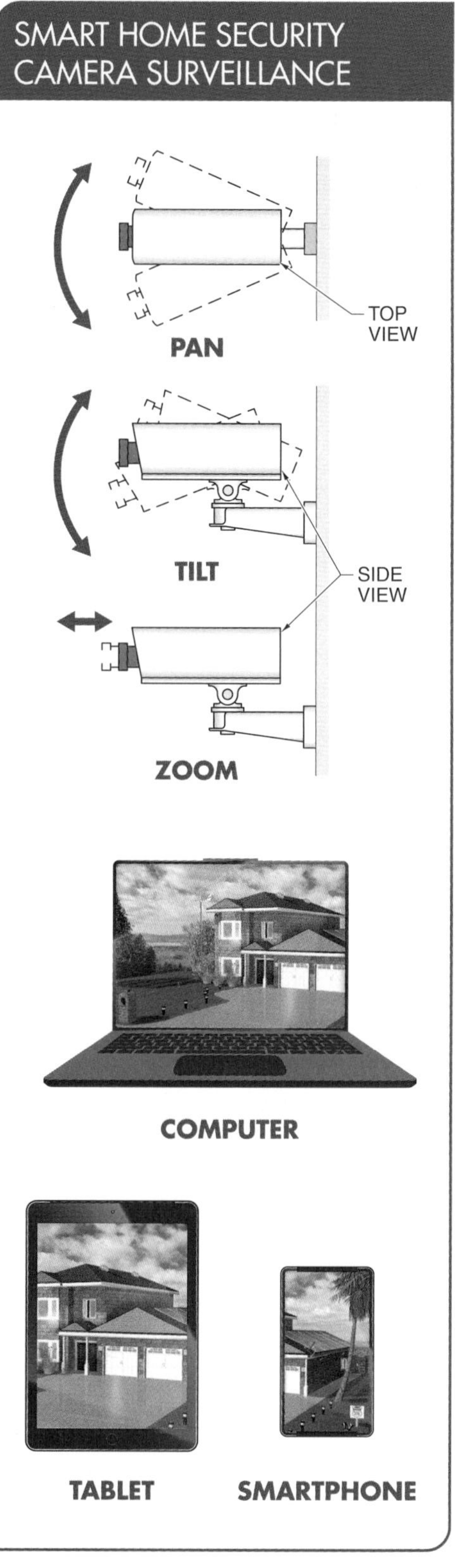

Figure 14-35. IP security cameras allow the viewing of an area through the cameras' PTZ capabilities using internet-connected devices such as computers, tablets, or smartphones.

Some cameras are available with a built-in motion sensor and microphone. This enables the camera to provide a notification alert when someone enters its detection zone. The microphone provides the ability to speak with the person remotely from a smartphone, tablet, or computer. This can create the effect that the homeowner is in the residence. These features can alert a homeowner that someone has approached their door, such as when a package is being delivered.

Some manufacturers have combined all these security camera features with a doorbell and security light with an optional siren activation capability. In such devices, the camera lens is in a fixed position but has an adjustable range. The smart doorbell may also have a stand-alone app and be part of a connected home. The area to be under video surveillance will determine which security camera features, as well as the type of camera, is best for the application.

Security Camera Types

Security cameras are designed for interior (indoor) or exterior (outdoor) environments and can be fixed or movable cameras mounted to walls, ceilings, or desktop surfaces. Common types of security cameras include pan-tilt-zoom (PTZ) cameras, dome cameras, and tabletop cameras. **See Figure 14-36.**

Outdoor security cameras must withstand harsher conditions than indoor security cameras. In cold environments, heaters may be needed. In warm environments, a fan may be needed. Outdoor security cameras should be placed in tamperproof housing to prevent attempts at disabling the camera.

Although fixed security cameras can make distance or range adjustments, they work best in areas where the viewing area does not change. In certain areas, a dome camera may be more useful for a larger view. A dome camera can cover 360° but the person in the viewing area may not know what the camera is viewing. Because dome cameras make it difficult to avoid detection, they are commonly used in retail stores and casinos. Dome cameras are also useful in large areas of a dwelling, such as a basement. Some security cameras offer full color, high resolution, and full motion. Some outdoor cameras provide IR night vision with a viewing range of 60′ to 150′.

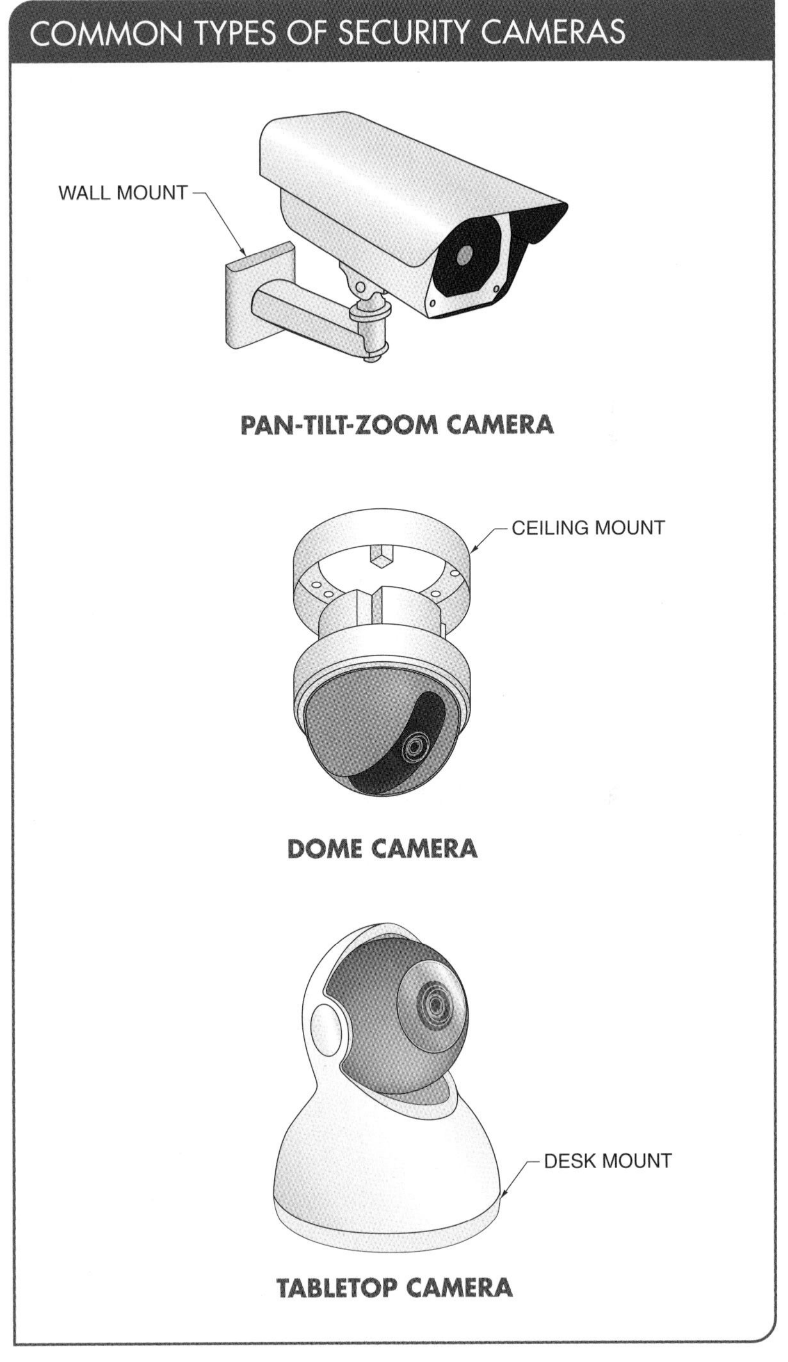

Figure 14-36. Common types of security cameras include PTZ cameras, dome cameras, and tabletop cameras. Security cameras can be wall-mounted, ceiling-mounted, or mounted to flat surfaces such as a desktop.

Security Camera Placement

Security cameras can be placed at all entry points and should be located high enough to provide a clear image of a person's face. Every security camera has a blind spot under the camera, which must be considered in the design of a perimeter security system. Narrowing a camera's field of view to increase its detection range further extends the blind spot. Effective perimeter security systems must be designed with a zero-blind spot approach, which requires that each camera's view overlap the blind spot of the camera in front of it.

Since many new home security systems are wireless, security cameras can be installed by the homeowner. However, if the installation will be part of a complete home security system or integrated into a whole house system, a certified professional may be necessary. A certified professional can provide additional information and technical support.

Outdoor Movement Filtering

Movement from wind, trees, leaves, small animals, and blowing trash can trigger alarms in outdoor video surveillance systems that include motion detection. If a system consistently produces nuisance alarms, a monitoring service or homeowner may begin to ignore them. Smart security cameras use image processing to filter out such movement. In that way, security cameras only trigger alarms for actual threats.

Digital Video Monitoring and Recording

Video images from security cameras are displayed on monitors. A monitor converts video signals from the security camera into monochrome or color images. Monitors are also used to view live video and video played from recording devices. Multiple video signals can be displayed on a single monitor.

It is too difficult for a person to have their full attention on multiple video images simultaneously, especially for hours at a time. In addition, situations may arise where it is necessary to have documented evidence of an event. Therefore, security systems with cameras should include some type of video recording device, such as a digital video recorder (DVR). **See Figure 14-37.**

A DVR for security purposes requires features not found in a typical DVR used for home entertainment. The sole purpose of a DVR in a home security system is to record images captured by the security cameras positioned around the home, either continuously or only when the camera detects motion. If a DVR has a built-in web server, recordings can be viewed using an internet-connected computer, tablet, or smartphone. A DVR requires approximately 4 terabytes (TB) of storage on average for four cameras.

The Cloud

Cloud storage, commonly referred to as "the Cloud," is a method for delivering services and storing data that can be retrieved from the internet through web-based tools and applications. **See Figure 14-38.** With the Cloud, a client pays for the services or storage space that they use, while maintenance of the storage and computing system is done without their involvement. The National Institute of Standards and Technology (NIST) has issued guidelines for the Cloud, although it also suggests that the federal government should protect the privacy of citizens.

Rather than video files being stored on a proprietary hard drive or local DVR, cloud storage makes it possible to save video files to a remote location or to access live footage from any web-enabled device. This service also saves storage space on desktop or laptop computers. Additionally, the structure of cloud storage allows users to upgrade software more quickly since software companies can offer their products for immediate download via the web rather than through traditional discs or flash drives. Other benefits of cloud storage include the following:

- Adaptability—Proprietary programs and applications allow owners control.
- Reliability—Technically competent third-party companies host the storage system.

- Scalability—Users can upgrade as needed.
- Security—Many companies offer encrypted connections.
- Performance—A service provider monitors system performance.

Cloud Security. The management of the Cloud is centralized and has the benefit of close supervision and control. It is, however, a prime target for cyber threats such as hacking. Other concerns include greater dependency on service providers, the possibility of being locked-in with one provider, thc possibility that providers will cease supporting a product, and a dependency on a reliable internet connection.

> TECH TIP
>
> *Many cellular telephone service providers also offer home security system packages that include security cameras and unlimited cloud storage of saved videos.*

All information going into the Cloud should have strong security measures embedded. Emergency back-up plans should be in place in the event of a security breach. Strong passwords are also a necessary measure for good security. Passwords should be used on every device in a residence that is installed on the network, including routers.

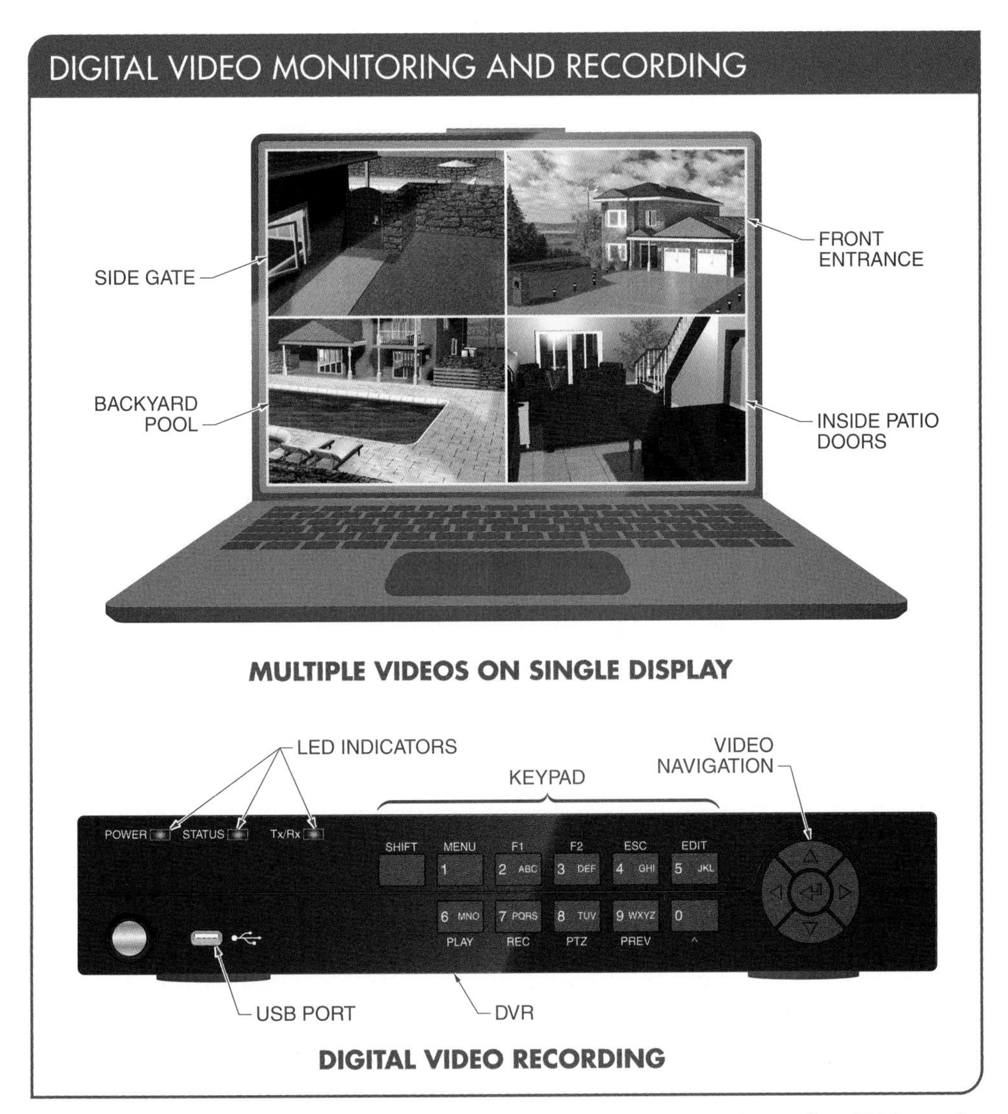

Figure 14-37. Since multiple video images can be displayed on a single monitor, DVRs can be used record video images and document events.

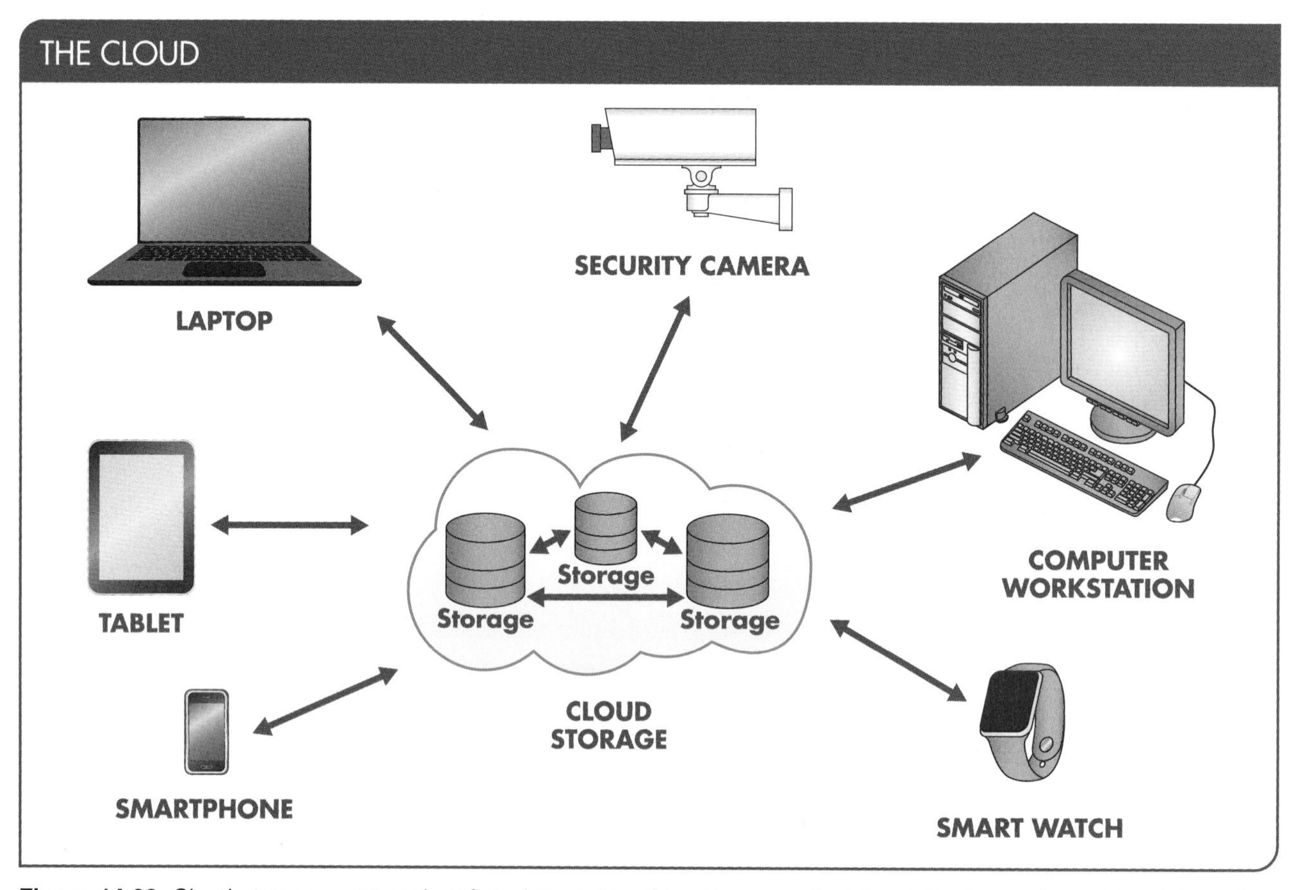

Figure 14-38. Cloud storage, commonly referred to as "the Cloud," is a method for delivering and receiving information through web-based applications.

> ***SECTION 14.5*** **CHECKPOINT**
>
> 1. What are the common types of security cameras?
> 2. What is an essential design consideration for an effective perimeter security system when mounting security cameras?
> 3. What is outdoor movement filtering?
> 4. What is the purpose of a DVR in a home security system?
> 5. What is cloud storage and how is it used for smart home security?

SECTION 14.6—SECURITY SYSTEM INSTALLATION

A permit is often required before a home security system can be installed. Permits help fire and police departments identify a residence when responding to an alert. The installation cost of a security system depends on the size of the system and who installs it. The cost may be less when a homeowner, rather than a professional, installs the system.

It is important to install a security system that has a battery backup (to provide power during outages), access code fluidity (to permit ease of adding or removing users), instant alerts, and rolling code technology that changes verification codes every few seconds to prevent eavesdropping and re-cording. Other security system installation considerations include security monitoring services, security system installation planning, and sensor mounting.

Security Monitoring Services

A 24/7 security monitoring service can be contracted monthly with an independent monitoring service or it may be provided as part of a professional installation. A security monitoring service should be selected based on whether they have 24/7 monitoring, good customer service, fast response times, and quality equipment.

A 24/7 monitoring service reduces the chance of a false alarm being sent to authorities. When a security breach is detected in a dwelling, the security system transmits a message to a central reporting station. When the message is received, the homeowner is called first to determine whether the alert is real. If the call is not answered or the keypad entry/exit password is incorrect, authorities are notified along with an alert sent to the homeowner. For these reasons, a security system control panel should be connected by a landline phone or wireless transmitter to a 24/7 monitoring service. **See Figure 14-39.**

Security System Installation Planning

An effective smart home security system requires detailed planning. The following steps can be used as a guide for the installation of an effective system:

1. Start with a floor plan or sketch to determine the entry access points. Have a separate floor plan for the main floor and each level of a multistory dwelling that can be reached without a ladder. Include access doors to a garage from inside a dwelling and to an outside garage entrance door.
2. Number all of the entry access points, beginning with the front door and moving in a clockwise direction, to determine the number of sensors required.

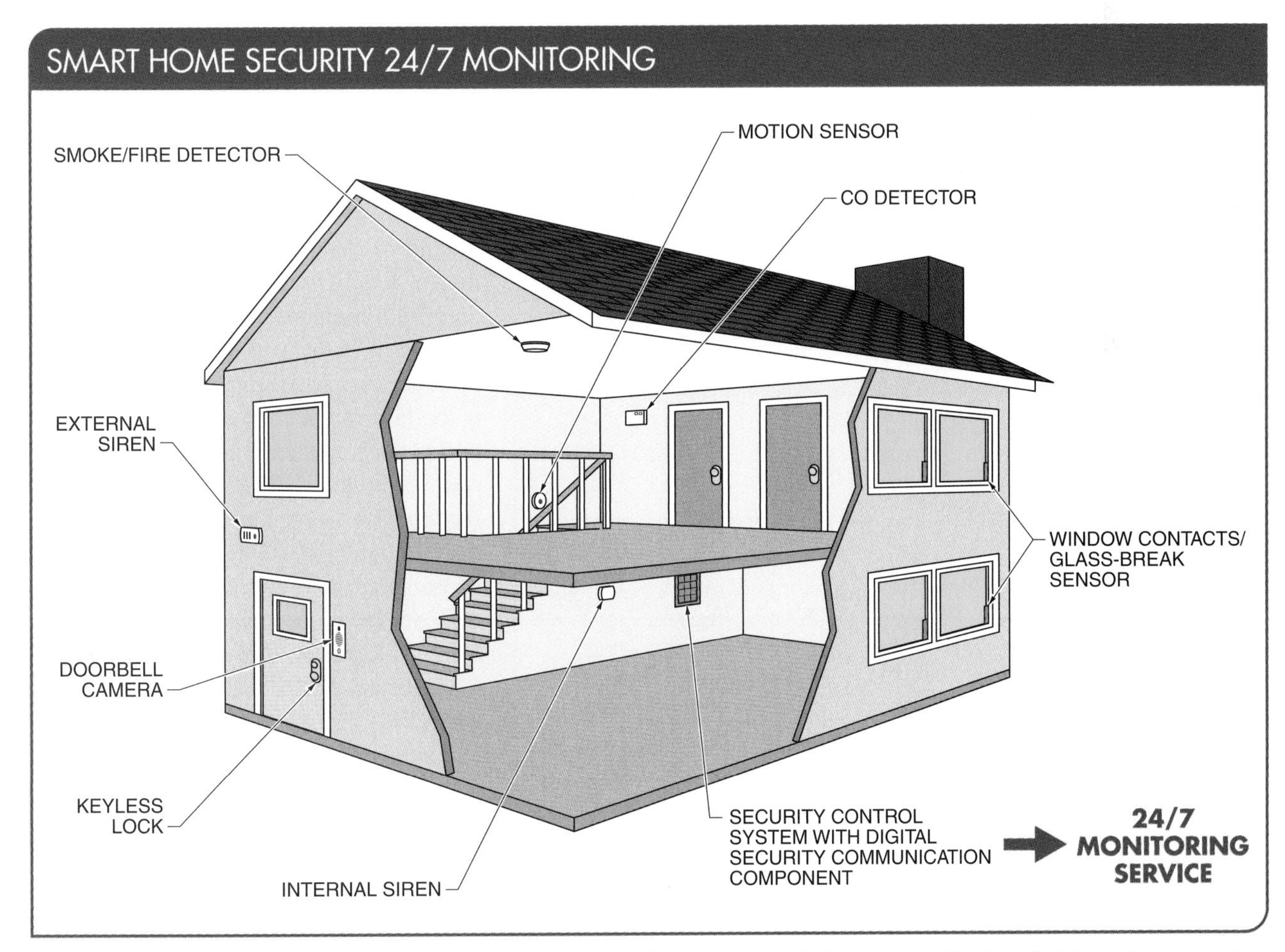

Figure 14-39. A control panel should be connected by a landline phone or wireless transmitter to a 24/7 monitoring service that is professionally staffed to handle security alerts.

3. Determine the type of security panel needed. The two common types of security panels are cabinet-style and self-contained security panels. A cabinet-style security panel consists of a keyboard that mounts near the door and a control panel (system control circuit board) that mounts near the phone interface. A cabinet-style security panel alerts a monitoring service even if the keypad is disconnected. A self-contained security panel contains both a control panel and a keypad in a single unit and is easy to install. However, the single unit could be knocked off the wall before an alert is sent to a monitoring service unless it has a built-in sensor. An RJ-31X phone jack is used to connect the control panel to the phone line for either type of security panel. **See Figure 14-40.**
4. Determine whether the security system will be connected to a 24/7 security monitoring service or if it will signal an alarm siren.
5. Ensure that the security monitoring service is equipped to handle a digital phone service Voice over Internet Protocol (VoIP).

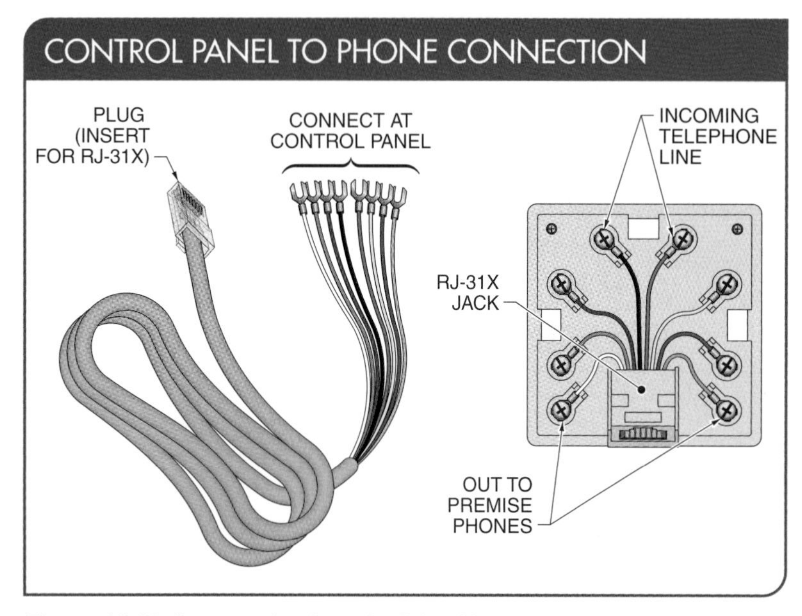

Figure 14-40. A connection from the RJ-31X phone jack to the control panel overrides the premise phone lines during a security breach and connects to a 24/7 security monitoring service.

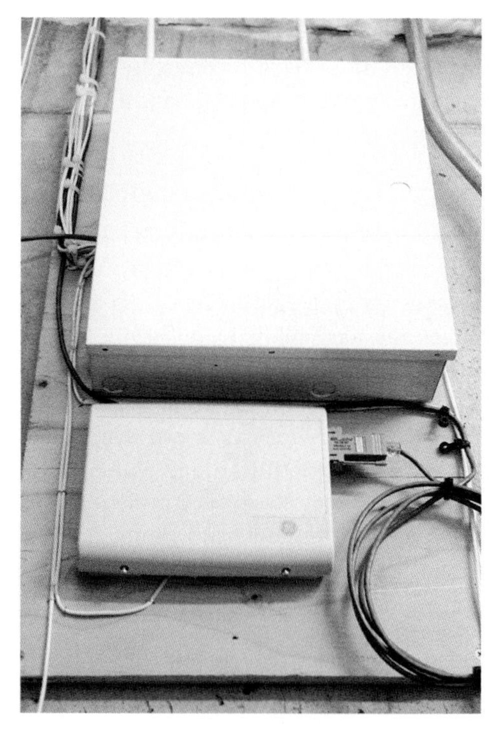

Security control panels are typically connected with a hardwired, RJ-31X-compatible communication cable or phone lines.

Sensor Mounting

The proper mounting of sensors is a fundamental requirement for an effective home security system. When a door or window sensor is installed, the sensor and the magnet must align with each other. If the magnet is too far away from the sensor, the sensor will not detect when the door or window is opened or closed. The transmitter is often mounted on a fixed structure, such as the door or window frame. The magnet can be mounted on the moving piece, such as the door or window. Some manufacturers include alignment dots on the sensors and magnets to make installation and alignment easier. **See Figure 14-41.**

Some sensors can be recessed into window casements or solid doors for eye appeal and to hide them and prevent tampering. Although some wireless devices are visible, newer units called vanishing sensors are now available. Vanishing sensors are thin, can be applied using adhesive tape, and blend into an environment. **See Figure 14-42.**

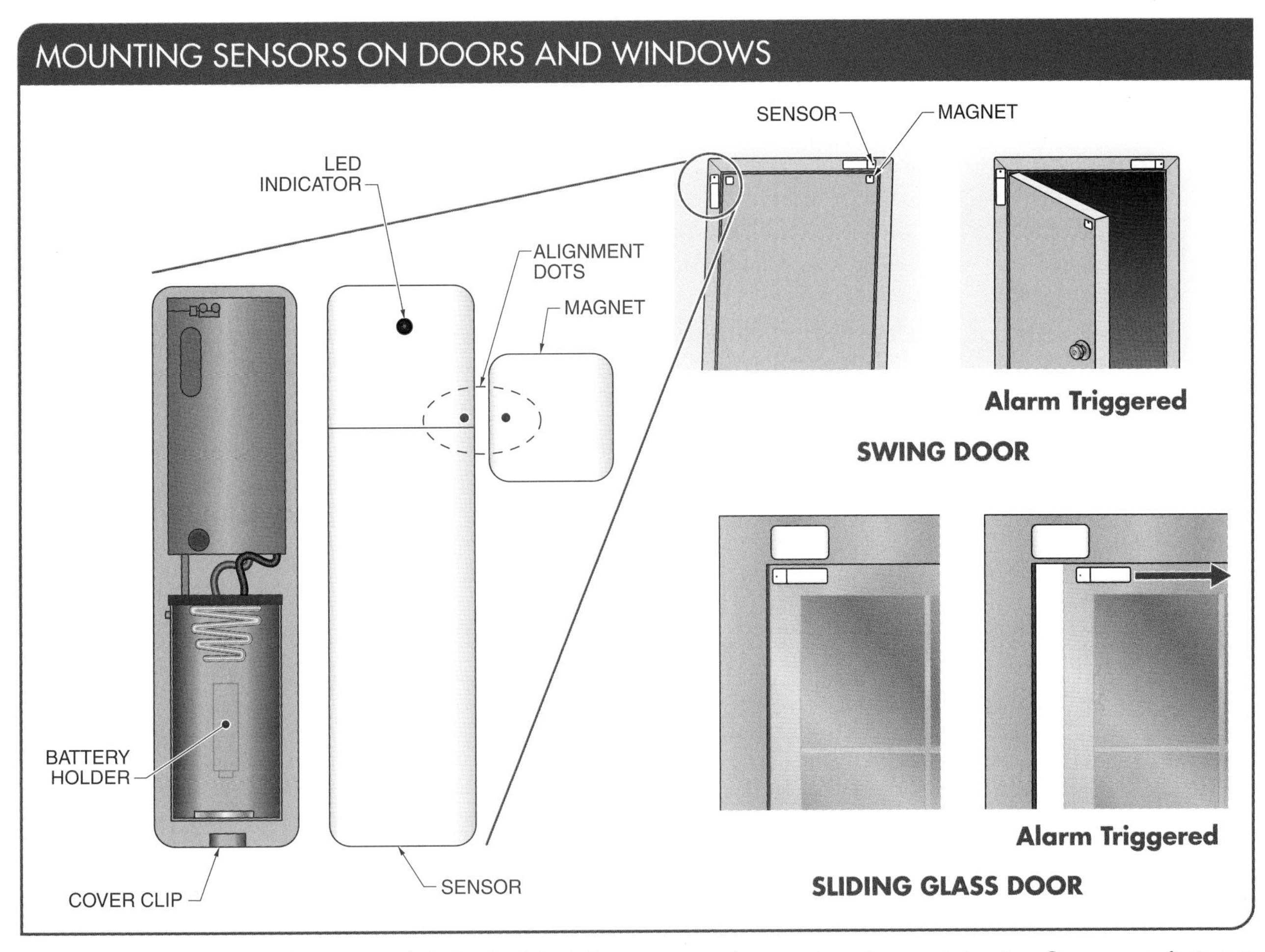

Figure 14-41. When a window sensor is being installed, the sensor and magnet must accurately align. Some manufacturers add alignment dots so that the units can be installed and checked before being activated.

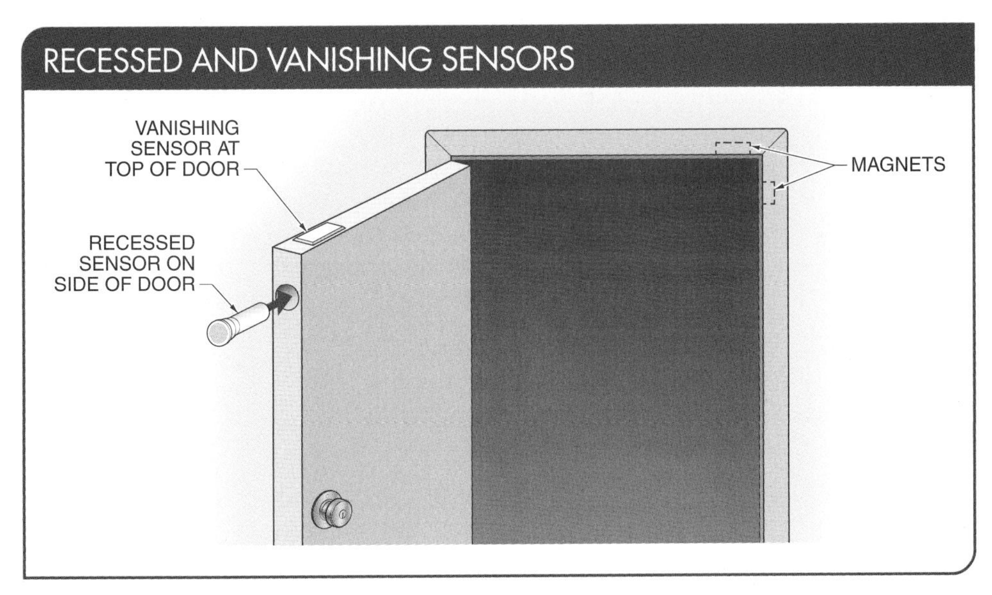

Figure 14-42. Sensors can be recessed into window casements or solid doors so as to be hidden from view. Vanishing sensors are thin and can be applied using adhesive tape.

Special-Purpose Sensor Locations. Anything that opens and closes, such as cabinets and drawers, can be protected using a door or a window sensor. For example, sensors can be attached to liquor, firearm, or medicine cabinets. Sensors can also be attached to outbuilding doors and windows (provided that the range of the sensor is not exceeded). Sensors can also be attached to in-window air conditioners to deter their unauthorized removal.

SECTION 14.6 CHECKPOINT

1. How should a security monitoring service be selected?
2. How is a control panel connected to a phone line?
3. What are the five basic steps for planning the installation of an effective home security system?
4. What is a fundamental installation requirement for a door or window sensor?

14 Security Systems and Smart Home Applications

Chapter Activities

Name ______________________________ Date ____________________

Common Types of Security Cameras

Identify the different types of security cameras shown.

_______________ **1.** Tabletop camera

_______________ **2.** Pan-tilt-zoom camera

_______________ **3.** Dome camera

A

B

C

Outdoor Lighting Devices

Identify the outdoor lighting devices shown.

_______________ **1.** Low-voltage lighting

_______________ **2.** Motion detectors

_______________ **3.** Solar lights

_______________ **4.** Photo cell

A

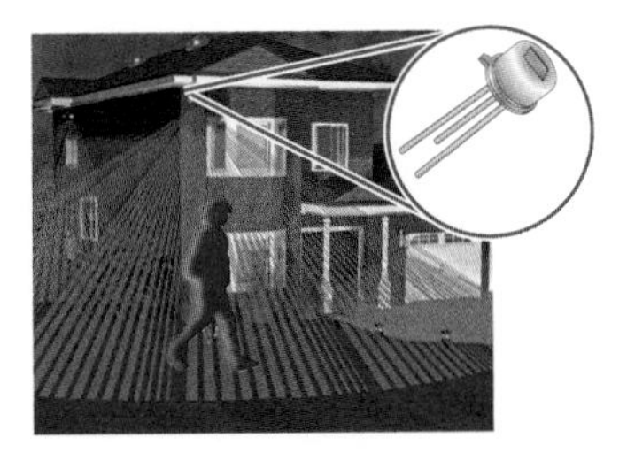

B

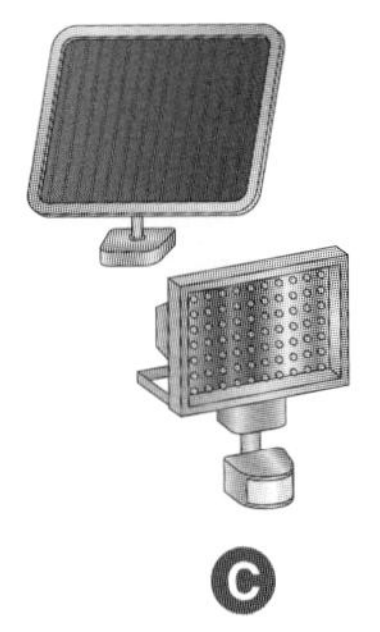

C

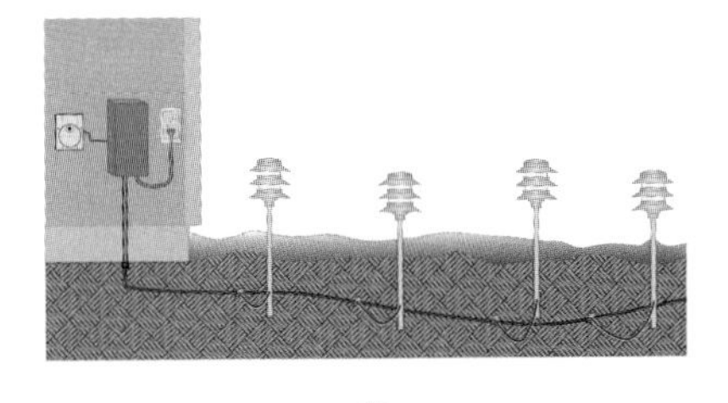

D

Garage Protection Sensors

Identify the different garage protection sensors shown.

_______________ **1.** Tilt switch sensors

_______________ **2.** CO sensors

_______________ **3.** IR sensors

_______________ **4.** Pressure sensors

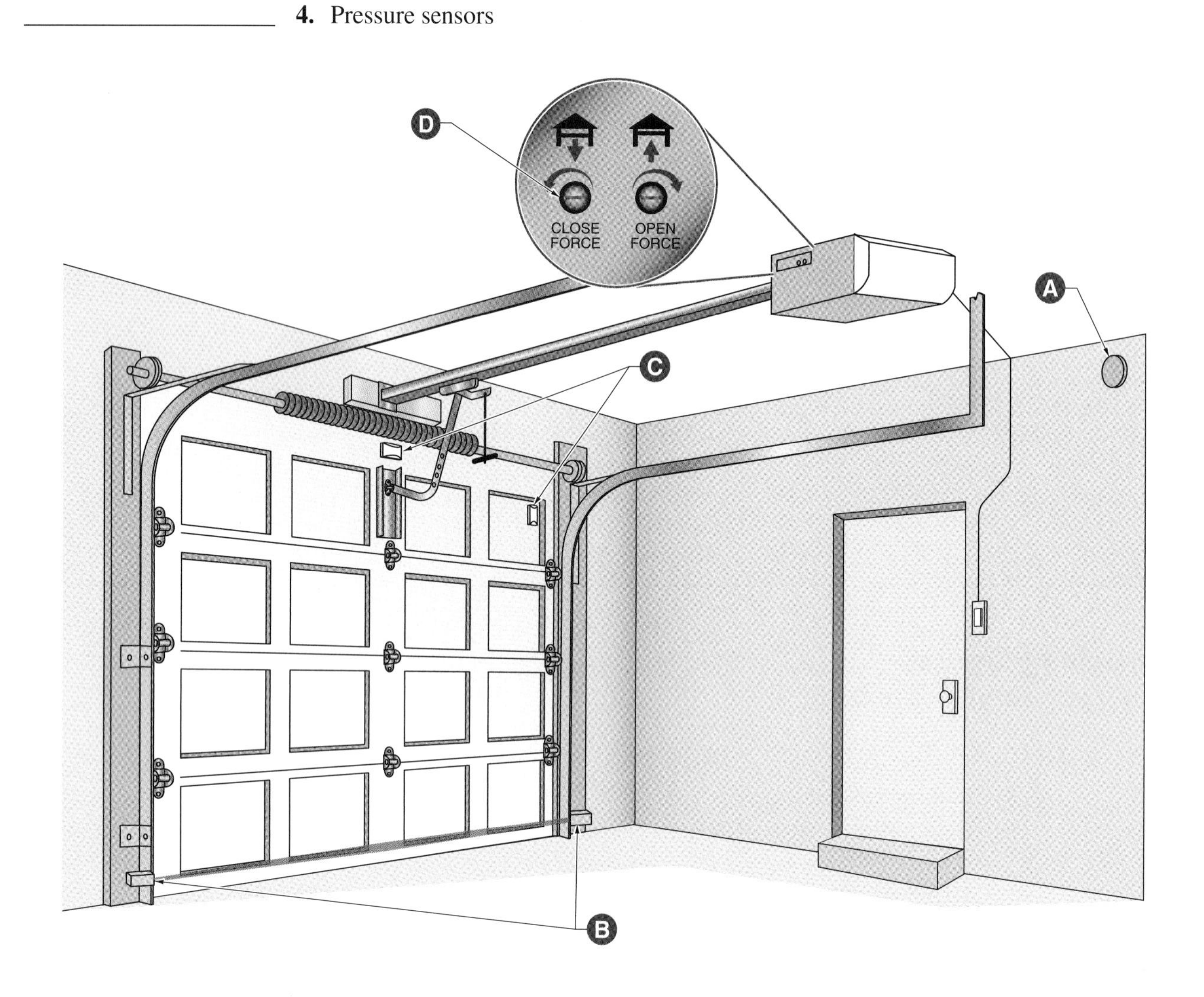

15 Fire Alarm Systems and Smart Home Applications

SECTION 15.1—FIRE ALARM SYSTEM DEVICES AND FIRE SAFETY

- Explain fire safety.
- Describe control panels.
- Describe annunciators.
- Compare smoke detectors, heat detectors, and carbon monoxide (CO) detectors.

SECTION 15.2—ALARM SYSTEM CONNECTIONS

- Describe registered jack (RJ) connectors.
- Compare and contrast electromechanical relays (EMRs) and solid-state relays (SSRs).
- Explain low-voltage wiring.
- Describe wireless and wired circuits.
- Explain series- and parallel-connected circuits.

Learner Resources

ATPeResources.com/QuickLinks
Access Code: 791712

A smart home fire alarm system uses devices to detect the presence smoke, fire, and carbon monoxide (CO) and to provide necessary warnings. A smart home fire alarm system can be a hardwired system or a wireless system, but both are governed by the NEC®. All installed fire alarm systems must be approved and inspected by the authority having jurisdiction (AHJ) to enforce code requirements. Some cities have codes in addition to the NEC® that must also be followed.

SECTION 15.1—FIRE ALARM SYSTEM DEVICES AND FIRE SAFETY

Fire safety includes fire prevention, installing smoke detectors and fire extinguishers on every level of the dwelling, and having an evacuation plan. Fire prevention includes good housekeeping, maintaining appliances and electrical wiring, and the safe use of extension cords, space heaters, and candles. An evacuation plan should include knowledge of all exits, crawling to exits to stay underneath rising smoke, and meeting at an agreed location outside the residence.

Fire alarm system devices include control panels, annunciators, and detectors to sound alarms and notify residents of a possible fire or emergency. The control panel receives input signals from smoke, heat, and CO detectors and outputs signals to annunciators that give visual and audible warnings. Fire alarm signals are sent from a control panel to a 24/7 monitoring service where the alarm signal is screened by professional operators to reduce false alarms. A verified fire alarm is then sent to the fire department. UL® and other inspecting agency certifications should be listed on all fire alarm system components.

Fire Extinguishers

Fire extinguishers that are labeled for Class A, B, and C fires should be installed on every level of a dwelling. **See Figure 15-1.** Fire extinguisher labels include class symbols for each type of fire as well as operating instructions for extinguishing fires. Each class is denoted by a letter indicating the type of material and how to extinguish it. A fire extinguisher can be used by aiming at the base of a small fire. If the fire is not extinguished, the dwelling must be evacuated immediately using a predetermined evacuation plan.

Control Panels

A *control panel* is an electronic device that receives input signals from detectors, processes information, turns output devices on and off, and sends notifications to a monitoring service. The control panel can determine which detector in a residence is transmitting the alarm signal and automatically notifies local authorities and the homeowner with a prerecorded voice message and/or text message in the event of a fire or emergency. **See Figure 15-2.**

Solid-state relays and the programming module are mounted on the control panel circuit board, which is inside the control panel enclosure. The programming module determines how relays open and close in response to information obtained from detectors, such as smoke detectors and heat detectors.

The operating electronics and user interface of a control panel can be powered by standard branch circuit wiring or a DC power source. The control panel enclosure usually includes a digital alphanumeric LCD that shows the status of the system and provides troubleshooting information. There is also a touch pad or keypad to enter codes so that the resident can silence the alarm and reset the system if the panel triggers a false alarm.

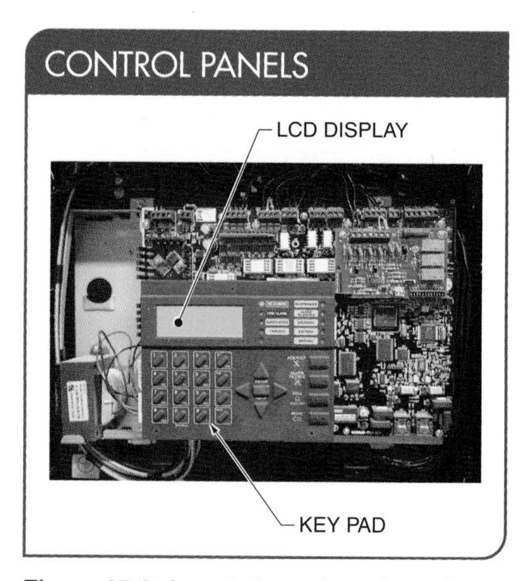

Figure 15-2. A control panel receives signals from detectors and turns on and off devices such as sirens and strobe lights.

Annunciators

When a control panel receives an input signal from a detector, an output signal is sent to an annunciator (an alarm). An *annunciator* is an audible or visual output

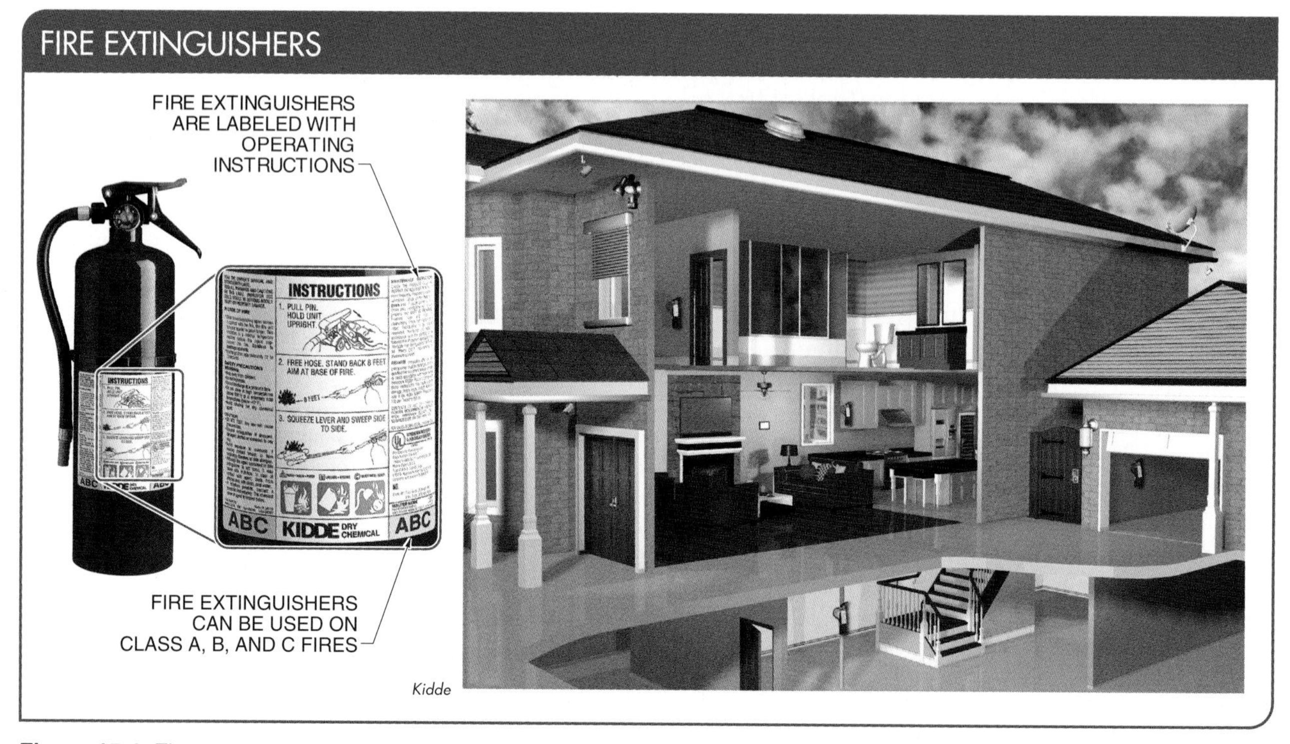

Figure 15-1. Fire extinguishers labeled for Class A, B, and C fires should be placed on every level of a dwelling.

device that notifies residents that one or more detectors have been activated. The different types of annunciators include sirens, strobe lights, and voice commands through speakers. Some annunciators combine audible and visual outputs in one device. **See Figure 15-3.**

ANNUNCIATORS

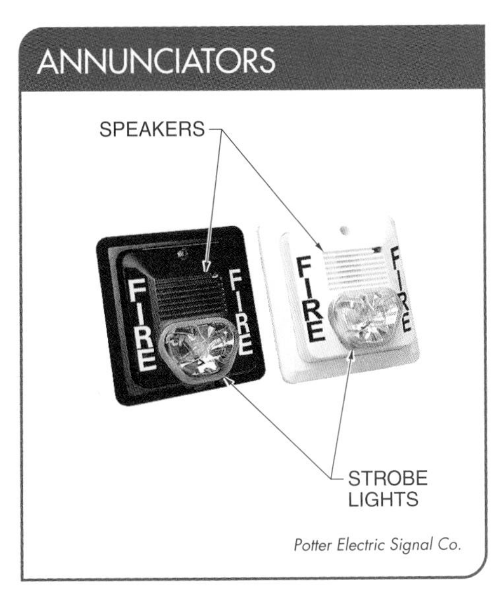

Figure 15-3. An annunciator is an audible or visual output device that includes sirens, strobe lights, and speakers for voice commands.

SMOKE DETECTORS

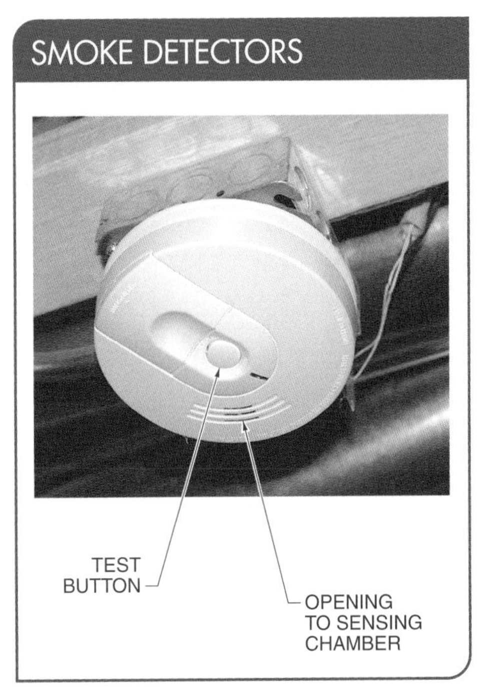

Figure 15-4. A smoke detector is a device that senses smoke optically (photoelectric) or by physical process (ionization).

Smoke Detectors

Per the NFPA, smoke detectors should be placed throughout a residence. A *smoke detector* is a device that senses smoke optically (photoelectric) or by physical process (ionization). Some smoke detectors use both types, photoelectric detectors and ionization detectors, in the same device. Smoke detectors come with several features and are available as battery-operated (DC) or AC-powered (hardwired) types.

Smoke detectors are activated by the presence of smoke particles. Smoke particles are composed of gases, solid particles, and liquid droplets suspended in the air. Smoke particles are often invisible to the unaided eye. Smoke detectors sense these particles as they enter a small sensing chamber within the device. **See Figure 15-4.**

Photoelectric Smoke Detectors. Photoelectric smoke detectors are the most common type of smoke detector. A *photoelectric smoke detector* is a device that has an infrared LED and a photocell sensor in the sensing chamber to detect smoke particles. When smoke particles enter the sensing chamber, they scatter the IR light from the LED onto the photocell sensor, which results in an alarm condition. When no smoke is present, the IR light from the LED does not reach the photocell sensor. The device resets once smoke concentrations fall below threshold levels. **See Figure 15-5.**

Ionization Smoke Detectors. An *ionization smoke detector* is a device that uses a radioactive source to ionize (electrically charge) the air to conduct a small current in the sensing chamber to detect smoke particles. Smoke particles entering the sensing chamber interfere with the normal current flow and the reduction in current is detected by the sensor, which initiates an alarm. **See Figure 15-6.**

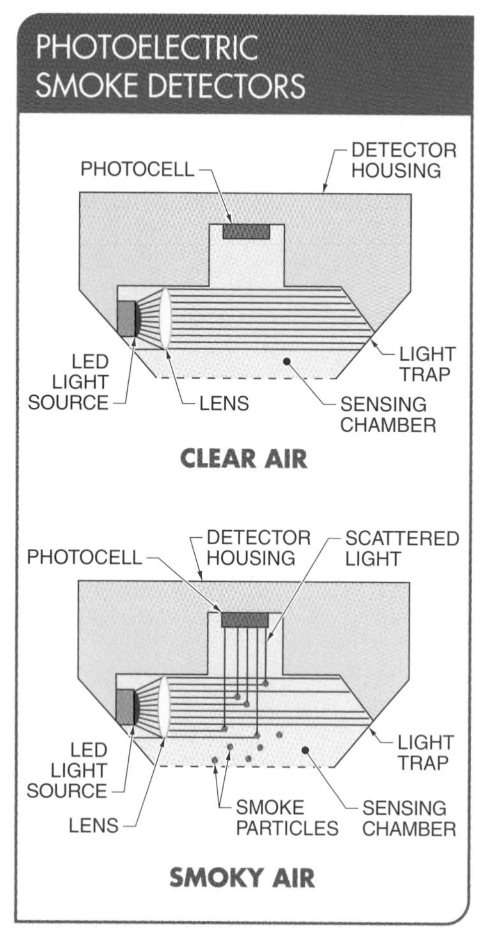

Figure 15-5. Photoelectric smoke detectors sense light scattered by smoke particles within the sensing chamber.

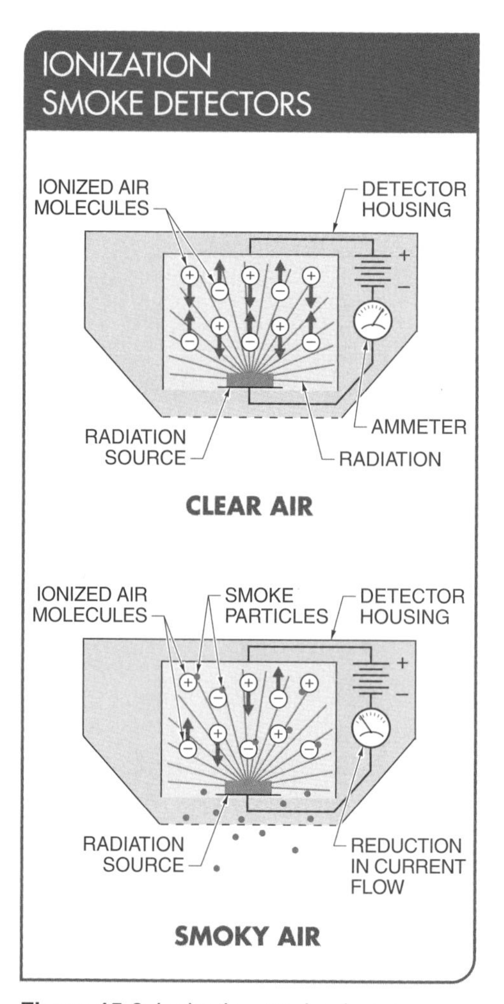

Figure 15-6. Ionization smoke detectors use a radioactive source to electrically charge the air in the sensing chamber. Smoke particles interfere with the normal current flow and trigger an alarm.

> TECH TIP
>
> *Battery-powered smoke detectors are recommended for residential installations because of their ability to operate in the event of a power outage. Batteries should be changed at least once a year for optimal operation.*

Ionization smoke alarms are generally more responsive to flaming fires, and photoelectric smoke alarms are generally more responsive to smoldering fires. Dual-sensor smoke detectors combine ionization detectors and photoelectric detectors in the same device.

Smoke Detector Features

Smoke detectors can have many features. Smart home smoke detectors can include LED status indicators, LED escape lights to illuminate hallways, speakers for voice alerts that announce "fire detected" or other instructions from recordings, deaf/hard-of-hearing alarms and accessories such as strobe lights or bed shakers, test/silence buttons, and hush buttons that will deactivate nuisance alarms for several minutes. **See Figure 15-7.** Stand-alone smoke detectors can be connected to other detectors or warning devices using wireless modules.

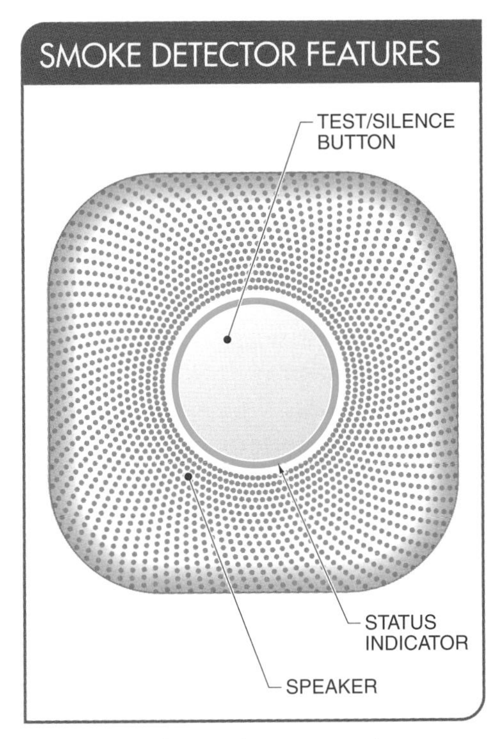

Figure 15-7. Smoke detectors can have many features, including lights, voice alerts, and recording capabilities for voice commands for children.

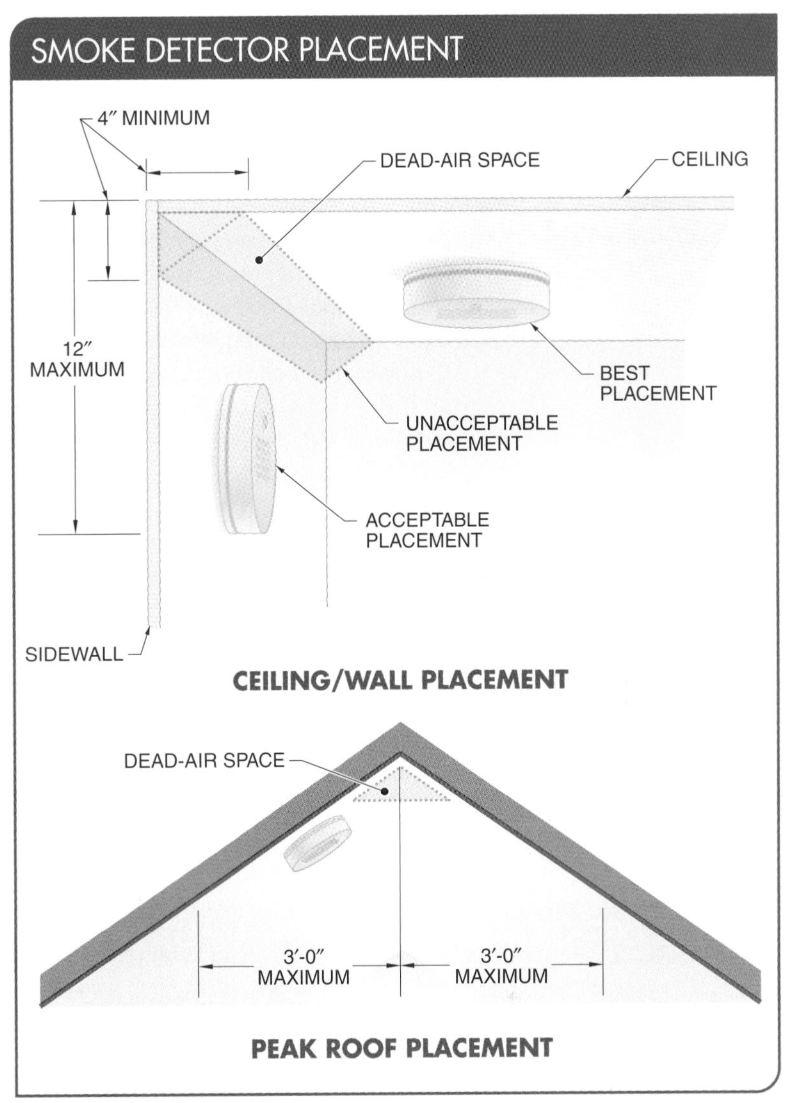

Figure 15-8. Smoke detectors should be placed on every level and in every sleeping area of a dwelling and away from dead-air spaces.

Smoke Detector Placement

The NFPA recommends that smoke detectors be placed on every level of a residence and in every sleeping area. It is important to place smoke detectors where they will activate at the earliest presence of smoke. Since smoke and heat rise to the ceiling and spread horizontally, smoke detectors should be placed on ceilings or high on walls to detect the first trace of smoke. Mounting a smoke detector on the ceiling in the center of a room places it closest to all the points in the room. Smoke detectors should not be placed in dead-air spaces, such as within 4″ of a roof peak or within 4″ of where ceilings and walls meet. **See Figure 15-8.**

Smoke detectors in basements should be placed near the bottom of the stairs and not at the top of the stairs were dead air may be trapped due to a closed door. Smoke detectors should also be placed in attics that have electrical equipment such as furnaces, air conditioners, or heaters.

Smoke detectors may not function properly if placed in certain locations. For example, smoke detectors should not be placed in garages or other spaces where the temperature can fall below 32°F or exceed 100°F. To avoid false alarms, smoke detectors should be placed at least 10′ away from cooking appliances, furnaces, water heaters, and high-humidity areas such as bathrooms, near dishwashers, or washing machines. In addition, smoke detectors should not be placed near windows, doors, or ducts where drafts might interfere with their operation. Smoke detectors should never be painted or affixed with stickers or decorations.

AC-Powered Smoke Detector Installation

AC-powered smoke detectors must meet NFPA code requirements. For example, AC-powered smoke detectors must have a visible "power on" indicator (to show that they have 120 VAC power), have a battery back-up source of power, and be capable to power the smoke detector for 7 days after the low-battery warning begins. In addition, AC-powered smoke detectors should be installed on a dedicated circuit and should never be installed on a switched outlet or on a GFCI-protected circuit.

It is also important to follow the NEC® and local codes when installing AC-powered smoke detectors. AC smoke detectors should be wired in parallel. A 14-2 cable is fed to the first smoke detector, then a 14-3 cable is used to connect any additional smoke detectors. The yellow wire from each smoke detector interconnects the entire system so that all detectors sound when one is triggered. **See Figure 15-9.**

> TECH TIP
>
> *Smoke detectors, including both AC and DC types, should be periodically inspected and cleaned of obstructions, such as dust, dirt, spider webs, and other debris, that can cause a smoke detector to fail or malfunction.*

Heat Detectors

A *heat detector* is a device that detects an increase in temperature of a heat-sensitive element. **See Figure 15-10.** Heat detectors can be used in kitchens, attics, and garages where smoke detectors are not practical due to dust or other particles that cause false alarms. Heat detectors are not meant to replace smoke detectors and are not recommended for use in bedrooms or sleeping areas. Heat detectors are typically classified as mechanical fixed-temperature heat detectors, mechanical rate-of-rise heat detectors, and electronic heat detectors.

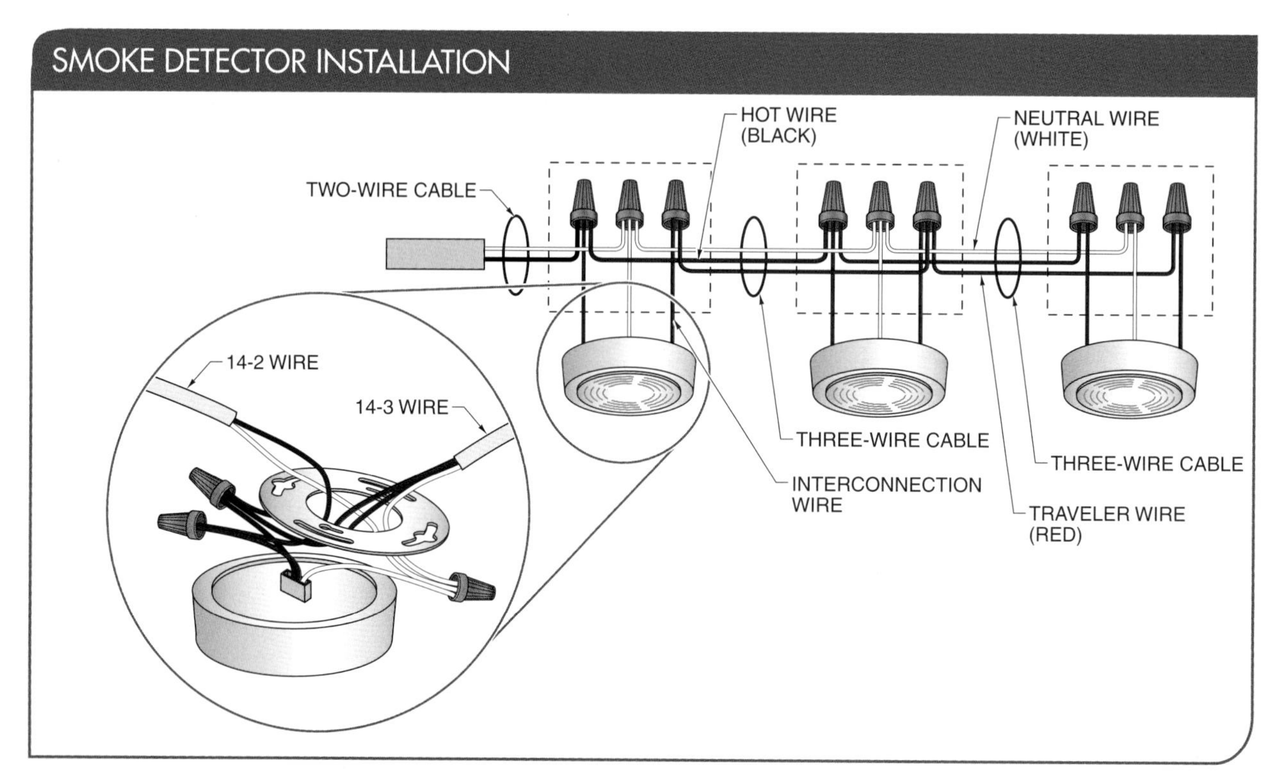

Figure 15-9. AC smoke detectors should be wired in parallel. The red lead from each smoke detector interconnects the entire system so that all detectors will sound when one is triggered.

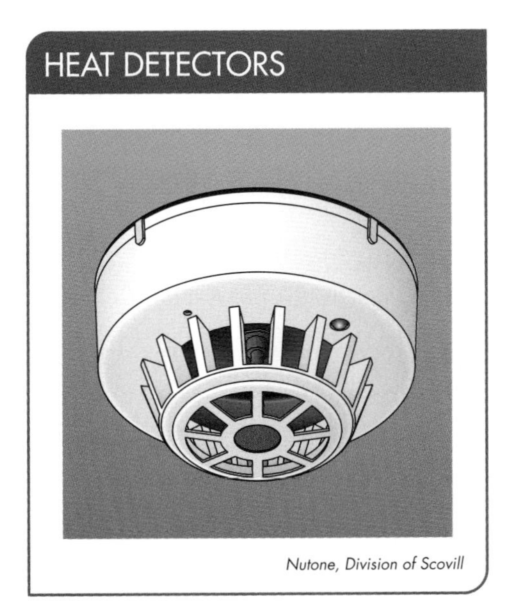

Figure 15-10. Heat detectors sense excess heat levels within a structure and are often used in conjunction with smoke detectors.

Mechanical Fixed-Temperature Heat Detectors. A *fixed-temperature heat detector* is a heat detector designed to respond when a room reaches a specific temperature. Once the specific temperature is reached, the contacts in the detector close and the alarm is activated. Fixed-temperature heat detectors may include fusible links or bimetallic strips for activating the alarm. **See Figure 15-11.**

Fusible-link fixed-temperature heat detectors include a small plunger and a compressed spring held by a small piece of material that melts at a setpoint temperature (such as 135°F). During a fire, the material melts, releasing the plunger and causing the spring to close the contacts to activate an alarm. Fusible-link fixed-temperature heat detectors are effective but cannot be restored. Meaning that once activated, these heat detectors must be replaced.

Bimetallic-strip fixed-temperature heat detectors include two bonded strips of different metals, each with a different coefficient of thermal expansion. When heated, one metal expands more than the other metal, causing the strip to bend in one direction and close a set of contacts. The advantage of bimetallic-strip fixed-temperature heat detectors is that they return to their original position when cool, reopening the contacts and restoring the function of the device. A disadvantage of these heat detectors is the time lag between when the ambient temperature reaches the setpoint temperature and when the device activates.

Mechanical Rate-of-Rise Heat Detectors. A *rate-of-rise heat detector* is a heat detector designed to respond to flash fires and possibly to slow-burning fires. The elements of a rate-of-rise heat detector sense any rapid change in temperature, such as a 100°F temperature change in 30 sec, to trigger an alarm. **See Figure 15-12.**

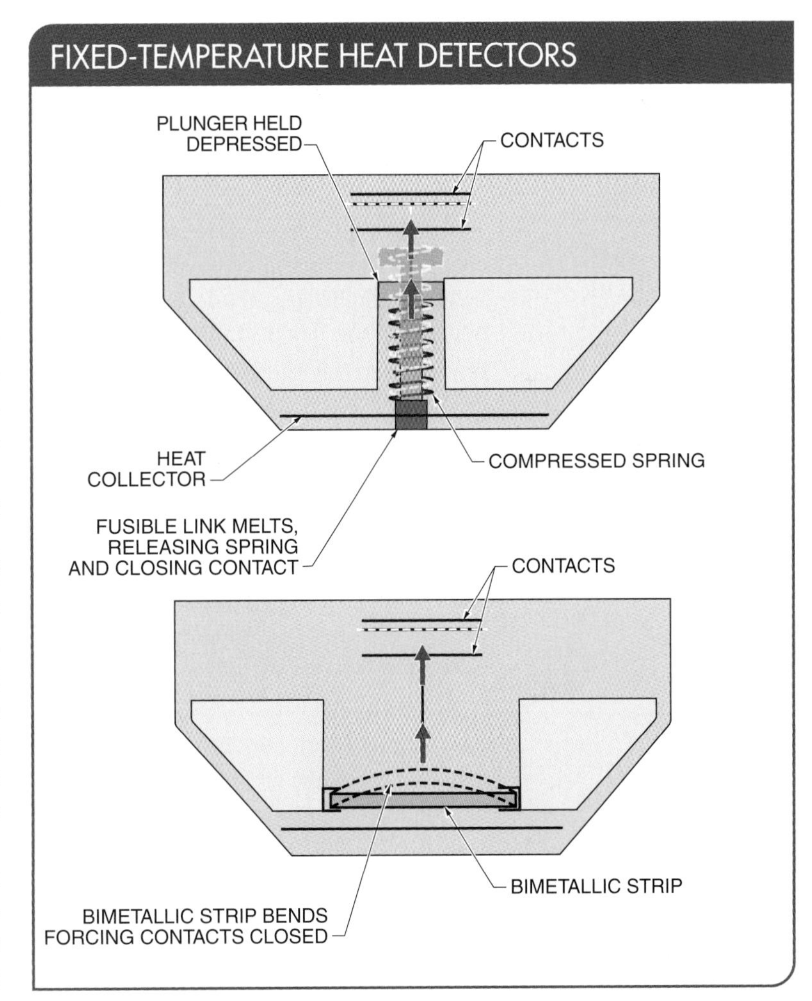

Figure 15-11. In fixed-temperature heat detectors, fusible links and bimetallic strips close a set of electrical contacts when the temperature reaches a certain setpoint.

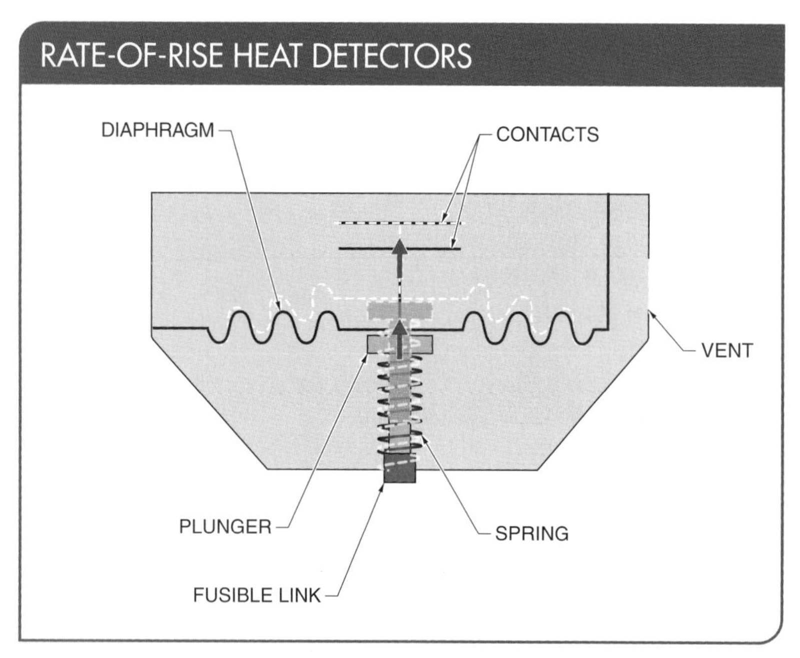

Figure 15-12. Rate-of-rise heat detectors use a diaphragm that flexes under high pressure to close electrical contacts. Many rate-of-rise heat detectors also include a fusible link as a back-up detector.

Rate-of-rise heat detectors react to changes in air pressure caused by extreme heat. A diaphragm inside a rate-of-rise heat detector flexes under sudden high pressure, which closes the contacts. As a back-up measure, many rate-of-rise heat detectors also include a fusible-link fixed-temperature heat detector. When the link melts, it releases a spring that pushes a plunger up on the diaphragm to close the contacts. The fusible-link backup ensures that the device can detect a slowly burning fire that does not raise the temperature fast enough to activate the rate-of-rise portion of the heat detector.

Electronic Heat Detectors. Electronic heat detectors are the most versatile type of heat detector because they are controlled by software and can be programmed to activate at a fixed temperature or a certain rate-of-rise temperature. Also, electronic heat detectors are restorable if they are not damaged by fire. The sensing element in electronic heat detectors is usually a thermocouple or thermistor. When the resulting output voltage (thermocouple) or resistance (thermistor) reaches the setpoint, the on-board electronics initiate an alarm signal.

Carbon Monoxide (CO) Detectors

A *carbon monoxide (CO) detector* is a device that detects the presence of CO gas to prevent CO poisoning. CO detectors should be placed in each bedroom as well as in the hallways leading to the bedrooms. CO detectors should be mounted to the ceiling where CO first rises and accumulates. CO detectors may be battery powered or hardwired and include a test button, siren, and LED indicators. **See Figure 15-13.** Some CO detectors have additional features such as a digital display that shows both the hazard level of the gas and the charge left in the battery.

CO Poisoning. CO is a poisonous gas that can cause sickness and death. CO can be produced from a malfunctioning gas stove, water heater, fireplace, or furnace or from a car left running in a closed garage. CO displaces oxygen in the blood, depriving the heart, brain, and other vital organs of the oxygen required to function properly. Prolonged exposure or large amounts of CO can overcome a person without warning.

Combination Smoke/CO Detectors

A combination smoke/CO detector provides both smoke and CO protection throughout a residence. A CO detector should be installed on each level of a residence, especially on any level with fuel-burning appliances. Wired combination smoke/CO detectors can satisfy the location requirements for smoke detectors and CO detectors in one device.

When either the smoke detector or the CO detector senses a potential hazard, one will communicate with the other. The alarm circuitry differentiates between a real fire/CO hazard and a false fire/CO hazard. The communication capabilities of combination smoke/CO detectors can enhance the performance of the fire alarm system.

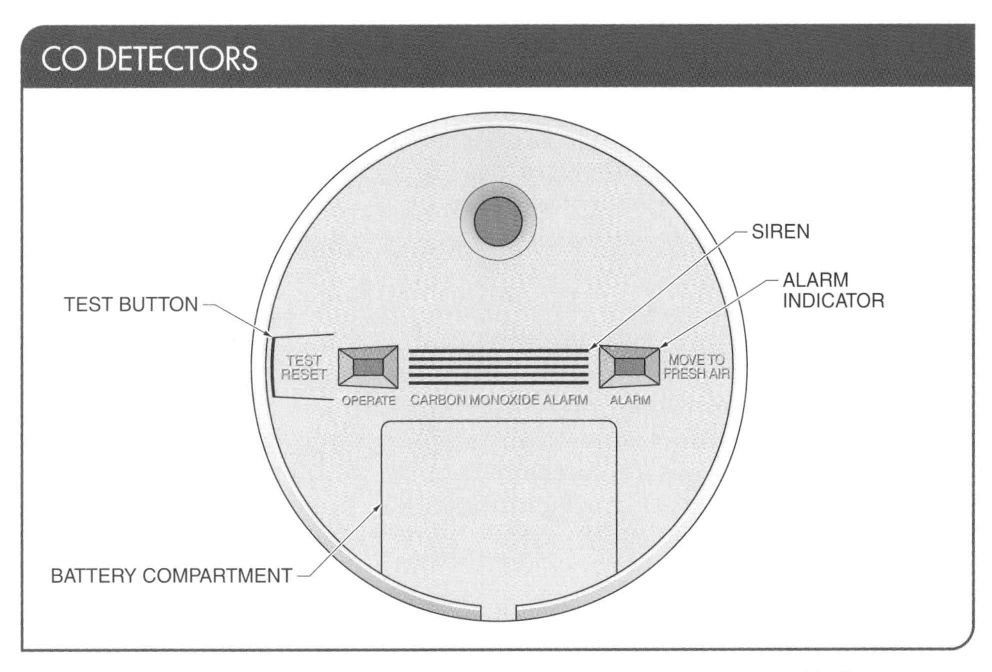

Figure 15-13. CO detectors should be mounted on ceilings in bedrooms and hallways.

A combination smoke/CO detector can also be interconnected wirelessly. This means that the detectors in a dwelling can be interconnected without the need to rewire. If one detector activates its alarm, it also sends a signal to all the interconnected detectors to respond. Therefore, if a fire starts upstairs, it can be made known immediately throughout the residence.

Detector Battery Power Supplies

Detectors that use batteries as power supplies should be checked and the batteries replaced according to the manufacturer's instructions or when there is a low-battery warning. The replacement batteries should be the same type of batteries listed by the manufacturer of the detector. Batteries should be replaced once every year. *Note:* AC-powered (hardwired) fire alarm systems are required to also have a battery backup, which should be tested monthly.

Some manufacturers have introduced detectors with nonreplaceable, long-life batteries that last up to 10 years. Detectors with nonreplaceable, long-life batteries, sometimes referred to as "10-year detectors," offer an advantage for people who may have difficulty replacing batteries. For a 10-year detector with a low battery, the entire detector must be replaced since the battery is manufactured sealed in the detector.

CODE CONNECT

NFPA 72®, National Fire Alarm and Signaling Code®, *requires, as a minimum, that smoke alarms be installed inside every sleeping room, including those in existing homes, as well as outside each sleeping area and on every level of the home.*

SECTION 15.1 CHECKPOINT

1. What does fire safety include?
2. What is an annunciator?
3. What are the two types of smoke detectors?
4. Where can heat detectors be used?
5. Where should CO detectors be placed?

SECTION 15.2—ALARM SYSTEM CONNECTIONS

Alarm systems are connected through modular connectors (jacks and plugs), relays, wires, and cables to the alarm system and are also interconnected to the telephone system. Alarm systems, such as fire alarm systems and security alarm systems, send out a signal from the control panel through an RJ-31X interface jack to a central monitoring service. Special tools are needed for residential alarm wiring configurations that include series-connected security alarm circuits and parallel-connected fire alarm circuits.

Modular Connectors

A *modular connector* is an electrical connector that was originally designed for use in telephone wiring but has since been also used for Ethernet applications. Modular connectors include plugs (male connectors) and jacks (female connectors). Plugs are used to terminate cables, and jacks are used for fixed locations such as walls, panels, and equipment.

Modular connectors are made in four sizes, 4P, 6P, 8P, and 10P, where *P* is a position or location for a contact. Not all positions may have contacts installed. If contacts are not needed, they are omitted from the outer positions, inward.

Modular connectors were mandated by the Federal Communications Commission (FCC) in 1976 as registered jacks (RJ). The RJ standard is no longer used, but RJ-11, RJ-31X, and RJ-45 remain common terms. The RJ connector is the standardized telecommunications network interface, consisting of plugs and jacks, used for connecting voice and data equipment to a service provider by a telephone company (telco). The letters *RJ* are followed by two digits that express the type of connector. The RJ-11, RJ-45, and RJ-31X are common types of connectors used in alarm system connectivity. **See Figure 15-14.**

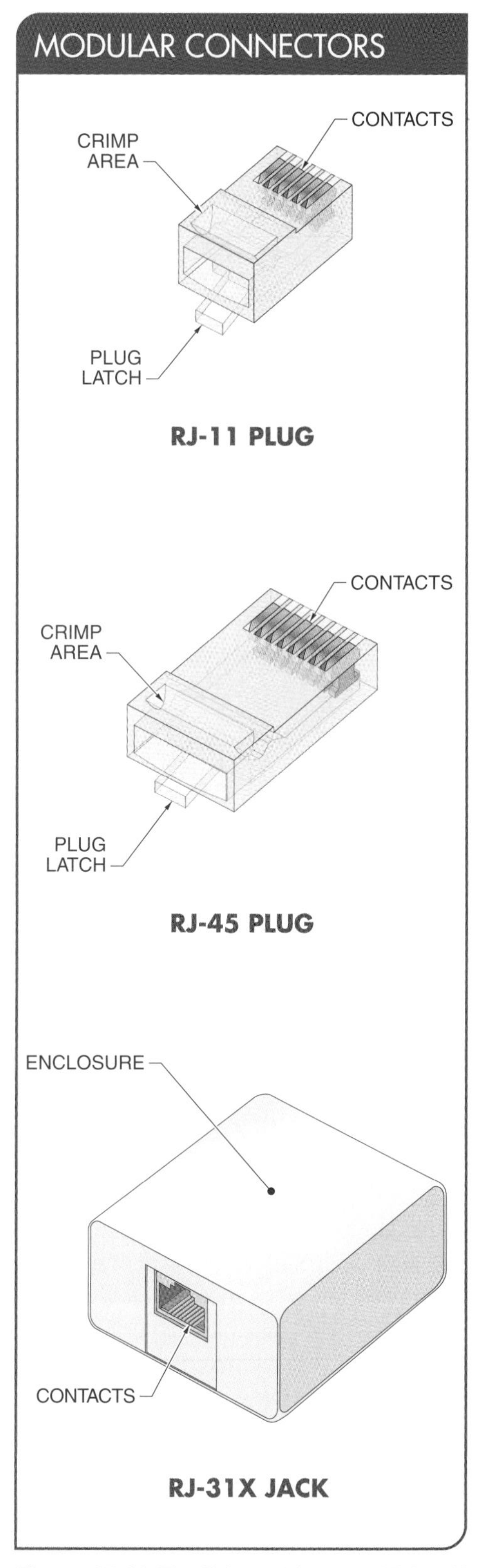

Figure 15-14. The RJ-11, RJ-45, and RJ-31X are three common types of modular connectors used in alarm systems.

RJ-11 Plugs. The RJ-11 plug is the most common type of telephone connection. RJ-11 plugs are used to connect telephones, modems, and fax machines to RJ-11 jacks located on walls and service panels. An RJ-11 plug usually has connection pins for six wires, though it is typically only used with four wires.

RJ-45 Plugs. The most common network connector used with copper wire systems is the RJ-45 plug. RJ-45 plugs are commonly used to terminate four-pair twisted Ethernet cable. RJ-45 plugs are sometimes referred to as 8P8C (eight position, eight contact). An installer must select an RJ-45 plug that is compatible with the four-pair twisted cable being used. Compatibility factors include whether the cable is solid or stranded, the AWG gauge of the individual conductors, whether the cable is round or flat, and the CAT rating of the cable. RJ-45 plugs are wider than RJ-11 plugs and are available in colors to help distinguish them from each other.

RJ-31X Alarm Interface Jacks. An RJ-31X jack is used as a quick disconnect point for control panels that communicate with a central monitoring service using copper telephone line. RJ-31X jacks are designed to provide preference to a fire alarm or security alarm control panel for dialing out in case of emergency. If any other terminal device (such as a telephone, fax, or modem) is using the line when the control panel needs to call, the control panel circuit breaks that connection and establishes its own.

Telephone Company Wiring Practices

Telephone cabling is terminated to a residence interface at what is called the "demarcation point" and is connected to the RJ-31X inside the dwelling. Different types of telephone interfaces have been installed over the years. Older dwellings may have a transient-voltage protector block with a lightning arrester.

However, telephone companies now install a network interface device (NID). An NID is usually an enclosure made of a strong tamper-resistant material. For security reasons, this enclosure is best located inside the residence to reduce the risk of tampering. Some homeowners also enclose outside telephone wiring in conduit to protect it from being cut. An NID furnished by a telephone company has two separate sections. One section is for telephone company access and the other section is for customer access. A telephone company will install protection clips or a lock to prevent customer access to the telco side of the box. **See Figure 15-15.**

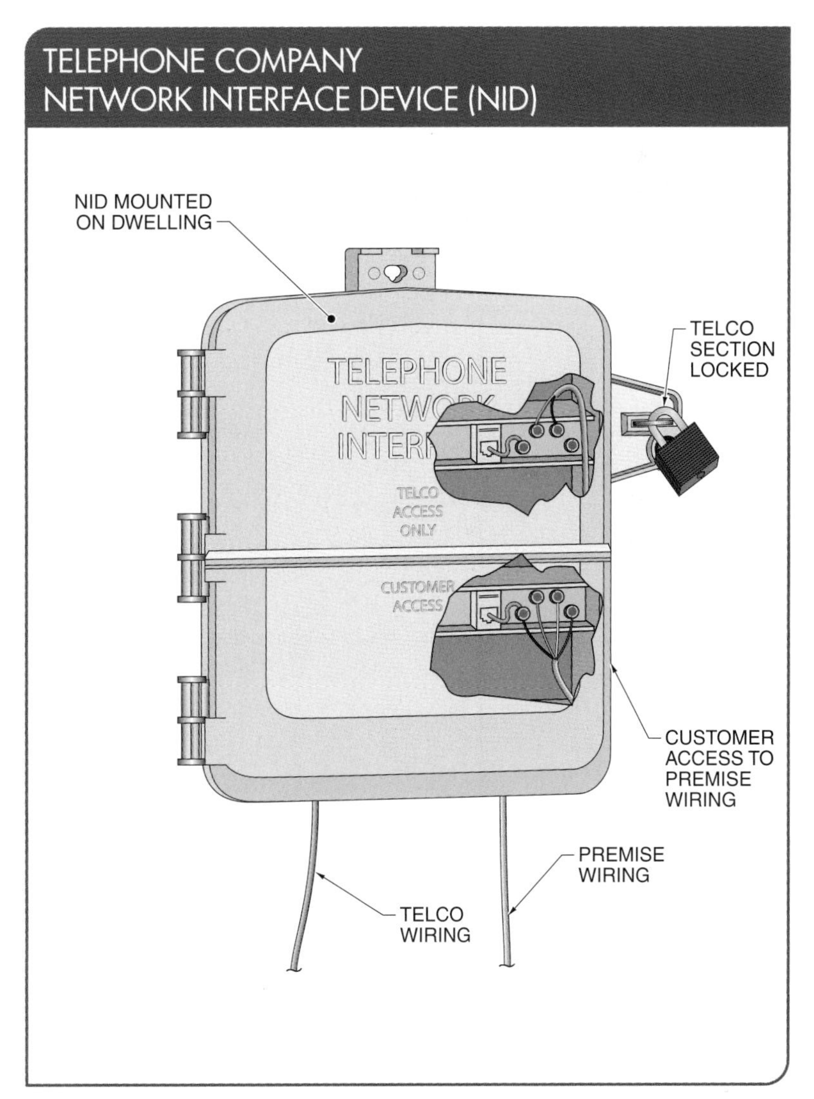

Figure 15-15. The telephone company NID is the connection point between the telco wires and the premise wires.

Tip and Ring Connections. Telephone wires at the NID are referred to as the tip and ring connections, which are the green and red wires connected to the RJ-31X jack. They are the equivalent of positive and negative designations. The terms "tip" and "ring" are a carryover from the early days of telephones when telephone operators used plugs to connect customer calls. They are named after the parts of the plug to which the wires were connected and are also often abbreviated with the letters *T* and *R*.

The "tip" was the tip of the plug and was the positive (+) side of the circuit. The "ring" was a conductive ring right behind the tip of the plug and was the negative (–) side of the circuit. Right behind the ring was the "sleeve," which was the ground connection. The ground is no longer used. **See Figure 15-16.**

The tip and ring carry 48 VDC when the phone is in the idle or on-hook condition. The tip and ring create a connection for a telephone line circuit. To make the telephone ring on an inbound call, around 90 V of 20 Hz AC is superimposed over the 48 VDC already available on the idle telephone line.

Relay Operation

Relays are used to interconnect control devices in fire alarm systems and security alarm systems. Relays are switches that open and close circuits using electromechanical contacts or electronic semiconductor contacts. Relays are normally activated by low voltages and current. Relay contacts are designed to control large voltages and currents. In an alarm circuit, a relay input

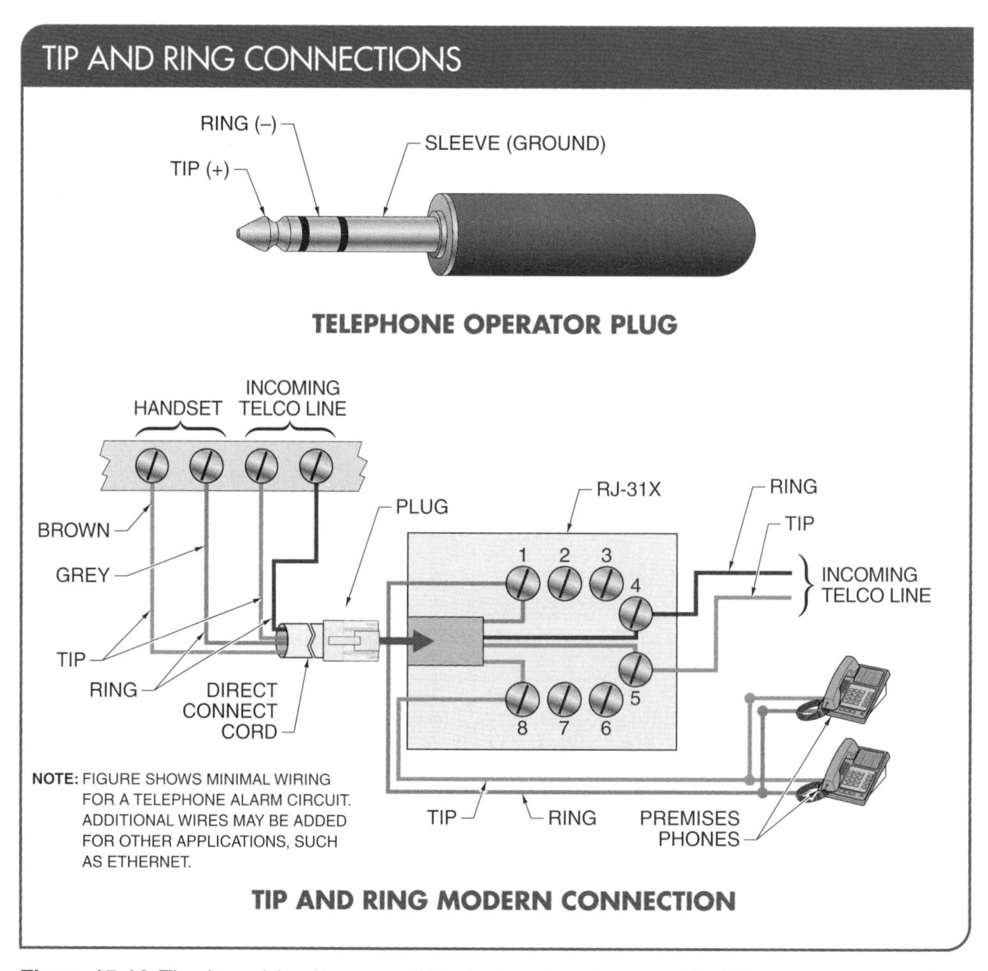

Figure 15-16. The tip and the ring carry 48 VDC when the phone is in the idle or on-hook condition.

voltage may be rated at 12 V, but the relay may be rated at 120 V (such as for turning on an outside light). A solid-state relay input voltage could be as low as 5 V.

Relay contacts are normally designated NO or NC, meaning they are normally open (NO) before being activated or normally closed (NC) before being activated. Once the relay is activated, the common (COM) moves the NO contacts closed or the NC contacts open. **See Figure 15-17.**

Electromechanical Relays vs. Solid-State Relays

Relays may be electromechanical (electric) relays or solid-state (electronic) relays. The schematic symbol and construction for each is slightly different. Electromechanical relays (EMRs) open or close contacts by magnetic force. When the relay coil is energized by a small current, it becomes an electromagnet moving a mechanical switch wired to contacts which control other devices. Solid-state relays (SSRs) switch contacts electronically using semiconductors.

EMRs and SSRs each have advantages and disadvantages. Less voltage is required to activate SSRs because they do not have a coil or open contacts. SSRs also turn on and turn off faster because there are no physical parts to move. SSRs are not subject to arcing like EMRs and do not wear out. However, the contacts on an EMR can be replaced, while an entire SSR must be replaced when it becomes defective. **See Figure 15-18.**

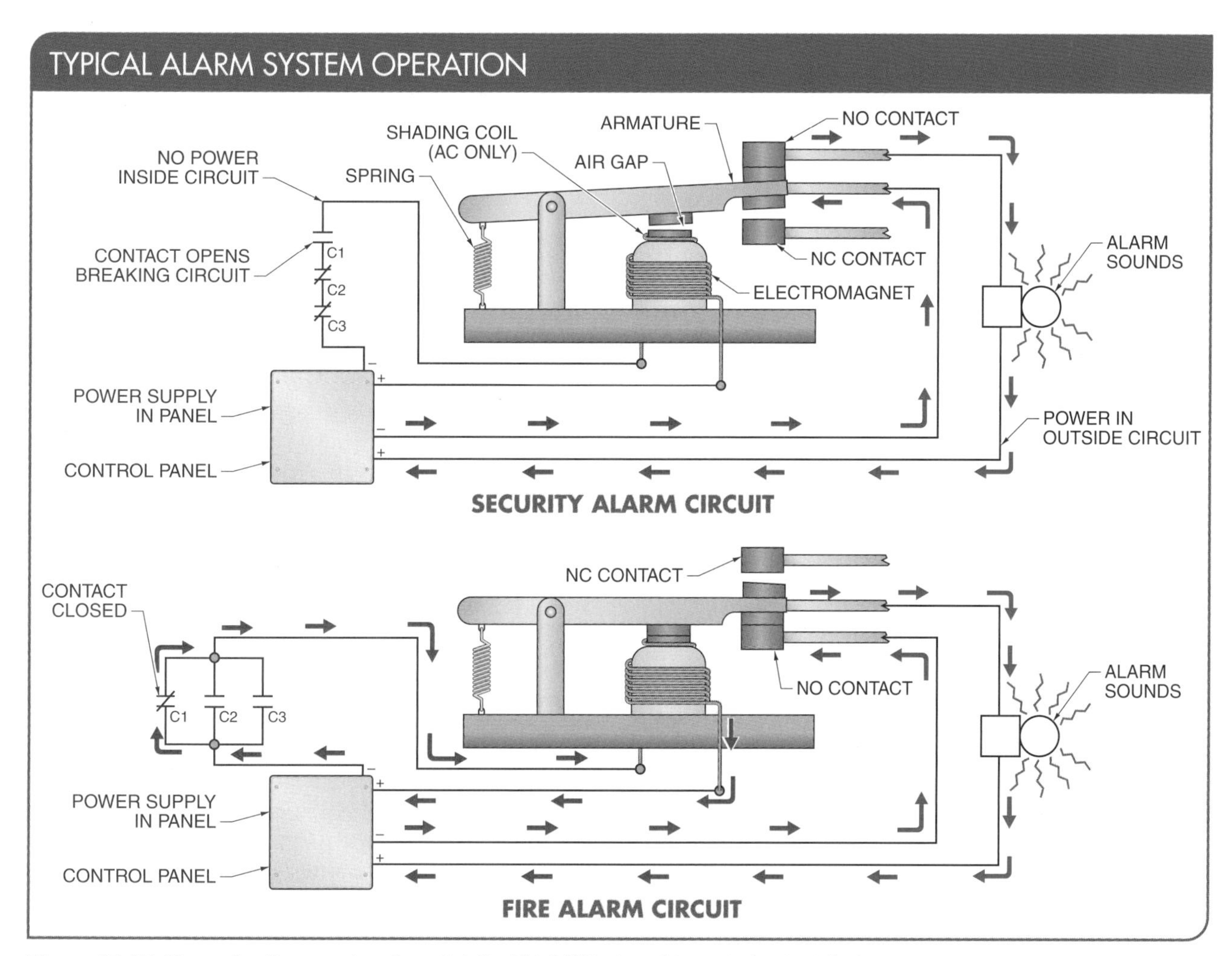

Figure 15-17. Alarm circuits use relays to switch the RJ-31X jack and to control output devices.

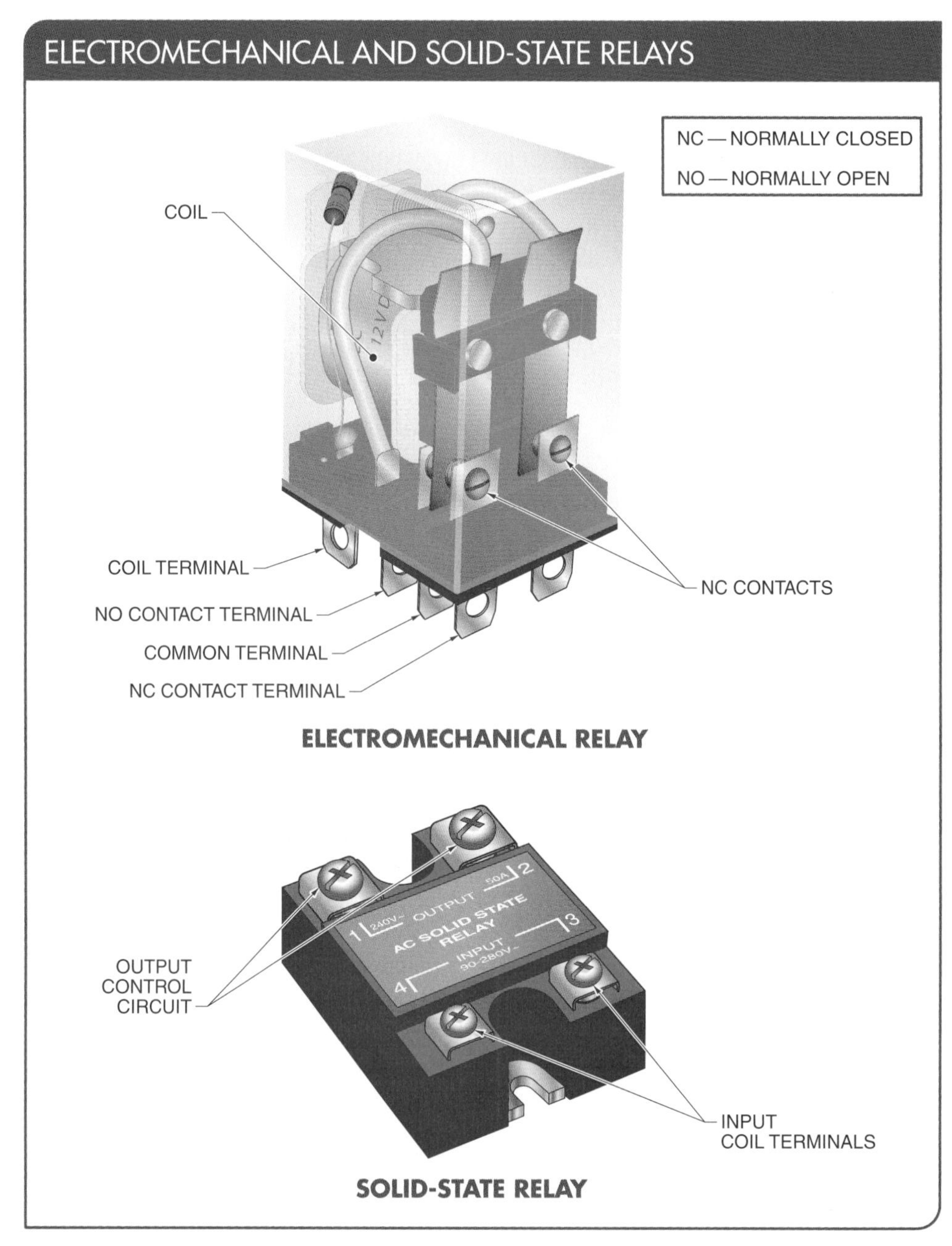

Figure 15-18. EMRs usually have normally open and normally closed contacts, while SSRs often have one normally open contact.

RJ-31X Jack Operation

An RJ-31X jack is a connection point between the telephone company and the phone-related devices inside a residence. When an alarm autodialer is plugged into the RJ-31X jack, the EMR or SSR in the autodialer connects R to R1 and T to T1 so that both the alarm system and the telephone devices in the residence are connected to the telephone company lines. In the event the alarm panel is activated by fire, smoke, or a trespasser, the relay in the autodialer is turned on and R1 and T1 are disconnected from the outside phone lines so they cannot interfere with the outgoing alarm signal. If a phone is left off the hook or the wires are disconnected, the alarm signal still goes through to the central monitoring service. **See Figure 15-19.**

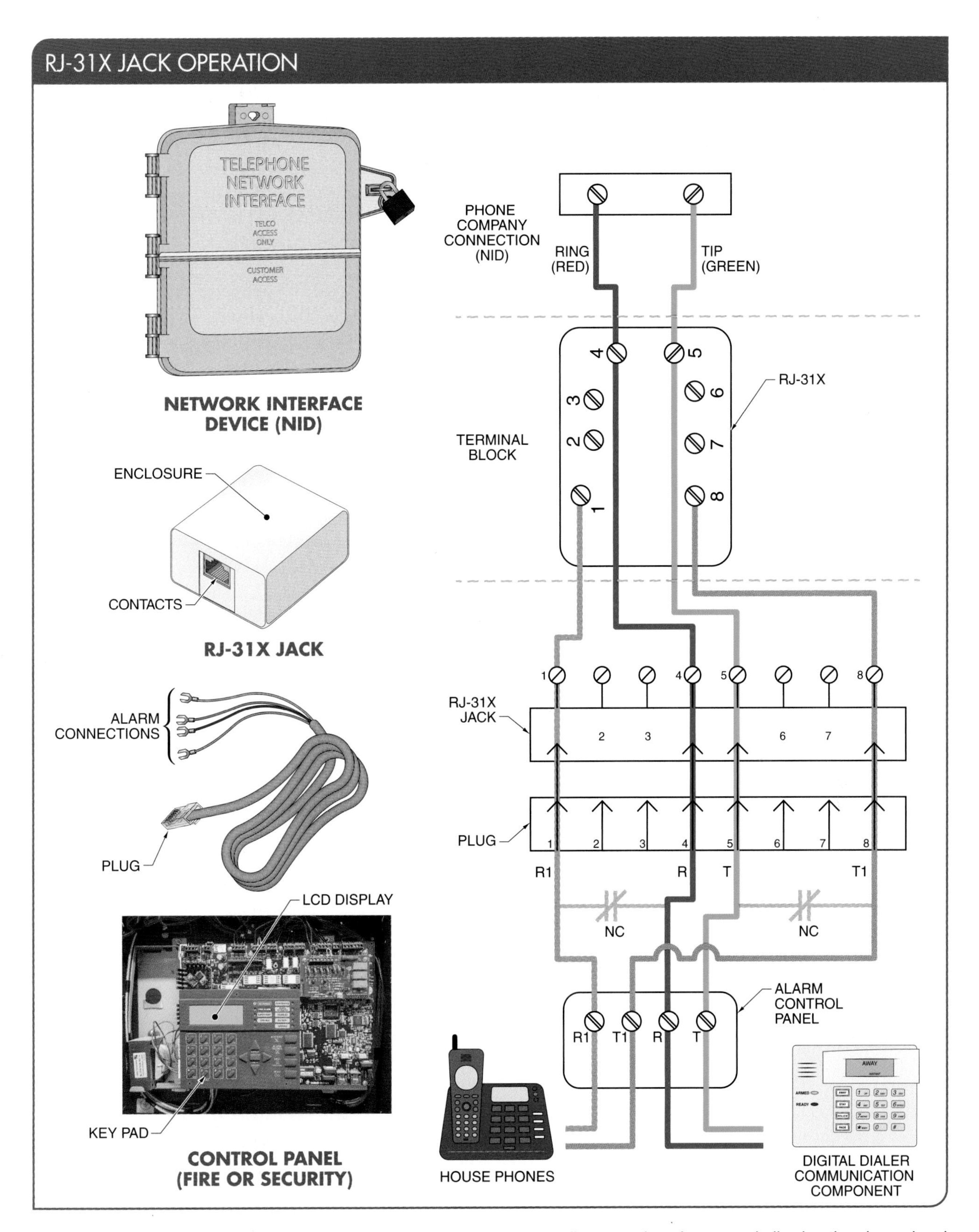

Figure 15-19. The control panel relay opens both R1 and T1 contacts, disconnecting phones and allowing the alarm signal to reach the central monitoring service.

Testing RJ-31X Jacks. The simplest method of testing an RJ-31X jack is as follows:

1. With no cable plugged into the jack, verify that every jack in the residence has a dial tone.
2. Carefully insert a standard phone into the RJ-31X jack. The jack is slightly wider than a standard phone jack. A standard phone plug will fit the wider jack if inserted carefully. Listen for a dial tone on the phone just inserted. The dial tone should be present. Hang up the phone, but leave it plugged in for the next step.
3. With the phone plug still inserted, recheck all other jacks in the residence. This must be done with the phone still inserted into the RJ-31X jack. All other jacks should be dead (no dial tone).
4. To restore normal phone service, remove the phone plugged into the RJ-31X jack. The dial tone should return on all the phones.

Low-Voltage Wire and Cable

Fire alarm systems and security alarm systems use low-voltage wires and cables. Wire for alarm systems may be two conductors or four conductors wrapped with a plastic cover to form a cable. Two conductor cables are used for windows and doors while four conductor cables are used for keypads and other powered devices. These conductors are designated with the AWG number first and the number of wires second, such as 22-2 or 22-4. **See Figure 15-20.** For smoke detectors and heat detectors, 22-2 or 22-4 fire-rated wire is often used. Heavier 18-gauge (such as 18-2 or 18-4) fire-rated wire can also be used and is sometimes easier to find.

Telephone Wiring. In the past, telephone wiring did not require a high-performance cable to work. The legacy cable often used for telephones was a four-conductor cable commonly referred to as plain old telephone standard (POTS). Now, the minimum wiring for a telephone circuit is CAT 5E or better. CAT-rated cables are not much more expensive than POTS and can be used for network cabling. An additional advantage of CAT-rated cables is that they carry four two-pair sets of wires, or eight conductors. The two-line pair capacity of POTS cable only provides four wires.

Also, CAT 5E cable has twisted pairs, whereas POTS cable does not. Twisted-pair cables provide a higher immunity from electrical interference. Twisted-pair cables may also be shielded for additional interference protection. However, for reasons of cost and cable space requirements, unshielded twisted-pair (UTP) cables are more commonly used.

Cable Color Coding. The outer coverings of cables, or cable jackets, are available in several colors. The color of a cable jacket can be used to identify the use of the cable. For example, one white cable for voice transmission and one blue cable for data transmission are commonly installed for each workstation in a work area. Cables with red jackets are commonly used for fire alarm system installations. Fiber-optic cables have orange, yellow, and teal jackets.

In addition to cable jacket color coding, the wire pairs inside the cable jacket also have standardized colors. In a four-pair cable, the colors of the insulation on the wire pairs are blue and white or white-striped; orange and white or white-striped; green and white or white-striped; and brown and white or white-striped. The striping pattern on white insulation can be vertical or horizontal. A standard color-coding system identifies each pair of wires in a cable. These colors must match the colors of the wires to ensure proper connections at both ends of the cable. **See Figure 15-21.**

CODE CONNECT

NEC® Subsection 760.24(A) states that fire alarm circuits shall be installed in a neat workmanlike manner. Cables and conductors exposed on the surface of ceilings and sidewalls shall be supported by the building structure, straps, hangers, staples, and cable ties without being damaged.

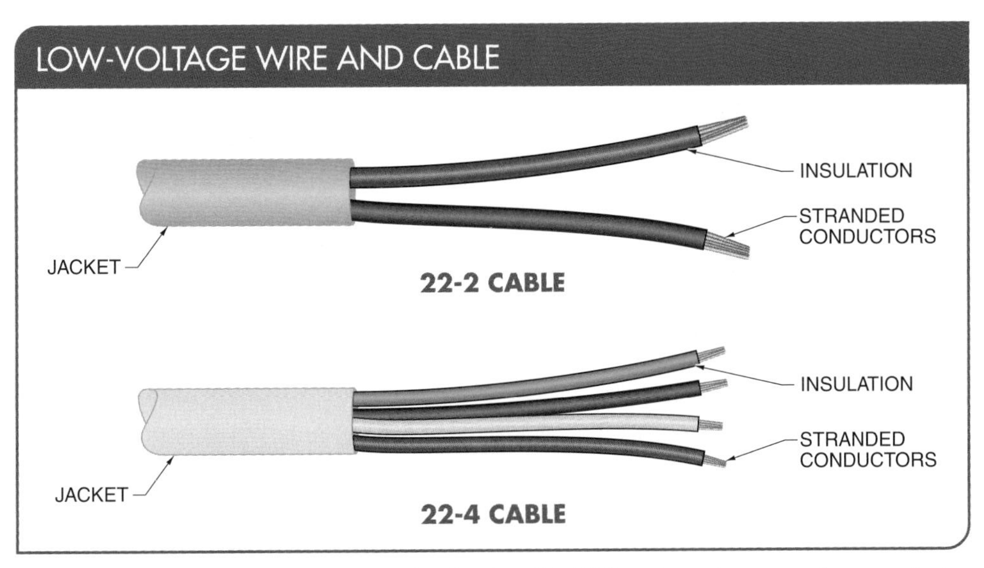

Figure 15-20. Wire for alarms systems is often No. 22 AWG and may be two- or four-conductor cables (22-2 or 22-4).

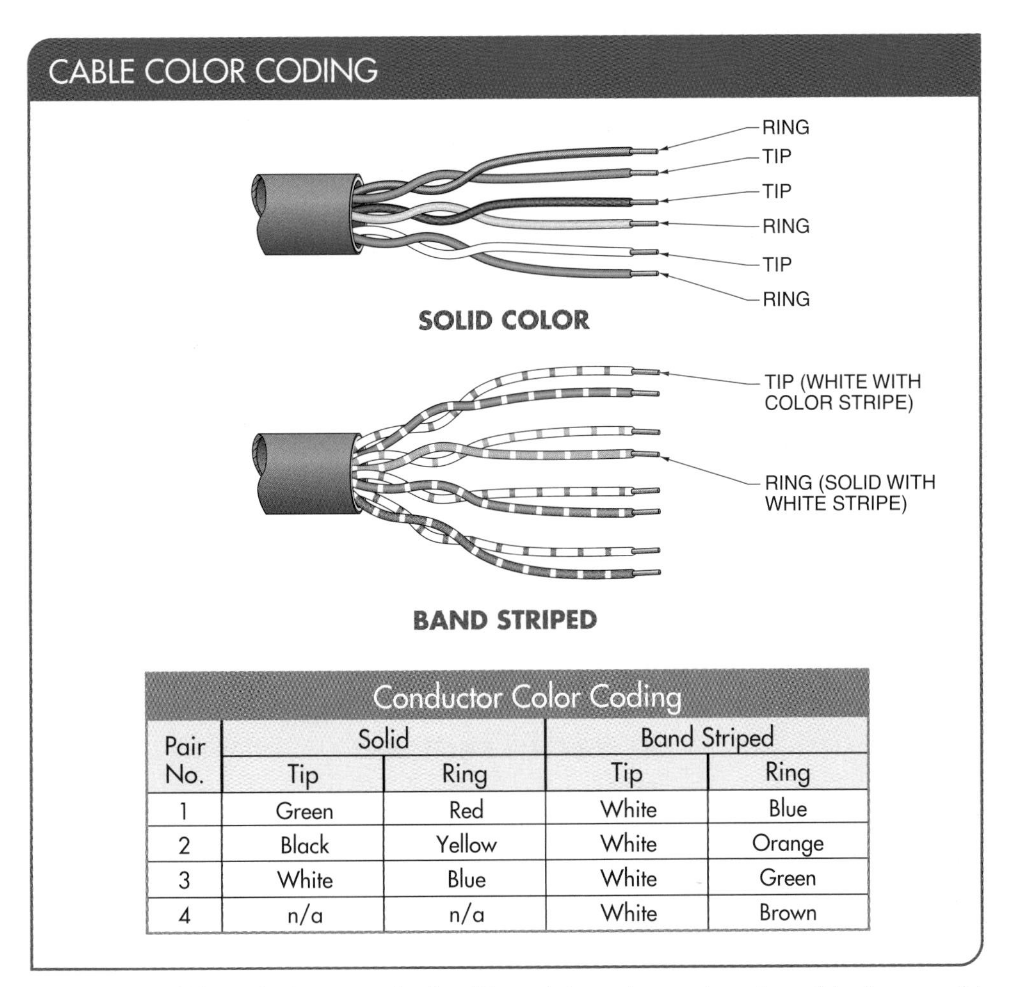

Conductor Color Coding				
Pair No.	Solid		Band Striped	
	Tip	Ring	Tip	Ring
1	Green	Red	White	Blue
2	Black	Yellow	White	Orange
3	White	Blue	White	Green
4	n/a	n/a	White	Brown

Figure 15-21. Wire pairs have standardized tip and ring colors and can be solid colors or solid colors with stripes.

Wire and Cable Stripping. There are specific tools needed when working with twisted-pair cable or coaxial cable. Common tools used for this work include UTP/coaxial wire strippers, an impact (punchdown) tool and combination blade, electrician's scissors, a wire spudger, and a toner. **See Figure 15-22.** A UTP/coaxial cable stripper is an all-in-one tool used to cut and strip CAT 5e twisted-pair and coaxial cables. Electrician's scissors can be used to cut open long pieces of cable or excess insulation. A wire spudger is used to loosen, release, poke, or adjust small wires in telecommunications equipment without cutting the wire or damaging nearby components in tight spaces.

Termination Procedures

Termination procedures ensure that devices are connected so the alarm system functions properly. A *termination* is the connection of the end of a wire to a contact or terminal. A crimp termination is performed when the device requires a contact or terminal. The wire insulation is stripped, and the contact or terminal is attached to the wire using a crimp tool. The tool crimps the contact or terminal onto the wire conductor. There are three common terminations that may be used for alarm systems: punchdown terminations, RJ-45 modular plug terminations, and legacy terminations.

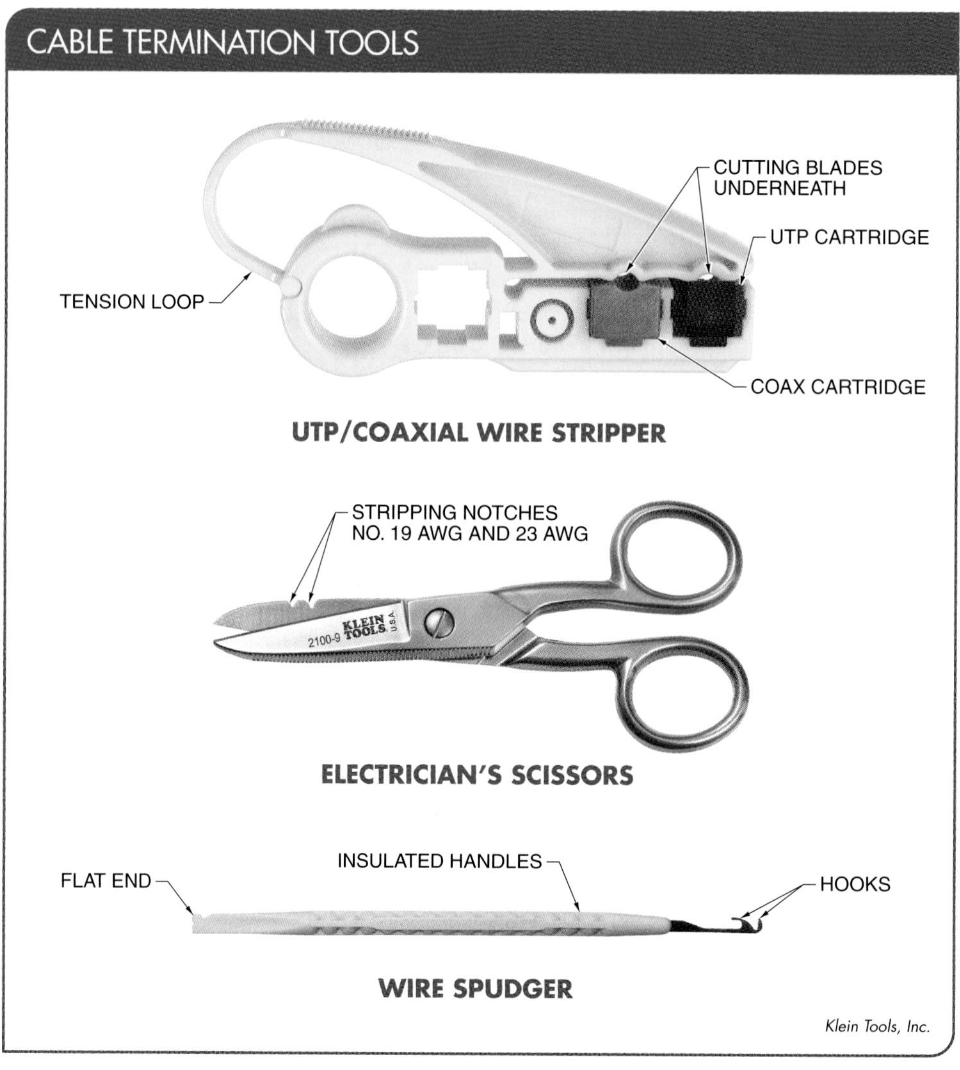

Figure 15-22. Specific tools are used to terminate and test cable when connecting alarm systems.

Punchdown Terminations. Punchdown tools, such as jack tools and impact (punchdown) tools, are available from several different manufacturers and have interchangeable blades intended for use with specific types of terminations, such as 110 blocks or RJ-45 jacks. A punchdown tool performs a fast termination method referred to as "insulation displacement."

The insulation displacement process increases the speed of making terminations. A punchdown tool works by placing an individual wire in a groove between two guide posts of a termination block and then piercing the bare section of wire. The process of making a connection with a punchdown tool provides a consistent, repeatable result. A punchdown tool seats the wire in place and trims off excess wire. The metal contacts are positioned to cut through the insulation and contact the copper wire inside. The conductor is evenly wedged between the contacts to hold it securely in place. **See Figure 15-23.**

RJ-45 Modular Plug Terminations. RJ-45 modular plugs are commonly used to terminate four-pair twisted cable. An installer must select a RJ-45 modular plug that is compatible with the four-pair twisted cable involved. Also, the RJ-45 modular plug must be compatible with the crimp tool. **See Figure 15-24.** The procedure for terminating an RJ-45 modular plug is as follows:

1. Position the RJ-45 plug with the tab down, the gold contacts up and away, and the pin numbers showing 1 through 8, left to right.
2. Strip approximately 2″ of jacket from the cable.
3. Determine which termination pattern to use, either T568A or T568B.
4. Fan out, separate, and straighten the wire from the cable into the proper termination pattern.
5. Trim the wire at 90° and strip the insulation as specified by the RJ-45 plug manufacturer.
6. Fully insert the wires into the RJ-45 plug.
7. Insert the RJ-45 plug into the crimp tool, making sure that the wires stay in place, and crimp.

Legacy Terminations. Legacy termination methods may be required for technicians working on older installations. Legacy terminations include the wire wrap, screw, spring compression, screw plate compression, crimp-on tap connector, and toolless insulation displacement contact (IDC) methods. There are applications involving newer systems, such as access security systems or fire alarm systems, which may interface with an older network and require the use of one of the legacy termination methods. **See Figure 15-25.**

TECH TIP

While many residential installations include legacy terminations such as coaxial cable and POTS systems, these systems are becoming obsolete since they are not compatible with the current technology used by service providers and electronics OEMs.

Toners

The most common instrument for testing wires and cable connections in telecommunications equipment is the toner. A *toner* is an instrument, consisting of a tone generator and a toner probe, used to insert a signal into individual conductors, twisted-pair cables, and coaxial cables to test for breaks in the line or the location of a connection. The tone generator creates a signal that when picked up by the probe is amplified and reproduced as an audible sound by the speaker in the probe. The tone generator can be connected to a wire using alligator clips or an RJ jack using the appropriate connector. Toners may also be used to trace lines connected to RJ jacks. **See Figure 15-26.**

A toner is not intended for use on conductors and cables that have any AC voltage, DC voltage, or active signals. Before using a specific toner, the operator must read and follow the manufacturer's instructions and all safety precautions, including using the required PPE.

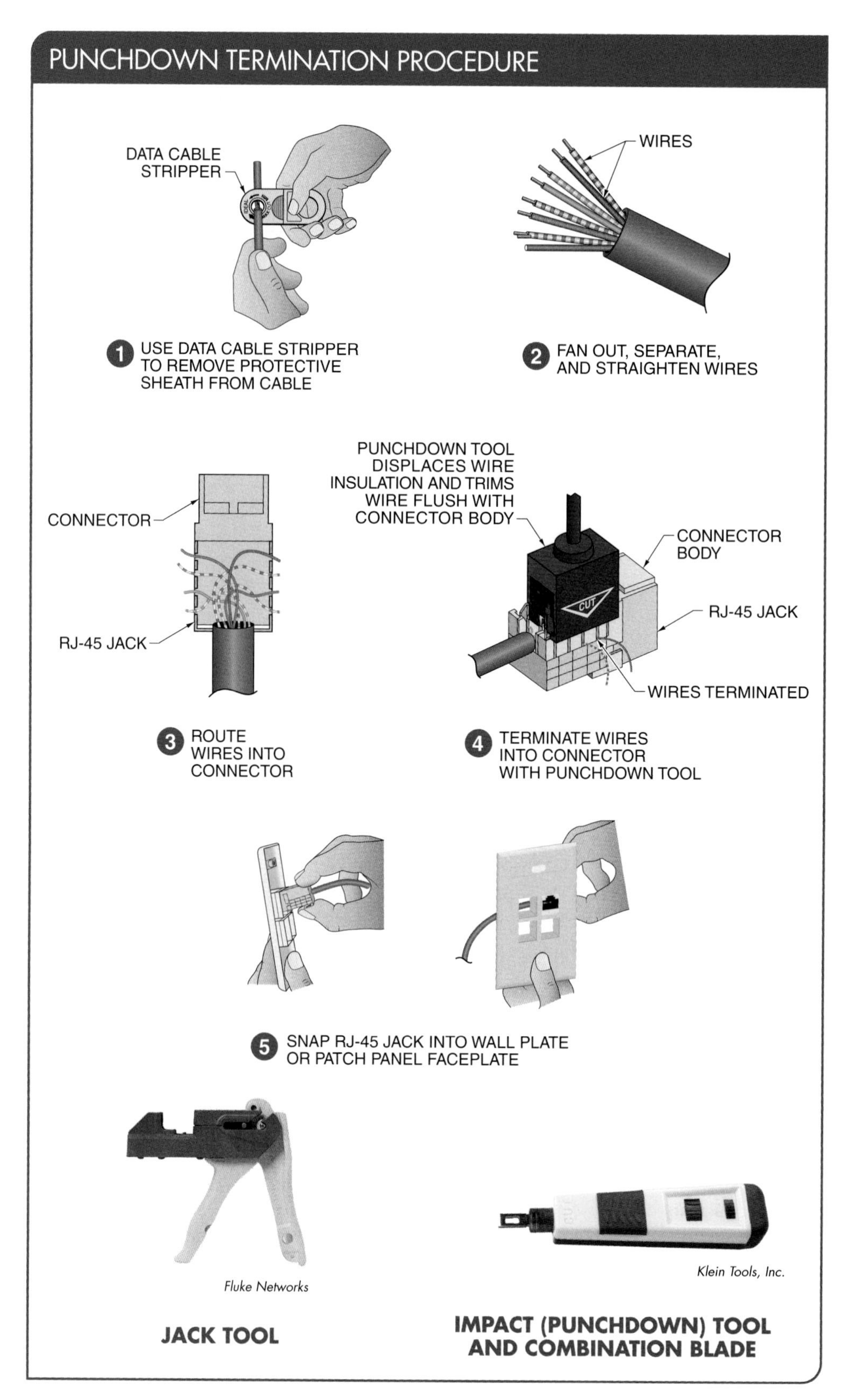

Figure 15-23. Jack tools and impact (punchdown) tools are used to secure copper wires to metal contacts and trim excess wire.

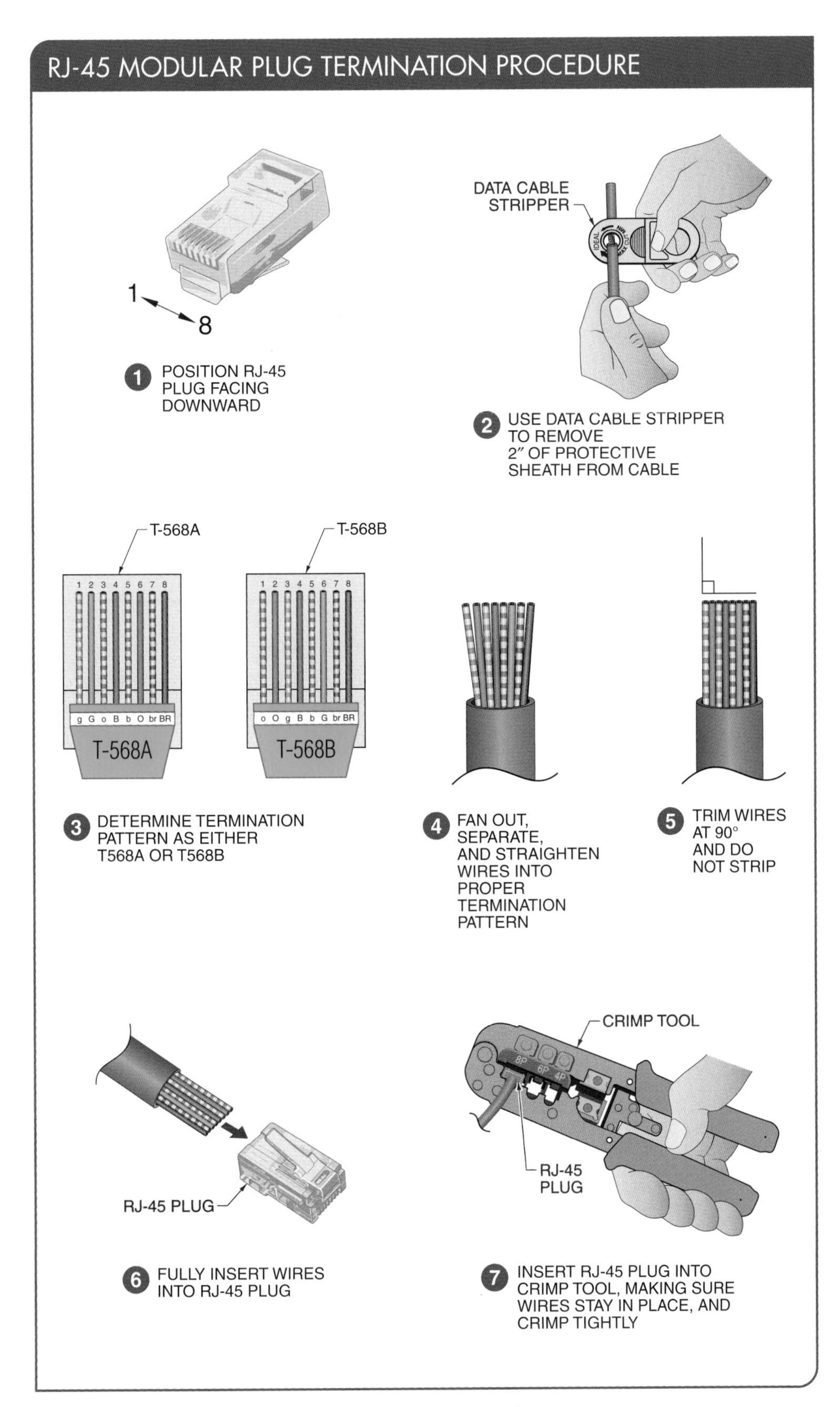

Figure 15-24. RJ-45 plugs are often used to terminate twisted-pair cable.

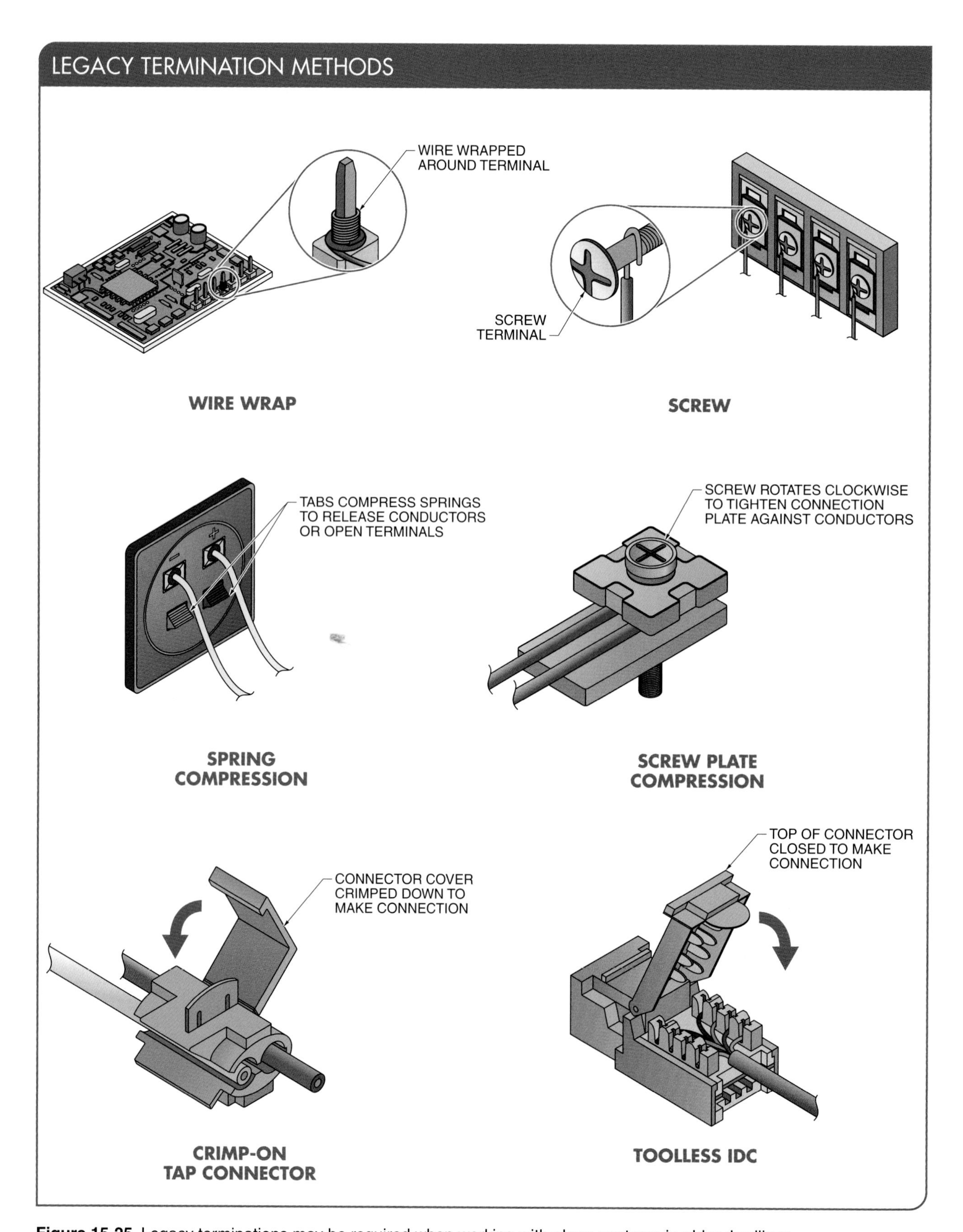

Figure 15-25. Legacy terminations may be required when working with alarm systems in older dwellings.

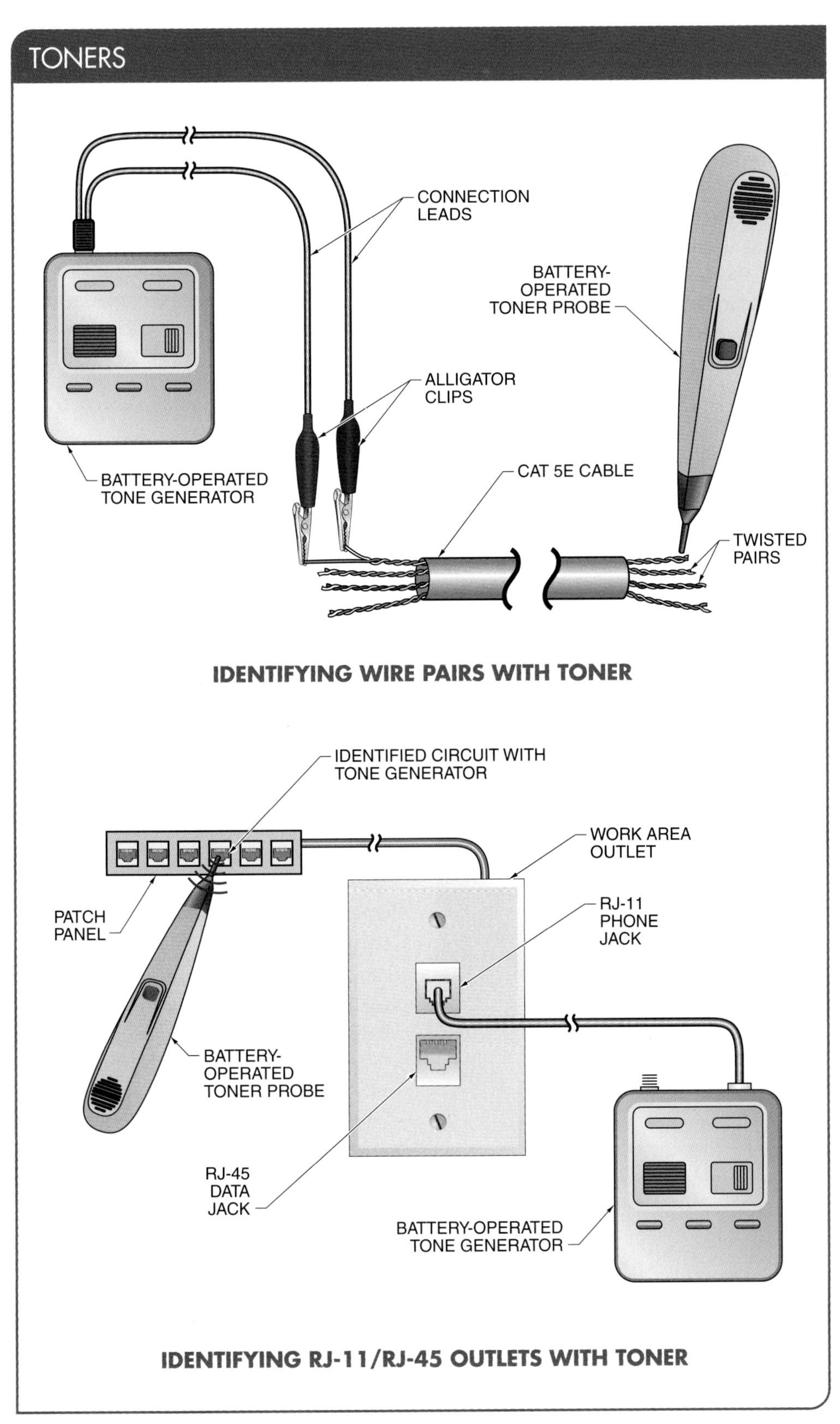

Figure 15-26. Toners are used to identify specific wire pairs in a cable and outlets, such as RJ-45 jacks.

Wired Circuits vs. Wireless Circuits

Previously, all wired security alarm systems were considered series circuits and all wired fire alarm systems that were considered parallel circuits. However, the introduction of wireless devices has changed that assumption. In a wireless system, the devices operate independently to form parallel circuits. Instead of wiring, each device has a transmitter that sends a radio signal to a receiver.

When security system components are wired, the devices must be physically connected to each other in the system to form a series circuit. Wired systems are very reliable and are easiest to install in new construction. Wireless systems can be used in existing construction where wiring would be difficult or expensive to install.

Hardwired security cameras are best installed in new construction projects rather than existing structures due to cost and ease of installation.

Series-Connected Security Alarm Circuits

Security alarm circuits are often series-loop circuits. A *series-loop circuit* is a circuit in which devices are connected in series and are normally closed (NC). Because the devices are connected in series, an alarm can be triggered anywhere a device opens in the circuit. This is based on the idea of protecting the entire circuit.

A series-loop circuit is used for door/window contacts and glass-break detectors. If a contact is opened or a wire is cut, the alarm would be set off. The alarm would be activated because current no longer flows in the relay circuit, causing the relay contacts to move from open to closed. Closed contacts allow current in the alarm circuit, activating the alarm. Once the sensor is reset and the alarm relay circuit is reconnected, the alarm shuts off. Although a series-loop circuit is quite simple, it is difficult for an unauthorized person to override. **See Figure 15-27.**

CODE CONNECT

When hardwired fire alarm circuits are installed, the type of conductors used should comply with Parts G and H of NEC® Section 760.179, which explain the fire alarm circuit integrity (CI) cable to be used with electrical circuit protective systems. CI cable is considered flame retardant.

Parallel-Connected Fire Alarm Circuits

Fire alarm circuits are often parallel circuits. A *parallel circuit* is a circuit in which devices are connected parallel and are normally open (NO). Normally open (open-circuit) wiring systems are typically used on the gas and smoke (fire) detection portions of a fire alarm system. An NO contact circuit consists of several NO sensors wired in parallel with an NO relay contact.

The principle behind a parallel circuit is that if the sensors remain in an open position, no current is available to circuit relay R1 and the annunciator circuit remains inoperative. If for any reason one of the sensors is activated, current passes through relay R1 and closes contacts R1 in the annunciator circuit. With contacts R1 closed, the bell, siren, or light turns on. **See Figure 15-28.**

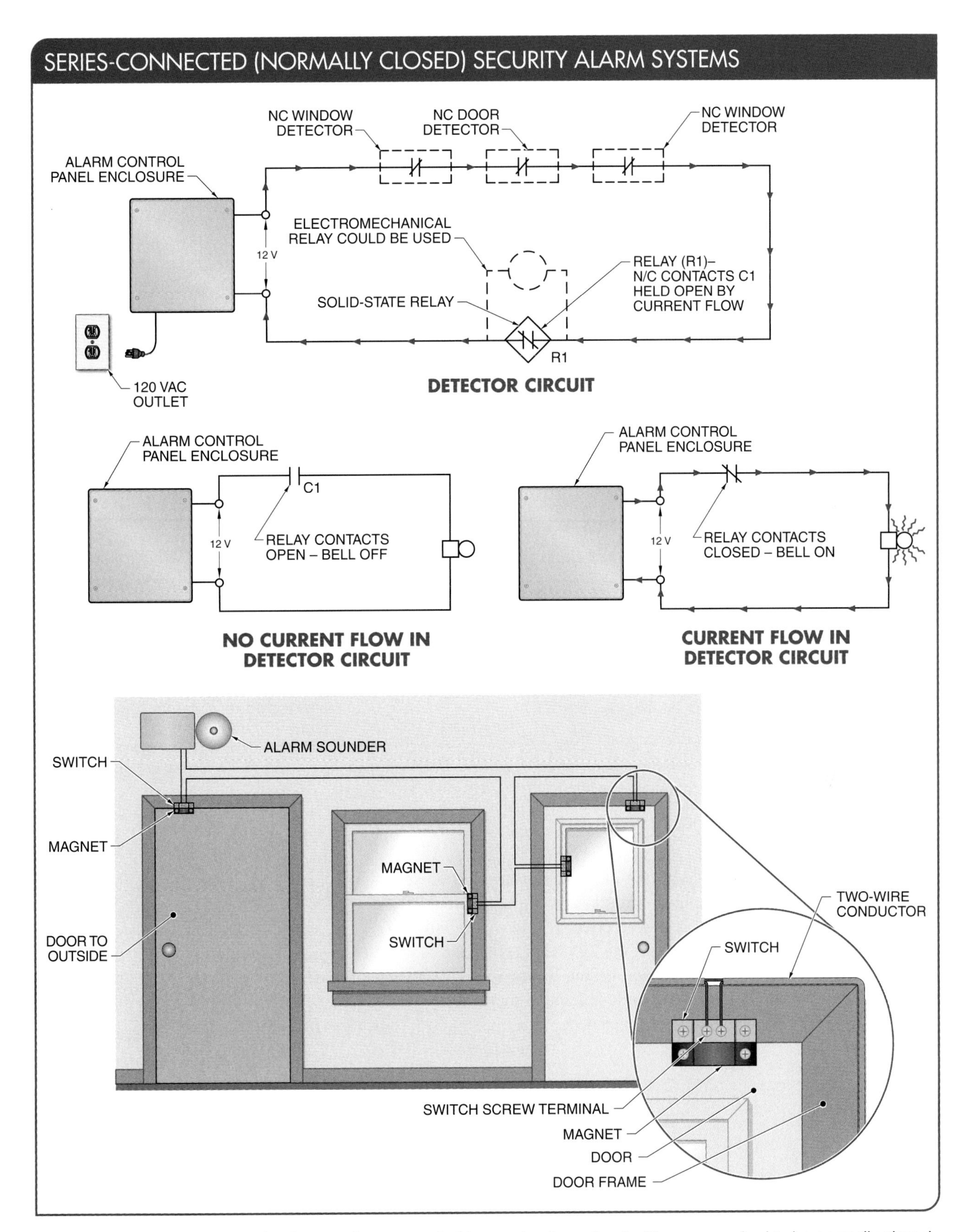

Figure 15-27. Security alarm circuits are often connected in a series-loop circuit with sensors wired to be normally closed.

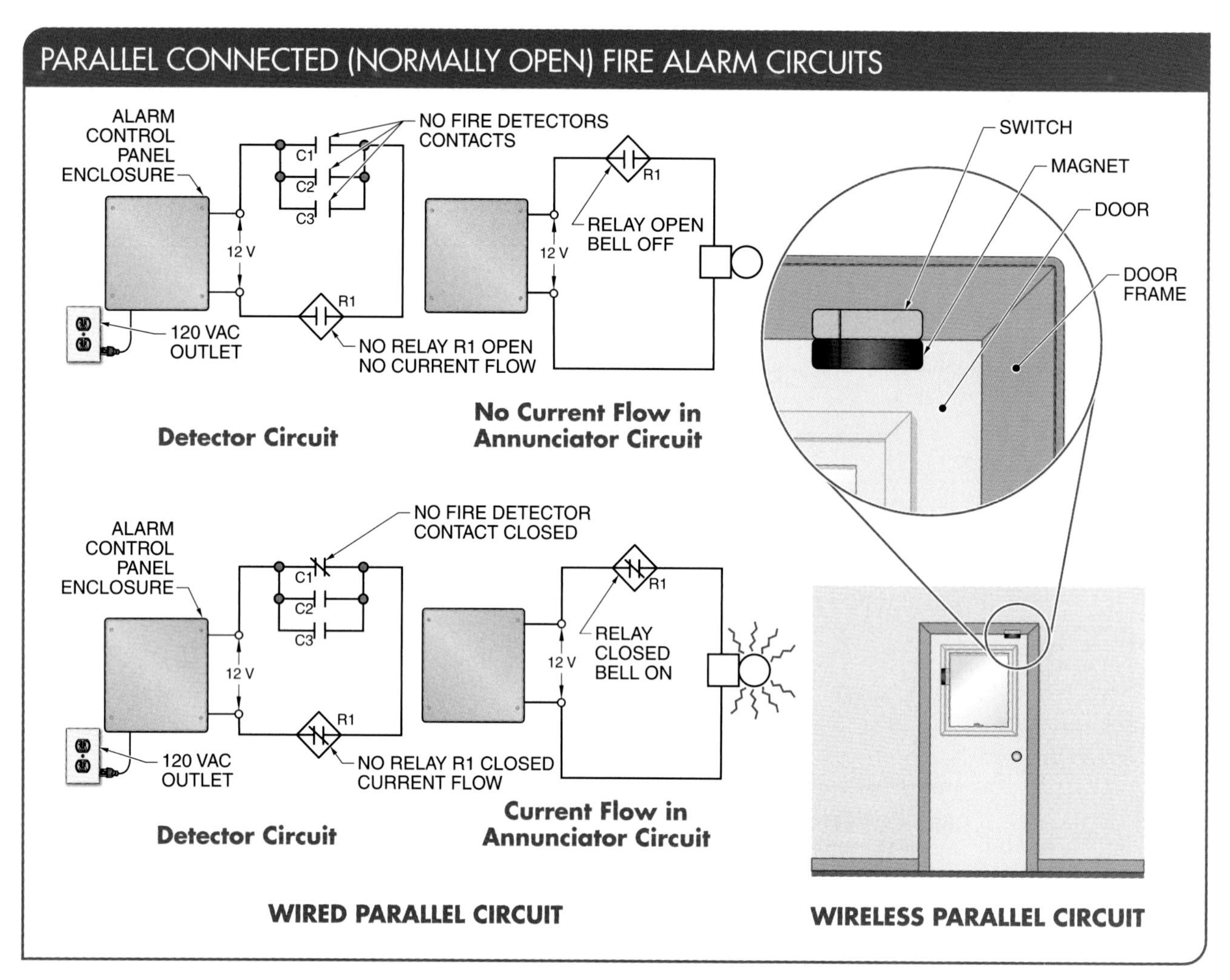

Figure 15-28. Fire alarm circuits are often connected in parallel with several sensors wired to be normally open. Wireless sensors are considered part of a parallel-loop circuit.

With the introduction of wireless sensors, all circuits are the equivalent of a parallel-loop circuit. With a parallel-loop circuit, all the devices are considered in parallel and normally open. Each sensor operates independently and generates a signal when closed.

TECH TIP

Since older smoke detectors have a single sensor that generally presents high false-alarm rates due to ambient temperature changes, modern smoke detectors have several different types of sensors built into their design.

SECTION 15.2 CHECKPOINT

1. What are three common types of RJ connectors?
2. What are RJ-31X jacks designed to do?
3. What are two types of relays?
4. What type of wire is often used to wire smoke detectors and heat detectors?
5. In what types of construction are wired and wireless systems usually installed?
6. What is a series-loop circuit?
7. What is a parallel circuit?

Chapter Activities

15 Fire Alarm Systems and Smart Home Applications

Name ______________________ Date ______________

Cable and Connector Termination Tools

Identify the cable and connector termination tools shown.

_______________ **1.** Crimp tool

_______________ **2.** Impact (punchdown) tool

_______________ **3.** Jack tool

_______________ **4.** Electrician's scissors

_______________ **5.** Wire spudger

_______________ **6.** UTP/coaxial wire stripper

Klein Tools, Inc.

A

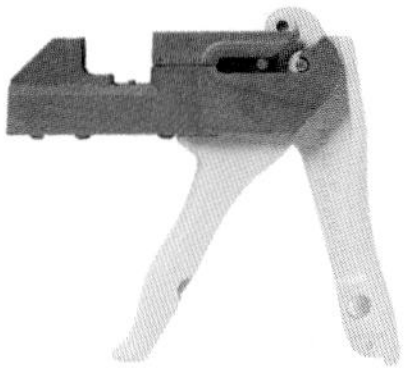

Fluke Networks

B

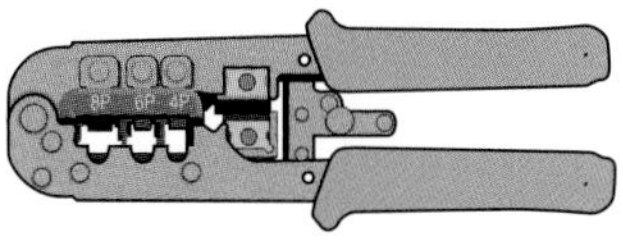

C

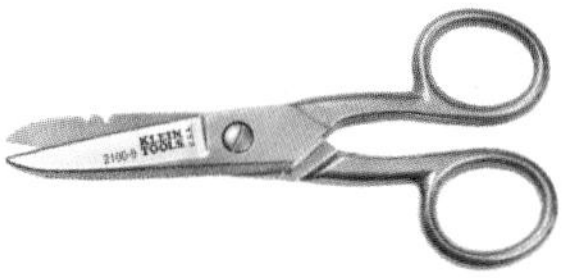

D

Klein Tools, Inc.

E

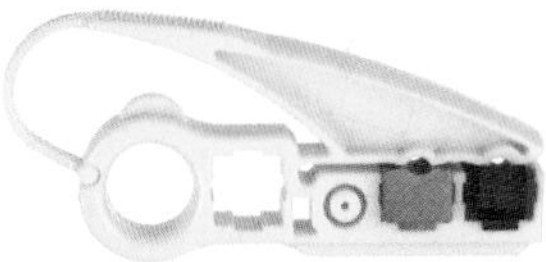

Klein Tools, Inc.

F

Relay Operation

Identify the parts of the relay shown.

_______________ **1.** Electromagnet

_______________ **2.** NC contact

_______________ **3.** Armature

_______________ **4.** NO contact

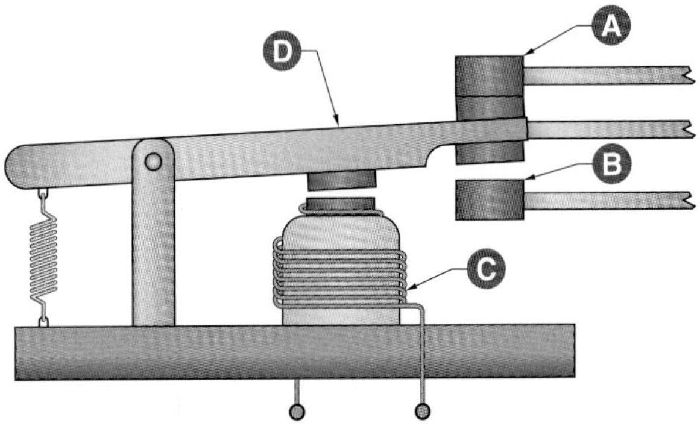

A

B

C

D

Smoke Detector Installation

Draw lines to connect the smoke detectors in parallel.

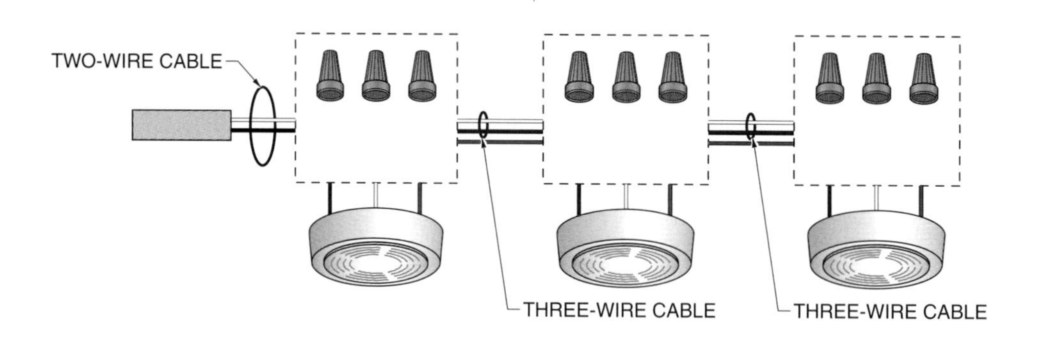

Series-Loop and Parallel Circuits

Identify the series-loop circuit and the parallel circuit.

_______________ **1.** Parallel circuit

_______________ **2.** Series-loop circuit

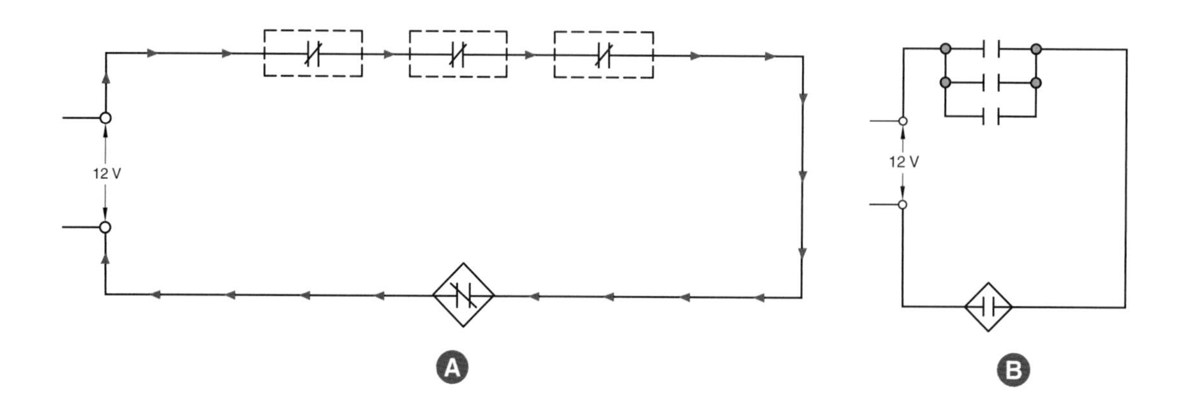

16 Home Energy Management and Smart Home Applications

SECTION 16.1—ENERGY AUDITS

- Explain energy audits and the major causes of structural energy loss.
- Explain continuous, intermittent, and phantom loads.
- Describe energy-monitoring devices.
- Describe smart power strips.

SECTION 16.2—LIGHTING MANAGEMENT

- Describe lamp types and lamp bulb shapes.
- Describe the Lighting Facts label.
- Explain how occupancy sensors and timers help reduce energy consumption.
- Describe the different types and styles of dimmer switches.
- Explain what heat sinks are and why they are used.

SECTION 16.3—INDOOR ENERGY MANAGEMENT

- Describe the different designs of thermostats and how they are wired.
- Describe the different types of fans and how they are controlled.
- Describe the application of duct dampers and HVAC control vents.
- Explain how window treatments can help reduce energy waste.

SECTION 16.4—ENERGY GENERATION AND STORAGE

- Compare grid-connected and off-grid (stand-alone) wind power systems.
- Describe solar photovoltaic power generation.
- Compare grid-connected and off-grid (stand-alone) PV systems.
- Describe building-integrated photovoltaics (BIPV).
- Describe PV system design and installation.

SECTION 16.5—HOME ENERGY MANAGEMENT

- Describe the fundamental components of a home energy management system (HEMS).
- Explain the collaborative energy management strategies of utilities and homeowners to reduce demand during peak hours.
- Explain how smart appliances play a part in energy conservation.
- Explain how residential and commercial microgrids help manage energy use.

Learner Resources

ATPeResources.com/QuickLinks
Access Code: 791712

Home energy management involves evaluating the energy consumption in a dwelling, identifying and reducing waste, and using smart home technology to improve energy efficiency. One of the many benefits of a smart home is the ability to manage energy consumption. Energy use can be monitored and controlled for appliances and other devices both inside a dwelling and outside on the property. Alternative energy sources can also be adopted by a homeowner and connected to a smart grid.

SECTION 16.1—ENERGY AUDITS

Home energy management should begin with an energy audit. An *energy audit* is an assessment of how much energy a dwelling consumes to determine what measures can be taken to make the dwelling more energy efficient. An energy audit can help locate where energy is being wasted in a dwelling and determine the priority of efficiency upgrades to stop or reduce energy waste. An energy audit should be performed before a home energy management (HEM) system is selected for a residence.

An energy audit consists of five steps that should result in energy cost savings. These steps include inspecting the dwelling, documenting potential energy losses, developing a home energy management plan, implementing the home energy management plan, and monitoring energy to retain the energy savings. **See Figure 16-1.** The energy audit should include a home energy management plan that shows which energy efficiency upgrades are most cost effective for the dwelling and the projected energy savings.

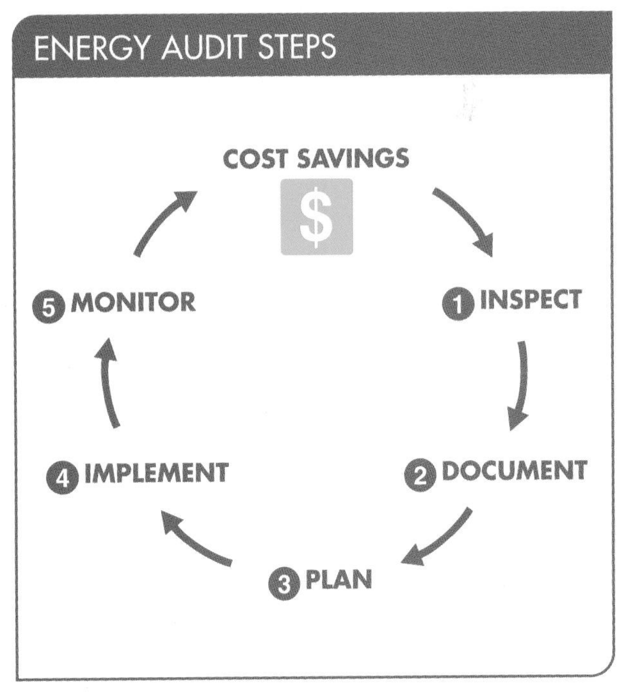

Figure 16-1. An energy audit includes five steps that should result in energy waste reduction and cost savings.

Home Construction Visual Inspections

A professional energy auditor may use special equipment, such as a thermal imaging camera, to detect structural sources of energy loss. **See Figure 16-2.** While not as thorough as a professional home energy audit, a homeowner can conduct their own energy audit by visually inspecting the home construction (building structure) and using energy-monitoring devices.

A residential energy audit begins with a visual inspection of the overall size, shape, orientation, and features of the dwelling and its surrounding. A sketch of the dwelling can be made to help pinpoint the energy losses that are due to the construction and age of the home. **See Figure 16-3.**

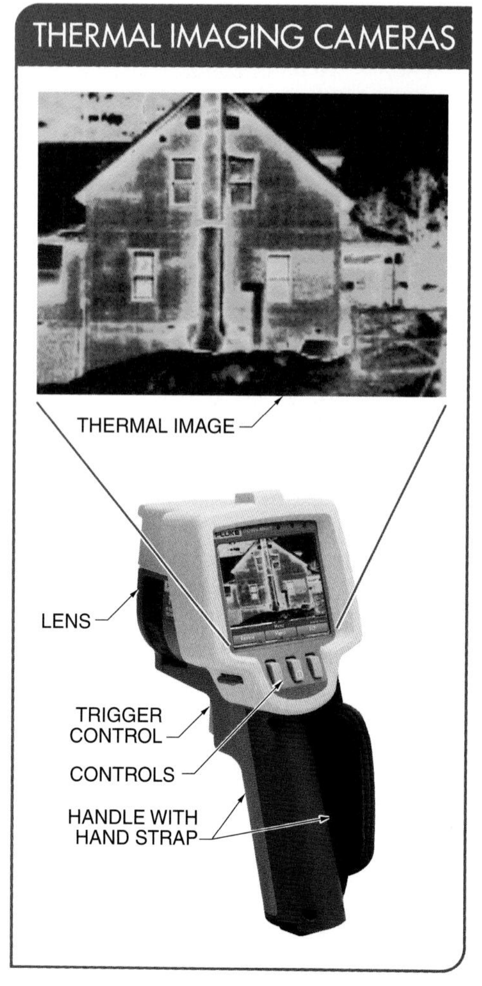

Figure 16-2. Thermal imaging cameras can be used to identify structural sources of energy loss.

Air leaks and lack of insulation are the major causes of structural energy loss. Air leakage can occur where there are gaps in the air barrier. Cold air from the outside enters a dwelling and warm air from the interior escapes. Since warm air rises, a heated home in winter acts like a chimney. As the warm air rises and escapes through ceiling openings, cold air is pulled in from the basement, garage, or crawl space.

Electrical Loads

The appliances and devices in a residence that consume electricity are referred to as loads. A *load* is anything that consumes electrical power. For the evaluation of the energy consumption of an appliance or device, the amount of power used by the appliance or device depends on whether the load is a continuous, intermittent, or phantom load. For a complete home energy management plan, all three load types must be considered and measured.

A *continuous load* is a load that operates all the time. Clocks and refrigerators are examples of continuous loads. An *intermittent load* is a load that does not operate all the time. Computers, microwaves, TVs, and entertainment equipment are examples of intermittent loads. A *phantom load*, also referred to as vampire load or standby power, is an appliance or device that uses power when plugged in despite technically being turned off. The US Department of Energy (DOE) estimates on average that 75% of the electricity used to power certain home appliances and entertainment electronics may be consumed while they are turned off.

Generally, the largest phantom loads are devices with remote controls, such as TVs or DVD players. Cell phone charging devices with external power supplies are also phantom loads. Many other common household devices and appliances, such as ovens, microwaves, and coffee makers with digital displays, are also phantom loads. The energy costs of these phantom loads can add up to almost 10% of residential electrical use. Energy-monitoring devices, such as service panel monitors, plug-in monitors, smart plugs, and smart power strips, can be used to monitor energy usage and identify phantoms loads.

Energy Monitors

An *energy monitor* is a device used inside a dwelling to provide residents with information on the electrical consumption of appliances and other devices. Energy monitors help identify opportunities to reduce electricity usage by providing real-time and historic information about when and how electricity is being consumed. Energy monitors detect which appliances are putting a heavy load on the energy bill, calculate how much electricity usage costs, and can be programmed to notify users what time of day electricity is most expensive. Service panel monitors and plug-in monitors are two configurations of basic energy monitors.

Service Panel Monitors. A *service panel monitor* is a device that is used to monitor the change in electrical usage as various appliances and other devices are switched on and off. This can be helpful for identifying the power consumption of devices such as electric stoves, air conditioning units, or appliances in which the plug is difficult to reach (such as a refrigerator). Most service panel monitors are made up of three parts: sensors, a transmitter, and a display panel. **See Figure 16-4.**

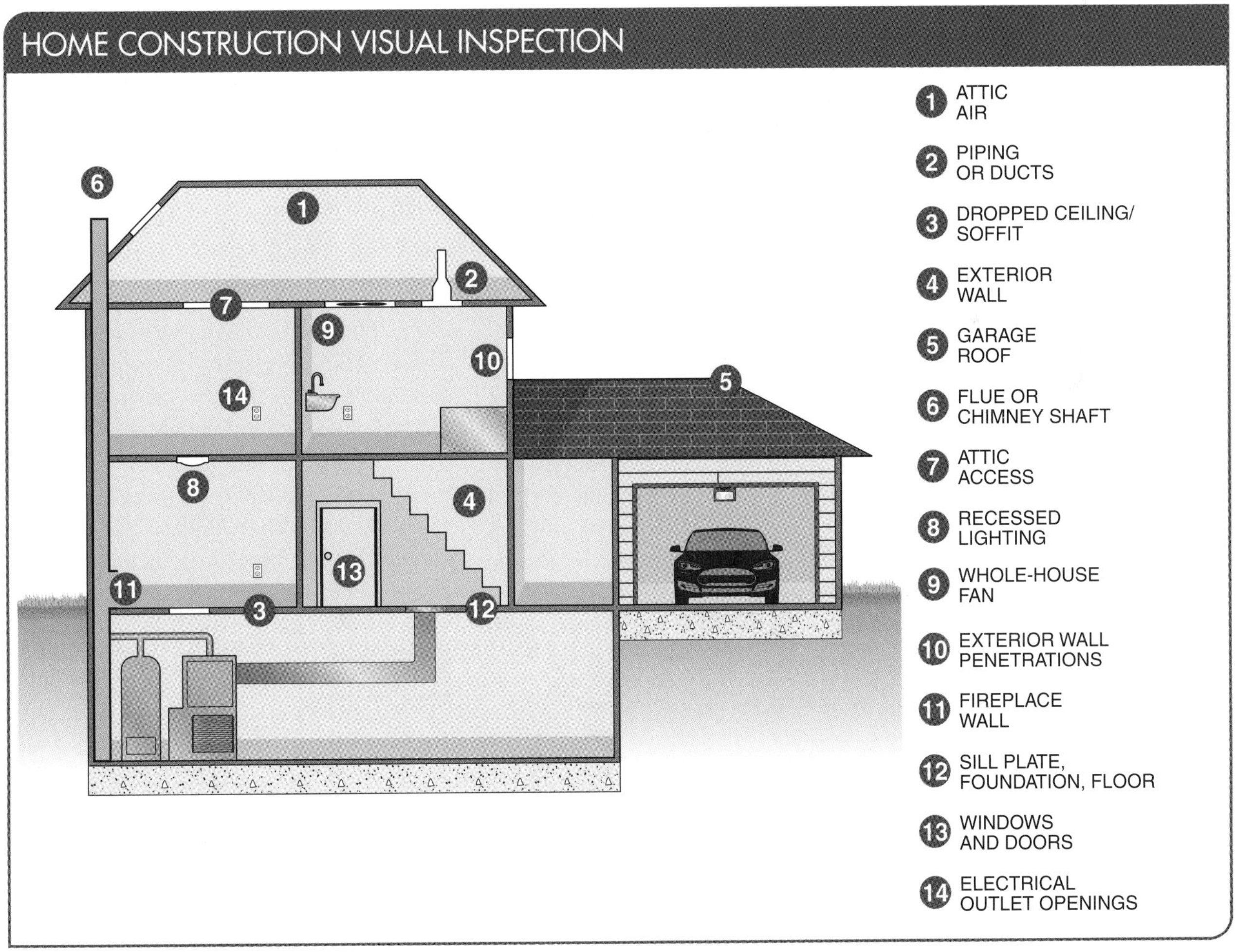

HOME CONSTRUCTION VISUAL INSPECTION

1 ATTIC AIR
2 PIPING OR DUCTS
3 DROPPED CEILING/ SOFFIT
4 EXTERIOR WALL
5 GARAGE ROOF
6 FLUE OR CHIMNEY SHAFT
7 ATTIC ACCESS
8 RECESSED LIGHTING
9 WHOLE-HOUSE FAN
10 EXTERIOR WALL PENETRATIONS
11 FIREPLACE WALL
12 SILL PLATE, FOUNDATION, FLOOR
13 WINDOWS AND DOORS
14 ELECTRICAL OUTLET OPENINGS

Figure 16-3. A visual home inspection can help identify where air leaks occur. Air leaks and lack of insulation are a major cause of structural energy loss.

SERVICE PANEL MONITORS

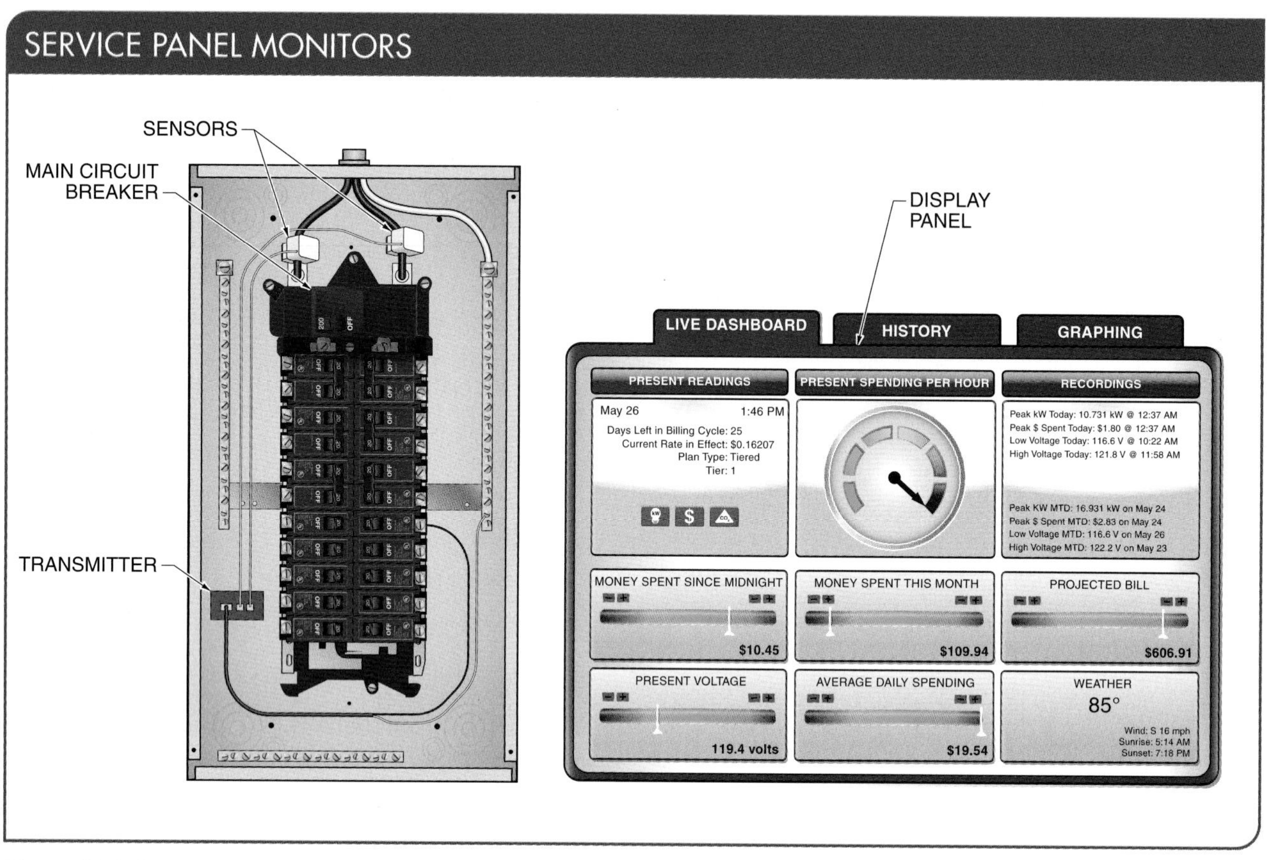

Figure 16-4. Service panel monitors typically use sensors, a transmitter, and a display to monitor electrical energy consumption.

The sensors, which are sometimes referred to as current transformers (CTs), clip onto the main power cables in the service panel enclosure. The sensors monitor the magnetic field around the incoming power cables to measure the electrical current passing through the cables. The sensors are attached to a transmitter that sends the information wirelessly to the display panel. The data is converted and displayed as real-time power usage in kilowatt-hours (kWh).

Plug-in Monitors. A *plug-in monitor* is a device that only measures the energy usage of one appliance at a time. A plug-in monitor is plugged in between the appliance and the electrical receptacle so that the monitor senses the amount of energy that an individual appliance or device consumes. To determine the amount of energy a group of appliances are using, the appliances could be plugged into a standard power strip that is then plugged into an energy monitor. **See Figure 16-5.** Typically, plug-in monitors include a portable battery-powered unit that can be moved around the home, a wireless connection to the unit attached to the electricity source, a display showing how much energy is currently being used, and an app to check on historical data, including daily, weekly, and monthly use.

Smart Plugs with Energy-Monitoring

Smart plugs are designed to fit into any standard electrical receptacle on a wall. The smart plug serves as an interface between the power source and an appliance or device. Smart plugs can be purchased with additional circuitry and software to measure the energy consumption of the appliances or devices plugged into them and to turn them on and off. The most important parameters measured by these smart energy-monitoring plugs are device running time, volts, amps, watts, and cumulative kilowatt-hours. This collected data can then be read on a smartphone. **See Figure 16-6.**

TECH TIP

Smartphones can be used to turn a device off when away from the home.

Electronics are used in smart plugs to provide special features. In addition to providing information on the energy consumption of the appliances, smart energy-monitoring plugs can be used to locate where and when the most energy is used. With this information, the homeowner can develop strategies to reduce energy use and costs. Smart energy-monitoring plugs can be moved around the dwelling every few days to find the phantom loads in a residence.

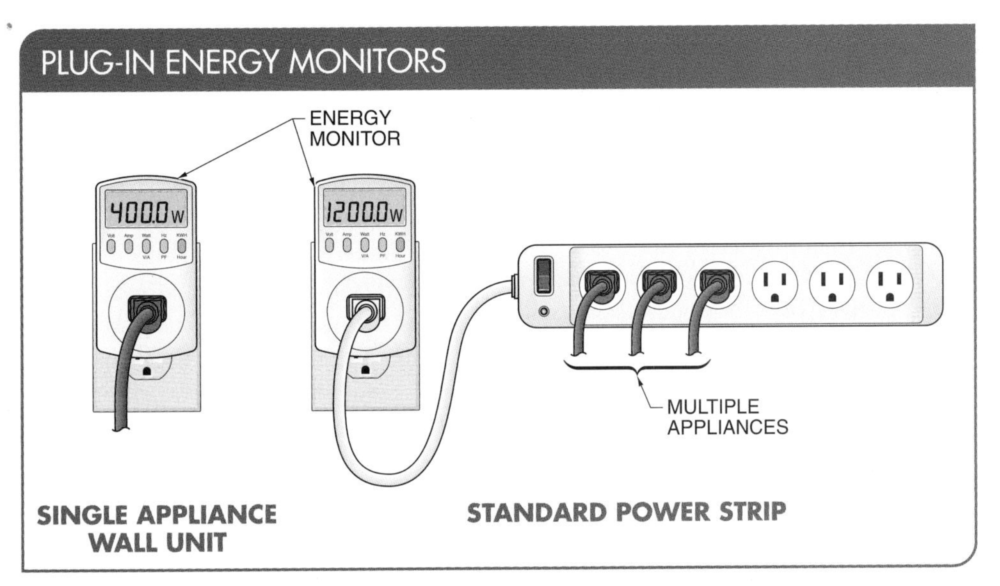

Figure 16-5. Plug-in monitors measure the energy of only one appliance or device at a time; although a group of appliances can be plugged into a standard power strip that is then plugged into the energy monitor to determine how much energy they consume.

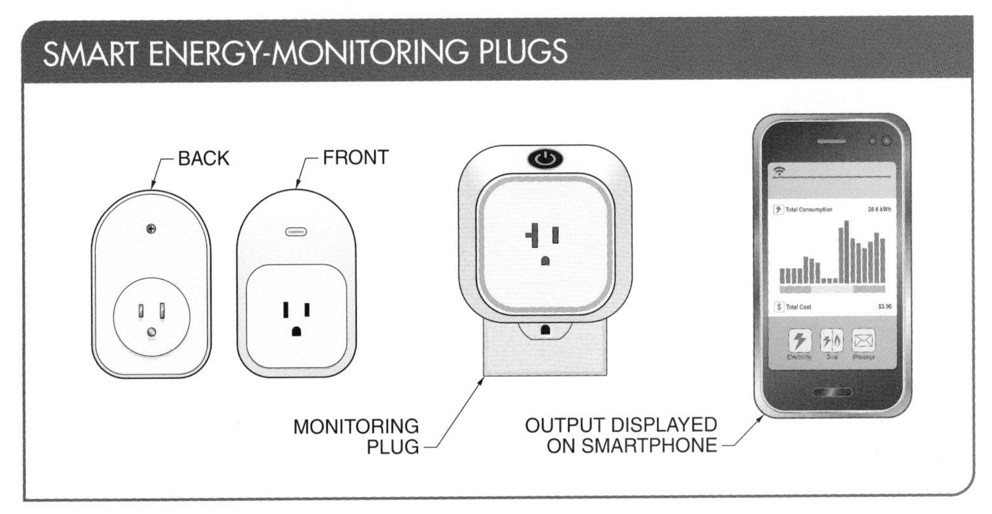

Figure 16-6. Smart energy-monitoring plugs can be used to monitor power consumption, turn a device on and off, and provide collected data to a smartphone.

Another special feature in a smart plug is the away mode, which enables the smart plug to be programmed to create the impression that someone who has left the residence is still home. This can be a useful security feature. When a light is plugged into a smart plug programmed in the away mode, it will be turned on and off in a random cycle by the smart plug. Smart plugs may be used to replace traditional mechanical timers currently used for many stand-alone lamps.

Geofencing is another special feature available in some smart plugs. *Geofencing* is the practice of using the global positioning system (GPS) to define a specific geographic boundary. Geofencing enables a smart plug to work with the GPS in a smartphone to pinpoint the homeowner's exact location at any time. This way, the smart plug can automatically turn devices (such as lights) on or off when the homeowner reaches a certain distance from their home (such as the driveway) as they return or leave.

Smart Power Strips

A smart power strip can provide all of the features of a smart plug as well as surge protection. A smart power strip is designed with several special-purpose outlets. The special-purpose outlets on a smart power strip include a control outlet, one or more switched outlets, and one or more constant on outlets. **See Figure 16-7.**

Smart Power Strip Controls. Smart power strips are controlled through electronic circuitry and smartphone and tablet apps. The four basic methods of controlling smart power strips consist of activity monitoring, timer-based controls, occupancy sensors, and app programs.

Activity monitoring is the process through which a smart power strip can determine when equipment is in use or in an idle state by monitoring equipment activity. If the equipment is idle for a length of time specified by the user, the smart power strip can shut off the device plugged into the control outlet and turn it on again only when activity resumes. The control circuit senses when power to a device drops below a preset threshold in the standby mode and turns the control outlet off. When someone turns the device back on, the circuit senses the power shift above the preset threshold and switches on power to the control outlet. However, any peripheral device, such as an internet router, plugged into the constant on outlet will remain on at all times. **See Figure 16-8.**

Smart power strips with timer-based control have outlets that are controlled by programmable timers. These outlets can be programmed to automatically turn on or off at designated times.

Occupancy-sensing smart power strips have outlets that can be controlled by a motion detector (occupancy sensor).

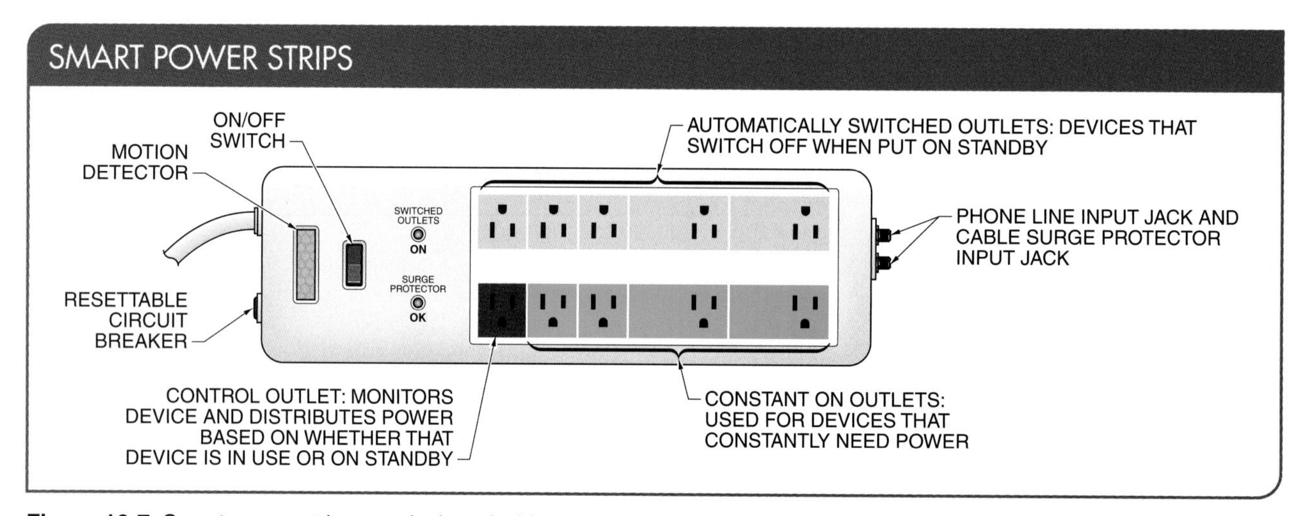

Figure 16-7. Smart power strips are designed with special-purpose outlets to control several devices at a time.

Devices plugged into outlets with occupancy sensors can automatically turn off or on in response to the presence of a person. The devices can also turn off once a defined period of time has passed after the person leaves the area. PIR sensors are the most common type of occupancy sensors used in smart power strips and can sense emitted heat up to a 15′ range.

Apps enable voice commands to be used to turn devices on and off through smart power strips. Apps allow the homeowner to monitor power and keep track of power consumption through their smart devices, such as smartphones and tablets. Apps may be added at any time to a smartphone or tablet to provide a variety of controls and information-gathering tools.

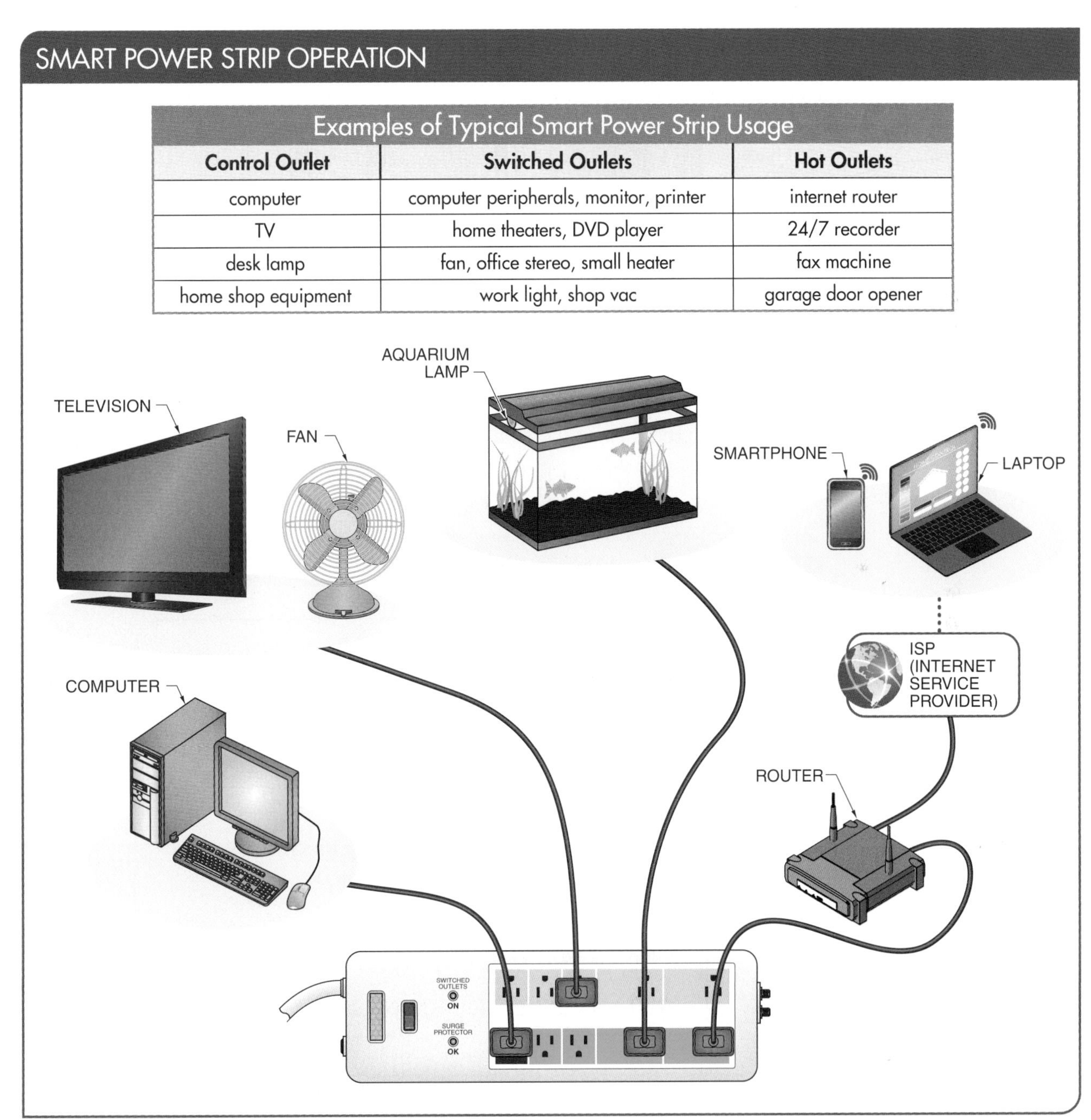

Examples of Typical Smart Power Strip Usage		
Control Outlet	**Switched Outlets**	**Hot Outlets**
computer	computer peripherals, monitor, printer	internet router
TV	home theaters, DVD player	24/7 recorder
desk lamp	fan, office stereo, small heater	fax machine
home shop equipment	work light, shop vac	garage door opener

Figure 16-8. Smart power strips can control devices through activity monitoring, timer-based controls, occupancy sensors, and app programs.

Smart Power Strip Surge Protection. There are different types of smart power strips with surge protection. Providing surge protection directly at an electrical outlet, called point-of-use protection, is common practice. Point-of-use surge protectors perform several layers of filtering to eliminate the noise on residential wiring and prevent damage to highly sensitive circuitry. They are also used to protect personal computers and home audio/video equipment. Some point-of-use surge protectors can electrically isolate their connected plug-ins so that the noise generated by a printer, for example, will not cross over to the computer.

A residence should be protected against power surges from outside lines, such as coaxial cable and telephone cable, entering the building. Surges on cable and telephone lines can do as much damage as surges over power lines due to lightning or other factors. A surge protector with a phone jack can protect a computer connected to phone lines via a modem. Cable surge protectors can be used to protect expensive equipment connected to coaxial cable.

Smart power strips with surge protection may have a UL® rating label on them. UL® rating labels indicate the level of protection the device is capable of providing. **See Figure 16-9.** Some important features on a UL®-listed surge protector include the following:

- Clamping voltage conveys the voltage level that causes the metal oxide varistors (MOVs) to conduct electricity safely to the ground line. MOVs are semiconductor devices that reduce their resistance when a higher-than-normal voltage is applied. There are three levels of voltage protection in the UL® rating: 330 V, 400 V, and 500 V. A lower clamping voltage indicates better protection.
- Surge protectors may include a light that indicates whether the protection components are functioning. MOVs may burn out after repeated power surges, but the unit will still function as a power strip.
- The energy absorption/dissipation rating, given in joules (J), indicates the amount of energy the surge protector can absorb before it fails. A high number indicates

Figure 16-9. Smart power strips with surge protection have UL® rating labels with operating voltage ratings.

greater protection. Many inexpensive surge protectors have a rating of 500 J or less. Better surge protectors have a rating of 1000 J or more.

- Surge protectors do not react immediately. There can be a slight delay as they respond to a power surge. A long response time indicates that the equipment will be exposed to the surge for a greater amount of time. Surge protectors that respond in less than one nanosecond are most effective.
- Smart power strips often include fuses and circuit breakers on the device for additional protection.

SECTION 16.1 CHECKPOINT

1. What can an energy audit determine?
2. What are the major causes of structural energy loss?
3. What three load types must be considered and measured for a home energy management plan?
4. What are four devices that can be used to monitor energy and help reduce energy waste?

SECTION 16.2—LIGHTING MANAGEMENT

Lighting is a major source of energy consumption in a residence. According to the US Energy Information Administration (EIA), the average US household has about 70 lamps (light bulbs) that account for approximately 14% of the average household electric bill. Only 10% of the energy used by incandescent lamps is converted into light, while the rest is wasted as heat. Proper indoor lighting can save energy, improve productivity, and increase comfort. For example, if lighting is too dark, it can cause eyestrain; and if lighting is too bright, energy is wasted. In addition, the proper lamp type, brightness level, and the installation of occupancy sensors, dimmers, and timers can reduce energy consumption.

Luminaries, Lamps, and Light Bulbs

Lighting technology has evolved since the early use of candles and oil lamps. Many of the lighting terms used today, such as "footcandles" and "lamps," have survived that evolution. Since some of these words have other definitions outside of lighting terminology, they may cause confusion.

In 2002, the NEC® began using the word "luminaire" to help clarify the meaning of lighting fixtures and their component parts. A *luminaire* is a lighting unit and all of its parts, which includes the internal and external parts as well as the lamp that actually provides illumination. A *lamp* is a device that converts electrical energy into light. The term "lamp" is considered the technically proper term for a traditional light source, rather than the term "light bulb." The term "light bulb" actually indicates the shape of the glass housing the light source.

Lamp Types

Lamp types are categorized by the method used to produce light. Common lamp types used in residences include incandescent, halogen, fluorescent, compact fluorescent, and LED. **See Figure 16-10.**

Incandescent and Halogen Lamps. An *incandescent lamp* is a lamp that produces light by the flow of electrical current through a tungsten filament inside a gas-filled, sealed glass bulb. The glowing filament produces light and a substantial amount of heat, making incandescent lamps very inefficient.

A special type of incandescent lamp is a halogen lamp. A *halogen lamp* is an incandescent lamp filled with a halogen gas. The halogen gas slows the deterioration of the tungsten filament, extending the life of the lamp. Halogens lamps are also very hot and inefficient.

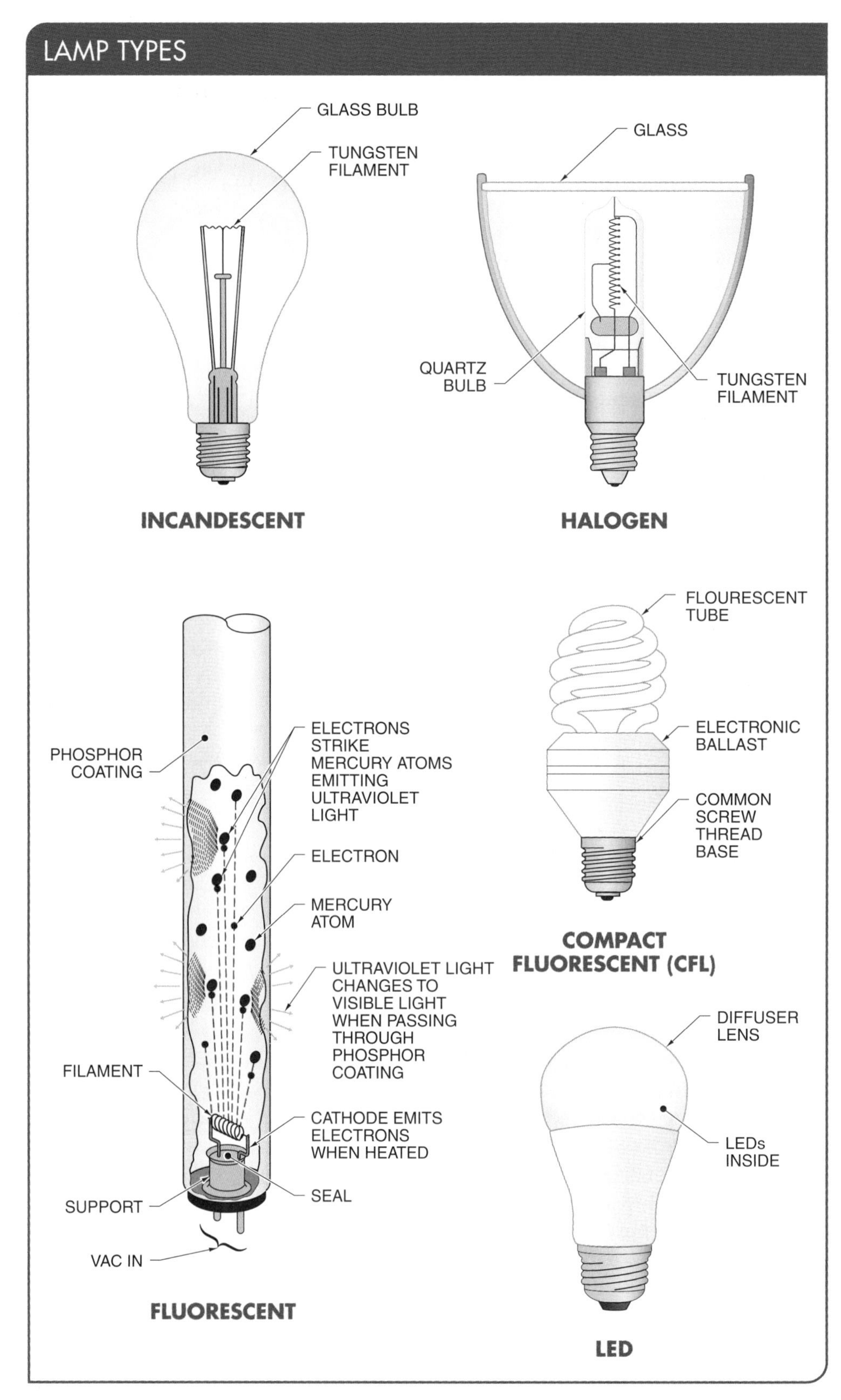

Figure 16-10. Common lamp types used in a residence include incandescent, halogen, fluorescent, compact fluorescent (CFL), and LED lamps.

Fluorescent Lamps. A *fluorescent lamp* is a low-pressure discharge lamp in which the ionization of mercury vapor transforms ultraviolet (UV) energy into light. The bulb contains a mixture of inert gas (normally argon and mercury vapor), which is bombarded by electrons from an electron-emitting cathode. This creates UV light that causes the fluorescent material on the inner surface of the bulb to emit visible light.

Fluorescent lamps require a ballast in the lighting circuit. A *ballast* is a device that controls the flow of current to a gas discharge lamp while providing a starting voltage. Most lamp ballast designs are based on either magnetic inductors or solid-state electronics. Generally, electronic ballasts are smaller, lighter, and more efficient than magnetic ballasts.

Compact fluorescent lamps (CFLs) are a design of fluorescent lamp that is compatible with traditional home light fixtures. The most common shape of a CFL is a spiral. The spiral shape mimics the shape of a conventional incandescent lamp.

LED Lights. An *LED light* is a solid-state semiconductor device used as an illumination source in a luminaire because it produces light when a DC current passes through it. LED lights produce bright light with minimal heat, making them very energy efficient. Therefore, one of the first upgrades for an energy-efficient lighting system can be to replace existing lamps with LED lights.

For example, LED tube lights are more efficient than fluorescent lamps and typically do not require a ballast that draws additional energy. LEDs are also more efficient than CFLs, last longer, and do not contain mercury, which is considered a hazardous material. Remote-controlled LED lights and dimmers are also becoming commonplace. The ability of LED lights to wirelessly communicate with other devices through an internet connection is a popular way to increase comfort and convenience in a residence.

Lamp Bulb Shapes

Lamp bulbs are available in a variety of shapes and sizes and are determined by the end use of the lamp. Lamp bulb shapes are designated by letters. The most common bulb shape is the A bulb, which is used for most residential lighting applications. **See Figure 16-11.**

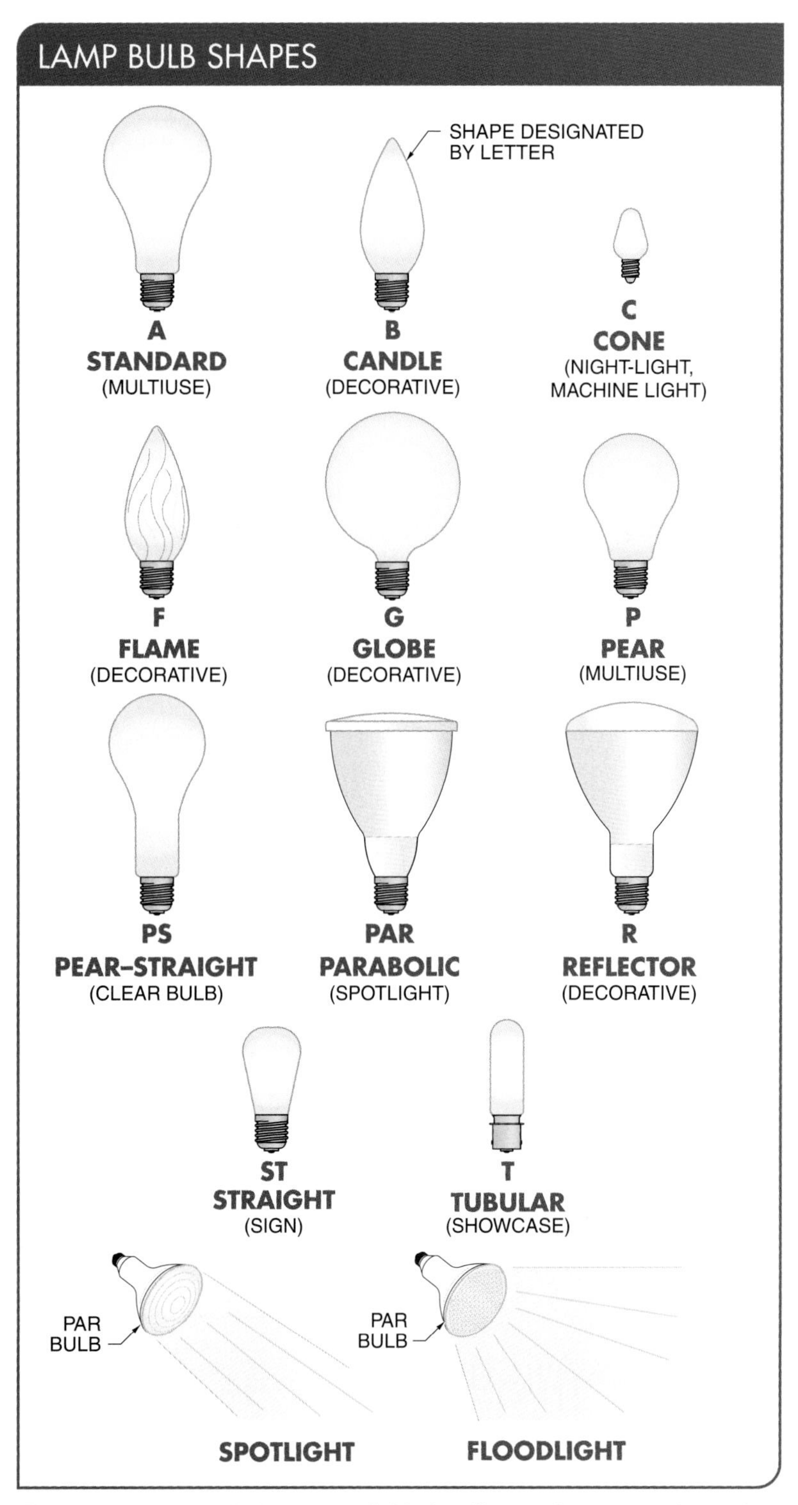

Figure 16-11. Lamp bulbs are available in different shapes, which are designated by a letter.

Other bulbs are used for decorative or special-purpose applications. The C bulb is used to withstand moderate vibrations found in appliances such as washing machines. The B, F, and G bulbs are used for decorative applications such as chandeliers, table lamps, and bathroom mirror lights.

A *floodlight* is a lamp that casts general light over a large area. Floodlights normally use R bulbs indoors or PAR bulbs outdoors. PAR bulbs are normally used as outdoor spotlights. A *spotlight* is a lamp that casts intensive light in a localized area. PAR bulbs are preferred outdoors because their shape allows for a more watertight seal than R bulbs when placed in fixtures.

Lighting Facts Labels

The Federal Trade Commission (FTC) requires that the Lighting Facts label be included on all lamp (light bulb) packages to help consumers compare the lighting information and energy efficiency characteristics of lamps. **See Figure 16-12.** The lighting information that is required to be provided on these labels includes brightness, appearance, and color temperature. Energy efficiency/efficacy characteristics are also included on some labels, especially LED lights.

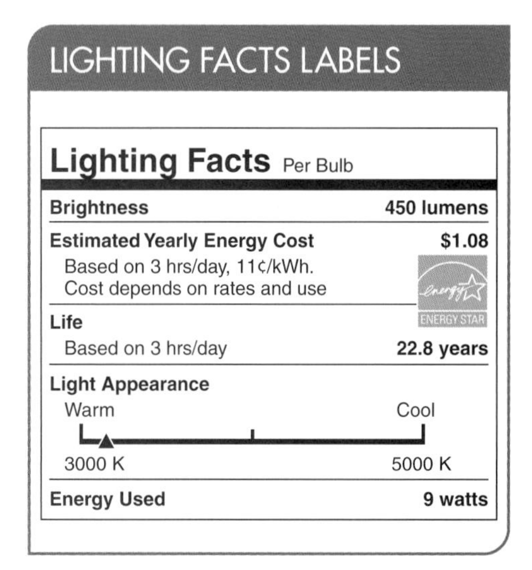

Figure 16-12. Lumens indicate the amount of light emitted by a lamp and are listed on the Lighting Facts label. The higher the lumen rating, the brighter the bulb.

Brightness. The brightness of the lamp is listed first on the label. While it may be customary to think about brightness in terms of watts, the brightness of an LED light is given in lumens. A *lumen (lm)* is a measure of the total amount of visible light from a light source. The higher the lumen rating, the brighter the lamp. The *lumens-per-watt rating* is an average rating over the lifetime of the lamp. However, a lamp becomes less efficient as it ages. Different rooms require different amounts of lumens per square foot to be effective. **See Figure 16-13.**

LIGHT BRIGHTNESS

Room	Lumens/sq ft
Kitchen	300 to 400
Living room	400 to 500
Hallway	300
Bedroom	300 to 400
Bathroom	500 to 600
Reading area	400
Working or task area	700 to 800

Figure 16-13. Different rooms require different amounts of lumens per square foot to be effective.

An incandescent lamp can be replaced with an LED light by determining the equivalent in lumens. For example, an inefficient 100 W incandescent lamps can be replaced with an equivalent energy-saving LED light rated around 1600 lm. A 60 W incandescent lamp can be replaced with an equivalent LED light with a rating around 800 lm. LED lights use far less energy (in watts) than incandescent lamps for equivalent lumens. Conversion charts are available to convert any wattage to lumens for most common lamp sizes. **See Figure 16-14.**

Appearance and Color Temperature. The color temperature of light is measured in Kelvin (K), which is commonly referred to as either warm or cool. Lamps with a low Kelvin value produce a warmer, yellowish light. Lamps with a higher Kelvin value produce a cool, blue light similar to daylight. Homeowners may prefer one color temperature throughout the home for uniformity, while others may use different color temperatures in each room for different tasks. **See Figure 16-15.**

LUMENS TO WATTAGE CONVERSION CHART		
Lumens	**Incandescent Lamps***	**LED Lights***
450	40	8
800	60	10
1100	75	17
1600	100	20

* approx. wattage (W)

Figure 16-14. Conversion charts compare the lumens and wattages of common bulb sizes.

Efficacy vs. Efficiency. The terms "efficacy" and "efficiency" are both used in describing lamp performance. Efficacy means the effectiveness or the ability to produce an output. Efficiency means how well something works. Lamp output light is measured in lumens, and electrical energy (which creates light) is measured in watts.

Efficacy is the number of lumens of light delivered for the energy (wattage) used, expressed in lumens per watt (lm/W). The efficacy of different lamp types falls within an approximate range of lumens per watt. For example, the efficacy of an incandescent lamp is approximately 14 lm/W, a CFL is approximately 65 lm/W, and an LED is approximately 85 lm/W.

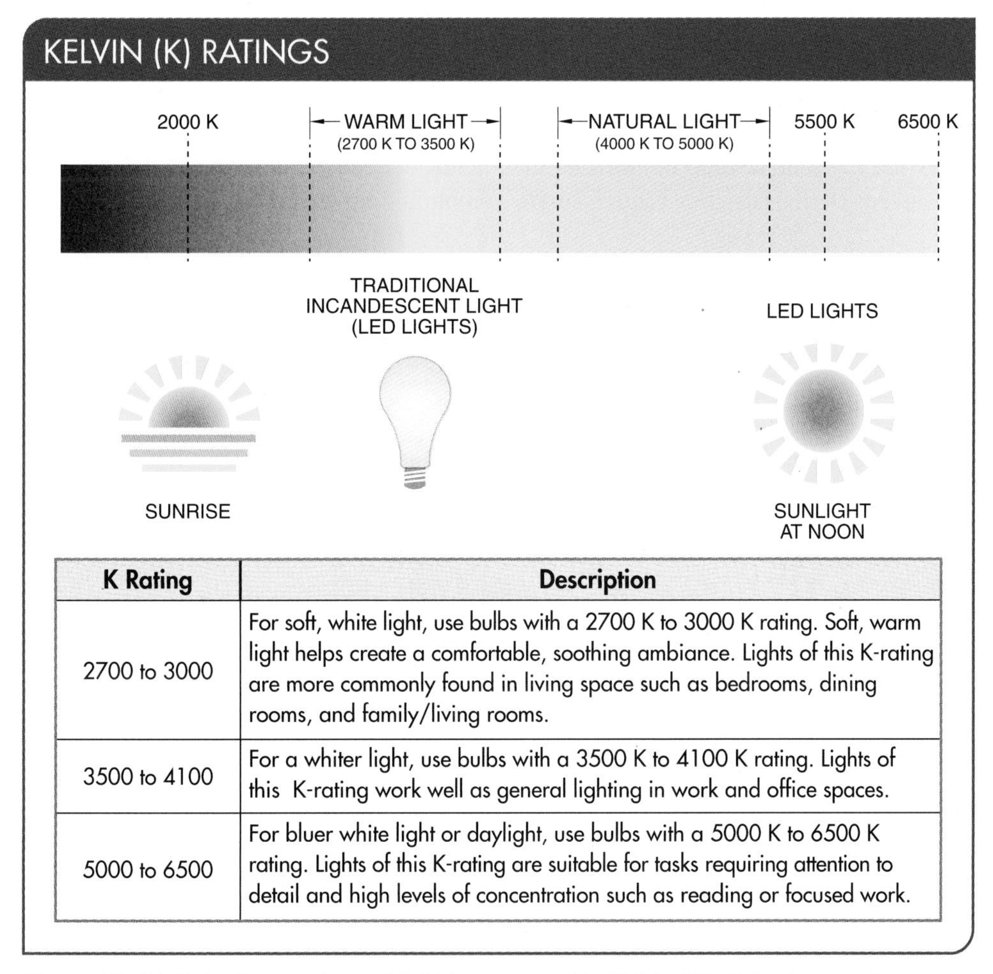

K Rating	Description
2700 to 3000	For soft, white light, use bulbs with a 2700 K to 3000 K rating. Soft, warm light helps create a comfortable, soothing ambiance. Lights of this K-rating are more commonly found in living space such as bedrooms, dining rooms, and family/living rooms.
3500 to 4100	For a whiter light, use bulbs with a 3500 K to 4100 K rating. Lights of this K-rating work well as general lighting in work and office spaces.
5000 to 6500	For bluer white light or daylight, use bulbs with a 5000 K to 6500 K rating. Lights of this K-rating are suitable for tasks requiring attention to detail and high levels of concentration such as reading or focused work.

Figure 16-15. Color temperature of light is measured in Kelvin (K). Color temperatures over 5000 K are called "cool colors" (bluish), while lower color temperatures (2700 K to 3000 K) are called "warm colors" (yellowish).

Reducing Lighting Energy Consumption

There are two basic ways to reduce energy consumption in a lamp: reduce the amount of time the light is on and/or reduce the amount of energy going to the lamp. Reducing the amount of time that the light is on can be accomplished through occupancy sensors and timers. Reducing energy going to the lamp can be accomplished through light dimmers.

Occupancy Sensors. Occupancy sensors are used to reduce energy use. The two most common types of occupancy sensors used are ultrasonic sensors and IR sensors. Ultrasonic sensors use sound to detect a moving object, while IR sensors detect heat and motion. Occupancy sensors turn lights or devices on or off as a person enters or leaves a room.

Wall switch occupancy sensors use the same advanced PIR and ultrasonic technology found in security motion detectors. The correct sensor must be installed to maximize energy savings while minimizing inconvenience to the user. **See Figure 16-16.**

Time-Based Control Systems. A *time-based control system* is an automated control system that uses the time of day to determine the desired operation of energy-consuming loads. Time-based control allows loads to be turned on and off at specific times. Time-based control strategies include seven-day programming, daily multiple-time-period scheduling, and timed overrides.

Seven-day programming allows homeowners to individually program on and off time functions for each day of the week. Daily multiple-time-period scheduling allows homeowners to schedule lighting, HVAC, and other loads to independently operate for multiple time periods during the day. A timed override allows homeowners to change a time period for temporary use. A timed override can be activated by a switch, a personal computer, tablet, or other portable device.

Time-based control systems can be used in many different parts of a home. For example, a time-based control system, such as a timer control, can be used for an exhaust fan in a

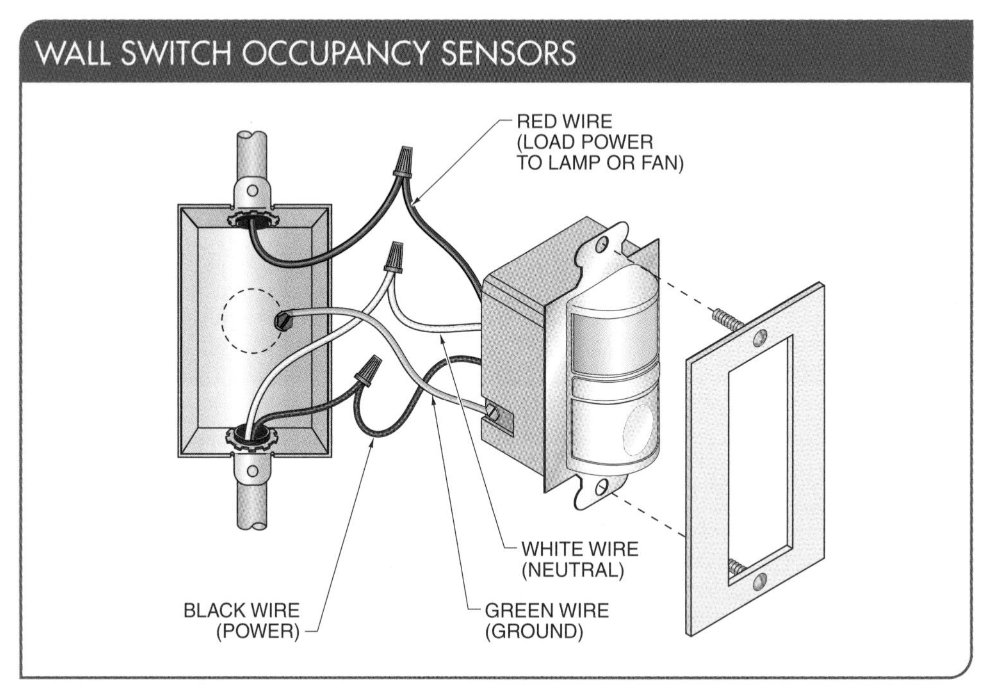

Figure 16-16. Wall switch occupancy sensors can be used to detect the presence of a person to automatically turn on or off lights.

bathroom. **See Figure 16-17.** This timer saves additional energy by reducing the length of time the fan blows warm air outside in the winter or cold air outside in the summer. Timers are also often used in hot tubs to manage the length of time that the jets are run.

Dimmer Switches. Dimmer switches can produce a softer lighting effect in a room and use less energy when full lamp brightness is not required. Since the characteristics of a lamp may vary, a specific type of dimmer may be required to match the type of lamp being dimmed. A dimmer that is designed, tested, and UL® listed for the specific lamp must be used.

For example, halogen and incandescent dimmers must only be used to control halogen and incandescent lamps. These dimmers should not be used to control CFLs and LEDs or dimmable CFLs and dimmable LEDs.

Fluorescent dimmers are only used to control fluorescent lights. All fluorescent lamps rely on ballasts inside of the fixture to operate properly. Fluorescent dimmers can control the illumination level produced by traditional tube-style fluorescent lights. The fixtures, however, must be equipped with dimming ballasts and rapid-start dimmable lamps.

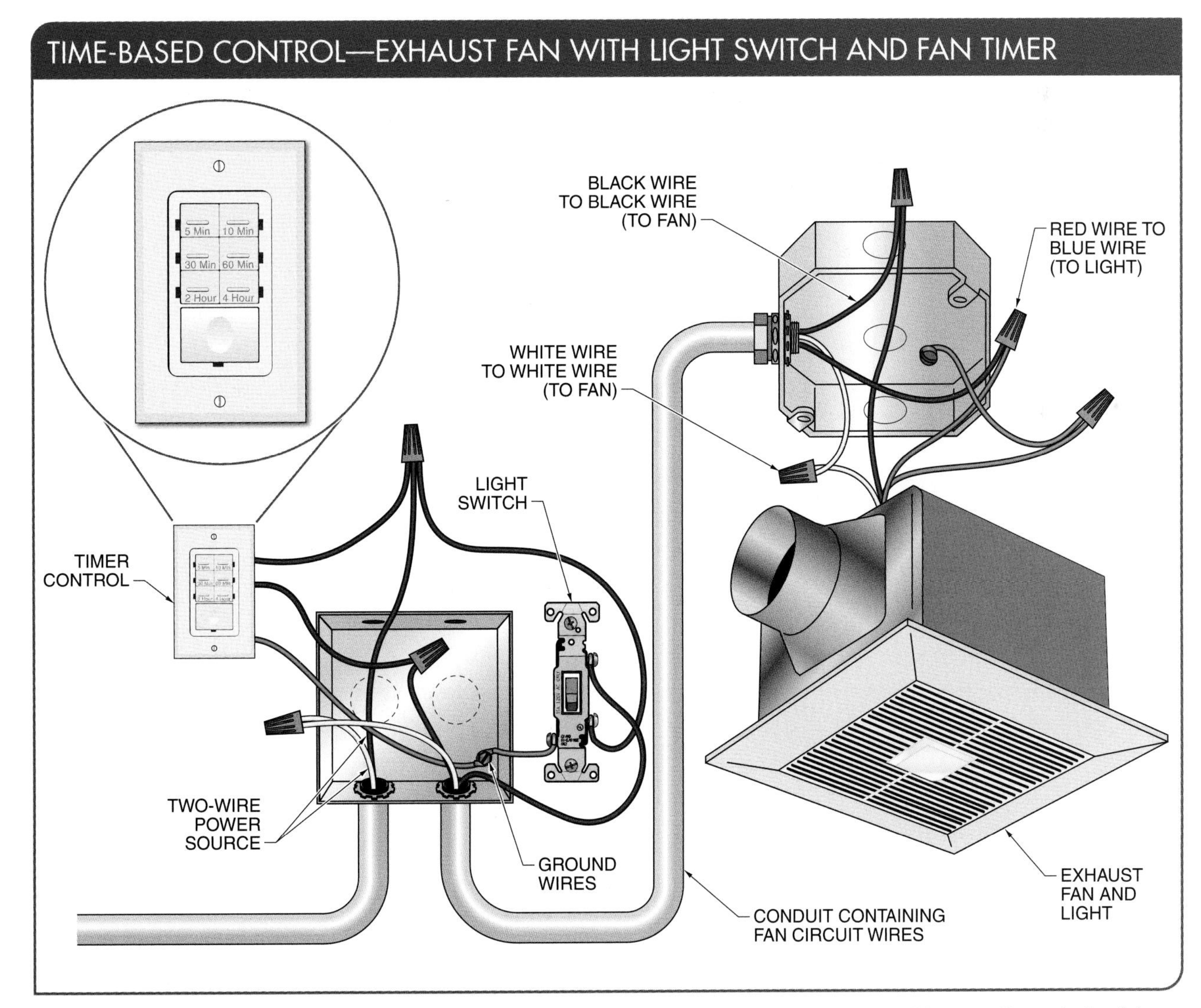

Figure 16-17. Time-based controls can turn devices, such as bathroom exhaust fans, on or off for specific periods of time.

Universal dimmers can be used to control halogen, incandescent, dimmable CFLs, and dimmable LEDs. Universal dimmers offer maximum-range dimming and soft startup. They also eliminate the flickering of lights. When CFLs or LEDs are used with a dimmer, they must be labeled to indicate that they are dimmable.

Dimmer Styles. Dimmers are available in a variety of different styles to meet all dimming applications. Dimmer styles include rotary, toggle, slide, rocker, and digital/touch dimmers. **See Figure 16-18.**

Rotary dimmers are a traditional, easy-to-use style of dimmer. Rotary dimmers use a knob that is rotated to adjust light levels. Twisting the knob all the way to the left turns off the connected fixtures.

Toggle dimmers use a switch similar to standard light switches to adjust light levels. However, toggle dimmers also combine the switch with a separate level control for dimming. Toggle switches allow the fixture to be turned on or off without changing the dimmer setting.

Slide dimmers use a slide that can be moved up and down to adjust light levels. Slide dimmers operate similarly to rotary dial dimmers except that they use a vertical or horizontal slide to adjust the light level. Slide dimmers often have a slide-to-off feature and some include multiple setting stops to quickly find preset light level settings.

Rocker dimmers are similar to toggle controls. Typically, rocker dimmers have a paddle-style switch that acts as an on/off

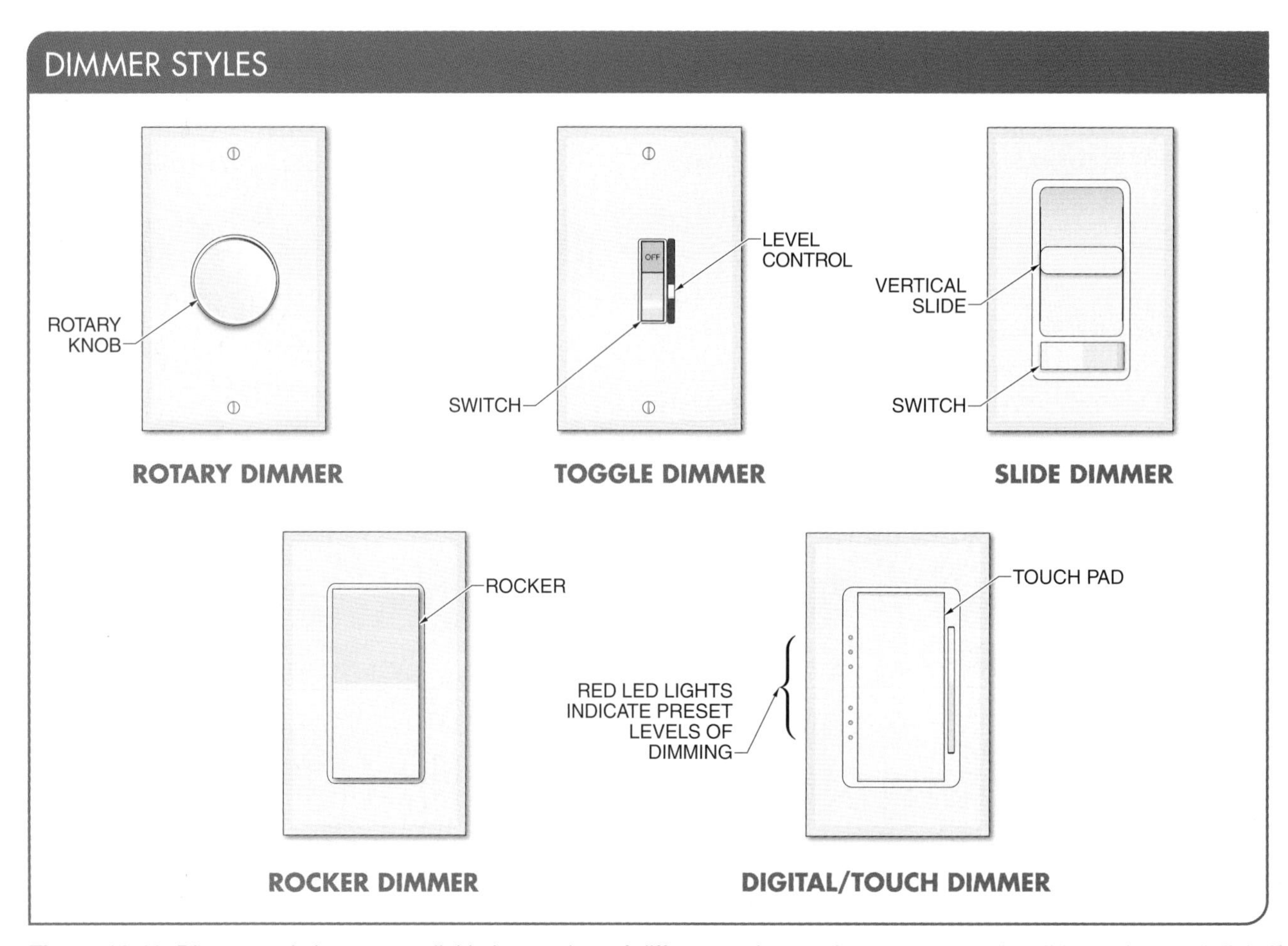

Figure 16-18. Dimmer switches are available in a variety of different styles, such as rotary, toggle, slide, rocker, and digital/touchpad dimmers.

control and a separate dimmer adjustment. Tapping or pressing and holding the up side of the switch increases illumination. Tapping or pressing and holding the down side of the switch reduces illumination.

Digital/touch dimmers feature touch pad control. The touch pad is used to control the light level through preset light level settings. Touch dimmers usually include a digital light display on the touch pad to provide users with visual indicators for increasing or decreasing dimming levels.

Dimmer Operation. Dimmers control the brightness of lights by controlling the amount of electrical energy supplied to the lamp. This is accomplished through the use of an electronic component such as a triac or a silicon-controlled rectifier (SCR). **See Figure 16-19.**

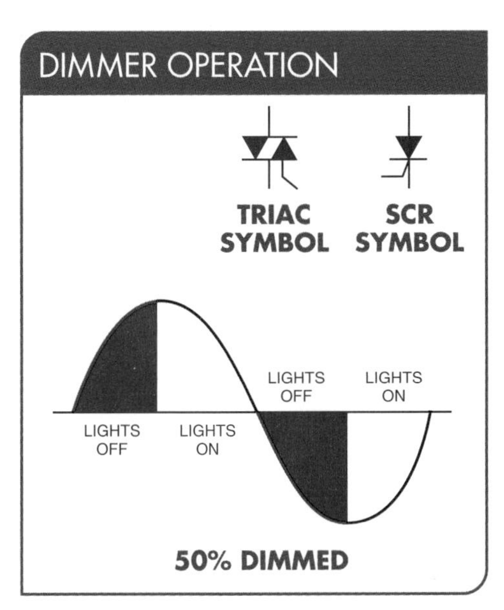

Figure 16-19. Dimmers control the brightness of a bulb using semiconductors such as triacs or SCRs.

Triacs and SCRs are fast-acting semiconductor switches that can turn on and off hundreds of times per second. When a dimmer is turned to full brightness, the semiconductors allow all of the electrical energy to flow through them to supply the light they control. When the light is dimmed, the semiconductors turn off the electrical energy for a specific period so that the lamp receives less energy and gives off less light.

Dimmers can be used to control light output from incandescent, halogen, CFL, and LED lights. Additional specialized equipment may be required to dim fluorescent tube lights. Some dimmers can be controlled by a wireless control system. These dimmers are now compatible with other internet control devices.

Dimmer Locations. Dimmer switches can be used in any single-pole, three-way or four-way location. Dimmers can be installed with standard wiring configurations as a direct replacement. **See Figure 16-20.**

When a dimming system is changed to be wireless, wireless dimmers can use the same electrical boxes as a direct replacement. However, the wiring configuration for wireless dimmer control is different. **See Figure 16-21.**

In a wireless dimming system, there will be only one control device that switches the lighting. The use of wireless remotes to activate that control device may eliminate some control wiring. Those wires can be removed from the circuit by using wire connectors and electrical tape to isolate them from the wireless circuitry.

TECH TIP

Lighting energy consumption can be reduced by turning off unused lights, installing motion sensors or timers, maximizing natural light from windows, replacing incandescent lamps with LED lights, and choosing small energy-efficient task lights for focused work.

Heat Dissipation

All electronic devices are affected by heat. Heat causes semiconductor materials to break down, which can result in reduced performance or failure over time. Heat can come from the surrounding air (the ambient temperature of a room) or from the electronic device itself. Heat can be dissipated by reducing ambient temperature or by using a heat sink.

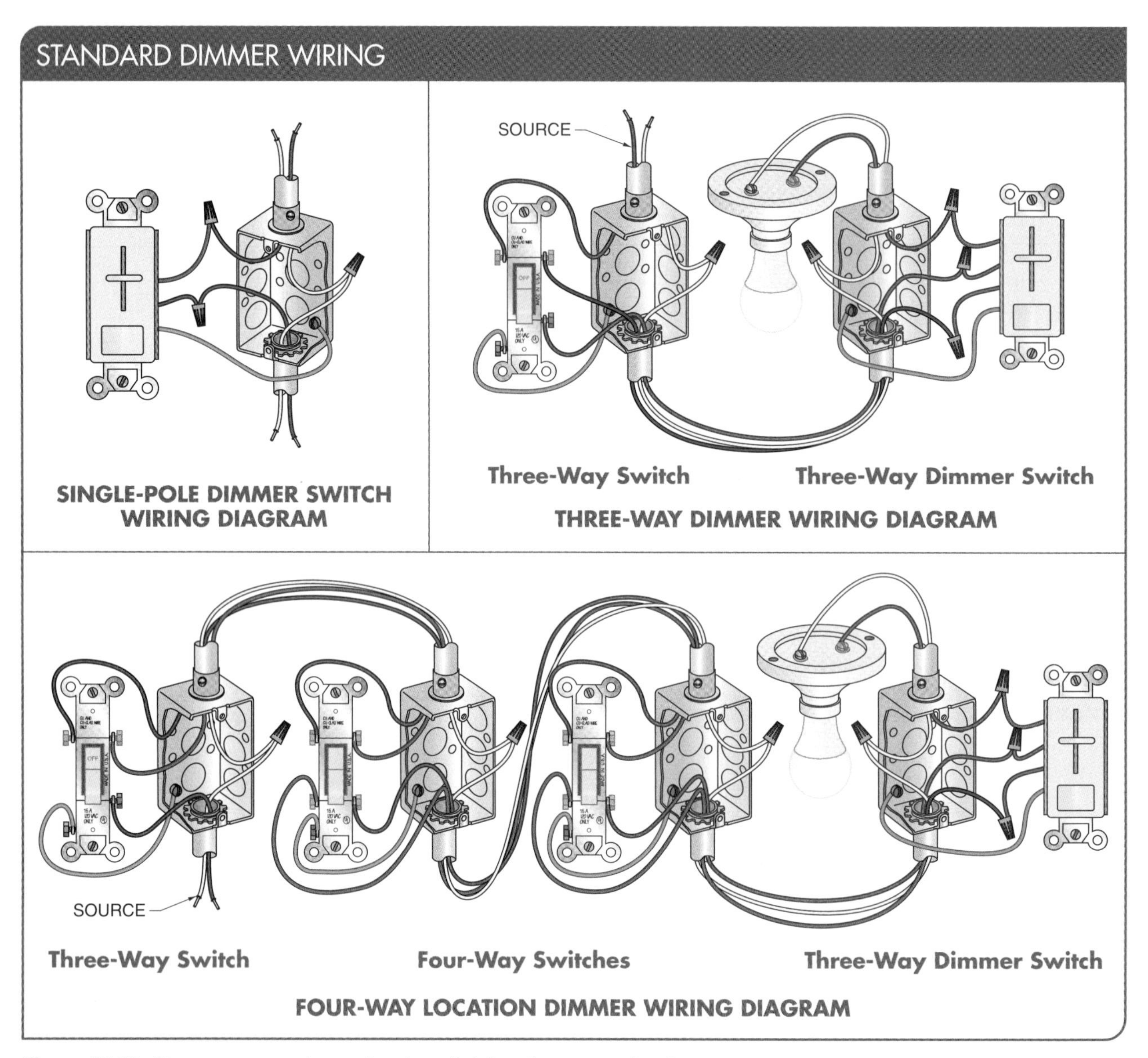

Figure 16-20. Dimmers can replace a toggle switch in a three-way circuit.

Reducing Ambient Temperature. A room's ambient temperature can be reduced by using air conditioning or installing electronic devices in a cooler location. For example, during normal operation, dimmers get warm to the touch. When properly installed per code and when operating on its rated load in a 77°F ambient temperature, the dimmer's accessible surfaces should stay below the UL® limits of 140°F. This is about the same temperature as a typical plug-in night-light.

Heat Sinks. Another way to dissipate heat is through the use of a heat sink. A *heat sink* is a device or material that absorbs and disperses heat away from an electronic device to keep it from overheating. Heat sinks are usually made of aluminum, which is a good conductor of heat. Heat sinks with removable aluminum fins are attached to standard light dimmers. High-wattage dimmers are mounted on aluminum heat sinks. **See Figure 16-22.**

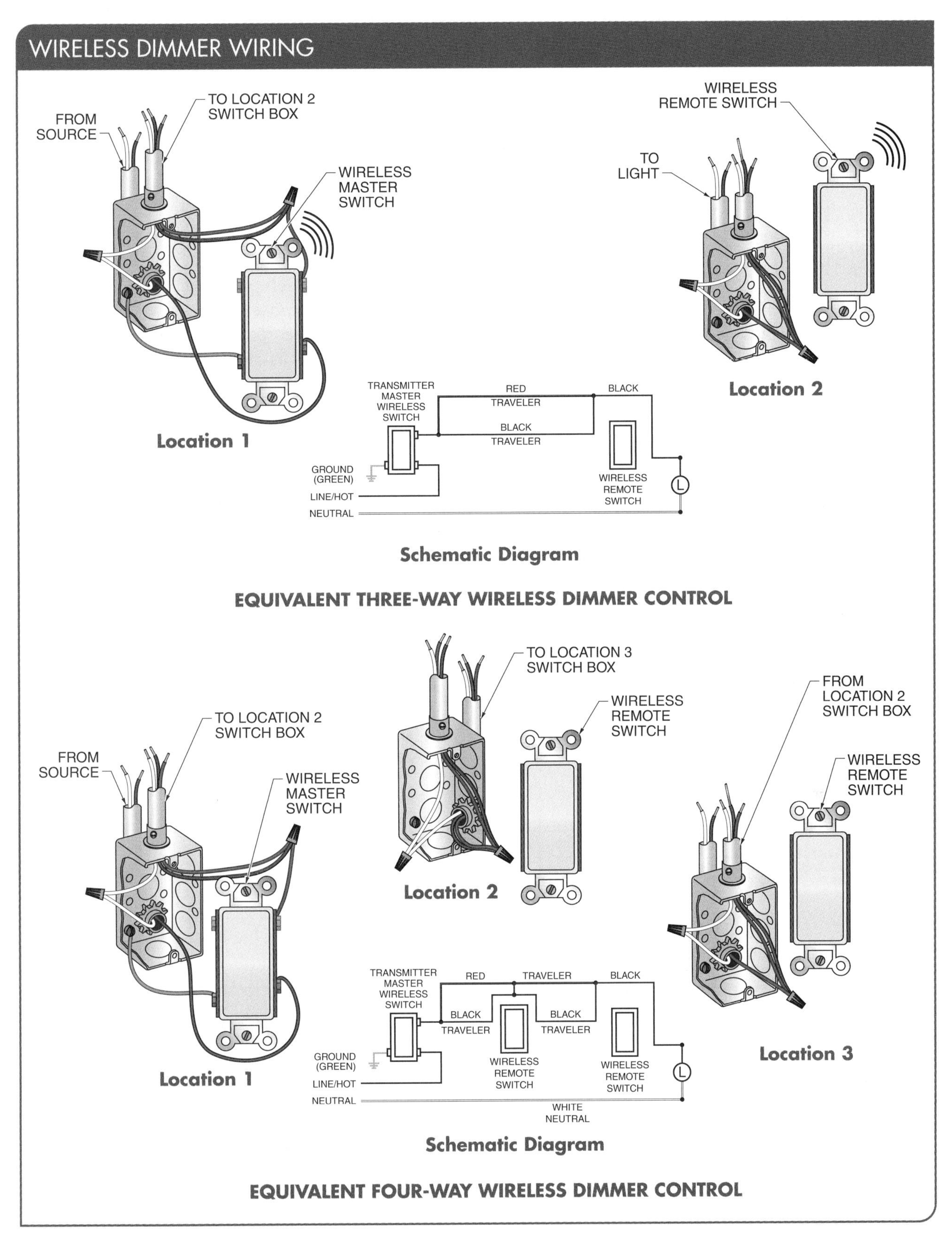

Figure 16-21. Wireless dimmers can be used when a wired lighting system is changed to a wireless configuration.

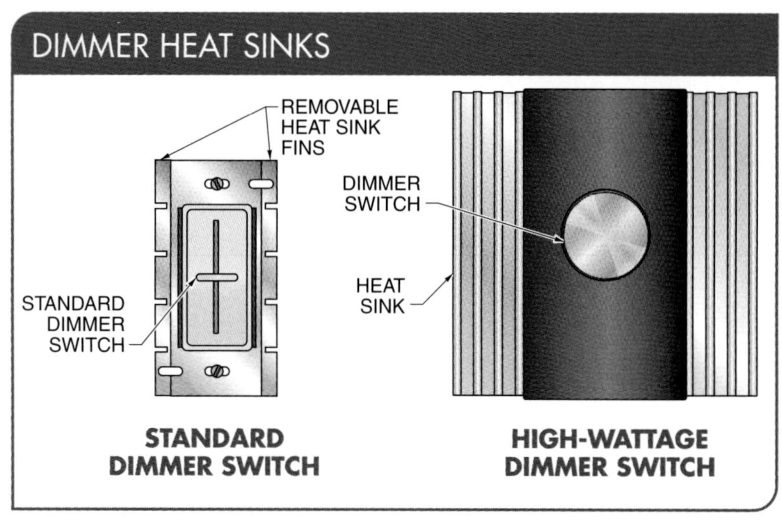

Figure 16-22. Heat sinks are used to absorb heat from light dimmers.

For example, most standard dimmers have heat sinks made up of individual fins (tabs) that can be removed using pliers. When more than one dimmer is installed side by side in an enclosure, the fins must be removed to enable them to fit. When two switches are installed side by side, these fins need to be removed from one side of each dimmer. When three or more dimmer switches are installed side-by-side, these fins must be removed on both sides of each dimmer, except for the dimmers at either end. The removal of fins from a dimmer reduces its capacity to dissipate heat and its maximum allowable wattage. This is called derating. **See Figure 16-23.**

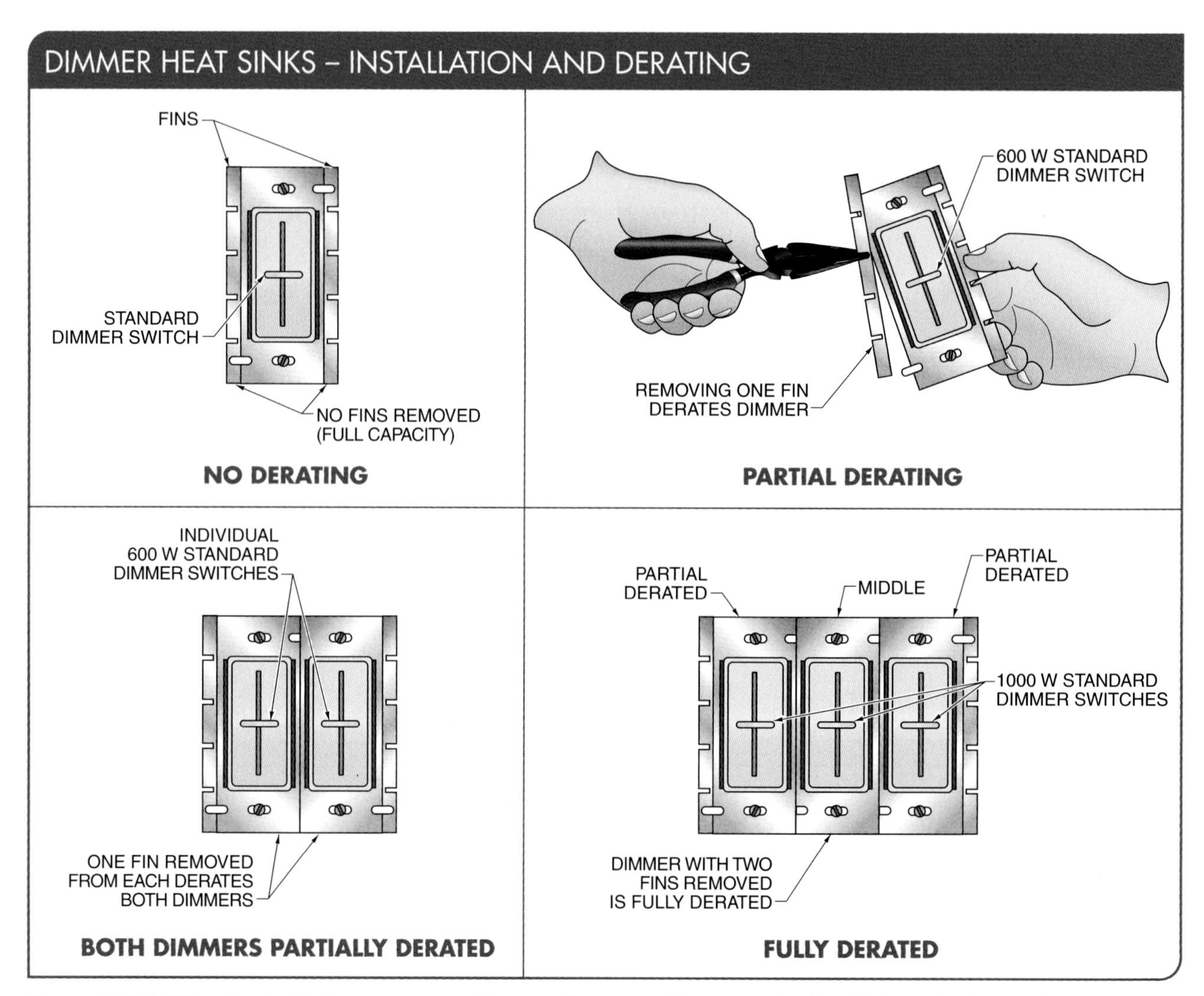

Figure 16-23. When heat sink fins are removed from a dimmer to allow space for installation, the dimmer is derated to a lower wattage. The amount of wattage derating is determined by the manufacturer.

TECH TIP

Each manufacturer may use a different derating system. Manufacturers' derating charts should always be followed, and consideration should be given to the ambient temperature of the room, airflow, direct sunlight exposure, and the size and type of mounting enclosure.

Heat sinks are also used to keep heat away from the electronic devices inside LED lights. Heat sink fins are often molded into the sides of LED lights to remove excess heat. In other types of LED lights, a thin layer of metal may be incorporated in the design to dissipate the heat. **See Figure 16-24.**

SECTION 16.2 CHECKPOINT

1. What are common types of lamps used in residences?
2. What is a lumen?
3. How is the color temperature of light measured?
4. What are the two basic ways to reduce energy consumption in a lamp?
5. What devices can be used to reduce the amount of time a light is on and to reduce the amount of energy going to a lamp?
6. What is derating?
7. What is a heat sink?

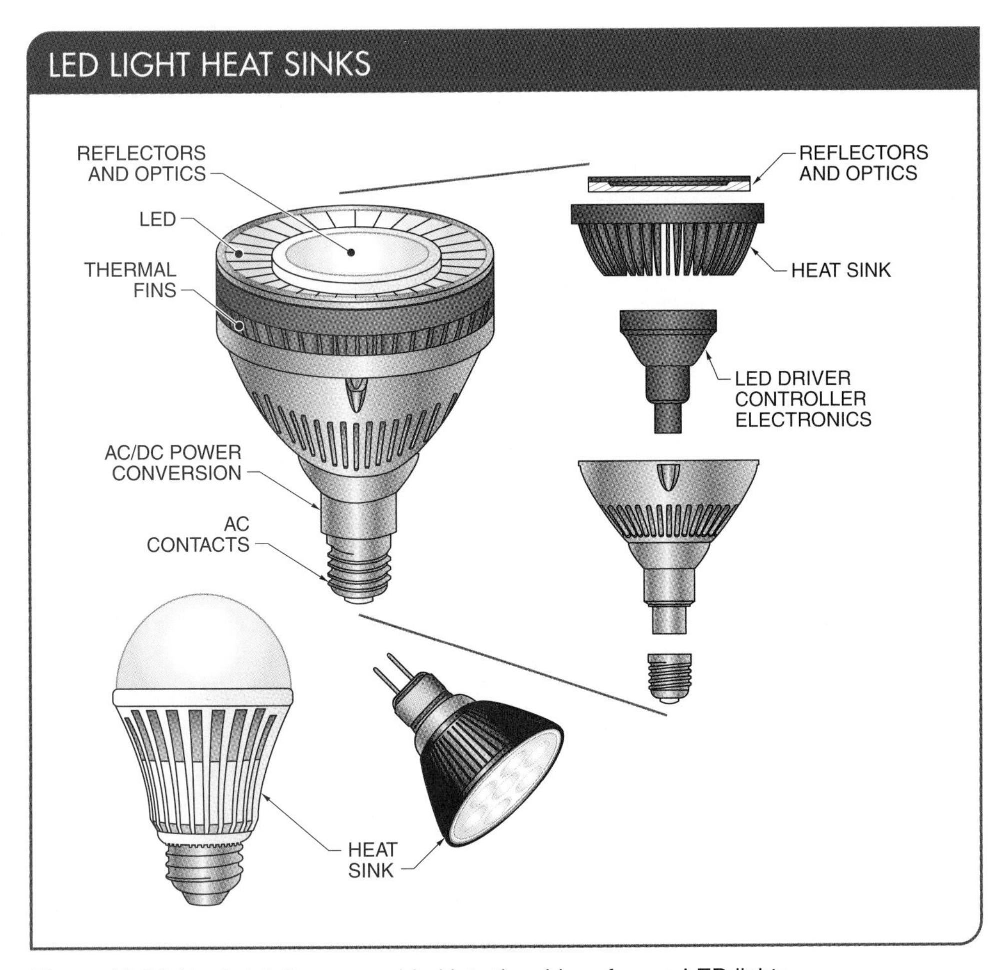

Figure 16-24. Heat sink fins are molded into the sides of some LED lights.

SECTION 16.3—INDOOR ENERGY MANAGEMENT

Indoor energy management involves improving HVAC system efficiencies, reducing energy waste, and lowering utility costs. This can be accomplished by using smart thermostats, ceiling mounted fans, whole-house fans, and attic fans. Remote-controlled electric dampers and HVAC vents can be added to existing homes for more precise control in individual rooms. New energy-efficient window treatments and electrochromic smart glass can help control heat gain or loss in a home.

Thermostats

A *thermostat* is a device that automatically regulates temperature by controlling the supply of energy, such as gas or electricity, to a heating or cooling apparatus. Thermostats for HVAC systems control the largest percentage, about 35% to 45%, of energy use in a residence.

There are two general categories of thermostat designs: nonprogrammable and programmable. **See Figure 16-25.** Nonprogrammable thermostats are the most affordable design, but the temperature must be set by hand to heat or cool a residence. Programmable thermostats operate by being programmed with specific instructions to automatically adjust heating or cooling based on multiple setpoints. Newer, wireless thermostat designs include connected thermostats and smart thermostats.

Nonprogrammable Thermostats. Most thermostats, and all nonprogrammable thermostats, are designed so that the setpoint temperature can be changed manually. *Setpoint temperature* is the temperature at which the switch in a thermostat opens and closes. The setpoint adjustor on a manual thermostat can be a lever or a dial that indicates the temperature on a visible scale.

Moving the setpoint increases or decreases the temperature at which the switch opens and closes. When a thermostat senses

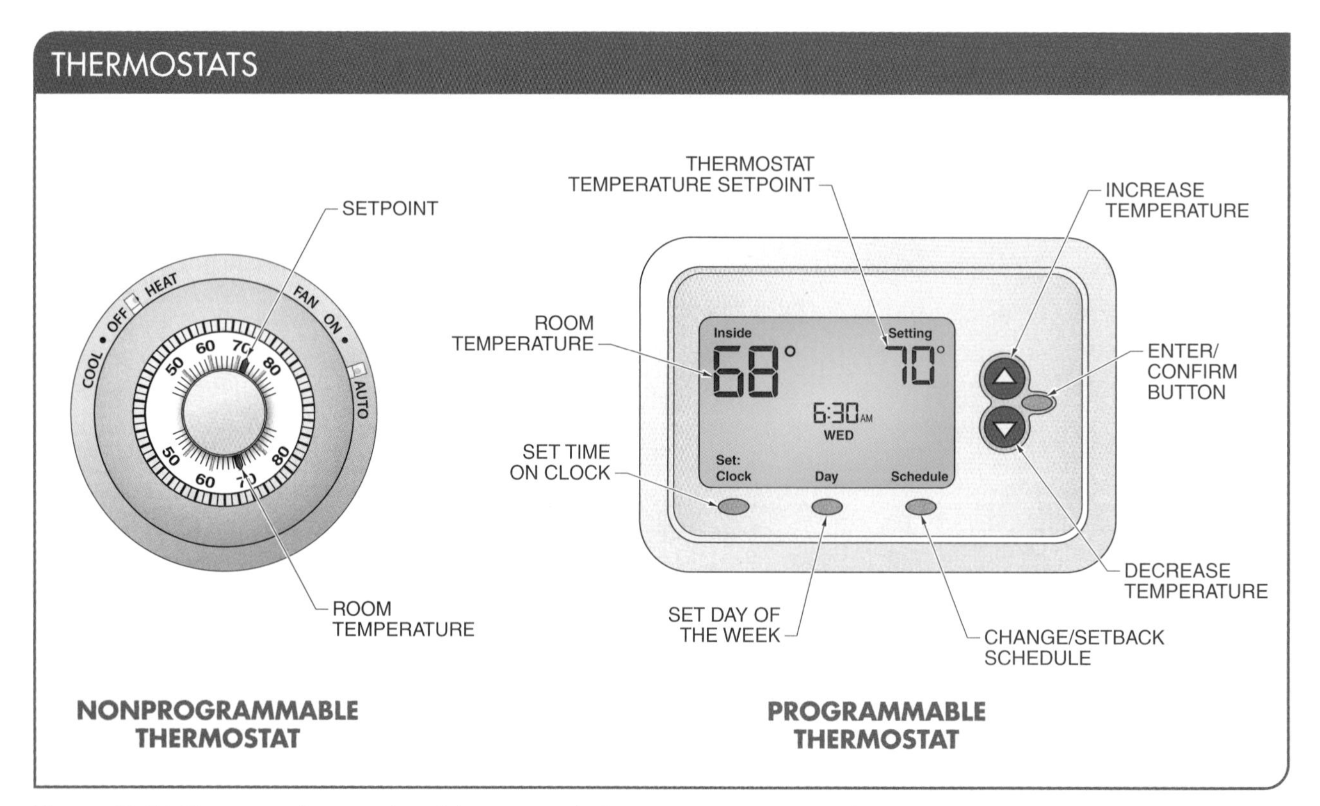

Figure 16-25. Two general categories of thermostat designs are nonprogrammable and programmable thermostats.

the setpoint temperature, it turns off the heating or cooling equipment. The switch in the thermostat turns the system on or off when the temperature varies approximately 2° above or below the setpoint, depending on whether heating or cooling is desired. The *differential* is the difference between the temperature at which the switch in the thermostat turns the system on and the temperature at which the thermostat turns the system off. The differential is necessary to prevent rapid cycling of the system.

Programmable Thermostats. A *programmable thermostat* is a thermostat designed to adjust temperature in a dwelling at different times of the day based on preprogrammed settings. Most programmable thermostats are digital. Digital thermostats offer the most features in terms of multiple setbacks, overrides, and adjustments. Some programmable thermostats include apps that allow a smartphone or Wi-Fi enabled device. Most programmable thermostats perform one or more of the following energy control functions:

- Store and repeat multiple daily settings, with a manual override that does not affect the rest of the daily or weekly program.
- Store six or more temperature settings a day.
- Adjust heating or air conditioning times as the outside temperature changes.

The four basic settings that are most used include wake, sleep, leave, and return. In addition, most programmable thermostats have settings for both weekdays and weekends. This means that these days can be programmed for different schedules, according to how much time someone is home.

Wireless Connected and Smart Thermostats. Wireless connected and smart thermostats use technology that allows them to connect to the internet. These new thermostats have more reporting and interactive features and are easier to use than programmable thermostats. The main difference between a connected thermostat and a smart thermostat is that the smart thermostat makes decisions by incorporating artificial intelligence (AI). **See Figure 16-26.**

AI uses a computer to perform tasks that represent human intelligence, such as vision, speech recognition, language translation, pattern tracking, and decision making. The primary focus of smart thermostat AI has been pattern recognition and decision making. Once a smart thermostat recognizes a certain pattern of use, the device can make decisions on when to adjust temperatures and send out alerts when equipment malfunctions.

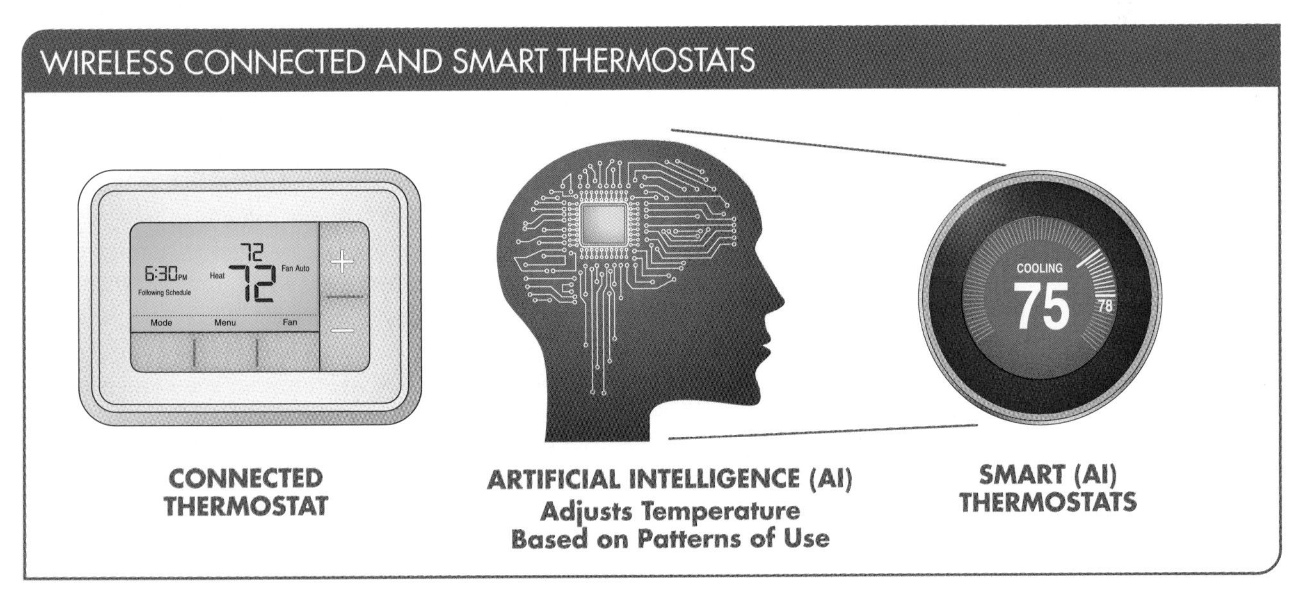

Figure 16-26. Wireless connected and smart thermostats connect to the internet, but smart thermostats also use an artificial intelligence (AI) that makes decisions by recognizing thermostat usage patterns.

Speech recognition and voice-activated assistance technology has become a very popular type of control. Other smart thermostat features include geofencing, the ability to change the temperature anytime Wi-Fi is available, multiple room control (with the addition of sensors), and energy history to show the amount of energy being used over time and its cost.

Temperature Settings. Energy Star recommends certain temperature setbacks at different times of the day for maximum energy savings. According to the US DOE, a home can save up to 1% of the energy costs for every degree Fahrenheit that a thermostat is turned down for heat or turned up for air conditioning in an 8-hour period each day. **See Figure 16-27.**

ENERGY STAR RECOMMENDED SETBACKS FOR MAXIMUM HEATING/COOLING SAVINGS				
Time of Day	Weekdays		Weekends	
	Heat (°F)	Cool (°F)	Heat (°F)	Cool (°F)
Wake (6 AM)	70°	65°	70°	75°
Leave (8 AM)	62°	83°	62°	83°
Return (6 PM)	70°	75°	70°	75°
Sleep (10 PM)	62°	83°	62°	83°

Figure 16-27. Turning a thermostat heat setting down 1°F and a thermostat air conditioning setting up 1°F saves up to 1% on energy costs.

Thermostat Location. The location of a thermostat can affect its performance and efficiency. A thermostat should be placed on an interior wall about 5′ above the floor in a room that is used often. A thermostat should be placed away from direct sunlight, drafts, doorways, skylights, windows, lamps, televisions, air vents (registers), and fireplaces. Damp areas should be avoided because they can lead to corrosion and shorten thermostat life. **See Figure 16-28.**

Thermostat Wiring. Until recently, only four wires were necessary for a thermostat to control the functions of an HVAC system. However, modern thermostats require a fifth wire referred to as the "C-wire," which stands for the common wire or return wire. The C-wire is used to provide a return connection to the 24 V power supply so that a continuous supply of power is available for thermostat digital screens and Wi-Fi transceivers. The C-wire serves the same functions as a neutral wire in a residential circuit.

The C-wire is located by removing the current thermostat to expose the wiring. The back of the thermostat has a series of letters on a terminal block, indicating connection points. If there are five wires connected to the thermostat, one should be the C-wire. Typically, the C-wire is a blue wire connected to the thermostat, although its color may vary. If the presence of a C-wire is not apparent, it might be recessed into the wall since some installers push the C-wire into the wall when it is not needed.

Another way to determine whether there is a C-wire is to look at the connections on the HVAC control board. It is necessary, however, before the C-wire is located on the control board that first power to the HVAC unit should be turned off and then the cover removed. Some control board covers may be secured with screws. If there is a terminal labeled with the letter C on the HVAC control board but no wire is connected, then there probably is no C-wire and one should be installed.

If there is no C-wire coming from the system, an additional wire can be run from the HVAC system to the thermostat. However, it may be infeasible to fish a line through a long run that is not fully accessible. When running a new wire is not feasible, an external transformer can be installed for a Wi-Fi thermostat instead. The transformer must be correctly wired to the Wi-Fi thermostat and then plugged into an electrical receptacle. **See Figure 16-29.**

Another option is to use an add-a-wire device when the thermostat being replaced is only a four-wire device and a C-wire is needed. By combining two of the relay wires, the add-a-wire device frees up a dedicated wire for a C-wire connection. The add-a-wire device provides a return wire for two devices, which are usually the furnace and air conditioner because they are usually turned on at the same time. Add-a-wire device kits are installed at the HVAC control board. **See Figure 16-30.**

Once the wires are pulled out from behind the wall, they should be labeled and then secured with a pencil or piece of wood to keep them from falling behind the wall. Many thermostat manufacturers provide labels as part of their installation kit. Wires should be left with 2½″ to 3″ of extra wire coming out of the wall opening. The ends of the wire should be stripped from about ⅛″ to ¼″ before being installed and screwed into the terminal block. **See Figure 16-31.**

New construction makes installing modern thermostats very easy. By running 18-5 thermostat wire from the HVAC equipment to the thermostat, all connecting wires, including the C-wire, are available for the thermostat.

Thermostat Wire Colors. There are no standards for wire colors used with thermostats. Although there are typical guidelines for wire color connections, these may vary by manufacturer. Regardless of color used, the connection lettering must match at the thermostat and the HVAC control board. **See Figure 16-32.**

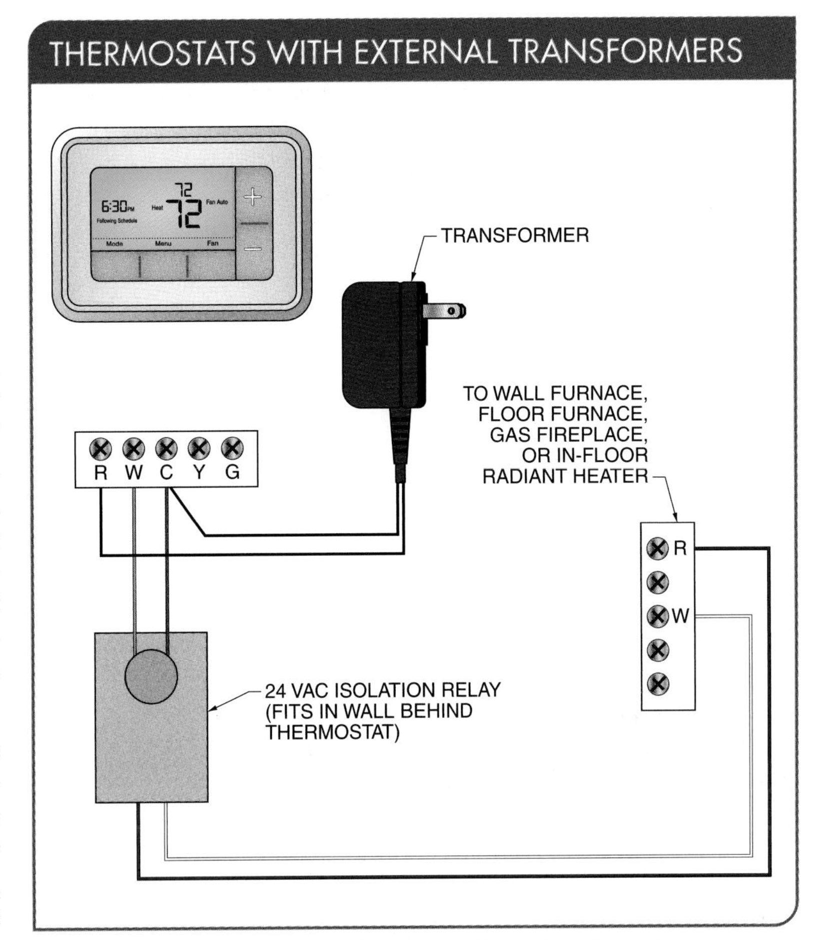

Figure 16-29. Thermostats can be powered by external transformers when there is no C-wire.

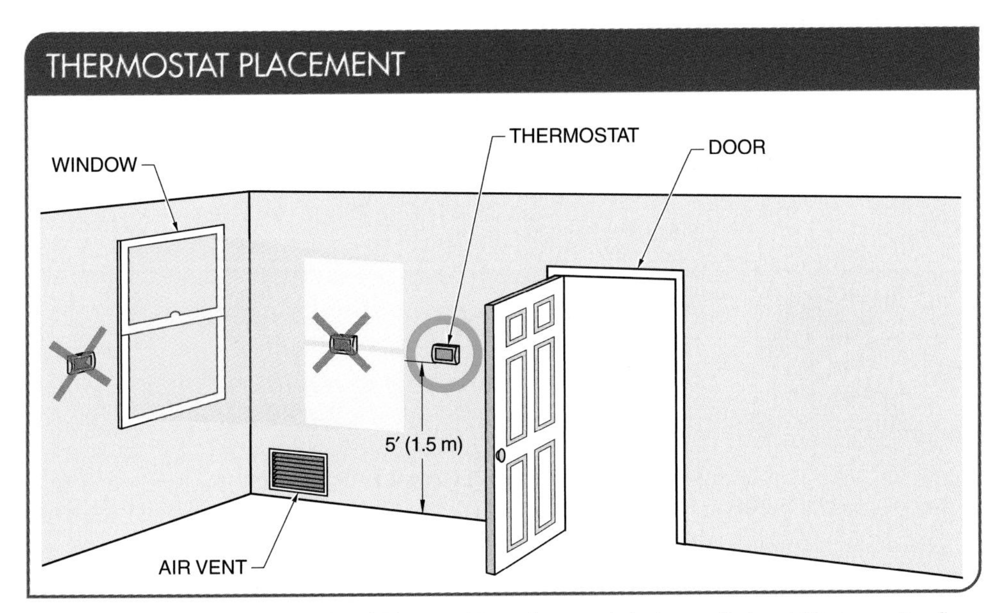

Figure 16-28. Thermostats should be positioned on an interior wall about 5′ above the floor and away from sunlight, air vents, windows, and doors.

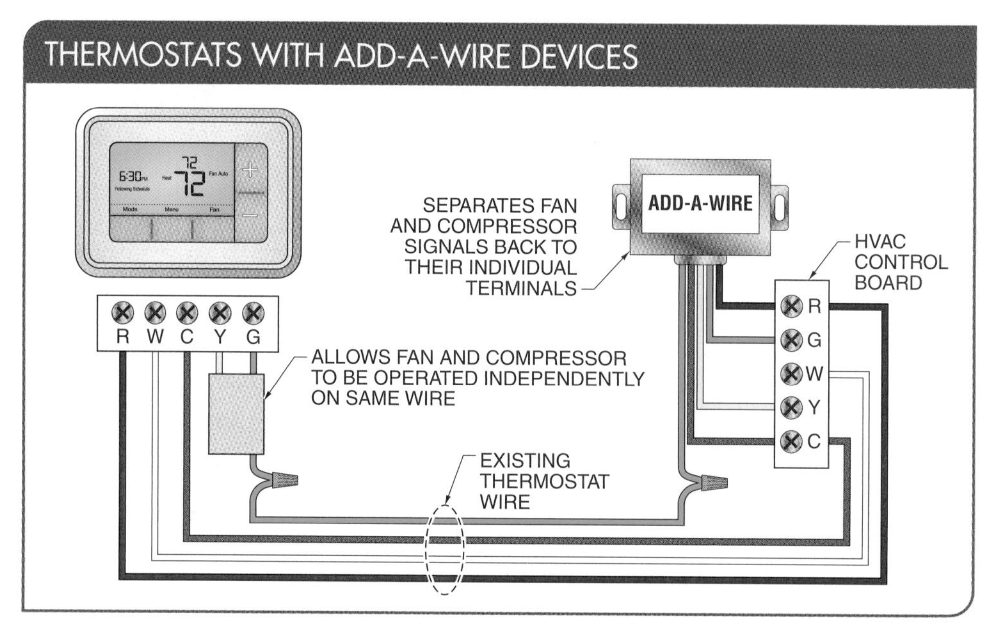

Figure 16-30. Add-a-wire devices can be used when there are only four wires but the new thermostat requires five wires.

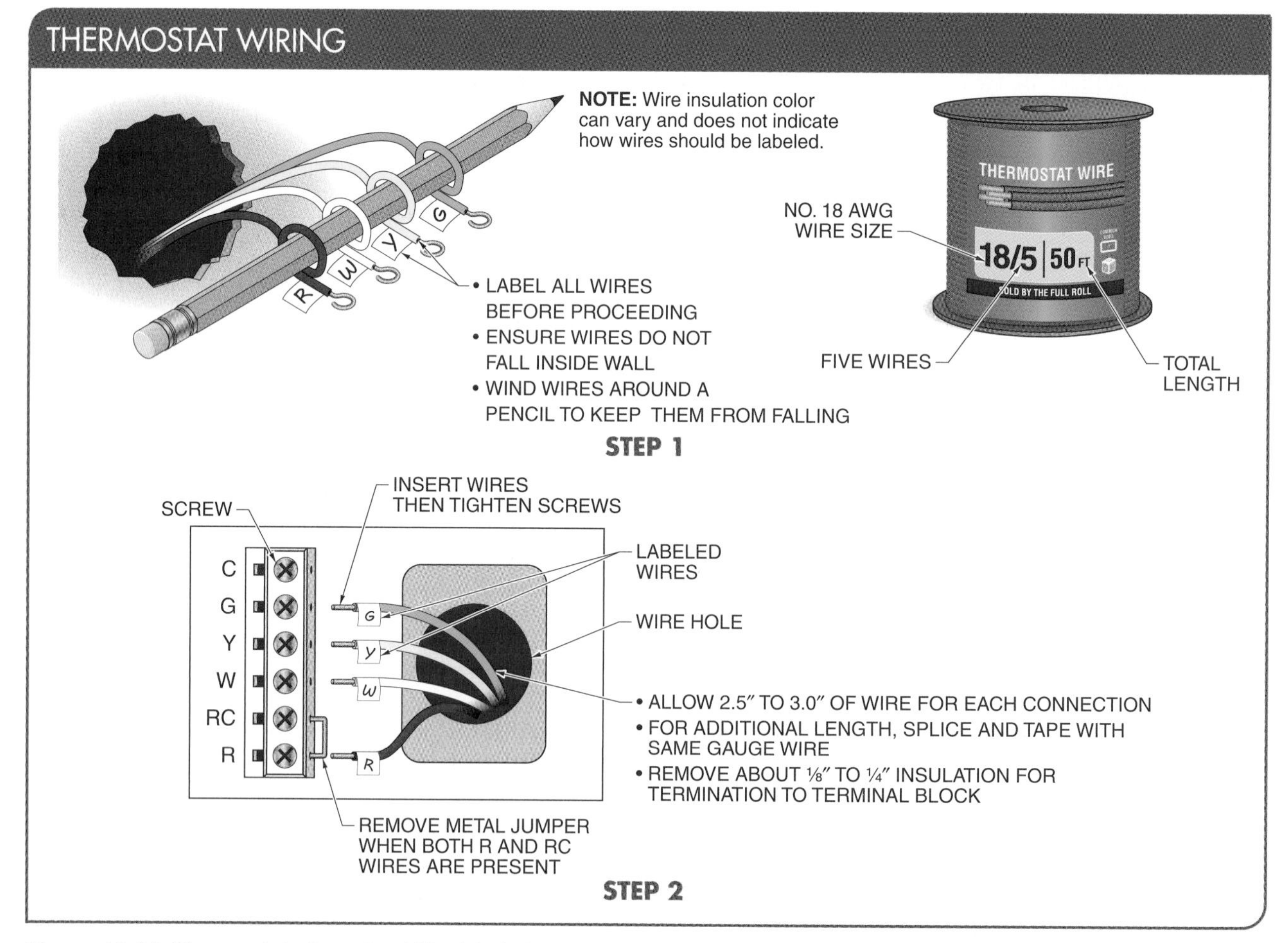

Figure 16-31. Thermostat wires should be labeled to match the terminal block letters where they terminate.

TYPICAL THERMOSTAT WIRE COLORS					
Connection	Description	Color			
C	Connects the common lead of the 24 VAC transformer. All HVAC switched components have one side of their power connected to this lead.	Brown	Blue	Purple	Black
G	Controls the fan blower. This fan blows air into the house. This fan is never completely off. It can be either set to automatic or on all the time. On automatic, the system is controlling the fan automatically when there is a need. If the fan is set to "on" it will circulate the air in the house all the time.	Green			
R-24 VAC (R and Rc)	Supplies the hot side of the 24 VAC line from the power transformer in an HVAC system.	Red			
Y (Y_1 – Y_2)	Controls the compressor, heat pump, or air conditioner. This controls the outdoor unit. If it is a two-stage unit, there will be two markings of these called Y_1 and Y_2.	Yellow			
W – Heat	In gas systems, this triggers the ignition process, opens the gas valve, turns on the ignitor, and lights the burners in the furnace. In electric heating furnaces, this lead turns on the heating elements. In a heat pump system, the white wire starts the compressor in the heating mode.	White			

Figure 16-32. While there are no standards for wire colors used with thermostats, typical guidelines for wire color connections may vary by manufacturer.

Ceiling Fans

Fans are used to ventilate, exhaust, and cool air in a dwelling. Temperature, humidity, and airflow are interrelated to the point that a change in one process can affect the other two. For example, if temperature or humidity is outside of the comfort range, changes in airflow may provide a better level of comfort.

The human body releases excess heat through perspiration and evaporation. Air movement speeds up this cooling process, creating what is known as the "wind chill factor." Wind chill factor can be created by a ceiling fan, and the breeze created can make a person feel several degrees cooler. A well-placed ceiling fan can keep most people comfortable without air conditioning in temperatures up to 80°F.

In cold weather, a built-in reverse switch can enable fan blades to spin clockwise, pulling air up toward the ceiling. This upward draft helps mix warm air at the ceiling with colder air near the floor without creating a noticeable wind chill. In a family room with a cathedral ceiling, temperatures can vary as much as 15°F from floor to ceiling. A properly sized ceiling fan can reduce this difference to only a few degrees.

Central air conditioning systems consume about 3000 W (3 kW) of power per hour, while a fan consumes only about 30 W. Fans, like smart thermostats, can be remotely controlled and programmed to turn on according to preset temperatures or programmed schedules.

Mounting Ceiling Fans. When a ceiling fan is mounted, the weight of the fan and the force created by the blades rotating must be considered. All ceiling fans are required to be mounted to an NEC®-listed fan box that is properly secured. Fans heavier than 35 lb must use the structural members of the house to support the fan. A common feature of ceiling fan installations is the use of lag screws. **See Figure 16-33.**

The greatest challenge of mounting a ceiling fan arises when structural members are not directly available to secure the fan. To aid in mounting ceiling fans, manufacturers may provide kits that are mounted to the nearest structural member. Typically, electricians use bar hangers that fit between ceiling joists to provide support for heavy ceiling fans.

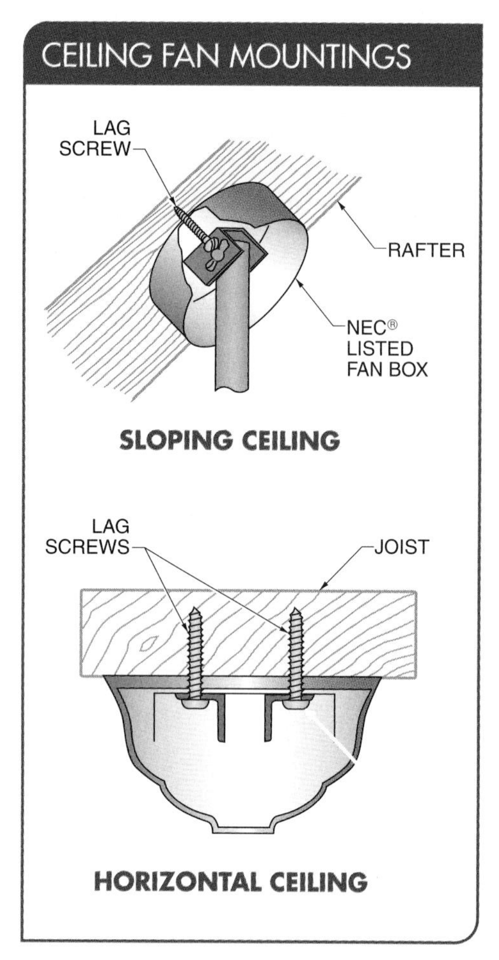

Figure 16-33. Ceiling fans must be securely mounted to support the weight of the fan and the rotating blades.

Downrod-mount ceiling fans must be grounded to the metal junction box and the downrod when they are mounted on the ceiling.

A bar hanger is installed by positioning it in place and forcing the spikes at the end of the bar into the joists to hold the bar in place. The bar flanges are then secured with screws. If access through the attic is not possible, there are fan brace kits that can be installed through the hole in the ceiling. **See Figure 16-34.**

Wiring Ceiling Fans. When ceiling fans are wired, they must be properly grounded. The grounding methods for flush-mount ceiling fans and for downrod-mount ceiling fans are different. Flush-mount ceiling fans are grounded directly to the metal junction box. Downrod-mount ceiling fans must be grounded to the metal junction box and the downrod.

Wiring a pull-chain ceiling fan only requires a two-wire cable with ground. A two-wire cable contains a black wire (hot), white wire (neutral), and a bare wire (ground). Two-wire cables are used when a fan and light will only be turned on and off. **See Figure 16-35.**

> CODE CONNECT
>
> *NEC® Section 422.19 states that canopies of ceiling-suspended (paddle) fans and outlet boxes taken together shall provide sufficient space so that conductors and their connecting devices are capable of being installed in accordance with Section 314.16.*

Sometimes a combined ceiling fan and light will be controlled from two separate switches. Power is supplied by a two-wire cable with ground. In the junction box, power is fed to the bottom of each of the single-pole switches. From there, one of the single-pole switches has a red wire going to the light and the other single-pole switch has a black wire going to the fan. Inside the fan junction box, the red wire is connected to the blue wire for the light and the black wires are connected for the fan. The neutral wires are connected and the ground wire is secured. **See Figure 16-36.**

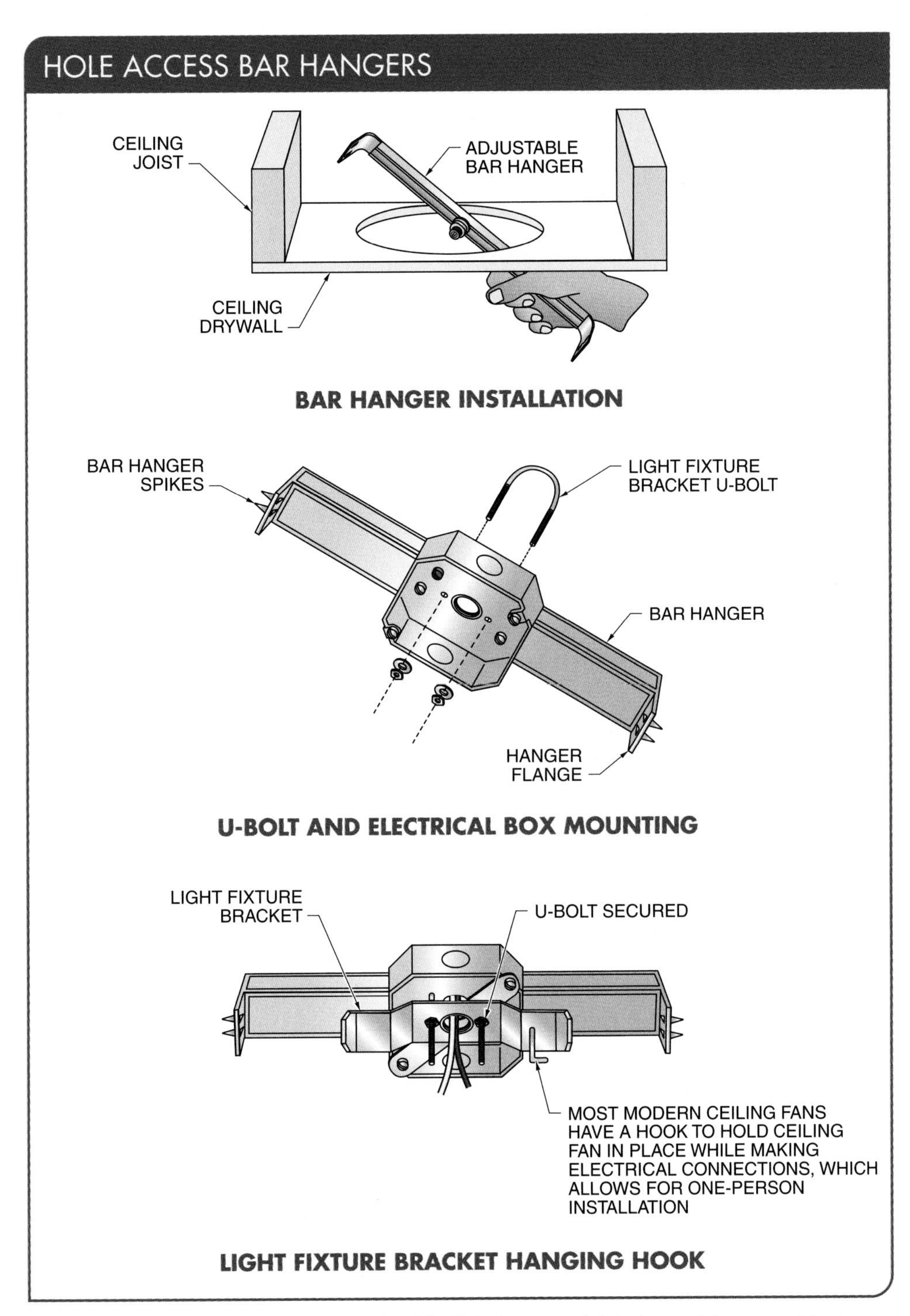

Figure 16-34. Bar hangers and mounting kits fit in between joists to provide support for heavy ceiling fans.

Ceiling Fan Remote Control. Fans and lights are typically controlled by wall switches. However, with smart home technology, fans and lights can also be remote controlled by smartphones, tablets, smart watches, or even voice control. Remote control is used to start, reverse, and speed control a fan, as well as to dim the lights.

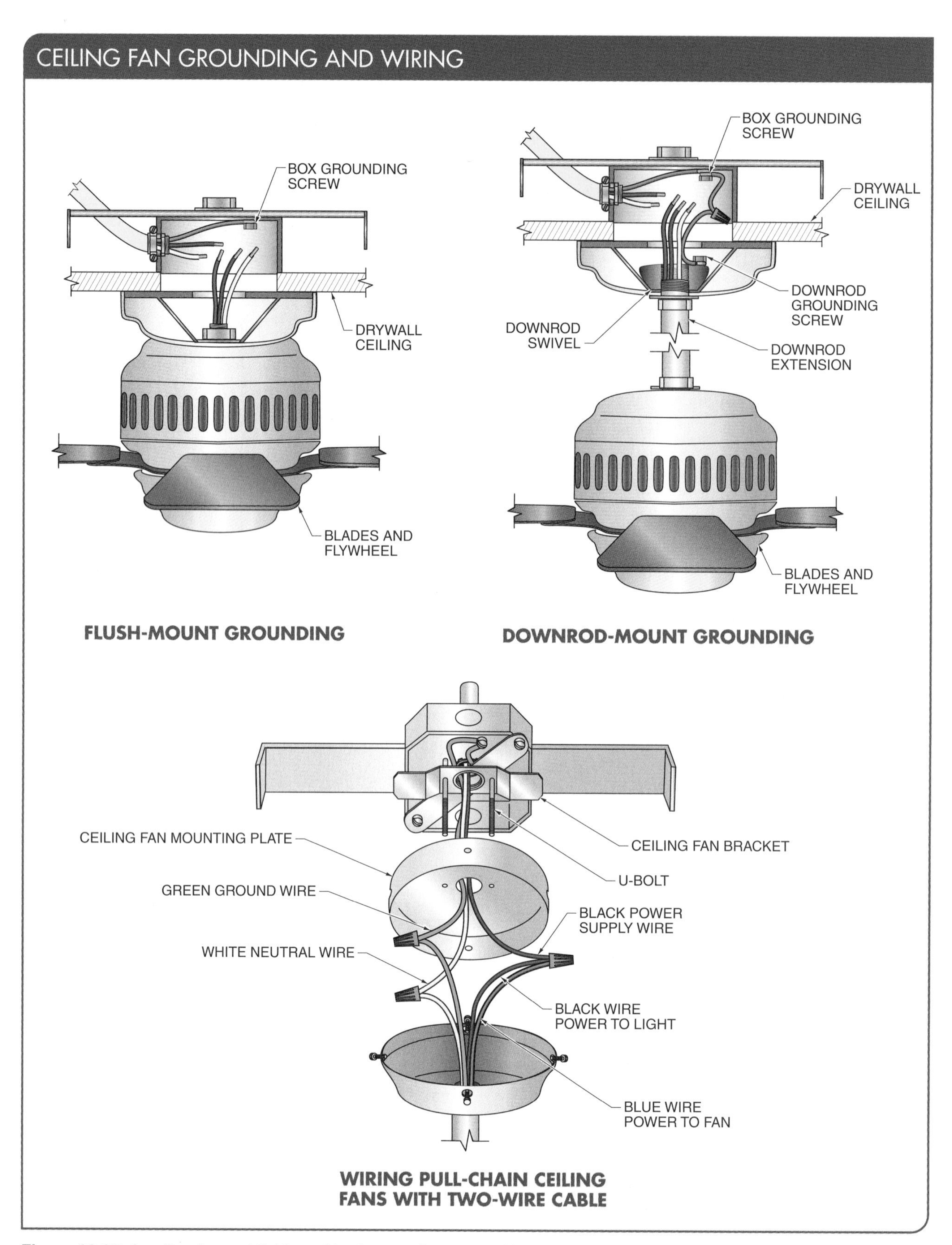

Figure 16-35. A ceiling fan and light combination can be powered by a two-wire cable with ground since only one hot wire needs to feed the two separate switches that control the fan and the light.

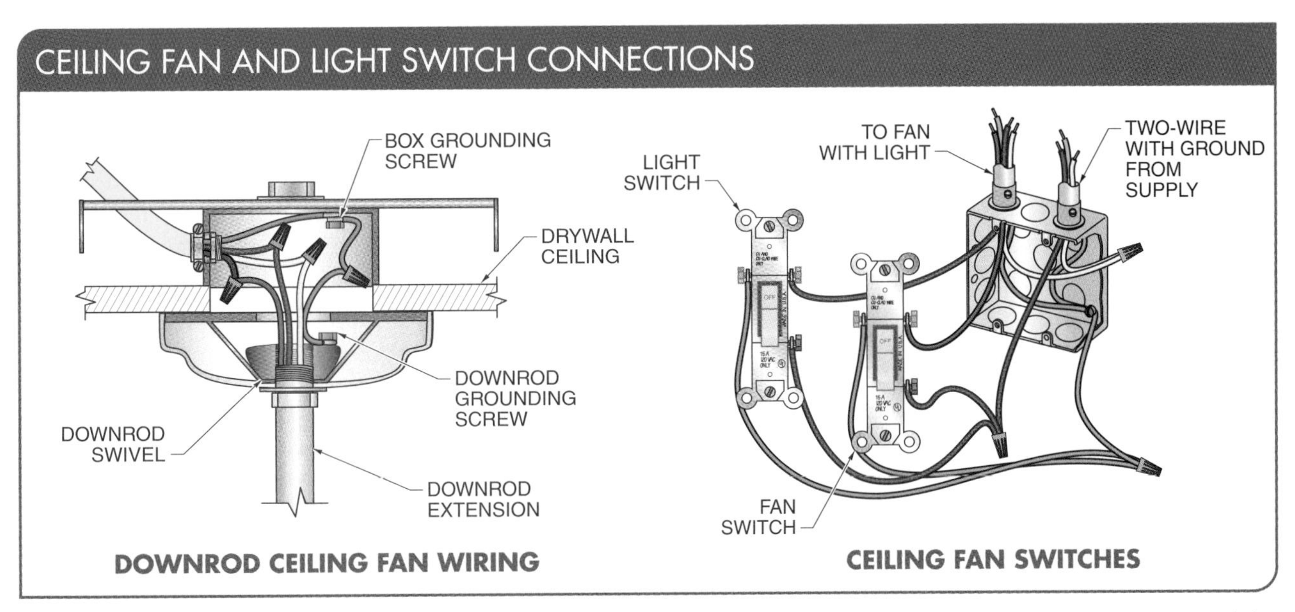

Figure 16-36. Downrod-mount ceiling fans require the ground wire to be attached to both the metal junction box and the downrod, but flush-mount ceiling fans are only grounded directly to the metal junction box.

Ceiling fan remote control kits allow old pull-chain or wall-switch ceiling fans (with on/off and fan speed control) to be controlled remotely. The use of a ceiling fan remote control kit depends on the amount of free space inside the fan canopy. To determine whether a specific style of fan has enough space in its canopy, the circuit breaker to the fan is shut off, the canopy is lowered, and the available space is compared to the size of the receiver. **See Figure 16-37.**

A ceiling fan remote control kit can then be installed by first determining how the fan is connected. Often, a diagram on the fixture shows the initial connections. The three-wire cable in the outlet box has a common neutral (white) wire, a red wire for the light, and a black wire for the fan. Only one of these switched hot wires is needed for power to the receiver unit. The black wire (along with the white neutral wire) is used to connect to the receiver, and the red wire can be removed or capped off permanently with a wire connecter. Then, to connect the remote receiver to the fan and light, the neutral-out from the receiver is connected to the common neutral for the fan and light, the black fan wire is connected to the black receiver wire, and the blue light wire is connected to the red receiver wire.

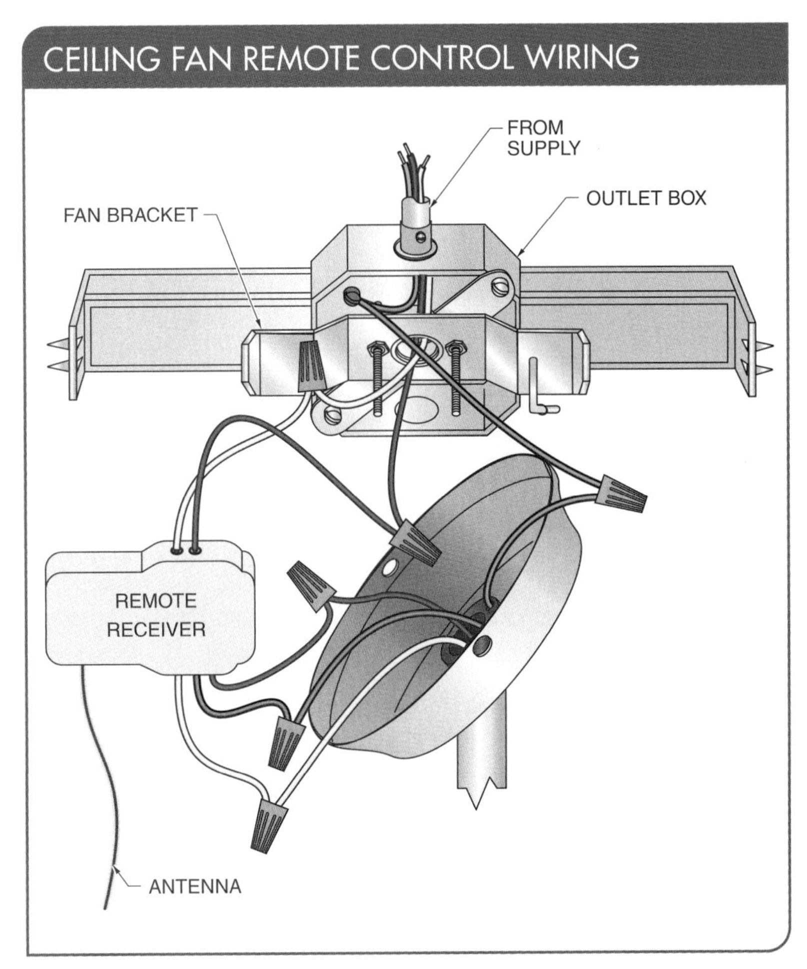

Figure 16-37. Ceiling fan remote controls can be installed in old pull-chain or wall-switch ceiling fans if there is space in the canopy for the remote receiver.

With all the connections made, the wires are tucked back into the box and the remote receiver is installed in the space available in the mounting bracket. It is important to ensure that no wires are pinched or that the receiver is not jammed in too tightly, causing damage to the wires. These conditions can potentially cause a short-circuit condition. When the remote receiver is installed and the canopy and trim ring are back in place, any changes in the multigang switch box can be made.

Bluetooth vs. Wi-Fi Fan Control. Bluetooth and Wi-Fi are two protocols used to control a fan or a combined fan and light. Bluetooth is used to wirelessly control the fan and light when the homeowner is in the same room. Wi-Fi, however, can be used to control a fan and light even when the homeowner is not at home.

Bluetooth can be convenient when the switch for the fan or light is located by the entrance to a room but the remote is kept near the homeowner in that room. Wi-Fi may be a better solution when the controller is tied into the smart home management or security system. Wi-Fi control allows the homeowner to set schedules that simulate activity when they are not at home. Wi-Fi also allows the fan to be connected to the thermostat so that it activates at certain temperature levels.

Attic Fans

An *attic fan* is a ventilation fan that exhausts hot air out of an attic. A thermostat is usually used to turn the fan on and off. Depending on the climate and the amount of insulation in the house, a roof vent attic fan can help reduce cooling costs. **See Figure 16-38.** The use of an air conditioner and fan in tandem better controls the cooling and keeps energy costs down.

During the summer, the air in an attic can reach temperatures well over 100°F, which radiates down into a dwelling. Thermostats can be used to turn attic fans on and off at preset temperatures, cooling the dwelling and lowering the utility bill by moving hot air out of the attic and drawing cooler air in through soffit vents. Even on a hot summer day, circulating outside air through the attic can lower the temperature in the attic. This keeps both the attic and the residence cooler, thus reducing the need for air conditioning.

Solar attic fans are easy to install since no wiring is needed, and they can help save money since they draw their power from the sun. To install a solar attic fan, a hole is cut in the roof to fit the solar panel and the fan. The manufacturer's installation instructions should be followed.

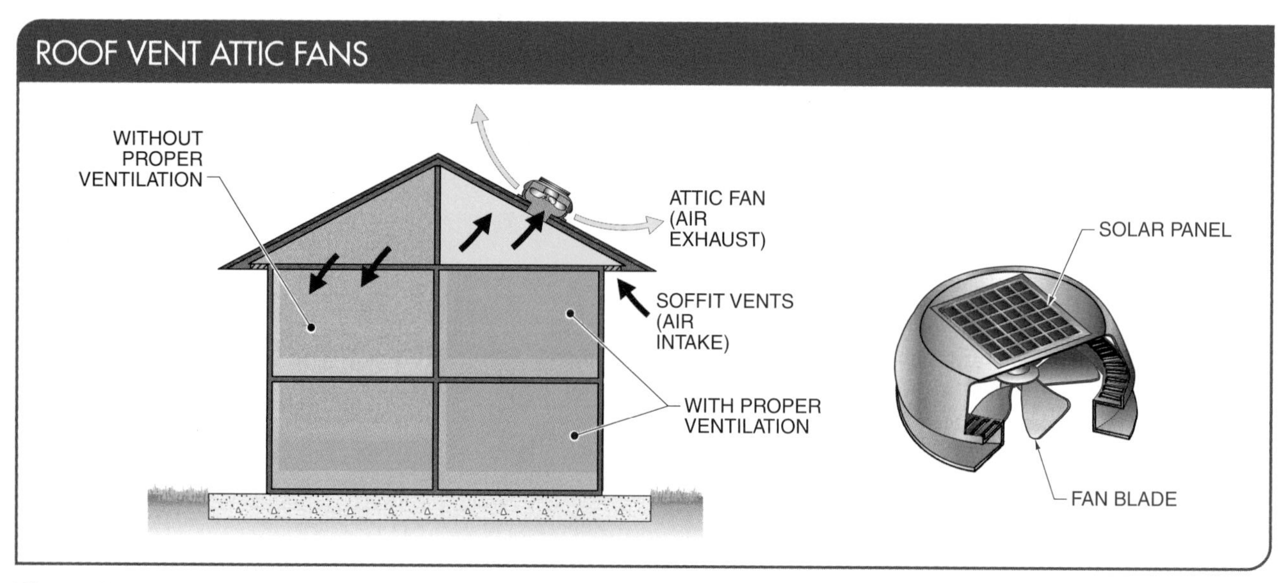

Figure 16-38. Attic fans circulate air by drawing outside air in through soffit vents and moving the hot air out of the attic through the roof.

Whole-House Exhaust Fan Installation. Whole-house exhaust fans can be used along with attic fans to help provide attic ventilation and cool a dwelling. A whole-house exhaust fan is often mounted in the ceiling of a hallway leading to bedrooms. The whole-house exhaust fan draws air in through open windows, creating a breeze through the dwelling. The whole-house exhaust fan then pulls that air into the attic where the attic fan blows it out through the roof and back outside.

A couple of considerations when purchasing and installing a whole-house exhaust fan include whether there is an adequate number of roof vents and whether a fan with a two-speed motor should be installed. The manufacturer specifications should be used to determine the fan size and amount of roof ventilation required. Generally, the airflow in cubic feet per minute (cfm) should be approximately three times the square footage of the dwelling.

Typically, a central hallway is the best location for a whole-house exhaust fan. To install a whole-house exhaust fan, the following procedure is applied:

1. Use a stud finder to locate the joists and find an opening between them. Verify that the fan will fit in the opening. The size of the opening is determined by the stud locations in the ceiling. **See Figure 16-39.**
2. Lay out and cut out the opening in the ceiling using a drywall or reciprocating saw. Drilling a hole in each corner will facilitate cutting and removing the drywall.
3. Set the fan in the ceiling opening and attach it to studs or joists with appropriate screws.
4. Connect the wires from the switch to wires for the fan in the junction box.
5. Install the louver panel to the fan with screws.
6. Install the switch in the wall box with it in the OFF position. Make the proper wire connections. Typically, a whole-house exhaust fan will operate off a single 120 V circuit with a two-speed switch. **See Figure 16-40.**
7. Turn on the circuit breaker and turn the switch to the ON position to test.

Electric Duct Dampers and Automatic HVAC Registers

Another way to control airflow is to install electric duct dampers or automatic HVAC registers. Electric duct dampers and automatic HVAC registers are used to automatically control and reduce airflow to unoccupied or unused rooms, providing shorter run times for heating and/or cooling systems. Through the addition of one or more electric duct dampers or automatic HVAC registers, zones can be created throughout a residence that can be remotely controlled. **See Figure 16-41.**

Ductwork should be inspected to determine the best location to install one or more electric duct dampers or automatic HVAC registers. When manual dampers and registers are replaced with electronically controlled designs, power is required. While electric duct dampers require an AC transformer, automatic HVAC registers can be battery operated. Once electric duct dampers and automatic HVAC registers have been installed, it may be necessary to obtain a properly balanced airflow throughout the residence. This may involve adjusting the registers and vents in other rooms since air that is blocked or reduced in one room may affect another room.

Electric duct dampers can be used to create zones throughout a dwelling that can be remotely controlled.

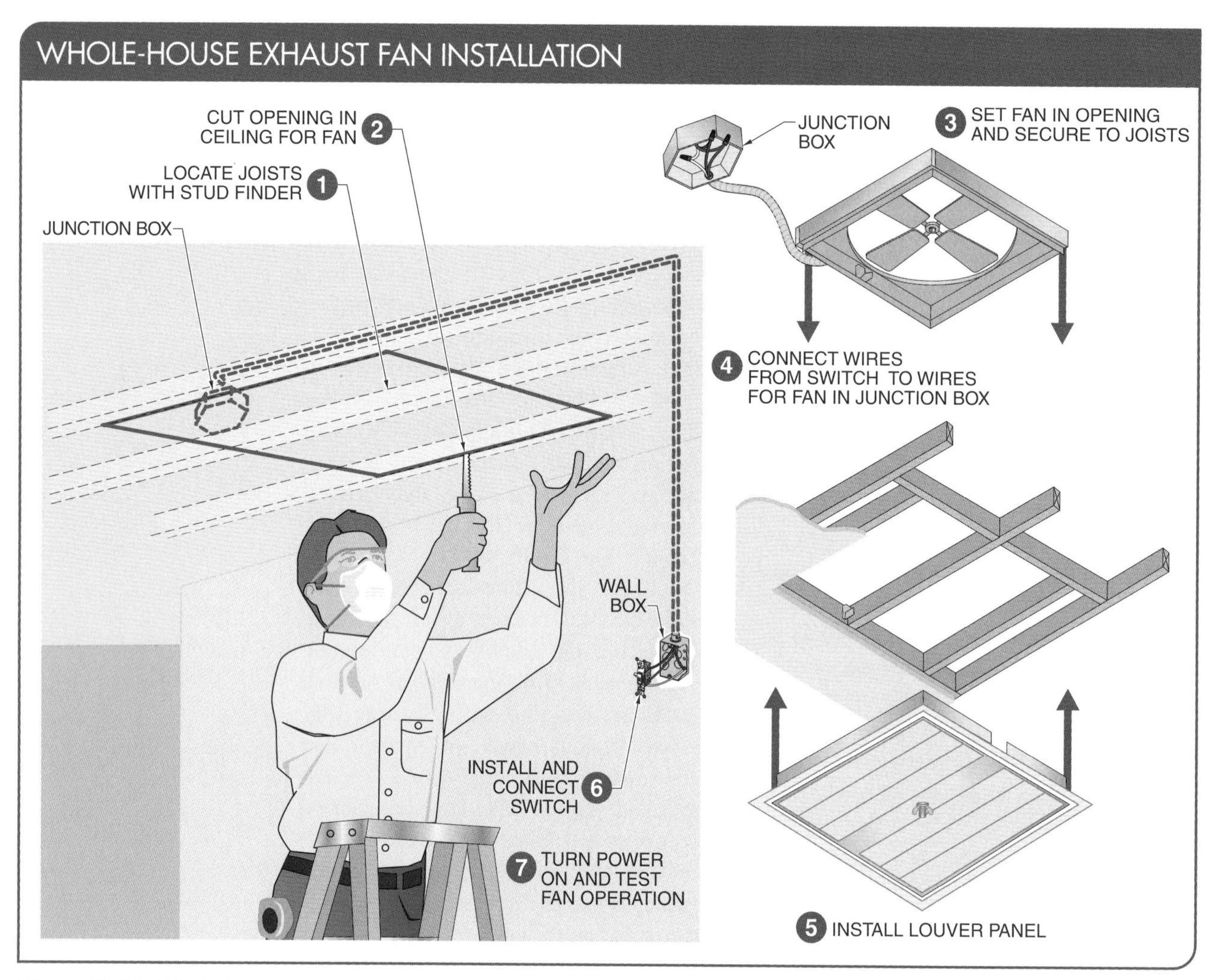

Figure 16-39. Whole-house exhaust fans are typically installed in a central hallway to help remove hot air from the dwelling.

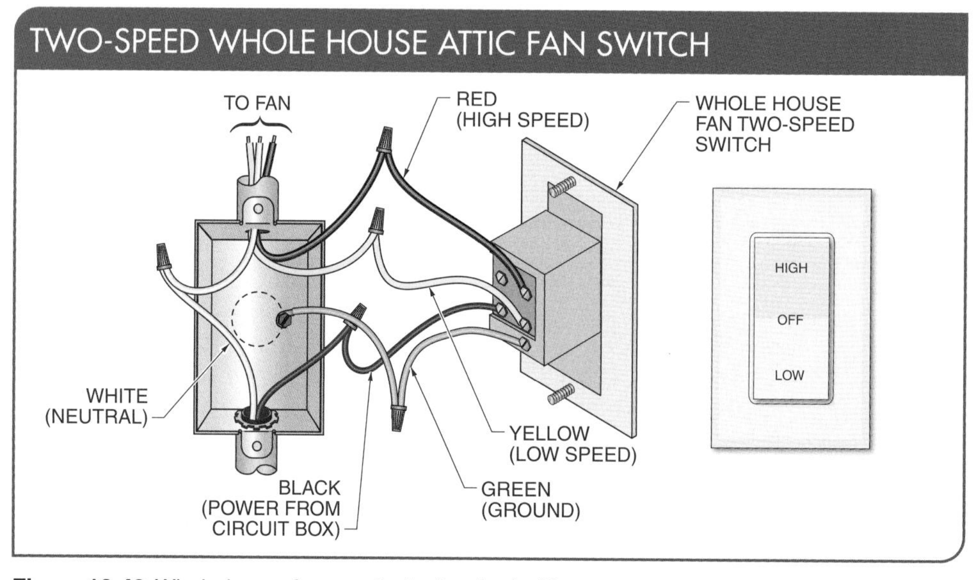

Figure 16-40. Whole-house fans are typically wired with a two-speed switch to control fan speed.

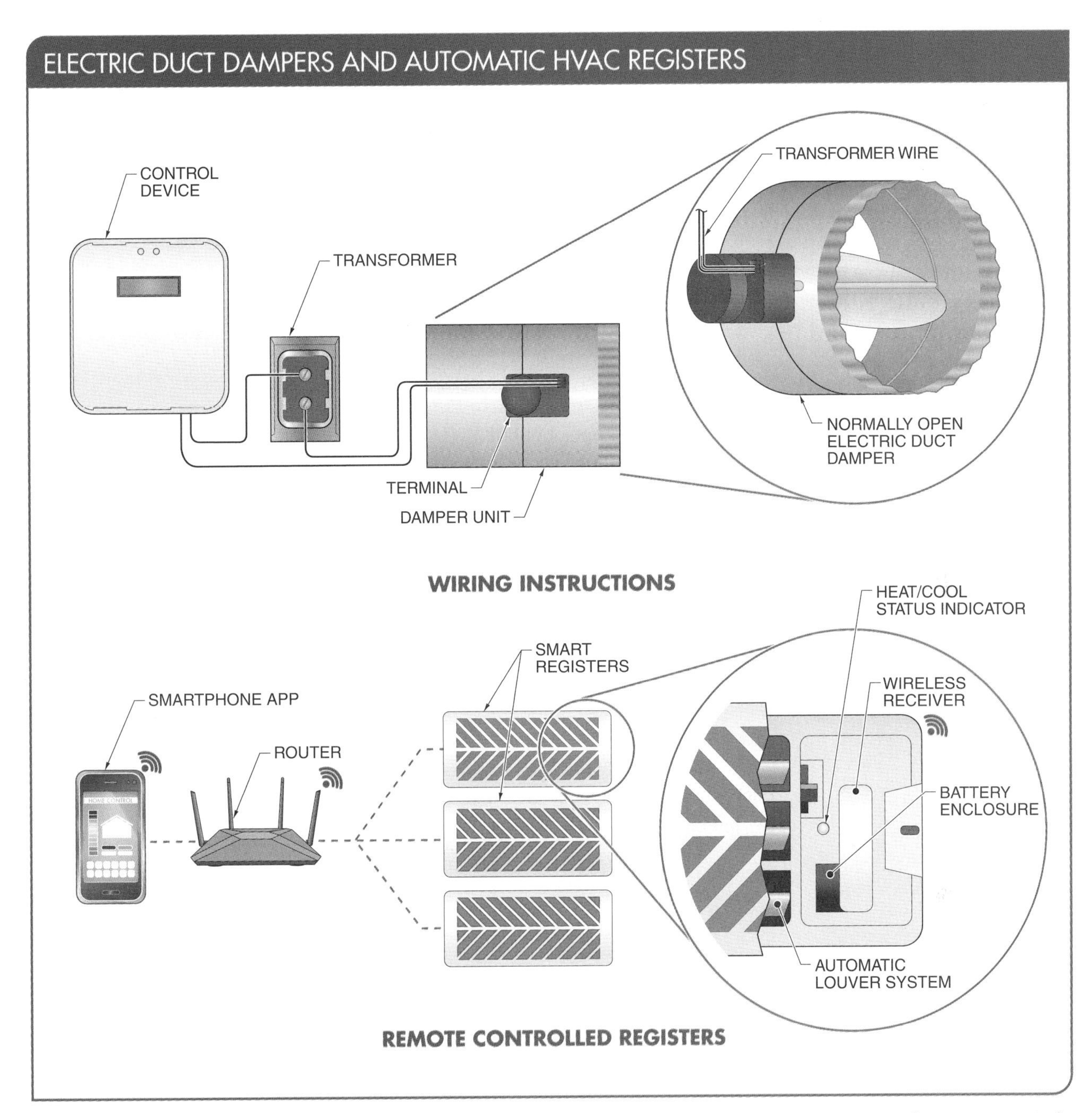

Figure 16-41. Electric duct dampers and automatic HVAC registers can adjust airflow in different rooms by remote control.

Window Treatments

Windows can account for more heat gain or loss than any other part of a dwelling. A homeowner can use window treatments to control how the sun's energy, in the form of heat and light, enters a dwelling. Window treatments can help reduce energy consumption, lower energy bills, lessen fading on interior furnishings, and provide a more consistent indoor temperature. Energy-efficient window treatments such as motorized shades and blinds as well as electrochromic smart glass can improve the home energy management of a residence.

Motorized Shades and Blinds. Automated window treatments include remote- or voice-controlled motorized shades and blinds. **See Figure 16-42.** As part of a smart home system, these automated window treatments can be programmed to take advantage of the sun's natural heat and light to reduce energy use. Motorized shades and blinds can also be integrated with door or motion sensors to operate automatically when the homeowner leaves, thereby contributing to security by closing certain windows.

One of the first considerations for the installation of motorized shades or blinds is the brightness in a room. Different materials, such as solar shades, thick fabrics, and wood, allow varying levels of brightness and visibility. Solar shades are made of a see-through fabric that retains the view to the outdoors while blocking the light and heat from the sun and reducing brightness. Thick fabrics and wood blinds block much more sunlight and visibility than solar shades.

Another consideration for the installation of motorized shades and blinds is their interoperability with lights, thermostats, and other energy-saving devices. Shades, drapes, and blind control systems must be compatible with smart home controllers. Motorized shades and blinds can be controlled through handheld remotes, keypads, or apps for smartphones, tablets, and smart watches. Motorized shades and blinds can also be battery operated or hardwired. Heavy window treatments, such as thick fabric or wood, require larger motors, meaning it is less likely that they can be battery operated. However, professional home installers often build custom soffits to completely hide the motorized hardware needed for hardwired systems.

Electrochromic Smart Glass. *Electrochromic smart glass* is a material that changes color when an electric current is applied to it. A smartphone app can be used to turn on the electrical supply to the glass. The voltage aligns the liquid crystal molecules in the glass, allowing ambient light to pass through. When the voltage is turned off, the liquid crystal molecules orientate randomly, scattering light and making the glass opaque to further block light. This process can also be set up through devices such as motion sensors.

The ability of electrochromic smart glass to change color allows it to remain transparent and minimize glare while helping to reduce the cost of energy use. A disadvantage of electrochromic smart glass is that it is more expensive than regular glass. As with any new technology, cost will be reduced as volume increases.

TECH TIP

Installing white window shades, blinds, or drapes helps to reflect heat away from a residence and can help keep a dwelling cool during warm weather. Closing curtains on the south- and west-facing windows and installing awnings can create shade and block heat.

Home Environment and Weather Monitoring

Weather monitoring is available as cloud-based software programs. These programs analyze and deliver weather information from a variety of sources, including home-based weather stations. Home environment and weather monitoring information may be as simple as measuring the outside temperature or it may be a sophisticated system that gathers information on humidity, wind speed and direction, rainfall, and barometric pressure. All of this information can be displayed on an in-home display, smartphone, tablet, or computer. **See Figure 16-43.**

With detailed information, both the homeowner and the HEMS can begin to make predictions and automatically change the temperature, humidity, and air quality of the HVAC system in the home through additional electronic sensors. Home environment and weather monitoring can also provide alerts for conditions such as approaching lightning storms, high winds, periods of varying rainfall and drought, and frost. These alerts provide a homeowner with time to act to protect their property.

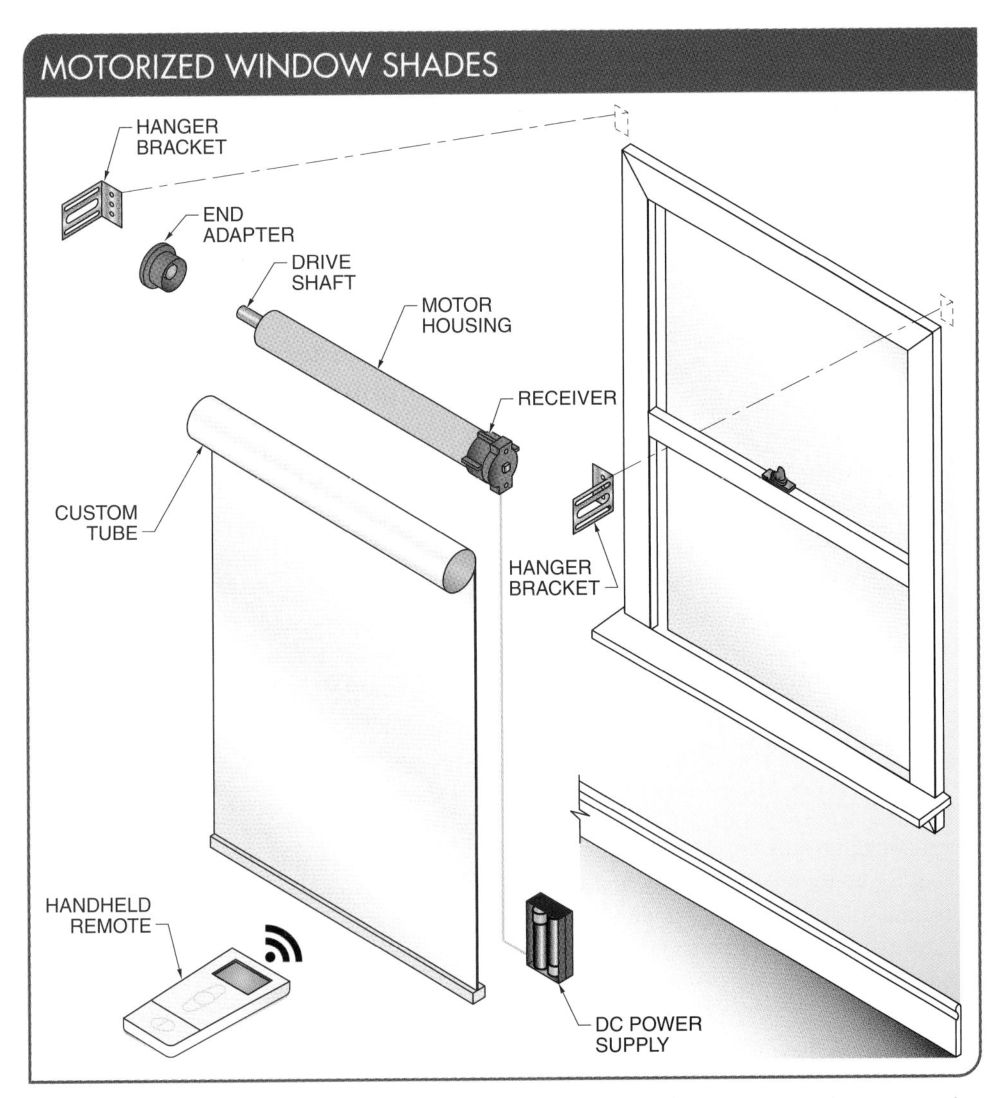

Figure 16-42. Motorized window shades can be controlled by remotes, smartphones, or voice commands.

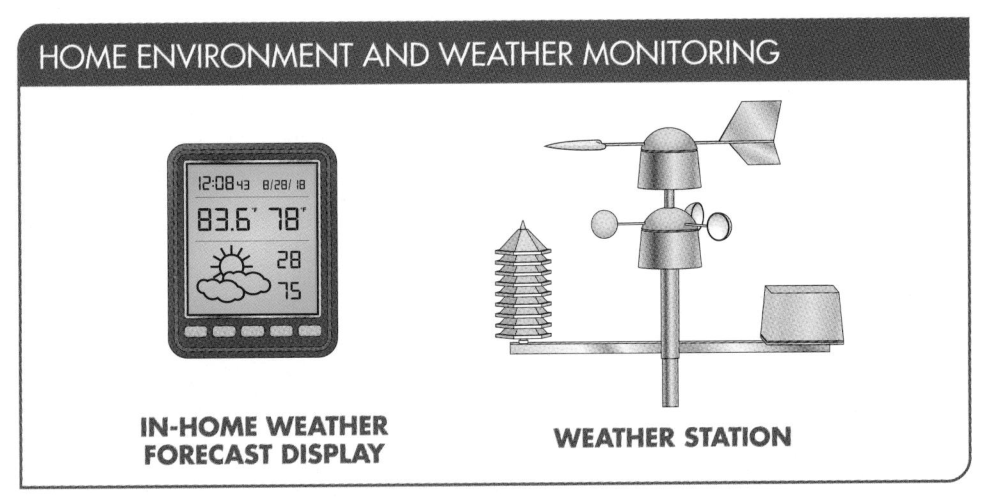

Figure 16-43. Home environment and weather monitoring information can be displayed on an in-home display, smartphone, tablet, or computer.

SECTION 16.3 CHECKPOINT

1. Where should thermostats be positioned?
2. What are the different designs of thermostats?
3. What is the difference between Bluetooth and Wi-Fi fan control?
4. How can attic fans cool a dwelling?
5. What are the uses for electric duct dampers and automatic HVAC registers?
6. What are automated window treatments?
7. What is electrochromic smart glass?

SECTION 16.4—ENERGY GENERATION AND STORAGE

For residential homes, energy generation primarily involves using wind energy and solar energy. Battery bank systems are used to store the generated electricity. When homeowners generate electricity on their premises, they can offset the cost of electricity consumed from the utility. In addition, homeowners may receive payment for any electricity they sell back to the utility.

Wind Power

Wind power is the use of airflow through turbines to provide energy to turn electric generators. A *small wind turbine* is a wind turbine that can be installed on properties as small as one acre in areas with sustained winds to create electricity. Small wind turbines typically have three propeller-like blades around a rotor connected to a shaft that spins a generator. **See Figure 16-44.** The two types of wind turbine systems are grid-connected wind turbine systems and off-grid (stand-alone) wind turbine systems.

Grid-Connected Wind Turbine Systems. Although small wind turbines are typically off-grid systems, they can also be connected to a utility's electrical distribution system (grid). These are called grid-connected wind turbine systems. To work effectively, a small wind turbine that is connected to the grid requires an average annual wind speed of about 10 mph to 15 mph.

Grid-connected wind turbines are only allowed to operate when the utility grid is on line. During power outages, the wind turbine is required to shut down due to safety concerns from islanding. *Islanding* is a condition in which a generator continues to power a location when electrical grid power is not present. Islanding can be dangerous to utility workers, who may not realize that a circuit is still powered.

A grid-connected wind turbine project requires working with the utility to make the interconnection. Utilities have developed interconnection standards for the equipment and special meters that need to be installed at the service. Also, an electrical inspector must sign off on the system before the utility will allow connection to the grid. The inspector will require that all electrical work be completed by a licensed electrician.

Off-Grid (Stand-Alone) Wind Turbine Systems. Small wind turbines that are not connected to the grid are called off-grid wind turbine systems, also known as stand-alone wind turbine systems. Off-grid wind systems can be installed to gain energy independence from the utility. However, a homeowner should be comfortable with uncertain power production due to fluctuations in wind speed. Off-grid wind turbine systems can be combined with solar PV systems to create a more reliable hybrid electric system. Wind and solar PV energy generation along with battery storage can offer enhanced improvements to an off-grid system.

Off-grid wind turbine systems are typically smaller and less expensive than grid-connected systems. Small wind turbines that are off-grid systems require annual maintenance. Annual maintenance usually requires that a person climb up the wind turbine tower. However, small wind turbines with tilt towers can be lowered to the ground for maintenance.

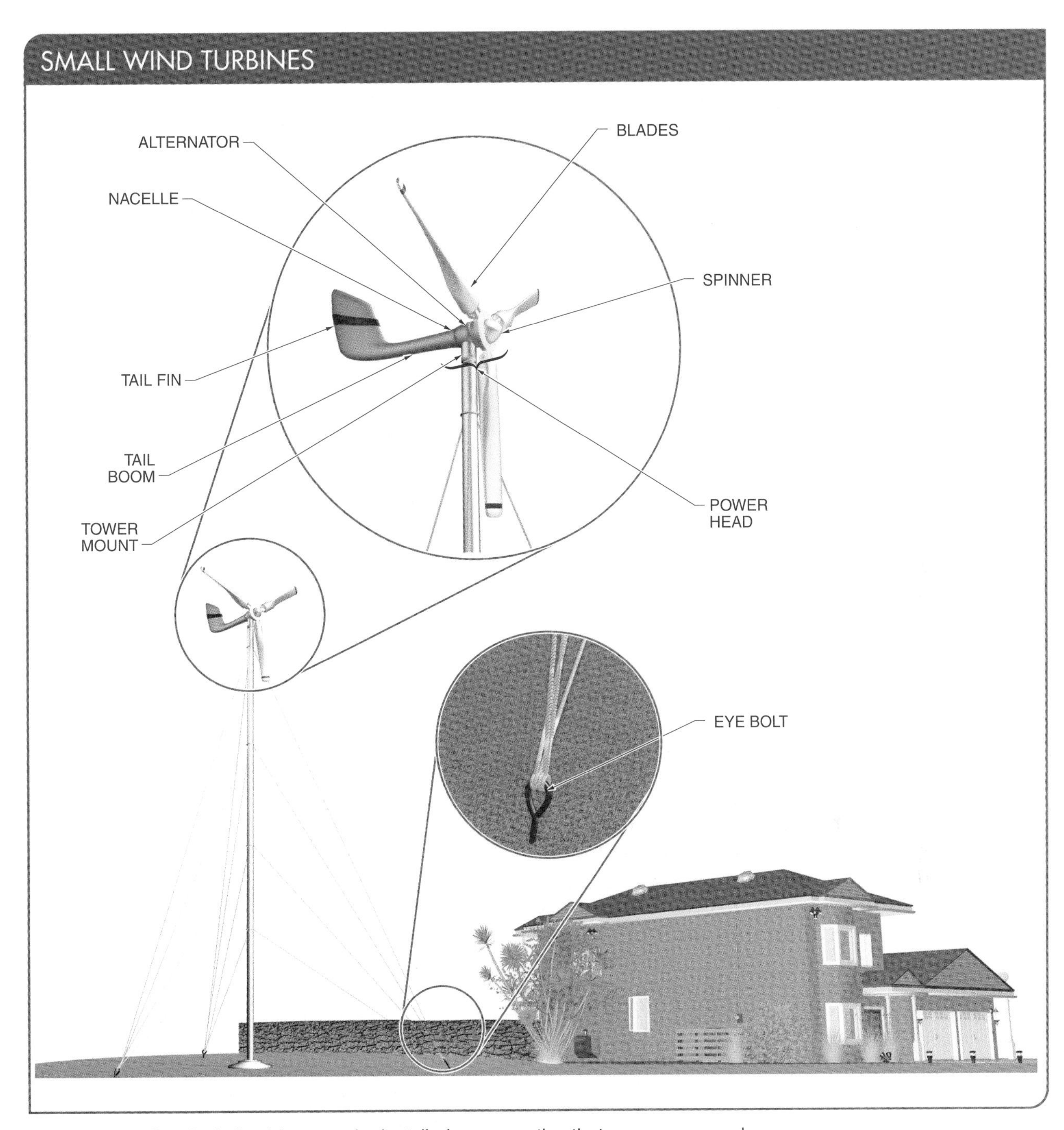

Figure 16-44. Small wind turbines can be installed on properties that are one acre or larger.

Vertical Axis Wind Turbines. A vertical axis wind turbine is a design of small wind turbine that does not require exact wind orientation and can still operate in unfavorable wind conditions. Unlike a traditional wind turbine on a horizontal axis, a vertical axis wind turbine does not have to track the wind to produce electricity. Some vertical axis wind turbines can also have solar panels embedded in their housings, which increases the energy output while using the same square footage of space. **See Figure 16-45.**

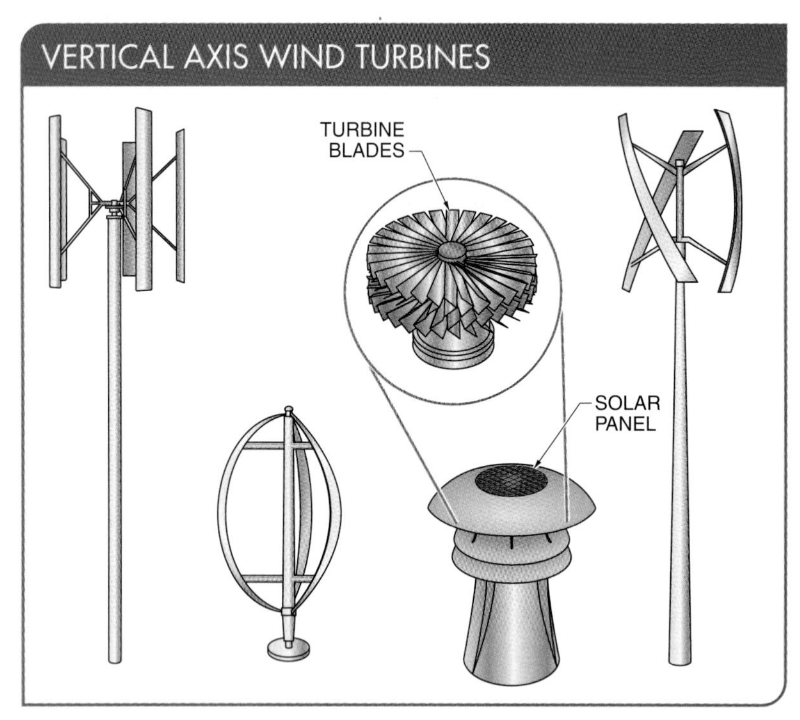

Figure 16-45. A vertical axis wind turbine does not require exact wind orientation and can operate in unfavorable wind conditions. Some units have solar panels embedded on top of their housing.

Purchasing Wind Energy Systems. To purchase a wind energy system, it is important to know the necessary tower height, the power required from the turbine, the installation cost, and the cost to maintain the system. There may be grants or incentives available to defer some costs. A homeowner should also purchase wind insurance for liability and damage to equipment.

The average height of a small wind turbine is about 80′, which is about twice the height of a residential telephone pole. However, small wind turbines can range in height from 30′ to 140′. The output needed to power a dwelling can range from 2 kW to 10 kW. A large, grid-connected system can range from $10,000 to $70,000, while the purchase and installation of an off-grid small wind turbine (less than 1 kW) generally costs $4,000 to $9,000. The ROI for a small wind turbine ranges from 6 years to 30 years. The ROI is based on the energy use of the dwelling, average wind speeds, and the turbine's height above ground.

Less than 1% of all small wind turbines are used in urban applications due to zoning restrictions and poor wind quality in densely built environments. Wind resource information can be found through the National Renewable Energy Laboratory (NREL), local airport wind data, and state guidelines through the DOE's Office of Energy Efficiency and Renewable Energy. There are incentives for the purchase of wind turbines and for the sale of excess electricity. The Public Utility Regulatory Policies Act of 1978 (PURPA) is a federal regulation that requires utilities to connect with and purchase power from small wind energy systems.

Solar Photovoltaic (PV) Power Generation

Solar photovoltaic (PV) power generation is the process of converting energy from the sun into electricity using solar panels. Solar panels, also called PV panels, are combined into arrays in a PV system. PV systems can also be installed in grid-connected or off-grid (stand-alone) configurations. The basic components of these two configurations of PV systems include solar panels, combiner boxes, inverters, optimizers, and disconnects. Grid-connected PV systems also may include meters, batteries, charge controllers, and battery disconnects. There are several advantages and disadvantages to solar PV power generation. **See Figure 16-46.**

Grid-Connected PV Systems. PV systems are most commonly in the grid-connected configuration because it is easier to design and typically less expensive compared to off-grid PV systems, which rely on batteries. Grid-connected PV systems allow homeowners to consume less power from the grid and supply unused or excess power back to the utility grid. **See Figure 16-47.** The application of the system will determine the system configuration and size. For example, residential grid-connected PV systems are rated less than 20 kW, commercial systems are rated from 20 kW to 1 MW, and utility energy-storage systems are rated at more than 1 MW.

SOLAR PHOTOVOLTAIC (PV) POWER GENERATION	
Advantages	**Disadvantages**
• Sunlight is free and readily available in many areas of the country • PV systems do not produce toxic-gas emissions, greenhouse gases, or noise • PV systems do not have moving parts • PV systems reduce dependence on oil • PV systems have the ability to generate electricity in remote locations that are not linked to a grid • Grid-connected PV systems can reduce electric bills	• PV systems have high initial investment • PV systems require large surface areas for electricity generation • Amount of sunlight can vary • PV systems require excess storage of energy or access to other sources, like the utility grid, when systems cannot provide full capacity

Figure 16-46. There are advantages and disadvantages to solar PV power generation.

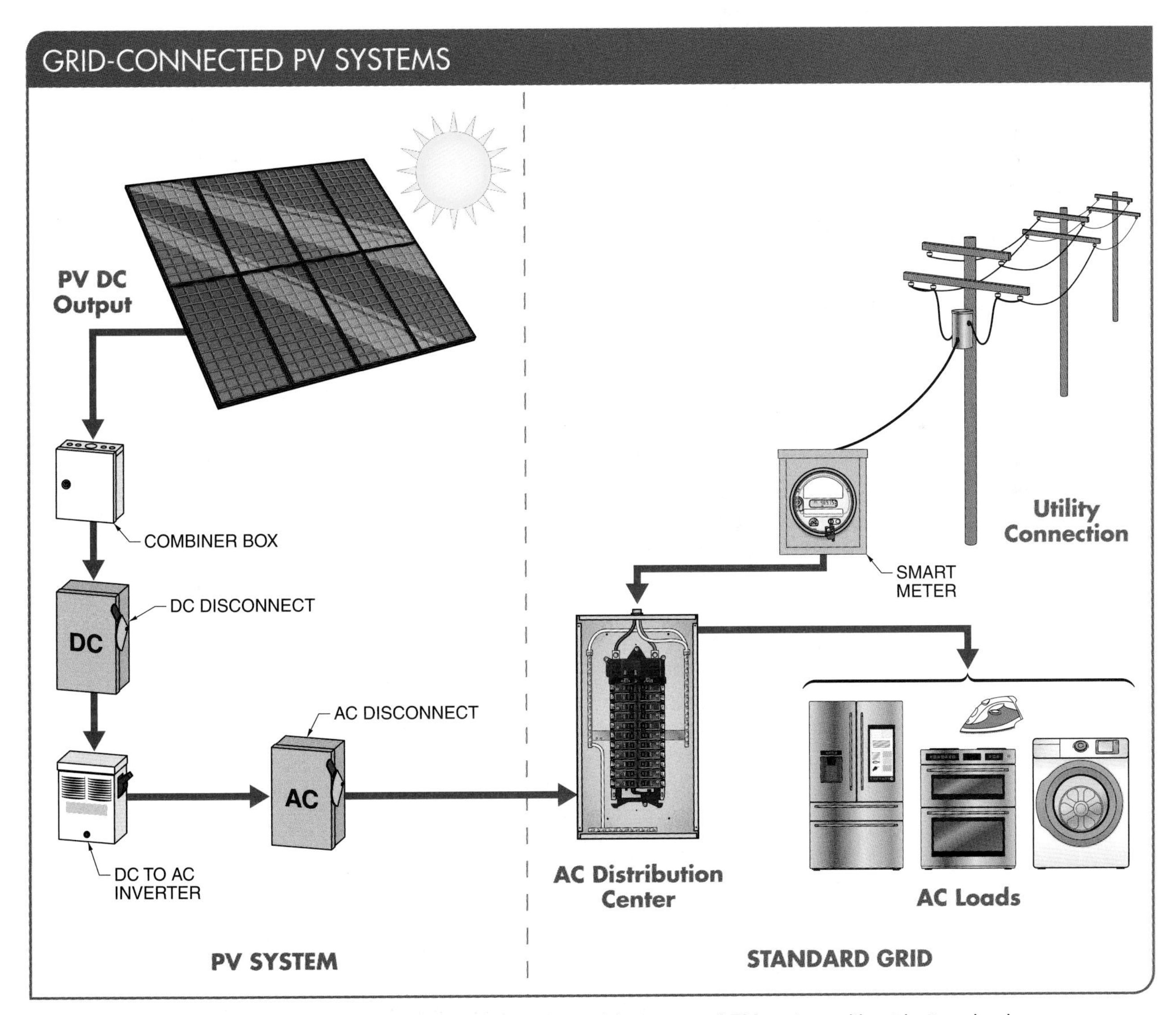

Figure 16-47. A common configuration for a PV system is a grid-connected PV system without battery backup.

Off-Grid (Stand-Alone) PV Systems. Off-grid (stand-alone) PV systems use arrays of solar panels to charge banks of rechargeable batteries during the day for use at night when energy from the sun is not available. The reasons for using an off-grid PV system include reduced energy costs and power outages, production of clean energy, and energy independence. Off-grid PV systems include battery banks, inverters, charge controllers, battery disconnects, and optional generators. **See Figure 16-48.**

Solar Panels. Solar panels used in PV systems are assemblies of solar cells, typically composed of silicon, and commonly mounted in a rigid flat frame. Solar panels are wired together in series to form strings, and strings of solar panels are wired in parallel to form arrays. Solar panels are rated by the amount of DC that they produce. Solar panels should be inspected periodically to remove dirt, debris, or snow, as well as to check electrical connections.

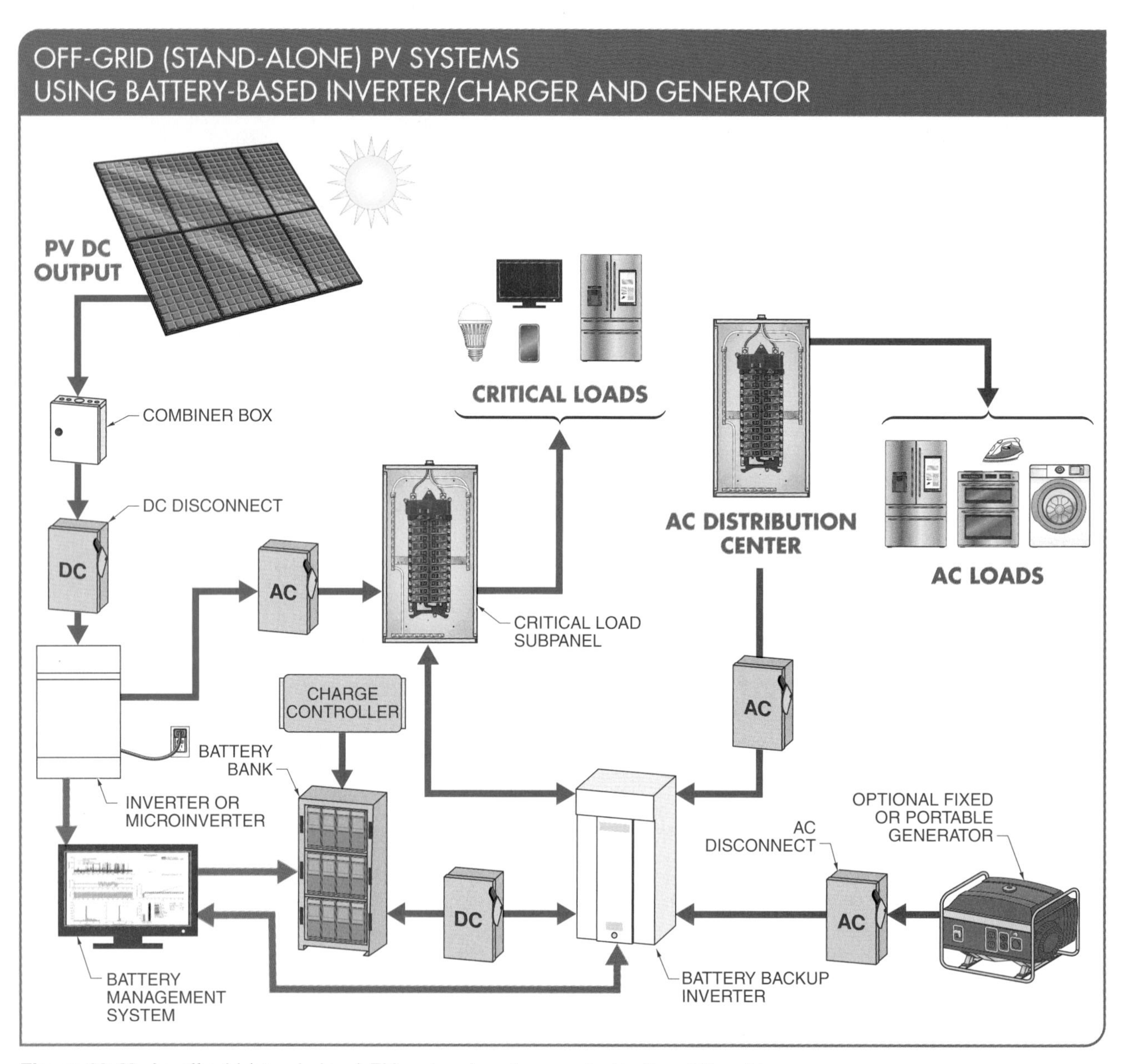

Figure 16-48. An off-grid (stand-alone) PV system is not connected to the utility grid.

Since photovoltaics are adversely affected by shade, any shadow can significantly reduce the power output of a solar panel. The performance of a solar panel will vary, but in most cases, guaranteed power output life expectancy is between 10 years and 25 years. Solar panel power output is measured in watts. Power output ratings range from 200 W to 350 W under ideal sunlight and temperature conditions.

Solar Arrays Construction and Mounting. When solar arrays are installed on a property, they must be mounted at an angle to best receive sunlight. Typical solar array mounts include roof, freestanding, and directional tracking mounts. **See Figure 16-49.**

Roof-mounted solar arrays can blend in with the architecture of a dwelling and will save yard space. Roof-mounted solar arrays attach to the roof rafters and are engineered to handle the same forces and climate conditions as the rooftop. Composition shingles are considered the easiest roofing on which to mount solar arrays, while slate and tile roofing materials are often considered the most difficult. The main drawback of roof-mounted solar arrays is that they require access for maintenance.

Freestanding solar arrays can be set at heights that allow convenient maintenance. However, freestanding solar arrays usually require a lot of space. Also, freestanding solar arrays should not be mounted on the ground in areas that receive a lot of snow.

Solar array mounts can also be either fixed or tracking. Fixed solar arrays, which are often roof-mounted or freestanding, are preset for height and angle and do not move with the sun. Directional tracking solar arrays move with the sun from east to west and adjust their angle to maintain the maximum exposure as the sun moves. Directional tracking solar arrays can increase the daily energy output of a PV system from 25% to 40%. However, despite the increased power output, directional tracking arrays may not justify the increased cost due to the complexity of the mounting system.

PV Combiner Boxes. A PV combiner box receives the output of several solar panel strings and consolidates this output into one main power feed that connects to an inverter. PV combiner boxes are normally installed close to solar panels and before inverters. PV combiner boxes can include overcurrent protection, surge protection, prewired fuse holders, and preconfigured connectors for ease of installation to the inverter. The use of prewired connectors saves running wires to the inverter. PV combiner boxes should be inspected periodically for leaks or loose connections.

PV combiner boxes are not required for every PV system installation. For example, when there are only two or three strings of solar panels, a combiner box may not be required. In these cases, the strings of solar panels are connected directly to the inverter.

Figure 16-49. Typical solar array mounts include roof, freestanding, and directional tracking mounts on the roof or on the ground.

PV Inverters. An *inverter* is a device that receives DC power and converts it to AC power. PV inverters serve three basic functions: they convert DC power from the PV panels to AC power, they ensure that the AC frequency produced remains at 60 cycles per second, and they minimize voltage fluctuations. The most common PV inverters are microinverters, string inverters, and power optimizers. **See Figure 16-50.**

A *microinverter* is a device that converts DC power to AC power and is mounted directly to individual solar panels. Because the DC to AC conversion happens at each solar panel, the microinverters maximize the potential output of a system. For example, if one solar panel is shaded by a tree, it will not affect the output of any other solar panels. Microinverters also eliminate the need for potentially hazardous high-voltage DC wiring.

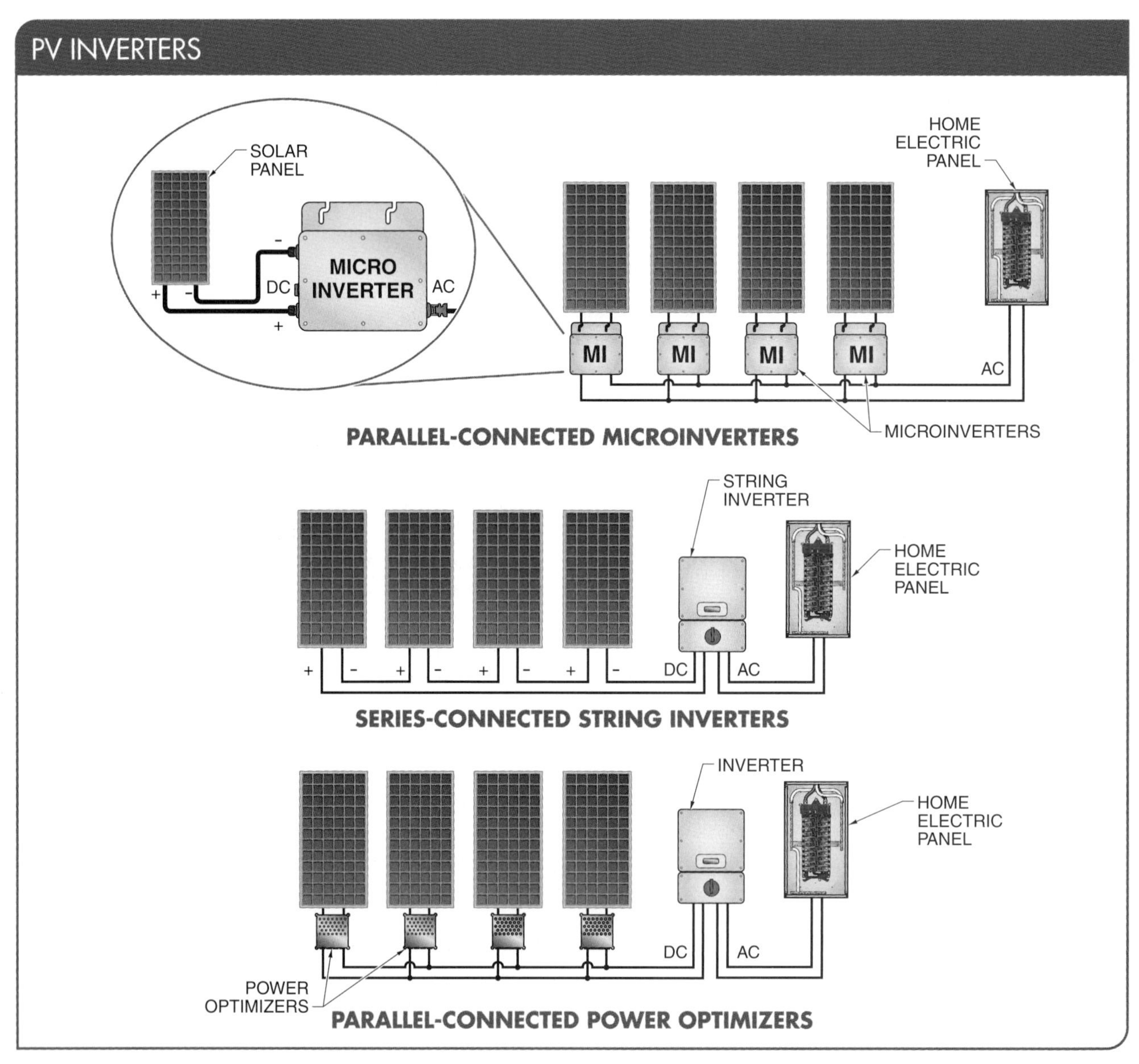

Figure 16-50. Microinverters are connected to each solar panel, which are connected in parallel, and convert DC directly to AC. String inverters are multiple solar panels connected in series. Power optimizers are installed on each solar panel, which are connected in parallel.

A *string inverter* is a device that converts DC power to AC power from several solar panels that are connected in series. However, in a series configuration, if one of the solar panels stops producing electricity, even due to temporary shading, it can decrease the performance of the whole system. String inverters are in the high-voltage range (600 V to 1000 V) and are used with large PV systems with no shading concerns. Usually only one string inverter is needed for a residential application.

A *power optimizer* (maximizer) is a hybrid microinverter system that conditions the DC power before sending it to a centralized inverter instead of converting the DC power from the solar panels directly into AC power. Power optimizers, like microinverters, still perform well when one or more panels are shaded or when panels are installed facing different directions. Power optimizer systems tend to cost more than string inverter systems, but less than microinverter systems.

PV Disconnects. Automatic and manual safety disconnects protect the wiring and components of PV systems from power surges and other equipment malfunctions. Disconnects ensure that the PV system can be safely shut down and system components can be removed for maintenance or repair. With grid-connected PV systems, safety disconnects ensure that the generating equipment is isolated from the grid for the safety of utility personnel. A disconnect is needed for each source of power or energy storage device in the PV system. An AC disconnect is typically installed inside the home before the main electrical panel. Utilities commonly require an exterior AC disconnect that is lockable and mounted next to the utility meter so that it is accessible to utility personnel.

PV Power Back-up Systems

Most PV systems include back-up power. In a grid-connected PV system, the utility grid is the back-up power. In off-grid (stand-alone) PV systems, a battery bank or a generator is the back-up power. A back-up generator is used to provide power during outages, cloudy weather, or when high household demand drains the battery bank. Some generators can run on alternative fuels, such as biodiesel.

Battery Banks. Battery banks store DC electrical energy for later use. Battery banks can be used to maintain power to critical loads such as lighting, refrigeration, and HVAC when a utility power outage occurs. However, there are some disadvantages to using batteries for back-up power, including expensive installation, maintenance requirements, and replacements needs that often occur before other parts of the PV system.

CODE CONNECT

NEC® Section 690.13 states that means shall be provided to disconnect a PV system from all wiring systems, including power systems, energy storage systems, and utilization equipment and its associated premises wiring.

Battery Charge Controllers. A PV system with a battery back-up system requires a battery charge controller. A *battery charge controller* is a device that regulates and maintains battery voltage, preventing batteries from being overcharged while allowing charging when it is needed. Battery charge controllers differ from battery chargers in that they detect the charge level to determine whether the battery needs a charge and how fast to charge it, while battery chargers typically charge at a constant rate.

A *battery charger* is a device used to put energy back into a rechargeable battery. Although many battery chargers charge at a constant rate, some models with microprocessors can switch off when the battery is fully charged. Microprocessor battery chargers can recognize how much charge was originally in the battery, and they can also indicate when adding water is necessary for liquid-based batteries. The rates at which batteries should be charged are determined by the manufacturer.

Battery Management Systems (BMSs). A *battery management system (BMS)* is an electronic system that protects and manages rechargeable batteries from operating outside their safe operating area. A BMS controls parameters such as battery temperature, depth of discharge (DoD), and state of charge (SoC), so that batteries are not overcharged or undercharged, which can affect battery life. A BMS is a software program that can be internal to a charge controller or on a separate circuit board installed at each battery. A BMS continuously measures battery voltages and currents and uses battery sensors to communicate abnormal measurements with external energy management systems.

A BMS is critical when using lithium-ion batteries because it regulates battery charging to obtain the best performance from the battery. However, when lead-acid batteries are used, a BMS is not critical. This is due to less risk of thermal issues and because the inverter or battery charge controller usually handles this charging process.

PV Storage Battery Types. A *PV storage battery* is a deep-cycle battery that provides energy storage for a PV system and is designed to be charged and discharged repeatedly. To maintain PV storage batteries in good condition and prolong their lifespan, manufacturers recommend limiting DoD to no more than 50%. PV storage battery cost is based on the battery material, capacity, depth of discharge, and lifespan.

SMA Technologie AG

Deep-cycle storage batteries are capable of being charged and discharged more often than other batteries.

There are several types of PV storage batteries that can provide energy daily and during peak demand. The common types of batteries used are lithium-ion batteries, lead-acid batteries, and flow batteries. Lithium-ion and lead-acid batteries are common in PV power back-up systems. There are also emerging battery technologies available, such as saltwater batteries, sodium–nickel chloride batteries, gel batteries, and absorbed glass mat (AGM) batteries. Each type of PV storage battery has advantages and disadvantages. **See Figure 16-51.**

Battery Capacity. The capacity of a storage battery is typically stated in ampere-hours (Ah). However, some PV manufacturers measure capacity in either watt-hours (Wh) or kilowatt-hours (kWh). When the capacity is known in ampere-hours, it is easy to convert to watt-hours. For example, the capacity in watt-hours of a 12 V battery rated at 100 Ah is determined by multiplying the voltage by the ampere-hours. Therefore, the capacity of this battery is 1200 Wh ($12 \times 100 = 1200$), or 1.2 kWh.

A battery's reserve capacity indicates how much time is available to keep equipment running when a charging system is not in place or has failed. New battery technology is leading to more powerful batteries with lower footprints. **See Figure 16-52.** Information needed to select the proper battery capacity includes average daily kilowatt-hours (kWh) used, energy used from summer to winter, and the appliances the batteries must supply power to and for how long.

Battery Lifetime. The lifetime of a battery must be specified in different ways depending on the application. For solar applications in which the battery is regularly charged and discharged, the most useful measure of lifetime is the number of charge/discharge cycles over which the battery maintains its capacity. This measure is called cycle life.

A battery's DoD refers to the amount of capacity that has been used. Because solar batteries are composed of electrochemical cells, they must always keep some charge since their useful life will be significantly shortened if a load uses 100% of their charge.

Most manufacturers will specify a maximum DoD for maximum performance. For example, if a 10 kWh battery has a DoD of 80%, it should not use more than 8 kWh of the battery before being recharged.

The age of a battery has a major impact on its capacity. The capacity will stay at or close to the battery's rated capacity for only a limited number of charge/discharge cycles, even when manufacturer's specifications on DoD are followed. Battery lifetime is typically affected by the gradual degradation in battery capacity, which is determined by charge/discharge cycles.

The safety regulations for each type of battery should be carefully checked. Battery systems used in renewable energy systems contain corrosive or dangerous chemicals and produce large currents. For example, lead-acid batteries produce hydrogen. Since most batteries contain toxic and/or corrosive material, it is important to safely dispose of the batteries to minimize any impact on the environment.

PV STORAGE BATTERY TYPES		
Battery Type	**Advantages**	**Disadvantages**
Lead-acid batteries	Low cost Proven technology Supply large surge currents Readily available	Emits explosive hydrogen Requires maintenance
Lithium-ion batteries	Low cost No explosive gases Lightweight Compact	Limited cycle life when deeply discharged Thermal runaway risk
Flow batteries	No explosive gas Replaceable cell Long cycle life Last 20 years	Large and heavy Solution can be toxic
Saltwater batteries	No explosive gas Fully recyclable Can add capacity Safest Battery	Large and heavy Low power availability
Sodium–nickel chloride batteries	No explosive gases Lightweight Compact Fully recyclable	Expensive Low production yield
Gel batteries	Does not release gases Can be used indoors Rugged Mounted in any position	Long charge time Cannot tolerate overcharging
Absorbed glass mat (AGM) batteries	Does not release gases Excellent performance Slow discharge Mounted in any position	Long charge time Cannot tolerate overcharging

Figure 16-51. Although lithium-ion and lead-acid batteries are the most commonly used types of batteries in PV systems, each type of PV storage battery has advantages and disadvantages.

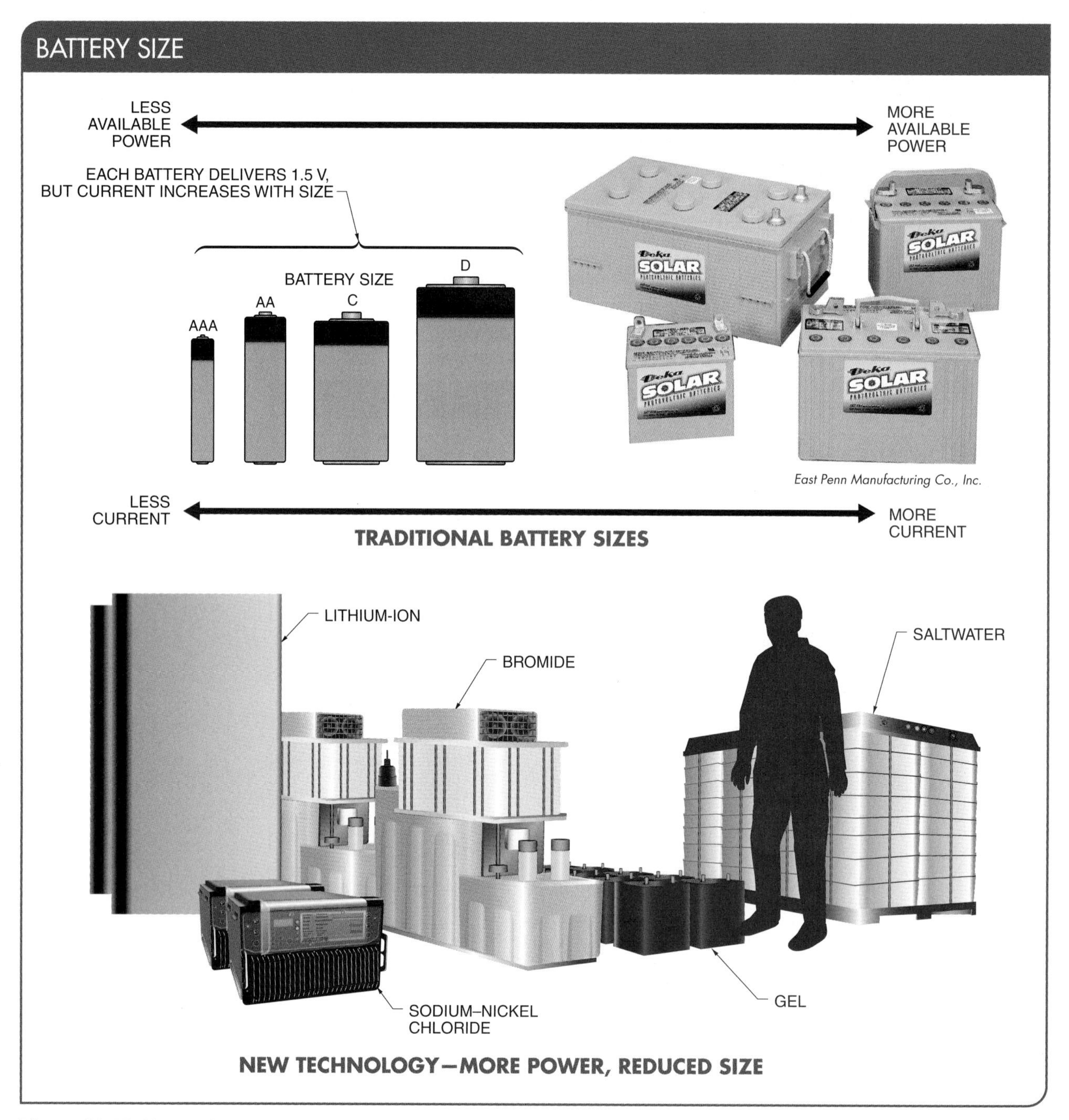

Figure 16-52. New technology enables more battery capacity in a smaller footprint.

Building-Integrated Photovoltaics (BIPV)

Building-integrated photovoltaics (BIPV) are PV materials that are used to replace conventional building materials in parts of the building envelope. Residential architects and builders are also beginning to integrate PV materials into the exterior of a dwelling. **See Figure 16-53.** BIPV can be attached to a residence as curtain walls, paneling, balconies, or sunshades. Also, PV vision glass can be used instead of traditional double-pane windows and skylights to provide both electricity and transparency.

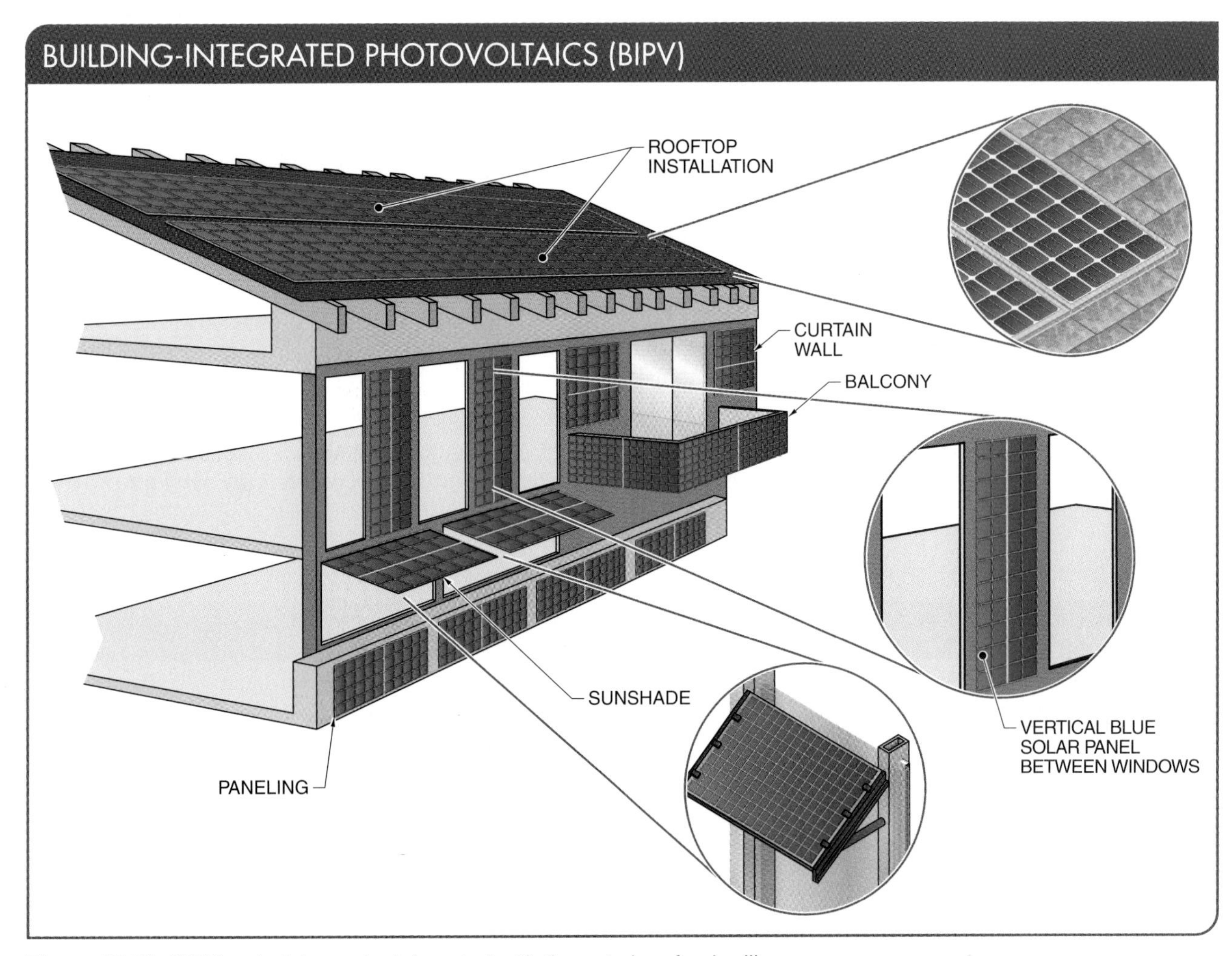

Figure 16-53. BIPV materials can be integrated with the exterior of a dwelling.

Several types of PV materials can be integrated into glass. For example, special solar PV glass blocks can be used to replace traditional glass blocks. These glass blocks contain solar cells with specialized optics that focus the light onto the PV material. **See Figure 16-54.**

The advantages of BIPV systems include that they do not require additional land, they reduce building energy consumption, and they can transfer energy with negligible transmission losses. Some barriers to BIPV systems can include the cost of BIPV products, maintenance, and a lack of knowledge to design with BIPV technology. The installation of BIPV also requires cooperation across multiple building trades, such as electricians, roofers, architects, and engineers.

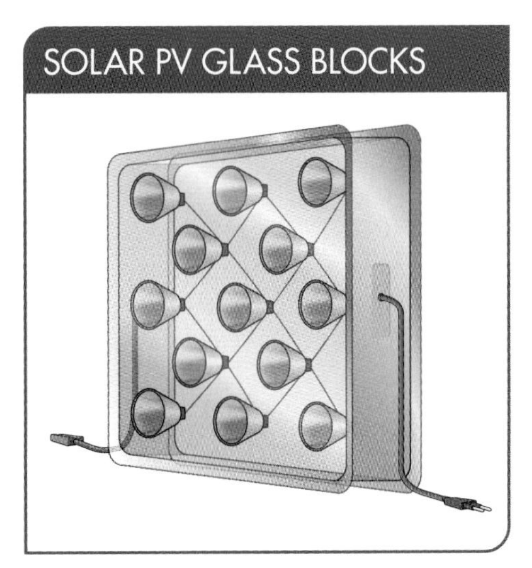

Figure 16-54. PV glass blocks can replace traditional glass blocks to harness the suns energy.

Solar Roofing. Solar roofs can be installed instead of traditional roofs using interconnecting solar sections or individual solar tiles. Some BIPV manufacturers, such as Tesla, have developed solar tiles that appear aesthetically similar to traditional roofing when viewed from street level. **See Figure 16-55.** Solar tiles are installed together to construct solar roofs with varying generation capacities. Solar tiles are made with tempered glass to make them stronger than standard roofing tiles. These materials tend not to degrade over time like asphalt or concrete tiles.

Solar Roofing Connections. Solar roofing requires the individual solar sections or tiles to be electrically connected to generate power for the dwelling. Some manufacturers design these solar sections or tiles to be electrically connected using wires and plug-in connectors. Tesla's solar tiles are unique because a polymer paste is used to connect one tile to another tile. The polymer paste bonds the two overlapping tiles, allowing electricity to flow from tile to tile. This type of connection is designed to be more resistant to temperature changes and weather conditions than soldered metal connections between cells. **See Figure 16-56.**

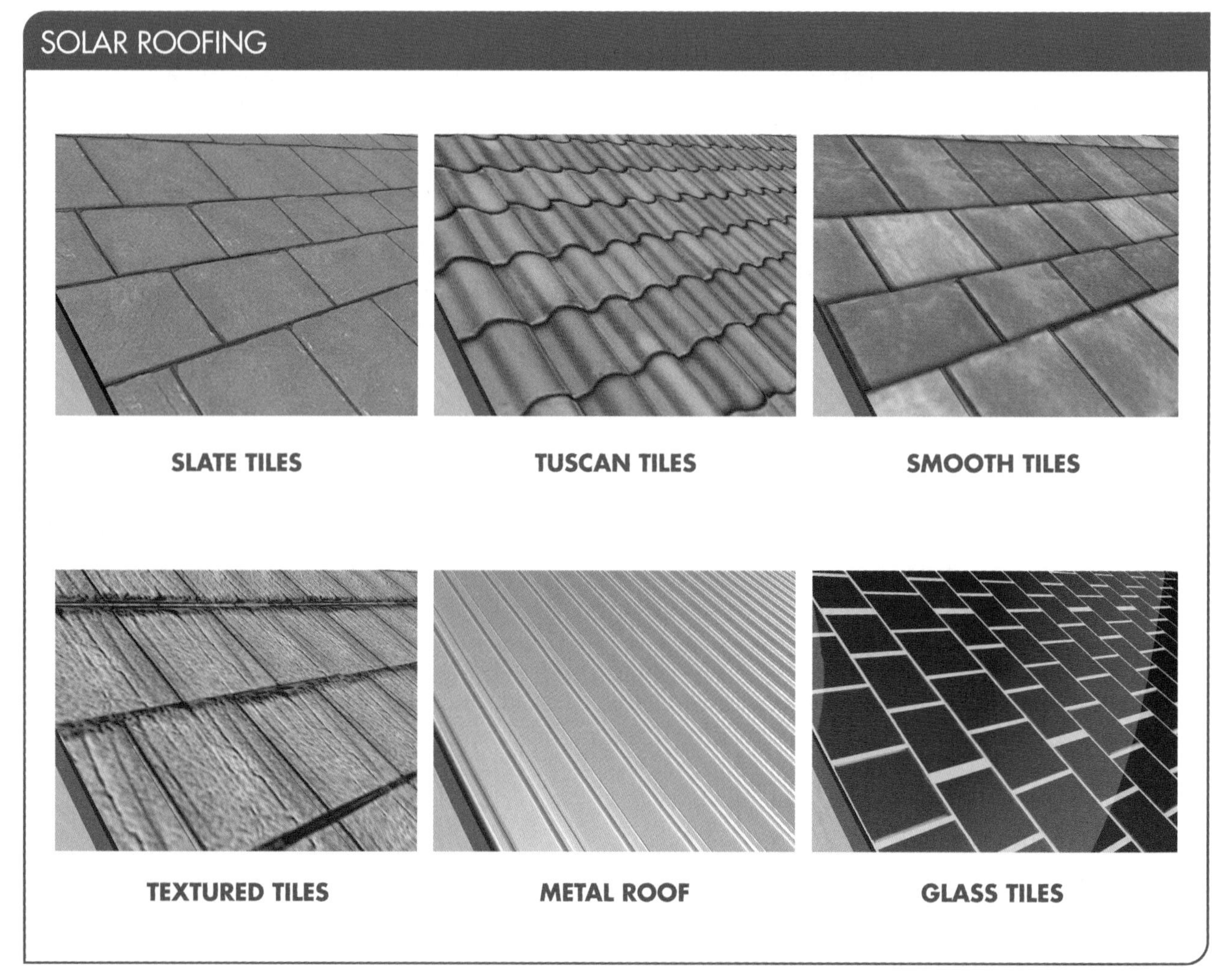

Figure 16-55. Instead of traditional roofs, aesthetically similar solar roofs can be installed using interconnecting solar sections or individual solar tiles.

Solar Roof Costs. The cost of a solar roof is based on the estimated square footage of the roof, which includes the cost of materials, installation, and the removal of the old roof. Additional costs can include gutter and skylight replacement. Other cost considerations include the number of sunlight hours per year where the dwelling is located, solar panel efficiency, and federal solar tax credits.

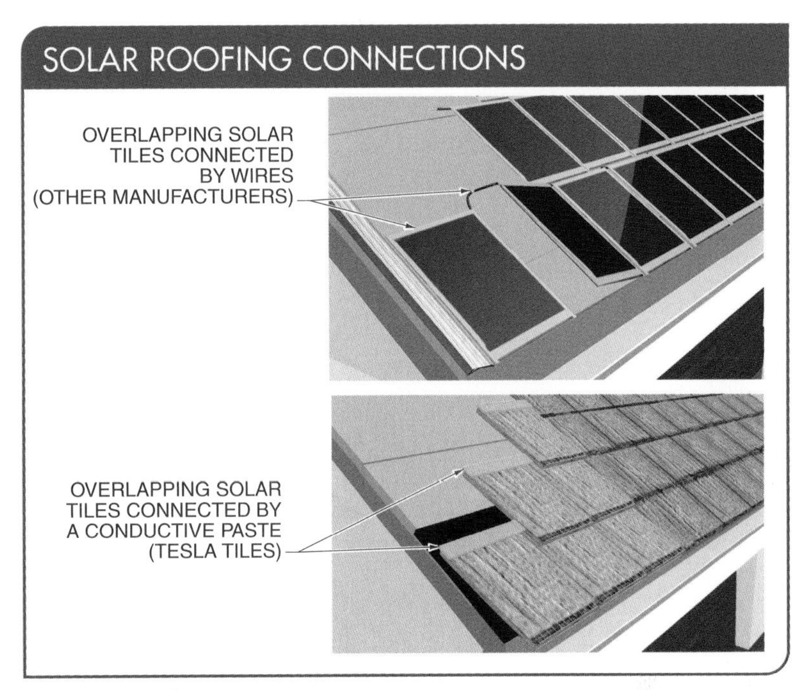

Figure 16-56. The individual solar sections or tiles of a solar roof are electrically connected together using wires and plug-in connectors or polymer paste to generate power for the dwelling.

Designing PV Systems

A homeowner can either design a PV system or buy a pre-engineered PV system that uses compatible devices to operate at maximum capacity. The first step in designing a PV system is to determine whether the site receives enough sunlight to make the system viable. The solar potential of a site can be calculated by consulting an insolation map. An *insolation map* is a map that indicates the average solar energy received in hours of peak sunlight per day on a specific area in a given month or year. **See Figure 16-57.** In the US, the summer months allow more solar energy to be produced than during the winter months.

On average, for every 1000 W of PV power required, a dwelling requires 100 sq ft of space to mount PV modules. The area around the PV modules must be left open for maintenance or repair access. If the location limits the physical size of the system, more efficient PV modules may be required. Each 1000 W of PV modules can generate about 1000 kWh per year depending upon location.

Solar Array Orientation. Solar arrays are ideally oriented south and produce 95% of their full power when within 20° of the sun. Therefore, the roof does not need to be perfectly tilted or oriented toward the sun. For flat roofs, solar arrays can be mounted on frames and tilted up toward true south. Less solar energy is received when a fixed array is oriented in a direction other than due south. This reduction increases with larger tilt angles, so orientation is more critical at higher latitudes. At lower latitudes with arrays installed at lower tilt angles, the effect of orientation is much less. **See Figure 16-58.**

The tilt angle is based on the latitude of the geographic location of the solar array, although it will vary by month. For example, Chicago, Illinois, at an approximate latitude of 42° N, requires a tilt angle of 48° in the spring and fall, but ranges from 24° in the winter to 72° in the summer. Online calculators are available for determining the proper tilt angle based on location.

Solar arrays need unobstructed access to sunlight from 9 AM to 3 PM throughout the year. Solar arrays can be optimized for either summer or winter gain by changing their tilt angle. To maximize summer energy production, the array should be mounted at a smaller tilt angle to take advantage of the sun's high arc through the summer sky. This may be desirable for a system with high energy requirements in the summer, such as for cooling loads. To maximize winter performance, the array should be mounted at a greater tilt angle to receive more energy when the sun is lower in the winter sky. This may be best for a system with greater electrical loads in the winter, such as for artificial lighting or heating.

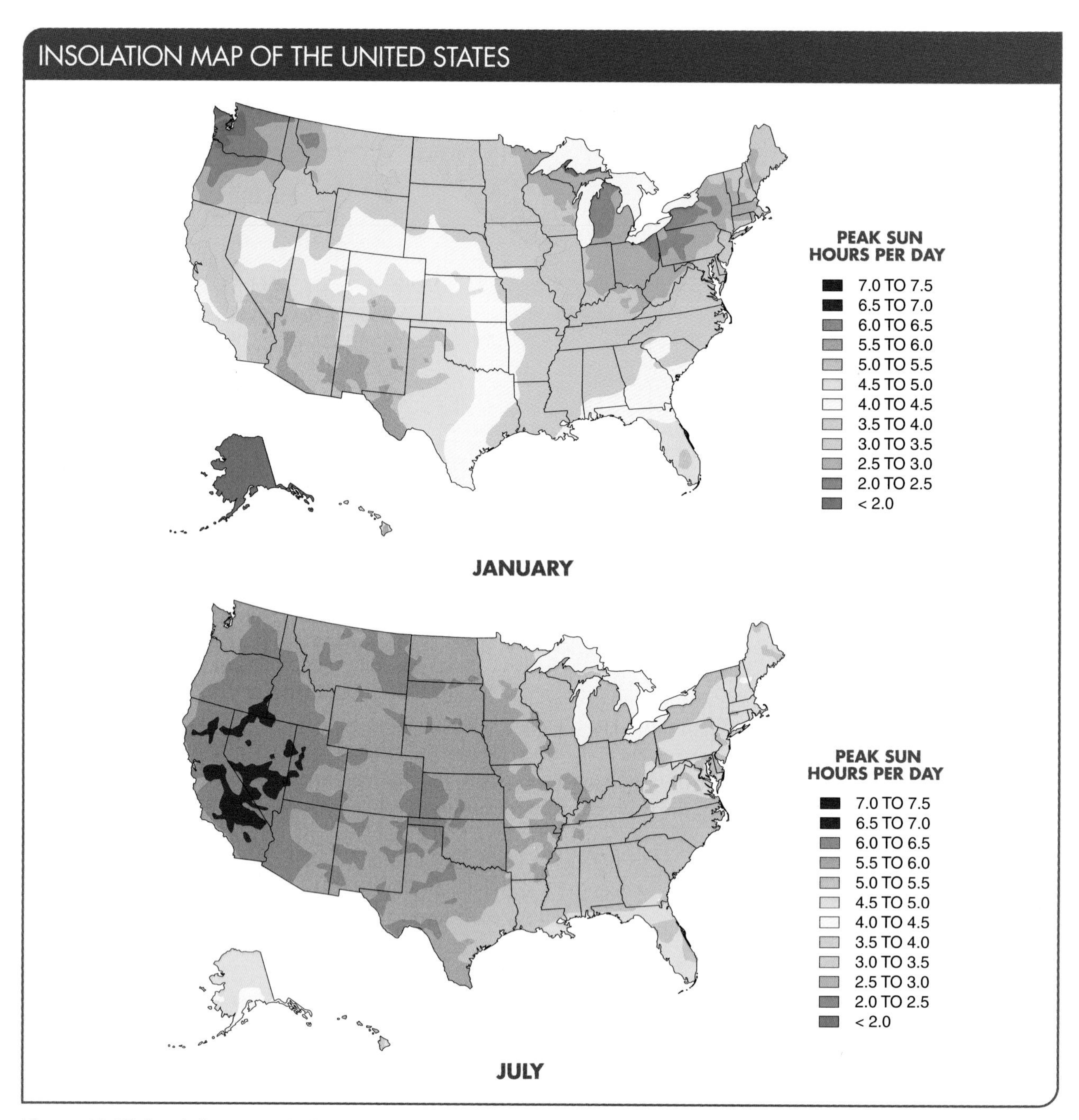

Figure 16-57. Insolation maps indicate the average solar energy received in hours of peak sunlight per day on a specific area in a given month or year.

PV System Installation Costs. The installation cost of a residential PV system in the US is based on the dollar amount per installed watt before government or utility incentives. For example, a PV system and the labor to install it may be $8 to $10 per watt. Some solar panels may have a lower cost per installed watt than higher efficiency panels, but they may also take up more space and require a larger mounting system. Other factors influencing the cost of a PV system include the size and type of the system, what type of mounting system is required, whether a battery backup is used.

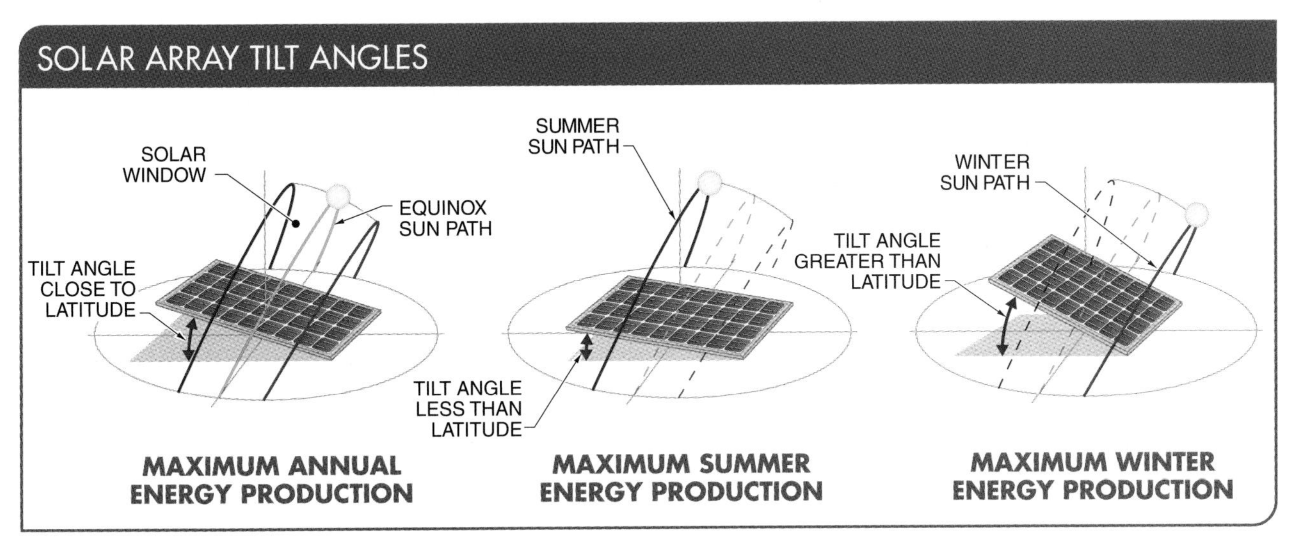

Figure 16-58. Solar array tilt angles can be adjusted in winter and summer to maximize system performance.

If batteries are included in the PV system, both material and installation costs will be greater. Batteries can add 50% or more to the time required for the installation. The factors that determine the cost of a battery back-up system include the price of components, cycle life, lifespan, and battery maintenance. A low installation cost may come at the expense of quality and battery life, and the need for frequent battery replacements will increase the cost over time.

The most important factor to consider in a battery back-up system may be cycle life. Design, materials, process, and quality influence how long the battery will cycle. The typical battery lifespan is approximately 10 years, with monthly and yearly maintenance.

Federal Solar Tax Credit. The federal solar tax credit, also known as the investment tax credit (ITC), allows homeowners to deduct 30% of the cost of installing a PV system from their federal taxes. There is no cap on the value of the purchase of a system. For example, if a homeowner spends $30,000 on a PV system, the 30% tax credit would be $9,000. If a homeowner spends $45,000 on a PV system, then the 30% tax credit would be $13,500. However, the federal solar tax credit for residential systems gradually steps down the value (percentage) of the credit and ends in 2022. **See Figure 16-59.**

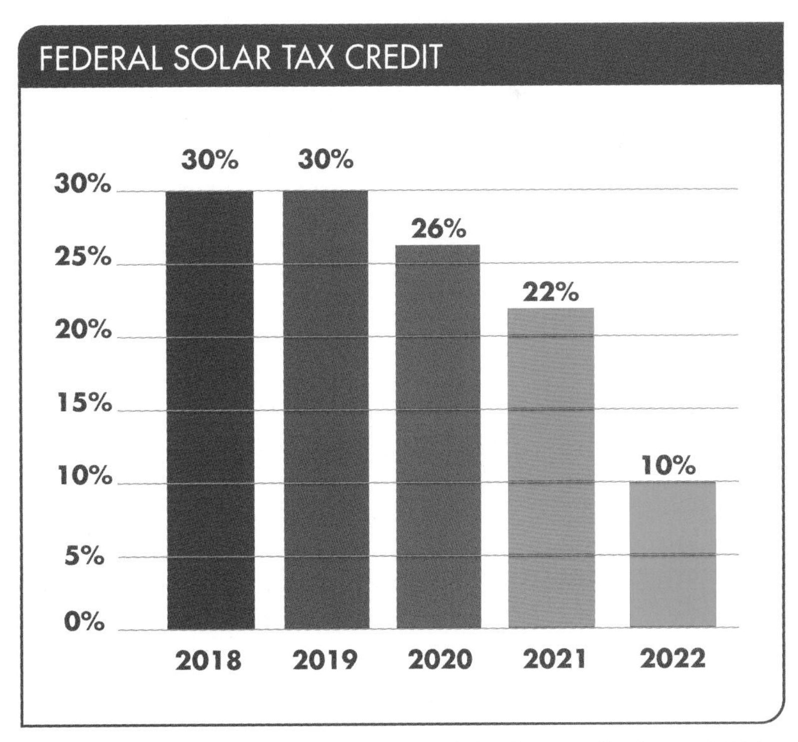

Figure 16-59. Homeowners can receive a federal tax credit of up to 30% for the cost of a PV system until 2022.

PV Applications—Pool Energy Management

Managing energy to clean, heat, and light a swimming pool can be part of a home energy management plan. This can be accomplished by using a PV system to supply energy to the pool, installing energy-efficient equipment, and using remote-control devices. **See Figure 16-60.** Pool heaters, pumps, and lighting systems can all be controlled remotely using a variety of handheld devices or home computers. By using Wi-Fi-enabled switches and app programs, a homeowner will be able to schedule, automate, and access all electrically connected devices to operate the pool.

Using properly sized, energy-efficient pumps and motors can lower the electric operating costs of a pool. A ¾ HP pump can generally operate a residential pool for 6 hr/day to 8 hr/day. A timer can run a pump for short cycles during the day, excluding peak load times, to keep the pool clean from debris.

The evaporation rate from an outdoor pool varies depending on the water temperature, air temperature and humidity, and the wind speed at the pool surface.

> CODE CONNECT
>
> *NEC® Section 680.33 states that an underwater luminaire, if installed, shall be installed in or on the wall of the storable pool, storage spa, or storable hot tub.*

Pool Water Temperature. Water temperature can be controlled to manage pool energy. Each degree the temperature of water is raised could cost an additional 10% in energy. The National Swimming Pool Foundation (NSPF) recommends pool water temperatures of 78° to 80° for active swimming and 82° to 84° for general use. Water heating costs in a shower area for the pool can be reduced by lowering its water temperature to 95°. The installation of low-flow showerheads and automatic shut-off valves can reduce water consumption and conserve energy. Insulating the water heater for the shower area will further reduce water heating costs.

A pool can be covered at night when it is not in use as an effective means of reducing heating losses. Pool covers also reduce evaporation, which is the largest energy loss for pools. Covers also provide many other benefits besides saving energy. Pool covers conserve makeup water by 30% to 50%, reduce chemical consumption, and reduce cleaning time by preventing dirt and other debris from entering the water.

High-Efficiency Pool Heating and Lighting Systems. There are three types of high-efficiency pool heating systems: gas pool heaters (with efficiencies as high as 97%), heat pump heaters (that are five to six times more efficient than electric resistance heaters), and solar pool heaters. The most cost-effective method to heat a swimming pool is using solar energy. Swimming pools require low-temperature heat, which is where solar panels are most efficient. Improving the efficiency of a pool lighting system is as simple as replacing incandescent lamps with LED lights.

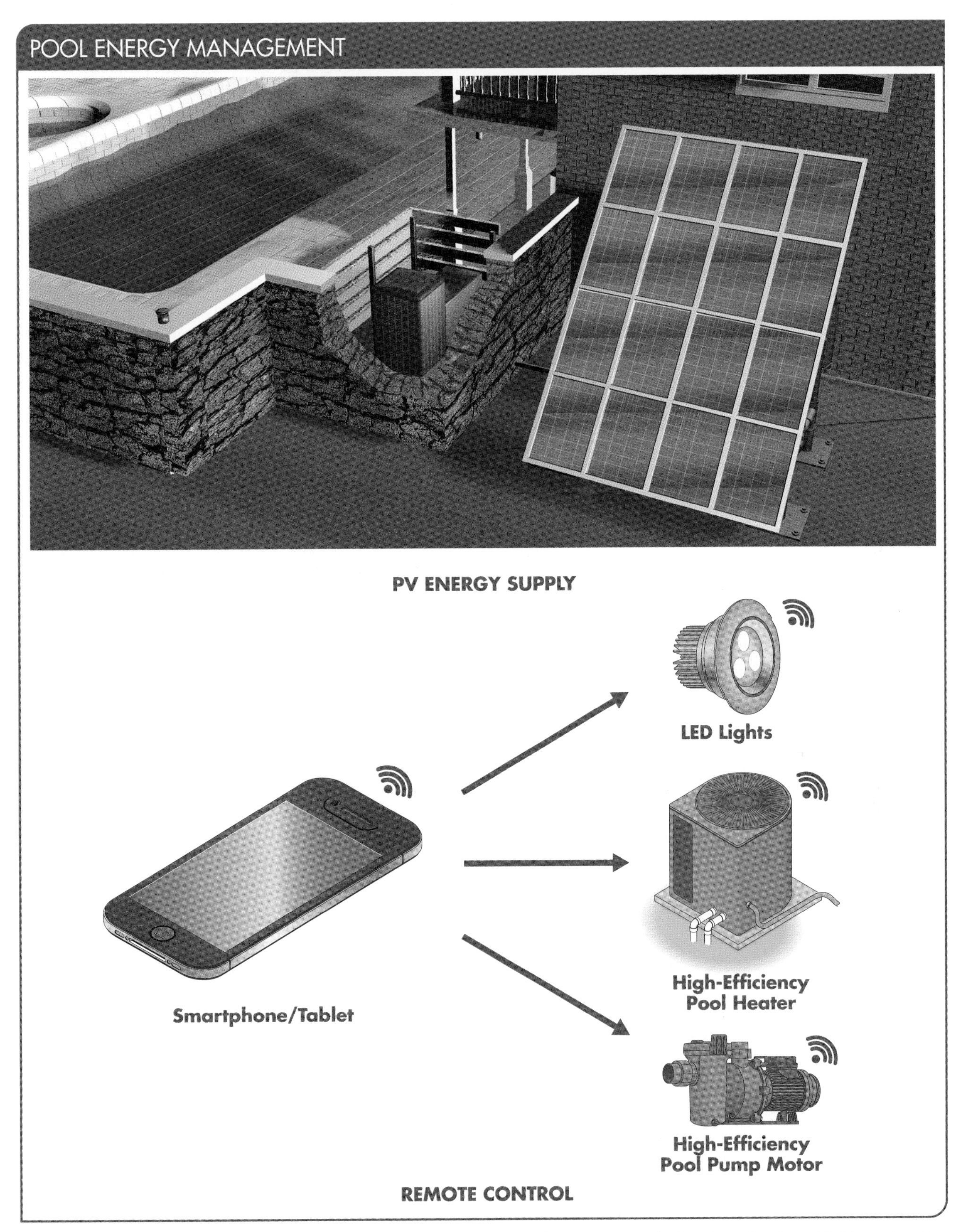

Figure 16-60. Part of a home energy management plan is to manage the energy needed to clean, heat, and light a swimming pool, which can be accomplished through the use of a PV system, energy-efficient equipment, and remote-control devices.

SECTION 16.4 CHECKPOINT

1. What is a small wind turbine?
2. What is solar photovoltaic (PV) power generation?
3. What is an inverter and what are three common PV inverters?
4. What are building-integrated photovoltaics (BIPV)?

SECTION 16.5 – HOME ENERGY MANAGEMENT

Home energy management is the process of monitoring and controlling energy use, while also ensuring the delivery of a reliable source of electricity. There are three distinct entities involved in energy management. The first entity is the utility, which is on the supply side of the smart meter. The second entity is the homeowner, who is on the consumer side of the smart meter. The third entity is the manufacturers, who produce the devices and appliances that allow the utility and the consumer to obtain and use the information on both sides of the smart meter for mutual benefit.

Home Energy Management Systems

A *home energy management system (HEMS)* is a group of products that processes and controls smart energy-using devices and their information on both sides of the smart meter to determine the best use of energy at any given time. The HEMS manages all of the information on both sides of the smart meter to determine the best use of energy at any given time. This involves alternative energy generation, energy storage, controlling smart appliances, coordinating with weather patterns, and, in some cases, linking to the home automation system. **See Figure 16-61.** This can be done with minimum interaction with the consumer and even when the consumer is not present.

Homeowners without some type of HEMS may experience high household energy use due to poor conservation and overuse. However, one of the barriers to using a HEMS includes making the initial investment in the hardware, such as displays, software, and sensors, needed to manage energy consumption. If a homeowner does not wish to purchase HEMS hardware, they can subscribe to an energy management service, along with smart home automation and security services if desired, for a monthly fee. Other barriers may include installation costs and difficulty interfacing with smart home automation systems.

The fundamental components of a HEMS include a HEMS controller, sensors, output devices and controls, and communication protocols. A HEMS controller typically has a touchscreen display that also serves as an input device. The HEMS controller is connected to both input devices and output devices that operate wirelessly through a common communications protocol.

HEMS Controllers. A HEMS controller can be programmed to establish the homeowner's priorities and manage pricing from the utility. By using utility data and the homeowner's consumption patterns, the HEMS controller can schedule optimal running time for various electrical loads in the most cost-effective manner. The HEMS controller utilizes the two-way flow of communication from the smart meter to achieve communication between the customer (homeowner) and utility. **See Figure 16-62.**

Sensors. Sensors are used in the home to monitor temperature, humidity, light, and other parameters affecting energy consumption. Many of these sensors are already built into devices such as wireless smart thermostats. Smart sensors send signals to the HEMS controller to control the operation of the devices connected to the HEMS.

CODE CONNECT

NEC® Subsection 750.30(C) states that an energy management system shall not cause a branch circuit, feeder, or service to be overloaded at any time.

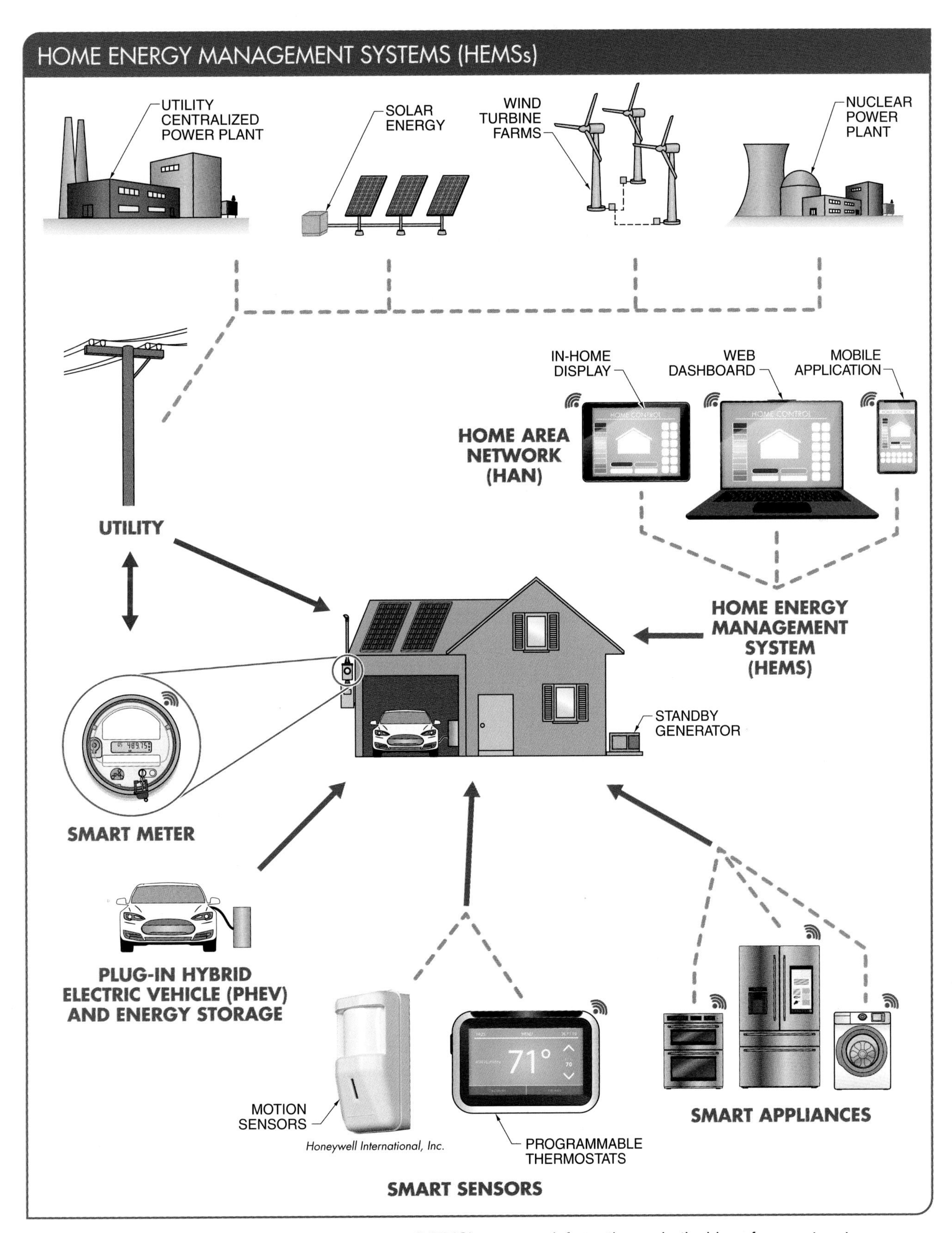

Figure 16-61. A home energy management system (HEMS) manages information on both sides of a smart meter.

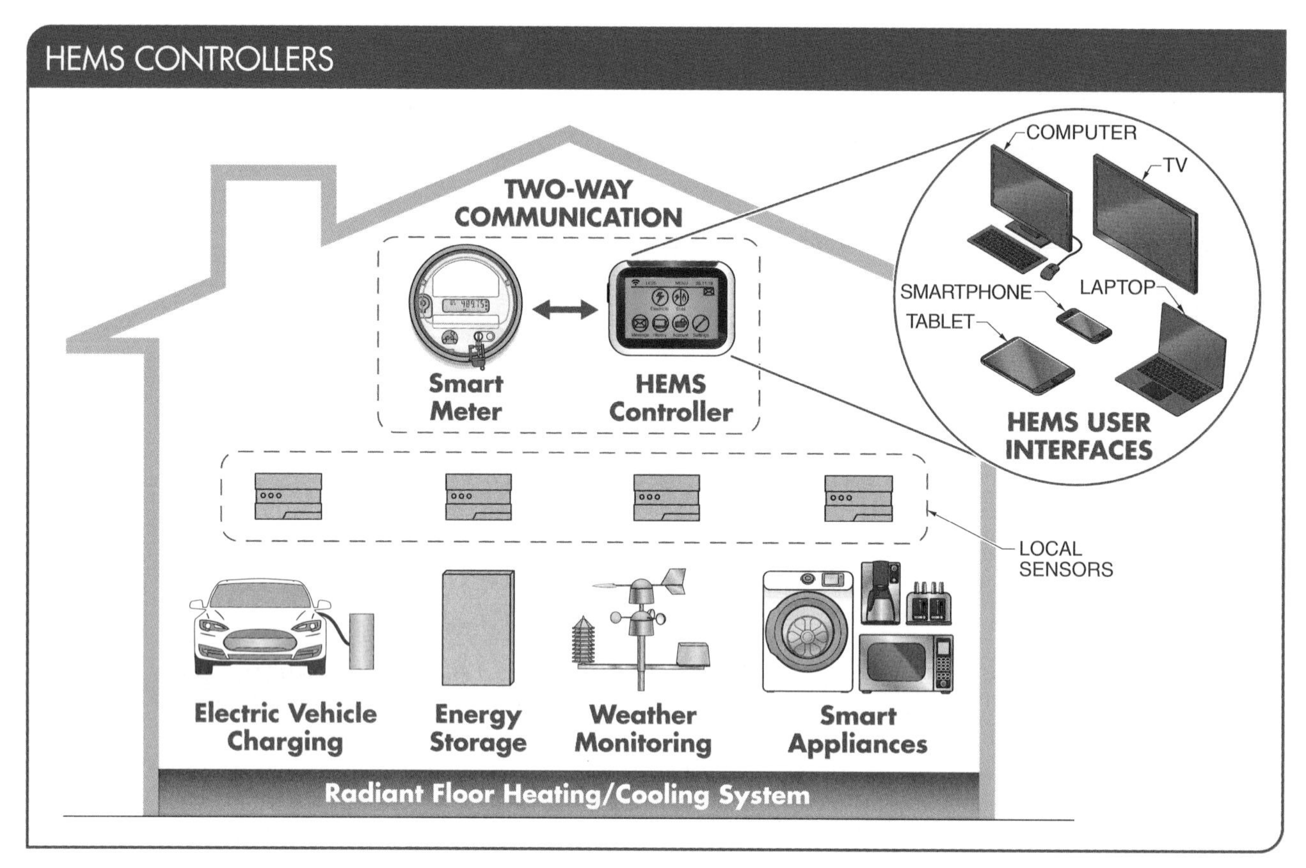

Figure 16-62. By using utility data and the homeowner's consumption patterns, a HEMS controller can schedule optimal running time for various electrical loads in the most cost-effective manner.

Output Devices and Controls. There are several levels of control associated with the HEMS. The HEMS itself is the centralized control. The homeowner also has control at the device level through the use of a smartphone or simple on/off switch. For example, smart appliances are built for either the HEMS to control or the homeowner to directly to control processes such as cooking, washing, drying, and refrigeration.

Communication Protocols. HEMS protocols are used to maintain communication between the utility and the smart sensor-based devices in the home. The objective of these communication protocols is to use the home area network (HAN) for controlling devices in the home. Quite often a computer sends and receives data between the utility and the smart home. HEMS rely on open and proprietary communication protocols such as Wi-Fi, Bluetooth, Zigbee, or Z-wave.

Energy Management Strategies

Residential smart meters allow the utility and the homeowner to work together to reduce energy demand during peak demand periods and save money. The utility saves money when demand is lowered during peak periods because this helps reduce the cost of new capital investments and minimizes the need for expensive peak power plants to operate during high demand. The homeowner can save money by managing energy use during periods of peak demand and taking advantage of reduced rates offered by the utility. This participation should also result in more reliable service during peak demand periods.

Demand Response Programs. Utilities may also offer demand response (DR) programs to encourage residential customers to reduce their electrical consumption. A *demand response (DR) program* is a utility incentive program for reducing electrical demand during peak hours. A DR program involves a contract between the utility and customer allowing the utility some control over the major appliances that use large amounts of electricity in the home. DR does not concern energy efficiency; it only aims at temporarily reducing loads to avoid overloading the grid's generation capacity.

For a load such as air conditioning or water heating, the temperature setting can be temporarily changed by the homeowner so that the electrical load is reduced or not used at all during hours of peak demand. The use of smart appliances and electric vehicle (EV) charging stations can also be rescheduled to a lower demand time. When these measures are applied to thousands of units/dwellings, the load on the grid can be significantly reduced.

Incentive-Based DR Programs. Incentive-based DR programs provide financial incentives to participating customers who reduce their energy consumption during peak hours. In return, the participants receive reduced bills or credit payments at the end of their billing cycles. Different utilities may offer several incentive programs, such as peak time rebates (PTRs), time-of-use rates, inclining block rates, and real-time pricing.

A *peak time rebate (PTR)* is a utility program that offers monetary rebates to customers who reduce energy consumption during periods of high-cost electricity (peak times). Customers who do not reduce usage during peak events are simply charged the normal rate. Customers tend to participate in higher amounts in PTR programs compared to many other DR programs. Utilities do not have to change their rate designs for PTR programs. PTR programs can benefit both customers and utilities when rates are monitored and set correctly.

A *time-of-use rate* is a utility program that can help in planning when to use major appliances. Energy rates are low when demand is low, but they will increase as times of peak demand approach. Energy costs can be reduced by operating major appliances during times of off-peak demand.

An *inclining block rate* is a utility program that charges a higher rate for each incremental block of consumption. Although this rate system does not encourage conservation, it does encourage the self-generation of electricity through alternative sources and the off-line storage of energy. Alternative sources of energy can provide electricity to dwellings at certain times of the day. The use of alternative sources of energy also puts the homeowner in the lower, less expensive tiers of the rate system. Inclining block rates can help individuals who traditionally aim to use less energy, such as low-income and elderly people, to save on their energy bills. **See Figure 16-63.**

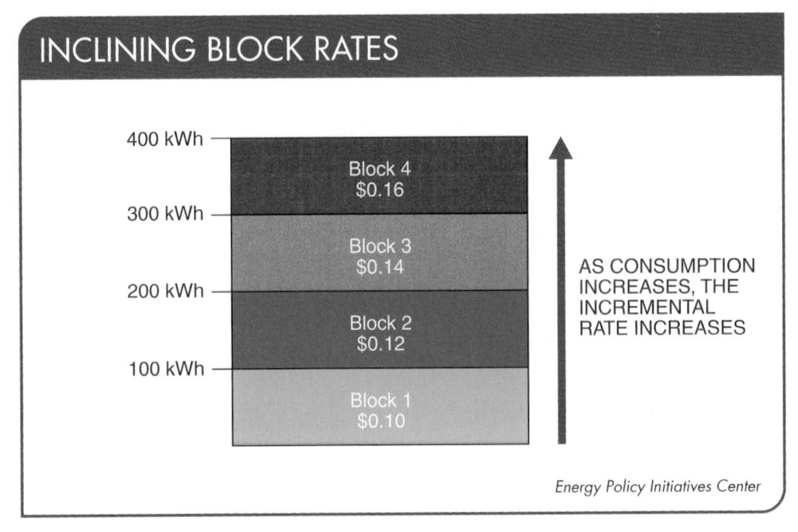

Figure 16-63. Inclining block rates increase incrementally with energy consumption.

Real-time pricing is a utility program that allows customers to pay for electricity at wholesale prices during a predetermined interval. This pricing is composed of the wholesale price of electricity plus a supplier margin. Utilities provide prices in advance and bills are calculated based on an hourly production and consumption rate. Some calculations can be as often as 15 minutes. Real-time pricing is used primarily with commercial and industrial customers who consume large amounts of electricity. Real-time pricing can be considered for residential homes where smart meters and an advanced metering infrastructure (AMI) is in place.

Net Metering. *Net metering* is an electric utility billing arrangement that credits customers for the electricity they add to the grid. Net metering enables customers who generate electricity on their premises to offset the cost of the electricity they consume by selling electricity back to the utility. Common generating systems suited for net metering include PV, wind energy, and hydroelectric systems. Utilities using interactive systems have electrical meters with bidirectional communication and metering capabilities to track the generation and consumption of electricity. If the generating system and the loads are perfectly balanced for annual energy production and demand, the result could possibly be an electric bill of zero. **See Figure 16-64.**

CODE CONNECT

NEC® Subsection 690.4(A) states that PV systems shall be permitted to supply a building or other structure in addition to any other electrical supply system(s).

Net metering programs provide an incentive for the consumer to invest in alternative distributed power generation. Currently, more than 35 states and the District of Columbia have net metering programs requiring utilities to purchase power from systems that qualify for the program. Each state has different rules and regulations. Additional information on net metering state programs can be obtained by visiting the Database of State Incentives for Renewables & Efficiency® (DSIRE®). Also, the Solar Energy Industries Association (SEIA®) promotes consumer advocacy programs for net metering and other pro-solar energy policies.

Smart Appliances

Smart appliances play a part in the energy conservation of the smart grid and home energy management. This is particularly true for large appliances such as refrigerators, electric ranges, washing machines, clothes dryers, and hot water heaters. Although small appliances contribute to energy consumption, major appliances have a much larger impact on electricity costs. Smart appliances with miniature sensors are capable of communicating with HEMSs and smart meters. **See Figure 16-65.**

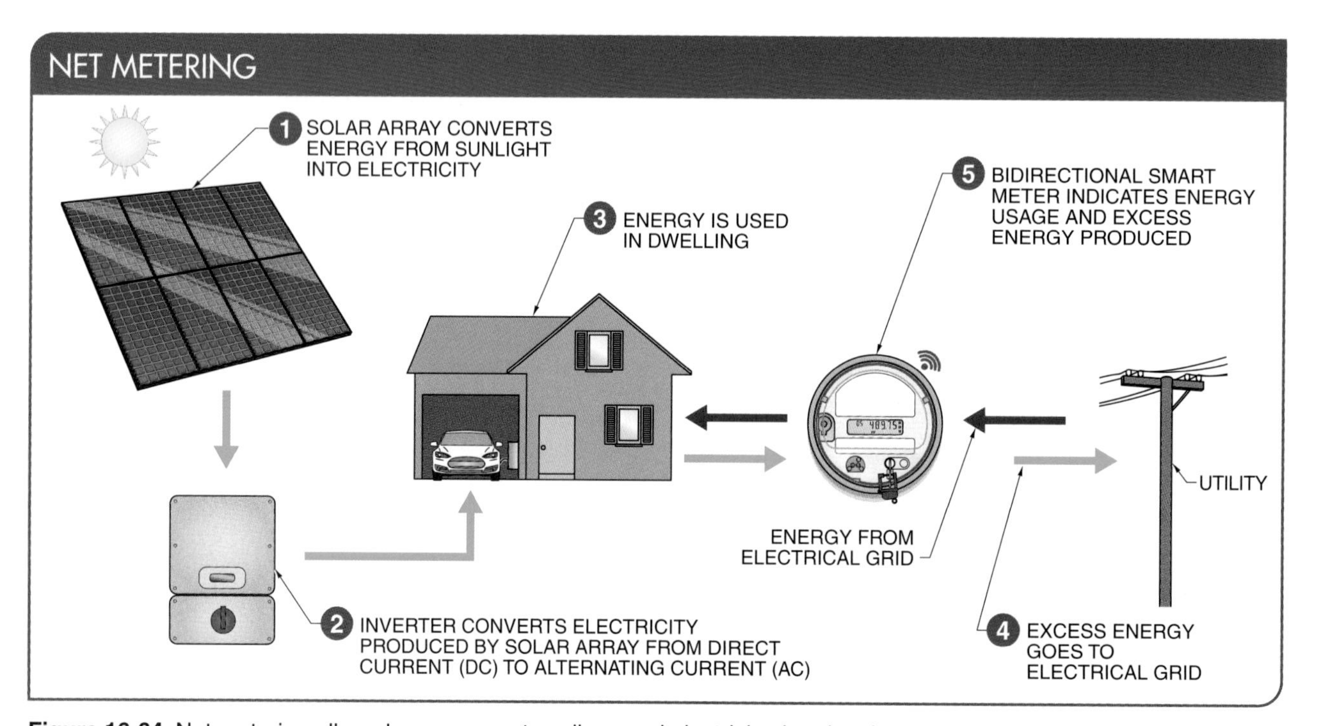

Figure 16-64. Net metering allows homeowners to sell unused electricity that they have generated back to the utility.

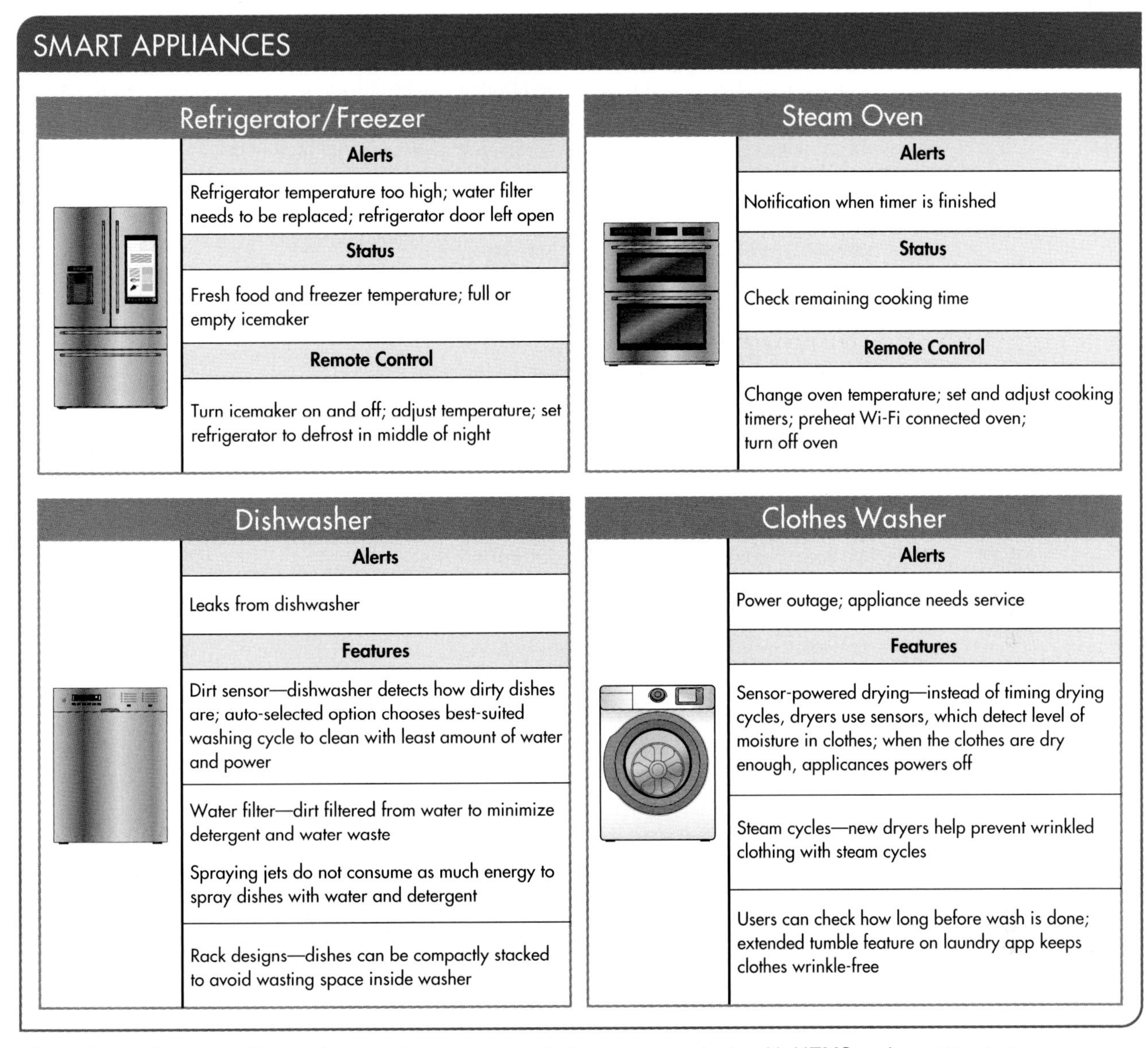

Figure 16-65. Smart appliances have miniature sensors that can communicate with HEMS and smart meters.

Smart Appliance Communication and Data Sharing. Both the utility and its customers (homeowners) benefit when smart appliances communicate with HEMSs and the smart grid. For example, customers are provided with the real-time energy use of smart appliances, allowed to remotely control their appliances, provided with outage information, and can plan for the most cost-effective time to use rates provided by the utility. The utility benefits because it is also provided with access to real-time energy usage. Also, the utility and its customers can agree to a contract for the utility to remotely control customers' appliances during periods of heavy electrical demand as part of a DR strategy.

Manufacturers also benefit from the data sharing of smart appliances. Provided that customers allow data sharing, smart appliances can create a revenue stream for manufacturers. For example, manufacturers can know when homeowners may need to purchase new refrigerators, washing machines, and air conditioners. Since many smart (AI) thermostats have motion detectors, patterns of behavior, such as sleeping and movement in the home, can be obtained and used for marketing purposes. Another way data can be used by manufacturers and other companies is with camera-equipped refrigerators and app-driven shopping lists. In these cases, customers may receive free coupons in a pop-up ad on a smartphone or computer or in the mailbox.

Smart Appliance Networks. Smart appliances are manufactured with multiple sensors and actuators. The sensors and actuators are connected to the main control board in the smart appliance through wired and wireless connections. These connections are used to control appliances and form a network of smart appliances in a dwelling.

Appliances can be controlled using Bluetooth or Wi-Fi protocols. For example, a smart clothes washer can be controlled through Bluetooth sensors or a Wi-Fi antenna. **See Figure 16-66.** A smartphone can be used to communicate with the Bluetooth sensors in the clothes washer for door locking/unlocking, liquid-level detection, and diagnostics. The smartphone can also control the clothes washer through the internet using the Wi-Fi antenna.

Wi-Fi connections and Bluetooth technology can also form a wireless network of smart appliances in the home. Smart appliances with Bluetooth can communicate with each other using the Bluetooth Low Energy (BLE) protocol, creating a BLE mesh network. **See Figure 16-67.** This extends the range of communication for Bluetooth sensors.

For example, if a user in one part of the house wants to know the ambient temperature in a different part, such as the living room, they can use a Bluetooth device to ask for this information. The data is sent from device to device through the BLE mesh network until it reaches the installed transceiver. This is similar to the type of mesh technology used with switches and receptacles. Although smart appliance mesh networking is relatively new, major appliance manufacturers are leading the way in its application.

CODE CONNECT

NEC® Subsection 422.60(A) states that each electrical appliance shall be provided with a nameplate giving the identifying name and the rating in volts and amperes or in volts and watts.

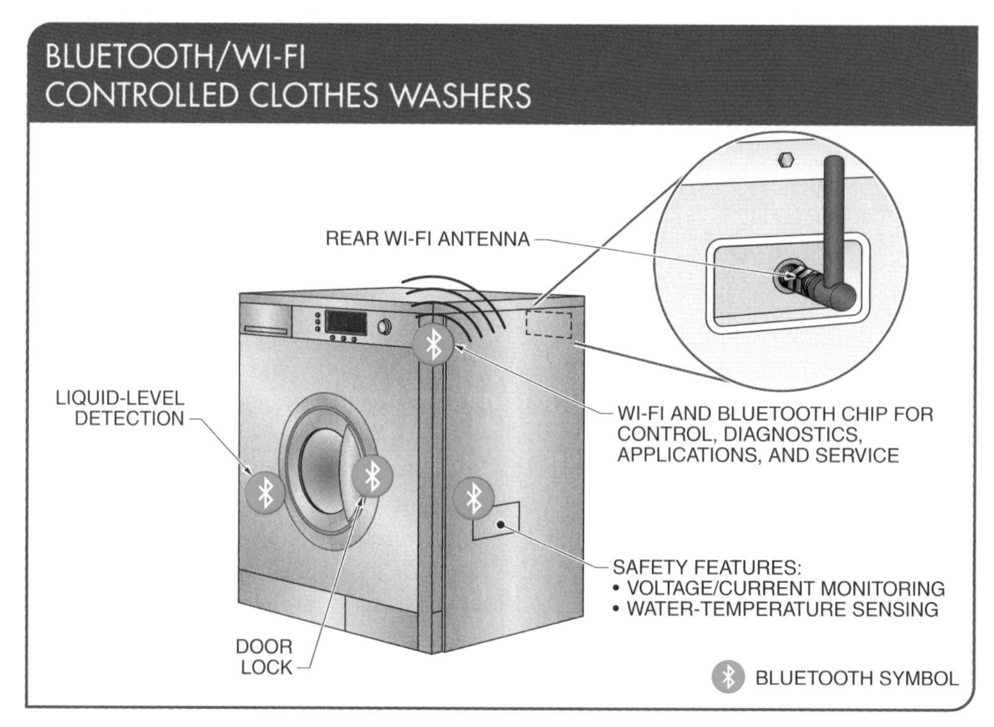

Figure 16-66. A smart clothes washer can be controlled with a smartphone through Bluetooth sensors or a Wi-Fi antenna.

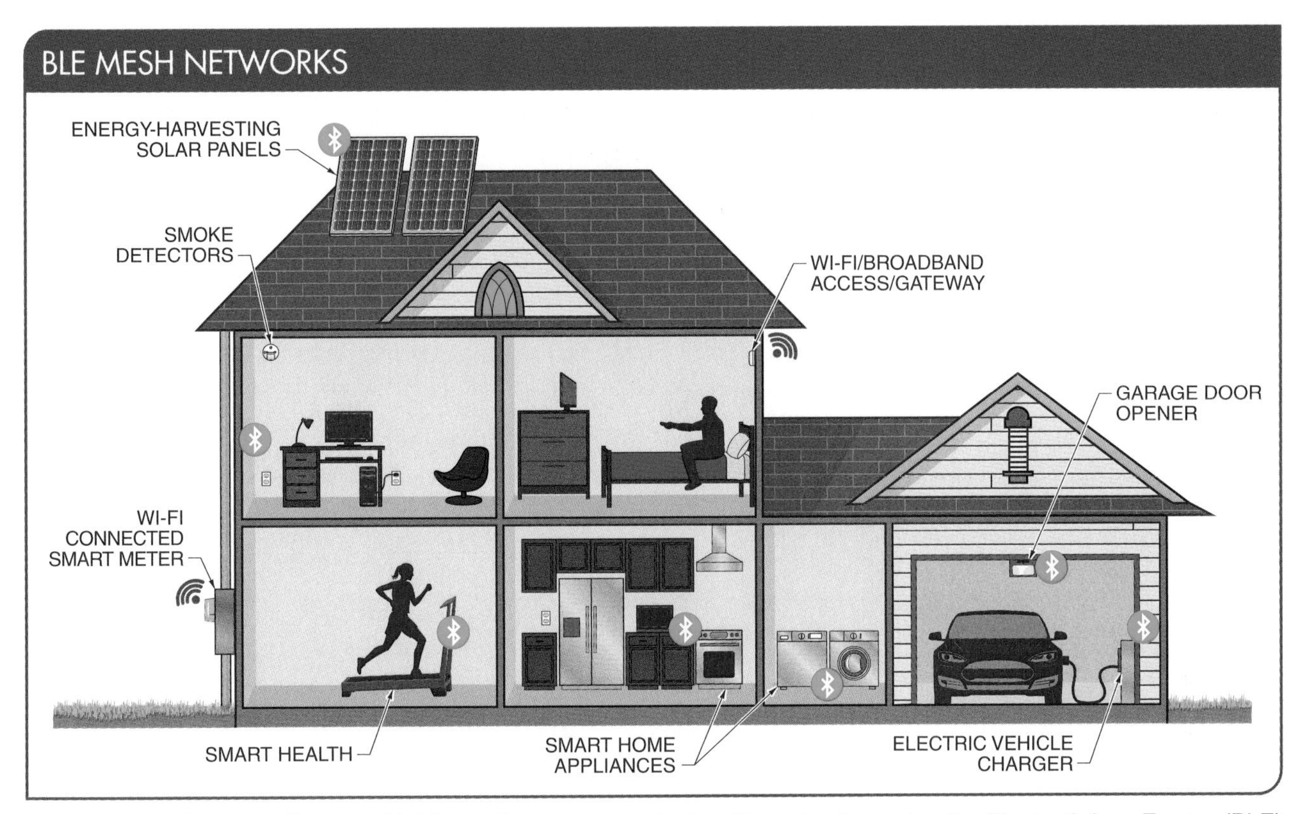

Figure 16-67. Smart appliances with Bluetooth can communicate with each other using the Bluetooth Low Energy (BLE) protocol, creating a BLE mesh network.

Energy Star® Appliances

Energy Star appliances have been identified and certified by the US Environmental Protection Agency (EPA) and US DOE as being the energy-efficient products in their class. The Energy Star logo is often included on the EnergyGuide label, which is found on appliances such as televisions, clothes washers, refrigerators, freezers, water heaters, air conditioners, furnaces, boilers, heat pumps, and pool heaters. However, although the EnergyGuide label does not appear on ranges, ovens, clothes dryers, humidifiers, and dehumidifiers, some of these appliances may be Energy Star rated. **See Figure 16-68.** The goal of the label to provide homeowners with energy usage information for a given product, pointing out both its strengths and weaknesses.

Many smart appliances are also Energy Star appliances and are integral to a smart home. According to the US DOE, Energy Star appliances can reduce home appliance energy usage and costs by as much as 10% to 50%. For example, Energy Star clothes washers and dryers use sensors to control the amount of water used for washing clothes and the amount of electricity for drying clothes to prevent energy waste. Energy Star clothes washers and dryers can also wash more clothes in a single load and dry them faster, which means less loads of laundry in less time. The long-term energy-saving benefits of Energy Star appliances outweigh their higher initial costs.

Water Heaters

Water heaters are used to provide a home with a hot water supply through the use of an electric heating element or natural gas burners. Water heating accounts for a large percentage of home energy usage. A home energy management plan should account for this energy usage. Energy savings can be maximized by installing a well-insulated hot water system set to the proper temperature and using low-flow faucets and showerheads.

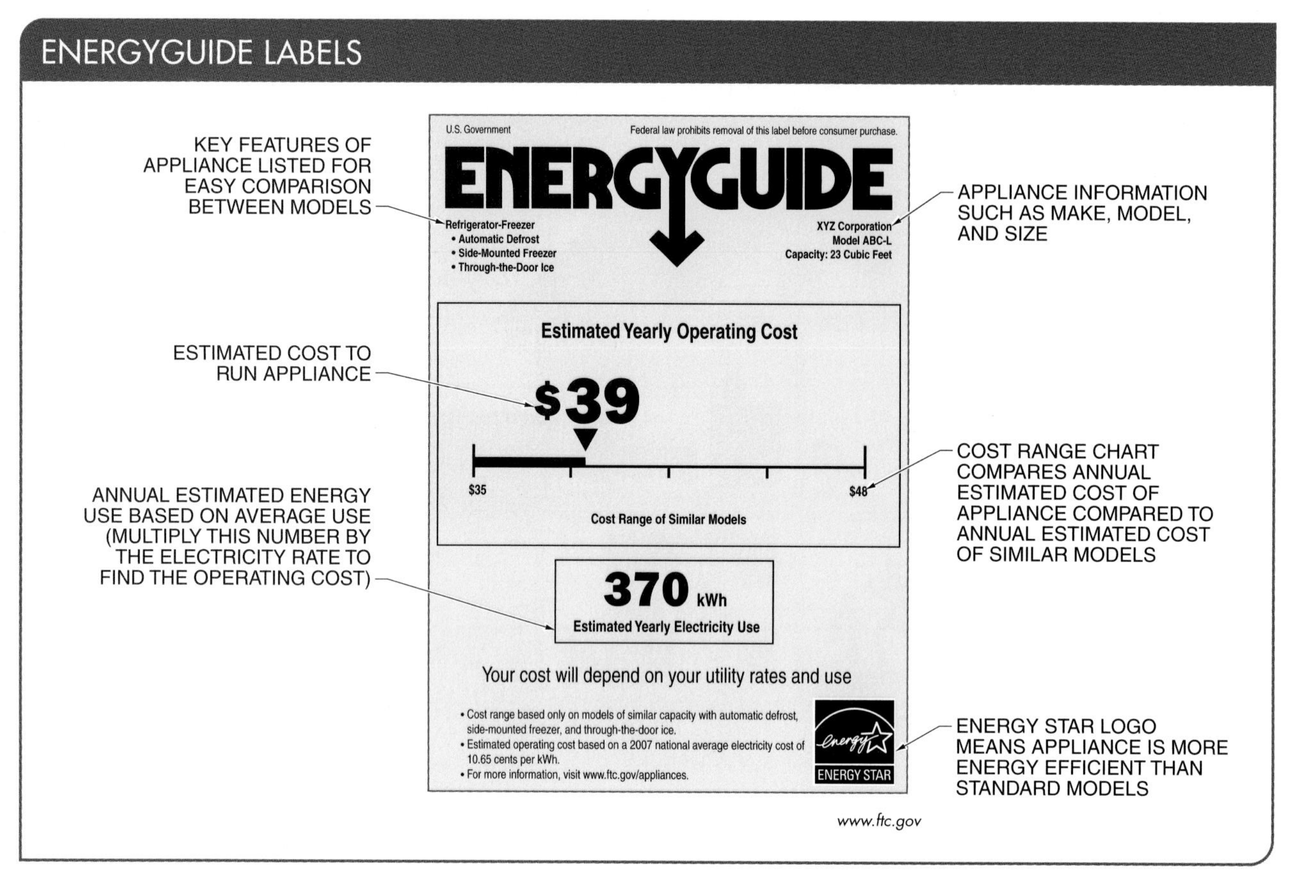

Figure 16-68. The Energy Star logo is often included on the EnergyGuide label, which is found on many home appliances.

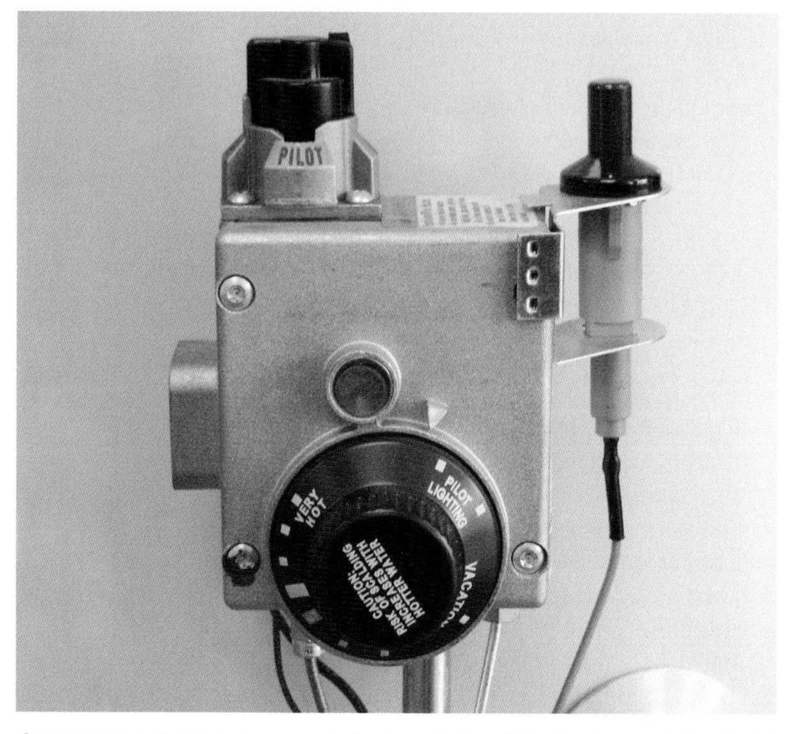

Gas water heaters have gas control valves that contain the thermostat and pilot light controls.

Residential water heaters usually have an easily accessible thermostat. Turning down the setpoint temperature of the water heater thermostat reduces the amount of electricity needed. Water heater thermostats should be set at approximately 120°F. **See Figure 16-69.** For water heaters that have two thermostats, both thermostats should be set at the same temperature. If the water heater thermostat is covered, the power should be turned off at circuit breaker before the cover is opened to set the thermostat.

Although newer water heaters are well insulated, many older models are not. Water heaters more than 15 years old may be replaced with new high-efficiency insulated water heaters for increased energy savings. Water heater insulation blankets are also available commercially from home supply stores or online vendors and are intended to be installed on electric water heaters. **See Figure 16-70.** In addition, insulating pipes with foam tubing, known as lagging, can help prevent heat loss.

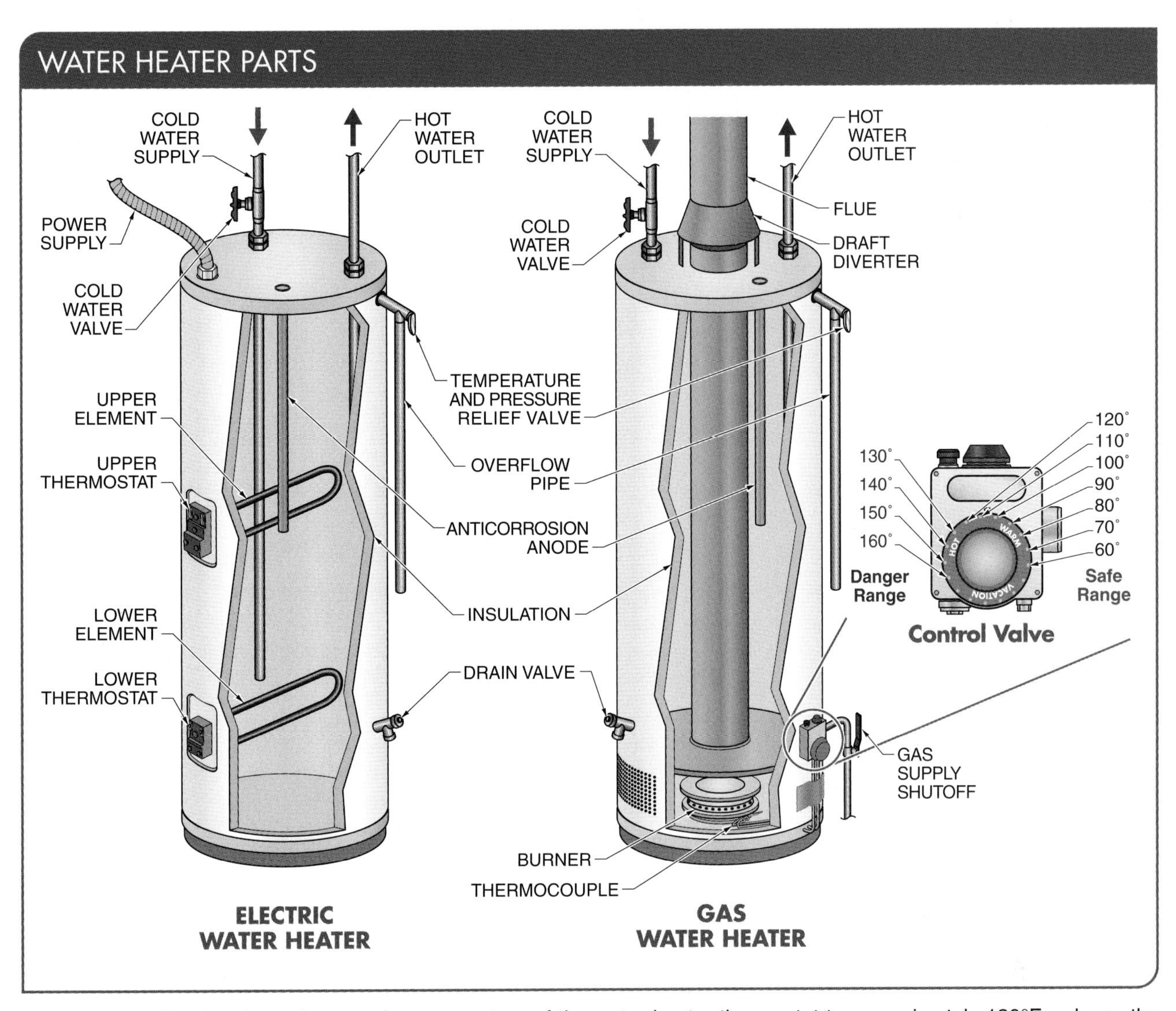

Figure 16-69. Turning down the setpoint temperature of the water heater thermostat to approximately 120°F reduces the amount of electricity needed.

Tankless On-Demand Water Heaters. A *tankless on-demand water heater* is a device that rapidly heats water by an electric heating element or a natural gas heat exchanger only when water is needed. For example, when a hot water tap is turned on, the water may be rapidly heated by a heating element, which stays on until the water stops flowing. This on-demand method of heating water saves energy because it only heats water at the time it is needed instead of keeping the water hot all of the time.

Tankless on-demand water heaters can be point-of-use water heaters and whole-house water heaters. Point-of-use tankless water heaters are smaller and are usually mounted closer to a fixture or appliance to avoid lag time and water waste. Whole-house tankless water heaters are larger and can supply multiple fixtures and appliances at a time. Tankless on-demand water heaters can heat the water through electricity or gas. **See Figure 16-71.**

CODE CONNECT

NEC® Section 422.13 states that a fixed storage-type water heater that has a capacity of 450 L (120 gal.) or less shall be considered a continuous load for the purposes of sizing branch circuits.

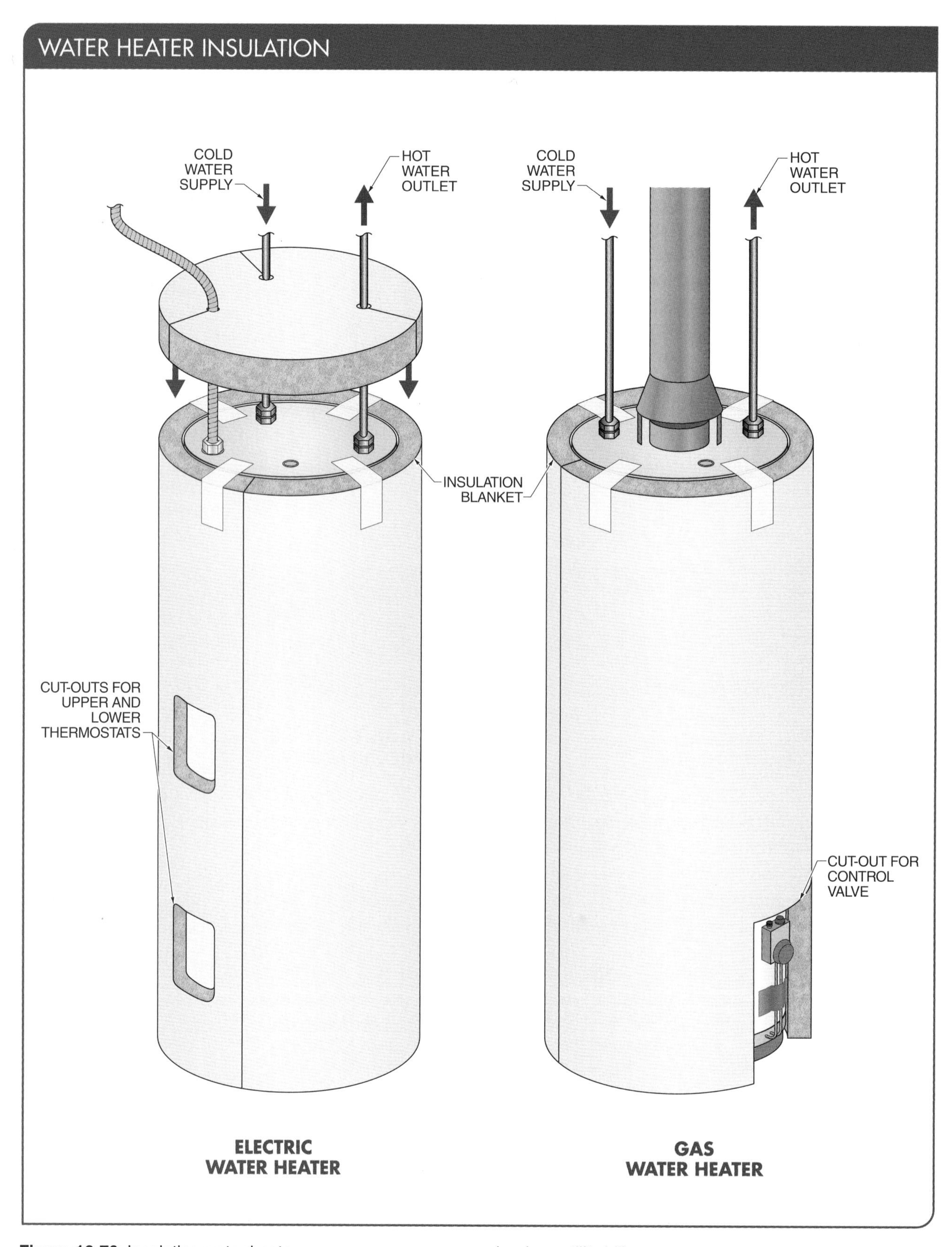

Figure 16-70. Insulating water heaters can conserve energy and reduce utility bills.

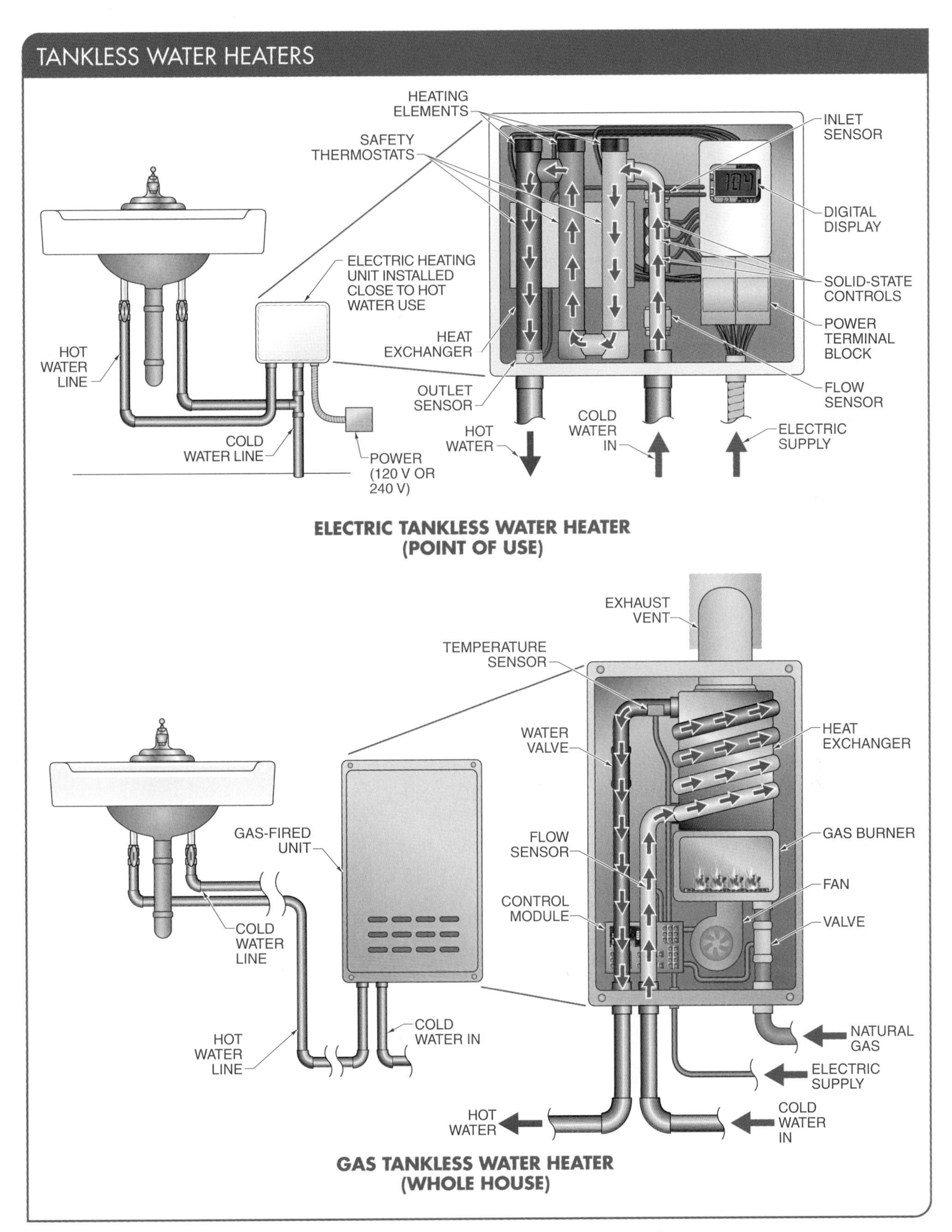

Figure 16-71. Tankless on-demand water heaters can heat the water through electricity or gas.

Intelligent Circuit Breakers

Circuit breakers serve as a distribution point for all of the electrical wiring in a house. Major appliances such as air conditioners and water heaters are served by separate circuit breakers. This makes the breaker box an ideal place from which to manage home energy use. The need for a wireless controller could be eliminated by using an intelligent circuit breaker. **See Figure 16-72.** With an intelligent circuit breaker, the water heater is controlled directly by the circuit breaker.

Most of the circuit breakers currently in service are simple electromechanical devices that sit idle most of the time. New intelligent circuit breaker technology includes features such as wireless connectivity and computing power, which allows the intelligent circuit breakers to operate more like smart meters. New intelligent circuit breaker technology being developed also includes microelectronics that are consolidated onto a single computer chip. This new technology allows circuit breakers to

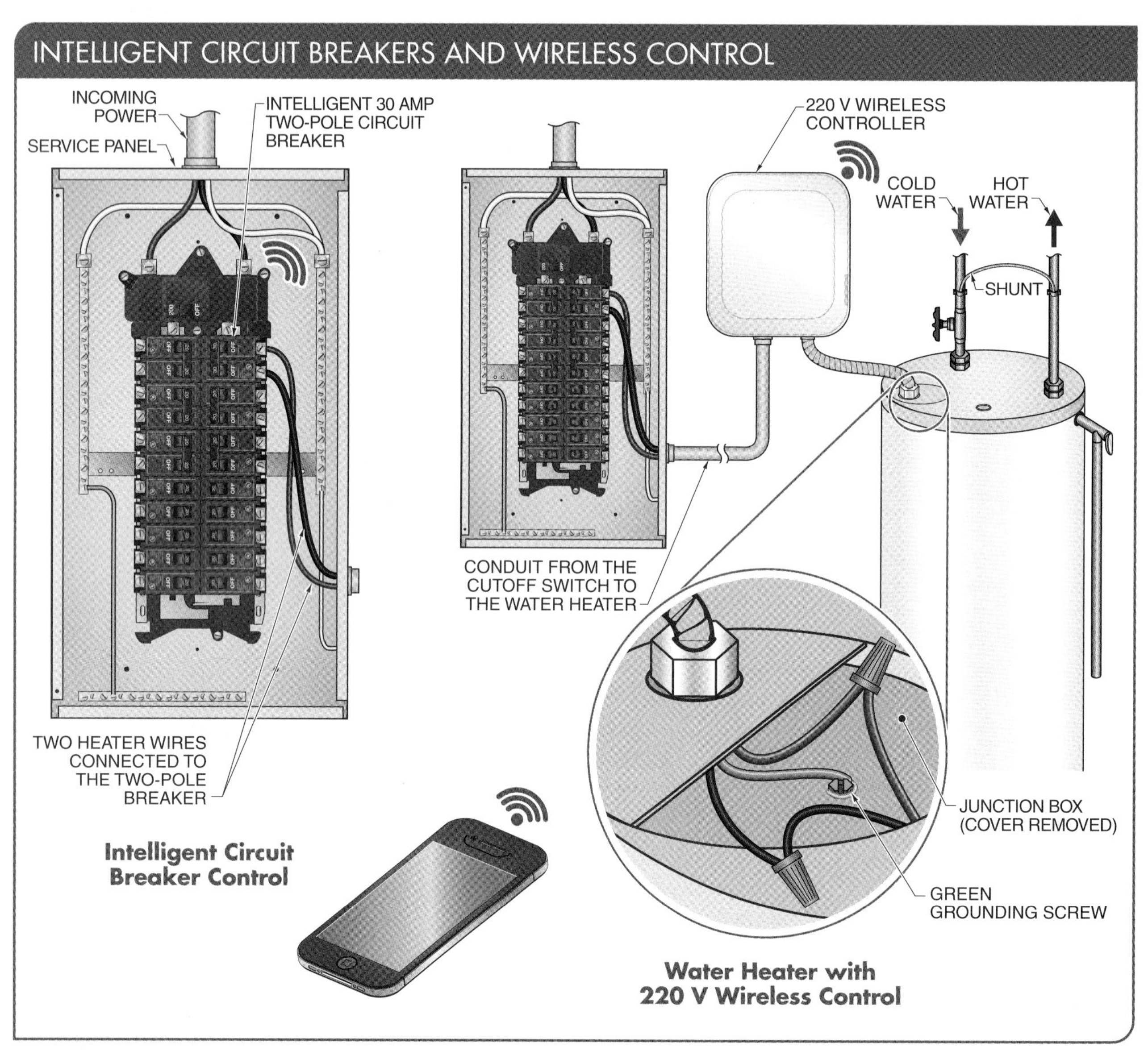

Figure 16-72. With an intelligent circuit breaker, an appliance such as a water heater is controlled directly by the circuit breaker, eliminating the need for a wireless controller.

become a powerful gateway to the smart home and smart grid.

Managing Energy. Intelligent circuit breakers can be a key component in effectively managing and maximizing a home's energy usage in real time. Cloud-based connectivity is incorporated into an intelligent circuit breaker, enabling advanced load monitoring and remote communication for an energy management system. Intelligent circuit breakers use their advanced technology to measure real-time energy consumption and work with the home automation system to determine which circuits to control in the home in the most efficient way possible.

Monitoring Capabilities. Intelligent circuit breakers with web-based applications can be accessed locally or via smartphones, tablets, or computers to display the power consumption of appliances. In this way, an intelligent circuit breaker provides an economic means of increasing a home's monitoring capabilities. By measuring voltage, current, power, energy use, and power quality, a homeowner can identify the cause of increased power consumption and reduce usage of certain appliances or equipment.

Residential Microgrids

A *residential microgrid* is an electrical network that has energy generation capability, storage capacity, and the ability for intelligent control that allows a dwelling to become a self-contained island completely independent of the main grid. A residential microgrid generally operates while connected to the main power grid, but it can break off and operate on its own using a local energy source in times of crisis, such as during storms or during power outages. Residential microgrids help home energy management by storing off-grid, alternative sources of energy and using it during high peak hours.

Self-Contained Energy Storage Systems. A new trend in residential microgrids is the use of self-contained energy storage systems without the implementation of any alternative energy generation, such as solar or wind power. Batteries are typically used as self-contained energy storage systems in a residential microgrid. **See Figure 16-73.**

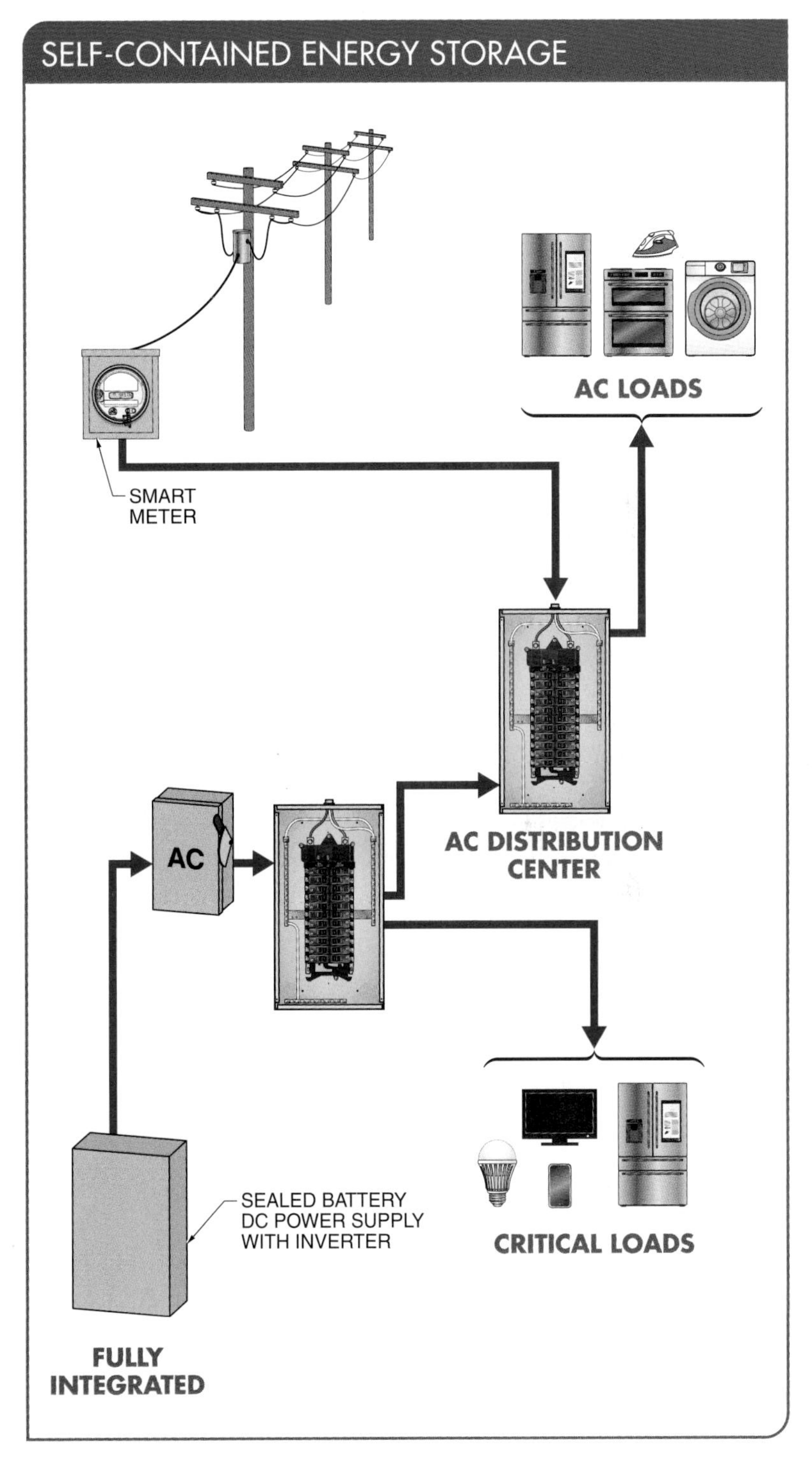

Figure 16-73. Batteries are typically used as self-contained energy storage systems in a residential microgrid, even without the implementation of alternative energy generation.

There are two major reasons for the use of self-contained energy storage systems. One reason is that these systems can serve as battery backups where outages are common. The second reason is that these systems can reduce the cost of electricity during peak demand hours. Batteries are charged during the late-night hours and early morning when the electricity is much cheaper than peak hours during the day. A self-contained energy storage system can supplement power or allow a dwelling to go completely off-grid during the day. The cheap energy stored from the previous evening can be used during the expensive peak hours the next day.

Installing Energy Storage Systems. Self-contained energy systems can be mounted on a wall inside a dwelling or in a garage. These wall-mounted units include batteries, an inverter, and all of the electronic controls to operate remotely. **See Figure 16-74.** Since all of the major components are included in the housing of the self-contained energy storage device, customers no longer need an additional box of electronics on the wall next to the battery box. The installation is simple, fast, cost effective, and more aesthetic.

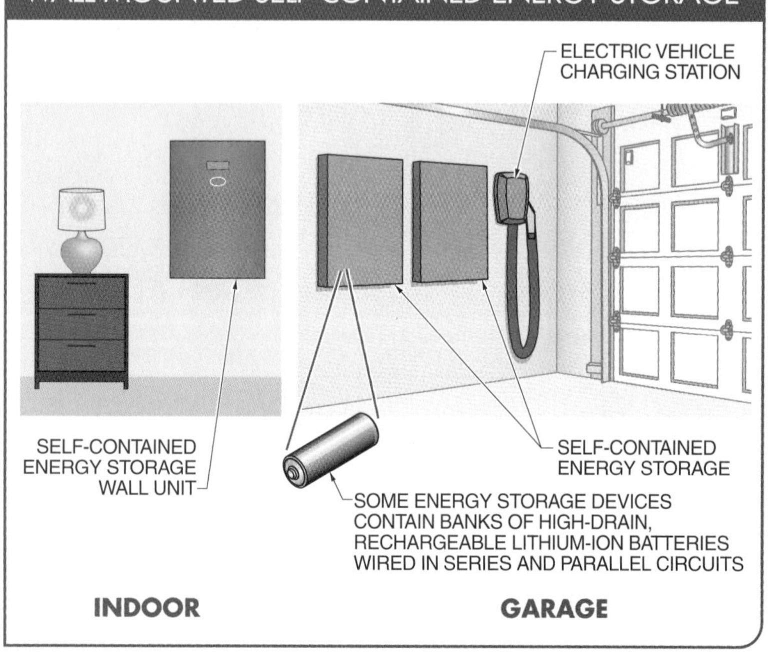

Figure 16-74. Self-contained energy systems can be mounted on a wall inside a dwelling or in a garage.

Wall-mounted units can be installed indoors or outdoors because they are able to operate within a wide range of temperatures, from –4°F to 122°F. Some units have built-in heating or cooling capacity for the high and low ends of their temperature ranges to improve battery life span. However, the units should not be installed in locations exposed to direct sunlight or subject to sustained low temperatures. Also, the units should not be installed in locations subject to flooding or near water sources, such as sprinkler systems. To maintain proper ventilation, the units should be kept clear of dense brush or accumulated snow.

Sizing Energy Storage Systems. When a home is using only battery storage for back-up purposes, it is important to decide what critical loads will need to be powered during outages and for how long. For example, homes in the US use about 30 kWh per day, although this can vary widely depending on location, climate, and the size of a household. A typical storage capacity is sized to provide electricity to critical loads for about one or two days in case of a power outage. It will be up to the homeowner to determine the loads that they consider to be critical or essential versus those that they consider to be conveniences.

Smart Meter Energy Measurement by Disaggregation. Smart meters can play an even bigger role in home energy management through energy disaggregation. *Energy disaggregation technology* is a new smart meter technology that can distinguish between various electrical loads in a consumer's home by monitoring the loads over time.

A smart meter that uses energy disaggregation technology can tell the difference between an air conditioner, refrigerator, and clothes dryer. Each appliance generates a distinct electronic waveform or fingerprint that identifies the appliance and how much power it draws. This information can be presented in the home by being visually displayed, sent to a HEMS, or sent back to the utility using a Wi-Fi connection. Utilities will be able to identify what loads are creating a larger demand on their side of the meter, allowing them to more effectively implement DR programs for heating and cooling and, eventually, things such as electric vehicle charging.

Grid Monitoring Subscriptions. For those consumers that do not have or want a smart meter, there is an alternative to obtaining this information. Some companies are providing a subscription service to gather information from the local power company if it cannot be obtained from a smart meter. The service plan is like a cell phone plan, where consumers have a choice between paying up front for equipment or a monthly fee for ongoing services.

Community Microgrids

A *community microgrid* is a group of electrical sources and loads that can be connected to the utility grid or operate by itself when the need arises. A community microgrid can consist of a few dwellings or a much larger community. A community microgrid uses systems of small battery-based community energy storage (CES) units to meet the energy storage needs at the edge of the grid, where it previously did not exist. **See Figure 16-75.**

The idea behind community microgrids is to use excess capacity stored in home systems and CES units to provide capacity to the grid and stability during times of crisis. Community microgrids help home energy management by collectively controlling and directing energy when the need arises for the benefit of the individual dwellings.

For example, power outages from hurricanes can impact large numbers of customers for weeks and even months. However, areas powered by community microgrids can maintain power even in such crises. To participate in a community microgrid storage system, the utility will contract with each individual homeowner on the amount of excess energy they may have available.

Security and Privacy Risks of Home Energy Management

Having the ability to monitor and control home energy consumption through a HEMS creates benefits as well as risks for the consumer if the information produced, or the control of equipment, falls into the wrong hands. These risks exist with every device that can connect to the internet. Manufacturers have recognized this problem and are developing methods to reduce these risks. Before a smart device or appliance for home energy management is purchased, a homeowner should understand any possible security concerns it may pose once connected to the internet.

As manufacturers of major appliances move appliance control to the internet, homeowners gain the ability to use their smartphones to activate appliances from a remote location. However, this benefit must be balanced against the risks associated with internet control. Consumers expect there to be a firewall that prevents outside access to the information but still allows communication with the utility. Consumers also expect that utilities and government agencies do not monitor how they use power without their permission.

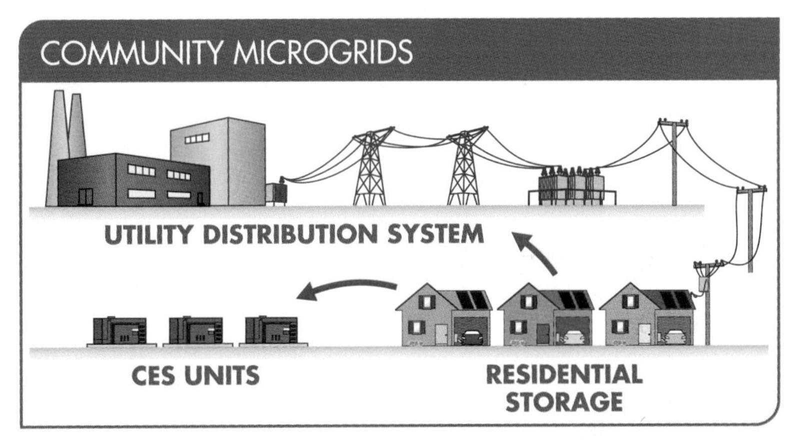

Figure 16-75. A community microgrid uses systems of small battery-based community energy storage (CES) units to meet the energy storage needs at the edge of the grid, where it previously did not exist.

SECTION 16.5 CHECKPOINT

1. What are the fundamental components of a HEMS?
2. What is net metering?
3. What smart appliances are playing a part in the energy conservation of the smart grid and home energy management?
4. How are intelligent circuit breakers used with web-based applications?
5. What is a residential microgrid?

Chapter Activities

16 Home Energy Management and Smart Home Applications

Name ______________________________ Date ______________

Lamp Types

Identify the lamp types shown.

______________ **1.** CFL

______________ **2.** Fluorescent

______________ **3.** LED

______________ **4.** Halogen

______________ **5.** Incandescent

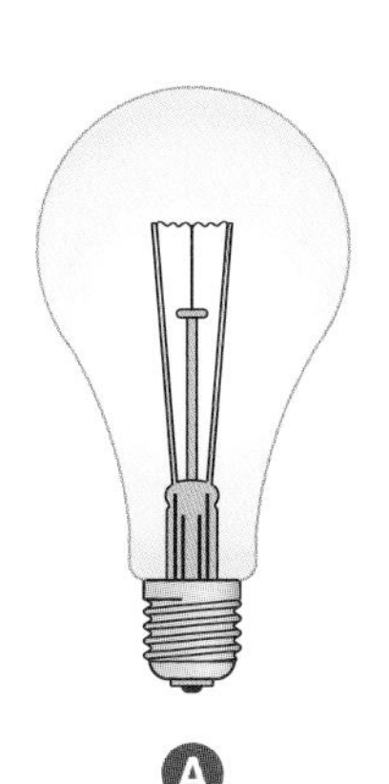

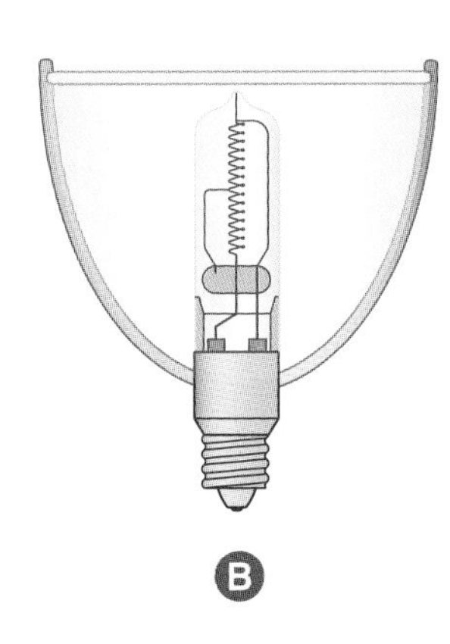

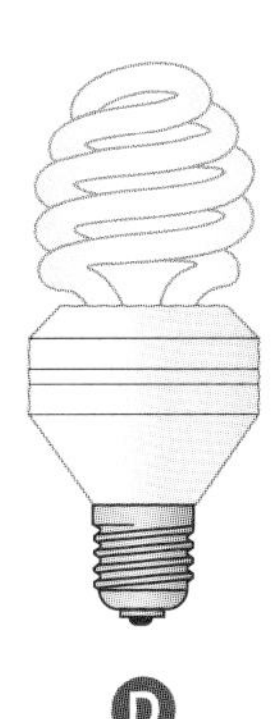

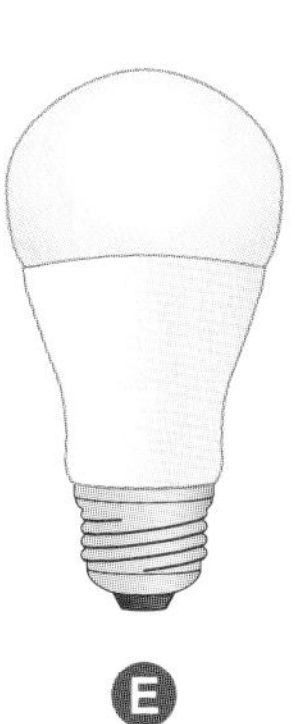

Dimmer Styles

Identify the dimmer styles shown.

_____________________ **1.** Slide dimmer

_____________________ **2.** Rotary dimmer

_____________________ **3.** Rocker dimmer

_____________________ **4.** Toggle dimmer

_____________________ **5.** Digital/touch dimmer

A

B

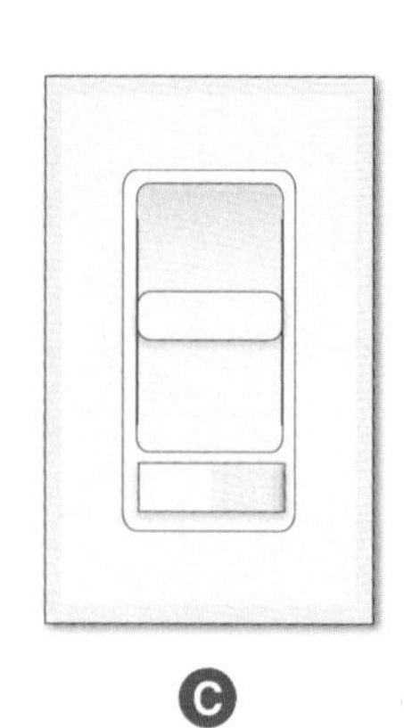

C

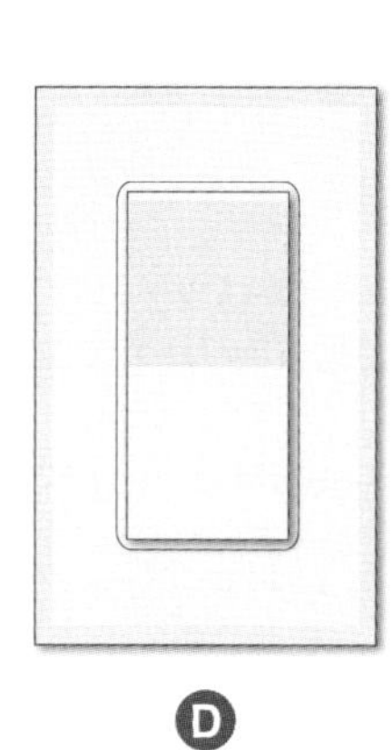

D

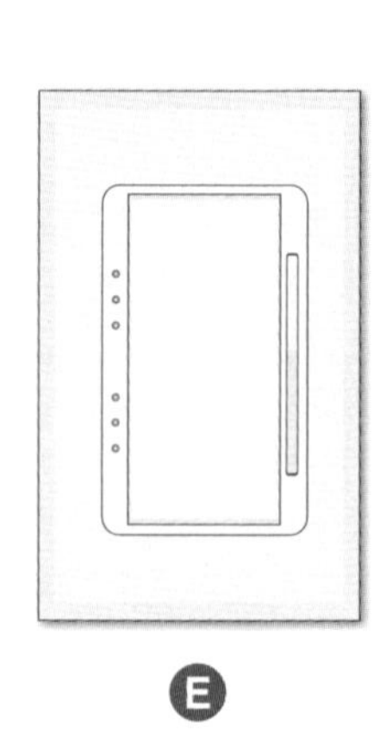

E

Solar Array Mounting

Identify the types of solar array mounting.

_____________________ **1.** Directional tracking

_____________________ **2.** Roof

_____________________ **3.** Freestanding

A

B

C

Ceiling Fan and Light Switch Connections

1. Draw lines from the neutral, hot, and ground wires to the wire nuts and ground screws to demonstrate how to wire a downrod ceiling fan.

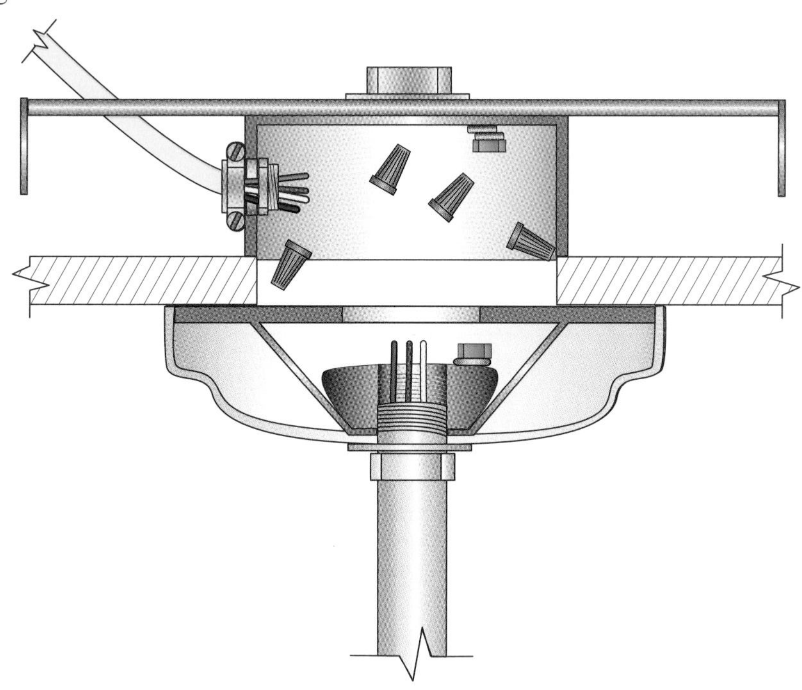

2. Draw lines from the neutral, hot, and ground wires to the wire nuts and screws terminals to demonstrate how to wire a fan switch and a light switch to a ceiling fan.

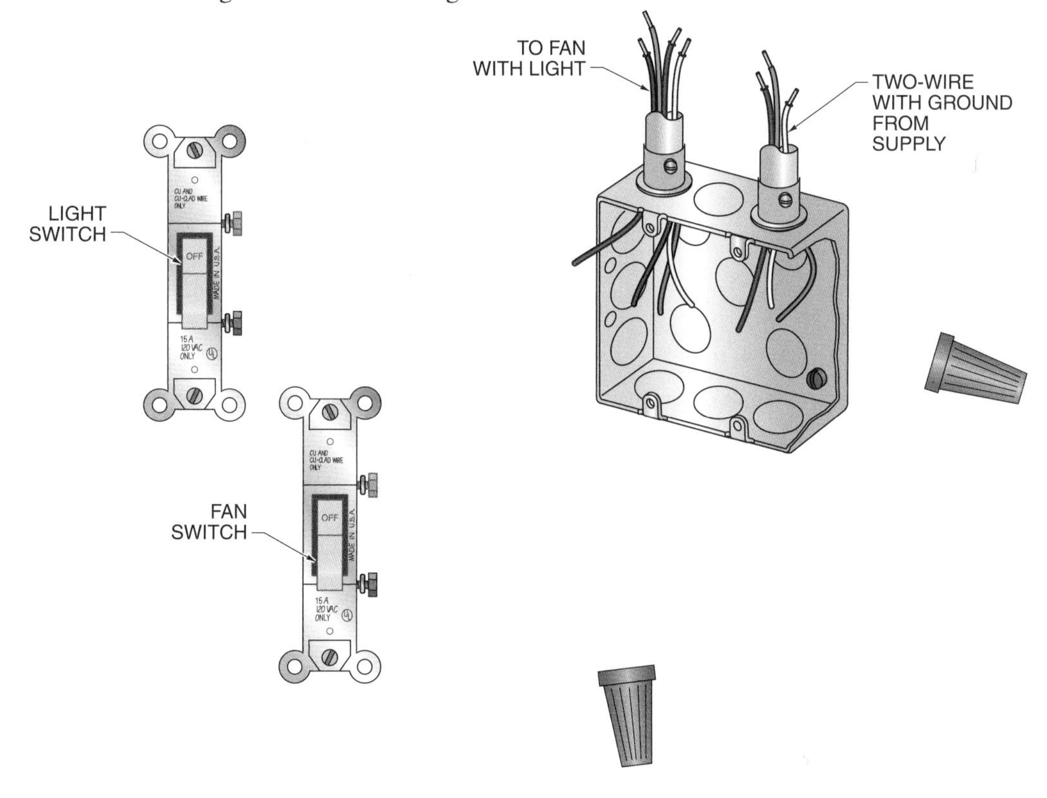

Ceiling Fan Remote Control Wiring

1. Draw lines from the neutral, hot, and ground wires to the wire nuts and screws terminals to demonstrate how to wire a ceiling fan remote control.

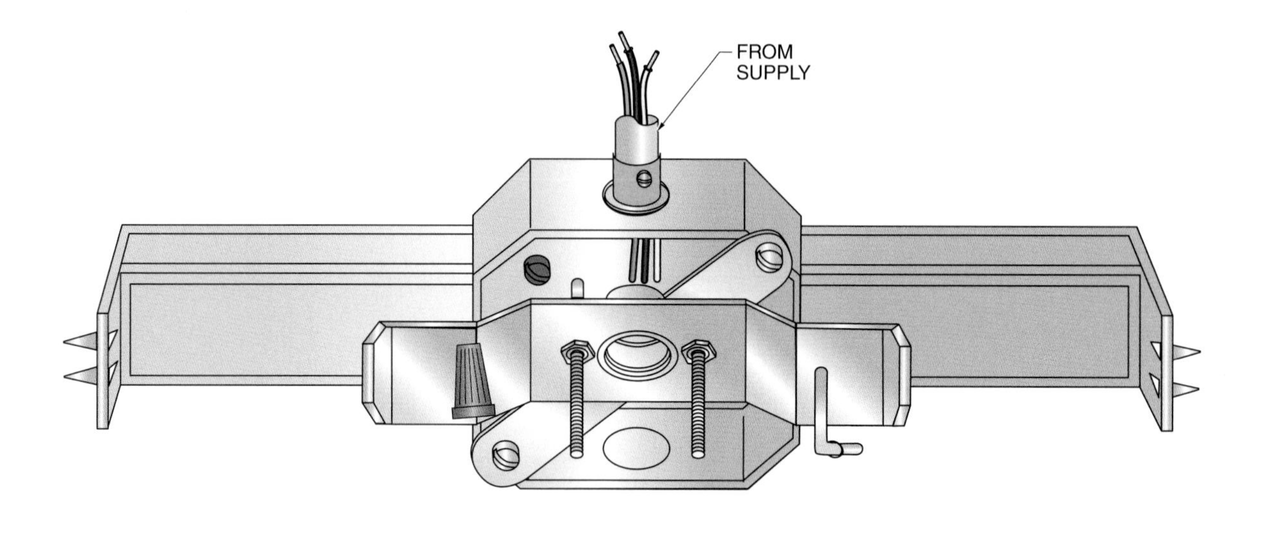

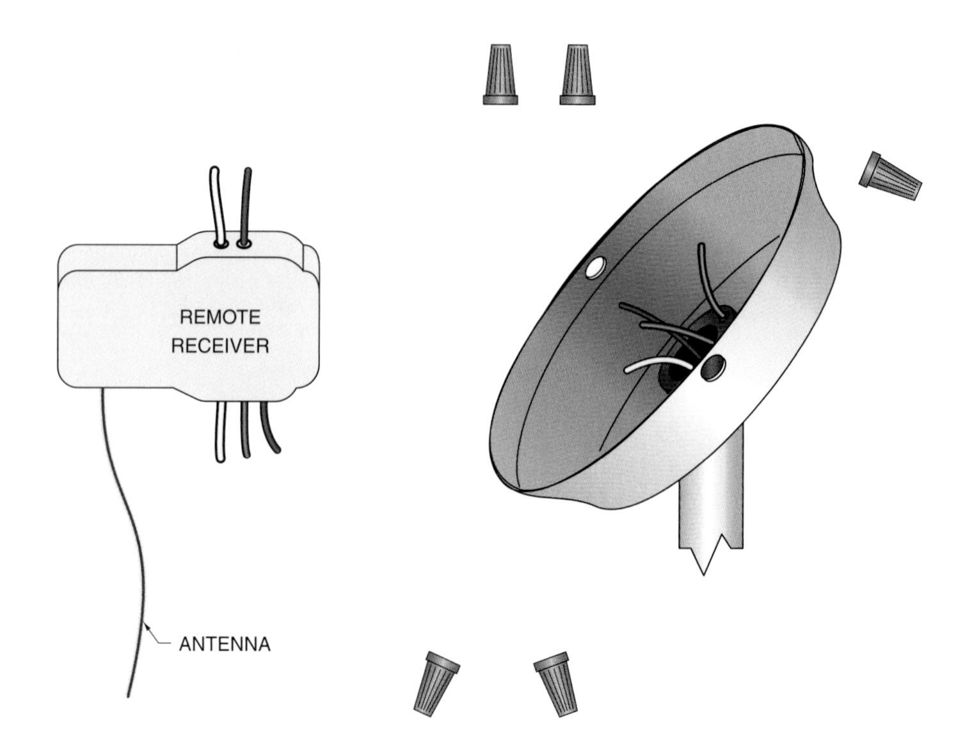

17 Lifestyle Applications of the Smart Home

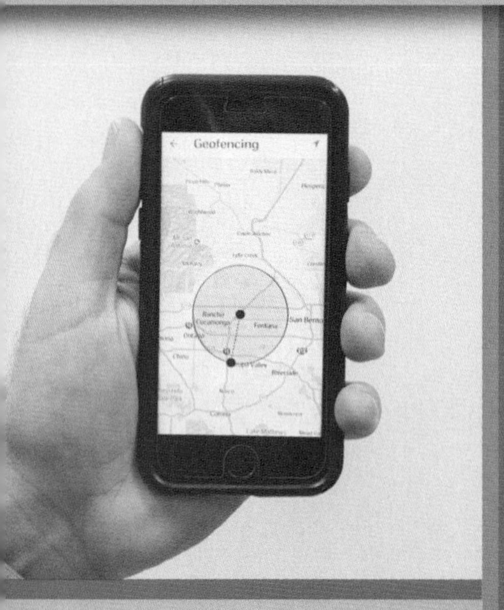

SECTION 17.1—SMART HOME LIFESTYLES
- Describe the different individual lifestyles.
- Describe the most common rooms to have smart technology.

SECTION 17.2—SMART BEDROOMS
- Describe how lighting can be used to create different scenes.
- Explain how smart devices can be used for lifestyle applications in a smart bedroom.
- Describe the monitoring devices that may be used in a smart nursery.

SECTION 17.3—SMART BATHROOMS
- Describe smart mirrors.
- Explain how a touchless sink operates.
- Describe the lifestyle applications of showers and tubs.

SECTION 17.4—SMART KITCHENS
- Explain general lighting and task lighting.
- Describe how the islands in a smart kitchen provide space and features for standing and sitting activities.
- Describe the features of a smart refrigerator.

SECTION 17.5—SMART FAMILY ROOMS AND LIVING ROOMS
- Describe how lighting can be used to create scenes in a smart family/living room.
- Explain how smart TVs can be used in a smart family/living room.
- Describe how a surround-sound experience is created in a smart family/living room.

SECTION 17.6—SMART GARAGES
- Describe electric vehicle supply equipment (EVSE).
- Explain electromagnetic inductive charging.

SECTION 17.7—OTHER LIFESTYLE APPLICATIONS
- Describe how robots can be used inside and outside of the smart home.
- Describe wearable technology (wearables).
- Explain power over ethernet (PoE).

Learner Resources
ATPeResources.com/QuickLinks
Access Code: 791712

Smart home technology can be used to keep dwellings safe and comfortable, to manage energy consumption, and to positively impact the lifestyle of the individuals who occupy a home. The smart home technology that is needed to fit the lifestyle of the individuals depends on their relationship and familial status, their age, and whether they have a disability. Smart home technology can also be used to care for family pets and other aspects of the home.

SECTION 17.1—SMART HOME LIFESTYLES

An individual's lifestyle, which may be active/functioning, in transition, or needing assistance, determines the types of smart devices selected for use in a smart home. Age, health, physical condition, and personal preferences, such as pet ownership, are factors that affect the lifestyles of individuals and families. **See Figure 17-1.** For example, an individual that fits the transition lifestyle may be someone rehabilitating from an injury, living with a chronic illness, or dealing with issues related to the aging process. Smart technology offers possible solutions to individuals in transition or needing assistance so that they can function at a level that allows them to remain in their home.

Types of Control

The type of control, as well as its cost and convenience, should be taken into consideration when choosing the smart home technology that best applies to an individual's lifestyle. In many cases, the type of control for the smart home technology will be the determining factor in the overall cost. Connected homes with a limited number of connected devices, smart homes, and intelligent systems represent the three general types of control.

A connected home with a limited number of connected devices is most often the least expensive choice. Typically, the connected devices in a connected home are easily controlled through a smartphone. For example, three apps can be installed on a smartphone to control three different connected devices. One app is used to control the HVAC through a smart thermostat, another app connects to a smart doorbell with a video camera, and the third app controls a few lights when an individual returns home. The rest of the wiring in the home remains the same.

If a smart home is the best fit for an individual's lifestyle, the costs will increase. A smart home hub (controller) may be added to allow an increasing number of smart devices to interconnect and communicate with each other. For example, a smart thermostat can communicate with ceiling fans and motorized window shades through the smart home hub to provide specific heating or cooling sequences to increase comfort. Though this type of control is often more expensive, it can also be more convenient.

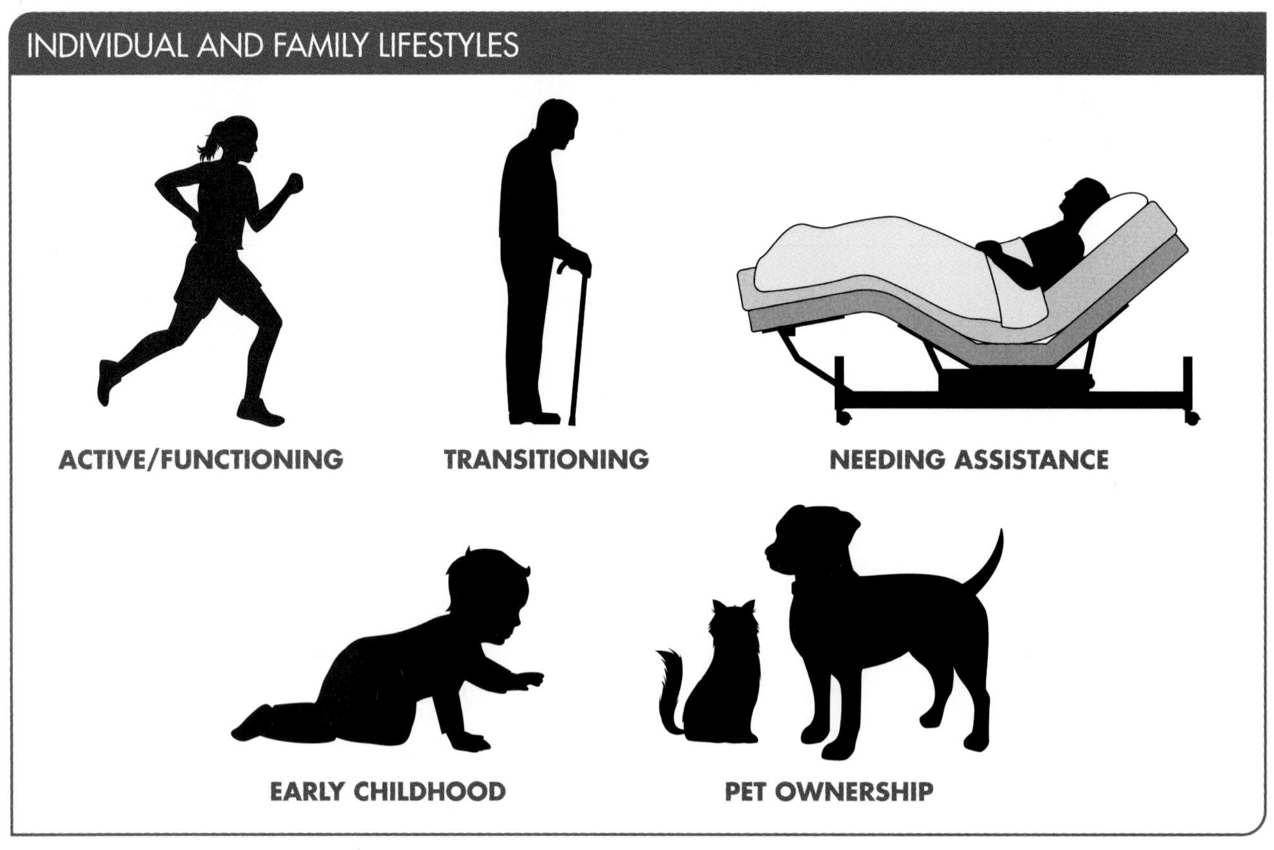

Figure 17-1. Age, health, physical condition, and personal preferences, such as pet ownership, are factors that affect the lifestyles of individuals and families.

The choice of an intelligent system for the control of smart home technology will have the highest costs. An intelligent system could potentially control almost everything that can be automated in a home. An intelligent home uses artificial intelligence where appropriate to learn patterns of behavior and provide automated responses to common needs. Typical examples of this level of control include coordinating events through geofencing when entering and leaving the home, setting lighting patterns, and determining HVAC settings. This type of system may require professional installation. Walking through each room can help determine what application is essential and how to control it.

Smart Rooms

The four most used rooms in a dwelling are the bedroom, bathroom, kitchen, and living/family room. Most of the lifestyle applications of a smart home will occur in these rooms. Typically, these rooms are also where most of the money for a smart home is invested.

As per NEC® code, GFCIs, AFCIs, smoke detectors, and security devices should already be installed in the bedrooms, bathrooms, kitchen, and living/family room of a dwelling. With smart home infrastructure in place, the necessary wiring to support smart devices keeps the dwelling safe, secure, and energy efficient. These four rooms can be further developed to fit an individual's lifestyle by adding unique smart devices and coordinating between them. Special consideration must be given to the addition of devices that enhance the capabilities of those individuals who have disabilities or wish to age in their own homes.

SECTION 17.1 CHECKPOINT

1. What are the different lifestyles and the factors that affect the lifestyles of individuals and families?
2. What four rooms are used the most in a dwelling?

SECTION 17.2—SMART BEDROOMS

Whether an individual is considered as functioning, in transition, or needing assistance, the bedroom is typically where the individual's day begins and ends. Although there are many smart devices that can enhance a smart room, space and accessibility remain important considerations. For example, space for a computer or television may need to be allocated. These devices will also require receptacles and internet access. For individuals with limited abilities, automatic doors activated by motion sensors or a remote device can be installed for easy access to the bathroom or other areas of the home. **See Figure 17-2.**

Comfort

Smart devices can be added to a bedroom to maintain the desired level of comfort. Depending upon the configuration of the HVAC system, a smart thermostat can be installed in the smart bedroom. Besides controlling the bedroom temperature, the smart thermostat may be programmed to work with a ceiling fan and control motorized window shades at certain times of the day. An air purifier and humidifier may also be added to the smart bedroom for additional comfort.

Lighting Scenes

With the installation of manual and wireless lighting systems, dimming technologies, and the use of smart plugs and smart bulbs, homeowners can create "scenes" in different rooms using smart technology. Scenes can be used for practical applications such as task lighting or for more cosmetic applications such as setting a mood for the room through different colors of light. Scenes that best apply to an individual's lifestyle can be programmed into the bedroom lighting system.

For example, a lighting scene for a smart bedroom can be programmed to be as simple as a light that turns on or off the when the individual enters or leaves the bedroom. Another scene, such as the activation of a night-light, can be programmed for when the bedroom light is turned off. Scenes can also be useful for security purposes when the homeowner is away. For example, a scene can be programmed to randomly turn on and off the bedroom light along with other lights in the dwelling to make it appear someone is home while away on vacation.

Scenes can be created using standard white lighting or by incorporating color. Innovations in smart LED lighting technology allow 16 million color combinations in a single light. These types of LED lights can be used in a smart bedroom to create "sleeping scenes" and "awakening scenes" that simulate sunset and sunrise. These scenes can be programmed to match an individual's sleep schedule.

Special Alarms

Depending on an individual's lifestyle needs, special alarms can be used in a smart bedroom. For example, lighting scenes that use series of colors to indicate emergencies can be developed for those individuals who may be hearing impaired. Blinking lights can be used to indicate a CO leak, fire alarm, or security alarm, which would also appear as alerts on a smartphone. A vibrating pad beneath a sleeping pillow can also be automated to activate in conjunction with these alarms. **See Figure 17-3.** Smart bulbs can also illuminate certain colors within their color palette to signify specific events, such as differentiating between the arrival of an unknown visitor versus that of a caregiver or family members.

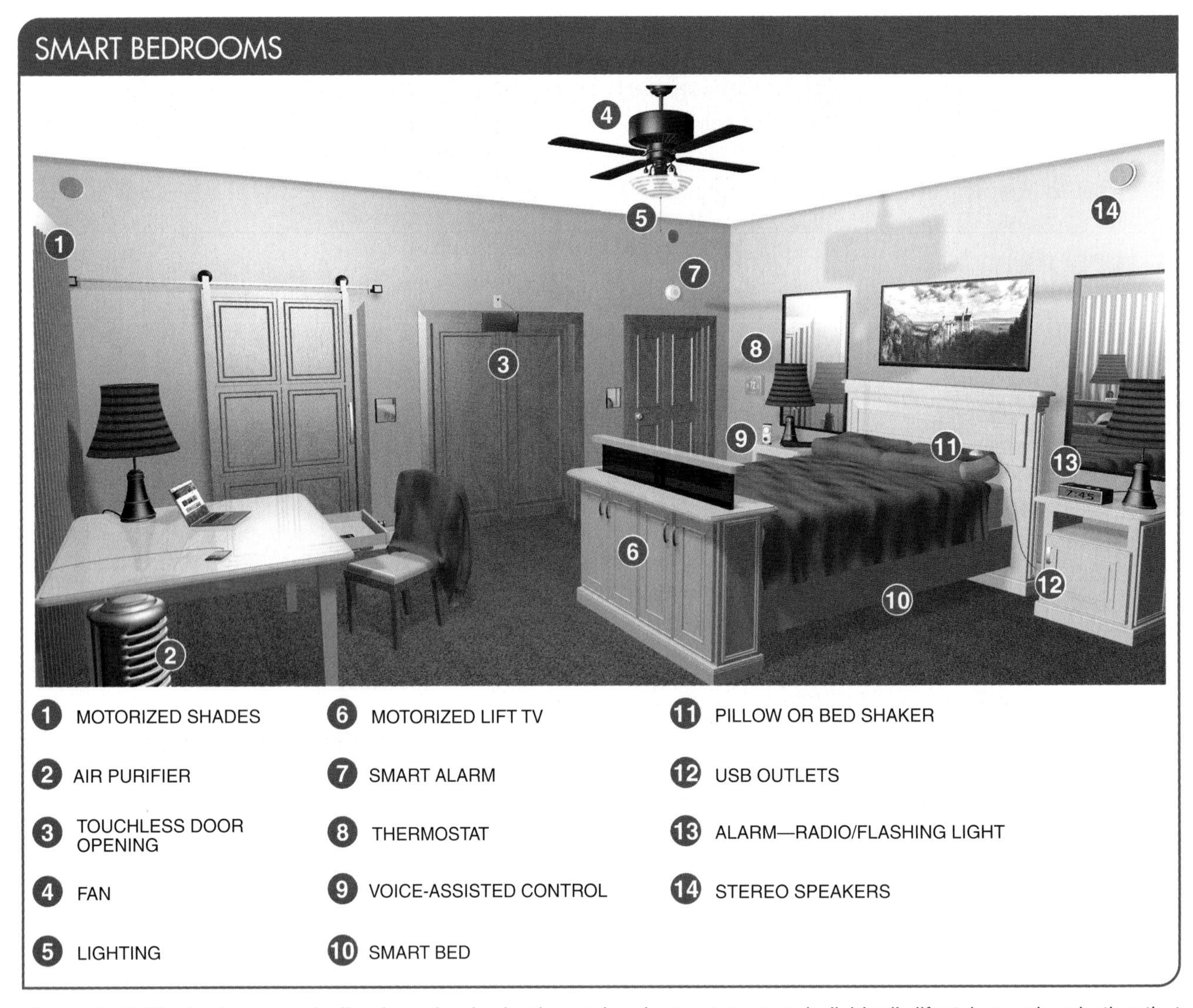

Figure 17-2. The bedroom, typically where the day begins and ends, must meet an individual's lifestyle needs, whether that is functioning, in transition, or needing assistance.

Safety

The National Electrical Code® (NEC®) requires that a bedroom be protected by an AFCI receptacle. This is required for all new construction, but it is also a good upgrade for a legacy home. Other receptacles can be replaced by charging receptacles with USB ports to eliminate bulky charger plugs where needed.

Daily Lifestyle Applications

Once the smart home technologies for the bedroom are established, an individual can decide how to use these devices to positively impact their lifestyle. For example, the smart thermostat can be programmed to begin warming up the bedroom and/or bathroom 10 min before the individual awakens. The lights can be programmed to gradually lighten the room for an awakening lighting scene. A program to begin playing soft music can also be set. The smart bedroom alarm clock can communicate with the coffee pot in the kitchen so that the coffee is brewed and ready when the individual enters the kitchen. A smart TV can be programmed to turn on in the bedroom, bathroom, or kitchen so that the individual can begin their daily routine.

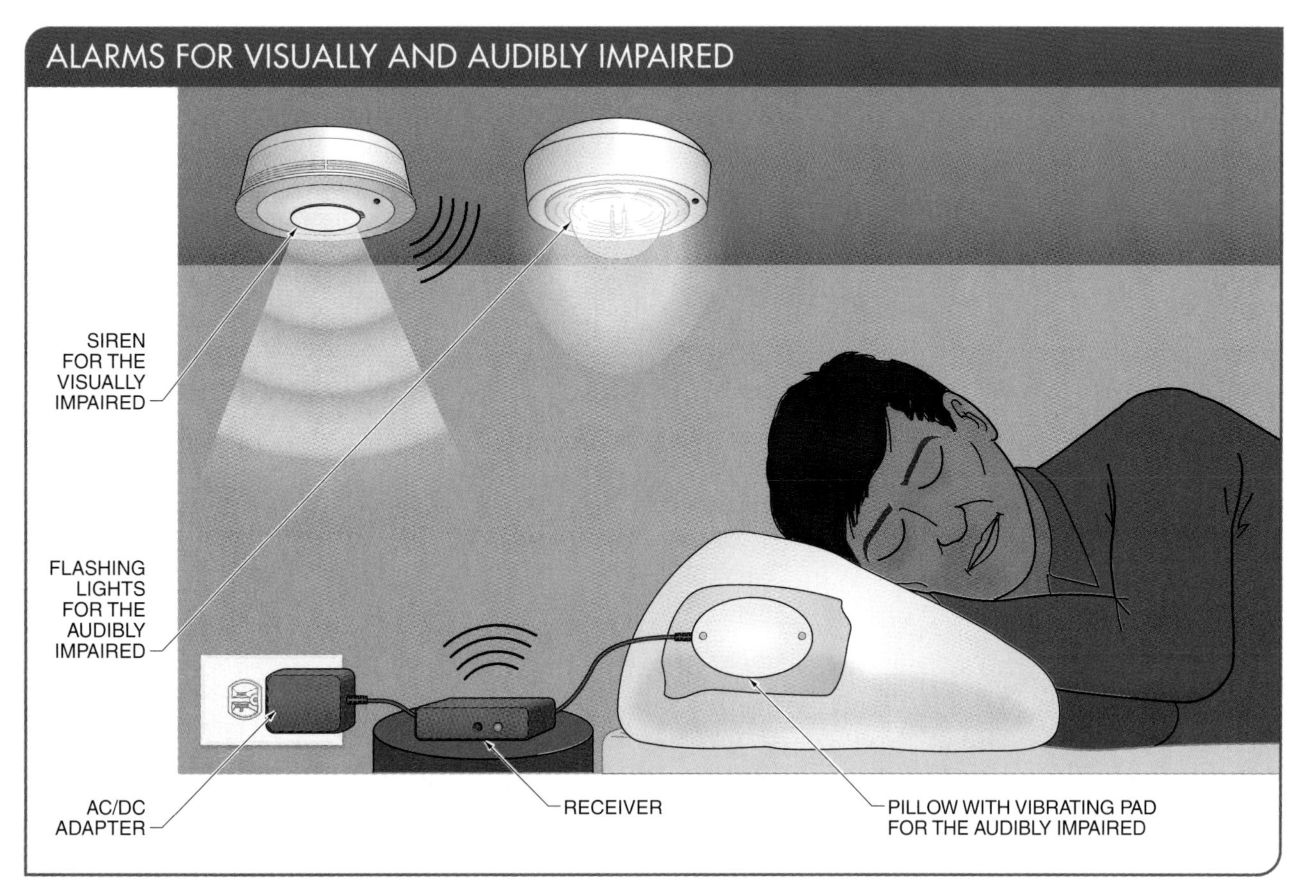

Figure 17-3. A vibrating pad beneath a sleeping pillow can also be automated to activate in conjunction with alarms.

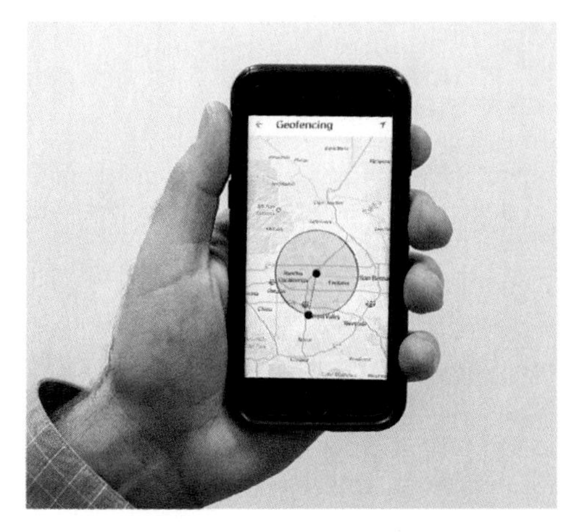

Geofencing can be used to trigger an alert on a smartphone when another device enters predetermined boundaries.

The assistance provided by smart bedrooms can help individuals whose lifestyles require that they receive specialized care throughout the day. For example, voice messages can be recorded by a virtual assistant to provide reminders to those individuals with memory issues when they awake in the morning. Caregivers can record multiple messages, such as when to take pills, have meals, or prepare for sleep, for broadcast at different times of the day. These messages can be linked to an audio or intercom system to provide this service throughout the home.

Children's Nurseries

While many traditional items such as cribs and rocking chairs are still popular accessories, smart devices are increasingly included in children's nurseries and some have become replacements for traditional household items. One direct replacement is the digital thermometer, which is used to monitor a child's temperature. Other smart devices commonly used in children's nurseries include video monitors and diaper sensors. **See Figure 17-4.**

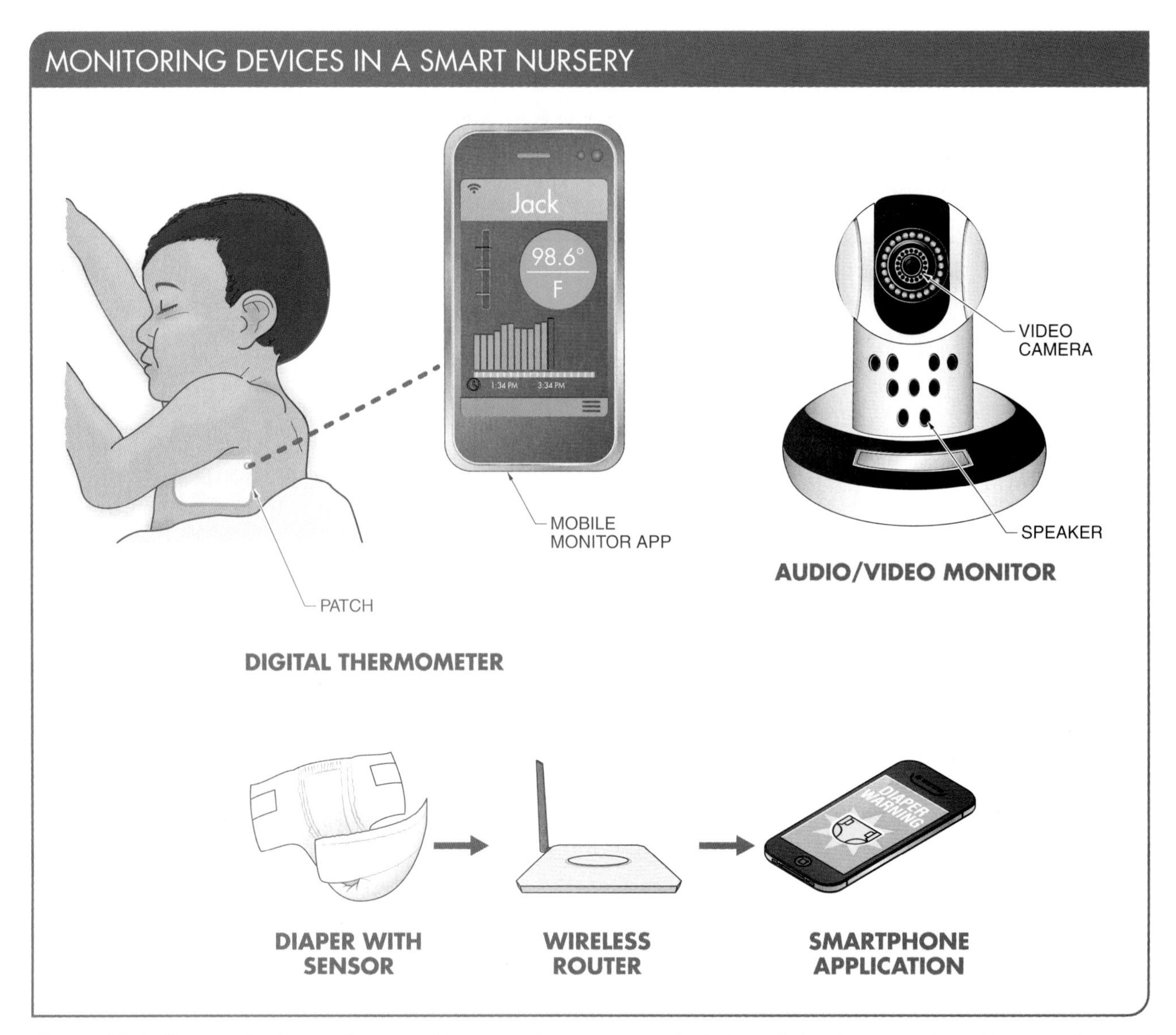

Figure 17-4. The monitoring devices used in a smart nursery may include a digital thermometer, video monitor, and diaper sensor.

Some monitoring devices used in a smart nursery have video monitoring and breathing monitoring capabilities. Video monitors may include two-way communication so that parents can both speak to and visually monitor the baby. Breathing monitors are used to see and monitor a baby's breathing pattern. Instant alerts provide information on breathing irregularities and can indicate when a baby is falling asleep or waking up. Some breathing monitors require a wearable accessory, others do not. This information is available at all times through the use of a smartphone app, no matter the location of the parents. Due to the private nature of video monitoring, it is essential that data be encrypted and stored securely.

For children and those with urination control issues, a diaper sensor can be placed on any disposable diaper. A single app can monitor multiple diapers. An alert is sent to a smartphone when the diaper needs to be changed. If needed, apps can provide daily and weekly records that may be useful when discussing urination issues with a physician.

SECTION 17.2 CHECKPOINT

1. How can a smart thermostat be programmed to maintain the desired level of comfort in a bedroom?
2. How many color combinations may be available in a single LED light and how can they be used to create lighting scenes?
3. How can voice messages be used be used in the smart bedroom?

SECTION 17.3—SMART BATHROOMS

The smart home technology that is used in a bathroom is based on the lifestyles of the individuals who occupy the home. The factors that affect the lifestyles of these individuals determines the application of smart devices in the bathroom. For example, a smart bathroom may be set up for single individuals or for those who are part of a family. The age of the individuals and their ability to move around and use the smart bathroom will also play a role in its setup. **See Figure 17-5.**

Figure 17-5. A smart bathroom should provide a positive impact on the lifestyle of an individual or family.

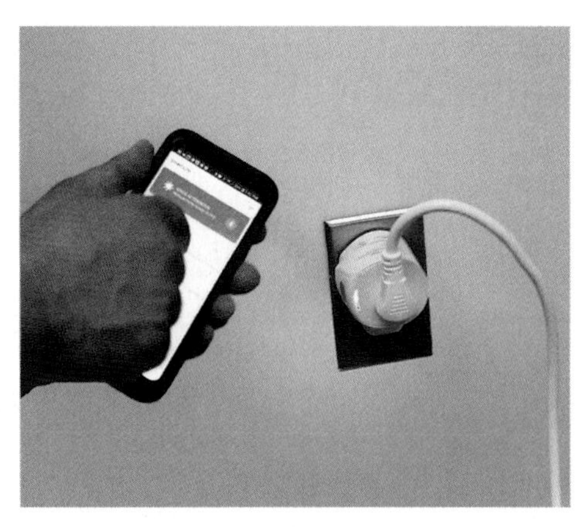

Smart plugs can be plugged into any receptacle to enable the control of a powered device from a smartphone.

Automated Sliding Doors

Accessing the bathroom door can be an issue for individuals in transition or needing assistance. In these cases, automated sliding doors can be installed to allow easier access. An automated sliding door can be opened, closed, locked, and unlocked by a remote on a key chain or by a motion detector.

Lighting

In a bathroom, a combination fan and light usually provides both lighting and ventilation. Proper lighting, such as lighting for the application of cosmetics, is often provided around the mirrors in a bathroom as well. In a smart bathroom, LED smart mirror lighting can help an individual prepare for different environments and times of day. The Kelvin value (K rating) of the LED lights can be adjusted to simulate the appropriate environment or time of day. A Kelvin value of 2700 K can be used to prepare for evening activities. A Kelvin value of 4600 K can be used to prepare for indoor day activities, while 6500 K can be used to prepare for outdoor day activities.

Smart Mirrors

A smart mirror is a mirror with an electronic display behind the glass that is connected to the internet. Smart mirrors enable individuals to adjust the lighting of the mirror and to start each day with personalized information displayed on the mirror. The information can be accessed through a menu on the touch-sensitive surface. Smart mirrors allow an individual to connect to calendars, weather information, TV programs, security cameras, dimmable lights, and smart devices in other rooms. For example, an individual can use a smart mirror in the morning to start a coffee pot in the kitchen if this was not initiated earlier in the smart bedroom.

Touchless Sinks

A touchless sink with a touchless faucet and soap dispenser may be installed in a bathroom to fit with an individual's lifestyle. Touchless faucets and soap dispensers operate by projecting a continuous IR beam from a sensor. When a user's hands enter the range of the beam, the sensor sends a signal to the solenoid, triggering the flow of water for as long as the user remains in range. When the user's hands move outside of the beam, the sensor shuts off the water until the beam is interrupted again. **See Figure 17-6.**

Smart Toilets and Bidets

A smart toilet and bidet are smart bathroom additions that will significantly impact an individual's lifestyle through features that provide cleanliness, comfort, and water conservation. Smart toilets and bidets commonly include touchless technology that uses motion detection to automatically raise and lower the seat or lid, flush, or illuminate a night-light. Both smart toilets and bidets can be wall mounted so that it is easy to clean underneath the fixtures. A bidet can help with personal cleanliness through features such as water-temperature settings, multiple spray settings, and an air dryer and deodorizer. Smart toilets can include several comfort features, such as a seat warmer, feet warmers, and Bluetooth-enabled speakers. Eco-friendly smart toilets repurpose stored water from washing to flush the toilet. **See Figure 17-7.**

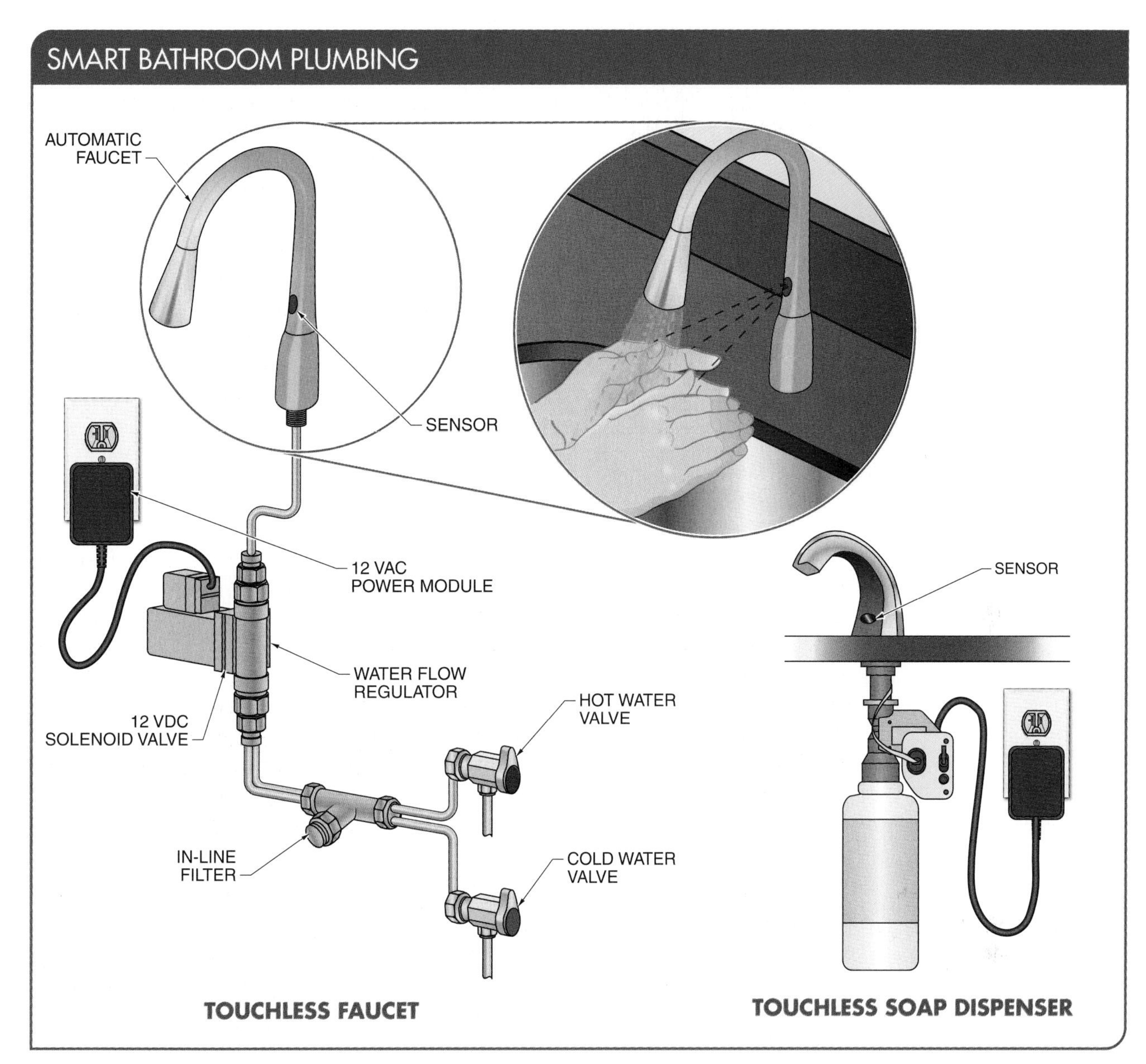

Figure 17-6. Touchless faucets and soap dispensers operate by projecting a continuous IR beam from a sensor.

> CODE CONNECT
>
> *Per NEC® Subsection 210.10(C)(3), at least one 120 V, 20 A branch circuit shall be provided to supply the bathroom(s) receptacle outlet(s).*

Showers and Tubs

There are several different features for showers and tubs in smart bathrooms that can be used to positively impact an individual's lifestyle. For example, chromotherapy (mood-enhanced lighting) and aromatherapy features in soaking tubs can be used to relieve stress. Showers can be equipped with multiple showerheads, handheld devices, lighting, and audio speakers. A showerhead with LED lights and audio capability can help relax the mind and body through colored-light therapy and music. **See Figure 17-8.** Safety devices such as tub walk-in doors and grip bars can help prevent slips and falls.

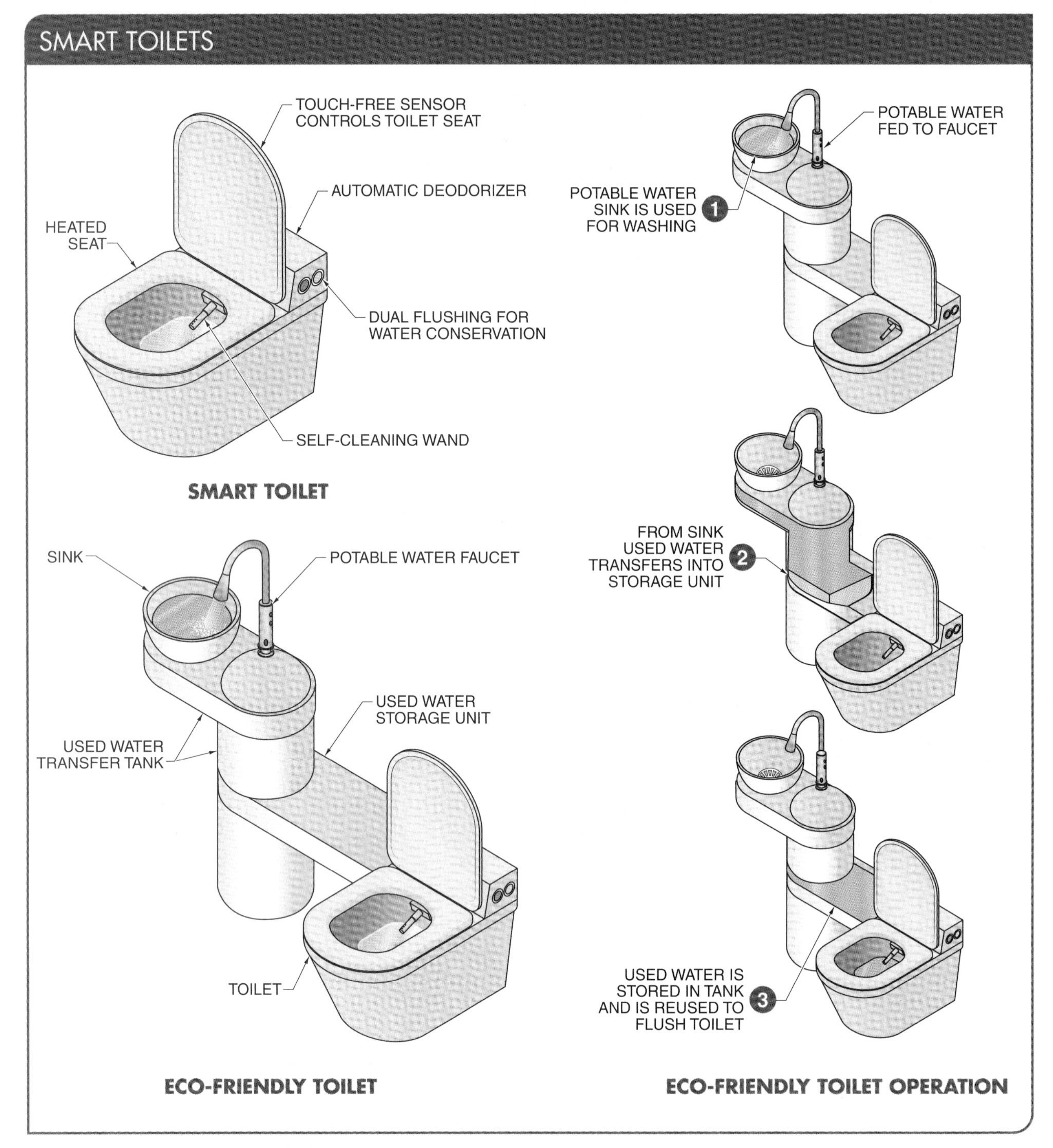

Figure 17-7. Eco-friendly smart toilets repurpose stored water from washing to flush the toilet.

Shower controls that connect to the Wi-Fi network allow control from an app on a smartphone or other device. A Wi-Fi-connected shower can be controlled from anywhere in the home to a preset water temperature. Showers may also have voice control that allows the user to vary water output from full power, mist, or steam settings. Shower privacy glass is available that can turn opaque or clear when plugged in and activated through an app.

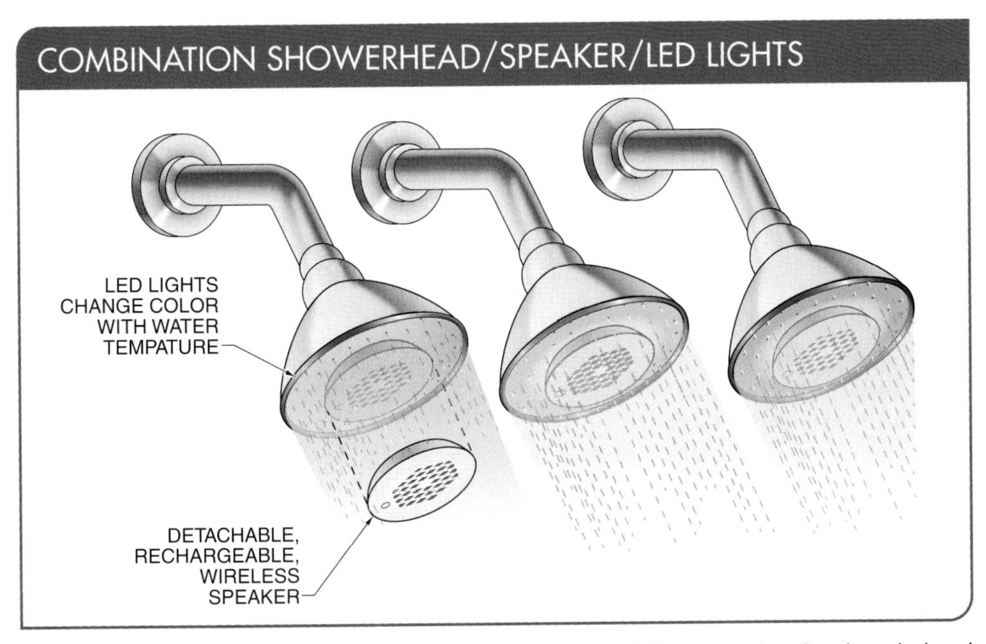

Figure 17-8. A showerhead with LED lights and audio capability can help relax the mind and body through colored-light therapy and music.

Heated Floors

Heated floors can provide comfort in a smart bathroom. A floor heating system is usually independently controlled. However, it may be possible to synchronize some floor heating systems to the house thermostat. A heated pad may also be used in specific locations in a smart bathroom, such as in front of a mirror and sink.

Smart Scales

It may be necessary for some individuals to track their overall health. Through the use of an app, an individual can track their overall health with a smart scale that records weight and other data and trends over time. Smart scales can also track multiple family members and keep report logs. Smart scales can be Wi-Fi or Bluetooth connected, or both. Smart scales can have features that measure or track the following:

- fat mass/percentage
- body mass index (BMI)
- water percentage
- bone mass
- resting heart rate
- pregnancy weight

SECTION 17.3 CHECKPOINT

1. At what Kelvin values should the LED lights in a smart bathroom be set to prepare for evening activities, indoor day activities, and outdoor day activities?
2. What is a smart mirror?
3. How do touchless faucets and soap dispensers operate?

SECTION 17.4—SMART KITCHENS

Numerous smart home technology is available for the kitchen. The types of smart devices incorporated into a kitchen depend on the lifestyle of the individual or family using the kitchen. A smart kitchen often includes both large and small smart appliances. Different methods of controlling smart devices in the kitchen and home can also be incorporated into the smart kitchen. The design of a smart kitchen is often centered around the main island and sometimes a service island.

Lighting

A kitchen must be properly lit for ease of use and safety. The lighting in a smart kitchen provides safe levels of illumination so that an individual can prepare meals without too much difficulty. Kitchen lighting falls into two categories: general lighting and task lighting.

General lighting provides enough natural and electrical lighting for safe movement about the kitchen during both the day and at night. Lighting can be provided by recessed can lights, surface mounted lights, and track lighting.

Task lighting is meant to provide a higher level of illumination at areas where kitchen tasks are completed. Task areas may include sinks, countertops, and islands. Task lighting can be provided by recessed can lights, pendant lights, and undercabinet lights.

The level of control considered for the lighting in a smart kitchen should be appropriate for the lifestyle of the individual using the kitchen. Controls for kitchen lighting can include touch screens, remotes, smartphones, and tablets. All systems should be compatible with each other and the types of lamps being used. It must be ensured that the system can handle the control of all lamp types.

Islands

A smart kitchen may include a main island or a service island, or both. The main island is often the center of most activity in the kitchen. The service island is often considered a preparation area and is supported by equipment on both walls. Both large and small appliances are close to the kitchen islands for ease of use.

A main island should provide space for standing and sitting activities. To accomplish this, the main island may be designed with upper and lower countertops. **See Figure 17-9.** The upper countertop can function as a space for guests to sit at during conversations or for family members to have a snack while homework is being supervised. The lower countertop should be able to accommodate a seating chair or a wheel chair and function as a workspace.

A touch screen (thin-film computer screen) can be built into the countertop's workspace, such as the lower countertop, for internet access. Individuals and families may find touch screens very useful if the screen is in a preparation area on the island or where workspace seating is available. If the built-in touch screen has voice capabilities, it can read the steps of a recipe. These types of screens can be scratch resistant and wiped clean after preparing a meal.

A pull-out drawer with USB ports can be installed in a main island to charge smartphones and tablets without taking up counter space. GFCI receptacles with USB ports can also be located along other countertop workspaces. Pop-up outlets can be used to plug in laptop computers or tablets, charge devices, or operate small DC appliances.

A main island may include a sink for convenience during food preparation and cleanup. For example, a main island sink with an instant hot water tap allows for the quick preparation of a hot beverage. A center-bowl garbage disposal with touchless faucets and soap dispensers provides cleanliness. A dual dishwasher can be installed at one end of the island for washing small loads or to be used at the same time as the main dishwasher after large meals.

A main island may also be designed with additional refrigeration space. For example, the other end of the island may include refrigerated drawers for storing vegetables. This makes it easy for items to be washed, cut, and stored until ready for use. The drawers also serve as extra refrigeration space when preparing for larger events.

A service island in a smart kitchen provides additional preparation areas and storage space. For example, a service island can be equipped with a refrigerated wine rack on one end and warming drawers on the opposite end. The warming drawers can be positioned close to the oven to receive baked goods until serving time. **See Figure 17-10.**

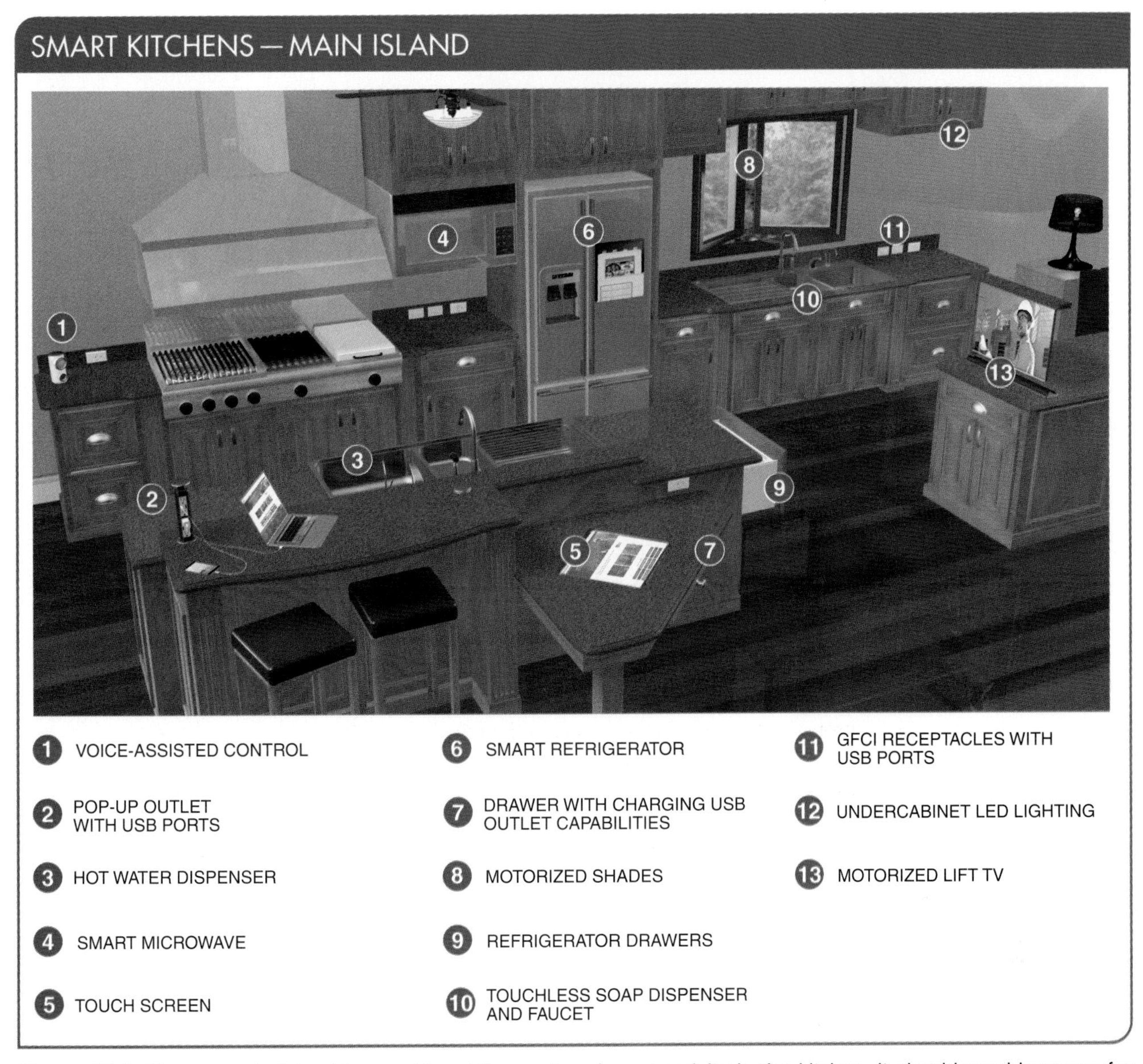

Figure 17-9. Since a main island is considered the center of most activity in the kitchen, it should provide space for standing and sitting activities, as well as features such as USB charging outlets, touch screens, touchless sinks, and refrigerator drawers.

Audio and Television

A smart kitchen can be equipped with wired or wireless audio as well as a television. A dedicated touch screen, voice assistant, or smartphone can connect to a variety of music systems, such as streaming services, local radio, or stored music. Speakers can be installed in the ceiling to save space.

Although televisions can be installed in several places in a smart kitchen, they require electrical receptacles. Many TVs also require internet connection. TV images can be displayed in a countertop, on the wall, or on the refrigerator. TVs can also be hidden in cabinets until needed and are also available as units that rise out of the counter on motorized lifts.

Figure 17-10. A service island in a smart kitchen provides additional preparation areas and storage space.

A voice assistant can be used to control other smart devices, play music, or search the internet for information.

Smart Kitchen Plumbing

There may be additional plumbing considerations for a smart kitchen. For example, appliances such as steam ovens require a water supply and drain. Sinks, water dispensers, and dishwashers installed in a main island also require a water supply and drain. If touchless faucets are installed, they may require additional space under countertops to account for digital controllers, sensor modules, cables, battery backups, and plug-in AC/DC adapters. Touchless faucets in a smart kitchen may have intelligent features that allow an individual to change water temperature and flow rates without the use of a lever or handle. **See Figure 17-11.**

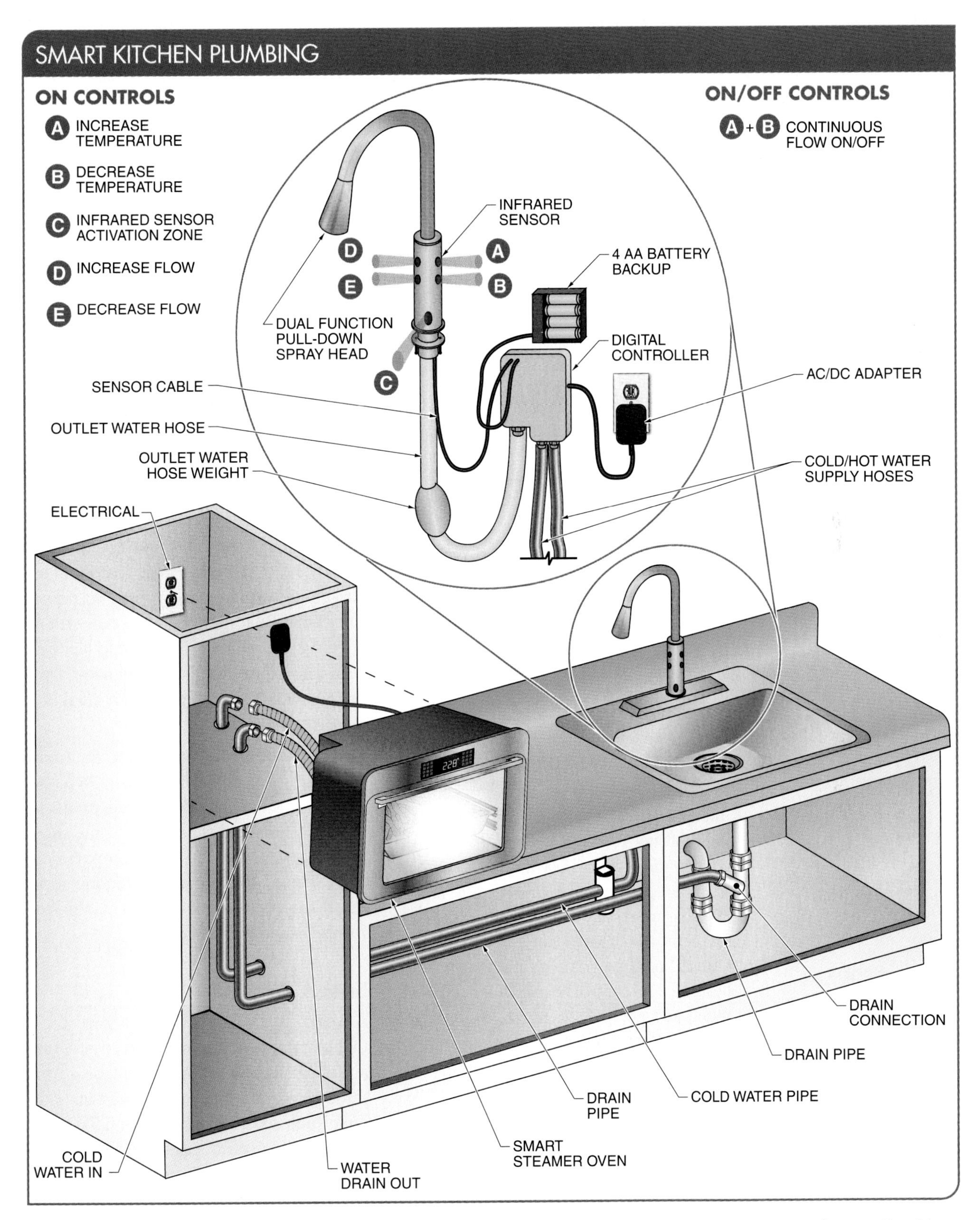

Figure 17-11. Additional plumbing considerations for a smart kitchen include steam ovens and touchless faucets. Touchless faucets in a smart kitchen may have intelligent features that allow an individual to change water temperature and flow rates without the use of a lever or handle.

Modular Components

One barrier to implementing smart technology may be its integration into the construction of a room. Prefabricated modular components may solve many of the issues that arise during integration. Many concerns about the delivery and quality of installation can be alleviated when the building process is moved off-site. For example, in the kitchen, countertops with touch-screen displays that are prefabricated off-site can be purchased and easily installed as prefabricated modular components.

Kitchen Controls

The kitchen is an ideal place to locate the main controller of a smart home. The kitchen is often centrally located and is a common gathering place for both the occupants and visitors in the home. From the kitchen, the energy, lighting, security, and comfort systems can be controlled from a main controller mounted on the wall. However, new thin film technology also allows these systems to be controlled from touch screens that are built into countertops or appliances such as refrigerators.

Smart Refrigerators

A *smart refrigerator* is a refrigerator with internet access that can sense and track the food items stored inside it through bar code reading or RFID scanning. Manufacturers are joining with third-party vendors to develop smart refrigerators that are the center of control for the entire smart home. These smart refrigerators are being designed to control the lighting, environment, and energy use in the home as well as access the security system so that the homeowner can view and speak to a visitor at the front door via the refrigerator touch-screen display.

Refrigerator Features. The features of a smart refrigerator include internet-connected touch-screen displays, water and ice dispensers, cameras, foot sensors, and connectivity to other appliances in the kitchen. **See Figure 17-12.** Water and ice dispensers are common features in many refrigerators, but smart refrigerators can be equipped with a hot water dispenser and an option for sparkling water when additional equipment is installed.

Internet-connected refrigerator touch screens typically offer many of the same capabilities as a tablet or countertop touch screen. For example, refrigerator touch screens can be used to access calendars, note postings, digital photos, recipes, grocery lists, and online shops. Bar-code readers track the dates that food items enter the refrigerator and send reminders about expiration dates. Touch screens may also become another means for advertising as manufacturers sign deals with food producers to offer coupons and discounts.

Besides bar-code readers, smart refrigerators also include cameras that allow an individual or family track the quantity and quality of their food items. Cameras are strategically placed inside the refrigerator so that food items can be seen without opening the door. A smartphone can also be used to access the cameras away from the home, such as when an individual is at a grocery store.

Cooling Options. Most smart refrigerators have several compartments with separate temperature controls for different products, such as meats and vegetables. Some smart refrigerators include two-compartment freezer units so that if not much food needs to be frozen, the other unit can be adjusted for use as a refrigerator.

Smart Ovens

A *smart oven* is an oven with internet access that can read food item labels to program and control temperature and cook time through a smartphone app. Smart technology, such as bar-code readers, is being included in electric, gas, and steam ovens. Bar-code readers scan food item bar-code labels to set the appropriate temperature and cook time. Smartphone or voice control can be used to preheat or adjust the temperature settings of a smart oven. Geofencing allows a smart oven to work with the GPS in an individual's smartphone to determine their location. When the individual leaves the home, they will receive an alert if the oven has been left on.

Small Smart Appliances

Larger smart appliances can communicate with the utility through the smart meters to conserve energy. However, small smart appliances communicate with each other for efficiency and convenience. The small appliances that may be included in a smart kitchen are smart microwaves, countertop ovens, coffee and tea pots, scales, plates, and garbage cans. These small appliances include the following features that positively impact an individual's lifestyle:

- Smart microwaves automatically set temperatures and times for the food being prepared. For example, when an individual prepares popcorn, an audio sensor determines when popping is complete.
- Smart countertop ovens include a Wi-Fi connection, a camera, and AI capabilities to weigh, monitor, and cook food with minimum input from the user.
- Smart coffee and tea pots select and brew coffee or tea portions on command and may be synchronized with an individual's morning or afternoon schedule.
- Smart kitchen scales track the exact amounts of ingredients when a recipe is chosen through an app.
- Smart plates are Wi-Fi- and Bluetooth-enabled, with three mini-cameras and a weight sensor that track what is eaten and advise on cutting back portions that are too large.
- Smart garbage cans include a bar-code reader that scans an item about to be discarded and adds it to the grocery list on a smartphone. When there is no bar code, a smart garbage can will still send an inquiry to the smartphone concerning what should be added to the grocery list.

Figure 17-12. Smart refrigerator features include internet-connected touch-screen displays, water and ice dispensers, cameras, foot sensors, and connectivity to other appliances in the kitchen.

SECTION 17.4 CHECKPOINT

1. How can internet access be incorporated into the countertop of a smart kitchen island?
2. What are some of the plumbing considerations for smart kitchens?
3. What is a smart refrigerator?
4. What is a smart oven?

SECTION 17.5—SMART FAMILY ROOMS AND LIVING ROOMS

A dwelling may have a family room, living room, or both, depending on the size of the dwelling and the purpose for which each room is intended to be used. Family rooms typically include television and audio systems to entertain all ages. Living rooms are often formal rooms used to entertain guests and display valued or collected items.

> TECH TIP
>
> *Multiple smart devices can often be controlled from a smartphone or a universal remote control.*

Smart home technology can be used in the family/living room to accommodate every type of lifestyle. Smart devices can be used in family/living rooms to enhance entertainment, security, and comfort. In a smart family/living room, wireless technology is used to control lighting, the HVAC system, ceiling fans, motorized drapes, and fireplaces. **See Figure 17-13.**

Lighting Scenes

A smart family/living room is a place in the home where lighting can be used to create different scenes, depending on the time of day or activity in the room. For example, a scene with even lighting can be created for

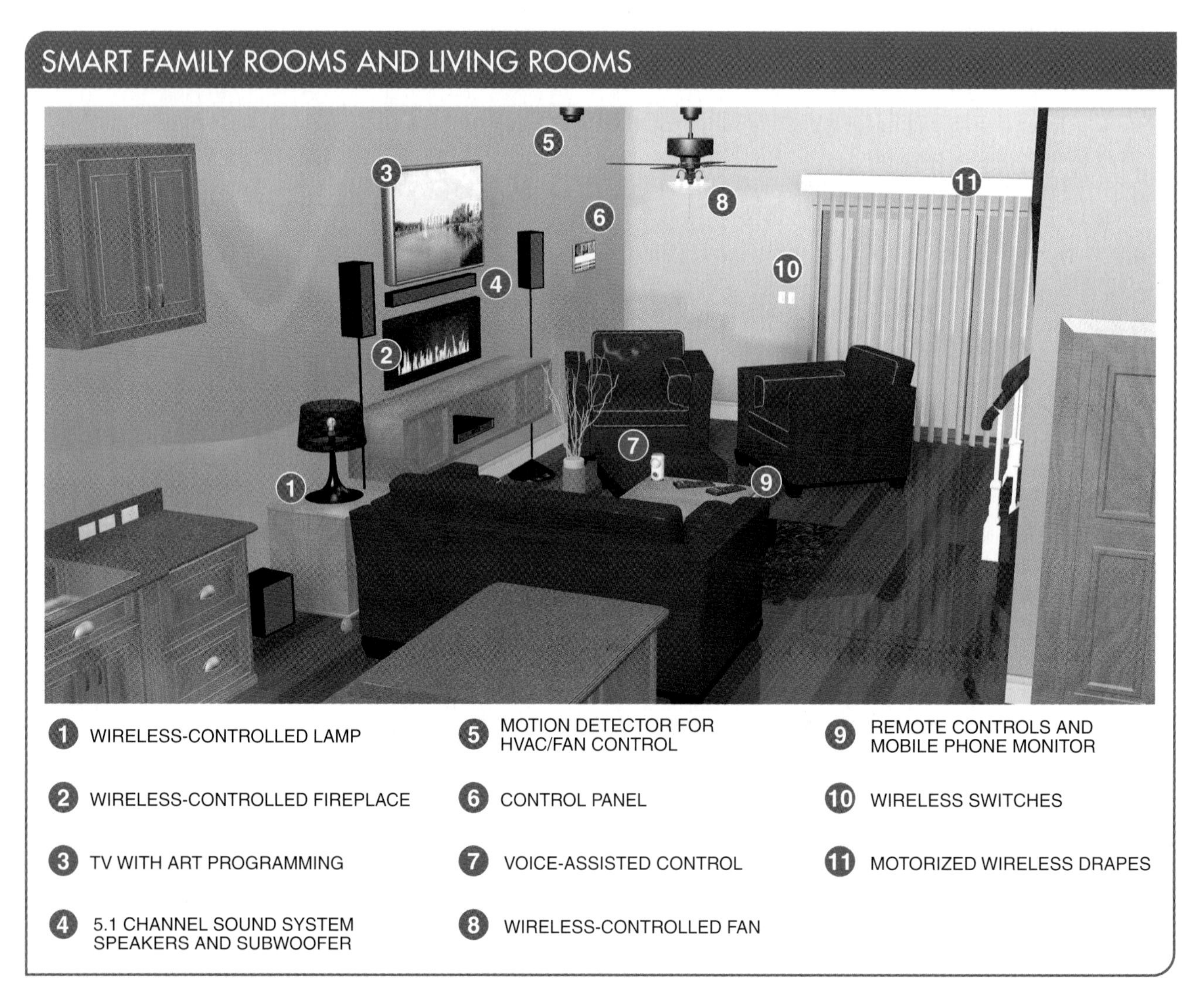

Figure 17-13. Smart devices can be used in family/living rooms to enhance entertainment, security, and comfort.

day or nighttime conversation. A scene for TV viewing can be created through dimmed lights and motorized drapes that have been wirelessly closed. Smart LED lights with millions of color combinations can be used to create different lighting moods depending upon the audience.

Smart TVs

Smart TVs can connect to the internet and allow individuals access to streaming video, social media, information, and interior views of rooms or the outside of the property. For example, a smart TV can be linked to a smart doorbell at the dwelling's entry point or to cameras in the bedroom of an individual who requires assistance. A smart TV can also be used to display information about other wireless connected smart devices and appliances in the home, such as smart washing machines.

A high-definition smart TV, such as a wide-screen 4K TV, provides the best performance when it is used for television viewing, regardless of an individual's lifestyle. A high-definition TV can also be programmed to display a static image of artwork for aesthetic purposes when not in use. A motorized lift TV can retract into the ceiling, exposing other artwork behind it. When possible, the TV should also be linked to surround-sound audio system for maximum performance.

Audio Systems

Since the family/living room is used to entertain guests or for viewing TV, the audio system must meet the lifestyle requirements of the individuals using it. For the best results in audio performance, a minimum of a 5.1 channel sound system with complimentary receivers may be necessary for a family/living room. A 5.1 channel sound system will support five speakers and one subwoofer to create a surround-sound experience. The receivers should have Bluetooth or Wi-Fi support for streaming music from web-based services.

Object-based audio adds to the surround-sound experience by adding a height dimension to the audio. This requires speakers to be installed in the ceiling. These ceiling speakers, along with specially designed wall speakers, are designed to provide overhead sound. Depending upon the size of the room and furniture configuration, a professional installer may be necessary for object-based audio, especially for systems that are a part of a home theater.

Controls

The most common type of control in the family/living room were remote controls for TVs and/or audio systems. With the introduction of smart home technology, the controls used in a family/living room may include touch screens and controls with programmable features instead of multiple remote controls. Many devices such as lamps, fireplaces, switches, motorized drapes, and ceiling fans can be wirelessly controlled to positively affect the lifestyle of individuals or families. An HVAC system can be controlled by motion sensors or a control panel. Depending on the smart technology used in the family room, a professional installer may be necessary.

Internet Access

An internet connection in the family/living room is essential to support all the resources that will be used by the audio-visual equipment and appliances in this room. For example, a smart TV can stream web-based content when connect to the internet. If there is not a wired connection available, a reliable wireless router should be located close to the family/living room.

SECTION 17.5 CHECKPOINT

1. How can a scene be created for TV viewing in a smart family/living room?
2. How can a smart TV be used to see images of other places in and around a property?
3. What is the minimum audio system configuration recommended for a surround-sound experience?

SECTION 17.6—SMART GARAGES

Smart home technology extends to the garage, where it is used to charge electric vehicles and to control the lighting and garage door operation. Charging systems for electric vehicles, remote-controlled overhead garage doors, and motion-controlled lights are common to a smart garage. Some smart garages may include multimedia equipment, such as in-wall speakers, ceiling speakers, and a TV, for entertainment purposes. GFCIs are required in a garage, although some receptacles can have USB ports for the low-current charging of smartphones and tablets.

Lighting

Motion sensors can be used to turn on the outside lights of a smart garage when a vehicle pulls in the driveway or a person walks up. A light (or multiple lights) in the garage can be connected to the remote garage door control so that it turns on when the garage door is opened using the remote control. Side lights can be connected to motion detectors for security and safety when a person must walk around the sides of the garage. Task lighting inside the garage should be available over workbenches or other areas that require specific lighting levels.

Electric Vehicle Types

There are different types of vehicles that use electric power. A battery electric vehicle (BEV) is a fully electric vehicle powered by an electric motor with no gas engine. A plug-in hybrid electric vehicle (PHEV) has a battery-powered electric motor and a gasoline engine to power the car. An extended range electric vehicle (EREV) has an electric motor that powers the vehicle with an onboard gas-powered generator to extend driving range when the battery's range is depleted. The difference between an EREV and a PHEV is that the EREV does not use the gasoline engine to provide mechanical energy to the drive train. These different types of electric vehicles can be charged in a smart garage with electric vehicle supply equipment (EVSE).

Electric Vehicle Supply Equipment (EVSE)

Electric vehicle supply equipment (EVSE) is the charging system for an electric vehicle and includes an electrical junction box and plug that connects the electric vehicle battery to an AC source. However, the actual charger for the electric vehicle battery is built into the vehicle. This onboard charger takes the AC voltage and converts it to DC voltage to charge the battery pack. An EVSE provides a safe connection for an electric vehicle to a 120/240 VAC source. **See Figure 17-14.** The Wi-Fi interface in an electric vehicle can be used to check the charging status.

There are three levels of EVSE charging systems for electric vehicles. The higher the level, the faster the electric vehicle battery is charged. Level 1 and Level 2 EVSEs can be installed in a dwelling's garage. A Level 1 EVSE is typically a 120 VAC, 20 A system. A Level 2 EVSE is typically a 240 VAC, 40 A system. Level 3 EVSEs are 480 VDC systems typically only installed as public charging systems.

EVSE Amperage Capacity. Many electric vehicles are often sold with a Level 1 EVSE charging cord so that they can be plugged into a standard 120 V receptacle. The 120 V receptacle should be on a separate circuit due to the current draw from charging. For a smart garage, an EVSE capable of delivering at least 20 A up to 40 A is recommended.

EVSE Plugs and Receptacles. Most electric vehicles and chargers use a receptacle rated up to 40 A and a charging cord with an SAE J1772 plug, the standard adopted from the Society of Automotive Engineers (SAE). An SAE J1772 plug can be used with a Level 1 or Level 2 EVSE. The service panel breaker(s) must be rated for the charging system that is being installed.

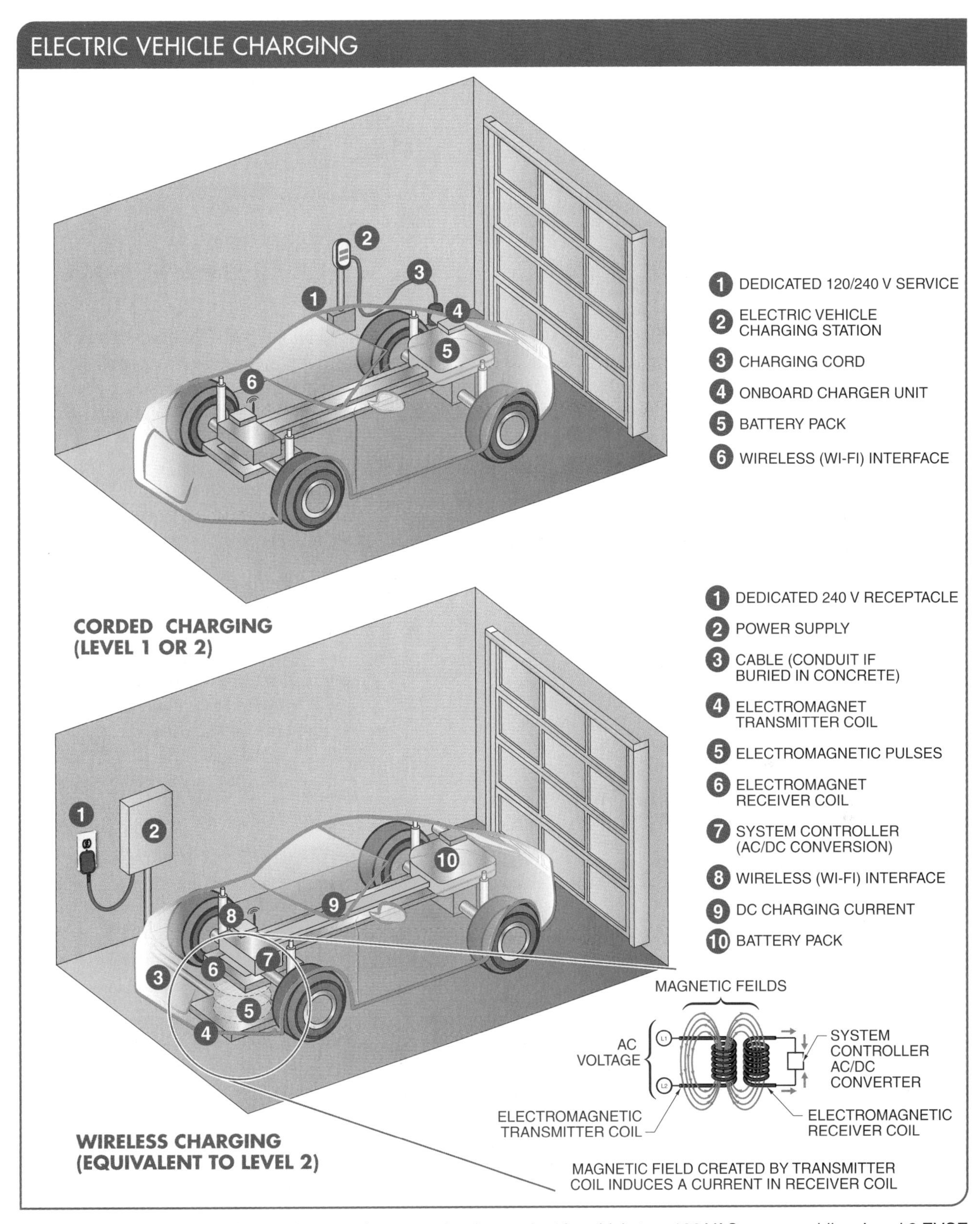

Figure 17-14. A Level 1 EVSE provides a safe connection for an electric vehicle to a 120 VAC source, while a Level 2 EVSE safely connects an electric vehicle to a 240 VAC source. An electric vehicle can also be charged wirelessly through electromagnetic inductive charging.

EVSE Installation. The first step in installing an EVSE is to determine the location where the electric vehicle will be parked. For a two-car garage, the EVSE can be centrally located since the cables usually run approximately 12′ to 25′. The cost of installation will vary depending on the distance from the dwelling's service panel.

The factory warranty should be reviewed before an EVSE is installed. Some manufacturers require the use of an EV installation specialist, but often any qualified electrician can handle the installation of an EVSE. A portable EVSE can be a good option for electric vehicle owners who move often because it is no more expensive than a permanent-mounted EVSE and is easily reinstalled in another location.

TECH TIP

The Transportation Electrification Accord encourages utilities and companies to participate in and facilitate the deployment of electric vehicle supply equipment (EVSE).

Electromagnetic Inductive Charging

Electromagnetic inductive charging is a wireless charging method that uses an electromagnetic field to transfer energy between coils. When an electric vehicle is charged wirelessly, a magnetic field is created by the EV charger in the transmitter coil, which sits on the garage floor. The magnetic field generated in the transmitter coil reaches the receiver coil in the electric vehicle and induces current to charge the batteries.

Wireless charging is convenient because the charging system is activated as soon as the vehicle is parked above the EV charger transmitter coil. It is not necessary to connect any cables for charging. Although there is always some energy lost during energy transfer, wireless charging units can reach efficiencies of 80% to 90%.

SECTION 17.6 **CHECKPOINT**

1. What is electric vehicle supply equipment (EVSE)?
2. What is electromagnetic inductive charging?

SECTION 17.7—OTHER LIFESTYLE APPLICATIONS

Smart devices can be used for other lifestyle applications both inside and outside of a dwelling. Smart home robots can be used for maintenance, entertainment, information, and security, as well as to aid individuals with disabilities. Wearable technology can be used for health monitoring, fitness monitoring, and tracking applications. In addition, smart devices can be used to aid with the responsibilities of pet ownership.

Smart Home Robots

Inside the smart home, robots and automation are merging. Voice control has been integrated into robots to provide easy access to information, music, and appliance controls that were once limited to stationary positions. Robots can be used all around the smart home to help with day-to-day tasks.

For example, indoor robotics can be used to provide help to individuals who may not be able to perform cleaning tasks around the home. A multipurpose vacuum robot can be used for automated cleaning and may include add-on components for air purification and security. Robotic window washers can be used for automated window cleaning. **See Figure 17-15.**

Advanced robots with AI capability, human-like sensory capability, emotional expression, and situational awareness are being developed for use around the home. These advanced robots will be capable of listening, talking, walking, providing information, and performing complex tasks. Advanced robots with onboard cameras will likely be incorporated into smart home security systems for mobile surveillance of a home. **See Figure 17-16.**

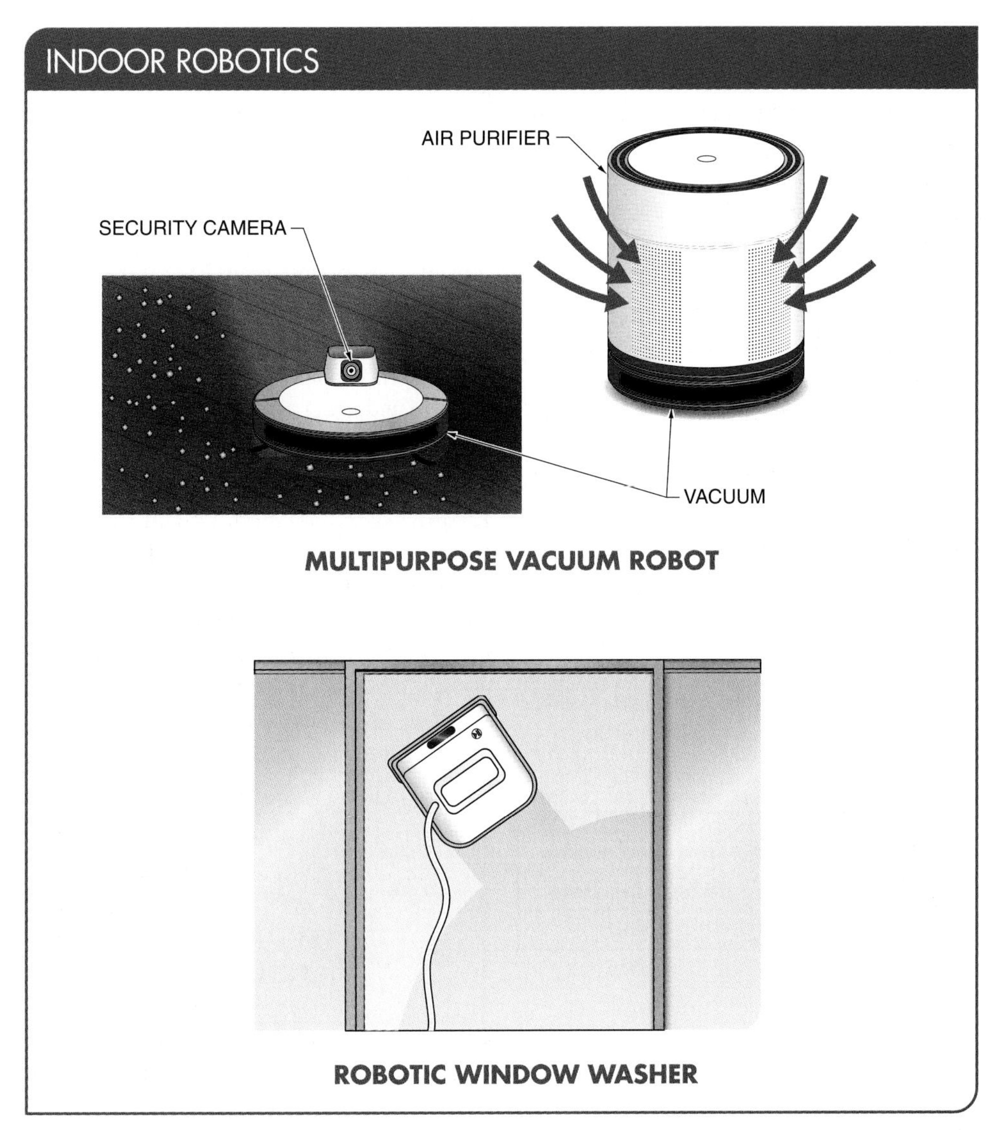

Figure 17-15. Indoor robotics, such as multipurpose vacuum robots and robotic window washers, can be used around a smart home to provide automated cleaning and other functions.

ADVANCED ROBOTS

Human-Like Sensory Capabilities	Advanced Robotics Technology
Hearing	Artificial intelligence
Speaking	
Object identification	Situational awareness system
Touching	
Emotional expression	Smart learning
Moving	

Figure 17-16. Advanced robots with AI capability, human-like sensory capability, emotional expression, and situational awareness are being developed for use around the home.

Children's Robotic Toys. Robotic toys that can sing, dance, and play music, videos, and games are commonly used to entertain children. Robotic toys with AI and speech capabilities are also commonly used to teach children. Parents with young children can use robotic toys for storytelling and nursery rhymes. For children with physical, developmental, or learning disabilities, a robot can also be a useful companion. Robotic toys vary in complexity and may be capable of voice recognition, facial recognition, speaking, walking, roaming, and obstacle avoidance. Most robotic toys are battery operated, although some may use solar power. **See Figure 17-17.**

CHILDREN'S ROBOTIC TOYS

Features
Voice recognition
Emotional expression
Roaming capability
Smartphone control
Speech
Video
Music
Games
Storytelling
Nursery rhymes

Figure 17-17. Robots can be useful companions for children with physical, developmental, or learning disabilities.

Robotic Pets and Pet Avatars. Robotic pets and pet avatars can be used to bring comfort to both young and elderly people needing assistance. **See Figure 17-18.** Robotic pets can be simple in design or they can have realistic features such as fur-like surfaces, barking or purring sounds, and the ability to run or roam on voice command.

Pet avatars are animal-like digital images that respond to the user. Pet avatars can carry on a conversation and be used in patient care for those suffering from memory issues and Alzheimer's disease. Because the representation on the screen is that of a pet animal, patients may respond better to the avatar image than an actual human face. A pet avatar can receive its personality from a caregiver in a remote location. Through the use smart devices with cameras such as tablets, a remote caregiver can see and hear the patient and respond appropriately. Pet avatars can also be used to send patients photos and other forms of encouragement from friends and family.

ROBOTIC PETS AND PET AVATARS

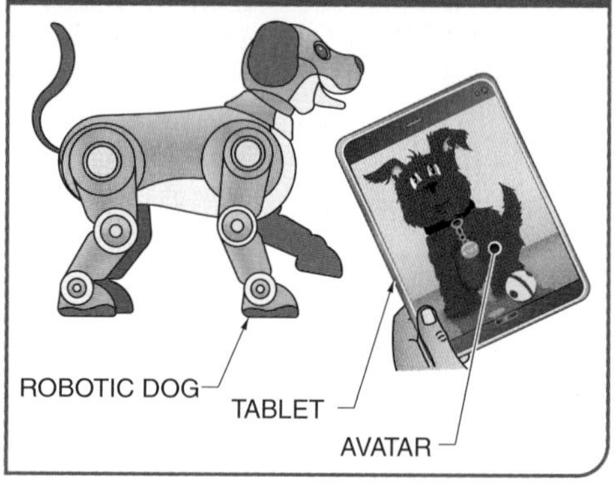

ROBOTIC DOG
TABLET
AVATAR

Figure 17-18. Robotic pets and avatars can be used to bring comfort to both young and elderly people needing assistance.

Outdoor Automation and Robotics

Smart home automation and robotics can be used for outside maintenance. Robotic lawn mowers can be combined with automated sprinkler systems and can be operated through a smartphone when occupants are home or away. **See Figure 17-19.**

Robotic Lawn Mowers. A robotic lawn mower has three basic components: a self-directed mower, a low-voltage perimeter limit wire, and a charging station. Navigation limits are provided primarily by the low-voltage wire buried along the perimeter of the area to be mowed. Some robotic lawn mowers also use IR or ultrasonic sensors to identify a corner or to avoid barriers and obstacles. Other features for a robotic lawn mower include GPS for more accurate movement, a theft-prevention alarm, and voice-activated controls. Safety features of robotic lawn mowers include the ability to stop the cutting blade when the mower is not on the ground or the blade is exposed and the ability to brake forward motion when a person steps in front of the mower.

Regardless of an individual's lifestyle, a robotic lawn mower offers several advantages. One advantage is that a robotic lawn mower saves time for a homeowner because the homeowner does not need to accompany the mower while it cuts the grass. Robotic lawn mowers are quiet and can operate 24/7 with zero emissions. When a robotic mower is set for frequent cutting, small grass clippings can be mulched to contribute to a healthy lawn. Robotic lawn mowers are also equipped with a rain sensor to avoid wet grass.

However, there are some disadvantages to robotic lawn mowers. A robotic lawn mower may face operational challenges when it does not or cannot avoid obstacles. For example, a robotic lawn mower may not be able to distinguish grass from a flower bed or planted garden. A robotic lawn mower may struggle on rugged terrain or steep slopes. Also, a robotic lawn mower may not be able to move through mud or large clumps of wet leaves.

Automated Sprinkler Systems. With sprinkler automation, a lawn or garden can be watered at any time from any location using a Wi-Fi connection and a smart garden hub. An automated sprinkler system offers many benefits to a homeowner regardless of their lifestyle. For example, an automated sprinkler system allows a homeowner to control one or more sprinkler zones at preset times and the ability to adjust water pressure for each zone. A moisture sensor in an automated sprinkler system can override the system on a rainy day. Also, an automated sprinkler system can connect to weather monitoring instruments or the internet to predict whether it should water the lawn as scheduled. When a flowmeter is added to the system, a homeowner can receive notifications when pipes break and can monitor water costs.

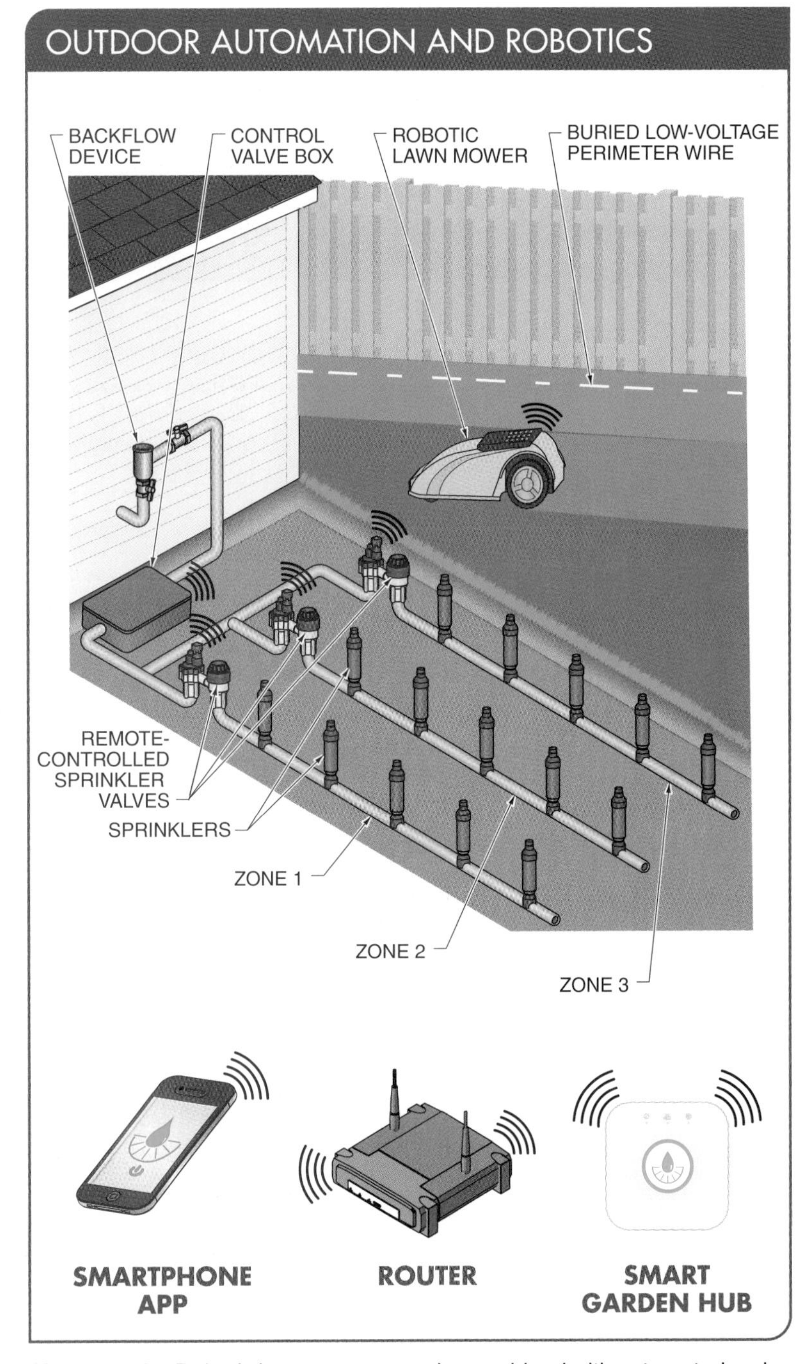

Figure 17-19. Robotic lawn mowers can be combined with automated sprinkler systems and can be operated through a smartphone when occupants are home or away.

Wearable Technology

Wearable technology (wearables) are electronic devices worn on the body or embedded in clothing for tracking information related to health and fitness, medical conditions, GPS location, as well as for playing music and making secure payments. Wearables monitor and enhance vision, sleep habits, exercise, and health. The miniaturization of sensors developed for medical technology is being applied to wearables for almost every surface of the body. **See Figure 17-20.**

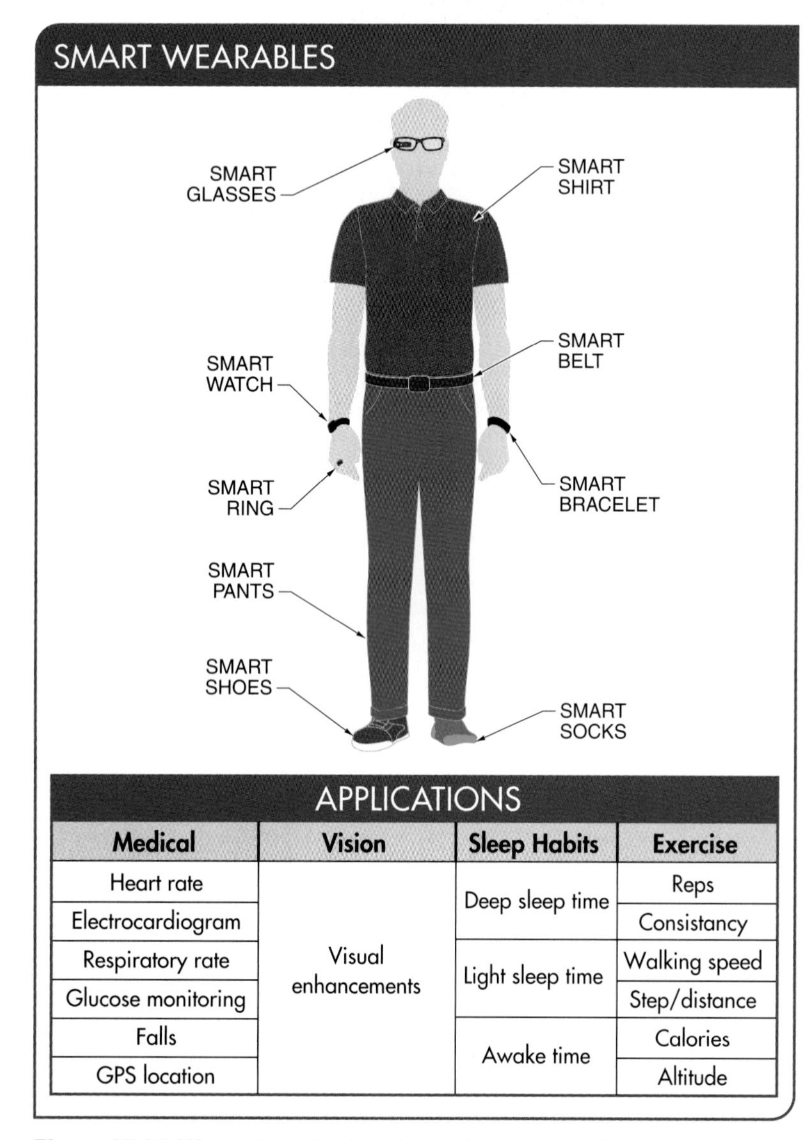

APPLICATIONS			
Medical	**Vision**	**Sleep Habits**	**Exercise**
Heart rate	Visual enhancements	Deep sleep time	Reps
Electrocardiogram			Consistancy
Respiratory rate		Light sleep time	Walking speed
Glucose monitoring			Step/distance
Falls		Awake time	Calories
GPS location			Altitude

Figure 17-20. Wearable technology (wearables) are electronic devices worn on the body or embedded in clothing for tracking information related to health and fitness, medical conditions, GPS location, as well as for playing music and making secure payments.

Smart clothing, including smart shirts, smart belts, smart pants, smart socks, and smart shoes, monitors the physical condition of the user to transmit biometric data and physical movement to an app in real time. For example, smart shoes can measure the vertical jump of a user or vibrate to guide the user to their destination. In addition, smart glasses can display information like a smart mirror alongside where the user is looking.

Smart devices can monitor walking speed, which can be helpful for individuals recovering from surgery. Monitoring walking speed can also provide an early indication of balance problems. For individuals recovering from an illness or confined to a bed, smart clothing can monitor vital functions such as heart rate, respiration, temperature, and electrocardiogram results, alerting the patient, caregiver, or physician if there is a problem. In addition to smart clothing, smart watches and smart rings can be worn to track GPS location, provide health and fitness information, and send medical emergency alerts. **See Figure 17-21.**

> TECH TIP
>
> *Smart devices worn to track fitness may be compatible with nutrition apps or health apps and can also be used to send reminders to move throughout the day.*

Smart Wheelchairs with Assistive Robotics

Smart wheelchairs with assistive robotics are used to support disabled individuals or those who require assistance in their daily activities. The assistive robotics can also help with specific activities and allow individuals with injuries or disabilities return to certain types work environments. A smart wheelchair can be equipped with a robotic manipulator arm, 3D camera, chin joystick, touch screen, voice assistant, and a form of brain control interface (BCI). A smart wheelchair with assistive robotics can be used in the smart home to help an individual perform tasks such as pouring and serving a drink, preparing and eating a meal, and handling objects like books without a caregiver's assistance. **See Figure 17-22.**

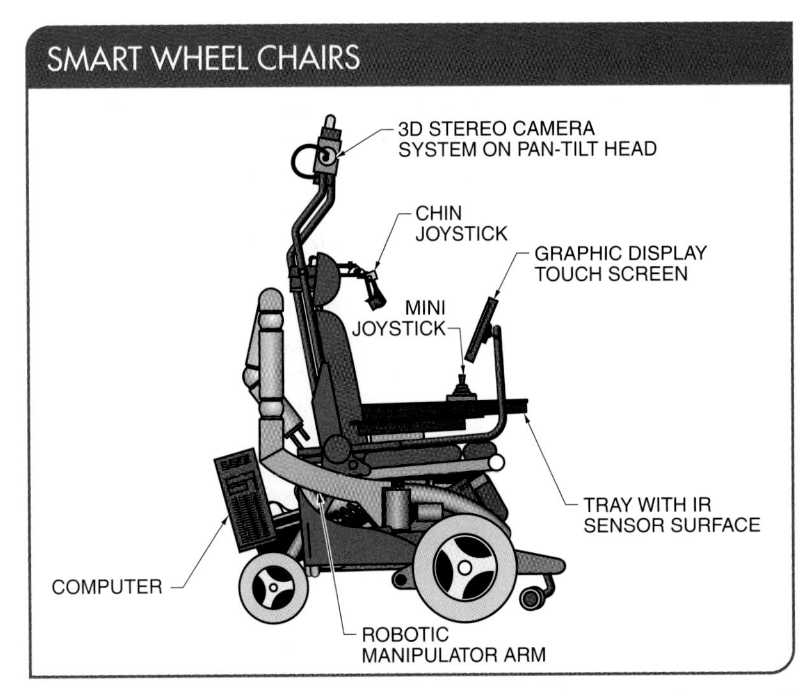

Figure 17-22. Smart wheelchairs with assistive robotics are used to support disabled individuals or those who require assistance in their daily activities.

Automated Pill Dispensers

An *automated pill dispenser* is an electronic device that stores medicine in compartments and can be programmed to sound alerts and dispense pills on schedule. Automated pill dispensers can be an important part of medication management for caregivers, parents monitoring a child's prescription drugs, and people with memory issues. Monitoring services that track and manage pill dispensing are also available for people who live alone or spend most of their day alone.

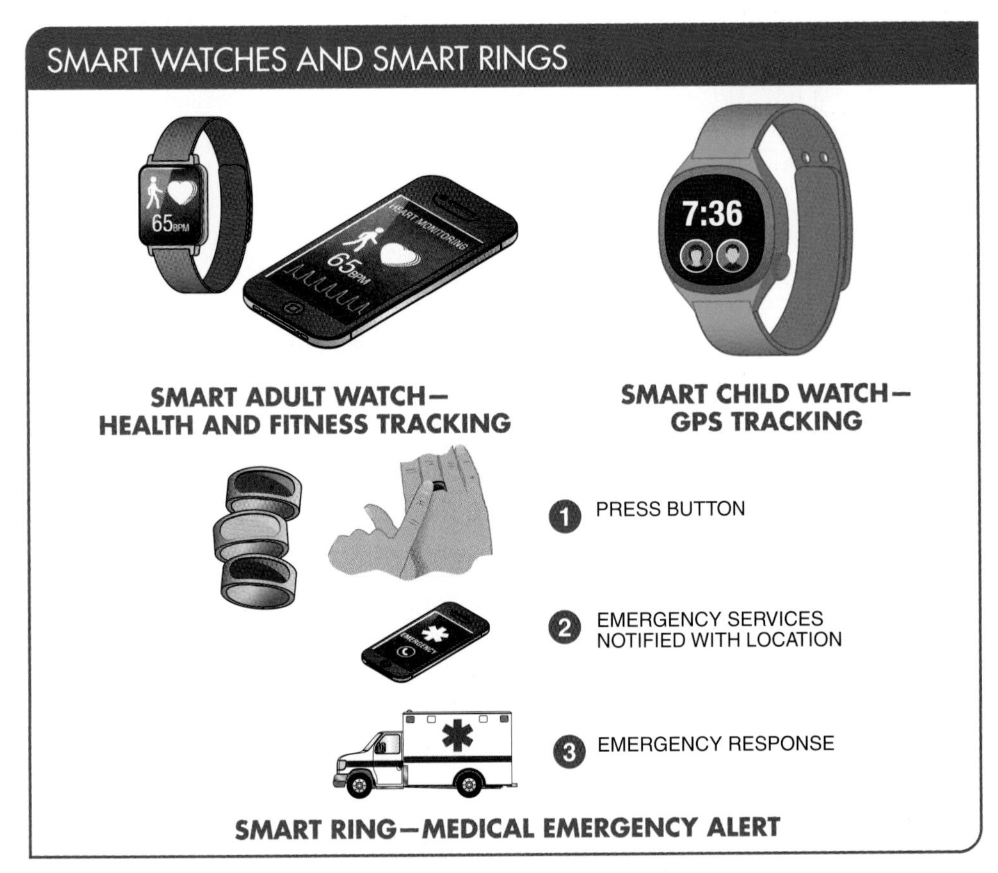

Figure 17-21. Smart watches and smart rings can be worn to track GPS location, provide health and fitness information, and send medical emergency alerts.

Automated pill dispensers are available with several different features. The templet for dosage scheduling should follow the days of the week. A user-friendly interface and Wi-Fi connection allow the dosage schedule to be set easily. The compartments can be deep so that several pills can be dispensed at once. The dispenser can have an alarm speaker and a flashing light. A charging station and rechargeable batteries provide power and will back up the dispenser during outages. **See Figure 17-23.**

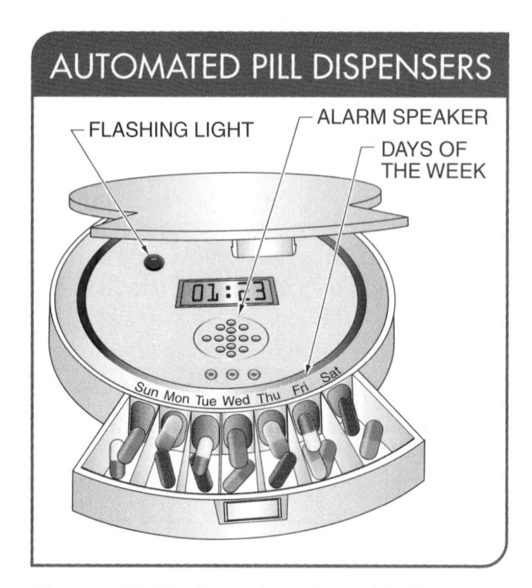

Figure 17-23. An automated pill dispenser is an electronic device that stores medicine in compartments and can be programmed to sound alerts and dispense pills on schedule.

Smart Devices for Pets

Smart devices can help people care for and protect their pets. These smart devices are also useful if the pet owners must be away from home for an extended time. Smart devices can be used to dispense food and water for pets, allow pets to enter or exit a dwelling, locate lost pets, and train pets.

Automated Pet Feeding/Watering Systems. Automated pet feeding/watering systems can be controlled remotely from a laptop, smartphone, or tablet. Feeding times, portion sizes, and food-dispensing speeds specifically for the type of pet being fed can be adjusted from anywhere in real-time through a Wi-Fi connection. Some automated pet feeding/watering systems may have temperature control to keep food and water from freezing when the feeders are placed outdoors. Other feeding/watering systems may have webcams that allow the pet owner to observe the pet from their smartphone. Automated pet feeding/watering systems with microphone recorders and speakers allow the owner to leave personalized voice messages to be played at feeding time. Automated pet feeding/watering systems are programmable 7 days a week, 365 days a year. **See Figure 17-24.**

Pet Access Doors. Smart pet doors allow pets to enter and exit a dwelling while preventing unwanted animals or intruders from entering the dwelling. Smart pet doors operate on AC or battery power and include features that address security concerns, pet safety, and weather conditions. The door is activated when an ultrasonic transmitter in the pet collar sends sound waves to the receiver. Access to and from the dwelling can be preset for the type of pets entering and leaving the home. A gravity-closing system with automatic retraction safely closes the door when the pet is clear of the door. A secure deadbolt can be automatically locked as well. A smart pet door can have an airtight weather seal and be thermally insulated. **See Figure 17-25.**

Pet Tracking Collars. A *pet tracking collar* is a GPS-equipped collar that tracks the location and movement of a pet in real time. A pet tracking collar can be useful for pet owners who allow their pets outdoors. A pet tracking collar can be used to set up a virtual fence so that the pet owner is notified when a roaming pet, such as cat or dog, leaves the designated area. An hour-by-hour log of the pet's location can be viewed through a smartphone app. Additional safety features include a heat alarm for when the surrounding area gets too hot and a light on the collar that can be turned on at night via an app to visually locate the pet. If the chip is detached from the pet tracking collar, a notification will be sent to the owner's smartphone. **See Figure 17-26.**

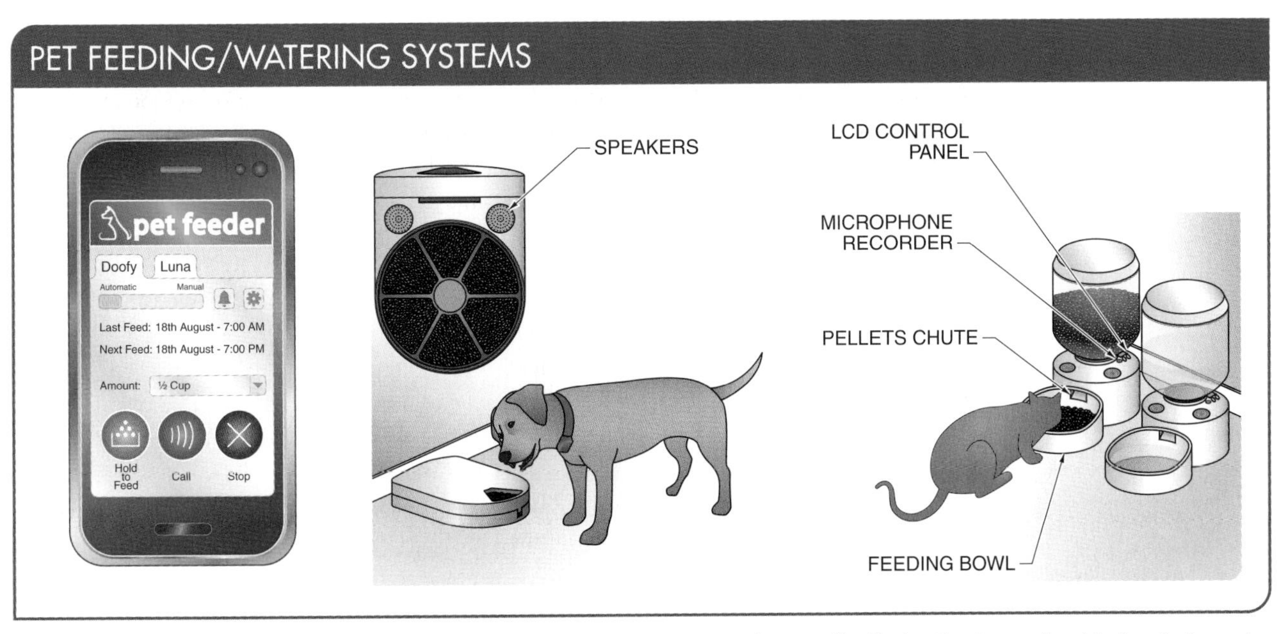

Figure 17-24. Feeding times, portion sizes, and food-dispensing speeds specifically for the type of pet being fed can be adjusted from anywhere in real-time through the Wi-Fi connection of automated feeding/watering systems.

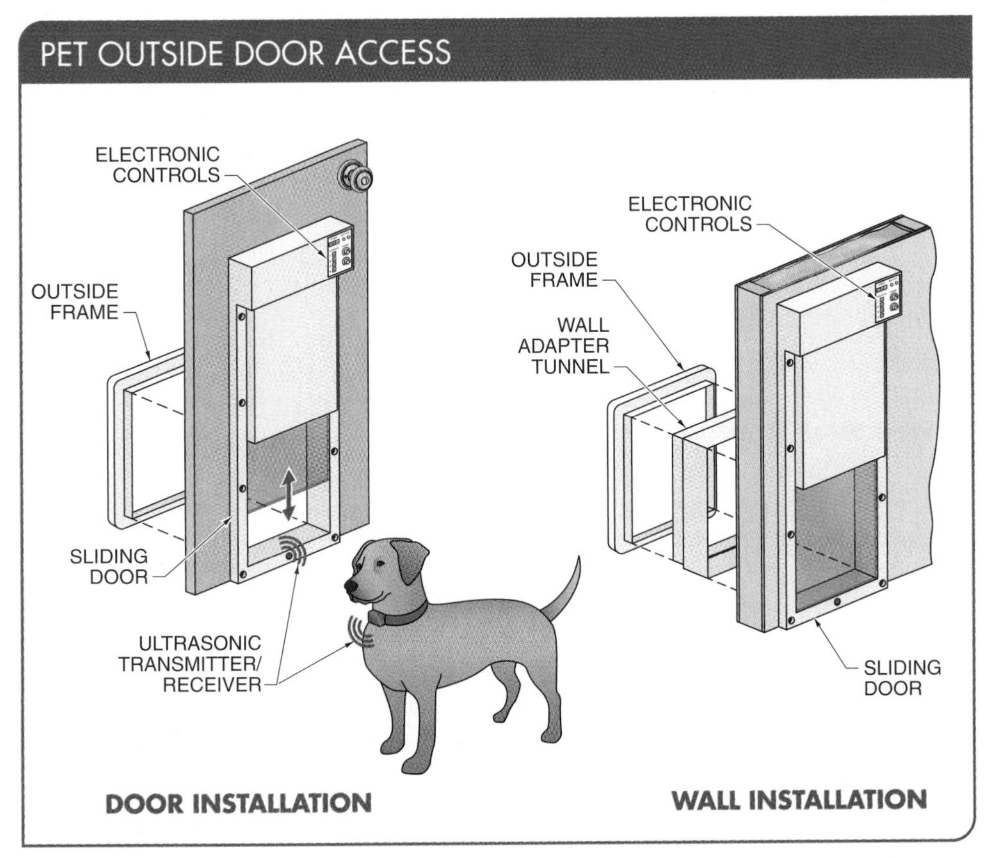

Figure 17-25. Smart pet doors allow pets to enter and exit a dwelling through an ultrasonic transmitter while preventing unwanted animals or intruders from entering the dwelling.

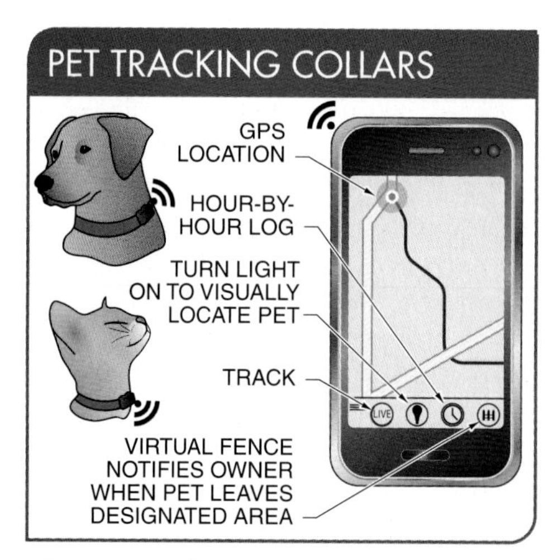

Figure 17-26. A pet tracking collar is a GPS-equipped collar that tracks the location and movement of a pet in real time.

Pet Training. Ultrasonic bark deterrents can help train pet dogs to stop nuisance barking. Ultrasonic bark deterrents are available as collars and portable, wall-mounted units. These smart devices allow 5 sec of natural barking before emitting ultrasonic sounds that are irritating to pets but inaudible to humans. The wall-mounted units can be used to deter barking in specific rooms of a dwelling, while a collar can be used to deter barking whenever the dog wears it. Ultrasonic bark deterrents can also have different sound levels (low, medium, and high) and manual buttons to deter other behaviors. **See Figure 17-27.**

Power over Ethernet (PoE)

Power over Ethernet (PoE) is a method of supplying power and data over a single Ethernet cable instead of having a separate power cord. PoE uses CAT 5E cable to connect DC-powered smart devices to the internet while also providing DC power to the devices without the need to run AC power. On one end of the cable in a PoE network is the power servicing equipment (PSE), and on the other end of the cable are the powered devices (PDs). Common examples of PDs are motorized window shades, webcams, and wireless access points. An advantage of PoE is that there is only one cable for both power and data, which is less expensive. PoE provides a solution when AC power may not be readily accessible and batteries are not an acceptable alternative.

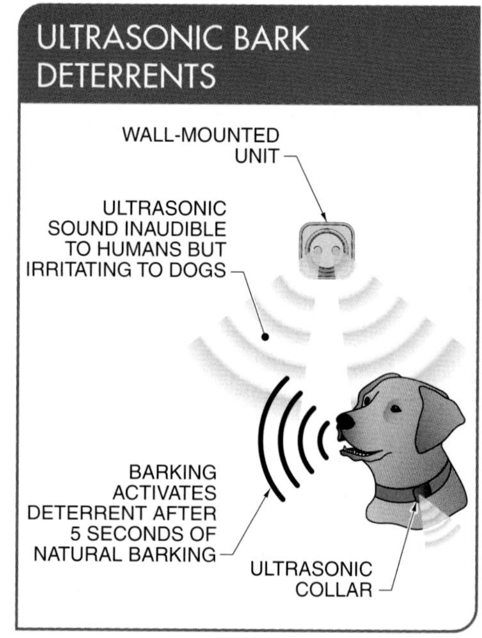

Figure 17-27. Ultrasonic bark deterrents are available as collars and portable, wall-mounted units.

There are two IEEE standards for PoE that might be used in a dwelling: IEEE 802.3af (PoE) and IEEE 802.3at (PoE+). The PoE standard provides for a maximum of 15.4 W per port, while PoE+ provides for a maximum of 30 W. Due to cable loss, the minimum guarantee is approximately 13 W per port for PoE and 25 W per for PoE+. The actual amount delivered depends on cable length. Therefore, if several devices are used in the system, the correct amount of power required must be calculated so that there is enough to properly operate the devices over long distances. Manufacturers' specifications must be carefully read or a professional installer should be hired when PoE will be used.

Do-it-yourself (DIY) vs. Professional Installation

The installation of smart home technology can be a do-it-yourself (DIY) project or it can be installed by a hired professional. Some smart devices are simple and can be installed with limited knowledge by following the manufacturers' directions. However, some manufacturers require that a professional install their products due to the specialized tools and access codes required for a proper installation. Improperly installed products may void the manufacturers' warranties.

When a professional is hired to install smart home technology, the homeowner should seek one who has the proper qualifications and is well trained. The Custom Electronic Design & Installation Association (CEDIA) is a professional association of member companies that prepares and certifies technicians to install smart systems and devices. The CEDIA certification process for home technology professionals (HTPs) includes a certification exam and continuing education for credential maintenance. It is recommended for homeowners to only hire CEDIA-certified or factory-authorized installers to install smart home technology.

Smart-Ready Homes and Smart Home Checklists

A survey by Coldwell Banker Real Estate LLC of 1250 adults (ages 18 and over) living in the US was conducted to determine what people desire in move-in ready homes. About 71% of the people surveyed had internet access and about 800 of the people surveyed owned at least one smart home product. According to the survey, many people want a move-in ready home to be "smart ready" as well. A smart-ready home has smart home technology and products already installed. For example, a smart-ready home is installed with products or tools that aid in controlling lighting, temperature, security, safety, and entertainment through the use of a smartphone, tablet, computer, or a separate autonomous system within the home.

Of the respondents who opted for a move-in ready home, 44% said that smart home technology should already be installed. Another 57% of the respondents would consider an older home to be updated if smart home technology was already installed in it. About 45% of the respondents agreed that a move-in ready home is new construction. Also, about 74% agree that a move-in ready home has new appliances, and 83% agree that a move-in ready home has updated kitchen and bathrooms. Another 85% agree that a move-in ready home has updated heating/cooling.

There are many considerations for both home buyers looking for a smart-ready home and homeowners looking to retrofit their dwelling with smart home technology. The following series of checklists can help an individual determine what is desired in, needed for, or already present in a smart-ready home:

Type of Installation

- DIY (do-it-yourself)
  - Wireless
  - Wired
  - Combination wired/wireless
- Professional
  - Wireless
  - Wired
  - Combination wired/wireless

Types of Controls

- Smartphones
  - Android
  - iOS
  - Windows Mobile
- Tablets
  - Android
  - iOS
  - Windows Mobile
- Remotes
- Touch screens
- Biometric
  - Voice
  - Gesture

Routers and Frequencies
- Single-band (2.4 GHz)
- Dual-band (2.4 GHz and 5.0 GHz)
- Tri-band (2.4 GHz and two 5.0 GHz)
- Mesh router network (900 MHz, 2.4 GHz, or 5.0 GHz)

Voice-Activated Virtual Assistants
- Amazon – Alexa
- Apple – Siri
- Google – Google Assistant
- Microsoft – Cortana

Smart Home Automation Systems
- Apple HomeKit
- Control4
- Crestron
- Insteon
- Google Home
- IFTTT
- Samsung SmartThings
- Vivint
- Wink
- ZigBee
- Z-Wave

Electrical Safety
- Ground-fault circuit interrupters (GFCIs)
- Arc-fault circuit interrupters (AFCIs)
- Standard grounded receptacles
- Isolated ground receptacles
- Tamper-resistant receptacles
- Universal serial bus (USB) receptacles
- Smart receptacles
- Double-insulated or grounded tools
- Circuit panel – 100 A or more
  - Properly rated circuit breakers/fuses
  - Intelligent circuit breakers

Perimeter and Entry/Exit Security
- Alarms
  - Audible
  - Visual
- Motion detectors
- Outdoor lighting (floodlights and solar lights)
  - Entry access points
  - Walkways
  - Landscaping
- Driveway sensor
- Wireless mailbox sensor
- Security signage
- Video surveillance security cameras
- Smart doorbell
- Smart locks (Wi-Fi/Bluetooth)
- Keyless fobs
- Biometric locks
  - Facial recognition
  - Fingerprint scanning
- Wireless gate sensors
- Pool safety systems
  - In-pool sensors
  - Pool water level sensors
  - Pool access door sensors

Interior Security
- Alarms
  - Audible
  - Visual
- Motion detectors
- Personal security fobs
- Panic buttons
- Keypads
- Smart pet door
- Door/window sensors
- Garage protection sensors
- Structural damage protection
  - Water detection system
  - Humidity sensors
  - Freeze sensors

Fire Safety
- Smoke detectors
- Heat detectors
- Carbon monoxide (CO) detectors
- Combination smoke/CO detectors
- Fire extinguishers

Electrical Surge Protection
- Whole-house surge protector
- Power strip surge protectors
- Smart power strips
- Uninterruptible power supplies (UPSs)

Energy Auditing/Monitoring Devices
- Smart meter
- Service panel monitor
- Plug-in energy monitors
- Smart energy-monitoring plugs

Climate Control
- Thermostats
  - Connected
  - Smart
- Dehumidifier
- Humidifier
- Air purifier
  - Stationary
  - Robotic
- Fans
  - Ceiling fans
  - Attic fans
  - Whole-house exhaust fans
- Roof vents
- Electric duct dampers
- Automatic HVAC registers
- Window treatments
  - Motorized shades and blinds
  - Electrochromic smart glass
- Weather station and in-home display

Lighting Control
- LED lights
- Motion sensor switches
- Timer switches
- Dimmer switches
- Smart switches
  - Smart plug wireless switches
  - Wireless ceiling fixture switches
  - Wireless wall switches

Electrical Test Instruments
- Voltage test light (neon tester)
- Voltage indicator
- Digital multimeter (DMM)
- Stud finder
- Branch circuit identifier

Basic DIY Tool Set
- Wire stripper
- Electrician's knife
- Pliers
  - Needle-nose
  - Diagonal-cutting
  - Side-cutting
- Screwdrivers
  - Flathead
  - Phillips
- Power drill and bits
- Electrician's hammer
- Drywall saw
- Adjustable wrench
- Tape rule (tape measure)
- Toolbox/tool pouch
- Torpedo level (pocket level)
- Fish tape
- Safety glasses
- Work gloves
- Work shoes

Specialty Tools
- Conduit bender
- PVC pipe cutter
- Offset drill
- Fuse puller
- Hacksaw

Basic DIY Supplies
- Wire connectors
- Wire markers
- Electrical tape
- NEC®-rated conductors (wire) and cables
  - Nonmetallic-sheathed
  - Metallic-sheathed
  - Single electrical conductors
- Switches
- Receptacles
- Electrical boxes/junction boxes

Specialty Supplies
- Conduit and connectors
- Heat-shrink tubing
- Soldering gun/soldering iron
- Propane torch

Smart Home Services
- 24/7 monitoring service
- Autodialer and RJ-31X jack
- Cloud-based storage (the Cloud)

SECTION 17.7 CHECKPOINT

1. How can indoor robots be used to clean the inside of a dwelling?
2. How can robots be used outside of a dwelling for outdoor maintenance?
3. What is wearable technology (wearables)?
4. What is power over ethernet (PoE)?

17 Lifestyle Applications of the Smart Home

Chapter Activities

Name ______________________ Date ______________

Smart Home Checklist

Use the following checklist to plan a smart-ready home or dwelling.

Type of Installation	
DIY (do-it-yourself)	
Wireless	
Wired	
Combination wired/wireless	
Professional	
Wireless	
Wired	
Combination wired/wireless	
Types of Controls	
Smartphones	
Android	
iOS	
Windows Mobile	
Tablets	
Android	
iOS	
Windows Mobile	
Remotes	
Touch screens	
Biometric	
Voice	
Gesture	

Routers and Frequencies	
Single-band (2.4 GHz)	
Dual-band (2.4 GHz and 5.0 GHz)	
Tri-band (2.4 GHz and two 5.0 GHz)	
Mesh router network (900 MHz, 2.4 GHz, or 5.0 GHz)	
Voice-Activated Virtual Assistants	
Amazon – Alexa	
Apple – Siri	
Google – Google Assistant	
Microsoft – Cortana	
Smart Home Automation Systems	
Apple HomeKit	
Control4	
Crestron	
Insteon	
Google Home	
IFTTT	
Samsung SmartThings	
Vivint	
Wink	
ZigBee	
Z-Wave	

Electrical Safety	
Ground-fault circuit interrupters (GFCIs)	
Arc-fault circuit interrupters (AFCIs)	
Standard grounded receptacles	
Isolated ground receptacles	
Tamper-resistant receptacles	
Universal serial bus (USB) receptacles	
Smart receptacles	
Double-insulated or grounded tools	
Circuit panel – 100 A or more	
Properly rated circuit breakers/fuses	
Intelligent circuit breakers	
Perimeter and Entry/Exit Security	
Alarms	
Audible	
Visual	
Motion detectors	
Outdoor lighting (floodlights and solar lights)	
Entry access points	
Walkways	
Landscaping	
Driveway sensor	
Wireless mailbox sensor	
Security signage	
Video surveillance security cameras	
Smart doorbell	
Smart locks (Wi-Fi/Bluetooth)	
Keyless fobs	
Biometric locks	
Facial recognition	
Fingerprint scanning	

Wireless gate sensors	
Pool safety systems	
In-pool sensors	
Pool water level sensors	
Pool access door sensors	
Interior Security	
Alarms	
Audible	
Visual	
Motion detectors	
Personal security fobs	
Panic buttons	
Keypads	
Smart pet door	
Door/window sensors	
Garage protection sensors	
Structural damage protection	
Water detection system	
Humidity sensors	
Freeze sensors	
Fire Safety	
Smoke detectors	
Heat detectors	
Carbon monoxide (CO) detectors	
Combination smoke/CO detectors	
Fire extinguishers	
Electrical Surge Protection	
Whole-house surge protector	
Power strip surge protectors	
Smart power strips	
Uninterruptible power supplies (UPSs)	

Energy Auditing/Monitoring Devices	
Smart meter	
Service panel monitor	
Plug-in energy monitors	
Smart energy-monitoring plugs	
Climate Control	
Thermostats	
Connected	
Smart	
Dehumidifier	
Humidifier	
Air purifier	
Stationary	
Robotic	
Fans	
Ceiling fans	
Attic fans	
Whole-house exhaust fans	
Roof vents	
Electric duct dampers	
Automatic HVAC registers	
Window treatments	
Motorized shades and blinds	
Electrochromic smart glass	
Weather station and in-home display	
Lighting Control	
LED lights	
Motion sensor switches	
Timer switches	
Dimmer switches	
Smart switches	
Smart plug wireless switches	
Wireless ceiling fixture switches	
Wireless wall switches	
Electrical Test Instruments	
Voltage test light (neon tester)	
Voltage indicator	
Digital multimeter (DMM)	
Stud finder	
Branch circuit identifier	
Basic DIY Tool Set	
Wire stripper	
Electrician's knife	
Pliers	
Needle-nose	
Diagonal-cutting	
Side-cutting	
Screwdrivers	
Flathead	
Phillips	
Power drill and bits	
Electrician's hammer	
Drywall saw	
Adjustable wrench	
Tape rule (tape measure)	
Toolbox/tool pouch	
Torpedo level (pocket level)	
Fish tape	
Safety glasses	
Work gloves	
Work shoes	

Specialty Tools	
Conduit bender	
PVC pipe cutter	
Offset drill	
Fuse puller	
Hacksaw	
Basic DIY Supplies	
Wire connectors	
Wire markers	
NEC®-rated conductors (wire) and cables	
Nonmetallic-sheathed	
Metallic-sheathed	
Single electrical conductors	
Switches	
Receptacles	
Electrical boxes/junction boxes	

Specialty Supplies	
Conduit and connectors	
Heat-shrink tubing	
Soldering gun/soldering iron	
Propane torch	
Smart Home Services	
24/7 monitoring service	
Autodialer and RJ-31X jack	
Cloud-based storage (the Cloud)	

Appendix

COMMON ELECTRICAL QUANTITIES AND PREFIXES

Name	Unit of Measure Abbreviations
1000 volts	1 kilovolt—1kV
One thousandth (1/1000) of an amp	1 milliamp—1 mA
47 thousand ohms	1 kiloohms—47 kΩ
1 millionth (1/1,000,000) of a farad	1 microfarad—1 μF
1000 watts	1 kilowatt—1 kW
One million (1,000,000) hertz	1 mega hertz—1 MHz

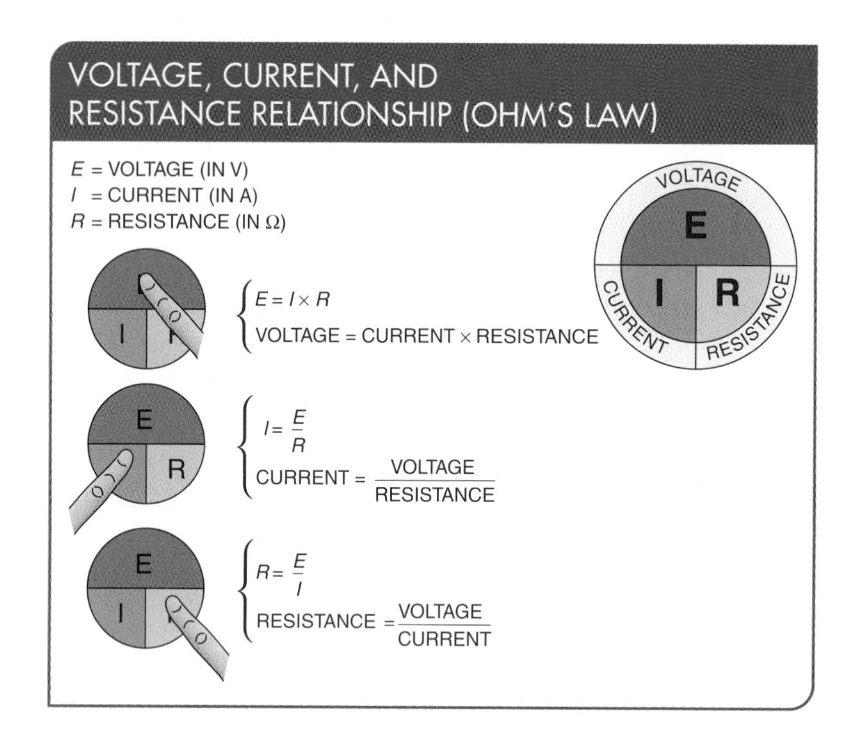

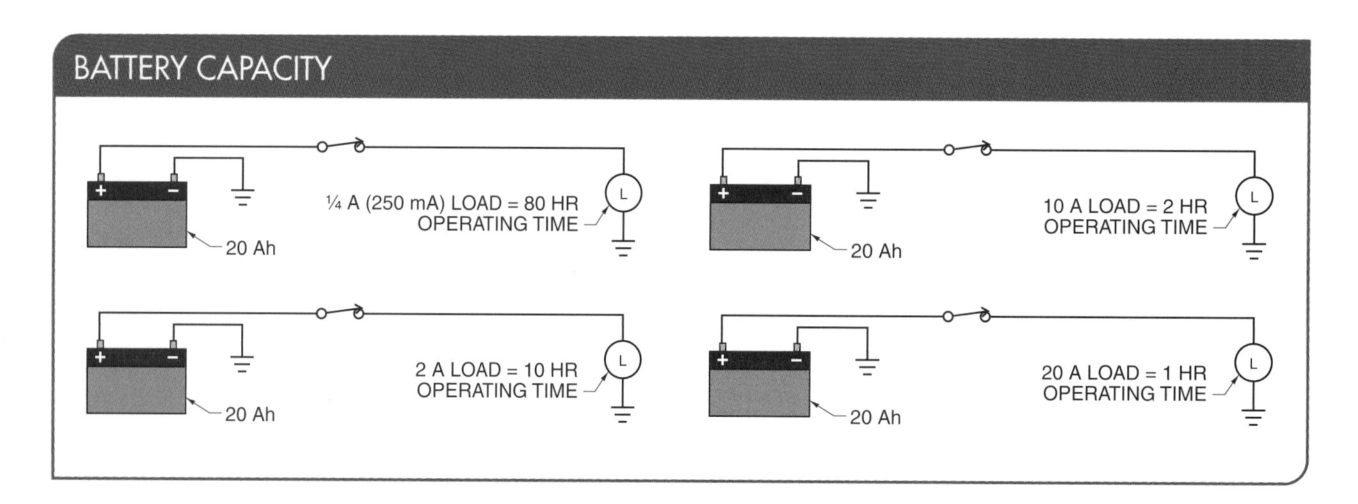

APPLIANCE STARTING AND RUNNING WATTAGES		
Appliance	**Wattage (Starting)**	**Wattage (Running)**
Refrigerator	2800	700
Freezer	2500	500
Well Pump –2 HP	6000	2000
Well Pump –3 HP	9000	3000
Gas Furnace Fan	500 to 2350	300 to 875

TYPICAL APPLIANCE WATTAGES	
Appliance	**Typical Wattage**
Radio	40 to 225
Toaster	200
Computer Monitor	200 to 800
Fan	275
Television	300 to 800
Printer	400 to 800
Microwave	600 to 1200
Dishwasher	700
Computer	700 to 1000
Electric Fry Pan	1000
Coffee Maker	1100
Washing Machine	1200
Hair Dryer	1500
Electric Cooktop Top – 8″	2100
Lights	4500
Electric Clothes Dryer	6000
Electric Oven	6500
Electric Water Heater	10,000

RUBBER INSULATING MATTING RATINGS					
Safety Standard	Material Thickness		Material Width (in in.)	Test Voltage	Maximum Working Voltage
	Inches	Millimeters			
BS921*	0.236	6	36	11,000	450
BS921*	0.236	6	48	11,000	450
BS921*	0.354	9	36	15,000	650
BS921*	0.354	9	48	15,000	650
VDE0680†	0.118	3	39	10,000	1000
ASTM D178‡	0.236	6	24	25,000	17,000
ASTM D178‡	0.236	6	30	25,000	17,000
ASTM D178‡	0.236	6	36	25,000	17,000
ASTM D178‡	0.236	6	48	25,000	17,000

* BSI–British Standards Institute
† VDE–Verband Deutscher Elektrotechniker Testing and Certification Institute
‡ ASTM International–American Society for Testing and Materials

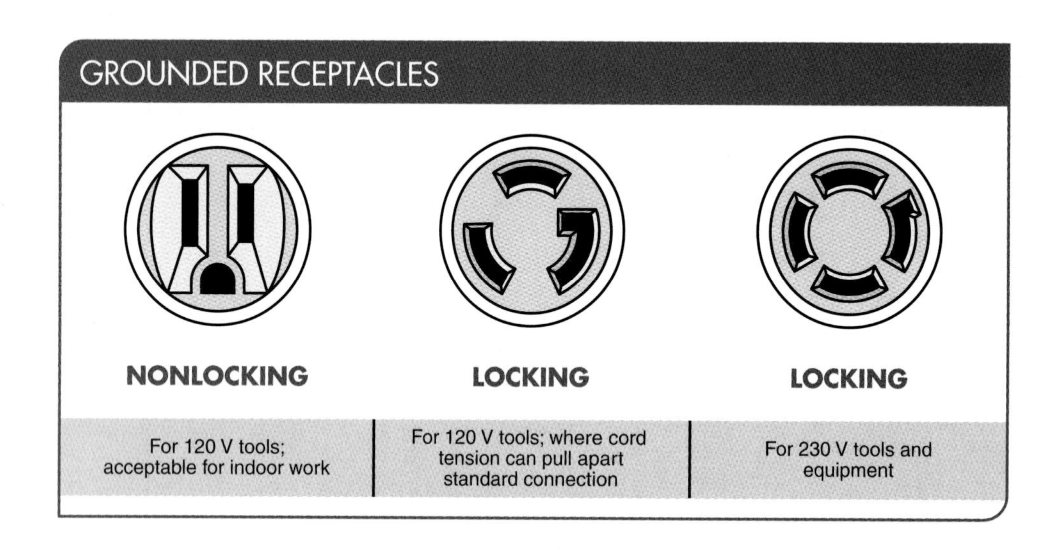

LIGHTING ABBREVIATIONS AND SYMBOLS			
Description	**Device**	**Abbrev**	**Symbol**
Wall Light	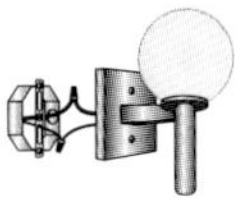	L	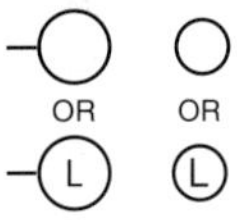 OR OR L L
Ceiling Light	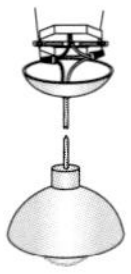	L	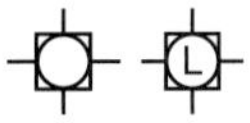 L
Recessed Ceiling Fixture Light	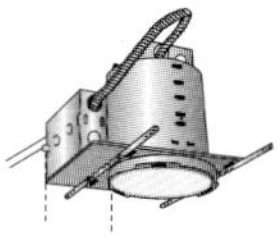	LAC	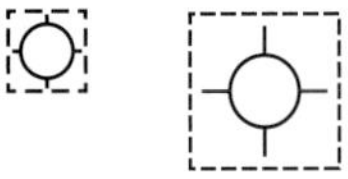 Outline indicates shape of fixture
Fluorescent Light		FLUOR	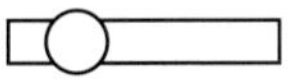 Extend rectangle for length of installation
Light with Lamp Holder		L	L
Light with Lamp Holder and Pull Switch	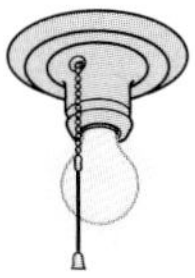	LPS	 L PS
Track Lighting	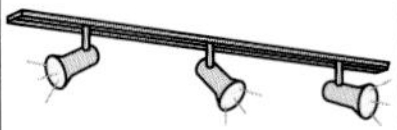	LTL	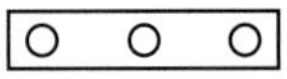
Fan Outlet	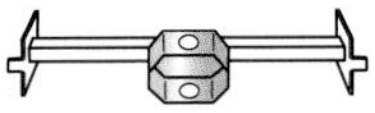	F	F Box listed as acceptable for fan support
Junction Box		J	J

RECEPTACLE OUTLET ABBREVIATIONS AND SYMBOLS			
Description	**Device**	**Abbrev**	**Symbol**
Single Receptacle (Outlet)	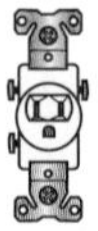	RCPT OR OUT	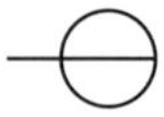
Duplex Receptacle		DX RCPT	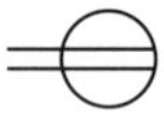
Split-Wired Duplex Receptacle	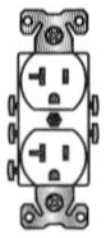	SPW RCPT	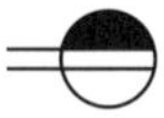
Triplex Receptacle	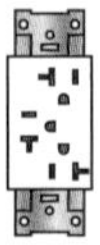	TRI RCPT	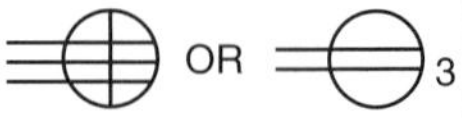 OR 3
Single Floor Outlet	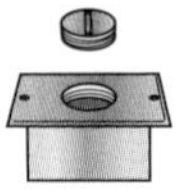	SF OUT	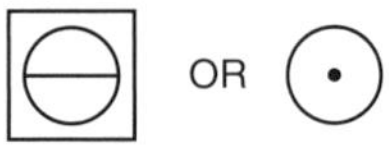 OR
Duplex Receptacle with Isolated Ground	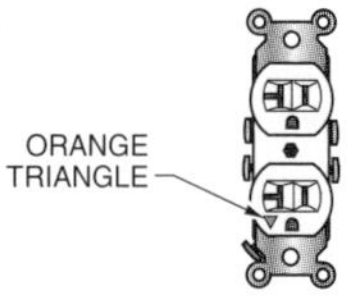 ORANGE TRIANGLE	DX IG	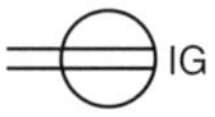 IG
Weatherproof Receptacle		IQ WP	 IG WP
Special-Purpose Outlet		OUT DW (DISHWASHER) OUT CD (CLOTHES DRYER)	 DW
Range Outlet		RNG OUT	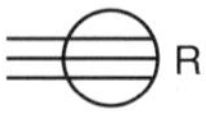 R
Combination Switch with Receptacle		SW & RCPT	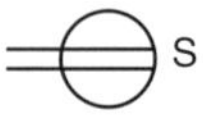 S
Ground-Fault Circuit Interuptor (GFCI)		GFCI	 GFCI

SWITCH ABBREVIATIONS AND SYMBOLS			
Description	**Device**	**Abbrev**	**Symbol**
Single-Pole Switch		SPST	$S - S_1$
Double-Pole Switch		DPST	S_2
Three-Way Switch		SPDT	S_3
Four-Way Switch		DPDT	S_4
Dimmer Switch		DMR SW	S_D
Switched Pilot Light		SPST	S_P
Weatherproof Switch		WP SW	S_{WP}

POWER ABBREVIATIONS AND SYMBOLS			
Description	Device	Abbrev	Symbol
Service Panel		SRV PNL OR PWR PNL	
Switch Wire Concealed in Wall, Ceiling, or Floor			
Motor	DC MOTOR	MTR	Ⓜ
Ground Connection		GR	
2-Wire Cable			
3-Wire Cable			
4-Wire Cable			
Cable Return to Service Panel		HR	
Wire Turned Up in Wall			
Wire Turned Down in Wall			
Exposed Wire			

CIRCUIT PROTECTION AND CONTROL DEVICE SYMBOLS			
Description	**Device**	**Abbrev**	**Symbol**
Fuses		FU	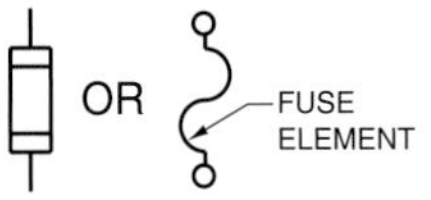 OR — FUSE ELEMENT **SINGLE FUSE**
Single-Pole Circuit Breaker	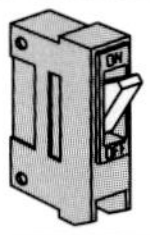	SPCB	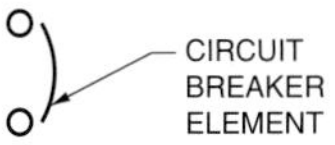 CIRCUIT BREAKER ELEMENT **SINGLE-POLE CIRCUIT BREAKER**
Double-Pole Circuit Breaker	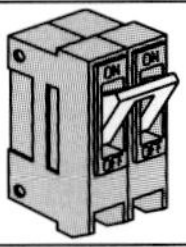	DPCB	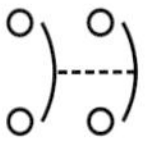 **DOUBLE-POLE CIRCUIT BREAKER**
Single-Pole Switch	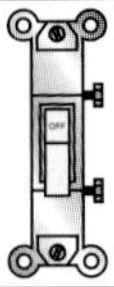	SPST	**SINGLE-POLE SINGLE-THROW**
Three-Way Switch	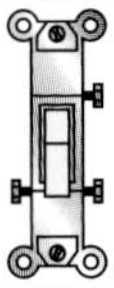	SPDT	**SINGLE-POLE DOUBLE-THROW**
Four-Way Switch	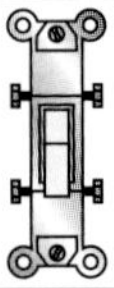	DPDT	**DOUBLE-POLE DOUBLE-THROW**
	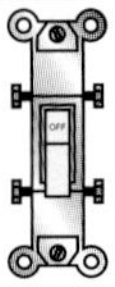	DPST	**DOUBLE-POLE SINGLE-THROW**
Dimmer Switch	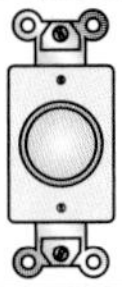	DS	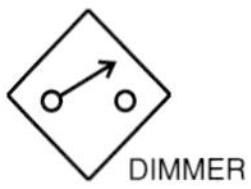 DIMMER
Float Switch		FS	 LEVEL OPERATOR **NORMALLY OPEN (NO)** **NORMALLY CLOSED (NC)**
Temperature Switch	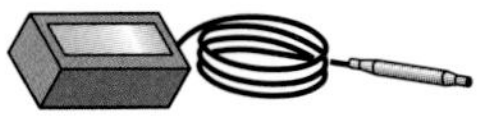	TEMP SW	 TEMPERATURE OPERATOR **NO** **NC**
Contacts	 **SMALL**  **LARGE** 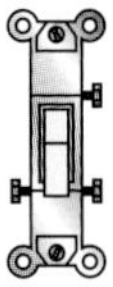 **SWITCH**	NO	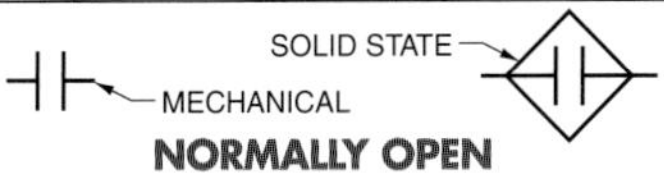 SOLID STATE MECHANICAL **NORMALLY OPEN**
		NC	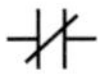 **NORMALLY CLOSED**

ELECTRONIC DEVICE AND LOAD SYMBOLS			
Description	Device	Abbrev	Symbol
AC Motors		1ϕ	T1 T2 T = TERMINAL SINGLE-PHASE, SINGLE- OR DUAL-VOLTAGE
Lights		L	L OR R LETTER INDICATES COLOR A = AMBER G = GREEN R = RED B = BLUE PILOT LIGHT
Alarms		AL or BELL	
		AL or HORN	
		BUZZ	
Diode		D	ANODE CATHODE
Light-Emitting Diode		LED	
Photodiode		D	

CONDUCTOR SIZES AND AMPACITIES							
Size AWG or kcmil	**Copper Conductor Ampacities**			**Aluminum Conductor Ampacities**			**Size AWG or kcmil**
	Temperature Rating of Conductor			Temperature Rating of Conductor			
	60°C	75°C	90°C	60°C	75°C	90°C	
	Types TW UF	Types RHW THHW THW XHHW THWN USE	Types RHH THHW RHW-2 THWN-2 XHHW THW-2 XHHW-2 THHN XHH USE-2	Types TW UF	Types RHW THHW THW XHHW THWN USE	Types RHH THHW RHW-2 THWN-2 XHHW THW-2 XHHW-2 THHN XHH USE-2	
14	20	20	25	—	—	—	—
12	25	25	30	20	20	25	12
10	30	35	40	25	30	35	10
8	40	50	55	30	40	45	8
6	55	65	75	40	50	60	6
4	70	85	95	55	65	75	4
3	85	100	110	65	75	85	3
2	95	115	130	75	90	100	2
1	110	130	150	85	100	115	1
1/0	125	150	170	100	120	135	1/0
2/0	145	175	195	115	135	150	2/0
3/0	165	200	225	130	155	175	3/0
4/0	195	230	260	150	180	205	4/0

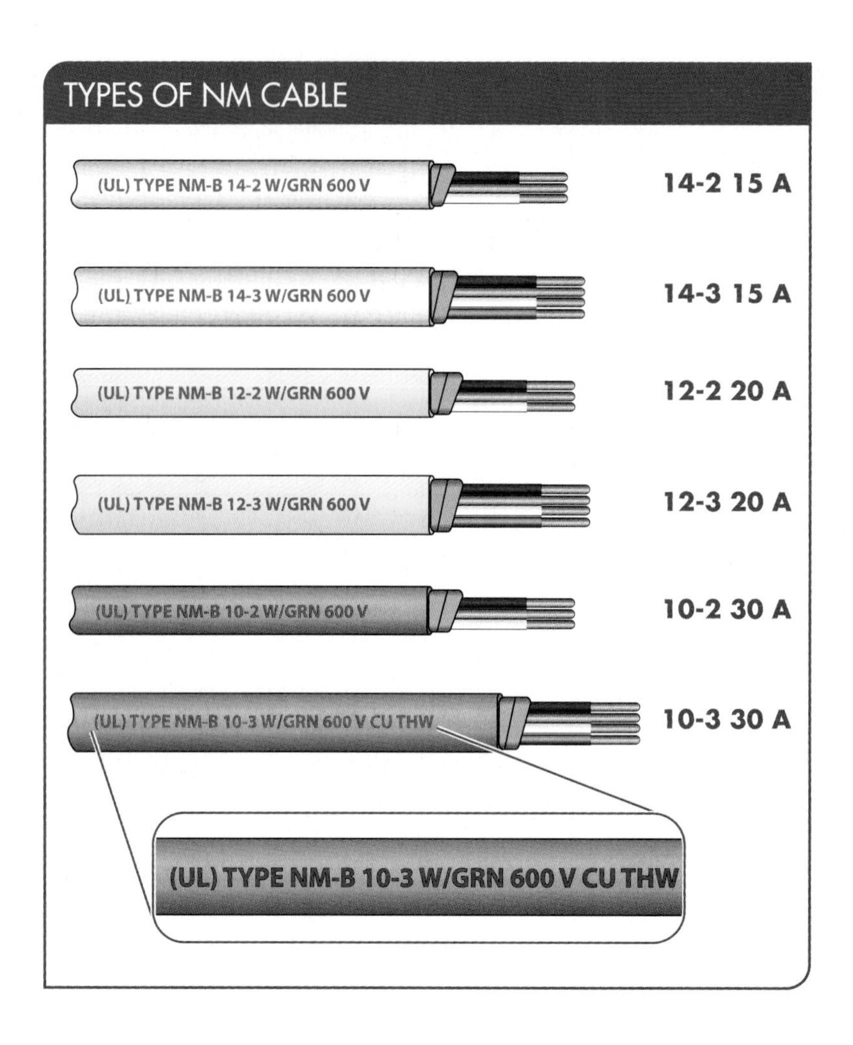
TYPES OF NM CABLE
(UL) TYPE NM-B 14-2 W/GRN 600 V
14-2 15 A
(UL) TYPE NM-B 14-3 W/GRN 600 V
14-3 15 A
(UL) TYPE NM-B 12-2 W/GRN 600 V
12-2 20 A
(UL) TYPE NM-B 12-3 W/GRN 600 V
12-3 20 A
(UL) TYPE NM-B 10-2 W/GRN 600 V
10-2 30 A
(UL) TYPE NM-B 10-3 W/GRN 600 V CU THW
10-3 30 A
(UL) TYPE NM-B 10-3 W/GRN 600 V CU THW

CODES AND STANDARDS ORGANIZATIONS

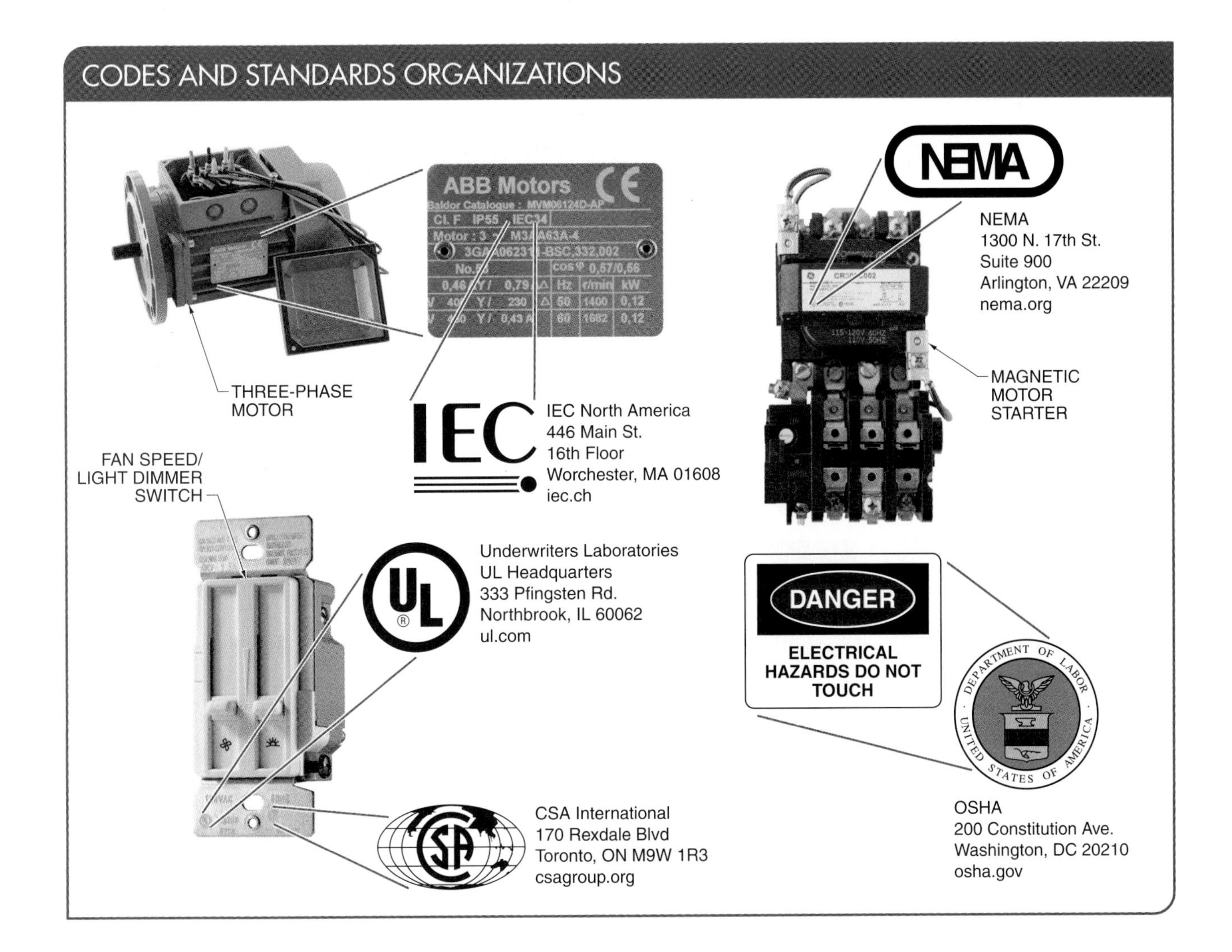

NATIONAL ELECTRICAL CODE® (NEC®)

Section	Description
Article 90	Introduction
Chapter 1	General
Chapter 2	Wiring and Protection
Chapter 3	Wiring Methods and Materials
Chapter 4	Equipment for General Use
Chapter 5	Special Occupancies
Chapter 6	Special Equipment
Chapter 7	Special Conditions
Chapter 8	Communication Systems
Chapter 9	Tables

Glossary

A

abbreviation: A letter or combination of letters that represents a word.

AC cycle: The complete positive and negative alternation of an AC cycle.

AC generator: A machine that converts mechanical energy into AC electricity.

activity monitoring: The process through which a smart power strip can determine when equipment is in use or in an idle state by monitoring equipment activity.

AC voltage: Voltage that causes current to reverse its direction of flow at regular intervals or cycles.

alternating current (AC): Current that reverses or alternates its direction of flow at regular intervals.

alternation: One half of an AC cycle.

ampere (A): The number of electrons passing a given point in one second.

annunciator: An audible or visual output device that notifies residents that one or more detectors have been activated.

anti-short bushing: A thermoplastic insulating device used to protect the wires of metallic-sheathed cable.

app: A specialized software program that can be downloaded to a device, such as smartphone or tablet, to perform a specific function.

appliance plug: A plug used to power an appliance that produces heat, such as an electric grill, roaster, broiler, waffle iron, or large coffeemaker. Also known as a heater plug.

arc-fault circuit interrupter (AFCI): A current-sensing device designed to detect a wide range of arcing electrical faults.

arc-fault circuit interrupter (AFCI) receptacle: A fast-acting electrical device that detects and opens a circuit in response to an electrical arc.

arc-rated face shield: An eye and face protection device that covers the entire face with a plastic shield.

armored cable (Type AC): A cable that consists of two, three, or four individually insulated wires and a bonding strip protected by a flexible metal outer jacket.

attic fan: A ventilation fan that exhausts hot air out of an attic.

automated pill dispenser: An electronic device that stores medicine in compartments and can be programmed to sound alerts and dispense pills on schedule.

automatically controlled circuit: Any circuit that uses a control device to initiate an action for a circuit to operate.

automatic iris: A motorized adjustable aperture that controls the amount of light passing through the lens of a camera to its CCD.

automatic transfer switch (ATS): A device that switches a load between two different sources of electric power.

B

back-feeding (islanding): The flow of electrical energy in the reverse direction from its normal direction of flow.

back-wired connector: A mechanical connection method used to secure wires to the backs of switches and receptacles. Also known as a quick connector.

ballast: A device that controls the flow of current to a gas discharge lamp while providing a starting voltage.

battery: A group of connected electrochemical cells.

battery charge controller: A device that regulates and maintains battery voltage, preventing batteries from being overcharged while allowing charging when it is needed.

battery charger: A device used to put energy back into a rechargeable battery.

battery management system (BMS): An electronic system that protects and manages rechargeable batteries from operating outside their safe operating area.

battery-powered cable cutter: A power tool designed to cut various diameters of electrical cables.

battery storage system: A network connected batteries that store electricity, typically generated from renewable energy sources, for use at a later time.

biometric verification: A means by which a person can be identified by evaluating, or reading, one or more distinguishing biological traits.

biometric reader: A specialized device that analyzes a person's physical or behavioral characteristics and compares these characteristics to a database for verification.

bit: *See* drill bit.

blade cartridge fuse: A snap-in OCPD with blade-like connectors at each end of the fuse that operates based on the heating effect of an element.

bonding conductor: A reliable conductor that ensures the electrical conductivity between two metal parts that must be connected electrically.

bonding strip: An uninsulated conductor inside armored cable that is used for grounding.

box-cutting knife: *See* utility knife.

branch circuit: An electrical circuit that spans from the service panel and throughout a dwelling to power electrical devices.

branch circuit identifier: A two-piece test instrument consisting of a transmitter that is plugged into a receptacle and a receiver that provides an audible indication when located near the circuit to which the transmitter is connected.

building-integrated photovoltaics (BIPV): PV materials that are used to replace conventional building materials in parts of the building envelope.

C

cable: Two or more insulated wires grouped together within a common protective cover.

cable run: A length of installed cable connecting two electrical devices that are not in immediate proximity to one another.

cable stripper: A tool used to strip insulation from cables with outside diameters from ¼″ to 1½″.

cable tie gun: A handheld device that is used to tighten plastic or steel cable ties around a bundle of wires or small-diameter cables.

Canadian Standards Association (CSA): A Canadian organization similar to the UL® that tests equipment and devices for conformity to meet national standards.

carbon monoxide (CO) detector: A device that detects the presence of CO gas to prevent CO poisoning.

carbon monoxide (CO) sensor: A device that detects CO gas.

cartridge fuse: A snap-in OCPD that operates on the same basic heating principle as a plug fuse.

charge-coupled device (CCD): A light-sensitive device that captures light and converts it to digital data.

circuit breaker: An OCPD designed to protect electrical devices and individuals from overcurrent conditions.

circuit breaker fuse: A screw-in OCPD that has the operating characteristics of a circuit breaker.

clamp-on ammeter adapter: A digital multimeter accessory that measures current in a circuit by measuring the strength of a magnetic field around a conductor.

clip grounding: A grounding method where a grounding clip is slipped over the grounding wire from the electrical device.

Cloud, the: *See* cloud storage.

cloud storage: A method for delivering services and storing data that can be retrieved from the internet through web-based tools and applications. Commonly referred to as "the Cloud."

code: A regulation or minimum requirement.

cold solder joint: A defective solder joint that results when the parts being joined do not exceed the liquid temperature of the solder wire.

community microgrid: A group of electrical sources and loads that can be connected to the utility grid or operate by itself when the need arises.

compass saw: A handsaw that is used to saw curves, holes, and other internal openings.

component grounding: A grounding method where the ground wire is attached directly to an electrical component, such as a receptacle.

component plan: A group of electrical component schedules that state the required locations for receptacles, lights, and switches.

compound-wound generator: A generator that includes series and shunt field windings.

compression connector: A type of box fitting that firmly secures conduit to a box by utilizing a nut that compresses a tapered metal ring (ferrule) into the conduit.

concentrator: A device that primarily functions to serve as the collection point for several smart meters within a NAN.

conductor: A material that has very little resistance and permits electrons to move easily through it.

conduit: A metallic tube that protects and routes electrical wiring.

conduit bender: A device used to bend certain types of conduit in order to follow walls or clear obstructions.

conduit coupling: A type of fitting used to join one length of conduit to another length of conduit and still maintain a smooth inner surface.

conduit screwdriver: A hand tool designed with a hooded blade to fasten conduit screws.

construction layout: An isometric drawing of a branch circuit in a dwelling.

contact arcing: An electrical arc that occurs when opening and closing circuit breakers.

continuous load: A load that operates all the time.

control panel: An electronic device that receives input signals from detectors, processes information, turns output devices on and off, and sends notifications to a monitoring service.

copper-clad aluminum: A conductor with copper bonded to the outside of an aluminum wire to counter the problem of oxidation.

crimp-type connector: An electrical device that is used to join wires together or to serve as terminal ends for screw connections.

CSA label: A stamped icon indicating that extensive tests have been conducted on an electrical device by the Canadian Standards Association (CSA).

current: The measure of the flow of electrons through an electrical circuit.

D

DC circuit breaker: An OCPD that protects electrical devices operating with DC and contains additional arc-extinguishing measures.

DC generator: A machine that converts mechanical energy into DC electricity.

DC voltage: Voltage that causes current to flow in one direction only.

decibel (dB): A unit of measure used to express the relative intensity of sound.

demand response (DR) program: A utility incentive program for reducing electrical demand during peak hours.

device layout: An exploded view of wires and electrical device connections.

differential: The difference between the temperature at which the switch in the thermostat turns the system on and the temperature at which the thermostat turns the system off.

digital multimeter (DMM): An electrical test instrument that can measure more than one electrical quantity and display the measured quantities as numerical values.

dimmer switch: An electrical control device that is used to control light brightness by adjusting the voltage level applied to the light.

direct current (DC): Current that flows in only one direction.

distributed power generation: The generation of power on site or in close proximity to where the power is used.

door/window sensor: A magnetic switch device that indicates whether a door or window is closed or open.

double-pole switch: A double-pole, single-throw (DPST) electrical control device.

drill bit: A rotary-end cutting tool that fits into the chuck of a power drill. Also known as a bit.

driveway sensor: A device that is designed to detect objects entering a driveway and transmit a signal to a receiver.

dry cell: Two electrodes in an electrolyte paste that are used to create a chemical reaction.

drywall saw: A handsaw that is used to saw cutouts in drywall for electrical boxes.

dual-band router: A wireless router that simultaneously broadcasts at 2.4 GHz and 5 GHz.

duplex receptacle: A standard receptacle that has two outlets for connecting two different plugs.

E

earplug: An ear protection device made of moldable rubber, foam, or plastic and inserted into the ear canal.

earmuff: An ear protection device worn over the ears.

efficacy: The number of lumens of light delivered for the energy (wattage) used, expressed in lumens per watt (lm/W).

electrical circuit: An assembly of conductors (wires), electrical devices (switches and receptacles), and electrical components/loads (lights and motors) through which current flows.

electrical component schedule: A list of electrical equipment that indicates manufacturer specifications and how many of each electrical device are required for a room or area in a component plan.

electrical consumption (usage): The total amount of electrical energy used during a billing period.

electrical floor plan: A print that shows all of the power circuits for a specific floor of a residence.

electrical grid: The entire network of power generation stations, transmission stations, distribution systems, and corresponding power lines.

electrical metallic tubing (EMT): Light-gauge metallic tube used to route wires.

electrical noise: Unwanted signals that are present on a power line.

electrical shock: A shock that results any time a body becomes part of an electrical circuit.

electrician's knife: A knife that is similar in design to a pocket knife and is used for cutting and scraping insulation from conductors.

electricity: A form of energy where electrons move from the outer shell of one atom to the outer shell of another atom when an electromotive force (EMF) is applied to a material.

electric vehicle supply equipment (EVSE): The charging system for an electric vehicle and includes an electrical junction box and plug that connects the electric vehicle battery to an AC source.

electrochromic smart glass: A material that changes color when an electric current is applied to it.

electrode: A long metal rod used to make contact with the earth for grounding purposes.

electromagnetic induction: The ability of a coil of conductor to induce a voltage in another circuit or conductor.

electromagnetic inductive charging: A wireless charging method that uses an electromagnetic field to transfer energy between coils.

electromagnetism: Magnetism that is produced when electric current passes through a conductor.

electromechanical power source: A device that uses chemical reactions to produce or store electrical energy.

energy: The capacity to do work.

energy audit: An assessment of how much energy a dwelling consumes to determine what measures can be taken to make the dwelling more energy efficient.

energy disaggregation technology: A new smart meter technology that can distinguish between various electrical loads in a consumer's home by monitoring the loads over time.

energy monitor: A device used inside a dwelling to provide residents with information on the electrical consumption of appliances and other devices.

equipment ground: A circuit designed to protect individual components of an electrical system.

eutectic alloy: An alloy that has one specific melting temperature with no intermediate stage.

F

ferrule cartridge fuse: A snap-in OCPD that has conductive ferrules at each end of the fuse.

fire block: A horizontal member between studs that slows down the passage of flames in case the structure catches fire.

firmware: A software program permanently embedded into a modem like a computer chip to give permanent instructions.

first aid: Help for a victim immediately after an injury and before professional medical help arrives.

fish tape: A retractable tape used to pull conductors through conduit, through inaccessible spaces, or around obstructions in walls.

fixed-temperature heat detector: A heat detector designed to respond when a room reaches a specific temperature.

flexible metal conduit (FMC): Metallic tube of interlocked metal strips that is bendable by hand.

floodlight: A lamp that casts general light over a large area.

floor mat: A sensing device that detects the weight of a person walking on it.

floor plan: A print that provides a plan view of a floor of a residence.

fluorescent lamp: A low-pressure discharge lamp in which the ionization of mercury vapor transforms ultraviolet (UV) energy into light.

flute: A spiral groove that runs along the length of a drill bit.

four-way switch: A double-pole, double-throw (DPDT) electrical control device.

frame rate: The number of images that are transmitted each second.

freeze sensor: A sensor that detects low-temperature conditions and provides a warning before freezing occurs.

fuse puller: A nonmetallic device that is used to safely remove fuses from service panels and subpanels.

flux: A chemical substance that cleans the soldering surface and promotes the melting of the solder wire.

fuel cell: An electrochemical device that combines hydrogen and oxygen to produce electricity.

fuse: An OCPD used to limit the rate of current flow in a circuit.

G

garage door opener: A motorized device that opens and closes a garage door.

generator: A machine that converts mechanical energy into electrical energy by means of electromagnetic induction.

geofencing: The practice of using the global positioning system (GPS) to define a specific geographic boundary.

geothermal energy: Renewable energy that is derived from heat contained within the earth.

GFCI receptacle tester: A test instrument that is plugged into a GFCI receptacle to determine whether the receptacle is properly wired and energized.

ground: A low-resistance conducting connection between electrical circuits, equipment, and the earth.

grounded conductor: A conductor that has been intentionally grounded.

ground fault: Any amount of current above the level that may deliver an electrical shock.

ground-fault circuit interrupter (GFCI): An electric device that protects personnel by detecting ground faults and quickly disconnecting power from the circuit.

ground-fault circuit interrupter (GFCI) receptacle: An electrical receptacle that offers protection by detecting ground faults and then quickly disconnecting power from the circuit.

H

hacksaw: A handsaw that has an adjustable steel frame designed to hold various lengths and types of blades for cutting different materials.

halogen lamp: An incandescent lamp filled with a halogen gas.

hammer: A striking or splitting hand tool with a hardened head fastened perpendicularly to a handle.

handsaw: A saw operated with one hand and used to cut and trim materials to their proper dimensions.

hand tool: A tool that uses no external power other than that supplied by the person using the tool.

heat detector: A device that detects an increase in temperature of a heat-sensitive element.

heater plug: *See* appliance plug.

heat pump: A mechanical compression refrigeration system containing devices and controls that reverse the flow of refrigerant to move heat from one area to another.

heat-shrink tubing: Plastic (polymer) tubing that is designed to shrink (contract) when heated.

heat sink: A device or material that absorbs and disperses heat away from an electronic device to keep it from overheating.

heavy-duty plug: A plug that is used on high-wattage appliances and equipment that operate on 230 V or 460 V.

home area network (HAN): A computer network that facilitates communication among devices within the close vicinity of a dwelling.

home energy management: The process of monitoring and controlling energy use, while also ensuring the delivery of a reliable source of electricity.

home energy management system (HEMS): A group of products that processes and controls smart energy-using devices and their information on both sides of the smart meter to determine the best use of energy at any given time.

home run: A line with an arrow that shows where circuit wiring should return to the service panel.

humidity sensor: A sensor that detects the amount of moisture that the air can hold at a given temperature.

hybrid AC/DC microgrid: A localized electrical grid or network that operates on AC and DC and can be disconnected from the traditional electrical grid or smart grid to operate on its own power source.

I

impact wrench: A corded or cordless wrench used to supply short, rapid impulses to sockets.

incandescent lamp: A lamp that produces light by the flow of electrical current through a tungsten filament inside a gas-filled, sealed glass bulb.

inclining block rate: A utility program that charges a higher rate for each incremental block of consumption.

indenter connector: A type of box fitting that secures conduit to a box with the use of a special indenting tool.

insolation map: A map that indicates the average solar energy received in hours of peak sunlight per day on a specific area in a given month or year.

interactive distributed generation: On-site power generation where switchgear can activate additional generation sources on demand.

interchangeability: The ability of an automated system to operate correctly when one device is substituted for another device.

intermittent load: A load that does not operate all the time.

International Electrotechnical Commission (IEC): An organization that develops international safety standards for electrical equipment.

internet service provider (ISP): A company that provides access to the internet.

inverter: A device that receives DC power and converts it to AC power.

ionization smoke detector: A device that uses a radioactive source to ionize (electrically charge) the air to conduct a small current in the sensing chamber to detect smoke particles.

isolated-ground receptacle: A special standard receptacle that minimizes electrical noise by providing its own grounding path.

islanding: A condition in which a generator continues to power a location when electrical grid power is not present.

J

joist: A horizontal wood member placed on edge to support a floor or ceiling.

junction box: An enclosure designed to house a termination of electrical wires.

K

kilowatt-hour (kWh): The amount of electricity used in an hour.

kinetic energy: The energy of motion.

knee pad: A rubber, leather, or plastic pad strapped onto the knees for protection.

knockout (KO): A round indentation stamped on a receptacle box that allows for quick installation of fittings to secure cables, conduit, or metallic cable.

L

lamp: A device that converts electrical energy into light.

LED light: A solid-state semiconductor device used as an illumination source in a luminaire because it produces light when a DC current passes through it.

light-emitting diode (LED): A device that emits a specific color of light when DC voltage is applied across a semiconductor junction.

light-emitting polymer (LEP): *See* organic light-emitting diode (OLED).

limited approach boundary: The distance from an exposed energized conductor or circuit part at which a person can get an electric shock and is the closest distance an unqualified electrical worker can approach.

line (ladder) diagram: A diagram that shows the logic of an electrical circuit or system using standard electrical symbols.

lineman's knife: A knife that is used for skinning and scraping insulation off of conductors and cables.

liquid crystal display (LCD): A flat, alphanumeric display that uses liquid crystals to display information without directly emitting light.

load: Anything that consumes electrical power.

lockout: The process of removing the source of electrical power and installing a lock that prevents the power from being turned on.

lockout device: A lightweight enclosure that allows the lockout of standard control devices.

low-voltage outdoor lighting: A system that uses 12 V or 24 V to power low-voltage lamps to illuminate walkways, landscaping, or dark areas on a property.

lubricant: A wet or dry compound that is applied to the exterior of wires to allow them to slide easier.

lumen (lm): A measure of the total amount of visible light from a light source.

lumens-per-watt rating: An average rating over the lifetime of the lamp.

luminaire: A lighting unit and all of its parts, which includes the internal and external parts as well as the lamp that actually provides illumination.

lux rating: The rating of the amount of light, or lux, that falls on an object being viewed by a camera.

M

magnetic circuit breaker: An OCPD that operates by using miniature electromagnets to open and close contacts.

manually controlled circuit: Any circuit that requires a person to initiate an action for the circuit to operate.

mechanical lock: A lock that includes a tumbler and matching key to permit the lock to be opened.

mesh network: A network that has no centralized access point but relies on all nodes to communicate with each other and form an interconnected network.

mesh technology: A system that extends the distance that a signal can travel within the system without increasing the power of each transmitter.

metal-clad (Type MC) cable: A cable that consists of two or more individually insulated wires and a separate grounding wire inside a flexible metal outer jacket.

metropolitan area network (MAN): *See* neighborhood area network (NAN).

microgrid: A mall local electrical network or grid that can operate completely separate from the main electrical grid.

microinverter: A device that converts DC power to AC power and is mounted directly to individual solar panels.

microwave detector: A device that detects a moving object by transmitting electromagnetic (EM) radio waves and receiving the reflected waves off of the object.

modem: A network device that transforms outgoing digital signals from a computer into analog signals that are transmitted over wires and that transforms incoming analog signals to digital signals for a computer.

modular connector: An electrical connector that was originally designed for use in telephone wiring but has since been also used for Ethernet applications.

momentary power interruption: A decrease to 0 V on one or more power lines lasting from 0.5 cycles up to 3 sec.

motion detector: A device that uses a sensor to detect movement and trigger a signal.

motion sensor switch: An electrical control device that automatically turns on or off based on the occupancy or vacancy of a room space.

N

National Electrical Code® (NEC®): A code book of electrical standards that indicate how electrical systems must be installed and how work must be performed.

National Electrical Manufacturers Association (NEMA): A national organization that assists with information and standards concerning proper selection, ratings, construction, testing, and performance of electrical equipment.

National Fire Protection Association® (NFPA®): A national organization that provides guidance in assessing the hazards of products of combustion.

neighborhood area network (NAN): The utility's outdoor network that collects data from multiple smart meters in a specific community or city location.

neon tester: *See* voltage test light.

net metering: An electric utility billing arrangement that credits customers for the electricity they add to the grid.

nonmetallic-sheathed (NM) cable: Electrical cable that has a set of insulated electrical conductors held together and protected by a strong plastic jacket.

O

Occupational Safety and Health Administration (OSHA): A federal agency that requires all employers to provide a safe environment for their employees.

offset: A double conduit bend with two equal angles bent in opposite directions in the same plane in a conduit run.

Ohm's law: The relationship between voltage, current, and resistance properties in an electrical circuit.

one-line diagram: A diagram that uses single lines and graphic symbols to show the current path, voltage values, circuit disconnect, OCPDs, transformers, and service panels for a residential electrical system.

open circuit transition switching: The process in which power is momentarily disconnected when switching a circuit from one voltage supply to another.

organic light-emitting diode (OLED): A thin film of carbon-based organic molecules that create light with the application of voltage. Also known as a light-emitting polymer (LEP).

outlet: *See* receptacle.

overcurrent protective device (OCPD): Electrical equipment that is used to protect service, feeder, and branch circuits and equipment from excess current by interrupting the flow of current.

P

panel display: A thin, lightweight screen used to project images.

parallel circuit: A circuit in which devices are connected parallel and are normally open (NO).

passive infrared (PIR) detector: A device that senses the difference in temperature between humans or animals and the background space.

peaker power plant: A back-up power plant that is reserved to run only during times of increased demand.

peak time rebate (PTR): A utility program that offers monetary rebates to customers who reduce energy consumption during periods of high-cost electricity (peak times).

pegboard: A hardboard that usually has a laminated finish on the front and is perforated with equally spaced holes for accepting hooks.

personal protective equipment (PPE): Clothing and/or equipment worn by individuals to reduce the possibility of injury in the work area.

pet tracking collar: A GPS-equipped collar that tracks the location and movement of a pet in real time.

phantom load: An appliance or device that uses power when plugged in despite technically being turned off. Also referred to as vampire load or standby power.

photocell (photoconductive cell): A light-activated control switch that varies resistance based on the intensity of the light striking it.

photoelectric smoke detector: A device that has an infrared LED and a photocell sensor in the sensing chamber to detect smoke particles.

photovoltaics: Solar energy technology that uses the unique properties of semi-conductors to convert solar radiation into electricity.

piconet: A network that allows a device to simultaneously communicate with several other devices.

pictorial drawing: A drawing that shows the length, height, and depth of an object in one view.

pigtail grounding: A grounding method where two grounding wires are used to connect an electrical device to a grounding screw in the box and then to system ground.

pigtail splice: A type of splice that is used to connect the ends of conductors (wires).

plug: A device at the end of a cord that connects equipment to an electrical power supply by means of a receptacle.

plug-in monitor: A device that only measures the energy usage of one appliance at a time.

pocket level: *See* torpedo level.

polarity: The positive (+) or negative (–) state of an object.

polarized plug: A plug in which one blade is wider than the other blade.

polarized receptacle: A standard receptacle that consists of one short (hot) slot and one long (neutral) slot.

pole: The number of completely separate circuits into which a switch can feed current.

potential energy: Stored energy due to position, chemical state, or physical condition.

power interruption: A loss of electric power for less than a second to a few minutes.

power optimizer (maximizer): A hybrid microinverter system that conditions the DC power before sending it to a centralized inverter instead of converting the DC power from the solar panels directly into AC power.

power outage: A short- or long-term sustained loss of electric power.

Power over Ethernet (PoE): A method of supplying power and data over a single Ethernet cable instead of having a separate power cord.

power quality: The condition of incoming power to a load.

power surge: A major increase in the electrical energy at any point in a power line.

pressure sensor: A device that measures pressure and converts it to an electrical signal.

primary cell: A cell that uses up its electrodes and electrolyte to the point where recharging is not practical.

primer: A chemical agent that cleans and softens a surface and allows solvent cement to penetrate more effectively into the surface.

programmable thermostat: A thermostat designed to adjust temperature in a dwelling at different times of the day based on preprogrammed settings.

protective helmet: A hard hat that is used in the workplace to prevent injury from the impact of falling and flying objects and from electrical shock.

protocol: An industry-standard machine language that enables the exchange of information between wireless-controlled devices.

pulling grip: A device that is attached to a fish tape to allow more leverage.

PV cell: *See* solar cell.

PVC pipe cutter: A handheld tool designed to quickly and accurately cut plastic pipe (PVC or PE) and rigid rubber hose up to 2″ diameter without the use of a vise.

PV panel (module): *See* solar panel.

PV storage battery: A deep-cycle battery that provides energy storage for a PV system and is designed to be charged and discharged repeatedly.

pyroelectric sensor: A sensor that generates a voltage in proportion to a change in temperature.

Q

qualified person: An individual with the necessary education and training who is familiar with the construction and operation of electrical equipment and devices.

quick connector: *See* back-wired connector.

R

rate-of-rise heat detector: A heat detector designed to respond to flash fires and possibly to slow-burning fires.

real-time pricing: A utility program that allows customers to pay for electricity at wholesale prices during a predetermined interval.

receiver: An electronic device that receives RF signals.

receptacle: An electrical contact device used to connect equipment with a cord and plug to an electrical system. Also known as an outlet.

receptacle box: An enclosure designed to house and protect switches, receptacles, and wiring connections.

receptacle tester: An electrical test instrument that is plugged into a standard receptacle to determine whether the receptacle is properly wired and energized.

reciprocating saw: A multipurpose cutting power tool in which the blade reciprocates to create a cutting action.

rectifier: A device that converts AC voltage to DC voltage by allowing the voltage and current to move in only one direction.

renewable energy: Energy from self-replenishing sources.

residential microgrid: An electrical network that has energy generation capability, storage capacity, and the ability for intelligent control that allows a dwelling to become a self-contained island completely independent of the main grid.

resistance: The opposition that a material offers to the flow of electric current.

resolution: A measure used to describe the sharpness and clarity of an image or picture.

restricted approach boundary: The distance from an exposed energized conductor or circuit equipment where an increased risk of electric shock exists due to the close proximity of the person.

rigid metal conduit (RMC): Heavy-duty galvanized metallic tube used to route service entrance conductors.

roughing-in: A phrase that refers to the placement of electrical boxes and wires before wall coverings and ceilings are installed.

rubber insulating matting: A floor covering that provides individuals protection from electrical shock when working on live electrical circuits.

S

safety glasses: An eye protection device with special impact-resistant glass or plastic lenses, a reinforced frame, and side shields.

schematic diagram: A drawing that indicates the electrical connections and functions of a specific circuit arrangement using graphic symbols.

screwdriver: A hand tool with a tip designed to fit into a screw head for fastening operations.

secondary cell: A cell that can be recharged.

security camera: A video surveillance camera that is used to record activity, which can help detect and prevent crime.

series-loop circuit: A circuit in which devices are connected in series and are normally closed (NC).

series-wound generator: A generator that has its field windings connected in series with the armature and the external circuit (load).

service drop entrance: A service entrance that has wires running from a utility pole to a service head.

service entrance: The wires installed between the meter fitting (socket) and the disconnecting means (main breaker, fuse box, or breaker panel) inside a dwelling.

service entrance cable: A cable that has a bare conductor wound around the insulated conductors and is often used for service drops.

service head (weatherhead): A weatherproof service point where overhead service drop conductors enter the service entrance riser pipe going to the meter socket.

service lateral entrance: A service entrance where wires are buried underground.

service panel: An electrical box designed to contain fuses or circuit breakers for protecting the individual circuits of a dwelling from the distribution system.

service panel monitor: A device that is used to monitor the change in electrical usage as various appliances and other devices are switched on and off.

service point: The connection between the electric utility and the premises wiring.

setpoint temperature: The temperature at which the switch in a thermostat opens and closes.

set screw connector: A type of box fitting that relies on the pressure of a screw against conduit to hold the conduit in place.

shrink constant: The reduction in distance that a conduit can run per inch of offset elevation.

shunt: A permanent conductor placed across a water meter to provide a continuous flow path for ground current.

shunt-wound generator: A generator that has its field windings connected in parallel (shunt) with the armature and the external circuit (load).

single-band router: A wireless router that only broadcasts one of the Wi-Fi frequencies.

single-pole switch: A single-pole, single-throw (SPST) electrical control device.

small wind turbine: A wind turbine that can be installed on properties as small as one acre in areas with sustained winds to create electricity.

smart bulb: A light bulb that contains a Wi-Fi receiver and switching device in the actual bulb.

smart light bulb socket: A Wi-Fi receiver and switching device installed in a light bulb socket that screws into standard lamps before a light bulb is installed in the smart socket.

smart doorbell: A Wi-Fi-enabled doorbell that includes video and audio capabilities and can send a signal to the homeowner when activated.

smart grid: An electrical grid in which electricity is delivered and monitored from a power source to an end user.

smart home: A residence that contains equipment, appliances, and software that are designed to provide security, energy management, and convenience.

smart home hub: A hardware device that connects multiple devices on a home automation network and controls communication between each device.

smart lock: An electromechanical device that locks and unlocks a door when it receives instructions from an authorized device using a wireless protocol and a cryptographic key to execute the authorization process.

smart meter: A meter that uses two-way digital communication to enable a utility to securely collect information regarding the electrical consumption (usage) of a consumer.

smart oven: An oven with internet access that can read food item labels to program and control temperature and cook time through a smartphone app.

smart receptacle: A receptacle controlled remotely by a control device using specific wireless signals.

smart refrigerator: A refrigerator with internet access that can sense and track the food items stored inside it through bar code reading or RFID scanning.

smoke detector: A device that senses smoke optically (photoelectric) or by physical process (ionization).

solar cell: An electrical device that converts sunlight (solar radiation) into electricity and stores it in a battery. Also known as a PV cell.

solar light: A light composed of an LED light, solar panel, battery, and a charge controller.

solar panel: An assembly of solar cells. Also known as a PV panel or module.

solar photovoltaic (PV) power generation: The process of converting energy from the sun into electricity using solar panels.

solar thermal energy system: A renewable energy system that collects and stores solar energy and is used to heat air and water in a residential structure.

solder: An alloy consisting of specific percentages of two or more metals.

soldering: A process of joining a base metal with a filler metal that has a melting point below that of the base metal.

solderless connector: A device used to firmly join wires without the help of solder.

solenoid: An electric actuator that consists of an iron plunger surrounded by an encased coil of wire.

solid wire: An insulated conductor composed of only one wire.

solvent cement: A chemical agent that penetrates and softens the surface of plastic pipe and fittings.

splice: The joining of two or more electrical wires by mechanically twisting the wires together or by using a special splicing device.

split-bolt connector: A solderless mechanical connector used for joining large cables.

split-wired receptacle: A standard duplex receptacle that has had the tab between the two brass-colored (hot) terminals removed, while the tab between the two silver-colored (neutral) terminals remains in place.

spotlight: A lamp that casts intensive light in a localized area.

standard: An accepted reference or practice.

standard plug fuse: A screw-in OCPD that contains a metal conducting element designed to melt when the current through the fuse exceeds the rated value.

standby power: *See* phantom load.

stranded wire: An insulated conductor composed of several smaller wires twisted together.

stress sensor: A sensing device that measures the weight of a load over a specific area.

string inverter: A device that converts DC power to AC power from several solar panels that are connected in series.

strip gauge: A short groove engraved in a switch to indicate the length of insulation that must be removed from a wire to fit it into a push-in fitting.

stud: An upright wood member that extends from the bottom to the top plates of a framed wall.

subpanel: A panel supplied by the main service panel that distributes circuits to a specific area of the dwelling or outbuilding.

sustained power interruption: A decrease to 0 V on all power lines for a period of more than 1 min.

switch: An electrical device used to control loads in an electrical circuit.

symbol: A graphic representation of a device, component, or object on a print.

system ground: A special circuit designed to protect the entire distribution system of a residence.

T

tagout: The process of placing a danger tag on the source of electrical power.

take-up: An adjustment made to a measurement when bending conduit.

tankless on-demand water heater: A device that rapidly heats water by an electric heating element or a natural gas heat exchanger only when water is needed.

tape measure: *See* tape rule.

tape rule: A measuring hand tool that consists of a long, continuous strip of fabric, plastic, or metal graduated in regular increments. Also known as a tape measure.

temporary power interruption: A decrease to 0 V on one or more power lines lasting between 3 sec and 1 min.

thermal circuit breaker: An OCPD that operates with a bimetallic strip that warps when overheated.

thermal imager: A device that detects heat patterns in the infrared-wavelength spectrum without making direct contact with equipment.

thermal-magnetic circuit breaker: An OCPD that combines the heating effect of a bimetallic strip with the pulling strength of a magnet to move a trip bar.

thermostat: A device that automatically regulates temperature by controlling the supply of energy, such as gas or electricity, to a heating or cooling apparatus.

three-way switch: A single-pole, double-throw (SPDT) electrical control device.

throw: The number of closed switch positions per pole.

termination: The connection of the end of a wire to a contact or terminal.

tilt switch sensor: A device that detects inclination or orientation.

time-based control system: An automated control system that uses the time of day to determine the desired operation of energy-consuming loads.

time-delay plug fuse: A screw-in OCPD with an internal dual element.

time-of-use rate: A utility program that can help in planning when to use major appliances.

timer switch: A programmable electrical control device that automatically turns off after an elapsed amount of time.

toner: An instrument, consisting of a tone generator and a toner probe, used to insert a signal into individual conductors, twisted-pair cables, and coaxial cables to test for breaks in the line or the location of a connection.

tool pouch: A small, open tool container (pouch) for storing electrical tools.

torpedo level: A small plumb and leveling tool. Also known as a pocket level.

transceiver: An electronic device that both transmits and receives RF signals.

transformer: An electric device that uses electromagnetism to change (step up or step down) AC voltage from one level to another.

transient voltage: A temporary, undesirable voltage in an electrical system.

transmission line: A conductor that carries large amounts of electrical power at high voltages over long distances.

transmitter: An electronic device that sends out RF signals.

tri-band router: A wireless router that simultaneously broadcasts three signals, which include one signal at 2.4 GHz and two separate 5 GHz signals.

T-tap splice: A type of splice that allows a connection to be made without cutting the main wire.

Type MCAP cable: Metal-clad cable manufactured with copper THHN insulated wires inside an interlocked flexible metal outer jacket.

U

UL® label: A stamped icon indicating that an electrical device has been approved for consumer use by Underwriters Laboratories Inc. (UL®).

ultrasonic detector: A device that detects a moving object by transmitting sound waves and receiving the reflected waves off of the object.

uninterruptible power supply (UPS): An electrical storage device that provides stable and reliable power during fluctuations or failures of the primary power source.

universal serial bus (USB): An industry-standard device composed of a connector, cable, and connector port for the communication of data and/or power supply between electronic devices.

USB receptacle: A receptacle that is constructed with USB connector ports for charging electronic devices.

utility knife: A knife with a blade that can be retracted into the knife body when not in use. Also known as a box-cutting knife.

V

vampire load: *See* phantom load.

verification: The process of identifying whether a person is qualified or allowed to enter a residence.

video surveillance: The monitoring and recording of activity in an area or building.

vinyl electrical tape: A type of plastic tape that has good insulation properties and can insulate up to 600 V per wrap.

voltage: The force or electrical potential between two points.

voltage indicator: An electrical device that indicates the presence of voltage when the test tip touches, or is near, an energized hot conductor or energized metal part.

voltage tester: An electrical test instrument that indicates an approximate amount of voltage and the voltage type (AC or DC) in an electrical circuit.

voltage test light: An electrical test instrument that is designed to illuminate in the presence of 120 V, 240 V, 277 V, or 480 V AC power. Also known as a neon tester.

W

watt: A unit of power used to quantify the rate of energy transfer produced by a current of one ampere across a potential difference of one volt.

wearable technology: Electronic devices worn on the body or embedded in clothing for tracking information related to health and fitness, medical conditions, GPS location, as well as for playing music and making secure payments.

Western Union splice: A type of splice that is used when the connection must be strong enough to support long lengths of heavy wire.

wet cell: Two electrodes in a liquid electrolyte solution that are used to create a chemical reaction.

wetting: The property of solder when it is molten.

whole-house generator: A permanently installed standby generator designed to supply temporary power to a dwelling during a power outage.

whole-house panel protector: A surge suppressor connected to the main service panel that diverts heavy electrical surges into the grounding system.

whole-house protection plan: The use of proper grounding methods and devices designed to suppress power surges and transient voltages to protect electrical and electronic circuits.

wide area network (WAN): The final communications link between the consumer and the utility.

wind power: The use of airflow through turbines to provide energy to turn electric generators.

wind turbine: A power generation system that converts the kinetic energy of wind into mechanical energy, which is used to rotate a generator that produces electrical energy.

wireless gate sensor: An accelerometer device that detects tilt or vibration movements and transmits a wireless signal to a receiver.

wireless mailbox sensor: A sensor that wirelessly transmits a signal to a receiver inside a dwelling to provide notification that mail is present in the mailbox.

wireless Wi-Fi router: A network device that routes data between internet-connected devices and the internet.

wire marker: A preprinted peel-off sticker designed to adhere to insulation when wrapped around a wire.

wire stripper: A hand tool that is designed to remove insulation from small-gauge wires.

wire stripper/crimper/cutter: A hand tool used to strip conductor insulation, crimp conductor terminals, and cut conductors.

wiring diagram: A diagram that shows the wired connections of all components in an electrical device.

wiring plan: An electrical floor plan that indicates the type, placement, and connection of all electrical devices required to wire an electrical circuit.

wrench: A hand tool with jaws at one or both ends that is designed to turn bolts, nuts, or other types of fasteners.

Index

Page numbers in italic refer to figures.

B

C

D

E

F

G

H

I

J

K

L

O

P

Q

R

S

T

U

V

W

X